Communications in Computer and Information Science 1257

Commenced Publication in 2007
Founding and Former Series Editors:
Simone Diniz Junqueira Barbosa, Phoebe Chen, Alfredo Cuzzocrea,
Xiaoyong Du, Orhun Kara, Ting Liu, Krishna M. Sivalingam,
Dominik Ślęzak, Takashi Washio, Xiaokang Yang, and Junsong Yuan

More information about this series at http://www.springer.com/series/7899

Jianchao Zeng · Weipeng Jing ·
Xianhua Song · Zeguang Lu (Eds.)

Data Science

6th International Conference
of Pioneering Computer Scientists,
Engineers and Educators, ICPCSEE 2020
Taiyuan, China, September 18–21, 2020
Proceedings, Part I

 Springer

Editors
Jianchao Zeng
North University of China
Taiyuan, China

Weipeng Jing
Northeast Forestry University
Harbin, China

Xianhua Song
Harbin University of Science
and Technology
Harbin, China

Zeguang Lu
National Academy of Guo Ding Institute
of Data Science
Beijing, China

ISSN 1865-0929   ISSN 1865-0937 (electronic)
Communications in Computer and Information Science
ISBN 978-981-15-7980-6   ISBN 978-981-15-7981-3 (eBook)
https://doi.org/10.1007/978-981-15-7981-3

This Springer imprint is published by the registered company Springer Nature Singapore Pte Ltd.
The registered company address is: 152 Beach Road, #21-01/04 Gateway East, Singapore 189721, Singapore

Preface

As the program chairs of the 6th International Conference of Pioneer Computer Scientists, Engineers and Educators 2020 (ICPCSEE 2020, originally ICYCSEE), it was our great pleasure to welcome you to the conference, which will be held in Taiyuan, China, September 18–21, 2020, hosted by North University of China and National Academy of Guo Ding Institute of Data Science, China. The goal of this conference is to provide a forum for computer scientists, engineers, and educators.

This conference attracted 392 paper submissions. After the hard work of the Program Committee, 98 papers were accepted to appear in the conference proceedings, with an acceptance rate of 25%. The major topic of this conference is data science. The accepted papers cover a wide range of areas related to basic theory and techniques for data science including mathematical issues in data science, computational theory for data science, big data management and applications, data quality and data preparation, evaluation and measurement in data science, data visualization, big data mining and knowledge management, infrastructure for data science, machine learning for data science, data security and privacy, applications of data science, case study of data science, multimedia data management and analysis, data-driven scientific research, data-driven bioinformatics, data-driven healthcare, data-driven management, data-driven e-government, data-driven smart city/planet, data marketing and economics, social media and recommendation systems, data-driven security, data-driven business model innovation, and social and/or organizational impacts of data science.

We would like to thank all the Program Committee members, 216 coming from 102 institutes, for their hard work in completing the review tasks. Their collective efforts made it possible to attain quality reviews for all the submissions within a few weeks. Their diverse expertise in each individual research area helped us to create an exciting program for the conference. Their comments and advice helped the authors to improve the quality of their papers and gain deeper insights.

We thank Dr. Lanlan Chang and Jane Li from Springer, whose professional assistance was invaluable in the production of the proceedings. A big thanks also to the authors and participants for their tremendous support in making the conference a success.

Besides the technical program, this year ICPCSEE offered different experiences to the participants. We hope you enjoyed the conference.

June 2020

Pinle Qin
Weipeng Jing

Organization

The 6th International Conference of Pioneering Computer Scientists, Engineers and Educators (ICPCSEE, originally ICYCSEE) 2020 (http://2020.icpcsee.org) was held in Taiyuan, China, during September 18–21, 2020, hosted by North University of China and National Academy of Guo Ding Institute of Data Science, China.

General Chair

Jianchao Zeng   North University of China, China

Program Chairs

Pinle Qin   North University of China, China
Weipeng Jing   Northeast Forestry University, China

Program Co-chairs

Yan Qiang   Taiyuan University of Technology, China
Yuhua Qian   Shanxi University, China
Peng Zhao   Taiyuan Normal University, China
Lihu Pan   Taiyuan University of Science and Technology, China
Alex kou   University of Victoria, Canada
Hongzhi Wang   Harbin Institute of Technology, China

Organization Chairs

Juanjuan Zhao   Taiyuan University of Technology, China
Fuyuan Cao   Shanxi University, China
Donglai Fu   North University of China, China
Xiaofang Mu   Taiyuan Normal University, China
Chang Song   Institute of Coal Chemistry, CAS, China

Organization Co-chairs

Rui Chai   North University of China, China
Yanbo Wang   North University of China, China
Haibo Yu   North University of China, China
Yi Yu   North University of China, China
Lifang Wang   North University of China, China

Hu Zhang	Shanxi University, China
Wei Wei	Shanxi University, China
Rui Zhang	Taiyuan University of Science and Technology, China

Publication Chair

| Guanglu Sun | Harbin University of Science and Technology, China |

Publication Co-chairs

| Xianhua Song | Harbin University of Science and Technology, China |
| Xie Wei | Harbin University of Science and Technology, China |

Forum Chairs

Haiwei Pan	Harbin Engineering University, China
Qiguang Miao	Xidian University, China
Fudong Liu	Information Engineering University, China
Feng Wang	RoarPanda Network Technology Co., Ltd., China

Oral Session and Post Chair

| Xia Liu | Sanya Aviation and Tourism college, China |

Mpetition Committee Chairs

| Peng Zhao | Taiyuan Normal University, China |
| Xiangfei Cai | Huiying Medical Technology Co., Ltd., China |

Registration and Financial Chairs

| Chunyan Hu | National Academy of Guo Ding Institute of Data Science, China |
| Yuanping Wang | Shanxi Jinyahui Culture Spreads Co., Ltd., China |

ICPCSEE Steering Committee

Jiajun Bu	Zhejiang University, China
Wanxiang Che	Harbin Institute of Technology, China
Jian Chen	ParaTera, China
Wenguang Chen	Tsinghua University, China
Xuebin Chen	North China University of Science and Technology, China
Xiaoju Dong	Shanghai Jiao Tong University, China
Qilong Han	Harbin Engineering University, China
Yiliang Han	Engineering University of CAPF, China

Yinhe Han	Institute of Computing Technology, CAS, China
Hai Jin	Huazhong University of Science and Technology, China
Weipeng Jing	Northeast Forestry University, China
Wei Li	Central Queensland University, China
Min Li	Central South University, China
Junyu Lin	Institute of Information Engineering, CAS, China
Yunhao Liu	Michigan State University, USA
Zeguang Lu	National Academy of Guo Ding Institute of Data Science, China
Rui Mao	Shenzhen University, China
Qiguang Miao	Xidian University, China
Haiwei Pan	Harbin Engineering University, China
Pinle Qin	North University of China, China
Zheng Shan	Information Engineering University, China
Guanglu Sun	Harbin University of Science and Technology, China
Jie Tang	Tsinghua University, China
Tian Feng	Institute of Software Chinese Academy of Sciences, China
Tao Wang	Peking University, China
Hongzhi Wang	Harbin Institute of Technology, China
Xiaohui Wei	Jilin University, China
lifang Wen	Beijing Huazhang Graphics & Information Co., Ltd., China
Liang Xiao	Nanjing University of Science and Technology, China
Yu Yao	Northeastern University, China
Xiaoru Yuan	Peking University, China
Yingtao Zhang	Harbin Institute of Technology, China
Yunquan Zhang	Institute of Computing Technology, CAS, China
Baokang Zhao	National University of Defense Technology, China
Min Zhu	Sichuan University, China
Liehuang Zhu	Beijing Institute of Technology, China

Program Committee Members

Witold Abramowicz	Poznań University of Economics and Business, Poland
Chunyu Ai	University of South Carolina Upstate, USA
Jiyao An	Hunan University, China
Ran Bi	Dalian University of Technology, China
Zhipeng Cai	Georgia State University, USA
Yi Cai	South China University of Technology, China
Zhipeng Cai	Georgia State University, USA
Zhao Cao	Beijing Institute of Technology, China
Richard Chbeir	LIUPPA Laboratory, France
Wanxiang Che	Harbin Institute of Technology, China
Wei Chen	Beijing Jiaotong University, China

Hao Chen	Hunan University, China
Xuebin Chen	North China University of Science and Technology, China
Chunyi Chen	Changchun University of Science and Technology, China
Yueguo Chen	Renmin University of China, China
Zhuang Chen	Guilin University of Electronic Technology, China
Siyao Cheng	Harbin Institute of Technology, China
Byron Choi	Hong Kong Baptist University, China
Vincenzo Deufemia	University of Salerno, Italy
Xiaofeng Ding	Huazhong University of Science and Technology, China
Jianrui Ding	Harbin Institute of Technology, China
Hongbin Dong	Harbin Engineering University, China
Minggang Dong	Guilin University of Technology, China
Longxu Dou	Harbin Institute of Technology, China
Pufeng Du	Tianjin University, China
Lei Duan	Sichuan University, China
Xiping Duan	Harbin Normal University, China
Zherui Fan	Xidian University, China
Xiaolin Fang	Southeast University, China
Ming Fang	Changchun University of Science and Technology, China
Jianlin Feng	Sun Yat-sen University, China
Yongkang Fu	Xidian University, China
Jing Gao	Dalian University of Technology, China
Shuolin Gao	Harbin Institute of Technology, China
Daohui Ge	Xidian University, China
Yu Gu	Northeastern University, China
Yingkai Guo	National University of Singapore, Singapore
Dianxuan Gong	North China University of Science and Technology, China
Qi Han	Harbin Institute of Technology, China
Meng Han	Georgia State University, USA
Qinglai He	Arizona State University, USA
Tieke He	Nanjing University, China
Zhixue He	North China Institute of Aerospace Engineering, China
Tao He	Harbin Institute of Technology, China
Leong Hou	University of Macau, China
Yutai Hou	Harbin Institute of Technology, China
Wei Hu	Nanjing University, China
Xu Hu	Xidian University, China
Lan Huang	Jilin University, China
Hao Huang	Wuhan University, China
Kuan Huang	Utah State University, USA
Hekai Huang	Harbin Institute of Technology, China

Cun Ji	Shandong Normal University, China
Feng Jiang	Harbin Institute of Technology, China
Bin Jiang	Hunan University, China
Xiaoyan Jiang	Shanghai University of Engineering Science, China
Wanchun Jiang	Central South University, China
Cheqing Jin	East China Normal University, China
Xin Jin	Beijing Electronic Science and Technology Institute, China
Chao Jing	Guilin University of Technology, China
Hanjiang Lai	Sun Yat-sen University, China
Shiyong Lan	Sichuan University, China
Wei Lan	Guangxi University, China
Hui Li	Xidian University, China
Zhixu Li	Soochow University, China
Mingzhao Li	RMIT University, Australia
Peng Li	Shaanxi Normal University, China
Jianjun Li	Huazhong University of Science and Technology, China
Xiaofeng Li	Sichuan University, China
Zheng Li	Sichuan University, China
Mohan Li	Jinan University, China
Min Li	South University, China
Zhixun Li	Nanchang University, China
Hua Li	Changchun University of Science and Technology, China
Rong-Hua Li	Shenzhen University, China
Cuiping Li	Renmin University of China, China
Qiong Li	Harbin Institute of Technology, China
Qingliang Li	Changchun University of Science and Technology, China
Wei Li	Georgia State University, USA
Yunan Li	Xidian University, China
Hongdong Li	Central South University, China
Xiangtao Li	Northeast Normal University, China
Xuwei Li	Computer Science College Sichuan University, China
Yanli Liu	Sichuan University, China
Hailong Liu	Northwestern Polytechnical University, China
Guanfeng Liu	Macquarie University, Australia
Yan Liu	Harbin Institute of Technology, China
Xia Liu	Sanya Aviation Tourism College, China
Yarong Liu	Guilin University of Technology, China
Shuaiqi Liu	Tianjin Normal University, China
Jin Liu	Central South University, China
Yijia Liu	Harbin Institute of Technology, China
Zeming Liu	Harbin Institute of Technology, China

Zeguang Lu	National Academy of Guo Ding Institute of Data Science, China
Binbin Lu	Sichuan University, China
Junling Lu	Shaanxi Normal University, China
Mingming Lu	Central South University, China
Jizhou Luo	Harbin Institute of Technology, China
Junwei Luo	Henan Polytechnic University, China
Zhiqiang Ma	Inner Mongolia University of Technology, China
Chenggang Mi	Northwestern Polytechnical University, China
Tiezheng Nie	Northeastern University, China
Haiwei Pan	Harbin Engineering University, China
Jialiang Peng	Norwegian University of Science and Technology, Norway
Fei Peng	Hunan University, China
Yuwei Peng	Wuhan University, China
Jianzhong Qi	The University of Melbourne, Australia
Xiangda Qi	Xidian University, China
Shaojie Qiao	Southwest Jiaotong University, China
Libo Qin	Research Center for Social Computing and Information Retrieval, China
Zhe Quan	Hunan University, China
Chang Ruan	Central South University of Sciences, China
Yingxia Shao	Peking University, China
Yingshan Shen	South China Normal University, China
Meng Shen	Xidian University, China
Feng Shi	Central South University, China
Yuanyuan Shi	Xi'an University of Electronic Science and Technology, China
Xiaoming Shi	Harbin Institute of Technology, China
Wei Song	North China University of Technology, China
Shoubao Su	Jinling Institute of Technology, China
Yanan Sun	Oklahoma State University, USA
Minghui Sun	Jilin University, China
Guanghua Tan	Hunan University, China
Dechuan Teng	Harbin Institute of Technology, China
Yongxin Tong	Beihang University, China
Xifeng Tong	Northeast Petroleum University, China
Vicenc Torra	Högskolan I Skövde, Sweden
Hongzhi Wang	Harbin Institute of Technology, China
Yingjie Wang	Yantai University, China
Dong Wang	Hunan University, China
Yongheng Wang	Hunan University, China
Chunnan Wang	Harbin Institute of Technology, China
Jinbao Wang	Harbin Institute of Technology, China
Xin Wang	Tianjin University, China
Peng Wang	Fudan University, China

Chaokun Wang   Tsinghua University, China
Xiaoling Wang   East China Normal University, China
Jiapeng Wang   Harbin Huade University, China
Qingshan Wang   Hefei University of Technology, China
Wenfeng Wang   CAS, China
Shaolei Wang   Harbin Institute of Technology, China
Yaqing Wang   Xidian University, China
Yuxuan Wang   Harbin Institute of Technology, China
Wei Wei   Xi'an Jiaotong University, China
Haoyang Wen   Harbin Institute of Technology, China
Huayu Wu   Institute for Infocomm Research, Singapore
Yan Wu   Changchun University of Science and Technology,
                        China
Huaming Wu   Tianjin University, China
Bin Wu   Institute of Information Engineering, CAS, China
Yue Wu   Xidian University, China
Min Xian   Utah State University, USA
Sheng Xiao   Hunan University, China
Wentian Xin   Xidian University, China
Ying Xu   Hunan University, China
Jing Xu   Changchun University of Science and Technology,
                        China
Jianqiu Xu   Nanjing University of Aeronautics and Astronautics,
                        China
Qingzheng Xu   National University of Defense Technology, China
Yang Xu   Harbin Institute of Technology, China
Yaohong Xue   Changchun University of Science and Technology,
                        China
Mingyuan Yan   University of North Georgia, USA
Yu Yan   Harbin Institute of Technology, China
Cheng Yan   Central South University, China
Yajun Yang   Tianjin University, China
Gaobo Yang   Hunan University, China
Lei Yang   Heilongjiang University, China
Ning Yang   Sichuan University, China
Xiaochun Yang   Northeastern University, China
Shiqin Yang   Xidian University, China
Bin Yao   Shanghai Jiao Tong University, China
Yuxin Ye   Jilin University, China
Xiufen Ye   Harbin Engineering University, China
Minghao Yin   Northeast Normal University, China
Dan Yin   Harbin Engineering University, China
Zhou Yong   China University of Mining and Technology, China
Jinguo You   Kunming University of Science and Technology, China
Xiaoyi Yu   Peking University, China
Ye Yuan   Northeastern University, China

Kun Yue	Yunnan University, China
Yue Yue	SUTD, Singapore
Xiaowang Zhang	Tianjin University, China
Lichen Zhang	Normal University, China
Yingtao Zhang	Harbin Institute of Technology, China
Yu Zhang	Harbin Institute of Technology, China
Wenjie Zhang	University of New South Wales, Australia
Dongxiang Zhang	University of Electronic Science and Technology of China, China
Xiao Zhang	Renmin University of China, China
Kejia Zhang	Harbin Engineering University, China
Yonggang Zhang	Jilin University, China
Huijie Zhang	Northeast Normal University, China
Boyu Zhang	Utah State University, USA
Jin Zhang	Beijing Normal University, China
Dejun Zhang	China University of Geosciences, China
Zhifei Zhang	Tongji University, China
Shigeng Zhang	Central South University, China
Mengyi Zhang	Harbin Institute of Technology, China
Yongqing Zhang	Chengdu University of Information Technology, China
Xiangxi Zhang	Harbin Institute of Technology, China
Meiyang Zhang	Southwest University, China
Zhen Zhang	Xidian University, China
Jian Zhao	Changchun University, China
Qijun Zhao	Sichuan University, China
Bihai Zhao	Changsha University, China
Xiaohui Zhao	University of Canberra, Australia
Peipei Zhao	Xidian University, China
Bo Zheng	Harbin Institute of Technology, China
Jiancheng Zhong	Hunan Normal University, China
Jiancheng Zhong	Central South University, China
Fucai Zhou	Northeastern University, China
Changjian Zhou	Northeast Agricultural University, China
Min Zhu	Sichuan University, China
Yuanyuan Zhu	Wuhan University, China
Yungang Zhu	Jilin University, China
Bing Zhu	Central South University, China
Wangmeng Zuo	Harbin Institute of Technology, China

Contents – Part I

Database

General Model for Index Recommendation Based on Convolutional
Neural Network 3
  Yu Yan and Hongzhi Wang

An Incremental Partitioning Graph Similarity Search
Based on Tree Structure Index 16
  Yuhang Li, Yan Yang, and Yingli Zhong

Data Cleaning About Student Information Based on Massive Open
Online Course System 33
  Shengjun Yin, Yaling Yi, and Hongzhi Wang

V2V Online Data Offloading Method Based on Vehicle Mobility 44
  Xianlang Hu, Dafang Zhang, Shizhao Zhao, Guangsheng Feng,
  and Hongwu Lv

Highly Parallel SPARQL Engine for RDF 61
  Fan Feng, Weikang Zhou, Ding Zhang, and Jinhui Pang

Identifying Propagation Source in Temporal Networks Based
on Label Propagation. 72
  Lilin Fan, Bingjie Li, Dong Liu, Huanhuan Dai, and Yan Ru

Analysis Method for Customer Value of Aviation Big Data Based
on LRFMC Model 89
  Yang Tao

Identifying Vital Nodes in Social Networks Using an Evidential
Methodology Combining with High-Order Analysis 101
  Meng Zhang, Guanghui Yan, Yishu Wang, and Ye Lv

Location Privacy-Preserving Method Based on Degree of Semantic
Distribution Similarity 118
  Rui Liu, Kaizhong Zuo, Yonglu Wang, and Jun Zhao

Supervisable Anonymous Management of Digital Certificates
for Blockchain PKI 130
  Shaozhuo Li, Na Wang, Xuehui Du, and Xuan Li

Product Customer Demand Mining and Its Functional Attribute
Configuration Driven by Big Data.   145
  Dianting Liu, Xia Huang, and Kangzheng Huang

Legal Effect of Smart Contracts Based on Blockchain   166
  Xichen Li

Machine Learning

Convex Reconstruction of Structured Matrix Signals from Linear
Measurements: Theoretical Results   189
  Yuan Tian

Design of Integrated Circuit Chip Fault Diagnosis System Based
on Neural Network   222
  Xinsheng Wang, Xiaoyao Qi, and Bin Sun

Improving Approximate Bayesian Computation with Pre-judgment Rule   230
  Yanbo Wang, Xiaoqing Yu, Pinle Qin, Rui Chai, and Gangzhu Qiao

Convolutional Neural Network Visualization in Adversarial
Example Attack   247
  Chenshuo Yu, Xiuli Wang, and Yang Li

Deeper Attention-Based Network for Structured Data   259
  Xiaohua Wu, Youping Fan, Wanwan Peng, Hong Pang, and Yu Luo

Roofline Model-Guided Compilation Optimization Parameter
Selection Method   268
  Qi Du, Hui Huang, and Chun Huang

Weighted Aggregator for the Open-World Knowledge Graph Completion . . .   283
  Yueyang Zhou, Shumin Shi, and Heyan Huang

Music Auto-Tagging with Capsule Network   292
  *Yongbin Yu, Yifan Tang, Minhui Qi, Feng Mai, Quanxin Deng,
  and Zhaxi Nima*

Complex-Valued Densely Connected Convolutional Networks   299
  Wenhan Li, Wenqing Xie, and Zhifang Wang

Representation Learning with Deconvolution for Multivariate Time
Series Classification and Visualization   310
  *Wei Song, Lu Liu, Minghao Liu, Wenxiang Wang, Xiao Wang,
  and Yu Song*

Learning Single-Shot Detector with Mask Prediction and Gate Mechanism. . .   327
  Jingyi Chen, Haiwei Pan, Qianna Cui, Yang Dong, and Shuning He

Network

Near-Optimal Transmission Optimization of Cache-Based SVC
Vehicle Video 341
   Xianlang Hu, He Dong, Xiangrui Kong, Ruibing Li, Guangsheng Feng,
   and Hongwu Lv

Improved Design of Energy Supply Unit of WSN Node Based on Boost
and BUCK Structure 354
   Shaojun Yu, Li Lin, Yujian Wang, Kaiguo Qian, and Shikai Shen

Disjoint-Path Routing Mechanism in Mobile Opportunistic Networks 368
   Peiyan Yuan, Hao Zhang, and Xiaoyan Huang

Double LSTM Structure for Network Traffic Flow Prediction. 380
   Lin Huang, Diangang Wang, Xiao Liu, Yongning Zhuo, and Yong Zeng

Super Peer-Based P2P VoD Architecture for Supporting
Multiple Terminals 389
   Pingshan Liu, Yaqing Fan, Kai Huang, and Guimin Huang

Localization Algorithm of Wireless Sensor Network Based on Concentric
Circle Distance Calculation 405
   KaiGuo Qian, Chunfen Pu, Yujian Wang, ShaoJun Yu, and Shikai Shen

Data Transmission Using HDD as Microphone. 416
   Yongyu Liang, Jinghua Zheng, and Guozheng Yang

Android Vault Application Behavior Analysis and Detection 428
   Nannan Xie, Hongpeng Bai, Rui Sun, and Xiaoqiang Di

Research on Detection and Identification of Dense Rebar
Based on Lightweight Network. 440
   Fang Qu, Caimao Li, Kai Peng, Cong Qu, and Chengrong Lin

Automatic Detection of Solar Radio Spectrum Based on Codebook Model. . . 447
   Guoliang Li, Guowu Yuan, Hao Zhou, Hao Wu, Chengming Tan,
   Liang Dong, Guannan Gao, and Ming Wang

Graphic Images

Self-service Behavior Recognition Algorithm Based on Improved Motion
History Image Network 463
   Liping Deng, Qingji Gao, and Da Xu

Supervised Deep Second-Order Covariance Hashing for Image Retrieval 476
   Qian Wang, Yue Wu, Jianxin Zhang, Hengbo Zhang, Chao Che,
   and Lin Shan

Using GMOSTNet for Tree Detection Under Complex Illumination
and Morphological Occlusion 488
    Zheng Qian, Hailin Feng, Yinhui Yang, Xiaochen Du, and Kai Xia

Improved YOLOv3 Infrared Image Pedestrian Detection Algorithm 506
    Jianting Shi, Guiqiang Zhang, Jie Yuan, and Yingtao Zhang

Facial Expression Recognition Based on PCD-CNN with Pose
and Expression 518
    Hongbin Dong, Jin Xu, and Qiang Fu

Regression-Based Face Pose Estimation with Deep Multi-modal
Feature Loss 534
    Yanqiu Wu, Chaoqun Hong, Liang Chen, and Zhiqiang Zeng

Breast Cancer Recognition Based on BP Neural Network Optimized
by Improved Genetic Algorithm 550
    Wenqing Xie, Wenhan Li, and Zhifang Wang

Amplification Method of Lung Nodule Data Based on DCGAN
Generation Algorithm 563
    Minghao Yu, Lei Cai, Liwei Gao, and Jingyang Gao

CP-Net: Channel Attention and Pixel Attention Network for Single
Image Dehazing 577
    Shunan Gao, Jinghua Zhu, and Yan Yang

Graphic Processor Unit Acceleration of Multi-exposure Image Fusion
with Median Filter 591
    Shijie Li, Yuan Yuan, Qiong Li, and Xuchao Xie

Improved PCA Face Recognition Algorithm 602
    Yang Tao and Yuanzi He

Classification of Remote Sensing Images Based on Band Selection
and Multi-mode Feature Fusion 612
    Xiaodong Yu, Hongbin Dong, Zihe Mu, and Yu Sun

System

Superpage-Friendly Page Table Design for Hybrid Memory Systems 623
    Xiaoyuan Wang, Haikun Liu, Xiaofei Liao, and Hai Jin

Recommendation Algorithm Based on Improved Convolutional Neural
Network and Matrix Factorization 642
    Shengbin Liang, Lulu Bai, and Hengming Zhang

Research on Service Recommendation Method Based on Cloud Model
Time Series Analysis.   655
  Zhiwu Zheng, Jing Yao, and Hua Zhang

Cloud Resource Allocation Based on Deep Q-Learning Network   666
  Zuocong Chen

Application of Polar Code-Based Scheme in Cloud Secure Storage   676
  Zhe Li, Yiliang Han, and Yu Li

Container Cluster Scheduling Strategy Based on Delay Decision Under
Multidimensional Constraints   690
  Yijun Xue, Ningjiang Chen, and Yongsheng Xie

Trusted Virtual Machine Model Based on High-Performance
Cipher Coprocessor.   705
  Haidong Liu, Songhui Guo, Lei Sun, and Song Guo

Author Index   717

Contents – Part II

Natural Language Processing

Content-Based Hybrid Deep Neural Network Citation
Recommendation Method. 3
    Leipeng Wang, Yuan Rao, Qinyu Bian, and Shuo Wang

Multi-factor Fusion POI Recommendation Model 21
    Xinxing Ma, Jinghua Zhu, Shuo Zhang, and Yingli Zhong

Effective Vietnamese Sentiment Analysis Model Using Sentiment Word
Embedding and Transfer Learning. 36
    Yong Huang, Siwei Liu, Liangdong Qu, and Yongsheng Li

Construction of Word Segmentation Model Based on HMM + BI-LSTM. . . . 47
    Hang Zhang and Bin Wen

POI Recommendations Using Self-attention Based on Side Information 62
    Chenbo Yue, Jinghua Zhu, Shuo Zhang, and Xinxing Ma

Feature Extraction by Using Attention Mechanism in Text Classification 77
    Yaling Wang and Yue Wang

Event Extraction via DMCNN in Open Domain Public
Sentiment Information 90
    Zhanghui Wang, Le Sun, Xiaoguang Li, and Linteng Wang

Deep Neural Semantic Network for Keywords Extraction on Short Text 101
    *Chundong She, Huanying You, Changhai Lin, Shaohua Liu,
    Boxiang Liang, Juan Jia, Xinglei Zhang, and Yanming Qi*

Application of News Features in News Recommendation Methods:
A Survey 113
    Jing Qin and Peng Lu

Security

New Lattice-Based Digital Multi-signature Scheme 129
    Chunyan Peng and Xiujuan Du

Research on Online Leakage Assessment 138
    *Zhengguang Shi, Fan Huang, Mengce Zheng, Wenlong Cao, Ruizhe Gu,
    Honggang Hu, and Nenghai Yu*

Differential Privacy Trajectory Protection Method Based
on Spatiotemporal Correlation.   151
   Kangkang Dou and Jian Liu

A New Lightweight Database Encryption and Security Scheme
for Internet-of-Things   167
   Jishun Liu, Yan Zhang, Zhengda Zhou, and Huabin Tang

Research on MLChecker Plagiarism Detection System.   176
   Haihao Yu, Chengzhe Huang, Leilei Kong, Xu Sun, Haoliang Qi,
   and Zhongyuan Han

Algorithm

Research on Global Competition and Acoustic Search Algorithm
Based on Random Attraction   185
   Xiaoming Zhang, Jiakang He, and Qing Shen

Over-Smoothing Algorithm and Its Application to GCN
Semi-supervised Classification   197
   Mingzhi Dai, Weibin Guo, and Xiang Feng

Improved Random Forest Algorithm Based on Adaptive Step Size Artificial
Bee Colony Optimization.   216
   Jiuyuan Huo, Xuan Qin, Hamzah Murad Mohammed Al-Neshmi,
   Lin Mu, and Tao Ju

Application

Visual Collaborative Maintenance Method for Urban Rail Vehicle Bogie
Based on Operation State Prediction   237
   Yi Liu, Qi Chang, Qinghai Gan, Guili Huang, Dehong Chen, and Lin Li

Occlusion Based Discriminative Feature Mining
for Vehicle Re-identification.   246
   Xianmin Lin, Shengwang Peng, Zhiqi Ma, Xiaoyi Zhou,
   and Aihua Zheng

StarIn: An Approach to Predict the Popularity of GitHub Repository.   258
   Leiming Ren, Shimin Shan, Xiujuan Xu, and Yu Liu

MCA-TFP Model: A Short-Term Traffic Flow Prediction Model Based
on Multi-characteristic Analysis   274
   Xiujuan Xu, Lu Xu, Yulin Bai, Zhenzhen Xu, Xiaowei Zhao, and Yu Liu

Research on Competition Game Algorithm Between High-Speed Railway
and Ordinary Railway System Under the Urban Corridor Background   290
  Lei Wang and Yao Lu

Stock Price Forecasting and Rule Extraction Based on L1-Orthogonal
Regularized GRU Decision Tree Interpretation Model   309
  Wenjun Wu, Yuechen Zhao, Yue Wang, and Xiuli Wang

Short-Term Predictions and LIME-Based Rule Extraction for Standard
and Poor's Index.   329
  Chunqi Qi, Yue Wang, Wenjun Wu, and Xiuli Wang

A Hybrid Data-Driven Approach for Predicting Remaining Useful Life
of Industrial Equipment   344
  Zheng Tan, Yiping Wen, and TianCai Li

Research on Block Storage and Analysis Model for Belt and Road Initiative
Ecological Carrying Capacity Evaluation System.   354
  Lihu Pan, Yunkai Li, Yu Dong, and Huimin Yan

Hybrid Optimization-Based GRU Neural Network for Software
Reliability Prediction   369
  Maochuan Wu, Junyu Lin, Shouchuang Shi, Long Ren,
  and Zhiwen Wang

Space Modeling Method for Ship Pipe Route Design   384
  Zongran Dong, Xuanyi Bian, and Song Yang

Simulation of Passenger Behavior in Virtual Subway Station Based
on Multi-Agent.   393
  Yan Li, Zequn Li, Fengting Yan, Yu Wei, Linghong Kong, Zhicai Shi,
  and Lei Qiao

Automatic Fault Diagnosis of Smart Water Meter Based
on BP Neural Network   409
  Jing Lin and Chunqiao Mi

Prediction of Primary Frequency Regulation Capability of Power System
Based on Deep Belief Network.   423
  Wei Cui, Wujing Li, Cong Wang, Nan Yang, Yunlong Zhu, Xin Bai,
  and Chen Xue

Simplified Programming Design Based on Automatic Test System
of Aeroengine.   436
  Shuang Xia

Intelligent Multi-objective Anomaly Detection Method Based on Robust
Sparse Flow   445
   Ke Wang, Weishan Huang, Yan Chen, Ningjiang Chen, and Ziqi He

Empirical Analysis of Tourism Revenues in Sanya   458
   Yuanhui Li and Haiyun Han

Study of CNN-Based News-Driven Stock Price Movement Prediction
in the A-Share Market   467
   Yimeng Shang and Yue Wang

Research on Equipment Fault Diagnosis Classification Model Based
on Integrated Incremental Dynamic Weight Combination   475
   Haipeng Ji, Xinduo Liu, Aoqi Tan, Zhijie Wang, and Bing Yu

Code Smell Detection Based on Multi-dimensional Software Data
and Complex Networks   490
   Heng Tong, Cheng Zhang, and Futian Wang

Improvement of Association Rule Algorithm Based on Hadoop
for Medical Data.   506
   Guangqian Kong, Huan Tian, Yun Wu, and Qiang Wei

An Interpretable Personal Credit Evaluation Model   521
   Bing Chen, Xiuli Wang, Yue Wang, and Weiyu Guo

Education

Research on TCSOL Master Course Setting   543
   Chunhong Qi and Xiao Zhang

Analysis of Learner's Behavior Characteristics Based on Open University
Online Teaching   561
   Yang Zhao

Empirical Study of College Teachers' Attitude Toward Introducing Mobile
Learning to Classroom Teaching.   574
   Dan Ren, Liang Liao, Yuanhui Li, and Xia Liu

Blended Teaching Mode Based on "Cloud Class + PAD Class"
in the Course of Computer Graphics and Image Design   588
   Yuhong Sun, Shengnan Pang, Baihui Yin, and Xiaofang Zhong

Application of 5G+VR Technology in Landscape Design Teaching   601
   Jun Liu and Tiejun Zhu

Feature Extraction of Chinese Characters Based on ASM Algorithm   620
   Xingchen Wei, Minhua Wu, and Liming Luo

Study of Priority Recommendation Method Based on Cognitive
Diagnosis Model.   638
    *Suojuan Zhang, Jiao Liu, Song Huang, Jinyu Song, Xiaohan Yu,
    and Xianglin Liao*

Author Index   649

Database

General Model for Index Recommendation Based on Convolutional Neural Network

Yu Yan[1] and Hongzhi Wang[1,2](✉)

[1] Harbin Institute of Technology, Harbin, China
{yuyan0618,wangzh}@hit.edu.cn
[2] Peng Cheng Laboratory, Shenzhen, China

Abstract. With the advent of big data, the cost of index recommendation (IR) increases exponentially, and the portability of IR model becomes an urgent problem to be solved. In this paper, a fine-grained classification model based on multi-core convolution neural network (CNNIR) is proposed to implement the transferable IR model. Using the strong knowledge representation ability of convolution network, CNNIR achieves the effective knowledge representation from data and workload, which greatly improves the classification accuracy. In the test set, the accuracy of model classification reaches over 95%. CNNIR has good robustness which can perform well under a series of different learning rate settings. Through experiments on MongoDB, the indexes recommended by CNNIR is effective and transferable.

Keywords: Document · Database · Index recommendation · CNN · Class model

1 Introduction

Database administrators (DBAs) usually need efforts to determine the best index for current applications [16]. It means that finding the best index for the current dataset and workload has always been a challenging task. To some extent, DBAs rely on so-called design consulting tools [2,7,10], which provide advice based on statistical data. However, due to the uncertainty and one sidedness of statistical data [9], DBAs usually find it difficult to make the best choice according to the experiences. With the development of informationization, the size of database increases explosively. As a result, DBAs need a lot of statistical data and efforts to index large-scale data. In order to solve these problems, scholars began to study automatic index recommendation (IR).

Nowadays, we are faced with two major challenges in IR. On the one hand, is that IR consumes a large amount of resources on the massive data, and thus it is impractical to solve IR problem in large scale data by using iterative method [3,5,14]. On the other hand, it is difficult to define the criteria of IR, sometimes IR lower the performance when the criteria selected are inappropriate.

© Springer Nature Singapore Pte Ltd. 2020
J. Zeng et al. (Eds.): ICPCSEE 2020, CCIS 1257, pp. 3–15, 2020.
https://doi.org/10.1007/978-981-15-7981-3_1

For example, only taking into account the workload without considering the original data, it is incomplete and sometimes lowers the performance.

IR has always been a challenge problem, which has been studied since 1970, such as literature [15,18] For the first time, reference [3] proposed the IR of RDBMS, which used the concept of "atomic configuration" mentioned in reference [13]. The system iterated from a single index to find all the best index configuration under the whole load. In reference [5,14], the candidate indexes were selected first, and then the indexes were iterated under these candidates. Through the above method, DBAs can find the best index by iterating on the original data set, without statistics and consultation. But it is impossible to find the most suitable index iteratively when there is a large amount of data, since we are unable to afford the cost of iterative recommendation for big data. What's more, the model is not transferable, and it needs to be re-trained when a new dataset is added, which is obviously inefficient.

In recent years, scholars have proposed some solutions based on machine learning [4,16], which can make IR more flexible. Reference [16] proposed a NoDBA system, which used reinforcement learning method for IR. It uses real measurement results or workload prediction results as reward function to determine whether a field should be indexed or not. Reinforcement learning method can automatically recommend index without a large number of data statistics and consultation. In reference [4], the author converted IR to a two-classification problem for improving recommendation precision. Since there is no need for a large number of iterations by using the classification model to determine whether the index is established, the model can greatly reduce the resource consumption. However, the above research still does not solve the problem of model migration. That is, when we want to recommend indexes for datasets, we must rely on the current datasets to train the model. E.g. We need to deal with IR on 10,000,000 datasets and 50,000,000 workloads. Using the model in [4–6,9,13–16] requires training at least 10,000,000 times to get the recommended index of all datasets. In the background of big data, each training will consume a lot of resources, so the mobility of the model has become an important problem to be solved.

To solve the above problems, this paper studies the transferability of the IR model for the first time and proposes a general multi-classification model (CNNIR) to solve the problem. The author considers the influence of dataset features (DF) and workload features (WF) which fundamentally affect IR results. Because the data in real life can always be divided into several kinds of distribution, we build CNNIR on the basis of DF distribution and WF distribution so that CNNIR can achieve better transferability. In summary, this paper firstly studies how to break the gap between different data sets from the perspective of data distribution, and really considers using one IR model to fit all data distribution.

CNNIR is a multi-classification model based on multi-core convolutional neural network. The input of CNNIR is data sequence (DS) and workload sequence (WS), and the output is index configuration containing some fine-grained characteristics of index. CNNIR uses convolution neural network to convolute sequence by multiple convolution kernels for learning features efficiently. The output of CNNIR contains a four-dimensional vector to illustrate the index configuration. It is a multi-classification model, more fine-grained IR, greatly improving the efficiency of the workload.

In summary, this paper makes the following contributions: (1) We propose a new idea from the perspective of data distribution to solve the problem of IR. (2) We propose a technique for characterizing data and workload into vectors. (3) We propose a transferable multi-classification model to solve IR problem. (4) We tested the model's portability on the classic NoSQL MongoDB, and evaluated the accuracy and robustness of CNNIR.

Our paper organized as follows. Section 2 defines the problem and introduces data vectorization. Then we illustrate CNNIR in details in Sect. 3. In Sect. 4, extensive experiments are conducted to test the accuracy and transferability of CNNIR. Finally, we make a conclusion for our contributions in Sect. 5.

2 Featurization

2.1 Problem Definition

With the development of network interconnection, the storage of semi-structured data becomes more and more important, and the IR approaches on semi-structured data could be applied on structured data. In this paper, we consider the IR problem on the semi-structured dataset. The input of this problem is document data (D) and workload (W). The output is the proper index configurations.

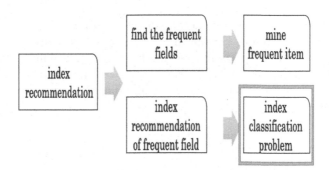

Fig. 1. Structure of IC problem

In this paper, the IR problem (see Fig. 1) is transformed into two subproblems. One is frequent field mining. The other is fine-grained index classification (IC) on frequent fields in the red box of Fig. 1. Since W in real life is usually huge, it will inevitably contain many fields with few queries. It is meaningless working on these fields. It can greatly speed up the recommendation by filtering out the fields with few queries in advance. Since the frequent itemset mining has been well studied, we attempt to use the classic frequent itemset

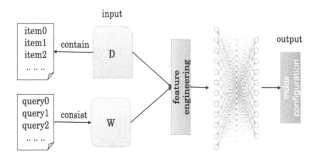

Fig. 2. Structure of IC problem

mining algorithm, FP growth [6] to discover the fields, which accomplish the mining by accessing data twice compared to Apriori [1].

Thus, we focus on the IC problem, as is demonstrated in Fig. 2 in this paper. The input of this problem is D which contains all values of a field, and W which consisted of queries. Then, in the next subsection, we will illustrate the vectorization for normalizing the input of classification task in the red box of Fig. 1.

2.2 Vectorization

We first describe how to vectorized D. Due to the large scale of the D, which is difficult to be directly used for model training, we consider using sampling method. Since stratified sampling can largely retain the features of D, and is widely used in various fields [11,12,19]. In this paper, we use the stratified sampling in the original data as the first step of the vectorization for reducing data size. Our stratified sampling (as shown in the Fig. 3) is divided into two steps: (1) sampling is conducted according to whether the data is empty, (2) sampling is conducted on non-empty data in lexicographic order. The two-level stratified sampling can retain the features of the original data greatly, which is the basis of the best IR.

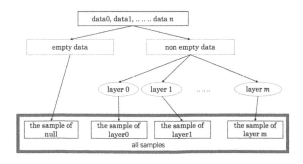

Fig. 3. Structure of stratified sampling

W consists of a series of queries. For the current classification, only the data related to the field affect the index configuration. Therefore, we extract all the data in the workload involved in the current classification as part of the model input. For example, $query0 = \{$name: "Amy"$\}$, $query1 = \{$name: "Tom"$\}$, $query2 = \{$age: 46$\}$. The result of extracting "name" field is ["Amy", "Tom"], and the result of extracting "age" field is [46].

For the result of preprocessing W and D, we employ a more fine-grained character (e.g. a to z, 1 to 9) level encoding method [20] as shown in the Fig. 4 to encoding them. As we know, one-hot coding is a common coding method in natural language processing. String-level encoding is very space-consuming. For example, to encode 1000 words, we need one-hot vector of 1000 dimensions in string-level encoding. In big data, data dimension is very high. Therefore, we consider more fine-grained encoding based on characters. The characters in the dictionary are limited, which greatly reduces the sparsity of encoding compared with the string-level one-hot coding.

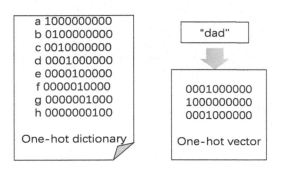

Fig. 4. Sample of coding

3　Classification Model

In this section, we introduce a classification model CNNIR to solve the IC problem defined in Sect. 2.1.

3.1　Classification Task

We transform IC problem in Sect. 2.1 to a classification task. The input of this task is the vector obtained in Sect. 2.2. The output is a four-dimensional vector which illustrates the index configuration of the field. The first dimension indicates whether the field is indexed, the second dimension indicates whether the index is sparse, the third dimension indicates whether the index is partial, and the fourth dimension indicates whether the hash or b-tree index is established. With the above four dimensions as output, we transform IC problem to a classification task, and then build a multi-classification model, as shown in Fig. 5, to solve this problem.

Existing IR method does not consider the transferability of the model. Whenever IR is needed for a new data set, these models always need to be trained. Our model considers from the perspective of data distribution, and studies whether the data with the same distribution will have the same index recommendation. And then, we introduce the CNNIR to achieve above targets. Our model learns the features of data distribution in order to achieve transferability.

For our CNNIR, when it is trained according to D and W under a certain distribution, any data with the same distribution can use this model to processing IR problem. For example, consider two data sets with four workloads, $D = [d_1, d_2]$, $W_{d1} = [w_1, w_2, w_3, w_4]$ and $W_{d2} = [w_1, w_2, w_3, w_4]$ with the same distribution. The index classification model can classify the remaining untrained but equally distributed data (d_2 and W_{d2}) after we use the d_1 and W_{d1} for training.

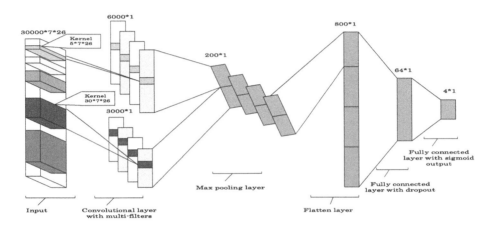

Fig. 5. Structure of CNNIR

3.2   Model Framework

The framework of our solution for the classification problem is shown in Fig. 5, and the input data is the vectorized data obtained in Sect. 2.2. The convolution layer is composed of several convolution kernels with different sizes, which can convolute the information under different steps to achieve better learning effectiveness. The max pooling layer is dynamic in order to make the convolutional output consistent. The length of pooling window is calculated by the blow formula.

$$pool_l = \frac{convshape0}{z}$$

The $convshape0$ is the length of first dimension in convolution output, and z is the normalization parameter. The flatten layer spreads the pool layer output into a one-dimensional vector to connect the full connection layer.

We list more detailed configuration information of models in Table 1. There are five convolution kernels, each of which can be convoluted in a different range and can learn more knowledge representation. Considering the amount of network parameters and representation effectiveness, it is inappropriate that our normalization parameter is too big or too small. Finally, we set $z = 100$ as normalization configuration. As we can see in Table 1, the convolution kernel $[5, 7, 26]$ corresponds to pooling kernel $[60, 1, 1]$ and the convolution kernel $[10, 7, 26]$ corresponds to pooling kernel $[30, 1, 1]$. Then, we can get the same output dimension ([None, 100, 2]) in different convolutional kernel. Additionally, due to the lack of training data, we use dropout [17] technology to avoid over fitting of neural network.

Table 1. Configuration of CNNIR framework

Layers	Configuration	Output shape
Input layer		[None, 30000, 7, 26]
Convolution layer	$[5, 7, 26] * 2$, $[10, 7, 26] * 2$, $[20, 7, 26] * 2$, $[30, 7, 26] * 2$, $[50, 7, 26] * 2$	[None, 6000, 1, 1, 2], [None, 3000, 1, 1, 2], [None, 1500, 1, 1, 2], [None, 1000, 1, 1, 2], [None, 600, 1, 1, 2]
Max-pooling layer	$[60, 1, 1]$, $[30, 1, 1]$ $[15, 1, 1]$, $[10, 1, 1]$ $[6, 1, 1]$	[None, 100, 2], [None, 100, 2] [None100, 2], [None, 100, 2] [None, 100, 2]
Flatten layer		[None, 1000]
Full connected layer	64, dropout = 0.1, relu	[None, 64]
Output layer	4, dropout = 0.1, sigmoid	[None, 4]

CNNIR with different kernels improve the accuracy of classification and has a good transferability. Convolutional neural network has been used in image classification field, which can improve the classification accuracy greatly. In this paper, multi-core convolution neural network is used to learn and represent the data and workload. By learning the features of data segments with different convolution steps, we can learn more data distribution features. The index configuration is determined by the original data distribution and workload data distribution of the field and the correlation between the two data sequences. For example, for a query on documents, building a sparse index on a sparse field is better than building a general index. When the query in a workload only involves a part of data, it is better to build a partial index on this field than to build a general index. CNNIR is able to distinguish the above situations by classification method.

3.3 Key Technical Points

In this section, we introduce the key technical points of CNNIR.

(1) Our model has transferability. By using multi-core convolution to learn data distribution features, we can achieve good representation learning for model transferability.

(2) We use character-level one-hot coding instead of string-level representation, which greatly reduces the dimension of input vector.

(3) We define more fine-grained classification output, which can make more detailed recommendation on the index, and greatly improve the efficiency of the workload execution.

(4) Our model has strong expansibility, and we can realize more diverse index classification by redefining the output vector.

4   Experimental Results

Our experiment is divided into two parts, one is to test the accuracy, robustness and convergence of CNNIR (in Sect. 4.2), and the other is to test the IR effectiveness and universality of the CNNIR (in Sect. 4.4). Our experiments run on machines with Intel Core i5 4210U, 8G of RAM and a 480G of disk.

4.1   Datasets

In this paper, we develop a benchmark to generate random document data with various configurations. We use it to generate various data distribution for neural network training and testing. The configurations are as follows: the number of documents, the data distribution (normal distribution or uniform distribution), the number of document fields, the sparsity of each field in the document (if the sparsity is 0.9, only 90% of documents have this field), the cardinality of each field (if the cardinality is 0.1, it means that the cardinality of this field is 0.1 by the number of fields in all documents), and the random seed (used to determine the seed generated by the random number and ensure the repeatability of the experiment). We generate dataset1 and dataset2 which is equally distributed to conduct our experiments.

4.2   Results of Model Performance

In this section we test the accuracy, robustness and convergence of the model. We divide dataset 1 into 80% training set and 20% test set. According to the training set and test set, we test the curve of model loss and the accuracy with the number of iterations under different learning rates (LR). CNNIR has a powerful classification ability through multi-core convolution learning. As shown in the Fig. 9, after 50 iterations, the classification accuracy of the model in the test set is over 95%. As shown in Fig. 6, 7, 8, 9, CNNIR has good robustness and can perform well under various learning rates. We use Adam Optimizer [8] for efficient stochastic optimization to train our model, which only requires first-order gradients with little memory requirement. Figure 9 clearly shows that the model can converge rapidly through several iterations under the learning rate of 0.01, 0.001 and 0.0001.

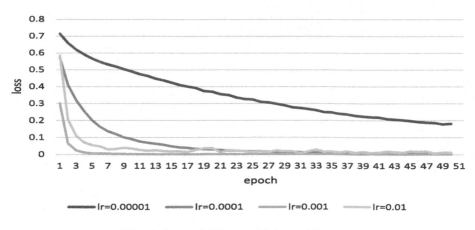

Fig. 6. Loss of different LR in training set

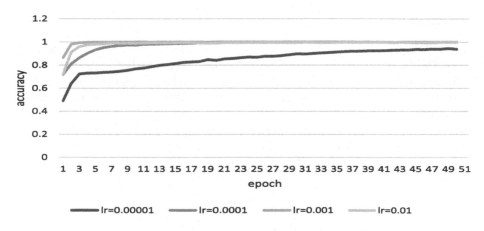

Fig. 7. Accuracy of different LR in training set

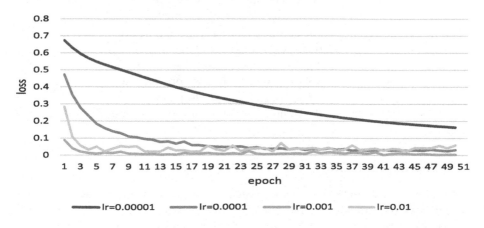

Fig. 8. Loss of different LR in test set

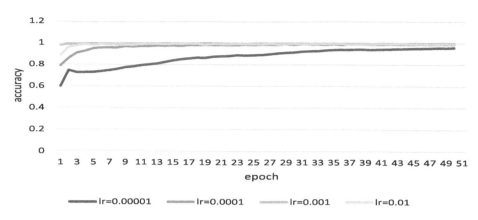

Fig. 9. Accuracy of different LR in test set

4.3   Workload

We use four types of workloads in Table 2 to test the IR efficiency of the model. Theoretically, two conditions should be taken into account when building indexes for these four types of workloads. The first is that when the query only involves a part of data, it is more proper to use a partial index instead of normal index. The second is that it is better to build a B-tree when the query data contains a large number of range queries, and it is better to build a hash index when the query data contains a large number of equivalent queries. We will discuss the IR performance of the model for these workloads in next section.

Table 2. Workloads

Workloads	Selections
W1	10,000 equivalent queries which are evenly distributed in the original data
W2	10,000 range queries which are evenly distributed in the original data
W3	10,000 equivalent queries which are distributed in part of the original data
W4	10,000 range queries which are distributed in part of the original data

4.4   The Experimental Results of Efficiency

In this section, we tested the efficiency on the workloads defined in Sect. 4.3. We use dataset1 to train the model, and then test it with dataset1 and dataset2. Since it takes too long to test all queries in each workload, we randomly select five queries from each workload. As shown in Table 3, through the index configurations from CNNIR, the average execution time of each query can be over 200 times faster than before. What's more, CNNIR only trained on dataset1 also performs well on dataset2, so it has been proved that CNNIR is a general model. That is to say, we only need to train once on CNNIR to make the model suitable for all the data with the same distribution, which can greatly reduce the resource consuming.

Table 3. Workload execution time

Dataset1			Dataset2		
Workloads	NoIndex [ms]	CNNIR [ms]	Workloads	NoIndex [ms]	CNNIR [ms]
W1:Q1	581.43	2.17	W1:Q1	798.80	2.45
W1:Q2	539.74	2.30	W1:Q2	541.10	2.32
W1:Q3	562.95	2.20	W1:Q3	542.76	2.27
W1:Q4	563.45	2.30	W1:Q4	574.98	2.27
W1:Q5	540.77	2.23	W1:Q5	550.04	2.16
W2:Q1	802.64	4.90	W2:Q1	724.54	3.43
W2:Q2	678.20	3.84	W2:Q2	716.98	5.56
W2:Q3	671.12	3.96	W2:Q3	713.30	3.80
W2:Q4	729.89	3.83	W2:Q4	770.02	3.86
W2:Q5	679.75	3.84	W2:Q5	690.09	3.51
W3:Q1	608.23	2.20	W3:Q1	567.71	2.34
W3:Q2	594.85	2.33	W3:Q2	561.83	2.91
W3:Q3	604.54	2.29	W3:Q3	563.07	2.39
W3:Q4	568.63	2.31	W3:Q4	542.59	3.48
W3:Q5	571.95	2.64	W3:Q5	528.28	2.55
W4:Q1	623.17	3.77	W4:Q1	638.14	2.85
W4:Q2	670.29	2.94	W4:Q2	609.24	3.42
W4:Q3	607.25	2.54	W4:Q3	611.63	3.00
W4:Q4	923.00	2.72	W4:Q4	614.00	3.36
W4:Q5	946.13	2.91	W4:Q5	596.52	2.45
Total	653.40	2.91	Total	622.78	3.02

5   Conclusion

This paper focuses on the migration of the IR model in big data. We transform the IR problem into a fine-grained multi-classification problem, and propose a model based on convolutional neural network to solve the classification problem. Two data sets and some workloads for model testing are generated by our benchmark. CNNIR uses the strong representation learning ability of convolutional neural network. As we can see in Sect. 4.2, CNNIR with a strong classification ability achieves over 95% accuracy on the test set. CNNIR is robust and performs well at several learning rates. According to the experimental results in Sect. 4.4, it is demonstrated that CNNIR has excellent universality. CNNIR only trained by dataset1 can achieve high efficiency on untrained dataset2 which has the same distributed dataset with dataset1. In summary, our migration model can reduce the resource consumption of IR by a big margin in big data. In the future, we will consider the case of multi-field IR problem by using Bayesian network to analyze the relationship between each field.

Acknowledgement. This paper was partially supported by NSFC grant U1866602, 61602129, 61772157.

References

1. Agrawal, R., Imieliński, T., Swami, A.: Mining association rules between sets of items in large databases. In: Proceedings of the 1993 ACM SIGMOD International Conference on Management of Data, pp. 207–216 (1993)
2. toolpark.de Alle Rechte vorbehalten: The website of powerdesigner (2016). http://powerdesigner.de
3. Chaudhuri, S., Narasayya, V.R.: An efficient, cost-driven index selection tool for Microsoft SQL server. In: VLDB, vol. 97, pp. 146–155. Citeseer (1997)
4. Ding, B., Das, S., Marcus, R., Wu, W., Chaudhuri, S., Narasayya, V.R.: AI meets AI: leveraging query executions to improve index recommendations. In: Proceedings of the 2019 International Conference on Management of Data, pp. 1241–1258 (2019)
5. Hammer, M., Chan, A.: Index selection in a self-adaptive data base management system. In: Proceedings of the 1976 ACM SIGMOD International Conference on Management of Data, Washington, D.C., 2–4 June 1976 (1976)
6. Han, J., Pei, J., Yin, Y., Mao, R.: Mining frequent patterns without candidate generation: a frequent-pattern tree approach. Data Min. Knowl. Disc. 8(1), 53–87 (2004). https://doi.org/10.1023/B:DAMI.0000005258.31418.83
7. IDERA, I.L.P.S.G.: The website of er/studio (2004–2020). https://www.idera.com
8. Kinga, D., Adam, J.B.: A method for stochastic optimization. In: International Conference on Learning Representations (ICLR) (2015)
9. Leis, V., Gubichev, A., Mirchev, A., Boncz, P., Kemper, A., Neumann, T.: How good are query optimizers, really? Proc. VLDB Endow. 9(3), 204–215 (2015)
10. Sparx Systems Pty Ltd.: The website of sparx enterprise architect (2000–2020). https://sparxsystems.com
11. Peng, J., Zhang, D., Wang, J., Jian, P.: AQP++: connecting approximate query processing with aggregate precomputation for interactive analytics. In: Proceedings of the 2018 International Conference on Management of Data (2018)
12. Quoc, D.L., et al.: ApproxJoin: approximate distributed joins. In: Proceedings of the ACM Symposium on Cloud Computing, pp. 426–438 (2018)
13. Finkelstein, S., Schkolnick, M., Tiberio, P.: Physical database design for relational databases. ACM Trans. Database Syst. (TODS) 13(1), 91–128 (1988)
14. Sattler, K.U., Geist, I., Schallehn, E.: Quiet: continuous query-driven index tuning. In: Proceedings 2003 VLDB Conference, pp. 1129–1132. Elsevier (2003)

15. Schkolnick, M.: The optimal selection of secondary indices for files. Inf. Syst. **1**(4), 141–146 (1975)
16. Sharma, A., Schuhknecht, F.M., Dittrich, J.: The case for automatic database administration using deep reinforcement learning. arXiv preprint arXiv:1801.05643 (2018)
17. Srivastava, N., Hinton, G.E., Krizhevsky, A., Sutskever, I., Salakhutdinov, R.: Dropout: a simple way to prevent neural networks from overfitting. J. Mach. Learn. Res. **15**(1), 1929–1958 (2014)
18. Stonebraker, M.: The choice of partial inversions and combined indices. Int. J. Comput. Inf. Sci. **3**(2), 167–188 (1974). https://doi.org/10.1007/BF00976642
19. Zhang, S., Yao, L., Sun, A., Tay, Y.: Deep learning based recommender system: a survey and new perspectives. ACM Comput. Surv. (CSUR) **52**(1), 1–38 (2019)
20. Zhang, X., Zhao, J., Lecun, Y.: Character-level convolutional networks for text classification (2015)

An Incremental Partitioning Graph Similarity Search Based on Tree Structure Index

Yuhang Li[1], Yan Yang[1,2(✉)], and Yingli Zhong[1]

[1] School of Computer Science Technology, Heilongjiang University,
Harbin, China
yangyan@hlju.edu.cn
[2] Key Laboratory of Database and Parallel Computing of Heilongjiang Province,
Harbin, China

Abstract. Graph similarity search is a common operation of graph database, and graph editing distance constraint is the most common similarity measure to solve graph similarity search problem. However, accurate calculation of graph editing distance is proved to be NP hard, and the filter and verification framework are adopted in current method. In this paper, a dictionary tree based clustering index structure is proposed to reduce the cost of candidate graph, and is verified in the filtering stage. An efficient incremental partition algorithm was designed. By calculating the distance between query graph and candidate graph partition, the filtering effect was further enhanced. Experiments on real large graph datasets show that the performance of this algorithm is significantly better than that of the existing algorithms.

Keywords: Graph similarity search · Similarity search · Graph partition

1 Introduction

In recent years, a large number of complex and interrelated data has grown. In order to ensure the integrity of data structure, it is modeled as a graph, and graph search operations are frequently used in database retrieval, so it is widely concerned. Due to the inevitable natural noise and human error input in real life, it is very necessary to search the similarity of graphs. It aims to search the set of graphs similar to the query graphs specified by users in large graph database. At present, there are many indicators to measure the similarity of graphs, such as the largest common subgraph [1, 2], edge missing [3]. Among them, the most commonly used similarity measure is the graph editing distance (GED), which can not only accurately capture the structural differences between graphs, but also can be applied in various fields, such as computer vision handwriting recognition [4], molecular analysis of compounds [5].

Therefore, this paper studies graph similarity search based on graph editing distance constraint: given a graph database and a query graph, the user gives GED threshold, the graph set in query graph database with graph editing distance within the threshold. The graph editing distance $GED(G_1, G_2)$ refers to two graphs G_1 and G_2. It is the minimum number of operations to convert G_1 to G_2 by inserting and deleting vertices/edges or relabel labels. But calculating the GED of two graphs has been proved to be NP hard [6].

© Springer Nature Singapore Pte Ltd. 2020
J. Zeng et al. (Eds.): ICPCSEE 2020, CCIS 1257, pp. 16–32, 2020.
https://doi.org/10.1007/978-981-15-7981-3_2

At present, the most widely used method to calculate the GED is A* algorithm [7], and literature [2] shows that the A* algorithm cannot calculate the editing distance of graphs with more than 12 vertices. Therefore, the solution of search problem in large graph database adopts the frame of filter and verification. Generally, a set of candidate graphs is generated under some constraints, and then GED is calculated for verification. At present, many different GED lower bounds and pruning technologies have been proposed, such as k-AT [8], GSimSearch [9], DISJOINT-PARTITION BASED FILTER [10], c-star [11], but they all share a lot of common substructures, resulting in the loose GED lower bounds. Inves [15] proposes a method of incremental graph partition, and considers the distance between partitions, but the filtering conditions of this method are too loose. Based on Inves, this paper makes the following contributions:

- We study the idea of incremental partition from a new perspective, and use the idea of q-gram to enhance the partition effect, so as to reduce the number of candidate sets.
- We propose a clustering index framework, which optimizes the K-Means clustering, uses the dictionary tree structure as the index, uses the dynamic algorithm to divide the incremental graph, and groups the similar graph according to the graph division structure to enhance the filtering ability.
- We have carried out a wide range of experiments on datasets, and the results show that the algorithm in this paper is significantly better than the existing algorithm in performance.

2 Preliminary

2.1 Problem Definition

For the convenience of illustration, this paper focuses on simple undirected label graph, and the method can also be extended to other types of graphs. Undirected label graph G is represented by triple (V_g, E_g, L_g), where V_g is the vertex set, $E_g \subseteq V_g \times V_g$ is the edge set, L_g is a function that maps vertices and edges to labels. $V_g(u)$ is the label of vertex u, $V_g(u)$ is the label of edge (u, v). In practical applications, labels can be expressed as properties of vertices and edges, such as chemical composition of compounds, chemical bonds, etc.

The editing operation of a graph refers to the editing operation of converting from one graph to another, including:

- Insert an isolated labeled vertex in the graph
- Remove an isolated labeled vertex from the graph
- Insert a labeled edge in the graph
- Delete a labeled edge in the graph
- Relabel a vertex label
- Relabel an edge label

The graph editing distance between graph G_1 and G_2 refers to the minimum number of graph editing operations converted from G_1 to G_2, expressed as $GED(G_1, G_2)$.

Example 1. Figure 1 shows two graphs G_1 and G_2, in which the vertex label is a chemical molecule and the edge label (single line and double line) is a chemical bond. G_1 can be converted to G_2 by changing u_3 label to 'S', u_4 label to 'N', deleting double key edge (u_6, u_7), and insert single key edge (u_6, u_7). Therefore, $GED(G_1, G_2) = 4$.

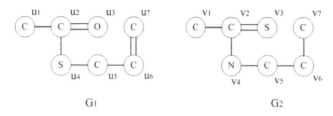

Fig. 1. Data graph G_1 and G_2

Definition 1. (graph similarity search problem) Given graph database $D = \{G_1, \ldots, G_n\}$, a query graph Q, and GED threshold τ, the problem of graph similarity search is to find the set of graphs satisfying $GED(G_i, Q) \leq \tau, G_i \in D$.

2.2   Related Work

Recently, much attention has been paid to graph similarity search. The previous work is to use the overlapping substructures of different graphs to filter, and then carry out expensive GED calculation.

k-AT [8] is inspired by the idea of q-gram of approximate string matching. By decomposing the graph into several subgraphs, the lower limit of editing distance is estimated by using the number of common subgraphs, and the inverted index is established by using the extracted features. However, k-AT algorithm is only suitable for sparse graphs. GSimSearch [9] proposes path based q-grams, which uses matching and mismatching features, and proposes a local label filtering method in its verification phase. Although it solves the limitation of k-AT, the paths overlap each other, and the partition size is fixed, resulting in poor pruning function.

c-star [10] uses 1-gram to construct star structure based on k-AT, and uses binary matching between star structures to filter. On the basis of traditional c-star, DISJOINT-PARTITION BASED FILTER [11] will tighten the filter lower bound of c-star by removing the leaf nodes, but this method still shares many common substructures, making the editing distance lower limit too loose. CSI_GED [2] based on the combination of edge mapping and backtracking algorithm, in the verification phase faster calculation of the graph GED. GBDA [12] proposes the graph branch distance and estimates the graph edit distance using probability method, but when calculating the prior distribution, the sampling of data is not accurate.

Pars [13] adopts random graph partition strategy, and proposes non overlapping and variable size graph partition. ML_index [14] uses vertex and edge tag frequency to define partition, and proposes selective graph partition algorithm, so as to improve filtering performance. Both Pars and ML_index use graph as index features, but large graph creation index costs a lot. In this paper, the dynamic algorithm is used to divide the graph incrementally, and a clustering index based on dictionary tree is proposed to enhance the filtering performance, so as to ensure its search performance in large-scale graph database.

3   IP-Tree Algorithm

3.1   Partition Based Filtering Scheme

Because it is very expensive to calculate the GED of each graph and query graph in a large graph database, this paper uses the filter and verification framework to calculate the GED lower bound between the candidate graph and query graph by using the partition method before calculating the GED accurately, so as to filter the candidate set and reduce the cost of calculating the GED accurately. Before we formally introduce the filter framework, we start with defining the induced subgraphs of graph partitions.

Definition 2. (induced subgraph isomorphism) Given a graph G_1 and a graph G_2, if there is a mapping $f : V_{G_1} \rightarrow V_{G_2}$ satisfying (1) $\forall u \in V_{G_1}, f(u) \in V_{G_2} \wedge L_{G_1}(u) = L_{G_2}(f(u))$ (2) $\forall u \in V_{G_1}, \forall v \in V_{G_2}, L_{V_{G_1}}(u,v) = L_{V_{G_2}}(f(u),f(v))$, it is said that the graph G_1 is the induced subgraph isomorphism of the graph G_2, expressed as $G_1 \subseteq G_2$.

Example 2. Figure 2 shows three graphs P_1, P_2 and G_1. It can be found that P_1 is not the induced subgraph isomorphism of G_1, because $L_{P_1}(u_2, u_3) \neq L_{G_1}(u_2, u_3)$, P_2 is the induced subgraph isomorphism of G_1.

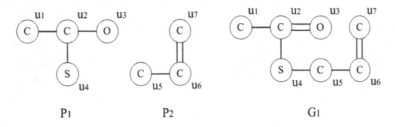

Fig. 2. Induced subgraph isomorphism

Definition 3. (graph division) The division of graph G is $P(G) = \{p_1, ..., p_k\}$, where p_k satisfies the condition: (1) $\forall i, p_i \subseteq G$ (2) $\forall i, j, V_{p_i} \cap V_{p_j} = \varnothing$ $s.t.$ $i \neq j$ (3) $V_G = \cup_{i=1}^{k} V_{p_i}$.

Definition 4. (partition matching) Given a graph G_1 and G_2, a partition $p \in P(G_1)$, if $p \subseteq G_2$, then partition p is the matching partition of graph G_2, otherwise partition p is the mismatching partition of graph G_2.

Lemma 1. Given graph G_1 and graph G_2 and partition $P(G_1)$ of graph G_1, $lb(G_1, G_2) = |p|p \subseteq P(G_1) \wedge p \nsubseteq G_2|$ is a lower bound of $GED(G_1, G_2)$.

Proof. Because G_1 partitions do not overlap and the editing operation of each partition does not affect other partitions, there is at least one editing operation from G_1 to G_2 for each mismatched partition.

Corollary 1. Given graph G_1, G_2, the division of graph G_1 and given GED threshold τ, if $lb(G_1, G_2) > \tau$, then graph G_1 is pruned and GED does not need to be calculated.

Example 3. Given two graphs G_1 and G_2 as shown in Fig. 1, the GED threshold τ. If we divide G_1 into $\{p_1, p_2, p_3\}$ according to the method shown in Fig. 3(a), we need to calculate GED of G_1 and G_2 according to corollary 1. If we divide G_1 into $\{p'_1, p'_2, p'_3\}$ according to the method shown in Fig. 3(b), according to corollary 1, we can prune it directly without calculating GED.

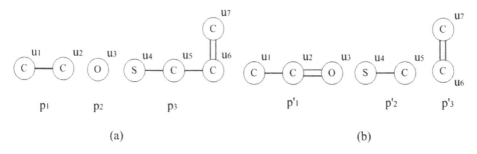

Fig. 3. Two divisions of figure G_1 in Fig. 1

It can be seen that the compactness of $lb(G_1, G_2)$ depends on the partition method of G_1, but the enumeration of all the partition methods is too time-consuming. Based on Inves [15], this paper studies the idea of incremental partition from a new perspective, and uses the idea of q-gram to enhance the partition effect and reduce the number of candidate sets.

3.2 Optimized Incremental Partition

In this section, we introduce an incremental partition strategy. The core idea is to minimize the editing of each mismatched partition p, that is, if some vertices of the partition p are deleted, the partition p will become a matching partition.

Definition 5. (incremental partition) Given two graphs G_1 and G_2, the incremental partition of G_1 is to extract the smallest mismatched partition from G_1. First, we obtain the vertex set $V_{G_1} = \{u_1, \ldots, u_n\}$ of graph G_1. We put the vertices in V_{G_1} into a partition p in order until we put a vertex u_{i+1} to make $p \nsubseteq G_2$ stop partition. At this time, G_1 is divided into $P(G_1) = \{p, G_1/p\}$, where G_1/p refers to the induced sub-graph composed of removing the vertices in the partition p. Repeat G_1/p division until $G_1/p \subseteq G_2$ or $G_1/p = \emptyset$.

The graph partition generated by Definition 5 satisfies the following properties, given two graphs G_1 and G_2, If G_2 uses incremental partition, the division is $P(g) = \{p_1, \ldots, p_k - 1, p_k\}$, then the $p_1, \ldots, p_k - 1$ does not match the G_2, p_k matches G_2. Therefore, $lb(G_1, G_2) = k - 1$.

Example 4. Given two graphs G_1 and G_2 as shown in Fig. 1, G_1 is divided incrementally. The process is shown in Fig. 4. First, select the vertex $\{u_1, u_2, u_3\}$ from G_1 to form the partition p_1 as shown in Fig. 4 (a). Because $(p_1 - \{u_3\}) \subseteq G_2 \wedge p_1 \nsubseteq G_2$, partition p_2' continues to be divided until it is shown in Fig. 4 (b). Because $p_4 = \emptyset$, the incremental division ends, $lb(G_1, G_2) = 3$.

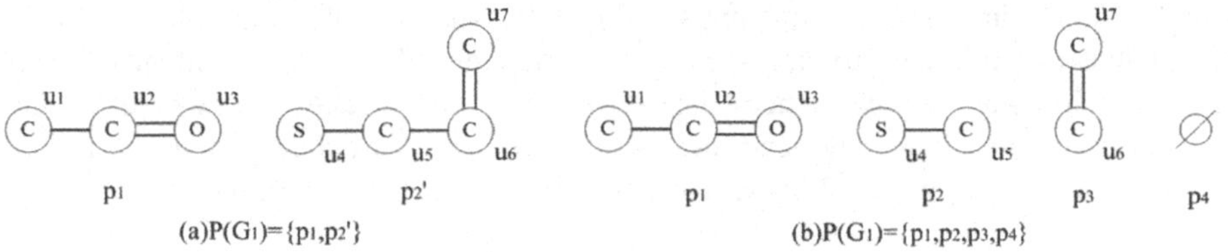

Fig. 4. Incremental division of G_1 in Fig. 1

After the incremental partition of the graph, it can be found that the last vertex put into the partition in each partition results in the mismatch between the partition and the target graph. As shown in the partition p_1 in Fig. 4 (a), the insertion of vertex u_3 makes $p_1 \nsubseteq G_2$, which indicates that the editing operation should be concentrated near the vertex u_3, not only that, the first selected vertex also has an impact on the subsequent entire incremental partition.

Therefore, the last vertex in the partition is taken as the starting vertex, and the partition is re divided in the same way. While maintaining connectivity, the points that may generate editing operations should be considered as early as possible, so as to reduce the size of mismatched partition.

Example 5. As shown in Fig. 5, suppose the vertex order of graph X is $\{u_1, u_2, u_3, u_4, u_5\}$. According to the incremental partition scheme, we choose the first partition p as $\{u_1, u_2, u_3\}$, $X/p \subseteq Y$, then $lb(X, Y) = 1$. We reorder the mismatched partition and get the new vertex order $\{u_3, u_2, u_1\}$ of the partition. Then we get the first partition p as $\{u_3, u_2\}$, $X/p \nsubseteq Y$. We continue to partition and get the second partition $\{u_1, u_4\}$, $X/p \subseteq Y$. We get a closer GED lower bound $lb(X, Y) = 2$.

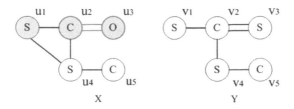

Fig. 5. Repartition mismatch partition

In this paper, the q-gram method is used to enhance the partition effect. Before incremental partition, the two graphs are decomposed into 1-gram to obtain the different structures of the two graphs. The vertices with the lowest frequency in the different structures are most likely to be relabel. The vertices are arranged in ascending order of frequency, and the incremental partition is carried out in this order.

Example 6. The graphs G_1 and G_2 in Fig. 1 are decomposed to get the 1-gram of G_1 and G_2 shown in Fig. 6. Comparing the different structures of G_1 and G_2, it can be found that the different structures are pink nodes, and the vertices with the lowest frequency in G_1 are yellow vertices. The vertices in G_1 are arranged in ascending order according to the frequency, and the vertex sequence $\{u_3, u_4, u_2, u_6, u_7, u_5, u_1\}$ is obtained, as shown in Fig. 7, incrementally partitions in this vertex order.

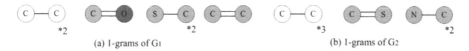

(a) 1-grams of G₁   (b) 1-grams of G₂

Fig. 6. 1-grams of graphs G_1 and G_2 in Fig. 1

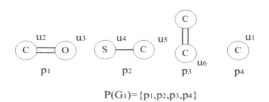

P(G₁)={p₁,p₂,p₃,p₄}

Fig. 7. Optimization increment division of G_1 in Fig. 1

The algorithm of incremental partitioning method is summarized in Algorithm 1.

Algorithm 1: IncremetalPartitioning(G, Q)

Data: Data graph G and Q
Result: A GED lower bound $lb(G, Q)$ based on incremental partition
  1: GetVertexOrder(G);
  2: $i \leftarrow$ SubgraphIsomorphism(G, Q);
  3: if $i > |V_G|$ then
  4:     return 0;
  5: end
  6: $p \leftarrow$ *the first vertex of the mismatched partition of G;*
  7: while $|V_p|$ *does not change* do
  8:     ReorderbByMismatchedVertices(p);
  9:     $i \leftarrow$ SubgraphIsomorphism(p, Q);
10:     $p \leftarrow$ *the first vertex of the mismatched partition of p;*
11: end
12: $G' \backslash \leftarrow G/p$;
13: return 1+IncremetalPartitioning(G', Q);

Given a pair of graphs G and Q, the algorithm uses the incremental partition method to partition, so as to calculate the lower limit $lb(G, Q)$.

Firstly, the q-grams algorithm is used to count the frequency to determine the vertex order of G (line 1), Then, the isomorphic test of induced subgraph of G is performed based on this order, and the unmatched vertex position of G in Q (line 2) is returned. If the number of unmatched vertex positions returned exceeds the number of vertex sets in G, then this pair of matching, return 0 (lines 3–5). Otherwise, the mismatched vertices are placed in partition p (line 6). Using 7–11 lines to reduce the size of the partition, reorder the vertices in the partition P (line 8), continue to perform the induced subgraph isomorphism test (line 9) on Q in this order, and return the unmatched vertex position (line 10) of G in Q until the partition p does not change. Then the partition p is separated from G to get G' (line 12). Iterate over the number of partitions in G' and return $lb(G, Q)$ (line 13).

3.3 Validation Algorithm

In the process of verification, A* algorithm is used to calculate GED. The performance of A* algorithm depends on the accuracy of the estimation of edit distance generated by unmapped vertices. In order to improve the accuracy of distance estimation, this paper uses the idea of bridge to predict the editing distance of unmapped vertices more accurately.

Definition 6. (bridge) Given a partition p, the bridge of vertex $u \in p$ is the edge of vertex u and v, where $v \notin p$.

Given a partition p in the graph partition of graph G_1, and a partition m matching the partition p in G_2, it is assumed that the vertex $u \in p$ of graph G_1 is mapped to the vertex $v \in q$ of graph G_2. Then, the edit distance between vertices u and v is $Be(u, v) = \Gamma(L_{br}(u), L_{br}(v))$, where $L_{br}(u)$ represents the set of labels of the bridge of vertices u, $\Gamma(X, Y) = max(|X - Y|, |Y - X|)$. The bridge editing distance from partition p to partition m is $B(p, m) = \sum_{u \to v \in M} Be(u, v)$, where m is the mapping of all vertices of partition p and partition M.

Example 7. Given the graphs G_1 and G_2 in Fig. 1, it is assumed that the matched partition is $p_1 = (u_1, u_2)$, and the partition p matches $m_1 = (v_5, v_6)$ in G_2, as shown in Fig. 8, $Be(u_1, v_5) = 1$, because there is no bridge in u_1, and there is a bridge in v_5, which is the same as $Be(u_2, v_6) = 1$, so $B(p_1, m_1) = 1 + 1 = 2$.

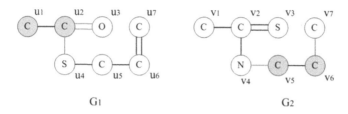

Fig. 8. Edit distance of matching partitioned bridges

When using the A* algorithm to calculate $GED(G_1, G_2)$, for the known vertex mapping set M, where $u \to v \in M, u \in V_{G_1}, v \in V_{G_2}$, we can get the predicted edit distance of unmapped vertices, as shown in Eq. 1.

$$h'(M) = B(M) + \Gamma(L_{G_1}(u'), L_{G_2}(v')) + \Gamma(L_{G_1}(u', v'), L_{G_2}(u', v')) \tag{1}$$

Where u', v' is the unmapped vertex, and the optimized A* algorithm is shown in Algorithm 2.

Algorithm 2: $\text{GED}(G, Q, \tau)$

Data: Data graph G and Q, The vertex set of graph G is $V_G = u_1, \ldots, u_n$ and the vertex set of graph Q is $V_q = v_1, \ldots, v_n$, GED threshold τ

Result: If the edit distance of data graph G and Q is less than τ, then the edit distance is returned, otherwise $\tau + 1$ is returned

1: $Map = \emptyset$; $M_{min} = \emptyset$;
2: foreach *node* $w \in V_Q \cup \varepsilon$ do
3:      $Map = Map \cup (u_1 \rightarrow w)$;
4: end
5: while *True* do
6:      $M_{min} = arg_{min}\{h'(M_{min}) + g(M_{min})$;
7:      $Map = Map/M_{min}$;
8:      if $Complete(M_{min})$ then
9:          if $existingDistance(M_{min}) > \tau + 1$ then
10:             return $\tau + 1$;
11:          end
12:          else
13:             return $existingDistance(M_{min})$;
14:          end
15:      end
16:      if *All vertices in V_G are mapped* then
17:          foreach *node $w \in$ Unmapped vertices V_Q* do
18:             $Map = Map \cup (\varepsilon \rightarrow w)$;
19:          end
20:      end
21:      else
22:          foreach *node $w \in$ Unmapped vertices $V_Q \cup \varepsilon$* do
23:             $Map = Map \cup next\ vertex\ currently\ matched\ to\ V_G \rightarrow w$;
24:          end
25:      end
26: end

First, initialize the set M_{min} storing the minimum overhead mapping and the candidate mapping set *Map* (line 1), which maps the first vertex of graph G to all vertices in Q (line 2–4). Select the *Map* with the minimum accumulated known overhead and prediction overhead as M_{min}, and delete it from the map (line 6–7). If M_{min} is already a complete *Map*, if the editing distance of the map exceeds τ, return $\tau + 1$, otherwise returns the edit distance of the map. (line 8–13) if all vertices in V_G are mapped, and it is not a complete mapping, it means that the remaining vertices in V_Q need to be mapped to null, otherwise, the mapping will continue in order (line 14–22).

3.4   Clustered Index Structure

In order to search in graph database faster, we propose a tree structure based on dictionary tree based on k-means. To enhance filtering, we try to assign more similar graphs to the same set. Then, a tree index is constructed for each set. In this paper, the i-th cluster is represented as C_i, and its corresponding tree index is expressed as T_i. When searching in the graph database, we search each index and merge the results of all searches. For a given query graph Q, if it shares a few graph partitions with C_i, most of the subtrees in T_i will be pruned to speed up the search time.

K-means algorithm is a common clustering algorithm. In this paper, the distance measure is defined as the GED lower bound of graph partition, and the center of clustering is expressed as C_i.

In order to solve the problem that k-means needs to specify the number of partitions, this paper proposes a method to keep the center of each cluster as far away from each other as possible. Firstly, the center of the first cluster is initialized randomly, and then the graph furthest from the current cluster center is selected as the new center iteratively until all graphs are close enough to the center of the cluster in which they are located. All partitions in the center of each cluster are counted, sorted according to the number of times that the graph in the cluster contains partitions, and a tree index is constructed based on the dictionary tree. The index has the following properties:

- The root node does not store data, and each path represents a partition of the cluster center map.
- Each node except the root node stores a set of graphs.
- From the root node to a node, the partition on the path belongs to every graph in the graph set.

Example 7. Given the $D = \{G_1, \ldots, G_n\}$ of graph database, first select G_1 as the center of the first cluster, calculate the distance between G_1 and other graphs, and obtain the division $P(G_1) = \{p_1, p_2, p_3, p_4\}$ of G_1, select the farthest graph G_k as the center of the second cluster, calculate the distance between G_k and other graphs, and update each cluster. Calculate the distance from the graph in each cluster to the cluster center, calculate the average distance, select the graph that is far away from both clusters as the new cluster center, iterate repeatedly until the distance between all vertices in the cluster and the cluster center is less than the average distance, or the cluster reaches the upper limit. Suppose that the cluster with G_1 as the center contains the graph G_2, G_4, G_5, G_8, where $p_1 \subseteq G_2 \wedge p_1 \subseteq G_4$, $p_2 \subseteq G_2 \wedge p_2 \subseteq G_5$, $p_3 \subseteq G_2 \wedge p_3 \subseteq G_5 \wedge p_3 \subseteq G_8, p_4 \subseteq G_5$ according to the number of times that the graph in the cluster contains the partition Row sorting gets the partition order $\{p_3, p_1, p_2, p_4\}$, and the tree index is constructed as shown in Fig. 9.

The algorithm for building the clustered index is shown in Algorithm 3.

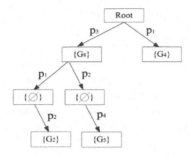

Fig. 9. Edit distance of matching partitioned bridges

Algorithm 3: Clustered_index(G, MAX)

Data: Graph database D and Maximum cluster number MAX

Result: Number of clusters n, set of tree index of clusters T

1: *Randomly select a graph $G \in D$ as the center c_1 of cluster C_1;*

2: IncremetalPartitioning(c_1, D);

3: *Find C_2, which is the farthest from C_1 editing distance, and initialize cluster C_2;*

4: The clusters C_1 and C_2 are stored in the cluster set C, n = 2;

5: while *True* do

6:      *Calculate the distance from the graph in each cluster to the cluster center;*

7:      Calculate the average distance L;

8:      if *Distance from $C_i \in D$ to the center of the current cluster longer than L and $n < MAX$* then

9:          *The new cluster C_{n+1} is initialized, and the cluster center is C_i;*

10:          $n = n + 1$;

11:      end

12:      else

13:          break;

14:      end

15: end

16: foreach $C_i \in C$ do

17:      Calculate the graph partition $P_i = \{p_{i1}, ..., p_{ik}\}$ at C_i of each cluster;

18: end

19: $T = \emptyset$;

20: foreach $c_i \in C$ and graph $g \in C_i$ do

21:      foreach *partition $p_{ij} \subseteq g, 1 \le j \le k$* do

22:          if *$T.child$ does not contain p_{ij}* then

23:              *Insert an empty set node into the tree and assign the edge as p_{ij};*

24:              *when p_{ij} is the last partition of g append g to the current node;*

25:          end

26:          $T = T.child$;

27:      end

28: end

Firstly, a graph is randomly selected as the center point C_1 (line 1) of the first cluster, and then the distance from the center point of the cluster to other graphs (line 2) is calculated. The graph C_2 with the farthest editing distance C_1 is found as the new cluster center vertex (line 3), calculate the distance from the graph in each cluster to the center of the cluster, find the average distance, select the graph that is far away from each cluster as the new cluster center, iterate repeatedly until the distance between all points in the cluster and the center of the cluster is less than the average distance, or the cluster reaches the upper limit (line 4–15). Get the partition in the center of each cluster (line 16–18).

For each cluster, the dictionary index is formed according to the graph containing partition in the cluster. The node in the tree is the set of graph, and the edge is the partition. It is constructed from the root node. If the partition P belongs to the current graph, when there is no partition p in the child of the root, an empty set node is inserted. If the partition P is the last structure belonging to the current graph, the current graph is inserted into the node (line 20–28).

4   Experiment

4.1   Data Set and Experimental Environment

In this paper, the real graph data set in graph similarity search field is selected.

1. AIDS: it is a data set of antiviral screening used in NCI/NIH development and treatment plan. It contains 42390 compounds, with an average of 25.4 vertices and 26.7 edges. It is a large graph database often used in the field of graph similarity search.
2. Protein: it comes from protein data bank and contains 600 protein structures, with an average of 32.63 vertices and 62.14 edges. It contains many dense graphs and has fewer vertex labels.

See Table 1 for data statistics, where LV and LE are vertex and edge labels.

Table 1. AIDS and Protein data set

Data set	$|G|$	$|V|_{arg}$	$|E|_{arg}$	$|LV|$	$|LE|$
AIDS	42390	25.40	26.70	62	3
Protein	600	32.63	62.14	3	5

For each data set, we use 100 random samples as query graphs. The index construction time, the size of the index, the query response time based on the GED threshold, and the size of the candidate set to be verified are compared in several aspects. All algorithms in this paper are implemented in Java. All experiments are run on MacBook Pro with inter Corei7 with 2.6 GHz and MacOS 10.14.6 (Mojave) system diagram with 16 GB main memory.

4.2 Comparison of Existing Technologies

In this paper, three commonly used algorithms are selected for comparison:

1. Pars: Pars [13] uses random partitioning strategy to propose a non overlapping, variable size graphics partitioning strategy and generate index. It is the most advanced partitioning method at present.
2. ML-index: ML-index [14] adopts a multi-layer graph index method. It contains L different layers, each layer represents a lower GED based on the partition, and a selection method is selected to generate the index. In this paper, the number of layers is defined as 3.
3. Inves: Inves [15] uses incremental partitioning method to divide the candidate graph accurately and gradually according to the query graph, so as to calculate the lower bound of their distance.

First of all, considering the time of index construction, as shown in Fig. 10. Notice that y-axis is log-scaled in all experiments. The Inves algorithm does not establish an index during the search process, and here it is modified using the tree index proposed in this article, so the index construction time is consistent with Ip-tree. Therefore, it is not shown in Fig. 10. Pars takes more time to build index, because it involves complex graph partition and sub graph isomorphism test in the index construction phase. In this paper, Ip-tree algorithm needs to cluster graph data, build index, and test some graph partition and subgraph isomorphism, so the time is slightly slower than ML-index.

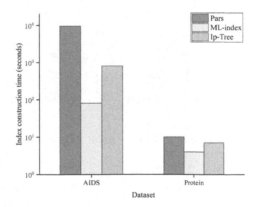

Fig. 10. Index construction time under AIDS and protein datasets

Figure 11 shows that under the AIDS data set, when the threshold changes, the processing time of different algorithms changes. Because the AIDS data set graph is large, we can clearly see that with the increase of threshold, the response time of all algorithms becomes longer. The algorithm in this paper is faster than other algorithms in processing time. Due to the small number of graphs in the protein data set, the GED threshold range is expanded to 1–8 in this paper. As shown in Fig. 12, the algorithm in this paper fluctuates slowly when the threshold rises.

Finally, this paper compares the number of candidate sets, as shown in Fig. 13 and Fig. 14. It can be found that the filtering effect of this algorithm is the best when the threshold value is in the range of 1–3.

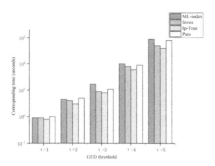

Fig. 11. Response time of AIDS

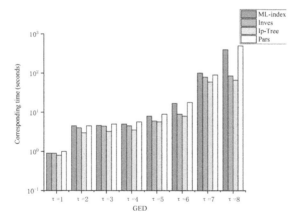

Fig. 12. Response time of Protein

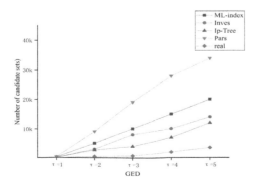

Fig. 13. AIDS candidate set size

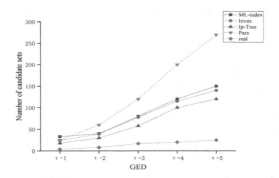

Fig. 14. Protein candidate set size

5 Conclusion

In this paper, we improve the existing algorithm Inves, and propose a dictionary tree based clustering index structure. The main idea is incrementally partition the candidate graph according to the query graph, and calculate its GED lower bound. Using clustering method, we cluster similar graphs and construct tree index, speed up the filtering of candidate graphs, and use the optimized A* algorithm to realize the accurate calculation of GED in the verification stage. Experiments on real large graph datasets show that the performance of algorithm proposed is significantly better than that of existing algorithms.

Acknowledgment. The Natural Science Foundation of Heilongjiang Province under Grant Nos. F2018028. Received 2000-00-00, Accepted 2000-00-00.

References

1. Shang, H., Lin, X., et al.: Connected substructure similarity search. In: SIGMOD 2010, pp. 903–914 (2010)
2. Gouda, K., Hassaan, M.: CSI_GED: an efficient approach for graph edit similarity computation. In: ICDE 2016, pp. 265–276 (2016)
3. Zhu, G., Lin, X., et al.: TreeSpan: efficiently computing similarity all-matching. In: SIGMOD 2012, pp. 529–540 (2012)
4. Maergner, P., Riesen, K., et al.: A structural approach to offline signature verification using graph edit distance. In: ICDAR 2017, pp. 1216–1222 (2017)
5. Geng, C., Jung, Y., et al.: iScore: a novel graph kernel-based function for scoring protein-protein docking models. Bioinformatics **36**(1), 112–121 (2020)
6. Zeng, Z., Tung, A.K.H., et al.: Comparing stars: on approximating graph edit distance. PVLDB **2**(1), 25–36 (2009)
7. Riesen, K., Emmenegger, S., Bunke, H.: A novel software toolkit for graph edit distance computation. In: Kropatsch, W.G., Artner, N.M., Haxhimusa, Y., Jiang, X. (eds.) GbRPR 2013. LNCS, vol. 7877, pp. 142–151. Springer, Heidelberg (2013). https://doi.org/10.1007/978-3-642-38221-5_15

8. Wang, G., Wang, B., et al.: Efficiently indexing large sparse graphs for similarity search. IEEE Trans. Knowl. Data Eng. **24**(3), 440–451 (2012)
9. Zhao, X., Xiao, C., Lin, X., Wang, W., Ishikawa, Y.: Efficient processing of graph similarity queries with edit distance constraints. VLDB J. **22**(6), 727–752 (2013). https://doi.org/10.1007/s00778-013-0306-1
10. Zheng, W., Zou, L., et al.: Efficient graph similarity search over large graph databases. IEEE Trans. Knowl. Data Eng. **27**(4), 964–978 (2015)
11. Ullmann, J.R.: Degree reduction in labeled graph retrieval. ACM J. Exp. Algorithmics **20**, 1.3:1.1–1.3:1.54 (2015)
12. Li, Z., Jian, X., et al.: An efficient probabilistic approach for graph similarity search. In: ICDE 2018, pp. 533–544 (2018)
13. Zhao, X., Xiao, C., et al.: A partition-based approach to structure similarity search. PVLDB **7**(3), 169–180 (2013)
14. Liang, Y., Zhao, P.: Similarity search in graph databases: a multi-layered indexing approach. In: ICDE 2017, pp. 783–794 (2017)
15. Kim, J., Choi, D.-H., Li, C.: Inves: incremental partitioning-based verification for graph similarity search. In: EDBT 2019, pp. 229–240 (2019)

Data Cleaning About Student Information Based on Massive Open Online Course System

Shengjun Yin, Yaling Yi, and Hongzhi Wang(⊠)

Harbin Institute of Technology, Harbin, China
441607987@qq.com, limeyiyaling@163.com, wangzh@hit.edu.cn

Abstract. Recently, Massive Open Online Courses (MOOCs) is a major way of online learning for millions of people around the world, which generates a large amount of data in the meantime. However, due to errors produced from collecting, system, and so on, these data have various inconsistencies and missing values. In order to support accurate analysis, this paper studies the data cleaning technology for online open curriculum system, including missing value-time filling for time series, and rule-based input error correction. The data cleaning algorithm designed in this paper is divided into six parts: pre-processing, missing data processing, format and content error processing, logical error processing, irrelevant data processing and correlation analysis. This paper designs and implements missing-value-filling algorithm based on time series in the missing data processing part. According to the large number of descriptive variables existing in the format and content error processing module, it proposed one-based and separability-based criteria Hot+J3+PCA. The online course data cleaning algorithm was analyzed in detail on algorithm design, implementation and testing. After a lot of rigorous testing, the function of each module performs normally, and the cleaning performance of the algorithm is of expectation.

Keywords: MOOC · Data cleaning · Time series · Intermittent missing · Dimension reduction

1 Introduction

Data cleaning [1–4] or data scrubbing aims to detect the errors or mismatches of data and then remove them out to improve the quality of data [5,6]. Traditional data cleaning methods are generally rule-based methods [7]. The data cleaning method based on deep learning has been a research hot-spot in recent years.

The data cleaning algorithm in this paper is aimed at the problems of inaccurate data analysis results caused by various errors (such as inconsistencies and missing values) in the input data due to input errors, system errors, and other reasons. In order to optimize the data to support accurate analysis, this project

© Springer Nature Singapore Pte Ltd. 2020
J. Zeng et al. (Eds.): ICPCSEE 2020, CCIS 1257, pp. 33–43, 2020.
https://doi.org/10.1007/978-981-15-7981-3_3

intends to study data cleaning techniques for online open course system students, including missing value filling for time series, and rule-based entry error correction [8]. The overall idea of the algorithm is similar to other data cleaning algorithms, but due to the particularity of online course data, this algorithm focuses on the problem of missing data filling on time series [9], encoding of category features, and feature extraction [10]. Time series-oriented missing value filling needs to establish a time series prediction model [11] based on the existing data, and predict and populate the results of missing values [9]. Category characteristics mainly refer to those discrete variables with special meanings. For example, the academic information of students includes primary school, junior high school, high school, and so on. Current machine learning algorithms struggle to handle these features. In the data cleaning phase, one-hot coding is needed to digitize these discrete features.

2   Technology

2.1   Data Cleaning Process

According to the modular design principle, we divide the data cleaning algorithm into 6 parts: preprocessing, missing data processing, format and content error processing, logical error processing, irrelevant data processing and correlation analysis as shown in Fig. 1.

Fig. 1. The overall process of the data cleaning algorithm in this paper

2.2   Preprocessing

The data of student information in the online open course system is huge, and it will take a lot of time to read the data using traditional methods. In order to solve this problem, we used a data structure "DataFrame" that supports the parallel reading of large amounts of data, and designed a data reading algorithm for this data structure. The DataFrame includes three parts: the header, the content, and the index. The internal implementation of the DateFrame is the nesting of multiple lists. We design two parallel processing methods for DataFrame:

1. Overall parallel processing method: For the reading of two-dimensional matrix data of DataFrame, we have designed an overall parallel processing algorithm that can perform input parallelization by performing overall feature conversion on the matrix. The implementation is similar to the FeatureUnion class in the pipeline package, thich improve the overall efficiency by calling multiple transform methods to read data from multiple parts at the same time.

2. Improved parallel processing method: we design a parallel processing algorithm for a large number of simple repetitive tasks. We have improved the FeatureUnion class by the following process. First, we add the attribute idx_list to record each column of the feature matrix that needs to be processed in parallel; then Modify the fit and fit_transform method so that it can read a part of the feature matrix; finally, modify the transform method so that it can output part of the feature matrix.

The above process can be summarized as follows,

- Step 2–1 = ('OneHoutEncoder', OneHotEncoder(sparse = False))
- Step 2–2 = ('Tolog', FunctionTransfer(log1p))
- Step 2–3 = ('ToBinary', Binarizer())
- Step 2 = ('FeatureUnionExt', FeatureUnionExt(tansformer_list = [Step2–1, Step2–2, Step2–3], idx_list=[[0], [1, 2, 3], [4]]))

2.3 Missing Data Processing

There are several ways to deal with missing value data:

- Feature discarding method: For the features with a higher degree of missing (the degree of missing is greater than 10
- Null value feature method: In some data sets, a null value can be used as a feature, but in order to facilitate processing, the null value needs to be specified as a special value.
- Mean filling method: If there are missing values in the training set but no missing values in the test set, the conditional mean or conditional median is usually taken for the missing values, because the training set data does not need to be particularly high quality.
- Upper and lower value or difference filling method: These two methods are filled using the feature information in the data table. The interpolation method needs to calculate the missing interval.
- Algorithm-Filled Filling: This is the most effective missing-value processing method currently in use. It learns from existing data to build an analysis model to predict missing values, including time-series prediction methods and logical prediction methods.

In this paper, we mainly use the time series prediction method in algorithm fitting and filling method to process missing values. The advantage is that it can analyze the time series, predict and fill the missing data at different time nodes, and the specific implementation of the algorithm.

2.4 Format and Content Error Handling

In this paper, due to format errors and content errors in the data, regular expressions are mainly used to reduce the format errors caused by user input, but this method is only for error prevention. If wrong data has been generated already, you need to use data binning, clustering, regression and other methods for processing. In this algorithm, data binning is mainly used to discretize continuous data. For example, the user's age data is distributed for small dispersion data in the range from 0 to 120, we use data binning to convert the data into large dispersion data for data analysis. For different data sets, we design supervised and unsupervised discretization algorithms. Supervised discretization algorithm: As for data with fixed category information, supervised discretization algorithm is required to discretize the data. Optimal criteria need to be formulated based on the category information. The determination of the optimal criteria needs to be judged based on the characteristics of the data set size Including but not limited to chi-square test, entropy discrimination, information gain, Gini coefficient and other methods. Unsupervised discretization algorithm: Unsupervised discretization algorithms need to be used to process data with unfixed class information. The main classification methods include dividing the value range of the feature's numerical attributes into K-divisions. When the data is approximately subject to uniform distribution, this kind of discretization algorithm has the best effect; another classification method is to classify the characteristic data by isoquantity by calculating the quantiles to ensure that the number of data samples in each interval is almost the same. After data binning, you need to perform data smoothing and other operations. This part of the algorithm involves processing category features.

2.5   Logical Error Handling

This part aims to judge the legality and validity of the data through simple logical reasoning. It mainly includes the three steps of removing duplicate values, removing irrational values, and correcting contradictions. Due to the intersection of the last two steps and formats with the content error processing algorithm, the data deduplication algorithm will be designed in this section. The deduplication algorithm is mainly divided into two parts: determining duplicate data and removing duplicate data. Determining duplicate data: The determination of duplicate data requires similarity analysis of the data. The algorithm processing flow is shown in Fig. 2. First, we perform pairwise similarity analysis on all the data in the data set; Calculate the threshold; finally use the threshold to eliminate duplicate data.

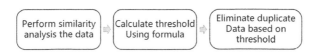

Fig. 2. Process of eliminating duplicate data

Using the operation pool to remove duplicate data: Due to the huge amount of data in the system, deleting one by one when deleting duplicate data will cause the DataFrame index to be re-write after each delete operation, which will cause serious performance problems. The solution to this problem is to use the operation pool to save the operation data to prevent performance loss caused by repeatedly calling and releasing the upper-level API of the DataFrame each time a delete operation is performed.

2.6 Irrelevant Data Processing

Irrelevant data refers to irrelevant features found during the analysis, these features or data will not affect the final results of the analysis, but deleting these data will adversely affect the readability and integrity of the analysis results. So you need to consider how to deal with this irrelevant data. As for redundant fields, direct deletion can be used for irrelevant data processing; for multi-index data features, it is necessary to determine whether to delete this feature based on the needs of the results.

2.7 Correlation Analysis

Association analysis uses PCA or LDA to integrate features to eliminate useless information. In this algorithm, PCA is mainly used to extract features. PCA (Principal Component Analysis), principal component analysis, is a data analysis technique. The main idea is to project high-dimensional data to a lower-dimensional space, extract the main factors of multiple things, and reveal its essential characteristics. Advantages of association analysis:

1. It can eliminate the relevant impact between the evaluation indicators, because the principal component analysis method transforms the original data indicator variables to form independent principal components. The practice proves that the higher the degree of correlation between indicators, the better the effect of principal component analysis.
2. It can reduce the workload of index selection. For other evaluation methods, it takes a lot of effort to select the index because it is difficult to eliminate the relevant impact between the evaluation indices. But the principal component analysis method can eliminate this related impact. As a result, index selection is relatively easy.
3. In the principal component analysis, each principal component is arranged in order according to the magnitude of the variance. When analyzing a problem, you can discard a part of the principal components and only take a few principal components with larger variances to represent the original variables, thereby reducing the calculation work. When using the principal component analysis method for comprehensive evaluation, the principle of selection is that the cumulative contribution rate is $geq85$.

3   Design and Implementation of Time Series Filling Algorithm

3.1   Data Smoothing

Time series data generally have periodic data fluctuations, which are the key characteristics of the entire time series. But these characteristics are too obvious. As a result, they will obscure other key characteristics to a certain extent, so it is necessary to retain the characteristics of the periodic data. Data difference algorithm: The original time series data is segmented by generating adaptive time series, and part of the data is intercepted every fixed length to divide the data into subsets. In the case where the intercept length is an integer multiple of the period T, each subset is data with the period characteristics removed, and features that are obscured by the period characteristics can be extracted. The implementation of the data difference algorithm is shown in Fig. 3.

Fig. 3. Main flow of data difference algorithm

And the comparison of the data distribution before and after the data difference processing is shown in Fig. 4.

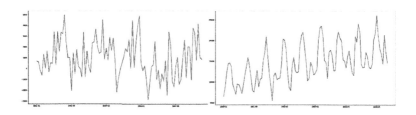

Fig. 4. Comparison before and after processing of the difference algorithm

3.2   Autocorrelation Graph Generation

By analyzing the correlation coefficients of the output variables of the time series to determine the characteristics of the time series, this method is called an autocorrelation method, also known as an autocorrelation map. The autocorrelation graph will show the correlation of each lagging feature and label these features by statistical feature analysis. The X-axis in the figure represents the value of the lagging feature, and the interval distribution of the Y-axis is $(-1, 1)$, showing the positive and negative correlations of these lag values. The points in the blue area are statistically significant. If the lag value of a feature is 0 but the correlation n is 1, it means that the observed value of the feature and itself 100% positive correlation. The implementation of the autocorrelation feature is shown in Fig. 5, which calls the DateFrame data structure we mentioned earlier.

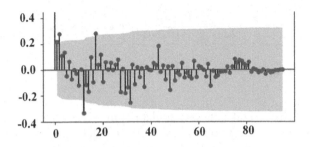

Fig. 5. Sample autocorrelation diagram

3.3 Optimization for Supervised Learning

If we take the lag variable as the input variable and the current observation variable as the output variable, we can transform the time series prediction problem into a supervised learning problem. The specific conversion steps are shown in Fig. 6. First, we use the shift function in the Pandas library to generate a new queue after the observations are transformed. Then we traverse the DataFrame to constrain the data specifications. Finally, we generate a prediction label by supervising the learning classifier.

Fig. 6. Supervised learning conversion process

3.4 Feature Selection for Lag Variables

Usually, a large number of features are weighted using decision tree algorithms (bagged trees, random forests, etc.). By calculating the weights of these features, the relative effectiveness of the features in the prediction model can be predicted. By using lag features, you can analyze other types of features based on time stamps, rolling statistics, and so on. Feature selection according to lag features needs to be considered from the following two aspects. Generation of feature weights for lag variables: In this algorithm, we use a random forest model to generate feature weights, as shown in Fig. 7, where RandomForestRegressor is a random forest model. To prevent accidental errors from affecting the experimental results, we use random seed initialization to randomly generate data.

Fig. 7. The process of fitting using random forest

The effect of the weight generation algorithm is shown in Fig. 8. The figure shows that the relative importance of $t-12$, $t-2$, and $t-4$ is high. In fact, similar results can be obtained by other design tree algorithms.

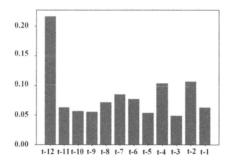

Fig. 8. Running effect of weight generation algorithm

Feature selection for lag variables: By using a recursive feature selection method (RFE) to create a predictive model that assigns different weights to each feature, and deletes those features with the smallest weight. Each time a feature is deleted, RFE is used to generate Model and repeat the above process until the number of features reaches the expected requirements. The implementation of the RFE algorithm is similar to Fig. 8. Running the algorithm found that the output results were $t-12$, $t-2$, and $t-4$. This result is the same as the highest weight result generated during the weight generation phase. The bar graph of the feature weight ranking is shown in Fig. 9.

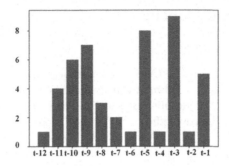

Fig. 9. Bar chart of feature weight ordering (smaller is better)

3.5 Selection of Core Algorithms

In order to solve the problems caused by vanishing gradients and local optimization, people find that local optimization can be solved by pre-training after years of exploration. At the same time, using maxout or RelU activation function instead of sigmoid can alleviate the problem of vanishing gradients. After this replacement, the structure is the basic structure of DNN. However, DNN is still a fully connected multi-layer perceptron in nature. As the number of layers increases, the exponential expansion of the number of parameters increase as well which mainly occurs in image recognition. CNN [12,13] uses convolution kernels and convolution layers to achieve the reuse of regional features, which perfectly solves many problems caused by the increase in the amount of calculations (such as the problem of overfitting). CNN uses the common features of the local structure to propagate forward. The execution efficiency of the algorithm depends on the size of the convolution kernel window. The time series is a very important feature in the user information data analyzed in this paper, so we select RNN-type algorithms as the core algorithm of the analysis model. However, to avoid the phenomenon of gradient disappearance, LSTM [14,15] needs to be used as the core algorithm for the amount of data is large.

4 Design and Implementation of Category Feature Processing Algorithm

There are so many data composed of category features in the data, such as: age, time, UID and so on, that the data corresponding to these features are often discrete and have large distribution intervals. In most cases, these data have no practical significance, so directly applying these features to model calculations affects the overall accuracy of the prediction model. We choose to use data binning and one-hot encoding to deal with these category features. Since one-hot encoding will cause the number of features to increase, we need to use feature selection and extraction algorithms to reduce the feature dimensions.

5 Evaluation and Conclusion

5.1 Parameter Tuning

Improved grid search algorithm: Optimize the repeated calculation steps of grid search. In the improved algorithm, we have modified the distance traveled by each search to reduce the number of searches, thereby reducing the number of cross-validations to improve efficiency. We use some data sets for grid search comparison tests, and use GBDT, Xgboost, RF, and SVM for testing. The model test results are shown in Fig. 10.

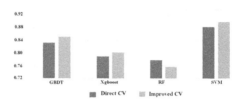

Fig. 10. Grid search comparison test results

5.2 Algorithm Test Results

The time series comparison of user behavior before and after the algorithm execution is shown in Fig. 11, and the data is successfully stabilized.

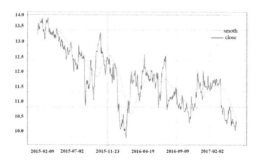

Fig. 11. Comparison of user behavior time series before and after algorithm execution

References

1. Li, W., Li, L., Li, Z., Cui, M.: Statistical relational learning based automatic data cleaning. Front. Comput. Sci. **13**(1), 215–217 (2019). https://doi.org/10.1007/s11704-018-7066-4
2. Zheng, Z., Quach, T.M., Jin, Z., Chiang, F., Milani, M.: Currentclean: interactive change exploration and cleaning of stale data. In: Proceedings of the 28th ACM International Conference on Information and Knowledge Management, CIKM 2019, pp. 2917–2920. ACM (2019)
3. Wang, T., Ke, H., Zheng, X., Wang, K., Sangaiah, A.K., Liu, A.: Big data cleaning based on mobile edge computing in industrial sensor-cloud. IEEE Trans. Industr. Inform. **16**, 1321–1329 (2019)
4. Lin, X., Peng, Y., Xu, J., Choi, B.: Human-powered data cleaning for probabilistic reachability queries on uncertain graphs. In: 34th IEEE International Conference on Data Engineering, ICDE 2018, Paris, France, 16–19 April 2018, pp. 1755–1756. IEEE Computer Society (2018)
5. Ilyas, I.F., Chu, X.: Data Cleaning. ACM, New York (2019)
6. Cichy, C., Rass, S.: An overview of data quality frameworks. IEEE Access **7**, 24634–24648 (2019)
7. Ganibardi, A., Ali, C.A.: Web usage data cleaning. In: Ordonez, C., Bellatreche, L. (eds.) DaWaK 2018. LNCS, vol. 11031, pp. 193–203. Springer, Cham (2018). https://doi.org/10.1007/978-3-319-98539-8_15
8. Polleres, A., Sobernig, S., Umbrich, J.: Data cleaning and preparation (basics) (2019)
9. Yi, X., Zheng, Y., Zhang, J., Li, T.: ST-MVL: filling missing values in geo-sensory time series data. In: Kambhampati, S. (ed.) Proceedings of the Twenty-Fifth International Joint Conference on Artificial Intelligence, IJCAI 2016, New York, NY, USA, 9–15 July 2016, pp. 2704–2710. IJCAI/AAAI Press (2016)
10. Joo, Y., Jeong, J.: Under sampling adaboosting shapelet transformation for time series feature extraction. In: Misra, S., et al. (eds.) ICCSA 2019. LNCS, vol. 11624, pp. 69–80. Springer, Cham (2019). https://doi.org/10.1007/978-3-030-24311-1_5
11. Mahalakshmi, G., Sridevi, S., Rajaram, S.: A survey on forecasting of time series data. In: International Conference on Computing Technologies and Intelligent Data Engineering (2016)
12. Barrow, E., Eastwood, M., Jayne, C.: Deep CNNs with rotational filters for rotation invariant character recognition. In: 2018 International Joint Conference on Neural Networks, IJCNN 2018, Rio de Janeiro, Brazil, 8–13 July 2018, pp. 1–8. IEEE (2018)
13. Cui, Y., Zhang, B., Yang, W., Yi, X., Tang, Y.: Deep CNN-based visual target tracking system relying on monocular image sensing. In: 2018 International Joint Conference on Neural Networks, IJCNN 2018, Rio de Janeiro, Brazil, 8–13 July 2018, pp. 1–8. IEEE (2018)
14. Wang, D., Fan, J., Fu, H., Zhang, B.: Research on optimization of big data construction engineering quality management based on RNN-LSTM. Complexity **2018**, 9691868:1–9691868:16 (2018)
15. Ma, Y., Peng, H., Cambria, E.: Targeted aspect-based sentiment analysis via embedding commonsense knowledge into an attentive LSTM. In: Proceedings of the Thirty-Second AAAI Conference on Artificial Intelligence, (AAAI 2018) (2018)

V2V Online Data Offloading Method Based on Vehicle Mobility

Xianlang Hu[1], Dafang Zhang[1], Shizhao Zhao[2],
Guangsheng Feng[2(✉)], and Hongwu Lv[2]

[1] Jiangsu Automation Research Institute, Lianyungang 222002, China
[2] College of Computer Science and Technology, Harbin Engineering University,
Harbin 150001, China
fengguangsheng@hrbeu.edu.cn

Abstract. As people are accustomed to getting information in the vehicles, mobile data offloading through Vehicular Ad Hoc Networks (VANETs) becomes prevalent nowadays. However, the impacts caused by the vehicle mobility (such as the relative speed and direction between vehicles) have great effects on mobile data offloading. In this paper, a V2V online data offloading method is proposed based on vehicle mobility. In this mechanism, the network service process was divided into continuous and equal-sized time slots. Data were transmitted in a multicast manner for the sake of fairness. The data offloading problem was formalized to maximize the overall satisfaction of the vehicle users. In each time slot, a genetic algorithm was used to solve the maximizing problem to obtain a mobile data offloading strategy. And then, the performance of the algorithm was enhanced by improving the algorithm. The experiment results show that vehicle mobility has a great effect on mobile data offloading, and the mobile data offloading method proposed in the paper is effective.

Keywords: Vehicular Ad Hoc Networks · Mobile data offloading · Vehicle mobility · Multicast manner · Genetic algorithm

1 Introduction

With the development of information technology, mobile devices have been widely and deeply applied to daily life. Due to vehicle users request network services anytime and anywhere, a large number of mobile data requests aggravate the load on the cellular network base station [1–3]. At the same time, some realistic conditions, such as the high-speed mobility of vehicles, affect the duration of the connection between vehicles and access points [4, 5]. With the increase of mobile data requests, network operators also continuously upgrade communication technologies [6], such as 5G technology, causing the cellular network hardly meet the data requests of vehicle users.

In vehicular networks, vehicles equipped with On Board Units (OBUs) can communicate with each other in their coverage ranges. Such vehicles can be applied to mobile data offloading. However, as a result of vehicle mobility, vehicles can not connect to each other continuously, which affects the Quality of Experience (QoE) of vehicle users greatly.

© Springer Nature Singapore Pte Ltd. 2020
J. Zeng et al. (Eds.): ICPCSEE 2020, CCIS 1257, pp. 44–60, 2020.
https://doi.org/10.1007/978-981-15-7981-3_4

The impacts of vehicle mobility on data offloading have been extensively studied in some researches [7–11]. By considering the vehicle mobility, the duration of the connection between vehicles is taken as a condition to make data offloading strategies more effective. However, most of the existing studies do not consider the impact of specific vehicle mobility (such as vehicle speed, direction, etc.) on the duration of the connection between vehicles. As to the way of data transmission through VANETs, most of the existing studies focus on the unicast manner without considering the multicast manner. From the perspective of users, most of the researches [12] did not take the differences of users into consideration.

Generally, the duration of the V2V connection is short when the relative speed between vehicles is large. As a result of this case, the data transmission can not be finished completely. In most instances of data offloading scenes, the vehicle users are more than seed vehicles. The factor results in the case that part of the vehicle users wait for service when the unicast manner is used in data transmission through V2V. Some vehicle users, who are sensitive about the network latency, are willing to buy the service from the cellular network while the others on the opposite side.

Therefore, we study the mobile data offloading through V2V in consideration of specific vehicle mobility as well as the QoE and the economic costs of users. Meanwhile, the multicast manner is applied to the data transformation. The major contributions of this work are summarized as follows.

(1) A vehicle movement model is built by referencing the Manhattan model. And the multicast manner is applied to the communication between vehicles for the sake of fairness.
(2) Considering the QoE and the economic costs of users, we formalize the problem to maximize the overall satisfaction of the vehicle users.
(3) A genetic algorithm based on Roulette and Fixed Cross Mutation Probability (GA-RFCMP) is proposed to solve the maximizing problem. Furthermore, to optimize the execution effect of the algorithm, we propose the genetic algorithm based on Roulette and Adaptive Cross Mutation Probability (GA-RACMP) and the genetic algorithm based on Elitist Retention and Adaptive Cross Mutation Probability (GA-ERACMP).

The rest of this paper is structured as follows. Section 2 introduces the system model. Section 3 formulates the problem to maximize the overall satisfaction of the vehicle users. In Sect. 4, we propose and design GA-RFCMP, GA-RACMP and GA-ERACMP to solve the maximizing problem, followed by the performance evaluation in Sect. 5. Finally, Sect. 6 concludes this paper.

2 System Model

In this paper, we assume that the cellular network covers all areas seamlessly. All vehicles can download data from the cellular network. And vehicles can also download data from other vehicles through a V2V connection by DSRC technology.

The set $R = \{u_1, u_2, \ldots, u_m\}$ denotes the vehicle users. The seed vehicles, which denoted by $V = \{v_1, v_2, \ldots, v_n\}$, are responsible for transmitting data to vehicle users.

The method of selecting seed vehicles is out of the scope of the paper. Access points refer to the cellular network and all seed vehicles.

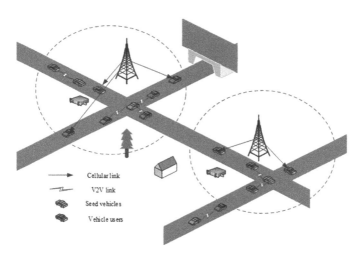

Fig. 1. Mobile data offloading through V2V

Seed vehicle v_i transmits data with a certain bandwidth resource r_i. Cellular network transmits data with a certain bandwidth resource r_c. As shown in Fig. 1, the number of vehicle users is usually bigger than the number of seed vehicles. In order to avoid the case that part of the vehicle users waiting for service, the multicast manner is applied to transmit data through V2V for the sake of fairness.

The network service process T is divided into continuous, equal-sized time slots. Due to the time slots is short enough, we assume the vehicle's moving status remain unchanged in each time slot. But the vehicles can move at different speeds in different time slots. Vehicle user u_j can download data from the only access point in each time slot, while u_j can download data from different access points in different time slots.

Each vehicle user requests one type of multimedia data from the cellular network, and the size of the multimedia data is constant. The system collects the information of vehicles and makes a mobile data offloading strategy once in each time slot. Due to the high efficiency of the genetic algorithm, the impact of the running time of the algorithm is out of the scope of the paper.

2.1 Vehicle Movement Model

We build vehicle movement model by referencing the Manhattan model [13–15]. The roads are composed of horizontal roads, vertical roads and the crossing area between roads. In each time slot, each vehicle maintains the speed of the previous time slot with 50% probability, and accelerates or decelerates with 25% probability. We assume that all of the probabilities of a vehicle going straight, turning left, and turning right at the crossing area are 33.33%. Due to the time slot is short, when a vehicle changes its moving direction in the crossing area, the vehicle has only completed the turning action during the rest time of the time slot, and its geographical position remains unchanged.

In time slot t, the tuple $(l_i(t), s_i(t), \theta_i(t))$ denotes the mobility of seed vehicle v_i. $l_i(t) = (x_i(t), y_i(t))$ denotes the position of v_i in the coordinate system A. $s_i(t)$ denotes the moving speed of v_i. $\theta_i(t)$ denotes the angle between the moving direction of v_i and the positive direction of the X axis in the coordinate system A, and $\theta_i(t) \in [0, 2\pi)$. The tuple $(l_j(t), s_j(t), \theta_j(t))$ denotes the mobility of vehicle user u_j. $l_j(t) = (x_j(t), y_j(t))$ denotes the position of u_j in the coordinate system A. $s_i(t)$ denotes the moving speed of u_j. $\theta_j(t)$ denotes the angle between the moving direction of u_j and the positive direction of the X axis in the coordinate system A, and $\theta_j(t) \in [0, 2\pi)$. R denotes the maximum communication distance between vehicles, and only when vehicles move to each other within R can they communicate with each other through V2V.

In time slot t, the relative speed between u_j and v_i is

$$s_{ij}(t) = \sqrt{s_i^2(t) + s_j^2(t) - 2 \cdot s_i(t) \cdot s_i(t) \cdot \cos(\theta_i(t) - \theta_j(t))} \tag{1}$$

The relative position between u_j and v_i are

$$\begin{cases} x_{ij}(t) = x_j(t) - x_i(t) \\ y_{ij}(t) = y_j(t) - y_i(t) \end{cases} \tag{2}$$

The horizontal relative speed $v_{ij}^x(t)$ and vertical relative speed $v_{ij}^y(t)$ are

$$\begin{cases} v_{ij}^x(t) = s_j(t) \cdot \cos \theta_j(t) - s_i(t) \cdot \cos \theta_i(t) \\ v_{ij}^y(t) = s_j(t) \cdot \sin \theta_j(t) - s_i(t) \cdot \sin \theta_i(t) \end{cases} \tag{3}$$

In the coordinate system A, $\varphi(t)$ denotes the angle between the direction of $s_{ij}(t)$ and the positive direction of the X axis, we have

$$\varphi(t) = \begin{cases} \arctan \frac{v_{ij}^y(t)}{v_{ij}^x(t)}, & \text{if } v_{ij}^y(t) \geq 0 \text{ and } v_{ij}^x(t) > 0 \\ \arctan \frac{v_{ij}^y(t)}{v_{ij}^x(t)} + \pi, & \text{if } v_{ij}^x(t) < 0 \\ \arctan \frac{v_{ij}^y(t)}{v_{ij}^x(t)} + 2\pi, & \text{if } v_{ij}^y(t) < 0 \text{ and } v_{ij}^x(t) > 0 \\ \text{undefined}, & \text{if } v_{ij}^x(t) = 0 \end{cases} \tag{4}$$

And the connect time between u_j and v_i is

$$t_{ij}(t) = \begin{cases} \dfrac{\sqrt{\left(\frac{-kb+\sqrt{k^2 \cdot R^2 - b^2 + R^2}}{k^2+1}-x_{ij}(t)\right)^2 + \left(\frac{b+k\sqrt{k^2 \cdot R^2 - b^2 + R^2}}{k^2+1}-y_{ij}(t)\right)^2}}{s_{ij}(t)}, & \text{if } v_{ij}^x(t) > 0 \text{ and } v_{ij}^y(t) \neq 0 \\[3ex] \dfrac{\sqrt{\left(\frac{-kb-\sqrt{k^2 \cdot R^2 - b^2 + R^2}}{k^2+1}-x_{ij}(t)\right)^2 + \left(\frac{b-k\sqrt{k^2 \cdot R^2 - b^2 + R^2}}{k^2+1}-y_{ij}(t)\right)^2}}{s_{ij}(t)}, & \text{if } v_{ij}^x(t) < 0 \text{ and } v_{ij}^y(t) \neq 0 \\[3ex] \dfrac{\sqrt{R^2 - x_{ij}^2(t)}-y_{ij}(t)}{s_{ij}(t)}, & \text{if } v_{ij}^x(t) = 0 \text{ and } v_{ij}^y(t) > 0 \\[3ex] \dfrac{y_{ij}-\sqrt{R^2 - x_{ij}^2(t)}}{s_{ij}(t)}, & \text{if } v_{ij}^x(t) = 0 \text{ and } v_{ij}^y(t) < 0 \end{cases} \tag{5}$$

And we can obtain that

$$t_{ij}(t) = \infty, \text{ if } v_{ij}^x(t) = 0 \text{ and } v_{ij}^x(t) = 0 \tag{6}$$

where $k = \tan \varphi(t)$ and $b = y_{ij}(t) - x_{ij}(t) \cdot \tan \varphi(t)$.

Due to the vehicle mobility, $y_{i,j}^k(t)$ denotes whether v_i is able to transmit data d_k to u_j or not. If v_i is able to transmit data d_k to u_j, $y_{i,j}^k(t) = 1$, on the other side $y_{i,j}^k(t) = 0$. We can obtain that

$$y_{i,j}^k(t) = \begin{cases} 1, & \text{if } t_{ij}(t) \geq t_l \\ 0, & \text{else} \end{cases} \tag{7}$$

2.2 Data Utility Model

We assume that u_j requests data d_k, and the amount of d_k is a_k. The remain data requests of u_j is ra_{jk}, In time slot t, the bandwidth allocated by system to u_j is $r_j(t)$. We have $ra_{jk} = ra_{jk} - r_j(t) \cdot t_l$, and if $ra_{jk} \leq 0$, we set $ra_{jk} = 0$, from then on, the system does not allocate bandwidth resources to u_j anymore. If $ra_{jk} = 0$, we have $rem_j = 0$, on the other hand, $rem_j = 1$. If v_i send data d_k to u_j, $h_{i,j}^k(t) = 1$, otherwise, $h_{i,j}^k(t) = 0$. If cellular network send data d_k to u_j, $h_{c,j}^k(t) = 1$, otherwise, $h_{c,j}^k(t) = 0$. We can obtain that

$$h_{i,j}^k(t) \leq rem_j \tag{8}$$

$$h_{c,j}^k(t) \leq rem_j \tag{9}$$

Considering the differences among vehicle users, we assume that the sensitivity of vehicle user u_j to network delay is s_j. In time slot t, the data utility that vehicle users obtain is U_t.

$$U_t = s_j \cdot t_l \left(\sum_{u_j \in U} \sum_{v_i \in V} \sum_{d_k \in D} h_{i,j}^k(t) \cdot r_{i,j}^k(t) + \sum_{u_j \in U} \sum_{d_k \in D} h_{c,j}^k(t) \cdot r_{c,j}^k(t) \right)$$

(10)

where $r_{i,j}^k(t)$ and $r_{c,j}^k(t)$ denote the bandwidth resources provided by seed vehicle v_i and cellular network, respectively.

2.3 Vehicle User Cost Model

The cost of vehicle users downloading multimedia data through cellular network is c_c/Mb, and the cost of vehicle users downloading multimedia data through seed vehicle is c_v/Mb. Considering the actuality, we assume that $c_v < c_c$ reasonably. In time slot t, the costs of all vehicle users are

$$C_t = t_l \cdot \left(\sum_{u_j \in U} \sum_{v_i \in V} \sum_{d_k \in D} h_{i,j}^k(t) \cdot r_{i,j}^k(t) \cdot c_v + \sum_{u_j \in U} \sum_{d_k \in D} h_{c,j}^k(t) \cdot r_{c,j}^k(t) \cdot c_c \right)$$

(11)

3 Problem Formulation

In each time slot, u_j can download data d_k from only one access point, which has cached data d_k and is able to transmit data to u_j, we can obtain that

$$\sum_{v_i \in V} h_{i,j}^k(t) + h_{c,j}^k(t) = 1$$

(12)

$$h_{i,j}^k(t) \le y_{i,j}^k(t)$$

(13)

$$h_{i,j}^k(t) \le p_i^k$$

(14)

where p_i^k denotes that whether v_i has cached data d_k or not. If v_i has cached data d_k, $p_i^k = 1$, otherwise, $p_i^k = 0$.

For each access point, the bandwidth resources used to transmit data can not exceed its total bandwidth resources, we can obtain that

$$\sum_{d_k \in D} \sum_{u_j \in U} r_{i,j}^k(t) \le r_i$$

(15)

$$\sum_{d_k \in D} \sum_{u_j \in U} r_{c,j}^k(t) \le r_c$$

(16)

Only when u_j download data d_k from an access point can system allocate the bandwidth resources of the access point for u_j. We can obtain that

$$h^k_{i,j}(t) \cdot r^k_{i,j}(t) = r^k_{i,j}(t) \tag{17}$$

$$h^k_{c,j}(t) \cdot r^k_{c,j}(t) = r^k_{c,j}(t) \tag{18}$$

Considering the difference of data utility sensitivity and economic cost sensitivity of different vehicle users, we assume that α_j and β_j represent the data utility sensitivity and economic cost sensitivity of vehicle user u_j, respectively. We assume that $\alpha_j, \beta_j \in (0, 1)$, and $\alpha_j + \beta_j = 1$. We formalize the data offloading problem as P0 to maximize the overall satisfaction of the vehicle users:

$$\max \left(\alpha_j \cdot U_t - \beta_j \cdot C_t \right) \tag{19}$$

$$\text{s.t. } h^k_{i,j}(t), h^k_{c,j}(t) \in \{0, 1\} \tag{20}$$

$$(10) - (11), (14) - (20)$$

In each time slot, the vehicle users download data right away. In order to maximize overall satisfaction of the vehicle users, for each vehicle user, there are two challenges, 1) which access point does each vehicle user download data from, 2) and how many bandwidth resources does each vehicle user obtain.

4   Problem Solution

In this section, we introduce GA-RFCMP to solve P0. And then, we propose GA-RACMP and GA-ERACMP to optimize the execution effect of GA-RFCMP.

The genetic algorithm can solve most of the resource allocation problems and obtain approximate optimal solutions; it can be executed at a fast rate; the algorithm is flexible to be designed as well as has strong expansibility to be optimized [16–18]. In each time slot, the genetic algorithm is used to solve P0. In this paper, we set the objective function in P0 as the fitness function of the genetic algorithm.

4.1   GA-RFCMP

Assume that there are n access points in total, and m vehicle users. The length of chromosome is $2m$. Then the first m bases represent the access points allocated by system for the vehicle users $1 \sim m$, whose ranges are $[1, n]$. The last m bases represent the bandwidth resources allocated by the system for the vehicle users $1 \sim m$. To simplify computation, we normalize the bandwidth of each accesses point to 1. So the value ranges of the last m bases are $[0, 1]$, which denote the ratios of the bandwidth resources of the corresponding access points at the first m bases.

Let *num* denote the size of the population, *testTime* denotes the iteration times. p and q represent the probability of cross and the probability of mutation, respectively. The best strategy of the solution denoted by *best_individual*.

Algorithm 1:GA-RFCMP

Input: m , n , num , $testTime$, p , q
Output: $best\_individual$
1 Init population
2 FOR 1 to $testTime$
3   Calculate population $fitness$
4   Get $best\_individual$
5   Select individuals according to Roulette
6   FOR $i=1:2:num$
          Randomly generate $PR \in [0,1]$, IF $PR < p$, Cross the two individuals
      END
7   FOR $i=1:num$
          Randomly generate $PR \in [0,1]$, IF $PR < q$, make the individual mutation
      END
8 END
9 return $best\_individual$

We define the proportion of each individual's fitness in the total fitness as the probability of the individual being selected. The process of GA-RFCMP is shown in Algorithm 1.

4.2 GA-RACMP

For genetic algorithm, cross and mutation are the main ways to generate new individuals, so an adaptive probability method is proposed for cross and mutation.

$$p = \begin{cases} p_1, & f_c < f_{avr} \\ \dfrac{p_2 \cdot (f_{max} - f_c)}{f_{max} - f_{avr}}, & f_c \geq f_{avr} \\ p_3, & p < p_3 \end{cases} \tag{21}$$

where f_{max} denotes the fitness of $best\_individual$, f_{avr} denotes the average fitness of the population. f_c denotes the higher fitness of the two individuals that participate in the cross activity. We assume that $p_1 > p_2 > p_3$.

$$q = \begin{cases} q_1, & f_m < f_{avr} \\ \dfrac{q_2 \cdot (f_{max} - f_m)}{f_{max} - f_{avr}}, & f_m \geq f_{avr} \\ q_3, & q < q_3 \end{cases} \tag{22}$$

where f_m denotes the fitness of the individual that participate in the mutation. We assume that $q_1 > q_2 > q_3$.

Based on GA-RFCMP, we propose GA-RACMP by calculating p and q according to Eq. (21) and Eq. (22), respectively.

4.3   GA-ERACMP

As the number of individuals in the population is large, resulting in a low probability of each individual to be selected. The individuals with higher fitness have low probability of being retained in the population through roulette method. Based on GA-RACMP, we propose elitist retention method to remain the individuals with higher fitness in the population. The process of GA-ERACMP is shown in Algorithm 2.

Algorithm 2:GA-ERACMP

Input: m , n , num , $testTime$, p_1 , p_2 , p_3 , q_1 , q_2 , q_3

Output: $best\_individual$

1 Init population
2 FOR 1 to $testTime$
3   Calculate population $fitness$
4   Get $best\_individual$
5   Select individuals according to elitist retention
6   FOR $i = 1 : 2 : num$
             Calculate p according to equation (21)
             Randomly generate PR $\in [0,1]$, IF PR $< p$, Cross the two individuals
         END
7   FOR $i = 1 : num$
             Calculate q according to equation (22)
             Randomly generate PR $\in [0,1]$, IF PR $< q$, make the individual mutation
         END
8 END
9 return $best\_individual$

---

5   Performance Evaluation

5.1   Experimental Simulation Scene and Parameter Settings

In this section, the convergence rate, the average fitness, the maximum fitness, the data offloading rate and the overall satisfaction of vehicle users are taken as the performance indicators of the algorithm.

There are 5 horizontal roads and 5 longitudinal roads in the scene, the distance between adjacent roads is 300 m, each road is a two-way lane, the width of the lane is 20 m, and the intersection area of the road is a square area of 20 m * 20 m.

The number of the seed vehicles is 6 and the number of the vehicle users is 18. There are 3 types of data, which denoted by d_1, d_2 and d_3, whose amounts are 100 Mb, 150 Mb and 180 Mb, respectively. Referring to [19–23], we set $R = 500$ m. The network service process duration is 60 s, and the time slot is 3 s. We set $r_c = 15$ Mbps, while $r_i \in [8\text{ Mbps}, 12\text{ Mbps}]$. In this section, we set $c_c = 0.05$ RMB/Mb and $c_v = 0.02$ RMB/Mb. The change of vehicle speed between adjacent time slots is 1 m/s. In this section, we assume that $s_j \in [1, 2]$.

5.2 Experimental Simulation Results and Analysis

Comparison of the Performances of 3 Genetic Algorithms. In the first time slot, Fig. 2 shows that GA-ERACMP, which can reach convergence at highest average fitness within 100 iteration times, performs better than GA-RACMP and GA-RFCMP.

In the whole network service process, Fig. 3 shows that the overall satisfactions of vehicle users are decreasing with time. This is caused by the amount of remaining data requests is decreasing, and the overall satisfactions are positively correlated with the amount of remaining data requests.

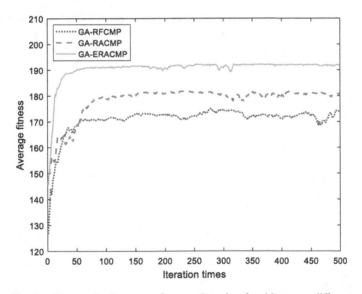

Fig. 2. The result of average fitness when the algorithms are different

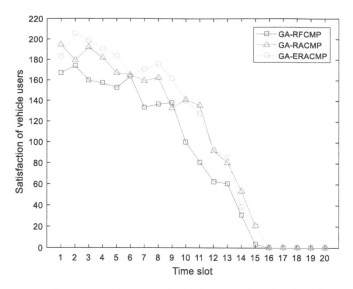

Fig. 3. The result of overall satisfactions of vehicle users when the algorithms are different

The Effects of Vehicles' Relative Moving Speed on Mobile Data Offloading.
Figure 4, Fig. 5 and Fig. 6 show that data offloading rates are basically stable with
vehicles' relative moving speed. When the model applies GA-ERACMP, which performs best, the data offloading rates of d_1 and d_2 keep above 84%.

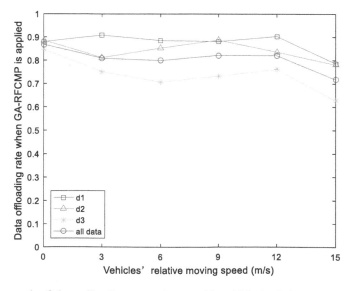

Fig. 4. The result of data offloading rates change with vehicles' relative moving speed when
GA-RFCMP is applied

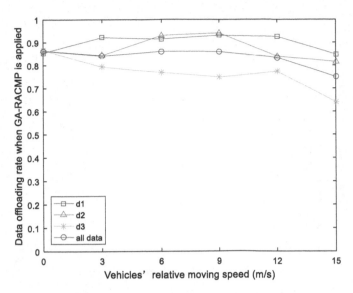

Fig. 5. The result of data offloading rates change with vehicles' relative moving speed when GA-RACMP is applied

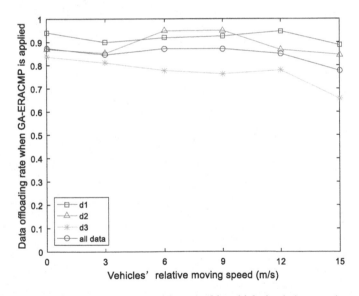

Fig. 6. The result of data offloading rates change with vehicles' relative moving speed when GA-ERACMP is applied

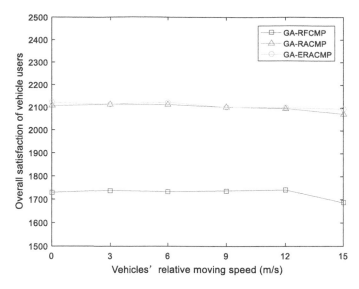

Fig. 7. The result of overall satisfactions change with vehicles' relative moving speed

Figure 7 shows that the overall satisfaction of vehicle users are basically stable. GA-RACMP and GA-ERACMP perform better than GA-RFCMP.

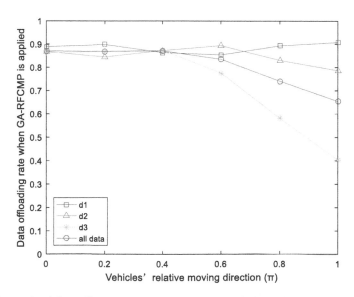

Fig. 8. The result of data offloading rates change with vehicles' relative moving direction when GA-RFCMP is applied

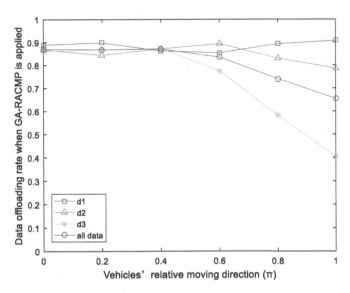

Fig. 9. The result of data offloading rates change with vehicles' relative moving direction when GA-RACMP is applied

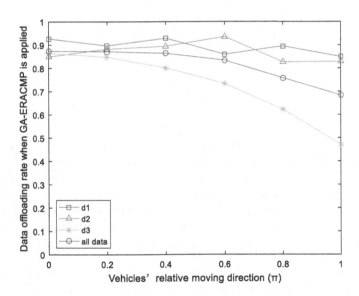

Fig. 10. The result of data offloading rates change with vehicles' relative moving direction when GA-ERACMP is applied

The Effects of Vehicles' Relative Moving Direction on Mobile Data Offloading. Figure 8, Fig. 9 and Fig. 10 show when the model applies GA-ERACMP, which performs best, the data offloading rates of d_1 and d_2 keep above 80%.

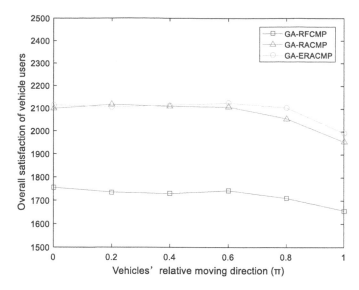

Fig. 11. The result of overall satisfactions of vehicle users change with vehicles' relative moving direction

Figure 11 shows that when the vehicles' relative moving direction below 0.6π, the overall satisfactions of vehicle users are basically stable. When the vehicles' relative moving direction above 0.6π, the overall satisfactions of vehicle users show a downward trend. GA-RACMP and GA-ERACMP perform better than GA-RFCMP.

6 Conclusion

In this paper, we propose a V2V online data offloading method based on vehicle mobility. Considering the vehicle mobility as well as the QoE and the economic costs of users, we formalize the problem to maximize the overall satisfaction of the vehicle users. And then, GA-RFCMP is designed to solve the problem. Meanwhile, GA-RACMP and GA-ERACMP are proposed to enhance the performance of the method. The experimental results show that the method proposed in the paper can relieve the load on the cellular network as well as guarantee the QoE of vehicle users.

Acknowledgement. This work is supported by the System Architecture Project (No. 614000 40503), the Natural Science Foundation of China (No. 61872104), the Natural Science Foundation of Heilongjiang Province in China (No. F2016028), the Fundamental Research Fund for the Central Universities in China, and Tianjin Key Laboratory of Advanced Networking (TANK) in College of Intelligence and Computing of Tianjin University.

References

1. Zhu, C.: Social sensor cloud: framework, greenness, issues, and outlook. IEEE Netw. **32**(5), 100–105 (2018)
2. Zhu, C.: Toward big data in green city. IEEE Commun. Mag. **55**(11), 14–18 (2017)
3. Khazraeian, S., Hadi, M.: Intelligent transportation systems in future smart cities. In: Amini, M.H., Boroojeni, K.G., Iyengar, S.S., Pardalos, P.M., Blaabjerg, F., Madni, A.M. (eds.) Sustainable Interdependent Networks II. SSDC, vol. 186, pp. 109–120. Springer, Cham (2019). https://doi.org/10.1007/978-3-319-98923-5_6
4. Zhou, H.: A time-ordered aggregation model-based centrality metric for mobile social networks. IEEE Access **6**, 25588–25599 (2018)
5. Chen, H.: Probabilistic detection of missing tags for anonymous multicategory RFID systems. IEEE Trans. Veh. Technol. **66**(12), 11295–11305 (2017)
6. Liu, D.: User association in 5G networks: a survey and an outlook. IEEE Commun. Surv. Tutor. **18**(2), 1018–1044 (2016)
7. Sun, Y.: Traffic offloading for online video service in vehicular networks: a cooperative approach. IEEE Trans. Veh. Technol. **67**(8), 7630–7642 (2018)
8. Zhu, X.: Contact-aware optimal resource allocation for mobile data offloading in opportunistic vehicular networks. IEEE Trans. Veh. Technol. **66**(8), 7384–7399 (2017)
9. Wang, N., Wu, J.: Optimal cellular traffic offloading through opportunistic mobile networks by data partitioning. In: 2018 IEEE International Conference on Communications (ICC), pp. 1–6 (2018)
10. Yuan, Q., Li, J., Liu, Z., et al.: Space and time constrained data offloading in vehicular networks. In: 2016 IEEE 18th International Conference on High Performance Computing and Communications; IEEE 14th International Conference on Smart City; IEEE 2nd International Conference on Data Science and Systems (HPCC/SmartCity/DSS), pp. 398–405 (2016)
11. Luoto, P., Bennis, M., Pirinen, P., et al.: Vehicle clustering for improving enhanced LTE-V2X network performance. In: 2017 European Conference on Networks and Communications (EuCNC), pp. 1–5 (2017)
12. Vigneri, L., Spyropoulos, T., Barakat, C.: Storage on wheels: offloading popular contents through a vehicular cloud. In: 2016 IEEE 17th International Symposium on A World of Wireless, Mobile and Multimedia Networks (WoWMoM), pp. 1–9 (2016)
13. Hanggoro, A., Sari, R.F.: Performance evaluation of the manhattan mobility model in vehicular ad-hoc networks for high mobility vehicle. In: 2013 IEEE International Conference on Communication, Networks and Satellite (COMNETSAT), pp. 31–36 (2013)
14. Gowrishankar, S., Sarkar, S., Basavaraju, T.: Simulation based performance comparison of community model, GFMM, RPGM, Manhattan model and RWP-SS mobility models in MANET. In: 2009 First International Conference on Networks & Communications, pp. 408–413 (2009)
15. Perdana, D., Nanda, M., Ode, R., et al.: Performance evaluation of PUMA routing protocol for Manhattan mobility model on vehicular ad-hoc network. In: 2015 22nd International Conference on Telecommunications (ICT), pp. 80–84 (2015)
16. Shrestha, A.P., Won, J., Yoo, S.-J., et al.: Genetic algorithm based sensing and channel allocation in cognitive ad-hoc networks. In: 2016 International Conference on Information and Communication Technology Convergence (ICTC), pp. 109–111 (2016)
17. Bhattacharjee, S., Konar, A., Nagar, A.K.: Channel allocation for a single cell cognitive radio network using genetic algorithm. In: 2011 Fifth International Conference on Innovative Mobile and Internet Services in Ubiquitous Computing, pp. 258–264 (2011)

18. Sun, W., Xie, W., He, J.: Data link network resource allocation method based on genetic algorithm. In: 2019 IEEE 3rd Information Technology, Networking, Electronic and Automation Control Conference (ITNEC), pp. 1875–1880 (2019)
19. Khan, A., Sadhu, S., Yeleswarapu, M.: A comparative analysis of DSRC and 802.11 over Vehicular Ad hoc Networks. Project Report, University of California, Santa Barbara, pp. 1–8 (2009)
20. Kaabi, F., Cataldi, P., Filali, F., et al.: Performance analysis of IEEE 802.11 p control channel. In: 2010 Sixth International Conference on Mobile Ad-hoc and Sensor Networks, pp. 211–214 (2010)
21. Neves, F., Cardote, A., Moreira, R., et al.: Real-world evaluation of IEEE 802.11 p for vehicular networks. In: Proceedings of the Eighth ACM International Workshop on Vehicular Inter-Networking, pp. 89–90 (2011)
22. Vinel, A., Belyaev, E., Lamotte, O., et al.: Video transmission over IEEE 802.11 p: real-world measurements. In: 2013 IEEE International Conference on Communications Workshops (ICC), pp. 505–509 (2013)
23. Gozálvez, J.: IEEE 802.11 p vehicle to infrastructure communications in urban environments. IEEE Commun. Mag. **50**(5), 176–183 (2012)

Highly Parallel SPARQL Engine for RDF

Fan Feng[1], Weikang Zhou[1], Ding Zhang[1], and Jinhui Pang[2(✉)]

[1] Tianjin University, Tianjin, China
[2] Beijing Institute of Technology, Beijing, China
pangjinhui@bit.edu.cn

Abstract. In this paper, a highly parallel batch processing engine is designed for SPARQL queries. Machine learning algorithms were applied to make time predictions of queries and reasonably group them, and further make reasonable estimates of the memory footprint of the queries to arrange the order of each group of queries. Finally, the query is processed in parallel by introducing pthreads. Based on the above three points, a spall time prediction algorithm was proposed, including data processing, to better deal with batch SPARQL queries, and the introduction of pthread can make our query processing faster. Since data processing was added to query time prediction, the method can be implemented in any set of data-queries. Experiments show that the engine can optimize time and maximize the use of memory when processing batch SPARQL queries.

Keywords: SPARQL · Pthread · Multithreading · Performance prediction

1 Introduction

With the explosion of RDF data, SPARQL query related technologies are also advancing by leaps and bounds. Over the years, many have been devoted to how to build SPARQL storage, and how to effectively answer the graphical query mode expressed in SPARQL. Currently, stand-alone RDF engines, such as RDF3X [1], gStore [4,5], can efficiently execute SPARQL queries serially. Their performance and stability are relatively high. With the continuous development of computer hardware, the memory of a single computer is getting larger and larger, and the parallel processing capability is getting higher and higher, but how to use the parallel capability to handle batch query problems has still not been solved well. But how to use the parallel ability to deal with the batch query problem has not been solved well.

The existing multi-query optimizations [MQO] [12] are based on finding common parts of batch queries and processing them to get answers. It uses query rewrite technology to achieve ideal and consistent MQO performance and guarantee its integrity on different RDF stores. However, the MQO on this SPARQL query is np-hard [11], and the equivalent relationship algebra that has been established between SPARQL and the relationship.

© The Author(s) 2020
J. Zeng et al. (Eds.): ICPCSEE 2020, CCIS 1257, pp. 61–71, 2020.
https://doi.org/10.1007/978-981-15-7981-3_5

Our method chose to avoid this np-hard problem and switched to using pthreads for parallel processing. We use multiple threads, which use the same address space with each other, share most of the data, and start a thread. Space is much smaller than the time it takes to start a process, and the time required to switch between threads is much less than the time required to switch between processes. It is fast to create and destroy. Naturally, this method is extremely useful for batch processing of saprql queries. Big advantage. At the same time, in order to make better use of computer performance, we also performed time prediction [2] and memory estimation on the query to do reasonable grouping and query. In a nutshell, the main contributions of this work are summarized as follows:

- We introduced query-time predictions to initially group batches of questions. For a problem, we use machine learning algorithms to predict its time consumption based on its algebraic tree structure, BGD size, and database format. In this way, a batch of problems is grouped, which is convenient to use multi-threading to process.
- We have also increased memory estimation to avoid unbalanced query memory usage. It is a reasonable method to evaluate the memory usage of the query by its intermediate result size. We make a cardinality estimation of the query and use its triple structure to find its position occupation in the data graph.
- We use pthread technology to process spall queries in batches in parallel. Through multi threading [16], we can improve the query speed very well, and also show exceptionally high practicality in batch processing queries.

2   Related Work

In relational databases [6–10], the problem of multi-query optimization has been well studied. The main idea is to identify common sub-expressions in a batch of queries. Construct a globally optimized query plan by reordering the join sequence and sharing intermediate results in the same set of queries, thereby minimizing the cost of evaluating common subexpressions. The same principle was also applied in [7], which proposed a set of heuristics based on dynamic programming to handle nested sub-expressions. The general MQO problem of a relational database is the NP problem. Even with the heuristic approach, the search space for a single candidate scheme and its combined hybrid scheme (i.e., the global scheme) is usually astronomical. Because of hardness, they proposed some heuristic methods, which have proven to be effective in practice.

All the above work is focused on MQO in a relational situation. To avoid this np problem, we chose to use pthread to process batches of SPARQL in parallel. At the same time, To make better optimization, we introduced query estimation to complete better distribution.

3 Preliminaries

In this section, we will discuss the main ideas for the design of this paper. We avoided MQO processing problems of np-hard difficulty and chose to use threads to process SPARQL in parallel in batches. However, since the usage time and memory usage of a single query are different, and even a large gap may occur, we must allocate it reasonably.

3.1 SPARQL

An RDF (Resource Description Framework) data model is proposed for modeling Web objects, which is part of the development of the Semantic Web. It has been used in various applications. For example, Yago and DBPedia automatically extract facts from Wikipedia and store the points in RDF format to support structured queries on Wikipedia [18]. In general, RDF data can be represented as a collection of triples represented by SPO (*object, attribute, object*). A running example G_D is given in follows.

> Barack_Obama $\langle bornIn \rangle$ Honolulu.
>
> Barack_Obama $\langle won \rangle$ Peace_Nobel_Prize.
>
> Barack_Obama $\langle won \rangle$ Grammy_Award.
>
> Honolulu $\langle locatedIn \rangle$ USA.

SPARQL is a query language and data acquisition protocol developed for RDF. It is defined by the RDF data model developed by the W3C [14] but can be used for any information resource that can be represented by RDF. A SPARQL query graph $G_Q(V_Q, E_Q, L, Vars)$ is a labeled, directed multi-graph where V_Q is the set of query nodes; E_Q is the set of edges connecting nodes in V_Q, L is the set of edge and node labels. An RDF example in triplet form (TTL/N3 format) and an example query (which retrieves all the people who are born in a city that is located in "USA" and who won some prize) is expressed in SPARQL as follows.

> SELECT   ?peron, ?city, ?prize WHERE {
>
> ?person $\langle bornIn \rangle$?city.
>
> ?city $\langle locatedIn \rangle$ USA.
>
> ?person $\langle won \rangle$?prize. }

Therefore, processing the SPARQL query graph G_Q for the RDF data graph G_D can solve the problem of finding the isomorphism of all subgraphs between G_Q and G_D. For example, the method of SPARQL query on our RDF data segment is as follows.

> Barack_Obama, Honolulu, Peace_Nobel_Prize.
>
> Barack_Obama, Honolulu, Grammy_Award.

3.2 Time Prediction

We use machine learning algorithms [2] and innovate existing time prediction algorithms. Wei Emma Zhang proposed the work of predicting SPARQL query execution time using machine learning techniques. In work, multiple regression using support vector regression (SVR) was adopted. The evaluation was performed on the open-source triple storage Jena TDB for benchmark queries. Feature models were extracted based on the graph edit distance (GED) [13] between each training query, and feature extraction was also performed on the query's algebraic tree structure [13]. Further, improve accuracy. But in fact, we found that its method does not perform relevant feature extraction on the data set. We believe that data also has a great influence on the query. The simplest example is that the larger the data, the slower the query will become.

In order to effectively predict SPARQL query time, we borrowed the ideas of Wei Emma Zhang. Still, we introduced a new feature vector of the data set to further introduce the characteristics of the data into the time prediction.

3.3 Memory Estimation and Thread Acceleration

We consider that the query memory usage of SPARQL lies in the storage of the intermediate results, so we have designed a cardinality estimation algorithm, which builds a b-tree index on the RDF data and maps the triples of the query to get the approximate answer range. Further, the connection cost of the triples is obtained to obtain the estimated cardinality of the query, which represents the memory usage of the query. Through the above two steps, the query is grouped, and the dynamic programming method is used for reasonable thread usage allocation to ensure that the query time is minimized and the occupied memory is minimized.

4 Time Prediction

We used Wei Emma Zhang to use machine learning technology to predict SPARQL query execution time. The method proposed that the prediction time mainly depends on the features in the training set, so how to extract the problem information is an important breakthrough point. The vector of data information is also introduced to improve the accuracy and applicability of problem prediction. In this study, we introduce algebraic features and BGP features, which are obtained by parsing query text (see Sects. 4.1 and 4.2). By applying algebraic and BGP feature-based selection algorithms to generate hybrid features (see Sect. 4.3), the dataset analyzes depth-width features (see Sect. 4.4).

4.1 Algebraic Features

Algebra refers to operators such as opt, union, etc. Algebra can be represented and extracted in the form of a tree:

Definition 1. (Algebra Tree) Given a SPARQL query Q, the algebra tree TAlgebra (Q) is a tree where the leaves are BGP, and the nodes are the algebraic operators of the hierarchical representation. The parent of each node is the parent operator of the current operator.

We obtain the SPARQL algebra tree and then traverse the tree to build an algebraic set by recording the occurrence and hierarchical information of each algebraic operator.

Definition 2. (Algebra Set) Given an algebra tree represented as $T_{Algebra}$ (Q), the algebra set is a set of tuples (opt$_i$, c$_i$, maxh$_i$, minh$_i$), where opt$_i$ is the name of the operator and c_i is the number of occurrences in $T_{Algebra}$ (Q) Opti, maxh$_i$ and minh$_i$ are the maximum and minimum heights of opt$_i$ in $T_{Algebra}$ (Q).

4.2   BGP Feature

Algebraic features can only represent part of the query information, and the specific structural features are still not obvious enough. Therefore we introduce the graph structure of BGP as a supplement. Obtain the features of the BGP structure and convert it into a vector representation.

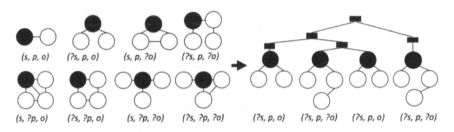

Fig. 1. Mapping triple patterns to graphs. Left: eight types of triple patterns are mapped to eight structurally different graphs. Right: mapping example query in Figure ?? to a graph

We map all eight types of ternary patterns to eight different graphs on the structure, and we convert the SPARQL in Figure ?? to Fig. 1 as an example. The black rectangles in the figure are the connection nodes. After converting the query into a graph representation, we use the graph edit distance as its BGP feature vector. The graph edit distance between two graphs is the minimum amount of editing operations required to convert one graph to another (i.e., delete, insert And replace nodes and edges). Figure 2 shows a GED calculation process. We calculated the target query and 18 query templates separately. These 18 query templates were filtered from the DBPSB benchmark [15] test. So we can get an eighteen-dimensional vector.

Fig. 2. Graph edit path

4.3   Data Feature

In view of the acquisition of query feature vectors, we have also derived the way of extracting data information. An RDF data set, which is linked in the form of a graph, must also have class and instance information, so we separate the class and instance from it.

Definition 3. (Data Information) Given a data set G, its class information, including the number of child classes and parent classes of each class, is used as the width of the data set. The information of the instance, including all the attributes of the instance, and the number of objects, is used as the depth of the data set.

Width calculation: We traverse the entire graph, and record the class when it meets the definition of the class, and traverse its parent and child classes for this class. If there is no parent and child class, the width of the graph is increased by one. If there is a parent and child class, the width of the graph is increased by Third, and record this category to prevent secondary statistics. Depth calculation, for a data set, traverse the graph, using a recursive algorithm, if the depth is already available, directly return the value depth of the instance plus 1, query the instance to get all its objects; traverse the object, perform the object Processing and then following the recursive algorithm. Each time an object is traversed, the depth is added to the depth of the object.

5   Memory Estimation

We introduce the estimate of the cardinality of the query to determine how much intermediate results the query has, and then determine its memory footprint. The challenge we face is how to estimate the cardinality of the query [17]. We graph the query and map each triplet pattern in BGP to node v. If there is a common variable between the triplet patterns, add an edge e between the nodes. After processing, you will get the SPARQL The connection graph, which is an undirected connected graph, is numbered according to the triples in the graph pattern entered by the user.

1. **Node estimation.** Build a B-tree index of RDF data. Use AllGroGraph to build three indexing strategies: *spo*, *pos*, and *osp*. These three include six patterns of triples, which can also save space. For example, in the triple pattern (*?spo*), if the subject's hash value range is set to [0-0xff], the hash

range of the entire triple can be determined. and set to [n1, n2]. For this value range in the B-tree, The query above calculates the number of results N with key values between n1 and n2, and the data in the range contains the results of triple matching. The estimate of its cardinality is:

$$cart(t) = 3 * N/4 \tag{1}$$

2. **Edge weight estimation.** The weights for the edges are calculated using an estimate of the size of the result set of the join operation on the node words. For the estimation of the size of the result set of the join operation, the estimated value of the cardinality of the node is used, and the formula is as follows:

$$T(R \bowtie S) = s * |R| * |S| \tag{2}$$

Among them, R and S respectively represent the result set after pattern matching of triples, and — R —, — S —, respectively represent the cardinalities of their triples results. s indicates the selection rate, which is divided into the following situations:

- No variables common to R and S. Then s is a constant coefficient of 1.

$$T(R \bowtie S) = |R| * |S| \tag{3}$$

- When left and right operands have common variables, suppose the set of public variables is the intersection of V_R and V_S:V_{RS}. $W(R, V_{RS})$ represents the number of different variables on R and V_{RS}. $W(S, V_{RS})$ represents the number of different variables on S and V_{RS}, take the maximum value between these two as *max*. And assuming that the variables are uniformly distributed, the formula becomes:

$$T(R \bowtie S) = \frac{|R| * |S|}{max} \tag{4}$$

6 Pthread Allocation

One of the reasons for using multi-threading is that it is a very "frugal" way of multitasking compared to processes. We know that starting a new process must allocate it an independent address space and create numerous data tables to maintain its code, stack, and data segments. This is an "expensive" multitasking way of working. And multiple threads running in a process use the same address space with each other and share most of the data. The space taken to start a thread is much less than the space taken to start a process. Moreover, the threads switch between each other. The time required is also much less than the time required for interprocess switching. According to statistics, in general, the cost of a process is about 30 times the cost of a thread. Of course, on a specific system, this data may vary greatly.

In our method, a batch of problems and their corresponding data are introduced. The problems are grouped by time estimation models into m groups, and the m groups of problems are processed in parallel using multiple threads while ensuring that the total time spent in each group reaches An average. Before querying, first, make a memory estimation of the problems in each group, and sort the problems in each group to ensure that when one problem is taken out from m groups for processing, the memory occupation of these m problems can be processed in this batch of problems. The total occupancy in the medium reaches an average.

7   Evaluation

In this section, we provide an assessment of our proposed method. We first introduce the experimental setup. We then report the results of various evaluations performed on different data sets and query sets.

7.1   Setup

Data Preparation. Data sets and queries use dbpedia3.9 data and query records. We first randomly selected ten datasets containing 100,000 triples from the DBPedia dataset, and then USEWOD 2014 provided log files for querying using the DPPedia SPARQL query engine. We decoded the log files and selected the A total of 10,000 questions were found and correctly returned in the selected data set. From this, we have ten pairs of data, including questions and answers.

System. The backing system of our local triple store was Jena TDB, installed on 64-bit Ubuntu 16.04 Linux operation system with 32 GB RAM and 16 CPU cores.

Evaluation Metric. For all experiments, we measure the number of optimized queries and their end-to-end evaluation time, including query rewrite, execution, and result distribution. We compare our Pthread algorithm with an evaluation without any optimization (i.e., No-MQO) and an algorithm with MQO (including various existing optimizations). At the same time, in order to ensure the effective use of memory, we have set up indicators that specifically detect memory usage to ensure that our memory allocation is indeed effective.

7.2   Experimental Results

We took ten sets of queries and data to train the time prediction model, and each group randomly extracted 600 queries, parsed out the query features and data features to build a prediction model, and used the remaining 400 queries for model detection. We directly give the results of the time prediction model trained by machine learning, as shown in Fig. 3. From this figure, we can see that we have made a good prediction of the query time, and this predicted time can be directly processed. We choose structure-based features that can be obtained directly from

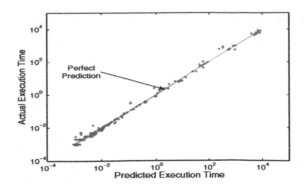

Fig. 3. Time predict effect

the query text. As stated in many works, similarly structured queries can have huge performance gaps due to the irregular distribution of values in the query data. However, based on our actual experience in this work, we have observed that although it may cause prediction distortion, based on the limited features we can obtain, the error rate is acceptable.

After the prediction model is processed, we randomly select a set of 1000 query data randomly, use the prediction model to group it, and then use memory estimation to sort within the group. Then use Pthread multi-thread batch processing to query our query performance. A comparison with other methods is shown in Table 1.

Table 1. Experimental results

Model	Processing time (s)
MQO	47
MQO-S	42
No-MQO	67
Pthread	**38**

8 Conclusion

We investigated batch query optimization in the context of RDF and SPARQL. We refer to the multi-query optimization processing method and choose to use Pthread to avoid its difficulties and handle these queries from a new perspective. Our method also introduces time prediction and memory estimation to make our query process faster, and more accurate and efficient. In addition, our technology is storage-independent, so it can be deployed on top of any RDF storage without modifying the query optimizer. The experimental results show that our method is effective.

Acknowledgments. This work is supported by Big Data Research Foundation of PICC.

References

1. Neumann, T., Weikum, G.: The RDF-3X engine for scalable management of RDF data. VLDB J. **19**(1), 91–113 (2010)
2. Zhang, W.E., Sheng, Q.Z., Qin, Y., Taylor, K., Yao, L.: Learning-based SPARQL query performance modeling and prediction. World Wide Web **21**(4), 1015–1035 (2017). https://doi.org/10.1007/s11280-017-0498-1
3. Le, W., Kementsietsidis, A., Duan, S., et al.: Scalable multi-query optimization for SPARQL. In: 2012 IEEE 28th International Conference on Data Engineering. IEEE Computer Society (2012)
4. Zou, L., Mo, J., Chen, L.: gStore: answering SPARQL queries via subgraph matching. Proc. VLDB Endow. **4**(8), 482–493 (2011)
5. Zou, L., Oezsu, M.T., Chen, L., et al.: gStore: a graph-based SPARQL query engine. VLDB J. **23**(4), 565–590 (2014)
6. Park, J., Segev, A.: Using common subexpressions to optimize multiple queries. In: International Conference on Data Engineering. IEEE (1988)
7. Roy, P., Seshadri, S., Sudarshan, S., et al.: Efficient and extensible algorithms for multi query optimization. ACM SIGMOD Rec. **29**(2), 249–260 (2000)
8. Sellis, T.K.: Multiple-query optimization. ACM Trans. Database Syst. **13**(1), 23–52 (1988)
9. Shim, K., Sellis, T.K., Nau, D.: Improvements on a heuristic algorithm for multiple-query optimization. Data Knowl. Eng. **12**(2), 197–222 (1994)
10. Zhao, Y., Deshpande, P., Naughton, J.F., Shukla, A.: Simultaneous optimization and evaluation of multiple dimensional queries. In: SIGMOD (1998)
11. Sellis, T., Ghosh, S.: On the multiple-query optimization problem. IEEE Trans. Knowl. Data Eng. **2**(2), 262–266 (1990)
12. Wang, M., Fu, H., Xu, F.: RDF multi-query optimization algorithm for query rewriting using common subgraphs. In: The 3rd International Conference (2019)
13. Hasan, R.: Predicting SPARQL query performance and explaining linked data. In: Presutti, V., d'Amato, C., Gandon, F., d'Aquin, M., Staab, S., Tordai, A. (eds.) ESWC 2014. LNCS, vol. 8465, pp. 795–805. Springer, Cham (2014). https://doi.org/10.1007/978-3-319-07443-6_53
14. Servidor web: World Wide Web Consortium (W3C) (2010)
15. Morsey, M., Lehmann, J., Auer, S., Ngomo, A.N.: Usage-centric benchmarking of RDF triple stores. In: Proceedings of the 26th AAAI Conference on Artificial Intelligence, Toronto, Canada (2012)
16. Filip, Z.: Parallel SPARQL query processing using bobox. Int. J. Adv. Intell. Syst. **5**, 302–314 (2012)
17. Gubichev, A., Neumann, T.: Exploiting the query structure for efficient join ordering in SPARQL queries. In: Proceedings of the 17th International Conference on Extending Database Technology (EDBT 2014), Athens, Greece, pp. 439–450 (2014)
18. Suchanek, F.M., Kasneci, G., Weikum, G.: Yago: a core of semantic knowledge. In: WWW, pp. 697–706 (2007)

Identifying Propagation Source in Temporal Networks Based on Label Propagation

Lilin Fan[1,2], Bingjie Li[1,2], Dong Liu[1,2(✉)], Huanhuan Dai[1,2], and Yan Ru[1,2]

[1] School of Computer and Information Engineering, Henan Normal University, Xinxiang, Henan, China
liudong@htu.cn
[2] Big Data Engineering Laboratory for Teaching Resources and Assessment of Education Quality, Xinxiang, Henan, China

Abstract. The spread of rumors and diseases threatens the development of society, it is of great practical significance to locate propagation source quickly and accurately when rumors or epidemic outbreaks occur. However, the topological structure of online social network changes with time, which makes it very difficult to locate the propagation source. There are few studies focus on propagation source identification in dynamic networks. However, it is usually necessary to know the propagation model in advance. In this paper the label propagation algorithm is proposed to locate propagation source in temporal network. Then the propagation source was identified by hierarchical processing of dynamic networks and label propagation backwards without any underlying information dissemination model. Different propagation models were applied for comparative experiments on static and dynamic networks. Experimental results verify the effectiveness of the algorithm on temporal networks.

Keywords: Source identification · Propagation model · Label propagation

1 Introduction

With the rapid development of social networks represented by Facebook, twitter, microblog and We-Chat, social network has become a very important information dissemination platform. At the same time users share information, they also have to face the adverse effects of rumors and other harmful information. In addition, respiratory infectious diseases such as tuberculosis and 2019-nCoV are highly infectious and seriously threaten human health [1–3]. Therefore, it is of great practical significance to locate source quickly and accurately when rumors or epidemic outbreaks occur.

The spread of rumors or diseases depends on specific networks. For example, the spread of rumors depends on social networks, and the spread of diseases depends on human interaction networks. If the infectious source of the disease is determined in a short period of time, it can better control the spread of the whole disease, narrow its transmission range, and reduce the threat of infectious diseases to people's life safety. If we can quickly locate the source of rumor in the critical period after spreading rumor, we can reduce the negative impact of rumor on politics, people's vital interests and

J. Zeng et al. (Eds.): ICPCSEE 2020, CCIS 1257, pp. 72–88, 2020.
https://doi.org/10.1007/978-981-15-7981-3_6

social stability. Generally, these networks have the characteristics of large scale and complex connections between nodes. Therefore, how to locate a source of infection or rumor in network has become a very meaningful and challenging problem.

The methods of propagation source location in static network are mainly divided into four categories: Propagation source centrality algorithm [4, 5, 9, 10], Belief propagation algorithm [11, 12], Monte Carlo algorithm [6], and Net-Sleuth algorithm [6]. The basic idea of propagation source centrality algorithm is directly obtaining central nodes as infection propagation source; Belief propagation algorithm identify the propagation source through message-passing on a factor graph; Monte Carlo algorithm identifies sources based on simulating the spread of disease; And the feature of the Net-Sleuth algorithm based on the minimum description length is that it does not need the number of sources. The above algorithms are all directed to solve the problem of identifying propagation source in static network. However, most real networks are dynamic and temporal.

In recent years, researchers have begun to focus on the problem of identifying propagation source in dynamic network. The algorithm based on the shortest temporal path effectively identifies the propagation source in temporal network [14]; A method of locating propagation sources based on reverse dissemination strategy was proposed by Jiang et al. [7]; The algorithm based on the time information of all nodes by Antulov-fantulin et al. [6]; and sparse signal reconstruction theories has been applied to locate propagation sources in dynamic network [19]. Although the above algorithm effectively solves the problem of source identification on dynamic networks, the pre-condition is to know the underlying propagation model in advance. Obviously, it is difficult to obtain model information in practice, and it is hard to obtain the true value of parameters in the preset basic propagation model.

As far as we know, only a few source identification studies have been carried out without any underlying information propagation model. The label propagation algorithm (LPSI) [8] is the first attempt to solve this problem, which is based on the idea of source prominence, namely the nodes surrounded by larger proportions of infected nodes are more likely to be source ones. The main idea of this algorithm is to identify the propagation source by propagating integer labels in the network. It effectively solves the problem that the propagation model cannot be known in advance in practice. However, it is only studied on static networks.

In this paper, we uses the label propagation algorithm to locate the source on the temporal network, the propagation source is identified by hierarchical processing of dynamic networks and iterative propagation of the label. Firstly, dividing dynamic network into multiple single layer static networks. Secondly, label propagation backwards from the last layer of the network to the first layer. Finally, the propagation source is derived based on the node labels of the first layer network. The major contributions of this paper can be summarized as follows:

- use the label propagation algorithm to locate the propagation source on the temporal network;
- identify propagation sources without any underlying propagation model information;
- comparative experiments on static and dynamic networks using different propagation models

The rest of this paper is organized as follows. In Sect. 2, we introduce the related work of propagation source identification in static and temporal networks. Section 3 gives the introduction of label propagation method. Section 4 analyzes the algorithm in detail. Section 5 compares the accuracy of locating propagation source on static network and temporal network under different propagation models. Section 6 summarizes this paper.

2 Related Work

In the past few years, researchers propose a series of methods to identify propagation sources in network. According to the time state of the network, they are divided into propagation source identification methods in static and dynamic networks.

For static networks, Shah et al. [4] put forward the first algorithm to solve the source identification problem for the SI propagation model on the regular tree network. Luo et al. [5] used Jordan center estimation (JCE) algorithm to solve the problem of single source estimation on tree network through Jordan infection center. Fioriti et al. [9] proposed a propagation source identification algorithm called DA by calculating the dynamic age of each node in the infection graph. Comin et al. [10] studied an unbiased betweenness algorithm for computing the centrality of unbiased mediations of all nodes in an infection graph. Lokhov et al. [11] established a dynamic message-passing algorithm for single source identification. Altarelli et al. [12] put forward a belief propagation algorithm (BP) based on factor graph on SIR propagation model. Antulov-fantulin et al. [6] proposed a source location algorithm based on Monte Carlo with single source and known states of all nodes. Prakash et al. [6, 13] recommended the Net-Sleuth algorithm based on the minimum description length in the source identification problem of multi-source and SI propagation model. Wang et al. [8] proposed a source identification method based on label propagation, which is the first attempt to solve the unknown problem of the propagation model in reality.

For temporal networks, Antulov-fantulin et al. [6] proposed a method to locate the source based on the time information of all nodes. Jiang et al. [7] established a method of locating propagation sources based on reverse dissemination strategy. Huang et al. [14] developed a reverse diffusion method to locate the propagation source by calculating the shortest time distance between nodes. Hu et al. [19] a general framework locate the source by bridging structural observability and sparse signal reconstruction theories.

The research goal of this paper is to locate the propagation sources on the dynamic network. However, the current methods of source identification on the dynamic network depend on the specific propagation model, which is not in line with the characteristics of real dissemination. Therefore, this paper presents a dynamic network source identification method based on label propagation. Firstly, dividing dynamic network into multiple single layer static networks. Secondly, label propagation backwards from the last layer of the network to the first layer. Finally, the propagation source is derived based on the node labels of the first layer network.

3 Methodology

The problem statement and algorithm description are shown in this section. Frequently used notations are summarized in Table 1.

Table 1. Notation summarization.

Notation Definition	Definition
G	The topological graph of temporal network
N	Number of nodes
L	Number of network layers
$T = (t_1, t_2, \ldots, t_L)$	Time series of temporal network G
$V = (v_1, v_2, \ldots, v_N)$	Set of vertexes in G
$E = (E_1, E_2, \ldots, E_L)$	Set of edge sets of L-layer temporal network
$Y = (Y_1, Y_2, \ldots, Y_L)$	Set of infection node vectors of L-layer temporal network
$Y_m = (y_1^m, y_2^m, \ldots, y_N^m)$	Infection node vector at layer m in G, $m \in \{1, 2, \ldots, L\}$
A^m	The adjacency matrix at layer m in G, $m \in \{1, 2, \ldots, L\}$
P^m	The label propagation probability at layer m in G, $m \in \{1, 2, \ldots, L\}$
$g_i^t\{m\}$	i-th node label value of layer m at iteration t, $m \in \{1, 2, \ldots, L\}$
S_m	Source estimate set at layer m in G, $m \in \{1, 2, \ldots, L\}$
s^*	The propagate source

3.1 Problem Statement

The problem of dynamic network source identification studied in this paper can be formulated as follows: Given a dynamic network $G = (V, E, T)$, According to the time series $T = (t_1, t_2, \ldots, t_L)$, the dynamic network is divided into m-layer static network. A node set $V = (v_1, v_2, \ldots, v_N)$ with a set E_m at layer m $\in \{1, 2, \ldots, L\}$. An infection node vector $Y_m = (y_1^m, y_2^m, \ldots, y_N^m)$ at layer m where $y_i^m = 1$ indicates node I is infected and $y_i^m = -1$ otherwise, the goal is to find the original infection source $s^* \subset V$. Because the algorithm used in this paper does not depend on the specific propagation model, it is not necessary to make assumptions on the propagation model.

3.2 Label Propagation Algorithm in Temporal Network

We formally present our method in Algorithm 1. One of the important characteristics of the label propagation algorithm is that it can identify the source independent of the propagation model. In this paper, the label propagation algorithm is used to identify the propagation source in dynamic network, which can be realized through the following three steps.

Assign Labels. We divide the dynamic network into L-layer static network. Known the infection node vector $Y = (Y_1, Y_2, \ldots, Y_L)$ when propagating to the last layer network. Notably, the underlying information propagation model for generating Y_L is unknown in our method. Based to the infected node vector Y_L to assign positive label (+1) and negative label (−1) to infected and uninfected nodes in the last layer network, and put the assigned label into $g_i^0\{L\}$.

Label Propagation. We need to obtain the adjacency matrix A^m at layer m, where $A_{ij}^m = 1$ if there is a connection between nodes i and j, otherwise $A_{ij}^m = 0$. Generating the label propagation probability P^m at layer m from node j to node i based on adjacency matrix. We start from the last network and reversely execute the label propagation algorithm. For each layer network, each node obtains a part of label information from its neighborhood, and retains some label information in its initial state. Therefore, the label value of node i of layer m at iteration $t + 1$ becomes:

$$g_i^{t+1}\{m\} = a \sum_{j:j\in N(i)} P_{ij}^m g_j^t\{m\} + (1 - a)y_i^m \tag{1}$$

where matrix $P^m = D^{(-1/2)}A^m D^{(-1/2)}$, matrix A^m is the adjacency matrix, D is a diagonal matrix with its (i, i)-element equal to the sum of the i-th row of A, and $0 < a < 1$ is the fraction of label information that node i gets from its neighbors. N(i) represents the set of neighbors of node i.

In addition to the first layer network, each layer carries out label propagation with iteration $t = 1$. For the label of nodes in (m − 1)-layer network, if node i belongs to the set of source estimation obtained from m-layer network, assign the positive label (+1), it inherits the label of m-layer network otherwise. Namely,

$$g_i^0\{m-1\} = \begin{cases} 1, & i \in S_m \\ g_i^0\{m\}, & otherwise \end{cases} \tag{2}$$

For each layer network, it is identified as a suspicious propagation source if a node i meets the following two conditions:

- Node i is initially an infected node;
- The final label value of node i is larger than those of its neighbors.

The first condition of the algorithm is to avoid the interference of uninfected nodes on the accuracy of source identification. The second condition is based on the principle of source prominence [10], namely the nodes surrounded by larger proportions of infected nodes are more likely to be propagation source.

Source Identification. When the label propagates back to the first layer network, the iterative propagation stops until the node label reaches convergence by Eq. 1. Here, "convergence" means that the label value of a node does not change in several consecutive iterations of the label propagation. It is worth mentioning that for a single-layer static network, only the node labels need to be iterated to convergence.

Finally, obtain the node with the highest label value from the suspect node as the propagation source s^*.

Algorithm 1 Input Generation Algorithm

Input: The infected temporal network G = (V, E, T), parameter a ; The initial infection node vector Y_L in the last layer network.

Output: The propagation source s^*

1: for each layer network do

2:    Form the weight matrix A^m defined by $A^m_{ij} = 1$ if there exists an edge connecting nodes i and j;

3:    Construct the matrix $P^m = D^{(-1/2)}A^m D^{(-1/2)}$, where D is a diagonal matrix with its (i, i) -element equal to the sum of the i -th row of Am;

4: endfor

5: $g_i^0\{L\} \leftarrow Y_L$

6: for m from L to 1 do

7:   if m!=1 then

8:     for each node i do

9:       $g_i^1\{m\} = a \sum_{j:j\in N(i)} P^m_{ij} g_j^0\{m\} + (1-a)y_i^m$

10:     endfor

11:     for each node i do

12:       if $g_i^1 >$ all i's neighbors' g_i^1 then

13:         $S_m = S_m \cup i$

14:       endif

15:     endfor

16:     $g_i^0\{m-1\} = \begin{cases} 1, & i \in S_m \\ g_i^0\{m\}, & otherwise \end{cases}$

17:   else

18:     while g_i^t dose not reach the convergence g^* do

19:       for each node i do

20:         $g_i^{t+1}\{m\} = a \sum_{j:j\in N(i)} P^m_{ij} g_j^t\{m\} + (1-a)y_i^m$

21:       endfor

22:       $t = t + 1$

23:     endwhile

24:   endif

25: endfor

26: for the first layer network (m=1) do

27:   for each node i do

28:     if $g_i^t\{1\} >$ all i's neighbors' g_i^t then

29:       $S_1 = S_1 \cup i$

30:     endif

31:   endfor

32: endfor

33: select node i with $max(g_{i\in S_1}^t\{1\})$

34: $s^* \leftarrow i$

35: return s^*

3.3   Algorithm Diagram

In order to further understand the algorithm in the section, Fig. 1 shows the process of propagation source identification in multi-layer dynamic network. The data set used in the example is a L-layer small world network with N = 10 nodes and average degree of $\langle k \rangle = 4$. Randomly select a source node to propagate through the SI model. We divide the dynamic network into L-layer static networks by chronological order. As shown in Algorithm 1, the green node is the suspect sources of each layer, the white point is the uninfected nodes and the orange one is the propagation source node. First, a set of source estimation nodes is obtained by Eq. 1 on the last layer network. Then, transfer labels to node i between layers in reverse by Eq. 2 and label propagation on the new network layer. Finally, in the first layer network, the label propagates stops until the label converges. The node with the largest label value is output as the propagation source.

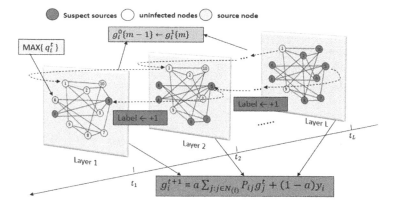

Fig. 1. The process of propagation source identification in multi-layer dynamic network.

4   Algorithm Analysis

Theoretical analysis [8] and time complexity analysis of the algorithm are shown in this section.

4.1   Convergence Analysis

The equation for calculating the label of each node at iteration t can be transformed into another writing method:

$$g^{t+1} = aPg^t + (1-a)y \tag{3}$$

By the initial condition that $g^0 = y$, we get:

$$g^t = (aP)^t + (1 - a) \sum_{i=0}^{t-1} (aP)^i y \tag{4}$$

According to $\lim_{t \to \infty} (aP)^t = 0$ and $\lim_{t \to \infty} \sum_{i=0}^{t-1} (aP)^i = (I - aP)^{-1}$, where I is an $n \times n$ identity matrix. Therefore, the label value will converge to:

$$g^* = (1 - a)(I - aP)^{-1} y \tag{5}$$

It can be seen that the label value will reach convergence after iterative propagation finally.

4.2 Source Prominence

Source Prominence, namely the node surrounded by larger proportions of infected nodes are more likely to be source ones. We can understand this idea in two ways, On the one hand, at the edge of the infected area, nodes tend to have fewer infected neighbors. On the other hand, in the center of the infected area, nodes tend to have more infected neighbors. This property has nothing to do with the propagation model, so we can apply this idea to source identification. The experimental results of this paper also verify this idea.

4.3 Time Complexity

For a single-layer network label propagation, the time complexity is $O(t * J * N)$, where J is the average number of neighbors per node and t is the number of iteration. Therefore, For a dynamic network, the time complexity is $O(L * t * J * N)$, where L is the number of network layer.

5 Experiments

5.1 Experiments Setup

Datasets. In this paper, two kinds of networks are selected for experiments, the first type are small-world networks, random networks, and scale-free networks. Number of nodes $N = 100, 500$, average degree $\langle k \rangle = 6, 8, 10$. The second type is the benchmark adopted by Lin et al. in [15], this data set [16] is a dynamic network of a fixed number of communities (named SYN-FIX), analogously to the classical benchmark proposed by Girvan and Newman in [17]. The network consists of 128 nodes divided into four communities of 32 nodes each.

Propagation Models. Existing propagation models can be divided into infection models and influence models. In order to verify that the algorithm can locate rumor sources without any underlying propagation model information, the experimental part of this paper mainly uses three models, namely SI model, SIR model and LT model [20].

For SI model, as shown in Fig. 2(a), each node has only two possible states: S and I. At the initial time of propagation, only one or a few nodes in the network are in state I, while other nodes are in state S. At each time step of the propagation process, any nodes in state I infects each neighbor in state S with the same probability p.

For SIR model, as shown in Fig. 2(b), each node has three possible states: susceptible (S), infected (I), and recovered (R). Every infected node tries to infect its susceptible (uninfected) neighbors independently with probability p. The infected node changes from probability q to R state, and the node will always be in R state, and will not be infected. The infection probability p is chosen uniformly from (0, 1) for the SI model and SIR model. For the SIR model, an extra recovery probability q is chosen uniformly from (0, p).

For LT model, there are only two possible states for each node: active state and inactive state, in which the active state represents that a node receives information in the propagation process, otherwise it is in the inactive state. As shown in Fig. 2(c), each node in the inactive state receives information only when the number of nodes in the active state around it exceeds a certain threshold. Until the sum of the influence of any active node in the network cannot activate the neighbor node in the inactive state, the propagation process ends. Assuming that nodes u, v have degrees du and dv, the infection weight between edges (u, v) is $1/dv$, and the infect weight between edges (v, u) is $1/du$. In addition, the threshold of each node is uniformly selected from a small interval [0, 0.5], in order to infect most of the network.

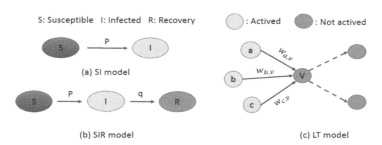

Fig. 2. Propagation Models. (a) SI model with the infection probability p; (b) SIR model with the infection probability p and the recovery probability q; (c) LT Models with a threshold θ.

Evaluation Metrics. We use F-score to evaluate the performance of algorithm.

F-score is a suitable metric for measuring multi-labeled classification problems, since it is a combination of precision and recall. The definition of F-score is following:

$$F - score = \frac{(1 + b^2) * Precision * Recall}{(b^2 * Precision) + Recall} \tag{6}$$

In this article, setting parameter b to 1 means that accuracy and recall are treated equally.

5.2 Results

In order to verify the effectiveness of the algorithm in locating propagation sources on dynamic networks, 100 independent experiments are carried out on four kinds of networks. We set the parameter a = 0.5. In each simulation, randomly select a node as the propagation source, and then use SI model, SIR model and LT model to simulate the propagation. In different propagation models, this paper studies the comparison of F-score between dynamic network and static network under different average degree ($\langle k \rangle = 6, 8, 10$) and different network layers (L = 4, 6, 8). In order to verify the effectiveness of the algorithm when the network scale is expanded, we increase the number of nodes to 500 for simulation experiments. In addition, the parameter a in Eq. 1 controls the influence of neighbor nodes in the process of label propagation. Therefore, this paper studies the optimal value of a by analyzing the influence of parameter a on the experimental results.

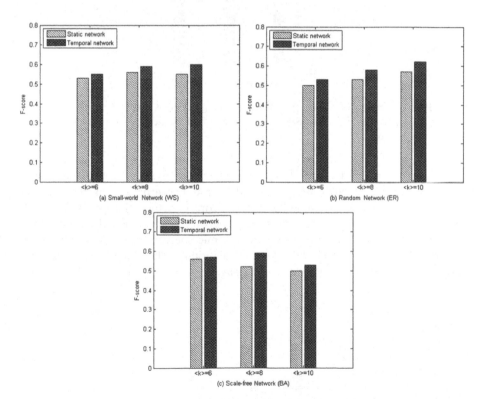

Fig. 3. The comparison of F-score of different average degrees ($\langle k \rangle = 6, 8, 10$) in dynamic networks and static networks under SI model. (a) Small-world network; (b) Random network; (c) Scale-free network.

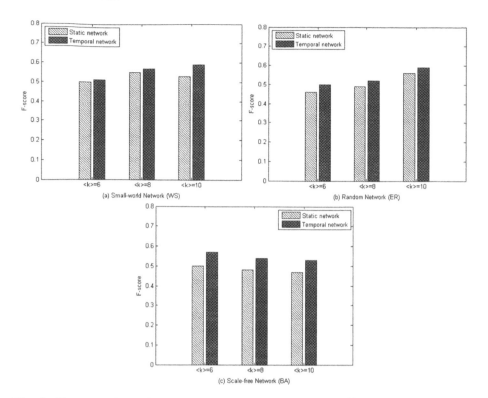

Fig. 4. The comparison of F-score of different average degrees ($\langle k \rangle = 6, 8, 10$) in dynamic networks and static networks under SIR model. (a) Small-world network; (b) Random network; (c) Scale-free network.

The comparison of F-score of different degrees of dynamic network and static network source in SI model, SIR model and LT model can be seen in Fig. 3, Fig. 4, Fig. 5. Figure 6 shows the comparison of F-score of temporal network and static network source under different propagation models for the data set called SYN-FIX. It can be seen that the location accuracy of dynamic source location algorithm is significantly higher than that of static location algorithm. The result proves the effectiveness of the algorithm to identify the propagation source on a temporal network.

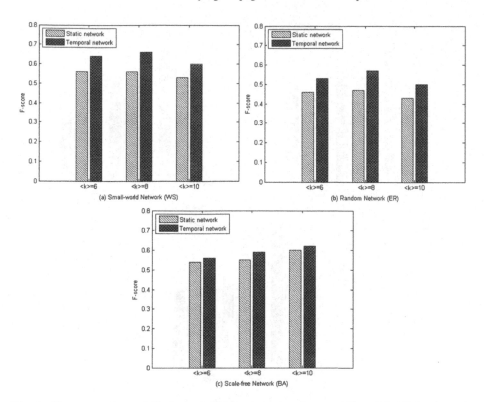

Fig. 5. The comparison of F-score of different average degrees ($\langle k \rangle = 6, 8, 10$) in dynamic networks and static networks under LT model. (a) Small-world network; (b) Random network; (c) Scale-free network.

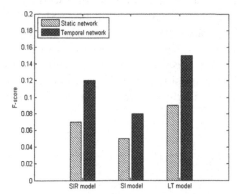

Fig. 6. The comparison of F-score of different propagation models in dynamic network and static network for SYN-FIX.

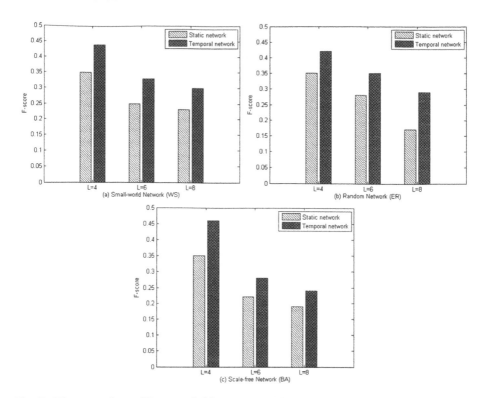

Fig. 7. The comparison of F-score of different network layers (L = 4, 6, 8) in dynamic networks and static networks under SI model. (a) Small-world network; (b) Random network; (c) Scale-free network.

Figure 7, Fig. 8, Fig. 9 shows the comparison of F-score of different network layers of dynamic network and static network source under SI model, SIR model and LT model respectively. The location probability of dynamic source location algorithm is significantly higher than that of static location algorithm. In addition, it can be seen that the accuracy of source location decreases with the increase of network layers. The main reason is that when the transmission time is long, information is widely spread in the network, and the topology will change greatly, which will reduce the accuracy.

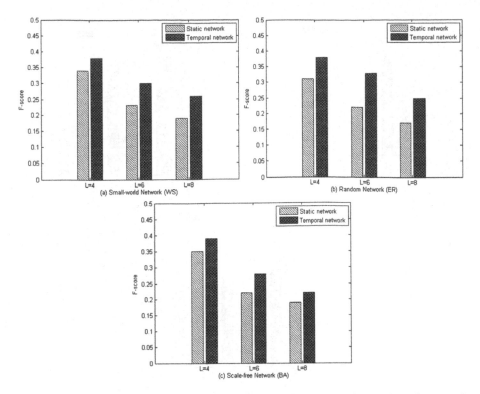

Fig. 8. The comparison of F-score of different network layers (L = 4, 6, 8) in dynamic networks and static networks under SIR model. (a) Small-world network; (b) Random network; (c) Scale-free network.

Figure 10 shows the comparison of F-score of different propagation models of dynamic networks and static networks with N = 500. The results show that when the network scale is expanded, the algorithm is still applicable.

Figure 11 is the result of the change of F-score on small world network with parameter a under SI model, SIR model and LT model. It is observed that the F-score decreases when parameter a approaches 0 or 1, regardless of the propagation model setting. On the other hand, the performances with a ∈ [0.2, 0.6] are always stable and preferred. The result confirm with the intuition that we should consider both the initial infection status and the effects from neighbors for source detection.

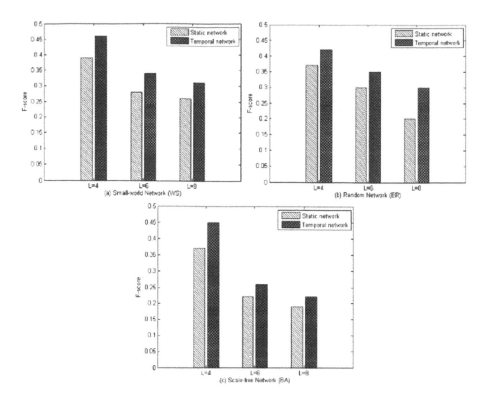

Fig. 9. The comparison of F-score of different network layers (L = 4, 6, 8) in dynamic networks and static networks under LT model. (a) Small-world network; (b) Random network; (c) Scale-free network.

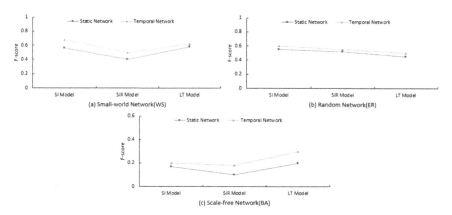

Fig. 10. The comparison of F-score of different propagation models of dynamic network and static network source with N = 500.

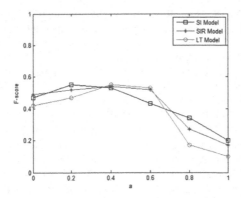

Fig. 11. The change of F-score with parameter a.

6 Conclusion

Locating the source in temporal network is a practical research issue. In this paper, label propagation algorithm is used to locate the source in dynamic network, and the simulation results on static and temporal network are given under SI model, SIR model and LT model. The experimental results show that the location accuracy of the temporal network is higher than that of the static network without considering the dynamic topology changes. In addition, the number of layers of temporal network has a certain impact on the accuracy of source location.

The algorithm in this paper has a premise that dynamic social networks only consider edge changes. However, for the online social network represented by microblog, it is in line with the actual situation in a short time. For the case of node and edge changing at the same time, it will be discussed in the future work.

References

1. Williams, B.G., Granich, R., Chauhan, L.S., et al.: The impact of HIV/AIDS on the control of tuberculosis in India. Proc. Nat. Acad. Sci. U.S.A. **102**(27), 9619–9624 (2005)
2. Smith, R.D.: Responding to global infectious disease outbreaks: lessons from SARS on the role of risk perception, communication and management. Soc. Sci. Med. **63**(12), 3113–3123 (2006)
3. World Health Organization: Global tuberculosis report 2013. World Health Organization, Geneva, Switzerland (2013)
4. Shah, D., Zaman, T.: Detecting sources of computer viruses in networks: theory and experiment. In: Proceedings of the ACM SIGMETRICS International Conference on Measurement and Modeling of Computer Systems, New York, USA, pp. 203–214 (2010)
5. Luo, W., Tay, W.P., Leng, M.: How to identify an infection source with limited observations. IEEE J. Sel. Top. Sign. Process. **8**(4), 586–597 (2014)
6. Antulovfantulin, N., Lancic, A., Šmuc, T., Štefančić, H., Šikić, M.: Identification of patient zero in static and temporal networks: robustness and limitations. Phys. Rev. Lett. **114**(24), 248701 (2015)

7. Jiang, J., Wen, S., Yu, S., et al.: Rumor source identification in social networks with time-varying topology. IEEE Trans. Dependable Secure Comput. **99**, 1 (2016)
8. Wang, Z., Wang, C., Pei, J., Ye, X.: Multiple source detection without knowing the underlying propagation model. In: AAAI. AAAI Press, pp. 217–223 (2017)
9. Fioriti, V., Chinnici, M.: Predicting the sources of an outbreak with a spectral technique. arXiv preprint arXiv:1211.2333 (2012)
10. Comin, C.H., Da Fontoura, C.L.: Identifying the starting point of a spreading process in complex networks. Phys. Rev. E **84**(5), 56105 (2011)
11. Lokhov, A.Y., Mzard, M., Ohta, H., et al.: Inferring the origin of an epidemic with a dynamic message-passing algorithm. Phys. Rev. E **90**(1), 12801 (2014)
12. Altarelli, F., Braunstein, A., Dall Asta, L., et al.: Bayesian inference of epidemics on networks via belief propagation. Phys. Rev. Lett. **112**(11), 118701 (2014)
13. Prakash, B.A., Vreeken, J., Faloutsos, C.: Spotting culprits in epidemics: how many and which ones? In: IEEE 12th International Conference on Data Mining (ICDM), Brussels, Belgium, vol. 2012, pp. 11–20 (2012)
14. Huang, Q.: Source locating of spreading dynamics in temporal networks. In: The 26th International Conference. International World Wide Web Conferences Steering Committee (2017)
15. Lin, Y.-R., Zhu, S., Sundaram, H., Tseng, B.L.: Analyzing communities and their evolutions in dynamic social networks. ACM Trans. Knowl. Discov. Data **3**(2), 18 (2009)
16. Folino, F., Pizzuti, C.: An evolutionary multiobjective approach for community discovery in dynamic networks. IEEE Trans. Knowl. Data Eng. **26**(8), 1838–1852 (2014)
17. Girvan, M., Newman, M.E.J.: Community structure in social and biological networks. Proc. Nat. Acad. Sci. U.S.A. **99**, 7821–7826 (2002)
18. Zhou, D., Bousquet, O., Lal, T.N., Weston, J., Scholkopf, B.: Learning with local and global consistency. Adv. Neural Inform. Process. Syst. **16**(16), 321–328 (2004)
19. Hu, Z.L., Shen, Z., Cao, S., et al.: Locating multiple diffusion sources in time varying networks from sparse observations. Sci. Rep. **8**(1) (2018)
20. Dong, M., Zheng, B., Hung, N., Su, H., Guohui, L.: Multiple rumor source detection with graph convolutional networks. 569–578 (2019). https://doi.org/10.1145/3357384.3357994

Analysis Method for Customer Value of Aviation Big Data Based on LRFMC Model

Yang Tao$^{(\boxtimes)}$ (iD)

Nanchang Institute of Science and Technology, Nanchang, China
taoyangxp@163.com

Abstract. In the era of Big Data, enterprise marketing focuses on customers instead of product center, and customer relationship management has become the core issue. In the aviation field, how to tap the high-quality customer base is more important. How to classify customers according to the characteristics of air passengers, and then make personalized marketing strategies for them, is the key problem to be solved. Aiming at optimizing resource allocation, with the help of aviation big data, a customer value analysis method is proposed based on the LRFMC model. First, Python was applied to clean, reduce and transform the data on the big data platform, and then to classify them. Moreover, characteristics of different customer categories were analyzed, and the customer value was evaluated. Finally, optimization methods based on K-Means algorithm were proposed, and the data were visualized, so that personalized services can be developed for different customers.

Keywords: LRFMC · Data analysis · Big data · Data visualization

1 Introduction

In modern society, China's civil aviation industry develops rapidly. With the continuous improvement of informatization, focusing on the civil aviation transportation industry, a large number of aviation customer data has been generated. With the continuous growth of airline network and air traffic volume, the competition among airlines is increasingly fierce, and the level of competition is also upgrading and expanding. In order to enhance competitiveness and attract aviation customers, airlines have put forward their own marketing plans, such as "Phoenix bosom friend" of Air China, "Oriental WanLiXing" of China Eastern Airlines, and "Pearl member" of China Southern Airlines. However, these alone cannot meet the needs of the explosive growth of the number of airline customers. The general membership level system has been difficult to assess the value and loyalty of airline passengers [1], nor to dig out the real value of massive customer data.

With the coming of the information age, the marketing focus of enterprises has changed from product center to customer center, and customer relationship management has become the core issue of enterprises. The key problem of customer relationship management is customer classification. Through customer classification, we can distinguish between valueless customers and high-value customers. Enterprises specify and optimize personalized services for different value customers, adopt

© Springer Nature Singapore Pte Ltd. 2020
J. Zeng et al. (Eds.): ICPCSEE 2020, CCIS 1257, pp. 89–100, 2020.
https://doi.org/10.1007/978-981-15-7981-3_7

different marketing strategies, concentrate the priority marketing resources with high-value customers, and achieve the goal of maximizing profits. Accurate customer classification result is an important basis for the distribution of marketing resources. The more and more customers, the more and more become one of the key problems to be solved in the customer relationship management system [2–4].

In the face of accumulated market competition, various airlines have launched more preferential marketing methods to attract more customers. Domestic capable airlines are faced with business crisis such as passenger loss, competitiveness decline and insufficient utilization of aviation resources. Just like the big data application system[5–7] in all walks of life, it is necessary and effective to provide personalized customer service for different customer groups by establishing a reasonable customer value evaluation model, grouping customers, analyzing and comparing customer values with different customer groups, and formulating corresponding marketing strategies. At present, the airline has accumulated a large number of member file information and flight records. Therefore, airlines usually need to adopt some models to effectively classify customers. However, the traditional processing method has low performance, long time-consuming, poor accuracy and poor interaction experience effect.

In order to solve the above problems and more effectively classify customers, this paper proposes a method to realize customer value analysis, which is based on LRFMC customer value evaluation model [8] and machine learning K-Means [9] clustering analysis algorithm. Through the use of Python data analysis module pandas as the main body, the whole process of data exploration, data preprocessing, modeling analysis and visualization is fully interpreted. Through data preprocessing, standardized feature vector and K-Means clustering analysis algorithm, rapid and efficient clustering analysis [10] is carried out to achieve the purpose of obtaining accurate customer value information.

2 Basic Theory and Models to Identify Customer Value

Dataology and Data Science [11] are the science of data, which are defined as the theory, method and technology to study and explore the mysteries of data in cyberspace. There are two main connotations: one is to study the data itself; the other is to provide a new method for natural science and social science research, called the data method of scientific research. It is widely used in multidimensional matrix and vector computation.

Our goal is customer value identification, which is to identify customers with different values through airline customer data. The most widely used model to identify customer value is to evaluate and analyze through three indicators, as follows:

- Recent Consumption Interval (Recency)
- Consumption frequency (Frequency)
- Consumption (Monetary)

This is our common model called the RFM model. By analyzing the three indicators of individual consumers, the consumer group customers can be segmented to identify high-value customers. In the RFM model, the consumption amount represents the total amount of the customer's purchase of the company's products over a period of time. As the air fare is affected by various factors such as transportation distance and class of cabin, different passengers with the same consumption amount have different values to the airline. Therefore, this single indicator is not applicable to the specific application scenarios of airline customer value analysis.

Instead of the amount of consumption, we choose two indicators: the customer's accumulated mileage M within a certain period of time and the average value C of the discount coefficient corresponding to the passenger's cabin class within a certain period of time. In addition, considering the length of the membership time of airline members can affect the customer value to a certain extent, the length of customer u relationship L is added to the model as another indicator to distinguish customers. The five indicators of customer relationship length L, consumption interval R, consumption frequency F, flight mileage M, and discount coefficient C are used as the indicator of airline customer identification (Table 1). Based on the simulation, the model adjusted to meet the special needs of the local scene is called the LRFMC model.

Table 1. LRFMC model (airline customer value model)

Model item	Meaning
L	The number of months since the member's joining time from the end of the observation time
R	Number of months since the member's last flight from the end of observation time
F	The total number of times the member has flown during the observation period
M	Miles accumulated during member observation time
C	The average value of the discount factor used by the member during the observation period

3   Customer Value Analysis

For the LRFMC model of the airline, if the attribute binning method analyzed by the traditional RFM model is used, as shown in Fig. 1 (it is divided according to the average value of the attribute, where the value greater than the average is represented as ↑, and the value less than the average is represented as ↓), although the customers with the best value can also be identified, but there are too many subdivided customer groups, which increases the cost of targeted marketing.

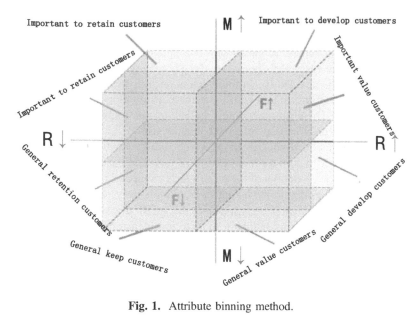

Fig. 1. Attribute binning method.

Therefore, the clustering method was used to identify customer value. By performing K-Means clustering on five indicators of the LRFMC model of airline customer value, the most valuable customers were identified.

3.1 Data Analysis Overall Process

After we have a preliminary understanding of the LRFMC model, we need to make a complete plan for the overall data mining and data analysis. According to the standard process of data mining analysis, we divide the whole process into the following five parts, as shown in Fig. 2.

Part 1: sorting and viewing data sources in the business system;
Part 2: data extraction according to business analysis requirements;
Part 3: data exploration and preprocessing;
Part 4: data modeling and application;
Part 5: result feedback and visual presentation.

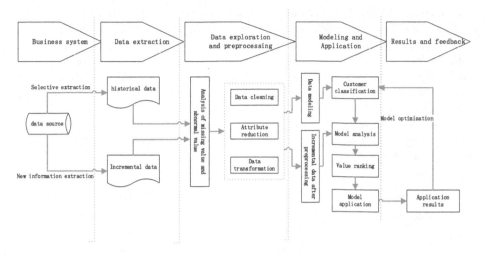

Fig. 2. Data analysis overall process.

Through the above five parts, we can complete the objective of data mining modeling and costumer value analysis of our project.

3.2 Data Mining Steps

The data mining of aviation customer value information mainly includes the following steps:

Step 1: selectively extract and add data from the data source of the airline to form historical data and incremental data respectively.

Step 2: carry out data exploration analysis and preprocessing for the two data sets formed in step 1, including data missing value and abnormal value exploration analysis, data attribute specification, cleaning and transformation.

Step 3: Based on the LRFMC model of customer value, the customer groups are divided and the characteristics of each customer group are analyzed.

Step 4: according to the customers who get different value from the model results, adopt different marketing methods to provide customized services.

After we have a preliminary understanding of the LRFMC model, we need to make a complete plan for the overall data mining and data analysis.

4   Application Case Analysis

We select a period of customer sample data of an airline as an example to analyze the application of the method. Take 2014-03-31 as the end time, select the two-year period as the analysis observation window, and extract the detailed data of all customers with boarding records in the observation window to form historical data. For the subsequent new customer details, the latest time point in the subsequent new data is taken as the end time, and the same method is used for extraction to form incremental data.

According to the last flight date, the detailed data of all passengers from April 1, 2012 to March 31, 2014 are extracted from the detailed data of basic customer information, flight information and integral information in the airline system, with a total of 62988 records. It includes 44 attributes, such as membership card number, joining time, gender, age, meeting card level, city of work, province of work, country of work, end time of observation window, integral points of observation window, kilometers of flight, times of flight, time of flight, interval between flights and average discount rate. Some of the aviation data are shown in Table 2.

Table 2. Aeronautical information data sheet (partial column slice data)

MEMBER_NO	FFP_DATE	FIRST_FLIGHT_DATE	FFP_TIER	WORK_COUNTRY
47229	2005/4/10	2005/4/10	6	CN
28474	2010/4/13	2010/4/13	6	US
58472	2010/2/14	2010/3/1	5	FR
13942	2010/10/14	2010/11/1	6	FR
45075	2007/2/1	2007/3/23	6	CN
47114	2005/1/15	2005/3/17	6	CN
54619	2006/1/7	2006/1/8	6	CN
12349	2008/6/16	2008/6/27	6	CN
35883	2006/4/11	2007/4/18	6	CN
56091	2004/11/25	2005/2/10	6	CN
2137	2005/4/11	2005/5/3	6	CN
27708	2006/3/20	2006/3/25	6	CN
28014	2006/12/1	2011/1/7	6	FR

In the data exploration and analysis, the missing value (null value) analysis and abnormal value analysis are carried out to analyze the data rules and abnormal values. Through the observation of the data, it is found that there are records in the original data that the ticket price is null, the minimum ticket price is zero, the minimum discount rate is zero, and the total flight kilometers are greater than zero. The data with null fare may be caused by the fact that the customer does not have a flight record, and other data may be generated by the customer taking a 0% discount ticket or exchanging points. Through reading and writing Python CSV file, data description and matrix transposition in Pandas module library, the number of empty values, maximum value and minimum value of each column of attribute observation value can be found. The results of data exploration and analysis are shown in Table 3.

Table 3. Data exploration and analysis data sheet (partial column slice data)

Attribute name	Null value number	Max	Min
MEMBER_NO	0	62988	1
FFP_DATE	0		
FIRST_FLIGHT_DATE	0		
GENDER	3		
FFP_TIER	0	6	4
WORK_CITY	2269		
WORK_PROVINCE	3248		
WORK_COUNTRY	26		
AGE	420	110	6
LOAD_TIME	0		
FLIGHT_COUNT	0	213	2
BP_SUM	0	505308	0
EP_SUM_YR_1	0	0	0

4.1 Data Cleaning

Through data exploration and analysis, it is found that there are missing values in the data, the minimum value of ticket price is zero, the minimum value of discount rate is zero, and the total flight kilometers are greater than zero. Due to the large amount of original data, the proportion of such data is small, which has little impact on the problem, so it is discarded. The specific rules and standards are as follows:

Discard the record with empty ticket price.
Discard the record that the fare is zero, the average discount rate is not zero, and the total flying kilometers are greater than zero.

The steps of data cleaning are as follows:

Step1: read the original sample CSV data.
Step2: set data filtering conditions according to standard requirements, and realize data clearing.
Step3: write the final result to the data file.

4.2 Attribute Reduction

There are too many attributes in the original data. According to the LRFMC model of airline customer value, six attributes related to the LRFMC model indexes are selected: FFP_DATE, LOAD_TIM, FLIGHT_COUNT, AVG_DISCOUNT, SEG_KM_SUM, and LAST_TO_END.

Delete attributes that are not related, weakly related or redundant, such as membership card number, gender, city of work, province of work, country of work and age. Through the data slicing function of pandas module, attribute reduction is realized. The data set after attribute selection is shown in Table 4.

Table 4. Data set after attribute selection (partial column slice data)

FFP_DATE	LOAD_TIME	FLIGHT_COUNT	AVG_DISCOUNT	SEG_KM_SUM
2006/11/2	2014/3/31	210	0.961639043	580717
2007/2/19	2014/3/31	140	1.25231444	293678
2007/2/1	2014/3/31	135	1.254675516	283712
2008/8/22	2014/3/31	23	1.090869565	281336
2009/4/10	2014/3/31	152	0.970657895	309928
2008/2/10	2014/3/31	92	0.967692483	294585
2006/3/22	2014/3/31	101	0.965346535	287042
2010/4/9	2014/3/31	73	0.962070222	287230
2011/6/7	2014/3/31	56	0.828478237	321489
2010/7/5	2014/3/31	64	0.708010153	375074
2010/11/18	2014/3/31	43	0.988658044	262013
2004/11/13	2014/3/31	145	0.95253487	271438
2006/11/23	2014/3/31	29	0.799126984	321529

4.3 Data Transformation

Data transformation is to transform data into "appropriate" format to meet the needs of mining tasks and algorithms. In this project, the main data transformation method is attribute construction. Because the original data does not directly give the five indicators of LRFMC, the five indicators need to be extracted from the original data. The specific calculation method is as follows:

- L = LOAD_TIME - FFP_DATE

The number of months between the time of membership and the end of observation window = the end time of observation window - the time of membership [unit: month].

- R = LAST_TO_END

The number of months from the last time the customer took the company's aircraft to the end of the observation window = the time from the last flight to the end of the observation window [unit: month].

- F = FLIGHT_COUNT

Number of times the customer takes the company's aircraft in the observation window = number of flights in the observation window [unit: Times].

- M = SEG_KM_SUM

Accumulated flight history of the customer in observation time = total flight kilometers of observation window [unit: km].

- C = AVG_DISCOUNT

Average value of the discount coefficient corresponding to the passenger space during the observation time = average discount rate [unit: none].

4.4 Data Normalization

At the same time, the standardized data is more conducive to the accuracy of model analysis. We use Z-score method to normalize the LRFMC index data. Z-score normalization is also known as standard deviation normalization. The normalized data is normally distributed, i.e. the mean value is zero, and the standard deviation is a formula as follows:

$$\chi^n = \frac{\chi - \mu}{\sigma} \tag{1}$$

Where μ is the mean value of all sample data and σ is the standard deviation of all sample data. Implemented in Python, the code is as follows:

$$\text{data} = (\text{data} - \text{data.mean (axis} = 0))/\text{data.std (axis} = 0)$$

The difference between the standard deviation and the standard deviation is that the standard deviation only reduces the variance and mean deviation of the original data by multiple, while the standard deviation standard deviation makes the standardized data variance one. This is more advantageous to many algorithms, but its disadvantage is that if the original data is not Gaussian distribution, the standardized data distribution effect is not good.

4.5 Modeling Analysis

The construction of customer value analysis model mainly consists of two parts. The first part: according to the data of five indicators of airline customers, cluster and group customers. The second part: combined with the business to analyze the characteristics of each customer group, analyze its customer value, and rank each customer group. K-Means clustering algorithm is used to classify customers into 5 categories (the number of customer categories needs to be determined by combining the understanding and analysis of business). In the specific implementation, the K-Means clustering algorithm we used is located in the clustering word library (sklearn.cluster) under the Scikit-Learn module library. We use SK-Learn module to create K-Means model object instance, use fit() function of K-Means object instance to complete model analysis, and finally get the model analysis result, as shown in Table 5.

Table 5. Cluster analysis results

Cluster No	ZL	ZR	ZF	ZM
4183	0.052674226	−0.002018209	−0.22675176	−0.231377928
24649	−0.700316003	−0.415473188	−0.160700141	−0.160404636
15746	1.160429765	−0.377379525	−0.087102357	−0.095201216
12139	−0.314170608	1.684986848	−0.573955288	−0.536720331
5334	0.483815811	−0.799389015	2.483818101	2.425261912

4.6   Data Visualization

According to the analysis results of the previous model, we want to visualize the distribution of customer clustering population by using histogram. First of all, we use the Pandas module to intercept the relevant data and generate the parameter data structure and type required for histogram. Then we build the basic structure of histogram based on python. Finally, we use Matplotlib module to generate histogram through data binding. Through the above steps, the distribution of customer clustering population is shown in Fig. 3.

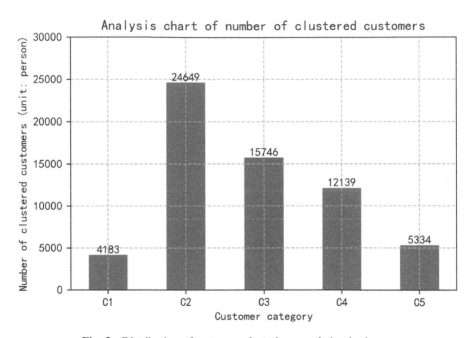

Fig. 3. Distribution of customer clustering population is shown.

Based on the above visualization of clustering population distribution, radar map can be constructed to visualize customer clustering eigenvalues, and the results are shown in Fig. 4.

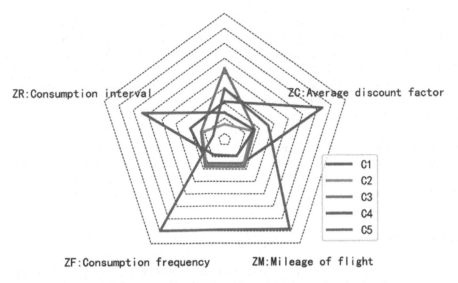

Fig. 4. Visualization of customer clustering eigenvalues is shown.

Based on the data and visual chart, the clustering results are analyzed. Through the chart observation, we can see that:

- customer group 1 is the largest in C attribute and relatively average in L, R, F and M;
- customer group 2 is relatively average in the five dimensions of LRFMC;
- customer group 3 is the largest in L attribute and relatively average in R, F, M and C, with no significant difference;
- customer group 4 is the largest in R attribute and relatively average in L, F, M and C;
- customer group 5 is the largest in F and M attributes, and the average in L, R and C attributes.

Combined with the business analysis, the paper evaluates and analyzes the characteristics of a group by comparing the size of each index among groups. There are maximum, minimum, sub maximum and sub minimum values in each indicator attribute, so we need to analyze and count them. For example, customer group 5 has the largest attribute of F and m, and the smallest attribute of R, so it can be said that f and m are the advantages of customer group 5. By analogy, F and m are inferior characteristics in customer group 4.

5 Conclusion

Starting from customer classification, based on the customer value analysis model of LRFMC, this paper analyzes customer resources and customer value mining to maximize customer value, and proposes a general standard process for big data analysis. Based on Python, according to the algorithm model of scientific computing and data analysis and machine learning, this paper implemented the application analysis based on aviation big data, in order to achieve the best customer classification results, and verified the effectiveness of the proposed method.

Acknowledgment. The project is provided with technical training guidance by China Soft International Education Technology Group. This paper is supported by the Ph.D. Research Initiation Fund of Nanchang Institute of Science and Technology with the Project (No. NGRCZX-18-01).

References

1. Ban, H.J., Joung, H.-W.: The text mining approach to understand seat comfort experience of airline passengers through online review. Culinary Sci. Hospitality Res. **25**(9), 38–46 (2019)
2. Kun, Q., Li, C.: Construction and research of airline customer relationship management system based on Python. J. Baotou Vocat. Tech. Coll. **20**(04), 9–14 (2019)
3. Xie, W.: Diagnostic Research on Customer Relationship Management System of China Eastern Airlines. China University of Geosciences, Beijing (2015)
4. Chiang, W.-Y.: Establishing high value markets for data-driven customer relationship management systems. Kybernetes **48**(3), 650–652 (2019)
5. Zhiliang, M., Mingkun, T., Yuan, R.: Building energy consumption information model for big data analysis. J. S. China Univ. Technol. (Natural Science Edition) **47**(12), 72–77 (2019)
6. Luo, X., Wen, X.: Student behavior analysis model based on artificial intelligence and big data. Educ. Obs. **8**(17), 11–13 (2019)
7. Sun, D., Wu, J., Wen, H., Xue, M.: Study on big data analysis model and application of mountain slope disaster resilience–taking Chengkou County as an example. J. Chongqing Norm. Univ. (Nat. Sci. Ed.) **36**(03), 64–71 (2019)
8. Tahanisaz, S., Sajjad, S.: Evaluation of passenger satisfaction with service quality: a consecutive method applied to the airline industry. J. Air Transp. Manage. **83**, 101764 (2020)
9. Li, J., Cao, Y., Wang, Z., Wang, G.: An algorithm for cloud simplification based on K-means clustering and Hausdorff distance. J. Wuhan Univ. (Inf. Sci. Ed.) **45**(02), 250–257 (2020)
10. Song, Y., Peng, G., Sun, D., Xie, X.: Active contours driven by Gaussian function and adaptive-scale local correntropy-based K-means clustering for fast image segmentation. Sig. Proc. **174**, 107625 (2020)
11. Data Science and Data Science. Tianjin Econ. (07), 63–64 (2017)

Identifying Vital Nodes in Social Networks Using an Evidential Methodology Combining with High-Order Analysis

Meng Zhang, Guanghui Yan[✉], Yishu Wang, and Ye Lv

School of Electronic and Information Engineering, Lanzhou Jiaotong University,
Lanzhou 730070, China
736336430@qq.com

Abstract. Identifying vital nodes is a basic problem in social network research. The existing theoretical framework mainly focuses on the lower-order structure of node-based and edge-based relations and often ignores important factors such as interactivity and transitivity between multiple nodes. To identify the vital nodes more accurately, a high-order structure, named as the motif, is introduced in this paper as the basic unit to evaluate the similarity among the node in the complex network. It proposes a notion of high-order degree of nodes in complex network and fused the effect of the high-order structure and the lower-order structure of nodes, using evidence theory to determine the vital nodes more efficiently and accurately. The algorithm was evaluated from the function of network structure. And the SIR model was adopted to examine the spreading influence of the nodes ranked. The results of experiments in different datasets demonstrate that the algorithm designed can identify vital nodes in the social network accurately.

Keywords: Vital nodes · High-order network · Evidence theory · SIR

1 Overview

In pace with the rapid development of information technology, the forms of communication and interaction have diversified. The resulting massive data can not only help us better understand the relationship between people, but also show the mode of information transmission between people [1–3]. Identifying vital nodes in the network helps us to guide the information dissemination better.

Centrality is a method which can measure the significance or importance of actors in social networks. Considerable centrality measures have been carried out previously for ranking the nodes based on network topology such as Degree Centrality (DC) [4], Closeness Centrality (CC) [5] and Betweenness Centrality

Supported by the Natural Science Foundation of China (No. 61662066, 61163010).

J. Zeng et al. (Eds.): ICPCSEE 2020, CCIS 1257, pp. 101–117, 2020.
https://doi.org/10.1007/978-981-15-7981-3_8

(BC) [6,7]. Although the DC is intuitive and simple, we just take into consideration the degree of nodes and ignores the global structure of the network. Whether
a node is important affects the importance of its neighbor nodes, which in the
social network its neighbor nodes would follow the behavior of the former. The
other two metrics based on the global structure of a network can better characterize the importance of nodes, but exhibit some serious drawbacks because
of the computational complexity in large-scale networks [8,9]. Considering the
operating efficiency and experimental results of the algorithm, Chen et al. [8]
proposed an effective Semi-local Centrality. These above methods take nodes
and edges as research objects. Despite the success of these methods in identifying vital nodes, they ignore the possible relationship between nodes, which
may lead us to deviate from the overall cognition of the network. So, an important issue is that how to describe the interaction, transitivity and other factors
between nodes.

A number of researches have shown that social networks contain abundant
subgraph structures, which are characterized by transitivity, interaction and so
on [10,11]. Usually, we describe this subgraph structure as network motif or
graphlet [11–13]. Compared with the method of researching from edges and
points, the network structure with small subgraph structure as the research
unit is called high-order network structure. In broader network analysis, high-
order structure is often described through the idea of a network motif. Since
the concept of motif was put forward in 2002 by Moli R et al., most of the
research had centered on how to count the number of motif efficiently in the
network [10,14,15]. Until 2016, Benson et al. proved that motif can be used for
graph clustering and community discovery, and proposed a series of theoretical
basis. The research of high-order network has become one of the important means
of current research [10,15].

The importance of nodes in the network is a vague and relative concept. As it
happens, the Dempster–Shafer evidence theory is a complete theory for dealing
with uncertainty information. It was first proposed by Dempster and then per-
fected by Shafer. Compared with traditional probability theory, D-S evidence
theory can not only express random uncertainty, but also express incomplete
information and subjective uncertainty information [8,16,17]. D-S evidence theory also provides a powerful Dempster combination rule for information fusion,
which can achieve the fusion of evidence without prior information, and can
effectively reduce the uncertainty of the system. On this basis, Wei et al. put
forward an algorithm to rank nodes in weighted networks [9].

In this paper, based on the high-order network analysis and Dempster-Shafer
evidence theory, we designed a high-order evidence semi-local centrality to identify vital users accurately in the social network. First, we designed a concept
based on high-order structure. And then the low-order information and high-
order information of nodes are regarded as two basic probability assignments
(BPAs). On the one hand, we verified the rationality of the proposed method
through the function of network structure. On the other hand, we adopted the
Susceptible-Infected-Recover (SIR) model to examine the spreading influence of

the top nodes by different centrality measures. The experiments on real social networks are applied to show the accuracy of the proposed method.

2   Related Work

High-order thinking has been shown to be useful in many applications such as social networks, biology, neuroscience, and so on [15]. Network motifs are the basic building blocks of networks and are also one of the important expressions of high-order network structure [10,11].

2.1   High-Order Network Structure

Combined with the basic theory of sociology, this paper takes the 3-order motif as the basic. Figure 1 shows the 13 connection modes of the 3-order motif.

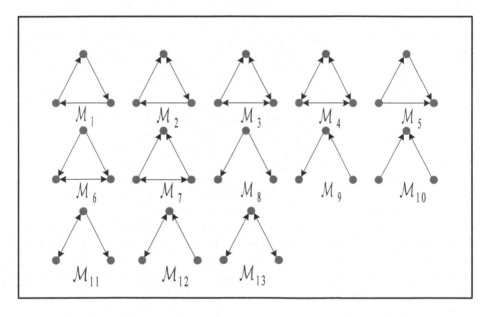

Fig. 1. All 13 connection modes of the 3-order motif $\mathcal{M}$ [15]

A motif $\mathcal{M}$ is usually defined as a tuple $(B, \mathcal{A})$ on k nodes, where B is a $k \times k$ binary matrix and $\mathcal{A} \subset \{1, 2, \cdots, k\}$ is a set of anchor nodes [10].

$$\mathcal{M}(B, \mathcal{A}) = \left\{ (set(\mathbf{v}), set(\mathcal{X}_\mathcal{A}(\mathbf{v}))) \, | \, \mathbf{v} \in V^k, v_1, \cdots, v_k, distinct, \mathbf{A}_\mathbf{v} = B \right\} \tag{1}$$

Given a motif $\mathcal{M}$, we define the motif-based adjacency matrix by $\mathbf{A}_\mathcal{M} = \{a_{ij}\}_{N \times N}$. In this paper, we use Benson et al. [11] improved algorithm to get the motif-based adjacency matrix. So, the algorithm is as follows [11,18]:

Algorithm 1 Algorithm for the motif-based adjacency matrix $\mathbf{A}_{\mathcal{M}}$.

Input: Directed network $G = (V, E)$ and selected motif $\mathcal{M}$
Output: $\mathbf{A}_{\mathcal{M}}$;
1: pro–processing: If $\mathcal{M}$ is $\mathcal{M}_4$, ignore all undirectional edges in G. If $\mathcal{M}$ is $\mathcal{M}_1$ or $\mathcal{M}_5$, ignore all bidirectional edges in G.
2: Obtain the undirected graph G_1 by getting rid of the direction of all edges in G.
3: d_u is the degree of node v_i in G_1. Sort the nodes in G_1 by ascending degree.
4: For every edge undirected edge u,v in G_1, if $d_u < d_v$, add directed edge (u,v) to G_2; otherwise, add directed edge (v,u) to G_2.
5: For every node in u in G_2 and every pair of directed edges (u,v) and (u,w), check to see if edge (v,w) or (w,v) is in G_2. If so, check whether these three nodes form motif $\mathcal{M}$ in G. If they do, increment the weights of edges $(\mathbf{A}_{\mathcal{M}})_{uv}$, $(\mathbf{A}_{\mathcal{M}})_{uw}$, and $(\mathbf{A}_{\mathcal{M}})_{uv}$ by 1.
        return $\mathbf{A}_{\mathcal{M}}$ as the motif-based adjacency matrix;

Let $G = (V, E, \mathbf{A}_{\mathcal{M}})$ be a directed and unweighted graph, where $|V| = \{v_i | i = 1, 2, 3, \ldots, n\}$ is the node set, and $|E| = \{e_{ij} | i, j = 1, 2, \ldots, n\}$ is the arc set, where e_{ij} is a directed edge from v_i to v_j. $\mathbf{A}_{\mathcal{M}}$ is the motif-based adjacency matrix.

2.2   Centrality Measures

Roughly speaking, there are two kinds of method about identifying vital nodes which are based on the number of neighbor and based on path in network. The former is characterized by the degree of nodes in the network as a measure of importance. This method is relatively intuitive and has good performance, such as DC. The latter measures the importance of nodes by controlling the information flow in the network, such as BC and CC. This kind of method is relatively complex and not suitable for large-scale networks [20].

The semi-local centrality is a node importance ranking method based on the number of neighbor in the networks. The method not only considered the neighbors of the nodes but also the neighbors and next neighbors of the neighbors. In other words, this method has low time complexity and is suitable for large-scale networks. Semi-local centrality of node v_i is defined as [9]:

$$SLC\,(i) = \sum_{j \in \Gamma(i)} \sum_{k \in \Gamma(j)} N^w\,(k) \tag{2}$$

where $N^w\,(k)$ is the number of the nearest and the nextest neighbors of node v_k, and $\Gamma\,(i)$ is the set of the nearest neighbors of node v_i.

2.3   Dempster-Shafer Theory of Evidence

The essence of D-S evidence theory is a generalization of probability theory. The basic event space in probability theory is extended into the power set space of the basic event, and the basic probability assignment function is established on it [13, 20].

Let $\Theta = \{\theta_1, \theta_2, \ldots, \theta_N\}$ be a finite complete set of N elements which is mutually exclusive. The frame of discernment is the set Θ. The power set of Θ is denoted as 2^Θ which is composed of 2^N elements.

$$2^\Theta = \{\varnothing, \theta_1, \theta_2, \ldots \theta_N, \theta_1 \cup \theta_2, \ldots, \theta_1 \cup \theta_2 \cup \theta_3, \ldots, \Theta\} \tag{3}$$

For a frame of discernment Θ, a basic probability assignment function is a mapping $m : 2^\Theta \to [0, 1]$, satisfying two conditions as follows:

$$m(\varnothing) = 0 \tag{4}$$

and

$$\sum_{A \subseteq \Theta} m(A) = 1 \tag{5}$$

where m is called the basic probability assignment (BPA), and $m(A)$ represents how strongly the evidence supports A.

In order to combine with information from multiple independent information sources, D-S evidence theory provides Dempster's Rules of Combination to achieve the fusion of multiple evidence. Its essence is the orthogonal sum of evidence.

$$\begin{cases} m(\varnothing) = 0 \\ m(A) = \frac{1}{1-k} \sum_{A_i \cap B_i = A} m_1(A_i) m_2(B_i) \end{cases} \tag{6}$$

where k is a normalization constant, called the conflict coefficient of BPAs.

$$k = \sum_{A_i \cap B_i = \varnothing} m_1(A_i) m_2(B_i) \tag{7}$$

3 High-Order Evidential Semi-local Centrality

In this section, some notations and some knowledge of high-order degree are given as follows. Two BPAs of a node are obtained Eq. 9 based on the high-order degree and degree of the node, respectively. An evaluation method of importance of the node is established by Dempster's rule of combination [9]. The influence of the node is identified by a new centrality measure, called the high-order evidential centrality. Inspired by semi-local centrality [English16], we propose high-order evidential semi centrality since it not only fuse the internal information of the network, but also considers the global structure information.

3.1 High-Order Degree

In the high-order network structure, the elements of motif-based adjacency matrix describe the local connection density of node pair (v_i, v_j). The higher the weight, the more the modal structures with the edge of node pairs are, the worse the anti-attack ability of the node pairs is, the higher the importance of the node pairs is. The higher the value of elements is, the more the number of motifs which contain the edge of node pairs are, that represent the worse the anti-attack ability of the node pair is, the higher the importance of the node pairs is [20].

The high-order degree of node v_i is the number of times node v_i appears in the given motif. The high-order degree of the node v_i denote as H_i. The H_i algorithm is as follows:

Algorithm 2 Algorithm for the H_i.

Input: Undirected network G_1, directed network G_2 and selected motif $\mathcal{M}$
Output: The number of $\mathcal{M}$ and high-order H_i
1: initialize: the H_i of all nodes in G_1.
2: for edge (u, v) node v_w in G_1 do:
3:        if node v_w is not node v_v:
4:            if edge (v, w) in G_2:
5:                $\mathcal{M}$ which consist of node v_u, node v_v and node v_w is isomorphic with subgraph which consist of node v_u, node v_v and node v_w in G_1
6:                    the value of H_v, H_u and H_w plus one respectively
7: end

3.2   BPAs of Degree and High-Order Degree

A node in the network is either important or not important, so we ascertain a frame of discernment Θ about each node, so a frame of discernment Θ is given as [9]:

$$\Theta = \{h, l\} \tag{8}$$

where h and l represent important and unimportant respectively which are two mutually exclusive elements.

The degree and the high-order degree are two indicators of importance about each node. And then we can obtain these two basic probability assignment functions from different independent sources.

So, two basic probability assignment functions are given as follows:

$$\begin{aligned} m_{d_i} : m_{d_i}(h), m_{d_i}(l), m_{d_i}(\theta) \\ m_{H_i} : m_{H_i}(h), m_{H_i}(l), m_{H_i}(\theta) \end{aligned} \tag{9}$$

where $m_{d_i}(\theta)$ and $m_{H_i}(\theta)$ represent the probability whether a node is important or not in the above two indicators. And their value are

$$\begin{aligned} m_{d_i}(\theta) = 1 - \left(m_{d_i}(h) + m_{d_i}(l) \right) \\ m_{H_i}(\theta) = 1 - \left(m_{H_i}(h) + m_{H_i}(l) \right) \end{aligned} \tag{10}$$

$$m_{d_i}(h) = \lambda_i \frac{|k_i - k_m|}{\sigma}$$
$$m_{d_i}(l) = (1 - \lambda_i) \frac{|k_i - k_M|}{\sigma}$$
$$m_{H_i}(h) = \frac{|H_i - H_m|}{\delta}$$
$$m_{H_i}(l) = \frac{|H_i - H_M|}{\delta}$$

(11)

where σ and δ are given as:

$$\sigma = k_M + \mu - (k_m - \mu) = k_M - k_m + 2\mu$$
$$\delta = w_M + \epsilon - (w_m - \epsilon) = w_M - w_m + 2\epsilon$$

(12)

μ and ϵ are given as 0.15. Because the value of μ and ϵ have no effect on the results. The influence value of node v_i is obtained by Dempster–Shafer theory of evidence, and is given by [9]:

$$M(i) = (m_i(h), m_i(l), m_i(\theta))$$

(13)

Normally, let $m_i(\theta)$ assign to $m_i(h)$ and $m_i(l)$ averagely, then

$$M_i(h) = m_i(h) + \frac{1}{2m_i(\theta)}$$
$$M_i(l) = m_i(l) + \frac{1}{2m_i(\theta)}$$

(14)

where $M_i(h)$ and $M_i(l)$ are the probability of importance and unimportance about node v_i, respectively. For node v_i, the higher the value of $M_i(h)$ is, the more important the node is. In other words, the lower the value of $M_i(l)$ is, the less important the node is [7,21].

The high-order evidential centrality $hec(i)$ of node v_i is defined as

$$hec(i) = M_i(h) - M_i(l) = m_i(h) - m_i(l)$$

(15)

where $hec(i)$ is a positive or negative number. In order to ensure $hec(i)$ is a positive number. The numerical treatment and normalization are denoted as follows,

$$HEC(i) = \frac{|\min(hec)| + hec(i)}{\sum_{i=1}^{N}\{|\min(hec)| + hec(i)\}}$$

(16)

The example is a directed network with 10 nodes, see Fig. 2, and k_i, H_i, $m_i(h)$, $m_i(h)$ and $HEC(i)$ for a single node is listed in Table 1.

3.3   High-Order Evidential Semi-local Centrality

We can calculate the value of HEC about each node though the above measure. Inspired by the semi-local centrality measure, we use HEC instead of degree of each node and then high-order evidential semi-local centrality (HESC) is defined. The HESC algorithm is as follows:

$$Q(j) = \sum_{k \in \Gamma(j)} N^w(k)$$

(17)

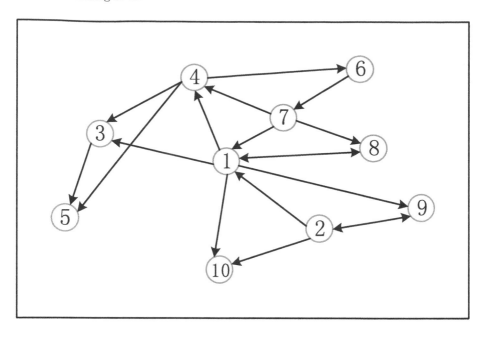

Fig. 2. High-order structures in network motifs $\mathcal{M}_{\triangledown}$

Table 1. An example of a HEC

Nodes	k_i	H_i	$m_i(h)$	$m_i(h)$	$HEC(i)$
1	5	3	0.9177	0	0.3033
2	3	1	0.2918	0.6452	0.0957
3	1	2	0.4817	0.4573	0.1574
4	3	3	0.9007	0.0316	0.2954
5	0	1	0.0964	0.8747	0.0263
6	1	0	0.0074	0.9466	0
7	3	1	0.2918	0.6452	0.0957
8	1	0	0.0074	0.9466	0
9	1	0	0.0074	0.9466	0
10	0	1	0.0964	0.8747	0.0263

$$HESC(i) = \sum_{j \epsilon \Gamma(i)} Q(j) \tag{18}$$

where $N^w(k)$ is the sum of HEC of nearest and next nearest neighbors of node v_k, and $\Gamma(i)$ is the set of the nearest neighbors of node v_i.

4   Example and Experimental Analysis

In this section, we will use the proposed HESC to obtain the ranking of nodes in three different social networks. Meanwhile, comparing with another four traditional centrality measures (DC, CC, BC and EC), we will show the difference between them.

4.1   Datum

We conducted experiments on three social network datasets. Specific Description about the three datasets is shown below:

Advogato. This is the trust network of Advogato. Advogato is an online community platform for developers of free software launched in 1999. Nodes represent Advogato users and a directed edge represent trust relationships called "certification". Advogato have three levels of "certification" corresponding to three edge weights, the weight of master is 1.0, the weight of journeyer is 0.8 and the weight of apprentice is 0.6. An observer without any trust certifications can only trust himself, and therefore the network contains loops [22].

Wiki-Vote. This is a social network that describes voting relationships among Wikipedia users. The network contains all the Wikipedia voting relationships from the inception of Wikipedia to January 2008. In this network, nodes can be regard as Wikipedia users and a directed edge from node v_i to node v_j is that user v_i voted on user v_j [23].

Table 2. Dataset statistics about three social networks

Dataset	Nodes	Edges	The number of $\mathcal{M}_1$ to $\mathcal{M}_7$
Advogato	6.5K	51.1K	18.3K
Wiki-Vote	7.1K	103.7K	608.4K
soc-Epinions1	75.9K	508.8K	1.6M

soc-Epinions1. This data set describes a who-trust-whom online social network of a general consumer review site Epinions.com. Whether to trust each other is decided by the members of the site. Reviews are shown to user based on the Web of Trust and review ratings. Nodes in the network represent consumers, and the directed edges are the trust relationship between consumers [24] (Table 2).

4.2 Relation Between Centrality Measures

In order to more intuitively characterize the relationship between different centrality measures, we compared the average centrality value of each node in the network by averaging over 100 independent runs under different centralities. The relationship between HESC and the other five centrality measures is shown in Fig. 4, Fig. 3 and Fig. 5, respectively. The Y-axis is the value of HESC. And the X-axis is the value of other centrality measures. From the Fig. 4, Fig. 3 and Fig. 5 we can see that the correlation between HESC and DC are the strongest as positively correlated (i.e., Fig. 4(a), Fig. 3(a) and Fig. 5(a)). And then because the HESC value of each node is obtained through the HODC value of each node and the local information of the network is considered at the same time, the correlation between HESC and DC is positive (Table 3).

4.3 Experimental Results Analysis

We remove the vital nodes in the network in turn, and compare the relative size of strong connected subgraphs to judge the network invulnerability under static attack. In this experiment, according to the order of above different centrality methods, 10 nodes are removed each time, and then the relative size of strongly connected subgraphs is calculated. Figure 6 shows the change of the relative size of the strong connected subgraphs corresponding to the five methods in different datum when removing the topn nodes from the network under static attack. X-axis represents the number of nodes removed from the network in order, and Y-axis represents the relative size of strongly connected component of the network.

Table 3. The number of various motifs in three datasets

$\mathcal{M}$	Advogato	Wiki-Vote	soc-Epinions1
$\mathcal{M}_1$	63	6795	7656
$\mathcal{M}_2$	2230	17667	84384
$\mathcal{M}_3$	3162	15275	328076
$\mathcal{M}_4$	1992	2119	160097
$\mathcal{M}_5$	4262	462715	531325
$\mathcal{M}_6$	3019	45559	281093
$\mathcal{M}_7$	3564	58259	231850

In contrast, the HESC method proposed on Advogato performs well. When the vital nodes are removed, the method has a strong destructive power to the network in Fig. 6(a). On Wiki-Vote, before removing Top50, the differences among the methods are small, and the node sorting performance of CC method is better. However, with the increase of the number of removed nodes, the advantages of HESC and HEC methods are shown, especially when the top 100 is

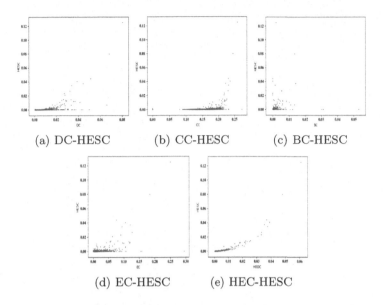

(a) DC-HESC   (b) CC-HESC   (c) BC-HESC

(d) EC-HESC   (e) HEC-HESC

Fig. 3. The relationship between HESC and others in Advogato

removed, the relative size of strongly connected subgraphs in HESC method is the smallest in Fig. 6(b). The HEC and HESC methods proposed are close to each other on soc-Epinions1, and the performance is only better than CC method. There are some differences in the methods of identifying vital nodes based on high-order structure in Fig. 6. But as a whole, with the vital nodes removed, the more seriously the network is destroyed.

In order to better evaluate our proposed methods, we carries out experiments on SIR model to test the propagation ability of nodes, and compares it with other algorithms. Nodes in SIR epidemic transmission model have three possible states at any time: susceptible, infected and recovered. At the time t, the proportion of these three groups of people in the crowd is used $S(t)$, $I(t)$ and $R(t)$ to express separately. $S(t)$ represents the proportion of nodes in a network that are vulnerable to infection. $I(t)$ represents the ability to transmit disease to other vulnerable nodes in an infected state. Each infected node can randomly transmit disease to its neighbor nodes through a certain probability. $R(t)$ represents the proportion of nodes that have been infected but have recovered and have immunity. In the SIR model of complex networks, we assume that all neighbor nodes around infected nodes have the chance to be infected.

We used the Top-10, Top-50 and Top-100 nodes ranked by various centralities as infected nodes in the initial network. Then we used the proportion of infected nodes and recovered nodes in the network to judge the influence of nodes when the network reaches steady state and compare the differences between different methods. Figure 7, Fig. 8 and Fig. 9 shows the propagation ability of infected

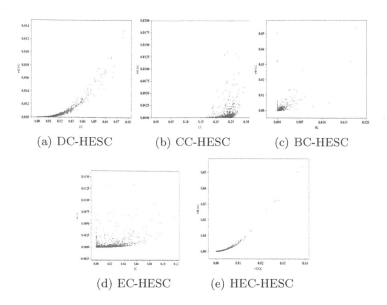

Fig. 4. The relationship between HESC and others in Wiki-Vote

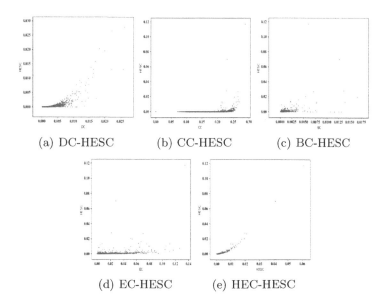

Fig. 5. The relationship between HESC and others in soc-Epinions1

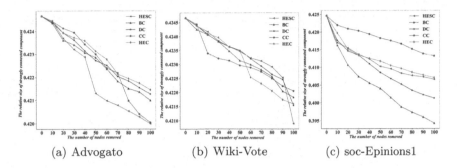

(a) Advogato   (b) Wiki-Vote   (c) soc-Epinions1

Fig. 6. The relative size of strongly connected component under static attack of different datum

nodes by the top-L nodes as ranked by six centrality measures under these three datasets.

Comparatively speaking, our proposed HESC is superior to the classical methods in both propagation range and propagation rate in Advogato dataset (see Fig. 7). HESC, BC and DC have almost the same performance on Advogato, because these two centralities are all positively correlated with HESC in this network (see Fig. 4). However, the propagation rate and range of HEC method are relatively poor, which may be due to the close number of various motifs in this data set. At this time, the advantage of HEC can not be reflected by taking the largest number of motifs in the network as the analysis object.

Figure 8(a) shows that HEC and HESC are basically similar to DC in terms of propagation range and propagation rate in Wiki-Vote. However, with the increase of infected nodes, HESC is better than traditional measurement methods in both transmission rate and transmission range (see Fig. 8(b) and 8(c)).

We can see that the transmission range and efficiency of HESC are better than other methods when Top10 node is used as a source of infection from Fig. 9.

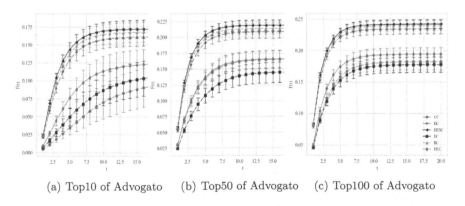

(a) Top10 of Advogato   (b) Top50 of Advogato   (c) Top100 of Advogato

Fig. 7. Experiment of TOP Nodes as initial infectious source node in Advogato

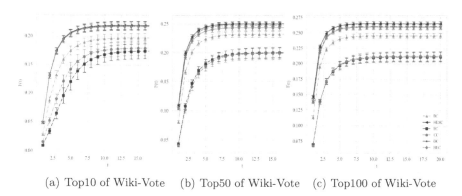

(a) Top10 of Wiki-Vote   (b) Top50 of Wiki-Vote   (c) Top100 of Wiki-Vote

Fig. 8. Experiment of TOP Nodes as initial infectious source node in Wiki-Vote

However, with the increase of infected nodes, the advantages of HESC gradually diminished. The propagation capability of Top nodes obtained by HEC is similar to that of HESC in Fig. 5. Furthermore, from the error bar of Fig. 9, we can see that the results are not sensitive to the dynamic process on networks.

Although HESC method can select more important nodes by the analysis of the experimental results of different data sets, there are some differences for different networks. In soc-Epinions1 network, only Top10 nodes have better propagation ability than other sorting results, which may be due to the large proportion of strongly connected subgraphs in the network. Therefore, when the number of selected source nodes is large, BC's advantages appear.

Table 4, Table 5 and Table 6 show the TOP5 nodes obtained by different methods in three datasets. Ego network is composed of a centered ego, direct contacts namely alters, and the iterations among them. We select the most influential node of the three networks by different centrality measures from Fig. 10, Fig. 11 and Fig. 12 to get its Ego networks. We try to explain why HESC

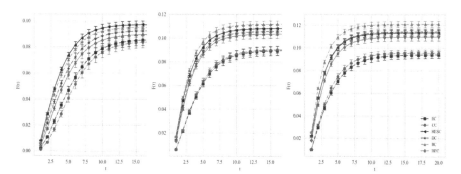

(a) Top10 of soc-Epinions1(b) Top50 of soc-Epinions(c) Top100 of soc-Epinions1

Fig. 9. Experiment of TOP Nodes as initial infectious source node in soc-Epinions 1

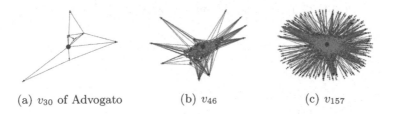

(a) v_{30} of Advogato   (b) v_{46} (c) v_{157}

Fig. 10. Ego-network of Top1 node by various methods in Advogato

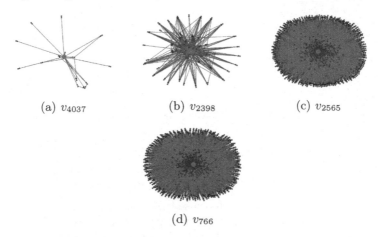

(a) v_{4037} (b) v_{2398} (c) v_{2565}

(d) v_{766}

Fig. 11. Ego-network of Top1 node by various methods in Wiki-Vote

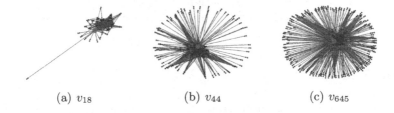

(a) v_{18} (b) v_{44} (c) v_{645}

Fig. 12. Ego-network of Top1 node by various methods in soc-Epinions1

Table 4. Top-5 nodes by different methods in Wiki-Vote

TOP	DC	BC	CC	EC	HEC	HESC
TOP1	2565	4037	2565	2398	2565	766
TOP2	1549	15	1549	4037	766	2565
TOP3	766	2398	15	15	11	1549
TOP4	11	1549	72	4191	2688	457
TOP5	1166	2535	737	2625	1549	11

Table 5. Top-5 nodes by different methods in Advogato

TOP	DC	BC	CC	EC	HEC	HESC
TOP1	157	46	157	46	30	157
TOP2	46	30	46	30	328	597
TOP3	597	328	597	328	126	232
TOP4	30	286	172	438	286	593
TOP5	328	719	328	719	172	793

Table 6. Top-5 nodes by different methods in soc-Epinions1

TOP	DC	BC	CC	EC	HEC	HESC
TOP1	18	18	44	18	18	645
TOP2	645	737	763	401	645	634
TOP3	634	136	634	550	634	44
TOP4	763	790	2066	737	143	71399
TOP5	143	143	645	34	790	763

outperforms others by Ego network in these three networks intuitively. The large solid circle in the center of Fig. 10, Fig. 11 and Fig. 12 is the most influential node obtained by various measurements. Clearly, the Ego network of the most influential node in HESC is more compact than other methods. To some extent, this reflects that when TOP nodes as initial infectious source nodes why the spreading rate of HESC is faster than that of others, and the total number of infected nodes of HESC is also larger than that of others.

5 Conclusions

This paper reconstructs the initial network by high-order structure. In order to describe the high-order information such as the interaction between nodes in the network, it defined the concept of high-order. By fusing the low-order information and high-order information of nodes, the HESC was proposed to identify the vital nodes. At the same time, the network topology and the propagation dynamic model was used to evaluate the node ranked. The results of experiments demonstrate that even though not all the nodes in initial network can form a high-order structure, the nodes that form a higher-order structure play a more influential role in the network. The influential nodes based on the high-order structure show slightly different performance in different networks. The propagation ability are better than that compared with influential nodes identified by other algorithms.

References

1. Zhang, J., Tang, J.: Survey of social influence analysis and modeling. Sci. Sin. Inform. **47**, 967–979 (2017) (in Chinese). https://doi.org/10.1360/N112017-00137
2. Han, Z.M., Chen, Y., Liu, W., Yuan, B.H., Li, M.Q., Duan, D.G.: Research on node influence analysis in social networks. Ruan Jian Xue Bao/J. Softw. **28**(1), 84–104 (2017). (in Chinese). http://www.jos.org.cn/1000-9825/5115.htm
3. Wasserman, S.: Social network analysis methods and applications. Contemp. Sociol. **91**(435), 219–220 (1995)
4. Bonacich, P.F.: Factoring and weighting approaches to status scores and clique identification. J. Math. Sociol. **2**(1), 113–120 (1972)
5. Freeman, L.C.: Centrality in social networks conceptual clarification. Soc. Netw. **1**(3), 215–239 (1978)
6. Freeman, L.C.: A set of measures of centrality based on betweenness. Sociometry **40**(1), 35–41 (1977)
7. Gao, C., Wei, D., Hu, Y., et al.: A modified evidential methodology of identifying influential nodes in weighted networks. Phys. A **392**(21), 5490–5500 (2013)
8. Chen, D., Lü, L., Shang, M., et al.: Identifying influential nodes in complex networks. Phys. A **391**(4), 1777–1787 (2012)
9. Wei, D., Deng, X., Zhang, X., et al.: Identifying influential nodes in weighted networks based on evidence theory. Phys. A **392**(10), 2564–2575 (2013)
10. Benson, A.R., Gleich, D.F., Leskovec, J.: Higher-order organization of complex networks. Science **353**(6295), 163–166 (2016)
11. Benson, A.R.: Tools for higher-order network analysis (2018)
12. Milo, R.: Network motifs: simple building blocks of complex networks. Science **298**(5594), 824–827 (2002)
13. Shen-Orr, S.S., Milo, R., Mangan, S., et al.: Network motifs in the transcriptional regulation network of Escherichia coli. Nat. Genet. **31**(1), 64–68 (2002)
14. Yin, H., Benson, A.R., Leskovec, J., et al.: Local higher-order graph clustering. In: KDD 2017, pp. 555–564 (2017)
15. Zhao, H., Xiaogang, X., Yangqiu, S., et al.: Ranking users in social networks with higher-order structures. In: AAAI (2018)
16. Shafer, G.: A Mathematical Theory of Evidence. by Glenn Shafer, vol. 73, no. 363, pp. 677–678 (1978)
17. Dempster, A.P.: Upper and lower probabilities induced by a multivalued mapping. Ann. Math. Stat. **38**(2), 325–339 (1967)
18. Schank, T., Wagner, D.: Finding, counting and listing all triangles in large graphs, an experimental study, vol. 3503, pp. 606–609. Springer, Heidelberg (2005)
19. Ren, X.L., Lü, L.Y.: Review of ranking nodes in complex networks. Chin. Sci. Bull. (Chin. Ver.) **59**(1197), 1175–1280 (2014) (in Chinese). https://doi.org/10.1360/72013-1280
20. Yan, G., Zhang, M., Luo, H., et al.: Identifying vital nodes algorithm in social networks fusing higher-order information. J. Commun. **40**(10), 109–118 (2019). (in Chinese)
21. Bian, T., Deng, Y.: A new evidential methodology of identifying influential nodes in complex networks. Chaos Solitons Fract. **103**, 101–110 (2017)
22. Kunegis, J.: KONECT - the koblenz network collection. In: Proceedings of the International Conference on World Wide Web Companion, pp. 1343–1350 (2013)
23. Leskovec, J., Huttenlocher, D., Kleinberg, J.: Signed networks in social media. In: CHI 2010 (2010)
24. Richardson, M., Agrawal, R., Domingos, P.: Trust management for the semantic web. In: ISWC (2003)

Location Privacy-Preserving Method Based on Degree of Semantic Distribution Similarity

Rui Liu[1,2], Kaizhong Zuo[1,2(✉)], Yonglu Wang[1,2], and Jun Zhao[1,2]

[1] College of Computer and Information, Anhui Normal University,
Wuhu 241002, Anhui, China
zuokz@ahnu.edu.cn
[2] Anhui Provincial Key Laboratory of Network and Information Security,
Anhui Normal University, Wuhu 241002, Anhui, China

Abstract. While enjoying the convenience brought by location-based services, mobile users also face the risk of leakage of location privacy. Therefore, it is necessary to protect location privacy. Most existing privacy-preserving methods are based on K-anonymous and L-segment diversity to construct an anonymous set, but lack consideration of the distribution of semantic location on the road segments. Thus, the number of various semantic location types in the anonymous set varies greatly, which leads to semantic inference attack and privacy disclosure. To solve this problem, a privacy-preserving method is proposed based on degree of semantic distribution similarity on the road segment, ensuring the privacy of the anonymous set. Finally, the feasibility and effectiveness of the method are proved by extensive experiments evaluations based on dataset of real road network.

Keywords: Location-based services · Road network · Semantic location · Privacy-preserving

1 Introduction

With the rapid development of communication technology and mobile positioning technology, users can use mobile devices such as in-vehicle terminals and mobile phones to obtain their locations anytime and anywhere, thereby application based on Location-based Services (LBS) have become more and more widespread [1–3]. If users wish to receive information from LBS, they have to share their exact location. For example, how to go to the nearest hospital? Meanwhile, users face the risk of leakage of location privacy [4]. More sensitive personal information can be stolen by the attacker. Therefore, how to solve the problem of leakage of location privacy in LBS has attracted the attention of scholars at home and abroad.

Currently, several schemes have been proposed by scholars [5–9] to protect the location privacy of users. For example, K-anonymous algorithm is usually used in Euclidean space [5, 6], where users can move freely. These algorithms construct anonymous set including k users instead of the exact location of the user, which makes it difficult for an attacker to distinguish the exact user from other anonymous users. However, the security of the K-anonymous algorithm is compromised when attackers

© Springer Nature Singapore Pte Ltd. 2020
J. Zeng et al. (Eds.): ICPCSEE 2020, CCIS 1257, pp. 118–129, 2020.
https://doi.org/10.1007/978-981-15-7981-3_9

mined road network information and semantic location information. Therefore, it is extremely important to propose location privacy-preserving scheme on the road network.

Location privacy-preserving schemes on the road network are mostly based on (K, L)-model, which means the anonymous set not only includes K anonymous users, but also includes L road segments. Chow et al. [10] designed a location privacy-preserving method on road network, which blurs the user's exact position into several adjacent road segments, and considers the query cost and query quality when constructing anonymous set. Pan et al. [11] constructed a unified set of anonymous users for users, and formed anonymous region through connected road segments. However, these methods do not take the information of semantic location on road network environment into consideration. Li et al. [8] used road network intersections and semantic locations to divide the road network into independent non-overlapping network Voronoi units, and optionally added neighboring Voronoi units to form anonymous region to meet semantic security. Xu et al. [12] proposed a global and local optimal incremental query method based on location semantic sensitivity rate. By reducing the number of unnecessary locations, the query overhead is reduced, and the user service quality is improved.

In the above methods guarantee some degree of privacy. Nevertheless, they all share a common drawback. The semantic location is represented by the nearest road network intersection, which is given semantic location information (semantic location type, popularity, sensitivity, etc.). If there are many different types of semantic locations in the intersection of the road network, it will be caused the problem that semantic locations cannot be represented in the road network environment. Chen et al. [13] proposed a privacy-preserving method based on location semantics, which fully considers the user's personalized privacy requirements. In this method, semantic location is directly distributed on the road segment, which makes the road network model more realistic. But the distribution of semantic location types on the road network is not considered, so the number of various semantic location types in the anonymous set varies greatly, which leads to semantic inference attack illustrated by the following example.

Example 1. Figure 1 shows the scenario that a user named Alice requests services through a mobile phone with GPS on road e_1. In order to prevent Alice's location leakage, the method based on K-anonymous and L-segment diversity protects Alice's location privacy. In our example, we assume $K = 45$, $L = 3$, then the anonymous set may be e_1, e_2, e_3. Unfortunately, it is easy for an attacker to infer that Alice is at school, and knows that Alice's identity information may be a student or a teacher. Because the user's semantic location type school accounts for a large proportion of anonymous sets. Therefore, even though Alice's location blur into several adjacent road segments, it's easy to a semantic inference attack.

The rest of the paper is organized as following. We introduce some necessary preliminaries such as fundamental concepts and models in Sect. 2, whereas Sect. 3 introduces the algorithms. The experimental settings and results of our experiments are illustrated in Sect. 4. Finally, Sect. 5 concludes this paper.

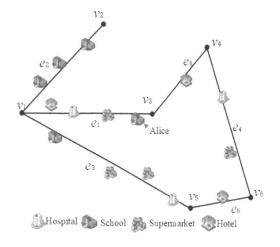

Edge Id	Number of users
e_1	10
e_2	15
e_3	25
e_4	20
e_5	10
e_6	15

(a)Simplified Road Network   (b) Number of users on the edge

Fig. 1. A snapshot of mobile users in road network

2   Preliminaries

2.1   Related Definition

Definition 1 (Semantic location). $loc = (slid, eid, x, y, type)$ is the semantic location on the road network, *slid* denotes the number of the semantic location, *eid* denotes the number of the road segment where the semantic location is located, x and y are the coordinate of the semantic location. And the *type* is the semantic location type, which contains m types and $Type = \{type_1, type_2, \ldots, type_m\}$ is types of the semantic location.

Definition 2 (Semantic location popularity). We use it to describe the popularity of a semantic location type. Every semantic location type $type_i \in Type$ has their own popularity pop_{type_i}. The set $POP = \{pop_{type_1}, pop_{type_2}, \ldots, pop_{type_m}\}$ indicates the popularity of all semantic location type.

Definition 3 (Vector of road segment popularity). It describes all the semantic location information on the road segment, which is consisted of the popularity and number of semantic locations. And it can be denotes as $\vec{e_i}$.

Definition 4 (Semantic Road network). A road network is an undirected graph $G = (V, E)$ with a node set V and an edge set E. A node $v_i \in V$ denotes the intersection of the road segment on the road network. While an edge $e_i = (eid, v_i, v_j, \vec{e_i}) \in E$ is the road segment, connects two nodes v_i and v_j, with *eid* is the road segment number, $\vec{e_i}$ denote the vector of the road segment popularity.

Example 2 (Semantic Road network). Figure 1 (a) displays a simplified semantic road network model, in which an edge is associated with the vector of the road segment popularity. For example, we assume that the popularity of hospitals, schools, super-markets, and hotels are 0.3, 0.3, 0.3, 0.1. The vector of road segment popularity on the road segment e_1 and e_2 is $\vec{e_1} = [0.3, 0.3, 0.3, 0.1]$ and $\vec{e_2} = [0, 0.9, 0, 0]$, respectively.

Definition 5 (Degree of semantic distribution similarity). It describes the degree of semantic distribution similarity between road segments, which is determined by the vector of road segment popularity and in a range of values [0, 1].

$$\delta_{\vec{e_i},\vec{e_j}} = \frac{\sum\limits_{k=1}^{n} |e_{ik} - e_{jk}|}{\sum\limits_{k=1}^{n} e_{ik} + \sum\limits_{k=1}^{n} e_{jk}}, \delta_{\vec{e_i},\vec{e_j}} \in [0, 1] \qquad (1)$$

Where $\vec{e_i}$ and $\vec{e_j}$ instead the vector of road segment popularity of the road segment e_i and e_j, respectively. The smaller $\delta_{\vec{e_i},\vec{e_j}}$ is, the higher degree of semantic distribution similarity of road segments is.

Definition 6 (Semantic location sensitivity). It denotes the sensitivity of semantic location type. Every user can set their own sen_{type_i} for each type of semantic location $type_i \in Type$ freely, and $Sen_u = \{sen_{type_1}, sen_{type_2}, \ldots, sen_{type_3}\}$ denotes sensitivity of all semantic location types.

Definition 7 (Anonymous set). It is a cloaked set of several adjacent road segments such that satisfies the user specified privacy requirements.

Definition 8 (Anonymous set popularity). The popularity Pop_{AS} of the anonymous set,

$$Pop_{AS} = \sum_{i=1}^{|Type|} \frac{|AS.locs.type = type_i|}{|AS.Locs|} pop_{type_i} \qquad (2)$$

Definition 9 (Anonymous set sensitivity). The sensitivity Sen_{AS} of the anonymous set,

$$Sen_{AS} = \sum_{i=1}^{|Type|} \frac{|AS.locs.type = type_i|}{|AS.Locs|} sen_{type_i} \qquad (3)$$

|Type| in above is the number of semantic location types contained in the anonymous set; |AS.Locs| is the number of semantic locations included in the anonymous set.

Definition 10 (Anonymous set privacy). The privacy PM_{AS} of the anonymous set,

$$PM_{AS} = \frac{Pop_{AS}}{Sen_{AS}} \tag{4}$$

Obviously, the popularity and sensitivity of anonymous set directly affect the privacy of anonymous set. The popularity of anonymous set is higher and the sensitivity is lower, the privacy of the anonymous set is higher.

Definition 11 (Privacy requirement). The user's privacy requirements is denote as $PR(UN, SN, \delta, Sen_u)$. UN and SN denotes the user-defined lowest number of mobile users and road segments, respectively; with δ is the highest value of degree of semantic distribution similarity on road segments; and Sen_u is sensitivity of different semantic location types.

2.2   System Architecture

Figure 2 shows the classic centralized server architecture [14], which mainly contains three components: user, anonymous server and LBS server. In this architecture, users can obtain their location information from the base station. Then, send it to the anonymous server together with query content and privacy requirements (step ①). Subsequently, the anonymous server uses the semantic location privacy-preserving module to blur the user's location into a set of road segment that meet the user's privacy requirement, and sends anonymous query to the LBS server (step ②). After the LBS server gets the candidate results for anonymous query and sends it to the anonymous server (step ③). Finally, the anonymous server computes the candidate results through the filter module and delivers exact result to the query user (step ④).

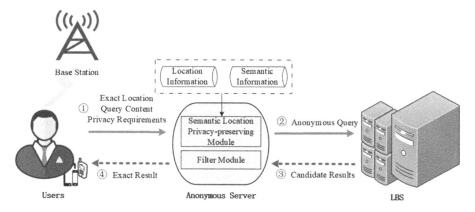

Fig. 2. System architecture

3   Semantic Location Privacy-Preserving Method

The method selects the appropriate adjacent road segments to construct an anonymous set according to user-defined privacy requirements. And it consists of two algorithms.

Algorithm 1 (ORSSA) is used to determine the optimal road segment in this paper. Calculate the popularity vector of the user's current road segment and adjacent road segments set, and take the road segments to candidate road segment sets which less than or equal to δ. Subsequently, select an optimal road segment from these road segment to join the anonymous set. Here, the optimal road segment means that the privacy level of the anonymous set composed of the road segment is highest. Detail process is depicted as following:

Algorithm 1 Optimal Road Segment Selection Algorithm (ORSSA)

Input: user u, current anonymous set CAS, adjacent road segments set $AdjacentEdges_{set}$, threshold δ, sensitivity set $Sens$, popularity set POP

Output: $OEdge$

1) $OEdge = \varnothing$; $CandEdges_{set} = \varnothing$; $PM_{set} = \varnothing$;

2) $MAX=0$;

3)   for each edge in $AdjacentEdges_{set}$

4)   assign the edge where user u located to e_1, calculate the vector of road segment popularity $\vec{e_1}$;

5)   assign edge to e_2, calculate the vector of road segment popularity $\vec{e_2}$;

6)   if $\sum_{k=1}^{n}|e_{1k} - e_{2k}| / (\sum_{k=1}^{n}e_{1k} + \sum_{k=1}^{n}e_{2k}) \leq \delta$ then

7) $CandEdges_{set} = CandEdges_{set} \cup edge$;

8)   end if

9)   end for

10) for each edge in $CandEdges_{set}$

11) $PM = Pop_{(CAS \cup edge)} / Sen_{(CAS \cup edge)}$;

12) $PM_{set} = PM_{set} \cup PM$;

13) $MAX = \{PM \mid \max\{PM \in PM_{set}\}\}$, and the corresponding edge assigned to $OEdge$;

14) end for

15) return $OEdge$

The next algorithm (DSDSPPA) is a privacy-preserving algorithm based on degree of semantic distribution similarity of road segments. The idea of our method is to start from the road segment where the user is located. If this road segment meets the user's privacy requirements, we return it as anonymous set to the LBS server. Otherwise, we pick the optimal road segment from all adjacent road segment until the user's privacy requirements are met. Detail process is depicted as following:

Algorithm 2 Privacy-preserving algorithm based on degree of semantic distribution similarity (DSDSPPA)

Input: user u, privacy requirements PR, popularity set POP
Output: Anonymous set AS

1) $AS = \varnothing$;
2) assign the edge where user u located to $OEdge$;
3) $AS = AS \cup OEdge$;
4) while $|AS.users| < PR.UN$ or $|AS.edges| < PR.SN$
5) $AdjacentEdges_{set} = GetEdges(AS)$;//find adjacent road segments of AS
6) $OEdge = ORSSA(AS, AdjacentEdges_{set}, PR.\delta, PR.Sen_u, POP)$;
7) $AS = AS \cup OEdge$;
8) $AdjacentEdges_{set} = \varnothing$;
9) *end while*
10) return AS;

---

4 Experimental Evaluation

Our algorithms are executed in java based on MyEclipse environment and all experiments are performed on Intel (R) Core(TM) i5-9400 CPU @ 2.90 GHz and 16 GB main memory with Microsoft Windows 10 Professional.

4.1 Datasets and Parameter

(1) Datasets

In this paper, we use the road network in California which includes 21048 vertices and 21693 edges. And real road network dataset contains semantic location of various categories, e.g., hospital, park, airport, bar, building [15]. The second experimental dataset used in this paper was collected from Gowalla, which has more than 6442890 check_ins made by users over the period of Feb.2009-Oct.2010 [16]. Then, we filter the user's check_ins data in California and calculate the popularity of different semantic location types.

(2) Query generator

We generate 10000 mobile users through by the Network Generator of moving objects [17] and choose 1,000 users send a query randomly. Table 1 depicts the parameters of our experiment.

Table 1. Parameter setting

Parameter	Default values	Range
PR.UN	25	[10,40]
PR.SN	6	[3,15]
PR.SN$_{max}$	20	
δ	0.7	[0.1,1]
The number of mobile users	10000	
The number of users that issue queries	1000	
The number of semantic location types	63	

4.2 Experimental Results

Assess the feasibility and effectiveness of the algorithms DSDSPPA, we compare algorithm LSBASC [13] and Enhance-LSBASC [18] from anonymous success rate, average anonymous execution time, relative anonymous and degree of privacy leakage.

(1) Effect of Number of Privacy Requirement Users

Figure 3 shows the effect of different number of privacy requirement users on three algorithms when PR.SN = 6, δ = 0.7, PR.SN$_{max}$ = 20. As can be seen from Fig. 3(a) that three algorithm show decreasing trend on the aspect of anonymous success rate. With the increase of number of privacy requirements users, more road segments are needed to anonymous set. So that the number of road segments is higher than PR.SN$_{max}$ lead to the failure of anonymity. However, the anonymous success rate of DSDSPPA algorithm is always higher than the other algorithms. In Fig. 3(b), the average anonymous execution time of the algorithm DSDSPPA is between the other algorithms, but it is always lower than the algorithm LSBASC. And the average anonymous execution time of all algorithms show increasing trend. The reason of this is that more users must be added in anonymous set in order to meet privacy requirement, so does more road segments. The algorithm DSDSPPA and LSBASC only add one road segment every time, while the algorithm Enhance-LSBASC adds several adjacent road segments, so the average anonymous execution time is less than the algorithm DSDSPPA and LSBASC.

Figure 3(c) shows that the relative anonymity of algorithm DSDSPPA is between the other algorithms. Because the algorithm Enhance-LSBASC chooses several adjacent road segments to the anonymous every time, the number of mobile users contained in the anonymous set is higher than algorithm DSDSPPA and LSBASC. Figure 3(d) is shown that the privacy leakage degree of the three algorithms. Algorithm DSDSPPA is always lower than the algorithm Enhanced-LSBASC and LSBASC, and the fluctuation degree is smallest. In order to satisfy the number of users with privacy requirements, new adjacent road segments need to be added to the anonymous set. Because the algorithm DSDSPPA selects the road segments with high degree of semantic distribution similarity as the candidate road segments, it balances the number of various semantic positions in the anonymous set, so that the attacker cannot infer the semantic position type of the user. However, the algorithm Enhanced-LSBASC and LSBASC only consider the privacy of the anonymous set, so they fluctuate greatly.

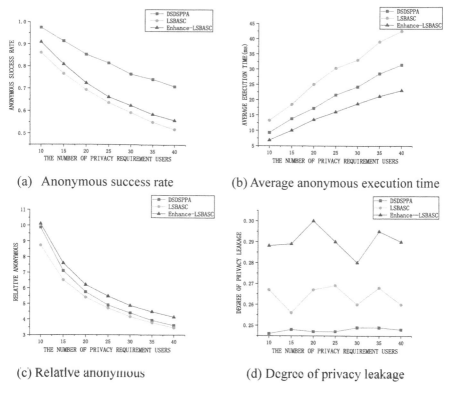

(a) Anonymous success rate

(b) Average anonymous execution time

(c) Relative anonymous

(d) Degree of privacy leakage

Fig. 3. Results of different number of privacy requirement users

(2) Effect of Number of Privacy Requirement Road Segments

Figure 4 shows the effect of different number of privacy requirement road segments on three algorithms when $PR.UN = 25$, $\delta = 0.7$, $PR.SN_{max} = 20$. Figure 4(a) shows that different number of privacy requirement road segments don't have an impact on anonymous success rate. In the case of the number of users with privacy requirements remains unchanged, as long as the number of road segments is within the allowed range, the anonymous success rate will not change. According to the experimental result in Fig. 4(b), the average anonymous execution time of all algorithms show increasing trend, but the execution time of DSDSPPA algorithm is between LSBASC and Enhance-LSBASC algorithm and always lower than LSBASC algorithm. In order to meet the number of privacy requirement road segments, more road segments should add to anonymous set. The algorithm DSDSPPA and LSBASC only add one road segment every time, while the algorithm Enhance-LSBASC adds several adjacent road segments, so the average anonymous execution time is less than the DSDSPPA algorithm and LSBASC.

According to the experimental result in Fig. 4(c), the relative anonymity of the three algorithms is increasing. And it can be known that the relative anonymity of the algorithm DSDSPPA is between the other algorithms. With the number of road segments added to the anonymous set to meet the privacy requirements, at the same time, the number of mobile users is also increasing. Figure 4(d) is shown that the privacy leakage degree of the algorithm DSDSPPA is always lower than the algorithm Enhanced-LSBASC and LSBASC, and the fluctuation degree is smallest. Because the algorithm DSDSPPA selects the road segments with high degree of semantic distribution similarity as the candidate road segments, it balances the number of various semantic positions in the anonymous set.

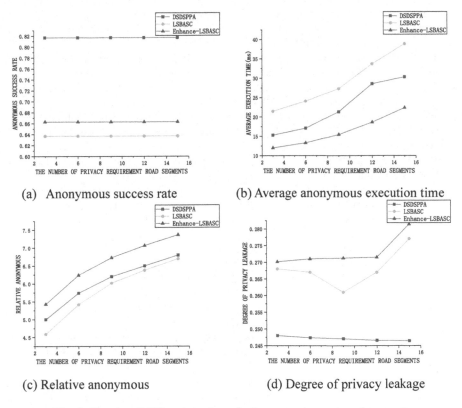

(a) Anonymous success rate

(b) Average anonymous execution time

(c) Relative anonymous

(d) Degree of privacy leakage

Fig. 4. Results of different number of privacy requirement road segments

5 Conclusion

As constructing anonymous set does not consider the distribution of semantic location types on the road network, the number of various semantic location types in the anonymous set varies greatly, which leads to semantic inference attack and privacy disclosure. Therefore, this paper proposes a location privacy-preserving method based on semantic location information on the road segment. It increases the indistinguishability of users' semantic location types and guarantees the privacy of anonymous sets. Finally, extensive experiments evaluations based on dataset of real road network show our algorithms is effective and feasible.

Acknowledgement. This paper was supported by the National Natural Science Foundation of China under Grant No. 61672039 and 61370050; and the Key Program of Universities Natural Science Research of the Anhui Provincial Department of Education under Grant No. KJ2019A1164.

References

1. Wan, S., Li, F.H., Niu, B., et al.: Research progress of location privacy protection technology. Chin. J. Commun. **37**(12), 124–141 (2016)
2. Zhang, X.J., Gui, X.L., Wu, Z.D.: Review of research on privacy protection of location services. Chin. J. Softw. **26**(9), 2373–2395 (2015)
3. Sun, Y., Chen, M., Hu, L., et al.: ASA: against statistical attacks for privacy-aware users in Location Based Service. Future Gener. Comput. Syst. **70**(70), 48–58 (2017)
4. Zhang, Y., Szabo, C., Sheng, Q.Z.: SNAF: observation filtering and location inference for event monitoring on twitter. World Wide Web **21**(2), 311–343 (2018)
5. Feng, Y., Xu, L., Bo, S.: (k, R, r)-anonymity: a light-weight and personalized location protection model for LBS query. In: ACM Turing Celebration Conference-China, pp. 1–7 (2017)
6. Ma, M., Du, Y.: USLD: a new approach for preserving location privacy in LBS. In: Workshop on Information Security Applications, pp. 181–186 (2017)
7. Cui, N., Yang, X., Wang, B.: A novel spatial cloaking scheme using hierarchical hilbert curve for Location-Based Services. In: Cui, B., Zhang, N., Xu, J., Lian, X., Liu, D. (eds.) WAIM 2016. LNCS, vol. 9659, pp. 15–27. Springer, Cham (2016). https://doi.org/10.1007/978-3-319-39958-4_2
8. Li, M., Qin, Z., Wang, C., et al.: Sensitive semantics-aware personality cloaking on road-network environment. Int. J. Secur. Appl. **8**(1), 133–146 (2014)
9. Li, H., Zhu, H., Du, S., et al.: Privacy leakage of location sharing in mobile social networks: attacks and defense. IEEE Trans. Dependable Secure Comput. **15**(4), 646–660 (2016)
10. Chow, C., Mokbel, M.F., Bao, J., et al.: Query-aware location anonymization for road networks. Geoinformatica **15**(3), 571–607 (2011)
11. Pan, X., Chen, W.Z., Sun, Y., et al.: Continuous queries privacy protection algorithm based on spatial-temporal similarity over road networks. Chin. J. Comput. Res. Dev. **54**(9), 2092–2101 (2017)
12. Xu, M., Xu, H., Xu, C.: Personalized semantic location privacy preservation algorithm based on query processing cost optimization. In: Wang, G., Atiquzzaman, M., Yan, Z., Choo, K.-K.R. (eds.) SpaCCS 2017. LNCS, vol. 10656, pp. 153–168. Springer, Cham (2017). https://doi.org/10.1007/978-3-319-72389-1_14
13. Chen, H., Qin, X.: Location-semantic-based location privacy protection for road network. Chin. J. Commun. **37**(8), 67–76 (2016)
14. Wang, Y., Zuo, K., Liu, R., Guo, L.: Semantic location privacy protection based on privacy preference for road network. In: Vaidya, J., Zhang, X., Li, J. (eds.) CSS 2019. LNCS, vol. 11983, pp. 330–342. Springer, Cham (2019). https://doi.org/10.1007/978-3-030-37352-8_30
15. Li, F., Cheng, D., Hadjieleftheriou, M., Kollios, G., Teng, S.-H.: On trip planning queries in spatial databases. In: Bauzer Medeiros, C., Egenhofer, M.J., Bertino, E. (eds.) SSTD 2005. LNCS, vol. 3633, pp. 273–290. Springer, Heidelberg (2005). https://doi.org/10.1007/11535331_16

16. Cho, E., Myers, S.A., Leskovec, J., et al.: Friendship and mobility: user movement in location-based social networks. In: ACM SIGKDD International Conference on Knowledge Discovery and Data Mining (KDD), pp. 1082–1090 (2011)
17. Brinkhoff, T.: A framework for generating network-based moving objects. GeoInformatica **6** (2), 153–180 (2002)
18. Lv, X., Shi, H., Wang, A., et al.: Semantic-based customizable location privacy protection scheme. In: International Symposium on Distributed Computing and Applications for Business Engineering and Science, pp. 148–154. IEEE Computer Society (2018)

Supervisable Anonymous Management of Digital Certificates for Blockchain PKI

Shaozhuo Li[1,2(✉)], Na Wang[1,2], Xuehui Du[1,2], and Xuan Li[3]

[1] National Digital Switching System Engineering and Technological Research Center, Zhengzhou 450000, China
494187944@qq.com
[2] Zhengzhou Science and Technology Institute, Zhengzhou 450000, China
[3] Jiuquan Satellite Launch Center, Jiuquan 732750, China

Abstract. Aiming at the requirement of anonymous supervision of digital certificates in blockchain public key infrastructure (PKI), this paper proposes a ring signature with multiple indirect verifications (RS-MIV). This mechanism can ensure multiple and indirect verification of certificate signer identity while preserving its anonymity. On this basis, a supervisable anonymous management scheme was designed based on smart contracts, which realizes the anonymity of certificate authority nodes, the anonymous issuance of digital certificates, the anonymous verification of digital certificates, and the traceability of illegal certificate issuers in the blockchain PKI. It is proved that the scheme can guarantee the anonymity and traceability of the certificate issuer's identity at an acceptable cost.

Keywords: Blockchain · Digital certificate · PKI · Ring signature

1 Introduction

Building a blockchain-based PKI and realizing open, transparent, and distributed management of digital certificates by uploading them to the blockchain can effectively solve the security problems caused by third-party CAs, which are being attacked or having weak security practices [1, 2]. These can also meet the cross-domain verification requirements of the digital certificates brought by the increasingly widespread application of distributed computing modes, such as Internet of Things, Big Data, and cloud computing [3–5].

Currently, blockchain-based PKI does not allow CA nodes (Node of CA user on blockchain) to be anonymous to ensure the credibility of the PKI. However, in some applications or scenarios where the commercial CAs are not willing to disclose their privacy, the blockchain-based PKI needs to ensure the anonymity of the CA nodes and realize anonymous management of the digital certificates. However, the anonymity of the CA nodes cannot be guaranteed without reducing the credibility of the PKI. For example, if an illegal digital certificate is detected in the blockchain, it must be accurately traced back to find the issuer of that certificate. As a result, this paper studies the supervisable anonymity management of digital certificates based on blockchain PKI.

© Springer Nature Singapore Pte Ltd. 2020
J. Zeng et al. (Eds.): ICPCSEE 2020, CCIS 1257, pp. 130–144, 2020.
https://doi.org/10.1007/978-981-15-7981-3_10

Verifiable ring signature mechanism can prove the signer's real identity by providing some relevant data when needed. However, it cannot be directly used to achieve the supervisable anonymity management of digital certificates as this needs multiple verification without destroying the anonymity. Thus, the indirect verification of the identity of the certificate issuer can be realized when it does not cooperate.

Based on this, this paper proposes a Ring Signature with Multiple Indirect Verifications (RS-MIV), which is based on RSA (Ron Rivest, Adi Shamir, Leonard Adleman algorithm) ring signature mechanism. By introducing one-to-one verification public and private keys corresponding to the digital certificates, the signer's signature of secret information is used instead of the initial secret value. In similar, to ensure the anonymity of the signer and realize multiple and indirect verification of the signer's identity when needed, binding ring signature is utilized. As a result, this paper designs a supervisable anonymity management scheme based on smart contracts, which can realize: i) the anonymity of CA (Certificate Authority) nodes; ii) the anonymity issuance of digital certificates; iii) the anonymity verification of digital certificates, and iv) the traceability of illegal certificate issuers in the blockchain PKI, and meet the actual demand of CA node Supervisable anonymity. It is proved that the proposed scheme can guarantee the anonymity and traceability of the identity of the certificate issuer at an acceptable cost.

2   Related Work

2.1   Current Literature on Blockchain PKI

Currently, the research on blockchain PKI mainly focuses on Certcoin, IKP, SCPKI, and Permor. Certcoin [6] uses the technical characteristics of blockchain decentralization to build a fully decentralized PKI by binding the user identity and the public keys in the blockchain. It takes Namecoin [7] as its underlying platform. IKP (instant karma PKI) [8] uses the characteristics of automatic and compulsory execution of smart contracts to detect the CA nodes that behave improperly or are under attack. It uses the economic incentive mechanism of Ethereum to reward the CA nodes that issue certificates correctly, and punish the nodes that issue illegal certificates. SCPKI (Smart Contract-based PKI and Identity System) [9] uses Ethereum smart contracts to implement PGP (Pretty Good Privacy), thereby building a fully distributed PKI system. Similar to PGP, SCPKI adopts a web of trust model to measure the credibility of public keys by using the trust relationship between the users. Pemcor [10] proposes to build two blockchains to store the hash value of the generated digital certificate and the hash value of the revoked digital certificate. These blockchains should be controlled by an authority, such as a bank or government. Therefore, if the hash value of the certificate is in the generated certificate blockchain, i.e., it is not in the revoked certificate, the certificate is valid. Otherwise, it is invalid.

This paper found out that the node identity should be completely open to ensure the credibility of the blockchain PKI. However, in some special applications or scenarios where the commercial CAs are not willing to disclose their privacy, blockchain PKI needs to realize the manageable anonymity of the CA nodes. That is to say, the CA nodes are allowed to issue certificates anonymously and upload the issued certificates to the blockchain. However, after discovering the illegal certificates, the identity of CA nodes that issued and uploaded the digital certificates must be confirmed for accountability to ensure the credibility of the blockchain PKI system.

2.2 Blockchain's Anonymous Mechanism

With the widespread adaptation of the blockchain in finance, the anonymity became compulsory. This mechanism can ensure the anonymity of transaction users to avoid the third parties getting the identity information of both parties from the transaction address and transaction content on the blockchain. For example, Dascoin introduces chain mixing and blinding technology to ensure user anonymity by mixing multiple users. In similar, Monroe coin hides address and ring signature. Zero Cash [11] uses zkSNARK (Zero-knowledge Succinct Non-interactive Arguments Of Knowledge) to achieve anonymity. This is by far the most secure and effective method; however, it is based on the NP (Non-deterministic Polynomial) problem, which limits its application, but makes it suitable for anonymity problems that are difficult to convert into NP problems (for example, the digital certificate Supervisable anonymity management problem solved in this paper). Furthermore, the initialization parameters of this method are complex.

Currently, the anonymity of blockchain is realized by changing the recording mode of transaction data, which hides transaction amount in the financial environment. However, these anonymous mechanisms achieve complete anonymity, and the node identity is not exposed from beginning to end, which does not meet the requirements of the CA node that can supervisable anonymity proposed in this paper.

2.3 Verifiable Ring Signature

Ring signature was first proposed by Rivest in 200112, which is mainly used in applications where the signer needs to be anonymous, such as anonymous voting and anonymous election.

A secure ring signature scheme should have the following properties:

(1) Correctness: the signature output by any member in the ring after executing the ring signature elaboration algorithm can pass the signature verification algorithm in the system;
(2) Anonymity: given a ring signature, any verifier will not identify the real signer with a probability greater than $1/n$, where n is the number of members in the ring.
(3) Unforgeability: any user who is not in the ring $U = \{U_1, U_2, \ldots U_n\}$ cannot effectively generate a message signature.

The concept of verifiable ring signature was proposed by LV [13] in 2003, which means that the real signer can prove their identity when necessary by presenting some relevant data. In 2004, Gan Zhi et al. proposed two verifiable ring signature schemes [14], based on: i) one-time identity verification; ii) zero knowledge verification. The latter adds some secret identity information to the initial parameters of the ring signature, and then confirms the identity of the signer by verifying the correctness of the secret identity information. This scheme can only achieve one-time verification of the signer's identity. In the former, the signer sets the initial value v as a product of two large prime numbers, and then uses zero knowledge proof to validate that it knows the decomposition of the initial value to prove its identity. In 2006, Zhang et al. proposed a verifiable ring signature based on Nyberg rueppel signature [15]. This scheme only used hash function to realize ring signature, which is suitable for small computation and signature scale. In 2007, Wang et al. put forward an extension scheme based on RSA ring signature [16], which realizes verifiable ring signature based on the designated confirmer signature and the verifier verification; however, the verification process needs multiple interactions, and the calculation is relatively complex. In 2008, I. Jeong et al. proposed a linkable ring signature scheme with strong anonymity and weak linkage [17], and proved that it can be used to construct an effective verifiable ring signature scheme, which is suitable for the ring signature scenarios with linkability requirements. In 2009, Luo Dawen et al. proposed a certificateless verifiable ring signature mechanism by combining certificateless Cryptosystem with verifiable ring signature mechanism [18]. This overcame the key escrow problem of the identity-based cryptosystem and avoided the storage and management problem of public key based on certificate cryptosystem. Because it is found that the scheme does not satisfy the non-repudiation, Li Xiaolin et al. proposed the corresponding improvement scheme [19]. Because it is found that Li Xiaolin's improvement scheme does not satisfy the non forgery, Zhang Jiao et al. proposed the corresponding improvement scheme [20]. In 2012, Qin et al. extended the verifiability based on RSA ring signature [21]. The initial value of the ring signature was replaced by private information and hash value related to signature, which allowed the signer to prove itself at any time, but only once. In 2013, based on a forward secure ring signature algorithm, Yang Xudong et al. proposed an improved verifiable strong forward secure ring signature scheme [22] by using the method of double private key update that guaranteed the forward and backward security of the ring signature. In 2017, bultel x et al. proposed a verifiable ring signature mechanism based on the DDH (decision-making Diffie Hellman hypothesis) and zero knowledge proof [23]. They proved the security of the mechanism under the random oracle model, which was applicable to the ring signature scenarios with non-linkable requirements.

In this paper, we find that the current verifiable ring signature mechanisms cannot be directly used in the supervisable anonymous management of digital certificates. The reasons are as follows: (1) some mechanisms, such as Gan Zhi's one-time authentication mechanism [14], Zhang's verifiable ring signature mechanism based on Nyberg rueppel signature [15], Qin's verifiable extension mechanism based on RSA ring signature [21], can verify the real identity of the signer only once, and after one-time verification, the signer's identity will be completely exposed. This means that the anonymity of the signer can no longer be guaranteed. However, in the scenario where the digital certificate can be managed anonymously, it is required to verify the signer's identity many times without affecting the anonymity of the signer; (2) all mechanisms require the signer to actively expose the evidence to verify the signer's identity. However, in the case of the digital certificate's supervisable aonymous management, there is the possibility that the CA node with malicious signature will not actively expose its identity. Therefore, it is necessary to confirm the signer's identity when the digital certificate issuer does not actively expose its identity.

2.4   RSA-Based Ring Signature Mechanism

Since the PKI mainly uses RSA algorithm to sign digital certificates, this paper focuses on improving the RSA based ring signature mechanism. Below, the principles of the RSA-based ring signature mechanism is described.

By supposing that the ring size is r, ring member is $A_1, A_2, A_3 \ldots A_r$, and each member has an RSA public key $P_i = (n_i, e_i)$, the one-way threshold replacement is: $f_i(x) = x^{e_i} (\bmod n_i), f_i \in Z_n$.

It is concluded that only A_i knows how to use the threshold information to effectively calculate inverse permutation $f_i^{-1}(x)$. E is a publicly defined symmetric encryption algorithm, so that for any length l of key k, the function E_k is a replacement on the bit string b. Define the composite function as $C_{k,v}(y_1, y_2, \ldots y_r)$, input as key k, initialization variable as v, and random number set as $\{0, 1\}^b$.

The ring signature process of the message m to be signed is as follows:

(1) The signer calculates hash for the signed message m, then symmetric key $k = h(m)$. The signer chooses a random value from $\{0, 1\}^b$ as v.
(2) The signer chooses x_i uniformly and independently from $\{0, 1\}^b$, and calculate $y_i = f_i(x_i)$.
(3) The signer solves y_s from $C_{k,v}(y_1, y_2, \ldots y_r) = v$.
(4) The signer uses its threshold knowledge to solve $x_s = f_s^{-1}(y_s)$.
(5) The output ring signature is: $(P_1, P_2, \ldots P_r; v; x_1, x_2, \ldots x_r)$.

The verifier verifies the ring signature $(P_1, P_2, \ldots P_r; v; x_1, x_2, \ldots x_r)$ as follows:

(1) The verifier calculates $y_i = f_i(x_i)$ for $i = 1, 2, \ldots, r$.
(2) The verifier calculates $k = h(m)$ for the encrypted message.
(3) The verifier calculates whether y_i satisfies $C_{k,v}(y_1, y_2, \ldots y_r) = v$. If so, the signature is legal. Otherwise, the signature is rejected.

3 Ring Signature Mechanism with Multiple Indirect Verifications

To verify the identity of the anonymous signer many times without exposing its identity, this paper improves the RSA-based ring signature mechanism and proposes a ring signature with multiple indirect verifications (RS-MIV).

In RS-MIV mechanism, before signing the digital certificate subject, the issuer synchronously generates a pair of public and private keys that are uniquely bound to the certificate. Then the issuer signs the digital certificate and the public key together. To verify the signer of a certificate many times, the signer only needs to show the signature of a message with the verification private key. To resist replay attack, the verification of certificate issuer is performed by challenge-response. To be able to confirm the identity of the issuer without revealing its identity, the RS-MIV uses the issuer's digital signature of the random number r and its own digital certificate serial number as secret information to generate parameters in the RSA ring signature

$v = h(sig(r, Cid))$, where the random number r is public. Thus, the legal certificate issuer can prove its validity to the verifier by generating its own digital signature of random number r and the certificate serial number. By excluding legal nodes, malicious nodes can be detected indirectly, i.e., the indirect verification of the identity of the digital certificate issuer can be realized.

The mechanism of the RS-MIV is detailed below:

(1) Generation of the ring signature:

1) The signer generates a pair of public and private keys for the certificate to be signed, i.e., the verification public and private keys. The public key is (n_c, e_c) and the private key is d_c. The private key is saved by the signer, which is not disclosed.

2) Generate symmetric key as $k = h(m, n_c, e_c)$ and initial value as $v = h(sig(r, Cid))$, where m is the certificate information to be signed, r is a random number, Cid is the serial number of the digital certificate of the signer, and $sig(r, Cid)$ is the digital signature of the signer to r and Cid.

3) The signer chooses x_i uniformly and independently from $\{0, 1\}^b$ and calculates $y_i = f_i(x_i)$.

4) The signer solves y_s from $C_{k,v}(y_1, y_2, \ldots y_r) = v$.

5) The signer uses its threshold knowledge to solve $x_s = f_s^{-1}(y_s)$.

6) The output ring signature is $(P_1, P_2, \ldots P_l; v; x_1, x_2, \ldots x_l; n_c, e_c, r)$.

(2) Ring signature verification:

1) For $i = 1, 2, \ldots, r$, calculate $y_i = f_i(x_i)$.

2) The verifier calculates $k = h(m, n_c, e_c)$ for the encrypted message.

3) The verifier calculates whether y_i satisfies $C_{k,v}(y_1, y_2, \ldots y_r) = v$. If so, the signature is legal. Otherwise, the signature is rejected.

(3) Ring signer authentication based on challenge-response:

1) The verifier generates a random number a and the digital certificate number to be verified, then it sends both to the signer.

2) The signer obtains the corresponding verification private key of the certificate according to the number of the digital certificate to be verified. Then it signs the random number with the verification private key of the certificate to be verified. Finally, it outputs $sig_c(a)$.

3) The verifier decrypts $sig_c(a)$ with the verification public key (n_c, e_c) in the certificate ring signature. If $sig_c^{e_c} = h(a) \bmod n_c$, the verification succeeds, and signer's identity is confirmed. Otherwise, the verification fails.

(4) Indirect verification of the signer identity:

1) The contract send a message to all members in the ring, requiring all members to send $(Cid_i, sig_i(r, Cid_i))$, where r is the random number in the signature of the ring to be verified, Cid_i is the certificate number of the member sending the message, and $sig_i(r, Cid_i)$ is the signature of the member sending the message. The members who do not send $(Cid_i, sig_i(r, Cid_i))$ are recorded as malicious.

2) The verifier finds the public key (n_i, e_i) of the member according to Cid_i, and then uses it to verify the authenticity of signature S. If the verification succeedstrue, it

means that the information sent by the member is true. If the verification is false, the member is listed as a malicious member.

3) Verify whether $v = h(sig_i(r, Cid_i))$ is valid. If true, the member is the issuer. Otherwise, the member is not the issuer, which results in member's exlusion.

The RSA-based ring signature mechanism has been proved to be secure under random oracle model in [12]. The RS-MIV mechanism proposed in this paper only replaces the initial value of the RSA-based ring signature, and does not improve its structure. Thus, it still has all the characteristics of the RSA-based ring signature mechanism. At the same time, when generating the symmetric key k, the RS-MIV added the verification public key to it. If the verification public key is tampered with, in the process of ring signature verification, when k is used to calculate $C_{k,v}(y_1, y_2, \ldots y_r) = v$, the correct result cannot be obtained, and the illegal signature result is obtained. Thus, the authenticity of the verification public key is guaranteed.

4   Supervisable Anonymous Management Scheme of Digital Certificates Based on Smart Contracts

This article uses smart contracts to implement the supervisable and anonymous management of digital certificates, including the anonymity issuance, the anonymity verification, and the traceability of the illegal digital certificate issuer in the blockchain PKI. Smart contracts can ensure the automatic execution and security of the three functions.

4.1   Anonymity Issuance of the Digital Certificate

The anonymity issuance of digital certificates means that the CA node issues the digital certificate with the RS-MIV mechanism after receiving a service request and then uploads the digital certificate to the blockchain.

In our scheme, the digital certificate adopts the standard format of X.509 (see Fig. 1); however, the following modifications need to be made: (1) the digital signature part of the certificate is the ring signature by CA using RS-MIV mechanism; (2) due to the characteristics of open consensus of the blockchain, hash algorithm is used to calculate the user's private information to be protected. The hash value of the user's private information is stored in the certificate. The method of obtaining the user's private information off-blockchain and comparing it with the hash value of the private information in the certificate are for ensuring the correctness of the obtained user's private information. (3) To ensure the anonymity of the CA node, the issuer's identity information is not kept in the certificate.

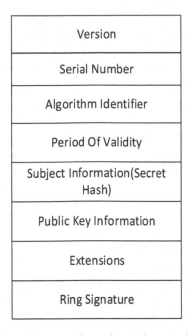

Fig. 1. Digital certificate format in our scheme.

4.2 Anonymity Verification of the Digital Certificate

When users get a digital certificate from the blockchain, they need to check the validity of the ring signature of that certificate first. In the verification process, the ring signature verification method in the RS-MIV mechanism is used. It should be noted that the current methods need to build a certificate chain from the root CA to the certificate issuing Ca, and realize one-to-one verification of the digital certificates in the certificate chain. However, in our scheme, due to the anonymity of the certificate issuing CA nodes, the certificate chain cannot be built. In this regard, the blockchain PKI adopts a node trust enhancement technology by default [2]. Under the premise that there are several root CAs with initial trust based on blockchain PKI, when a CA node wants to join the blockchain, the technology establishes the trust of CA node in the chain by verifying the certificate chain from the root CA to the CA node. Therefore, even if the scheme cannot verify the digital certificate, the node trust enhancement technology can guarantee the credibility of the CA node that issues the digital certificate anonymously.

The pseudo code of the smart contract used for anonymous authentication of digital certificate is as follows:

Algorithm 1: Anonymous certificate verification contract

Input: Verified certificate serial number :Serial number;Verified certificate ring signature: Ring Signature;

Output: judgment result: flag;

1. Certification cert=null;int flag=null;

2. **for** i=0 **to** addcert.length **do**

3.   **if**(addcert[i].Serial number=Serial number)**then**

4.   {cert=addcert[i];

5.   **End for;**}

6.   **if**(date<cert.Period Of Validity)**then**{

7.   **if**(revokelistquery(cert.Serial number)=1&&RS-MIVverification(cert. Serial number,ce rt.Ring Signature)=1)**then**{

8.   flag=1;

9.   }**else** flag=0;

10. }**else** flag=0;

11. **return** flag;

4.3   Traceability of the Issuer of Illegal Digital Certificates

After finding the illegal digital certificates in the blockchain, it is necessary to trace the issuer of these certificates.

Under normal circumstances, the first thing to do is checking the ring signature of the certificate and finding the ring signed for it. Then, the serial number of the certificate is sent to the ring group member, and the CA node that issues the certificate claims it through the ring signature authentication method based on challenge-response in the RS-MIV mechanism. When the verification is successful, the CA node that issued the illegal certificate needs to revoke the certificate. To encourage CA nodes to actively report illegal certificates, the economic incentive feature of the blockchain can be used as a reward mechanism.

In the case that CA node does not report illegal certificates, the ring signature function must be suspended. This is followed by asking all nodes in the ring to show proof $sig(Cid, r)$ and Cid. Then, the indirect verification method in the RS-MIV mechanism is used to confirm the identity of the node presenting the proof, excluding the legal CA node. The CA nodes that fail to pass identity verification are considered as malicious nodes. For malicious nodes, economic punishment measures based on blockchain can be taken. If the node is no longer trusted, all certificates issued by the node will be revoked, and the node will be removed from the blockchain PKI. By following, a new ring will be formed with the remaining nodes in the ring.

The smart contract pseudo code is as follows:

Algorithm 2: Anonymous certificate traceability contract

Input: illegal certificate serial number: Serial number;

Output: illegal certificate issuer:cert ; malicious node:illegalnode;

1. Certification cert=null;Certification[] illegalnode=null;

2. **if**(RS-MIVactivelytraceability(Serial numbert)=null)**then**{

3.   illegalnode=RS-MIVtraceability(Serial number);

4.   **return** illegalnode;

5. }**else**{

6.   cert=RS-MIVactivelytraceability(Serial number);

7.   **return** cert;

8. }

5 Security Analysis

5.1 Anonymity

Conclusion 1: If the hash function and the RSA algorithm are secure and the RS-MIV mechanism satisfies anonymity, the certificate issuer satisfies anonymity.

Prove: $A_{IBAnony}$ is defined as the adversary to attack the anonymity in the simulation attack game, A_{Hash} is the adversary to attack the hash function, A_{RRS} is the adversary to attack the anonymity of the RS-MIV mechanism, and A_{RSA} is the adversary to attack the RSA algorithm. A polynomial time algorithm $A \in (A_{Hash}, A_{RRS}, A_{RSA})$, which contains the ability of all the above attackers, is defined and A through the interaction of $A_{IBAnony}$ and A in anonymous simulation attack game is constructed. Thus, it can perform the above-explained attacks. If $A_{IBAnony}$ successfully attacks the anonymity of this scheme, then A can successfully attack the other parts, including the hash function, the RS-MIV mechanism, and the RSA algorithm, under a certain probability.

(1) Initialization: Algorithm A initializes the system, runs the anonymous certificate issuance and verification process, gives the certificate public key PK to the attacker $A_{IBAnony}$, and keeps the certificate private key SK and Ca for the certificate signature S.

(2) Query: Opponent $A_{IBAnony}$ queries algorithm A with polynomial bounded degree:

1) Ask for the private key SK of the corresponding anonymous certificate. The algorithm A attacks the ring signature scheme and the RSA algorithm by running A_{Hash}, A_{RRS}, A_{RSA}. Then it hands the private key SK to the attacker $A_{IBAnony}$.
2) Ask for the secret information $sig(Cid, r)$ of the issuing CA corresponding to the anonymous certificate. Algorithm A attacks by running A_{Hash}, A_{RRS}, A_{RSA}, and returns the secret information to $A_{IBAnony}$.

(3) Challenge: When the attacker $A_{IBAnony}$ finishes asking, A selects two nodes i_0, i_1, and generates corresponding private keys SK_{i_0}, SK_{i_1} according to the RSA algorithm. Then, it randomly selects a bit $\mu \in \{0, 1\}$, executes the above parts, and obtains the certificate $Cert$ and secret information $sig(Cid, r)$, Finally, it extracts the challenge certificate and returns $clCert = CLCert(SK_{i_b}, PK, Cert, sig(Cid, r))$ to A.

(4) Guess: The attacker $A_{IBAnony}$ conducts polynomial bounded query on A as before, but it is not allowed to query the private key of i_0 and i_1 and the secret information of the issuing CA.

(5) Output: Finally, the attacker $A_{IBAnony}$ outputs a guess $\mu' \in \{0, 1\}$. If $\mu' = \mu$, it means that the attacker $A_{IBAnony}$ wins the game. The probability of the opponent $A_{IBAnony}$ success is:

$$
\begin{aligned}
Adv_{A_{IBAnony}}(k) &= \Pr\left[Exp_{A_{IBAnony}}(k) = 1\right] \\
&= \Pr\left[A_{IBAnony}(guess) = 1 | \mu = 1\right] \cdot \Pr[\mu = 1] + \Pr\left[A_{IBAnony}(guess) = 0 | \mu = 0\right] \cdot \Pr[\mu = 0] \\
&= \frac{1}{2}\left(\Pr\begin{bmatrix} A_{Hash}(guess) = 1 \\ A_{RRS}(guess) = 1 \\ A_{RSA}(guess) = 1 \end{bmatrix} | \mu = 1 + \Pr\begin{bmatrix} A_{Hash}(guess) = 0 \\ A_{RRS}(guess) = 0 \\ A_{RSA}(guess) = 0 \end{bmatrix} | \mu = 0 \right) \\
&< \frac{1}{2}\left(\Pr[A_{Hash}(guess) = 1 | \mu = 1] \cdot \Pr[\mu = 1] + \Pr[A_{Hash}(guess) = 0 | \mu = 0] \cdot \Pr[\mu = 0]\right) + \\
&\quad \frac{1}{2}\left(\Pr[A_{RRS}(guess) = 1 | \mu = 1] \cdot \Pr[\mu = 1] + \Pr[A_{RRS}(guess) = 0 | \mu = 0] \cdot \Pr[\mu = 0]\right) + \\
&\quad \frac{1}{2}\left(\Pr[A_{RSA}(guess) = 1 | \mu = 1] \cdot \Pr[\mu = 1] + \Pr[A_{RSA}(guess) = 0 | \mu = 0] \cdot \Pr[\mu = 0]\right) \\
&= \Pr[Exp_{A_{Hash}}(k) = 1] + \Pr[Exp_{A_{RRS}}(k) = 1] + \Pr[Exp_{A_{RSA}}(k) = 1] \\
&= Adv_{A_{Hash}}(k) + Adv_{A_{RRS}}(k) + Adv_{A_{RSA}}(k)
\end{aligned}
$$

As a result, if the attacker A_{Hash} successfully attacks the hash function, A_{RRS} successfully attacks the anonymity of the RS-MIV mechanism, and A_{RSA} successfully attacks the RSA algorithm. Thus, $A_{IBAnony}$ will win the anonymity simulation attack game of this scheme. However, thanks to the above-given algorithm and the security of the RS-MIV scheme, the probability of successful attack of the opponent $A_{IBAnony}$ can be ignored, so the scheme satisfies anonymity.

5.2   Traceability

Conclusion 2: When the CA node is trusted and if the blockchain meets the requirement of non-tamperability, the RS-MIV mechanism meets the requirements of non-forgery and verifiability. This means that the RSA algorithm is secure, and the identity of certificate issuer can be traced when necessary.

Prove: $A_{IBAnony}$ is defined as the adversary who attacks the anonymity simulation attack of the scheme, A_{Blc} as the adversary against the non-tamperable blockchain, A_{RRS} as the adversary against the unforgeability and verifiability of the RS-MIV case, and A_{RSA} as the opponent against RSA. A polynomial time algorithm $A \in (A_{Blc}, A_{RRS}, A_{RSA})$, which comprises the abilities of all the above-defined attackers, and constructs A through the interaction of $A_{IBAnony}$ and A in the anonymous simulation attack game to perform the attacks. If $A_{IBAnony}$ successfully attacks the traceability, A can successfully attack other parts with a certain probability, including the blockchain non-tamperability, the RS-MIV scheme, and the RSA algorithm.

(1) Initialization: Algorithm A initializes the system, runs the anonymous certificate issuance and verification process in the scheme, hands the certificate public key PK to the attacker $A_{IBAnony}$, and keeps the certificate private key SK and the CA's ring signature S for the certificate.

(2) Query: Opponent $A_{IBAnony}$ queries algorithm A with polynomial bounded degree:

1) The adversary $A_{IBAnony}$ requests the private key SK corresponding to the certificate owned by node i, and algorithm A sends the obtained private key SK to $A_{IBAnony}$ by running the simulated attack games of A_{Blc}, A_{RRS}, and A_{RSA}, respectively.
2) The adversary $A_{IBAnony}$ requests the ring signature threshold knowledge Knl corresponding to the certificate owned by node I, And algorithm A sends the acquired threshold knowledge to $A_{IBAnony}$ by running the simulated attack games of A_{Blc}, A_{RRS}, and A_{RSA}, respectively.
3) The adversary $A_{IBAnony}$ requests the secret information of the CA that issued the certificate to node I, and algorithm A sends the obtained secret information to $A_{IBAnony}$ by running the simulated attack games of A_{Blc}, A_{RRS}, and A_{RSA}, respectively.

(3) Challenge: The adversary $A_{IBAnony}$ outputs the certificate $clCert = CLCert(SK, PK, Cert, Knl)$ and CA's secret message $sig(Cid, r)$ based on the information obtained.

(4) Output: If the certificate output or the CA's secret information is invalid, then the attack is considered as successful.

As a result, the successful attack probability of the adversary $A_{IBAnony}$ is:

$$
\begin{aligned}
Adv_{A_{IBAnony}}(k) &= \Pr\left[Exp_{A_{IBAnony}}(k) = 1\right] \\
&= \Pr[clCert = 1] \cdot \Pr[Address = 1] + \Pr[clCert = 0] \cdot \Pr[Address = 1] + \\
&\quad \Pr[clCert = 1] \cdot \Pr[Address = 0] + \Pr[clCert = 0] \cdot \Pr[Address = 0] \\
&= (\Pr[clCert = 1] + \Pr[clCert = 0]) \cdot (\Pr[Address = 1] + \Pr[Address = 0]) \\
&= (\Pr[Exp_{A_{Blc}}(k) = 1] + \Pr[Exp_{A_{RRS}}(k) = 1]) \cdot \Pr[Exp_{A_{RSA}}(k) = 1] \\
&= (Adv_{A_{Blc}}(k) + Adv_{A_{RRS}}(k)) \cdot Adv_{A_{RSA}}(k)
\end{aligned}
$$

If the attacker A_{Blc} successfully attacks the tamperability of the blockchain, the attacker A_{RRS} successfully attacks the RS-MIV scheme, and the attacker A_{RSA} successfully attacks the RSA algorithm. Thus, $A_{IBAnony}$ can win the traceability simulation attack game of this scheme. However, according to the security of known components, the successful attack probability of $A_{IBAnony}$ is ignored, and thus the scheme meets the traceability.

6 Performance Analysis

We selected the RSA algorithm as 1024 bit, defined E as the exponential operation cost, H as the hash operation cost, the ring size as r, and ignored the cost of multiplication and addition. Table 1 illustrates the calculation cost of the RS-MIV mechanism.

Table 1. The calculation cost of the RS-MIV mechanism.

Process	Algorithm	Expenses
Anonymity issuance of the digital certificate	Hash	H
	RSA	$E+H$
	RS-MIV signature	$(3r+2)E+rH+H$
Anonymity verification of the digital certificate	RS-MIV signature	$rE+H$

From Table 1, we can see that the total calculation cost of our scheme is $(4r+3)E+(r+4)H$, where the highest cost belongs to the RS-MIV signature algorithm. The performance of RS-MIV mechanism is tested on PC with Win10 (64 bit), inter (R) core (TM) i7-7700 @ 3.6 GHz and 16 GB memory. The test results are shown in Fig. 2. When the r is close to 100, the RS-MIV signature algorithm takes 1.2 s, while the RS-MIV signature verification algorithm takes about 0.3 s. However, in practice, the signature algorithm is only used when the digital certificate is issued, which does not affect the performance of the digital certificate application. The efficiency of digital certificate application is only affected by signature verification algorithm. And in practice, the number of Ca nodes that need anonymous digital certificate management will not reach a very large number. When the number of nodes participating in the ring is limited, the time-consuming of the algorithm is acceptable. On the premise that digital certificates can be managed anonymously, this paper considers that the increased time is acceptable for users.

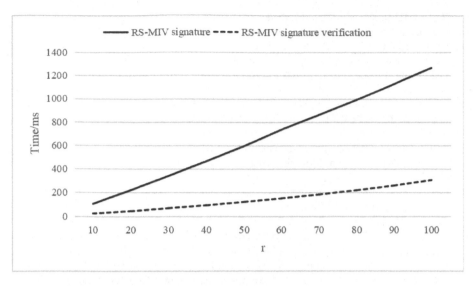

Fig. 2. The relationship between the main algorithm time cost and r.

7 Concluding Remarks

In this paper, we propose a ring signature mechanism that can be indirectly verified many times. We design a supervisable anonymous management scheme for digital certificate based on smart contracts, which can guarantees the anonymity of CA nodes in the blockchain PKI and realizes the supervisable anonymous management of digital certificate. However, our solution only uses hash encryption to protect the user's private information in the digital certificate, which cannot meet the on-demand disclosure requirements. Thus, this issue will be studied in future work.

Acknowledgements. This work was supported in part by the National Key Research and Development Program of China under Grant 2018YFB0803603 and Grant 2016YFB0501901, and in part by the National Natural Science Foundation of China under Grant 61502531, Grant 61702550, and Grant 61802436.

References

1. Liu, A., Du, X., Wang, N., Li, S.: Blockchain technology and its research progress in the field of information security. J. Softw. **7**, 2092–2115 (2018)
2. Li, S., Wang, N., Du, X., Liu, A.: Internet web trust system based on smart contract. In: Cheng, X., Jing, W., Song, X., Lu, Z. (eds.) ICPCSEE 2019. CCIS, vol. 1058, pp. 295–311. Springer, Singapore (2019). https://doi.org/10.1007/978-981-15-0118-0_23
3. Faisca, J.G., Rogado, J.Q.: Personal cloud interoperability. In: World of Wireless, Mobile and Multimedia Networks, pp. 1–3 (2016)
4. Zhu, J., Fu, Y.: Dynamic multi center collaborative authentication model of supply chain based on blockchain. J. Netw. Inf. Secur. **2**(1), 27–33 (2016)

5. Kuo, T.T., Hsu, C.N., Ohno-Machado. L.: ModelChain: decentralized privacy-preserving healthcare predictive modeling framework on private blockchain networks
6. Fromknecht, C., Velicanu, D., Yakoubov, S.A.: Decentralized public key infrastructure with identity retention. IACR Cryptology ePrint Archive, 2014: 803 (2014)
7. Wikipedia, "Namecoin," [EB/OL], 28 December 2018. https://en.wikipedia.org/wiki/Namecoin
8. Matsumoto, S., Reischuk, R.M.: IKP: Turning a PKI around with decentralized automated incentives. In: 2017 IEEE Symposium on Security and Privacy (SP), pp. 410–426. IEEE (2017)
9. Al-Bassam, M.: SCPKI: a smart contract-based PKI and identity system. In: Proceedings of the ACM Workshop on Blockchain, Cryptocurrencies and Contracts, pp. 35–40. ACM (2017)
10. Corella, F.: "Implementing a PKI on a Blockchain," Pomcor research in mobile and web technology, [EB/OL], 28 December 2018. https://pomcor.com/2016/10/25/implementing-a-pki-on-a-blockchain/
11. Ben Sasson, E., Chiesa, A., Garman, C.: Zerocash: Decentralized anonymous payments from bitcoin. In: 2014 IEEE Symposium on Security and Privacy (SP). IEEE (2014)
12. Rivest, R.L., Shamir, A., Tauman, Y.: How to leak a secret. In: Boyd, C. (ed.) ASIACRYPT 2001. LNCS, vol. 2248, pp. 552–565. Springer, Heidelberg (2001). https://doi.org/10.1007/3-540-45682-1_32
13. Lv, J., Wang, X.: Verifiable ring signature. In: Proceedings of DMS 2003-The 9th International Conference on Distribted Multimedia Systems, pp. 663–667 (2003)
14. Zhi, G., Ke-Fei, C.: A New verifiable ring signature scheme. Acta Scientiarum Naturalium Universitatis Sunyatseni, 43(2), 132–134 (2004)
15. Zhang, C., Liu, Y., He, D.: A new verifiable ring signature scheme based on Nyberg-Rueppel scheme. In: International Conference on Signal Processing. IEEE (2006)
16. Wang, C.H., Liu, C.Y.: A new ring signature scheme with signer-admission property. Inf. Sci. 177(3), 747–754 (2007)
17. Jeong, I., Kwon, J., Lee, D.: Ring signature with weak linkability and its applications. IEEE Trans. Knowl. Data Eng. 20(8), 1145–1148 (2008)
18. Wen, L.D., Xing, H., Yi, L.: Certificateless verifiable ring signature scheme. Comput. Eng. 15, 141–143
19. Xiaolin, L., Qianqian, L., Kui, L., et al.: Analysis and improvement of verifiable ring signature scheme. Comput. Appl. 32(12), 3466–3469 (2012)
20. Zhang, J., He, Y., Li, X.: Security analysis and improvement of two verifiable ring signature schemes. Comput. Eng. Appl. 8, 115–119 (2016)
21. Dong, Q., Li, X., Liu, Y.: Two extensions of the ring signature scheme of Rivest–Shamir–Taumann. Inf. Sci. 188, 338–345 (2012)
22. Yang, X.: Research on the strong forward security ring signature scheme based on improved verifiability. Comput. Appl. Softw. 4, 325–328
23. Bultel, X., Lafourcade, P.: Unlinkable and strongly accountable sanitizable signatures from verifiable ring signatures. In: Capkun, S., Chow, Sherman S.M. (eds.) CANS 2017. LNCS, vol. 11261, pp. 203–226. Springer, Cham (2018). https://doi.org/10.1007/978-3-030-02641-7_10

Product Customer Demand Mining and Its Functional Attribute Configuration Driven by Big Data

Dianting Liu[1,2(✉)], Xia Huang[2], and Kangzheng Huang[1]

[1] College of Mechanical and Control Engineering,
Guilin University of Technology, Guilin 541004, Guangxi, China
365565379@qq.com
[2] College of Information Science and Engineering,
Guilin University of Technology, Guilin 541004, Guangxi, China

Abstract. The maturity of big data analysis theory and its tools improve the efficiency and reduce the cost of massive data mining. This paper discusses the method of product customer demand mining based on big data, and further studies the configuration of product function attributes. Firstly, the Hadoop platform was used to perform product attribute data participle and feature word extraction based on Apriori algorithm was used to mine product customer demand information. And then the MapReduce model on the big data platform was applied into efficient parallel data processing, obtaining product attributes with research value, and their weights and attribute levels. After that, the cloud model and the MNL model were employed to construct the product function attribute configuration model, and the improved artificial bee colony algorithm was used to solve the model. The optimal solution of the product function attribute configuration model was got. Finally, an example was given to illustrate the feasibility of the proposed method in this paper.

Keywords: Big data · Customer demand · Product function attribute configuration · Apriori · MNL model · Artificial bee colony algorithm

1 Introduction

At present, in traditional methods such as market survey questionnaires, household interviews, observation method, etc., consumer demand for products is obtained manually, which is not only time-consuming and laborious, but it is difficult for users to express their actual views on the product also objectively and rationally. It causes that the workload of text processing is large, which affects the validity and real-time of information in the process of requirement analysis.

With the popularization and in-depth application of the Internet, more and more users like to express their opinions on products on the network; It has broken through the time and space restrictions, and is generally non-interfering and spontaneous, which can be regarded as true and reliable. These product demand information can be automatically and efficiently collected and processed by computer systems, and then the actual needs of users can be analyzed timely and effectively to assist enterprises in product innovation.

© Springer Nature Singapore Pte Ltd. 2020
J. Zeng et al. (Eds.): ICPCSEE 2020, CCIS 1257, pp. 145–165, 2020.
https://doi.org/10.1007/978-981-15-7981-3_11

In the e-commerce website and WeChat, blog, QQ social network space, people's comments on products have the features of big data [1] in Volume, Velocity, Variety, Value, Veracity. This type of data set is rich and huge, and cannot be collected, managed, and analyzed using traditional data processing methods. At present, big data platforms and analysis tools such as Hadoop platform and MapReduce are commonly used to perform parallel processing calculations for solving the hardware bottleneck and software performance constraints in mining algorithms [2–4].

The product function attribute configuration [5–7] can be considered as a decision process to determine the attribute level value of a new product under the premise of knowing customer preferences, competitive products and market information. Before developing a product, an enterprise can investigate customer needs and preferences, establish an optimization model to obtain the optimal product attribute level value and attribute combination, and at the same time combine product engineering performance and market performance to optimize the design of the product.

This article discusses the use of big data platforms and analysis tools to collect customers' comments on products on the network, and then to mine customer needs for products; On this basis, the theory and method of combining cloud models and MNL models to configure product function attributes is researched.

2   Product Customer Demand Mining Based on Big Data

Modern products are becoming more and more complex and with numerous product attributes. When an enterprise's product is to be positioned, the product attributes and attribute levels must be firstly determined. If the data in this step is inaccurate, it will lead to the inaccuracy of subsequent data. Because the data collected by traditional methods such as sampling surveys and questionnaires on potential customers or experts is too random, and with incomplete data types and high blindness, it cannot be used as the decisive data to determine product attributes. Therefore, big data collection platform and analysis tools should be used to collect data on a large scale, so that the collected data is sufficiently comprehensive and complete to play a decision-making role.

2.1   Data Collection

The Jingdong online shopping platform is selected as the object for crawling data in this paper. The process of crawling the data has two steps: the first step is to crawl the product comments, the product comments data segmentation is done under the Hadoop platform, and then the product attributes with research value is to be mined by Apriori algorithm; In the second step, the product attributes excavated in the first step are used to calculate the relevant product attribute level and sales, so that the product attribute level with research value will be determined.

(1) Crawling data. The required data is by Scrapy. Scrapy is a fast, high-level web crawling Web Spider framework which is developed using Python language, and used to crawl structured data from web site pages. The steps to crawl product comments are as follows: First step analyzed the page and defined the fields that

need to crawl; The second step is to analyze the interface url and parse the crawling content field through xpath and json; In the third step, write the storage method in the pipelines.py file; The fourth step is to start crawling; Finally stored in ElasticSearch database.

(2) Data preprocessing. While scraping the product information, the data is simply denoised, such as removing duplicate data and deleting blank lines. And then the data must be further processed, such as removing emoji, special characters, stop words in Chinese and English and other junk data.

2.2   Product Attribute Mining

The first part of the data is obtained through the crawling program, which is the comments data of the product on the Jingdong Mall, and then the data segmentation and Apriori feature extraction are performed.

Data Segmentation. The Chinese word segmentation tool used in this article is jieba, which is currently the most used Chinese word segmentation tool in China. The captured data stored in MySQL is imported into HDFS using Sqoop and the jieba package is imported on the Hadoop project, and the word segmentation is calculated on the mapreduce program. The process is shown in Fig. 1:

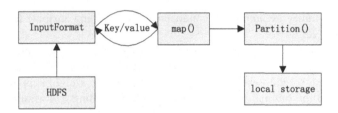

Fig. 1. Map take execution process

Only simple word segmentation is performed here, so there is no need for multiple mappers and no reducer. There is no control on the Inputformat here. If there are many files, in order to ensure the highest computational efficiency of mapreduce, the Inputformat must be controlled to limit the size of the slices. To set 'Key' for the offset of each line of text and 'value' for the content of text, and the main function is realized by the function of map().

Feature Words Extraction Based on Hadoop and Apriori Algorithm. Introduction of Apriori algorithm [8–12] to extract feature words.

Use the comments data and feature set A to construct 0-1 matrix M:

$$
M = \left\{
\begin{matrix}
a_{11} & a_{12} & \cdots & a_{1n} \\
a_{21} & a_{22} & \cdots & a_{2n} \\
\vdots & \vdots & \vdots & \vdots \\
a_{m1} & a_{m2} & \cdots & a_{mn}
\end{matrix}
\right\}
\tag{1}
$$

Where a_{ij} is equal to 0 or 1, this comment has this attribute feature represented by 1, if not, it is represented by 0, $i = 1, 2, ..., m$; $j = 1, 2, ..., n$; N attribute feature sets are represented by $I = \{I_1, I_2, I_3, ..., I_N\}$. According to formula (2), calculate the probability of I_j appearing in the transaction database is $p(I_j)$; calculate the weight $w(I_j)$ of I_j by formula (3).

$$p(I_j) = k/m \tag{2}$$

$$w(I_j) = 1/p(I_j) \tag{3}$$

Where the frequency of I_j appearing in the transaction set is represented by k, which is the number of 1 in column j of matrix M, and the total number of comments in matrix M is represented by m.

In formula (4), the l-th comment in the dataset is represented by R_l. Take the average weight of all attribute features in this comment and record it as $wr(R_l)$, which is the $w(I_j)$ sum of all aij = 1 in line i is averaged. The weight calculation method of Rl is calculated according to formula (4).

$$wr(R_l) = \sum_{j=1}^{I_j \in R_l} w(I_j)/|R_l| \tag{4}$$

In the above formula, the number of commented R_l containing attribute feature items is represented by $|R_l|$.

The weight support of attribute is denoted as *wsupport*, the weight represents the proportion of transaction weights containing attribute features to all transaction weights, and then set the lowest threshold according to the weight support of attribute features to form the optimal feature set. Calculated according to formula (5).

$$w\sup port(S) = \sum_{l=1}^{S \subseteq R_l} wr(R_l)/\sum_{l=1}^{m} wr(R_l) \tag{5}$$

In the above formula, any attribute characteristic item in the transaction database is denoted by S.

The Apriori algorithm steps are as follows:

Step 1: Scan the comment database, construct a Boolean matrix of attribute features, and calculate $p(I_j)$ according to the comment transaction matrix, which is the probability of each attribute feature appearing in the transaction database, and then calculate the weight $w(I_j)$ of each attribute feature and $wr(R_l)$ of each comment transaction.

Step 2: Calculate the weight support *wsupport(S)* of the attribute items to obtain the candidate 1-item set. Generate frequent 1-item sets according to the weight support threshold of the minimum item, and finally get the result.

Implementation of Apriori algorithm to extract feature words under MapReduce.

When the Apriori algorithm is being calculated, a large number of candidate sets will be generated, and the database needs to be scanned repeatedly. Because there is a lot of data, MapReduce built on Hadoop for parallel calculation can improve the performance and efficiency of the algorithm. Figure 2 is the MapReduce model implementation process of the Apriori algorithm under the Hadoop framework.

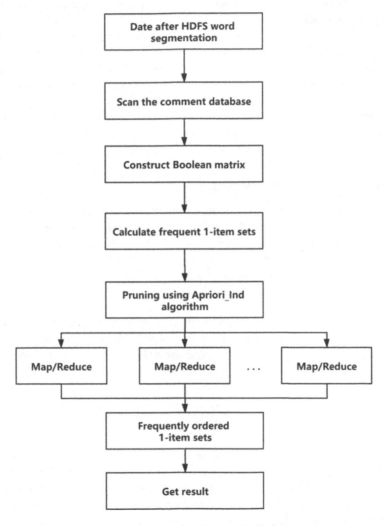

Fig. 2. Apriori algorithm under Hadoop

The MapReduce function is used to perform distributed processing on data, and a frequent 1-item set that meets the weight support threshold of the minimum item is retrieved. The traditional Apriori algorithm has the disadvantage of generating a large number of redundant candidate item sets and frequently scanned data. Therefore, in this paper, combined with the Apriori_Ind algorithm in the Reference [12], which improves the representation of each node's block strategy and frequent item set. The feature of different items of each node realizes that each node will generate a unique candidate set and pruning based on frequent item sets, which effectively reduces the time of generating, pruning and support statistics of each node's candidate set. At the same time, the representation of frequent item sets is improved to <previous, post>, which can reduce the size of data transmission between Hadoop clusters and accelerate the generation of pruning and candidate sets.

The first-level candidate set is generated by calculating each set during the process of reading the dataset after HDFS word segmentation, and then the first-level candidate set is pruned by the Apriori_Ind algorithm. For this candidate set, the Map function under the Hadoop framework is used to divide the entire original dataset into several subsets (first-level candidate set), and then it is distributed in parallel to the Reduce function for reduction, filter the frequent 1-item sets by defining the weight support threshold of the minimum item. Finally, sorted out the attribute feature words, which are product attributes with research value.

2.3   Product Attribute Level Mining

On the basis of the above product attribute determination, the current product sales and product attribute level data are collected and analyzed, so as to determine the product attribute level with more research value, and lay the foundation for the calculation of product attribute preference in the following. The main steps are as follows: First to crawl the product attributes and sales of Jingdong Mall by crawl program, and then preprocess the data, after that get the product attribute level of more research value by using liner statistical analysis.

To select the most valuable research attribute level is equivalent to finding the sum of sales of all products containing this attribute level. Higher sales indicate that this attribute level is more popular, and the more valuable it is for research. Calculated as follows:

$$sum(u_{ij}) = \sum_{n=1}^{6000} salevolume(u_{ij})_n * x_{ij} \tag{6}$$

In the above formula, the j-th attribute level of the i-th attribute of the product is represented by u_{ij}; x_{ij} represents the value of 0 or 1, the j-th attribute level with i-th attribute is 1, and 0 if not included; The n-th product contains the sales of the j-th attribute level of the i-th attribute is represented by $salevolume(u_{ij})_n$, and the total sales of products including the j-th attribute level of i-th attribute is represented by $sum(u_{ij})$.

3   Product Function Attribute Configuration

In general, the combination of product attribute levels is effective, and the product outline can be represented by a combination of attribute levels. For example, the product has I product attributes, and each attribute i has J attribute levels. The goal of the new product's functional attribute configuration is to find the optimal product profile of the new product on the basis of maximizing the company's total profit.

3.1   Customer Preferences

When describing customer preferences, linearly superimpose the horizontal utility value of the product's own attributes is the most common method. Product attribute weights and product price attributes are considered in this article. In the product market, product price attributes have decisive relationship with whether customers choose to purchase this product. Some customers will only purchase the price that they can accept. As product prices rise, customer preferences will decrease and sales will decline. The price factor p is introduced when calculating U_k for customer preferences; customers pay different attention to different attributes of products, so when calculating the U_k, the product attribute weight q_i will be introduced.

$$U_k = \sum_{i=1}^{I} \sum_{j=1}^{J} u_{ij} q_i x_{ij} - \gamma p \tag{7}$$

Where the total utility value of product is represented by U_k, the partial utility value of the j-th attribute level selected in the i-th attribute of the product is represented by u_{ij}, which is calculated by combing the cloud model and the discrete selection model. $x_{ij}(i = 1, 2, .., I; j = 1, 2, ..., J)$ is a variable of 0 and 1, when the j-th attribute level of the i-th attribute is selected, $x_{ij} = 1$, otherwise $x_{ij} = 0$. The customer's attention weight for the i-th attribute of the product is q_i. γ is the weight of price influence, where $\sum_{i=1}^{I} q_i + \gamma = 1$; The sales price of product k is p.

3.2   Product Attribute Utility Value

(1) Language evaluation of product attributes and conversion of cloud model. Suppose customers use n-level scale to evaluate product attributes, where n = 2t + 1, $t \in N$; The evaluation value is recorded as H: $H = \{h_v | v = -t, ..., 0, ..., t, t \in N\}$ in the natural language set. For example, according to the 7-level evaluation scale, the language evaluation set for product attributes is $H = \{h_{-3} = worst, h_{-2} = worse, h_{-1} = bad, h_0 = general, h_1 = good, h_2 = better, h_3 = best\}$. As the show in Table 1, each attribute level of the product is represented by the first line, and a customer's natural language evaluation of each attribute level is represented by the second line. The natural language evaluation of the j-th attribute level selected in the i-th attribute is $(h_v)ij$.

Table 1. Qualitative evaluation of product attributes

The attribute level of the selected product k	k_{1j}	k_{2j}	...	k_{ij}	...	k_{Ij}
Customer evaluation set h_v	$(h_v)_{1j}$	$(h_v)_{2j}$	...	$(h_v)_{ij}$	...	$(h_v)_{Ij}$

Definition [13] given a language evaluation set $H = \{h_v | v = -t, \ldots, 0, \ldots, t, t \in N\}$, there is a function that coverts h_v to the corresponding value θ_v, where $\theta_v \in [0, 1]$:

$$\theta_v = \begin{cases} \frac{a^t - a^{-v}}{2a^t - 2}, & -t \leq v \leq 0 \\ \frac{a^t + a^{-v} - 2}{2a^t - 2}, & 0 \leq v \leq t \end{cases} \tag{8}$$

The value of a is in the interval [1.36, 1.4], which can be obtained from the experiment [14]. In this paper, $a \approx 1.37$.

The conversion process from qualitative to quantitative is shown in the following Algorithm 1:

Algorithm 1: Standard evaluation cloud generator.

Input: The attribute evaluation scale is n-level and the effective domain of attribute comment value is $[D_{min}, D_{max}]$.

Output: Standard evaluation cloud $A_v = Y(Ex_v, En_v, He_v)$, where $v = 1, 2, \ldots, n$.

The algorithm steps are as follows:

Step1: Calculated θ_v by formula (8)

Step2: Calculated the expected value Ex_v based on the upper and lower limits of the domain $[D_{min}, D_{max}]$.

$$Ex_v = D_{min} + \theta_v (D_{max} - D_{min}) \tag{9}$$

Step3: Calculated the entropy En_v according to Step2.

$$En_v = \begin{cases} \frac{(1-\theta_v)(D_{max} - D_{min})}{3}, & -t \leq v \leq 0 \\ \frac{\theta_v (D_{max} - D_{min})}{3}, & 0 \leq v \leq t \end{cases} \tag{10}$$

$$En_{-v} = En_v = \begin{cases} \frac{(\theta_{|v|-1} + \theta_{|v|} + \theta_{|v|+1})(D_{max} - D_{min})}{9}, & 0 < |v| \leq t - 1 \\ \frac{(\theta_{|v|-1} + \theta_{|v|})(D_{max} - D_{min})}{6}, & |v| = t \\ \frac{(\theta_v + 2\theta_{v+1})(D_{max} - D_{min})}{9}, & v = 0 \end{cases} \tag{11}$$

Step4: Calculated the super-entropy He_v according to Step3.

$$He_{-v} = He_v = \frac{En'^+ - En}{3} \tag{12}$$

$$En'^+ = \max_k \left\{ En'_k \right\} \tag{13}$$

After Algorithm 1, the qualitative natural language evaluation can be converted into a quantitative value, which is the characteristic $Y(Ex_v, En_v, He_v)$ of cloud number that used standard evaluation cloud to represent each language evaluation interval. As shown in Table 2, the level value k_{ij} of each attribute of the product k is represented by the first line; The customer's natural language evaluation $(h_v)_{ij}$ of each attribute level is represented by the second line; And the converted cloud $Y((Ex_v)_{ij}, (En_v)_{ij}, (He_v)_{ij})$ of customer evaluation is represented by the last line.

Table 2. Product attribute evaluation cloud

Product k attribute level	k_{1j}	k_{2j}	...	k_{Ij}
Customer natural language evaluation h_v	$(h_v)_{1j}$	$(h_v)_{2j}$	...	$(h_v)_{Ij}$
Converted cloud	$Y((Ex_v)_{1j},$ $(En_v)_{1j}, (He_v)_{1j})$	$Y((Ex_v)_{2j},$ $(En_v)_{2j}, (He_v)_{2j})$	...	$Y((Ex_v)_{Ij},$ $(En_v)_{Ij}, (He_v)_{Ij})$

(2) Product attribute utility. The attribute level utility value u_{ij} is calculated based to the customer's evaluation cloud $Y((Ex_v)_{ij}, (En_v)_{ij}, (He_v)_{ij})$ for each attribute level of the product. Generally speaking, products have multiple customers, and customers have different characteristics in real life, such as gender, age, position, etc., so they need to be classified. However, the proportion of each type of customer is different, and the importance of product evaluation is also different. A weighting factor β is introduced to the calculation of attribute utility value u_{ij}. The value of β can be adjusted according to the proportion of the customer's characteristic attributes, and then combined the evaluation cloud's Ex_v calculation to enhance its rationality and credibility. Suppose that product k has L types of customers, and each type of customer has M individuals. The calculation formula of u_{ij} is as follows:

$$u_{ij} = \frac{e^{R_{ij}}}{e^{R_{ij}} + \sum_{j=1}^{J} e^{R_{ij}}} \tag{14}$$

$$R_{ij} = \frac{\sum_{m=1}^{M} \beta_l (Ex_v)_{ij}^{lm}}{\sum_{l=1}^{L} M_1} \tag{15}$$

Where the expectation of the type l customer m for the evaluation of the j-th attribute level of the i-th attribute of the product is represented by $(Ex_v)_{ij}^{lm}$, which can be obtained by formula (9). The weight of type l customer evaluation is β_l, where $\sum_{l=1}^{L} \beta_l = 1$; The number of customers of type l is denoted by M_l; The total expectation of the j-th attribute level of the i-th attribute of the product is represented by R_{ij}.

3.3   Product Selected Probability

According to the MNL model, the probability C_k of a customer choosing a new product k among many competing products is calculated by formula (16):

$$C_k = \frac{e^{\chi U_k}}{e^{\chi U_k} + \sum_{r=1}^{r-1} e^{\chi U_r}} \tag{16}$$

In the formula, the overall utility value of product k is represented by U_k; The utility value of the r-th competitive product is U_r; The proportionality parameter is represented by χ, if χ has a tendency to approach infinity, then this model approximates deterministic choice, which means that the customer is absolutely rational when making a choice, and the final product performance preference is the best one. If χ is closed to zero, so this model approximates random selection, and the selection probability tends to be randomly and uniformly distributed. In this paper, the MNL proportionality parameter χ is calibrated to 0.5.

3.4   Product Function Attribute Configuration Model

Based on the customer purchase selection rules, the expected number of customers who purchase a new product k is Q_k, $Q_k = QC_k$. The product profitability index EP:

$$EP = Q_k(p - W) = QC_k(p - W) = QC_k$$
$$= \frac{e^{\chi U_k}}{e^{\chi U_k} + \sum_{r=1}^{r-1} e^{\chi U_r}}(p - \sum_{i=1}^{I}\sum_{j=1}^{J} f_{ij}x_{ij}) \qquad (17)$$

In the above formula, the meaning of C_k and U_k have been discussed in the foregoing, f_{ij} is unit cost of the j-th attribute level of i-th attribute, the number of potential customers is Q, the product cost is W, the product price is p.

3.5   Selection of Algorithm for Solving the Model

The product positioning optimization model is a discrete nonlinear model. In this model, x_{ij} is discrete variable, the price p is continuous variable. The determination of the product profile is a combination of product multiple attributes and attribute levels. Because of the variety and complexity of product attributes and attribute levels, it can be classified as NP (Non-Deterministic Polynomial) in combination optimization. Compared with GA (Genetic Algorithm), PSO (Particle Swarm Optimization Algorithm) and DE (Differential Evolution Algorithm), ABC (Artificial Bee Colony Algorithm) has the advantages of less parameter setting, fast convergence speed, and high convergence accuracy. In this paper, the improved ABC algorithm is used to solve the above product positioning design optimization model.

In the ABC algorithm, the initial solution is randomly generated twice. Once is when the population is initialized; And the other is when a food source is not updated within the maximum limit times, then the initial solution is generated by the detective bee. Therefore, the initialization will be improved separately in this article. It also proposes improved methods for search strategies. The specific improvement methods are as follows:

(1) Improvement of population initialization. Because the initial solution is randomly generated, there may be excessive concentration of individuals in the random solution, and reducing the global search performance of the algorithm and relying more on the detection bee. In this paper, the reverse learning strategy is combined to improve the initialization, the improvement ideals are as follows:

Randomly generate N/2 food sources within the search space (set the population to $N = r * M$), and set the space solution to $g_{i,j} \in (g_{i,min}, g_{i,max})$, and calculation formula of reverse solution is as follows:

$$g'_{i,j} = (g_{i,\min} + g_{i,\max} - g_{i,j}) \tag{18}$$

Calculated the fitness of all food sources, including reverse food sources, and the best r food source are selected and used as the center points of the subpopulations in thinking evolution. The distribution is based on r center points, and each generates M random food sources that obey the normal distribution.

(2) Improvement of detection Bee initialization. The traditional ABC algorithm is too random for the position of the food source generated by the initial detection of the detection bee, which leads to slow convergence and easy to fall into the local optimal. However, the Gaussian distribution has strong perturbation. In the Gaussian distribution, the application of random perturbation terms can solve the problem of individuals falling into local optimality, and can improve the accuracy of the solution. The improved formula used is shown in (19).

$$g_{i,j} = g_{best,j} + g_{best,i} \cdot N(0,1) \tag{19}$$

(3) Improvement of search strategy. In the original artificial bee colony algorithm, when detection bees and following bees to search, the search strategy adopted is better ability for global search, but it ignores the ability for local search. Therefore, by referring to the PSO (Particle Swarm Optimization Algorithm) and introducing the current optimal and suboptimal solutions, and a new search method is proposed:

$$v_{i,j} = g_{best,j} + \varphi_{i,j}(g_{w,j} - g_{k,j}) + \delta_{i,j}(g_{\sec ondbest,j} - g_{i,j}) \tag{20}$$

In the formula, the candidate food sources are represented by $v_{i,j}$, $g_{w,j}$ and $g_{k,j}$ are randomly generated unequal known solutions, $\varphi_{i,j}$ and $\delta_{i,j}$ are random value on $[-1, 1]$, the current optimal food source position is $g_{best,j}$, the second best food source position is $g_{secondbest,j}$. After introducing the current optimal and suboptimal solutions, the local search ability of the algorithm is improved to a certain extent, and the convergence speed is accelerated.

4   Examples and Analysis

4.1   Software and Hardware Environment

The experimental cluster is composed of 5 PCs, one of which is a computer with a higher CPU frequency, configured as a Master and used as a Slave at the same time, and the remaining 4 computers are isomorphic only as Slave. The configuration is shown in Table 3, using a Windows64 system, Using Hadoop 2.6.0-cdh 5.7.0 version, jdk is using version 1.7.0.

Table 3. Experimental equipment configuration

Node type	Node name	CPU	RAM
Master/Slave	Namenode	i5-8250U 3.4 GHz	8G
Slave	Datanode	I5-3210 M 2.50 GHz	4G

4.2 Example Application

Suppose an enterprise performs the configuration design of the functional attributes of a certain model of smartphone. After investigation and statistics, this model of smartphone has more than 20 attributes and more than 60 attribute levels. The product attributes and attribute levels with research value are analyzed through big data mining, and then the customer preferences of these attributes are obtained through questionnaires, and then the product function attribute configuration model is solved using an improved ABC (Artificial Bee Colony) algorithm to obtain the optimal product function property configuration.

Determination of Phone Attributes. By crawling the mobile phone comments of Jingdong Mall, and then performing word segmentation and Apriori algorithm feature extraction on the Hadoop platform. The weight support of each attribute feature item is calculated by formula (5), and then top 9 phone attribute features are extracted, which are the mobile phone attributes selected in this paper. As shown in Table 4.

Table 4. Mobile phone attribute feature extraction results

Attributes	wsupport
CPU	0.3462
RAM	0.3211
Price	0.2886
Mobile phone pixel	0.2412
Fingerprint unlock	0.2133
Screen size	0.1739
Battery	0.1621
ROM	0.1380
Resolution	0.1258

After the mobile phone attribute feature item is selected, the weight of each attribute feature item is calculated according to formula (3). As shown in Table 5.

Table 5. Weights of mobile phone attributes

Attributes	Weights
CPU	0.1713
RAM	0.1640
Price	0.1370
Mobile phone pixel	0.1127
Fingerprint unlock	0.1055
Screen size	0.0981
Battery	0.0909
ROM	0.0662
Resolution	0.0542

Through the feature item extraction of the Apriori algorithm, the top 9 phone attribute features are extracted, indicating that these 9 attributes are also the mobile phone attributes that users are most concerned about. Among them, the other 8 attributes except the price belong to the hardware attributes of the phone, which are included in the next section of the attribute level research. The weight of the influence of price is 0.137 from Table 5, that is, $\gamma = 0.137$ in the previous chapter.

Determination of Mobile Phone Attribute Level. On the basis of the determination of the above product attributes, the current mobile phone product sales and product attribute level data are collected and analyzed, and then to determine the product attribute level that is more valuable for research.

The relevant attributes and sales of mobile phone products are used by the spider program to crawl in Jingdong Mall, and then the 23 product attribute levels with the most research value are finally determined according to formula (6). The specific product attributes and attribute levels are shown in Table 6.

Phone attribute Preferences. (1) Questionnaire of mobile phone attribute preferences. The content of this questionnaire is to set the preference of 23 attribute levels of the mobile phone to be studied above. The answers included seven levels: best, better, good, general, bad, worse and worst. This questionnaire is for students at school, and 200 students are randomly selected as the object of the survey. According to the investigation, the number of valid questionnaires among the 200 statistical results obtained is 186. Among the valid questionnaires, there are 100 men and 86 women. After analyzing 186 valid questionnaires, the 23 attribute level preference values of the mobile phone were obtained.

Table 6. Smartphone product attributes and attribute levels

Attribute	Attribute level
Screen resolution	1920 * 1080
Screen resolution	2340 * 1080
Screen resolution	1440 * 720
CPU core + RAM (running memory)	Eight core+6g
CPU core + RAM (running memory)	Eight core+4g
CPU core + RAM (running memory)	Eight core+3g
ROM (body memory)	128g
ROM (body memory)	64g
ROM (body memory)	32g
Screen size	5–5.5 in.
Screen size	5.5–6 in.
Screen size	6–6.5 in.
Front camera pixels	5–10 million
Front camera pixels	10–16 million
Front camera pixels	20–25 million
Rear camera pixels	800 or less
Rear camera pixels	1200–1900
Rear camera pixels	2000–2400
Battery capacity	3000–3500 mAh
Battery capacity	3500–4000 mAh
Battery capacity	4000–5000 mAh
Fingerprint recognition	Support
Fingerprint recognition	Unsupport

Table 7. Survey results show

Research objects	CPU			Body memory			...
	Eight core+3g	Eight core+4g	Eight core+6g	128g	64g	32g	...
1	h_{-2}	h_1	h_2	h_{-1}	h_3	h_{-2}	...
2	h_{-3}	h_0	h_2	h_3	h_2	h_0	...
3	h_1	h_2	h_{-1}	h_0	h_1	h_{-1}	...
⋮	⋮	⋮	⋮	⋮	⋮	⋮	⋮
186	h_0	h_3	h_{-1}	h_{-2}	h_3	h_{-3}	...

(2) Analysis of questionnaire results. Due to too much data, part of the survey data is shown in Table 7. The language evaluation set of product attributes is $H = \{h\text{-}3 = worst, h\text{-}2 = worse, h\text{-}1 = bad, h0 = general, h1 = good, h2 = better, h3 = best\}$, after sorting, Table 7 is obtained:

① According to formula (8), θ_i can be obtained, and the qualitative evaluation language is converted into a cloud model, when $t = 3$, then: $\theta_{-3} = 0$, $\theta_{-2} = 0.221$, $\theta_{-1} = 0.382$, $\theta_0 = 0.5$, $\theta_1 = 0.618$, $\theta_2 = 0.779$, $\theta_3 = 1$.

② The three digital features are calculated using formulas (9)–(13), assuming that the domain: $[D_{min}, D_{max}] = [2, 8]$, then: $Ex_{-3} = 2$, $Ex_{-2} = 3.326$, $Ex_{-1} = 4.292$, $Ex_0 = 5$, $Ex_1 = 5.708$, $Ex_2 = 6.674$, $Ex_3 = 8$. $En_{-3} = 1.779$, $En_{-2} = En_2 = 1.598$, $En_1 = En_{-1} = 1.265$, $En_0 = 1.157$, $En_3 = 1$. $He_{-3} = He_3 = 0.074$, $He_{-2} = He_2 = 0.134$, $He_{-1} = He_1 = 0.245$, $He_0 = 0.281$.

③ The seven-scale language value is converted to seven clouds: $Y_{-3}(2, 1.779, 0.074)$, $Y_{-2}(3.326, 1.589, 0.134)$, $Y_{-1}(4.292, 1.265, 0.245)$, $Y_0(5, 1.157, 0.281)$, $Y_1(5.708, 1.265, 0.245)$, $Y_2(6.674, 1.598, 0.134)$, $Y_3(8, 1.779, 0.074)$.

④ Calculated u_{ij} by formula (14)–(15), the results are shown in Table 8.

Table 8. Partial utility value result at attribute level

Attributes i	Attribute level j	Partial utility value u_{ij}
CPU	Eight core+3g	0.1030
CPU	Eight core+4g	0.1778
CPU	Eight core+6g	0.4008
Body memory	128g	0.4163
Body memory	32g	0.0527
Body memory	64g	0.1878
battery capacity/mAh	3000–3500	0.1725
battery capacity/mAh	3500–4000	0.3862
battery capacity/mAh	4000–5000	0.1396
Screen size/inch	5–5.5	0.2075
Screen size/inch	5.5–6	0.3533
Screen size/inch	6–6.5	0.1610
Front camera pixels/10,000 pixels	500–1000	0.1652
Front camera pixels/10,000 pixels	1000–1600	0.3775
Front camera pixels/10,000 pixels	2000–2500	0.1636
Rear camera pixels/10,000 pixels	800 or less	0.0341
Rear camera pixels/10,000 pixels	1200–1900	0.4609
Rear camera pixels/10,000 pixels	2000–2400	0.0990
Fingerprint recognition	Support	0.4962
Fingerprint recognition	Unsupport	0.0149
Resolution	1920 * 1080	0.2797
Resolution	2340 * 1080	0.3334
Resolution	1440 * 720	0.1003

Cost of Mobile Phone Attribute Level. The research in this paper only considers the cost of mobile phone hardware. Different brands of mobile phones use different devices, and it is difficult to uniformly demarcate their attribute levels. Therefore, it is assumed that all types of mobile phones use the same accessories, such as speakers, from the same manufacturer. Through the investigation and analysis of the mobile phone bill of materials, the cost price of the hardware of different attribute levels of the mobile phone can be known. The statistical results are shown in Table 9.

Table 9. Cost of attribute level hardware

Attributes i	Attribute level j	Attribute level cost f_{ij}/thousand yuan
CPU	Eight core+3g	0.2
CPU	Eight core+4g	0.25
CPU	Eight core+6g	0.3
Body memory	128g	0.2
Body memory	32g	0.05
Body memory	64g	0.1
battery capacity/mAh	3000–3500	0.03
battery capacity/mAh	3500–4000	0.04
battery capacity/mAh	4000–5000	0.08
Screen size/inch	5–5.5	0.12
Screen size/inch	5.5–6	0.15
Screen size/inch	6–6.5	0.2
Front camera pixels/10,000 pixels	500–1000	0.08
Front camera pixels/10,000 pixels	1000–1600	0.1
Front camera pixels/10,000 pixels	2000–2500	0.12
Rear camera pixels/10,000 pixels	800 or less	0.09
Rear camera pixels/10,000 pixels	1200–1900	0.13
Rear camera pixels/10,000 pixels	2000–2400	0.15
Fingerprint recognition	Support	0.08
Fingerprint recognition	Unsupport	0
Resolution	1920 * 1080	0.1
Resolution	2340 * 1080	0.14
Resolution	1440 * 720	0.08

Phone Price. In this paper, the price p of the mobile phone is set to 120%, 150%, 180%, 210%, and 230% of the cost price, and the weight of price influence is set to $\gamma = 0.1372$ according to Table 5.

4.3   Experimental Results and Analysis

In this paper, the improved ABC (Artificial Bee Colony) algorithm is used in MATLAB, and the test function is used to make a detailed comparative analysis of the results before and after the improvement. It is concluded that the improved ABC (Artificial Bee Colony) algorithm can effectively compensate for the shortcomings of the original algorithm local optimization, and the convergence speed has also increased to a certain extent.

Algorithm Comparison Results and Analysis. The original version of the ABC algorithm, the Reference [15] algorithm (IABC) and the improved ABC algorithm in this paper were simulated using MATLAB in this section. The initial solutions of the three algorithms are randomly generated. Set the size of the population $N = 100$, the search spatial dimension $Dim = 50$, the maximum number of iterations $MCN = 2000$, and the number of cycles $limit = 100$. The original ABC algorithm, the Reference [15] algorithm (IABC) and the improved ABC algorithm in this paper were tested on the Rosenbrock function and the Rastrigin function, and the test results were compared one by one. The parameters of each test function are shown in Table 10. The test results of the three algorithms are shown in Fig. 3 and Fig. 4.

Table 10. Expressions, search interval, and minimum values of the four test functions

Function name	Function expression	Search space	Minimum value
Rosenbrock	$f_1(x) = \sum_{i=1}^{n} 100(x_{i+1} - x_i^2)^2 + (1 - x_i)^2$	$[-100, 100]$	0
Rastrigin	$f_2(x) = \sum_{i=1}^{n} (x_i^2 - 10(\cos(2\pi xi)) + 10)$	$[-5.12, 5.12]$	0

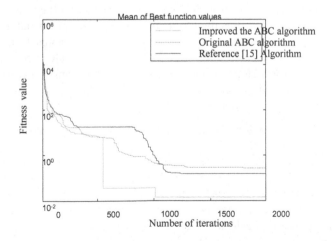

Fig. 3. Fitness changes of Rosenbrock function

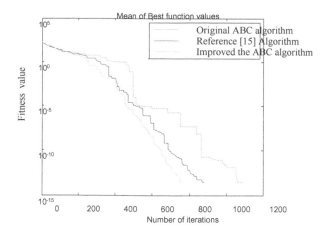

Fig. 4. Fitness changes of Rastrigin function

From Fig. 3 and Fig. 4, the original ABC algorithm will fall into local optimum and slow convergence rate in the test function can be obtained; Although the algorithm of Reference [15] is reduced in the number of iterations and the convergence rate is improved compared with the original algorithm, but it still lacks in the global search ability; The improved ABC algorithm in this paper combined with the reverse learning strategy to improve the population initialization, and has a good global search ability. In terms of search strategy, from Fig. 4, after introducing the current optimal and suboptimal solution, the local search capability of the algorithm has been improved to a certain extent, the convergence rate has also been accelerated, and the number of iterations has been reduced. When the detection bee is initialized, form Fig. 3, after introducing the Gaussian distribution factor, it can help individuals jump out of the local optimal solution, thereby improving the accuracy of the solution.

According to the above analysis, it can be obtained that through the experimental comparison of the two test functions, the improved ABC (Artificial Bee Colony) algorithm has improved resolution and convergence rate compared to the Reference [15] algorithm and the original ABC (Artificial Bee Colony) algorithm. To a certain extent. The defect that the ABC (Artificial Bee Colony) algorithm is easy to fall into the local optimal solution and the shortcoming of the later convergence rate is relatively slow are solved to a certain extent.

Comparison of Solution Results of Product Function Attribute Configuration Model

Figure 5 shows an iterative graph of product line profits. As shown in the figure, when the original ABC algorithm is used to solve the model, it takes 51 iterations to find the optimal solution, and the improved ABC algorithm in this paper finds the optimal solution after 29 iterations, indicating that the algorithm in this paper is better and the speed of convergence is accelerated, and the global search ability is also improved. Through continuous iteration, the value of the objective function is increasing until the optimal solution is found, that is, the value of the objective function is the largest, and

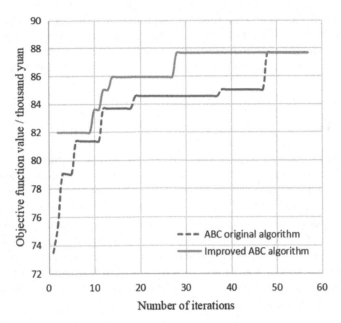

Fig. 5. Iterative graph of product profit

Table 11. Optimal solutions for product positioning

Attribute	Attribute level
CPU and running memory	8 cores+6g
Body memory	128g
Screen resolution	2340 * 1080
Front camera pixels	10–16 million
Rear camera pixels	12–19 million
Fingerprint recognition	Support
Battery capacity	3500–4000 mAh
Screen size	5.5–6 in. screen
Cost	1.140 thousand yuan
Price	2.070 thousand yuan

the product profit is also the largest. According to the operation results of the algorithm, when the expected number of customers is $Q = 200$, the maximum total profit value of the new product generated is $EP = 87.64$ thousand yuan, and the profit of each new mobile phone product is 0.93 thousand yuan. In Table 11, the optimal solution of the ABC (Artificial Bee Colony) algorithm is given, and the optimized configuration and price of the new product are obtained.

The above table shows that the positioning of the product is not a combination of all the optimal attributes, but to re-match the attribute levels of these attributes, which reduces the product attribute configuration that some users do not pay much attention to, it also reduces the price of the product, and finally the goal of maximizing product profit is achieved. In this article, the es database is mainly used for massive data storage, and the big data hadhoop platform mapreduce parallel computing for comment word segmentation mining, which speeds up the running speed and calculation accuracy.

5 Conclusion

In order to improve the accuracy of product function attribute configuration, this paper proposes a method of mining product customer demand and function attribute configuration driven by big data. The mapreduce parallel computing mining was used on the Hadoop platform to determine the product attributes, attribute levels, etc., and the efficiency of calculation was greatly improved. Then customer preferences were obtained through the questionnaire, and the customer preference function is improved. The product function attribute configuration model was established based on the discrete selection model MNL, and the improved ABC (Artificial Bee Colony) algorithm was used to solve the model. Finally, an empirical analysis was carried out on the case of mobile phone products.

Acknowledgments. This work was supported by the National Natural Science Foundation of China granted 71961005 and the Guangxi Science and Technology Program granted 1598007-15.

References

1. David, L., Ryan, K., Gary, K., Alessandro, V.: Big data. The parable of Google Flu: traps in big data analysis. Science (New York, N.Y.) **343**(6176), 1203–1205 (2014)
2. Jin, S.: Based on Hadoop Practice. Machinery Industry Press, Beijing (2011)
3. Alyass, A., Turcotte, M., Meyre, D.: From big data analysis to personalized medicine for all: challenges and opportunities. BMC Med. Genomics **8**(1), 1–12 (2015)
4. Yang, J.: Data mining technology in the view of cloud computing. Electron. Technol. Softw. Eng. (05), 151 (2019)
5. Zhai, G., Cao, Y.: Research on brand choice of mobile phone consumers in China: empirical analysis based on discrete choice model. Mod. Commer. Ind. **20**(01), 55–56 (2008)
6. Zhai, X., Cai, L., Kwong, C.K.: Multi-objective optimization method for product family design. Comput. Integr. Manuf. Syst. **17**(07), 1345–1355 (2011)
7. Luo, X., Cao, Y., Kuang, Z.: New product positioning model and algorithm considering negative utility. J. Syst. Eng. **28**(06), 820–829 (2013)
8. Li, C., Pang, C., Li, M.: Application of weight-based Apriori algorithm in text statistical feature extraction method. Data Anal. Knowl. Discov. **1**(09), 83–89 (2017)
9. Wang, Q.S., Jiang, F.S., Li, F.: Multi-label learning algorithm based on association rules in big data environment. J. Comput. Sci. **47**(05), 90–95 (2020)
10. Yang, Q., Zhang, Y.W., Zhang, Q., Yuan, P.L.: Research and application of multi-dimensional association rule mining algorithm based on hadoop. Comput. Eng. Sci. **41**(12), 2127–2133 (2019)
11. Luo, Z.H., Che, Y., Yang, Z.W.: Research and analysis of massive data mining algorithm based on hadoop platform. Digit. Commun. World (07), 67–68 (2019)

12. Wang, Q.S., Jiang, F.S.: An improved Apriori algorithm under Hadoop framework. J. Liaoning Univ. (Nat. Sci. Ed.) **46**(03), 257–264 (2019)
13. Wang, J.Q., Peng, L., Zhang, H.Y., et al.: Method of multi-criteria group decision-making based on cloud aggregation operators with linguistic information. Inf. Sci. **274**(274), 177–191 (2014)
14. Bao, G., Lian, X., He, M., et al.: A two-dimensional semantic improvement model based on new language evaluation scale. Control and Decis. **25**(05), 780–784 (2010)
15. Yu, J., Zheng, J., Mei, H.: K-means clustering algorithm based on improved artificial bee colony algorithm. J. Comput. Appl. **34**(04), 1065–1069+1088 (2014)

Legal Effect of Smart Contracts Based on Blockchain

Xichen Li[(✉)] [iD]

Xihua University, Chengdu 610039, Sichuan, China
xichenli@mail.xhu.edu.cn

Abstract. Smart contracts are increasingly used in financial innovation area. In view of smart contracts' legal effects, a study is implemented in this paper. First, Smart contracts appear as a set of computer codes, and carry the mutual consensus of the transaction parties and under the principle of freedom of contract form. Thus, smart contracts can be understood as a type of contractual written form in the high-tech context. Then the agreement automatic enforcement was made by the five-element structure, and the traditional situation of enforcement uncertainty was avoided. The results show that it is necessary to examine whether the meaning of the machine matches the meaning of the party, and whether the machine meaning is in conformity with the legal provisions. Finally, the conclusion was drawn that the legal effect of smart contract can be clarified in response to the booming financial technology.

Keywords: Smart contracts · Legal effect · Automatic enforcement

Driven by big data, cloud computing, artificial intelligence, blockchain and mobile internet, the Fintech industry has become a high-level development form of modern finance with national strategic significance. The financial industry enabled by science and technology is rapidly becoming a new engine of innovative economic development. In 2016 and 2018, the official White Paper "China Blockchain Technology and Application Development" issued by the Ministry of Industry and Information Technology of China, emphasized that smart contracts are one of the six core technologies of the blockchain and an important cornerstone of future digital civilization. Smart contracts have received extensive attention from the financial practice community. More and more companies and investment focus on the field of smart contracts. Smart contracts are different from traditional contracts, and their widespread use in the field of practice faces questions about whether they have legal effect and what kind of legal effect. If the legal validity of smart contracts is not studied in a timely and in-depth manner, it will not only place the current smart contracts application in the risk of legal uncertainty, but also delay and reduce the development of Fintech and international competitiveness.

Supported by National Social Science Foundation of China (Grant No. 19BFX120).

© Springer Nature Singapore Pte Ltd. 2020
J. Zeng et al. (Eds.): ICPCSEE 2020, CCIS 1257, pp. 166–186, 2020.
https://doi.org/10.1007/978-981-15-7981-3_12

1   The Essence of Expression Will in Smart Contracts

Blockchain-based smart contracts can form trust among strangers. At present, they have been widely used in finance and other fields. In sharp contrast to the booming smart contract practice, the "legal effect of smart contract" and "what kind of legal effect" are rarely studied. To answer these questions, first of all, it is necessary to analyze what is the essence behind the smart contract: is it simply computer codes, machine meaning, or is it carrying the meaning of human being? If it is a machine expression of human meaning, can it be the consensus of both parties to the transaction? Analyzing the legal relationship behind the smart contract and the formation process of the smart contract is undoubtedly the key of the problem solving.

1.1   Legal Relationship Behind Smart Contracts

Blockchain-based smart contracts are a high-tech form of innovative transactions that bring together the participation of multiple parties. From the perspective of the core subjects, smart contract transactions include blockchain platform providers, transaction parties, smart contract producers, and transaction validator [1].

– **Blockchain platform provider.** Blockchain platform providers provide users with a complete set of blockchain systems to ensure that the platform has features such as decentralization, openness, autonomy, immutability, and traceability. The platform has the function of sending and storing data, verifying transactions, and running programs intelligently. Blockchain platform establishes a mechanism of providing trust among strangers. Technically, the blockchain platform consists of a data layer, a network layer, a consensus layer, an incentive layer, and a smart contract layer [2]. The legal relationship between the blockchain platform provider and the smart contract trader is the relation of platform services supplying and accepting. The platform provider should technically guarantee the stability, security, and continuity of the features mentioned above.
– **Parties of the transactions.** Both parties of the transaction are users of the blockchain platform, and use the trust and intelligent mechanism created by the blockchain technology to carry out various transactions. There is a contractual relationship between the two parties. They agree on the subject matter, price, mode, performance method, time limit, etc. The two parties enjoy rights and perform obligations in accordance with the agreement.
– **Smart contract makers.** A smart contract maker is a very special type of subjects in smart contract transactions. They do not exist in traditional contract transactions. Such subjects accept the commission of the transaction party to provide technical language interpretation services that help to convert the commercial contracts expressed by the two parties in natural language into Smart contracts edited in computer language. Smart contract producers compile codes because transacting parties believe their technical skills,

so they have the obligation to correctly understand the contractual meaning of the transacting parties, accurately translate it into computer codes, and ensure that the codes can be executed correctly.

- **Transaction validator.** Effective nodes in the blockchain network play the role of smart contract transaction verifiers. They compete for bookkeeping rights, receive remuneration, witness transactions, build blocks, and record them in the blockchain in accordance with the unified rules of the entire network. For smart contract transactions, the owner of an effective node provides transaction witness, verification, and bookkeeping services for a fee. Its core obligation is to correctly witness the transaction and ensure that the verified transaction is authentic and legal.

The various parties involved in blockchain smart contracts are in a legal relationship network. Smart contracts are agreements in which transaction parties arrange transaction content, their respective rights and obligations, and are written in machine language; smart contracts can only be executed in a reliable and automatic fulfillment function offered by the blockchain platform provider; blockchain as a peer-to-peer network community, witnessing and recording transactions are inseparable from the participation of node users. The smart contract itself and the blockchain environment in which it resides bring together multiple parties' rights and interests. Although the appearance of the smart contract is computer codes, it is actually a reflection of the expression will of the transaction parties; the normal execution of the smart contract must also get the clear cooperation of other parties.

1.2 The Formation of Smart Contracts

Transactions through smart contracts can be roughly divided into two types: one is that the two parties negotiate first, enter into a traditional contract, and then form a smart contract; the other is that one party to the transaction drafts the transaction terms and translates them into computer codes deployed on the blockchain and the counterparty "Click" to enter into the smart contract.

In the first type, the two parties to the transaction agree on the transaction itself, form a traditional written or oral contract, and agree to use the blockchain-based smart contract for transactions. At this time, the transaction party needs to download the blockchain application to become a node in the blockchain network. The client of a full node usually includes functions such as account creation, wallet management, mining management, data management, and deployment of smart contracts. The verbal or written contract (usually an electronic contract) which is already formed by the two parties of the transaction is written in natural language and cannot be identified and run even if it is uploaded to the blockchain as an electronic record. To make a smart contract, you need to convert the contract terms into computer codes and upload it to the blockchain [3]. Therefore, it is necessary to convert the formed natural language into computer codes. The transaction party entrusts the smart contract producer to use a computer language such as Solidity to edit the smart contract, edits the metadata through an editor, and finally publishes it to the blockchain platform [4].

Every node scattered the smart contract on the blockchain [5]. Smart contract will be executed in the operating environment provided by the blockchain platform (such as the "Ethernet virtual machine"). Once specific conditions are met, they are automatically executed. Smart contracts and automatic execution are witnessed on the entire network. The rights and obligations deployed in the smart contract are realized.

One party to the transaction drafts the contract terms and translates them into computer codes, which reflects the transaction intentions of the drafter. What needs to be discussed here is whether the later party agrees on the rights and obligations arrangements in the smart contract, and whether the parties to the transaction have reached an agreement on the content of the smart contract. The legal requirement for consensus protects the contractual freedom of the parties. Each party should have the opportunity to understand and understand the terms of the draft contract and have the right to choose whether to agree. The law should make it not easy but also not difficult for the parties to form consensus [6]. 112 (a) of the United States Uniform Computer Information Transaction Act stipulates that "click" usually has the legal effect of "consent", and "click" is regarded by all rational persons as consent, just as the meaning of consent is expressed orally [7]. In addition, the US "Electronic Signatures in Global and National Act" also recognized the legal effect of electronic signatures. Article 48 of "E-Commerce Law of the People's Republic of China" stipulates that: "E-commerce parties' use of automatic information systems to enter into or perform contracts has legal effect on the parties using the system." Article 49 states: "Commodities issued by e-commerce operators or if the service information meets the conditions of the offer, the user selects the product or service and submits the order successfully, and the contract is established. If the parties agree otherwise, the agreement will be adopted." It can be seen that China and the United States hold that reading the contract terms offered unilaterally and clicking by the counterparty are usually regarded as accepting the terms of the contract and reaching agreement.

Regardless of whether the two parties of the transaction first form an agreement and then convert it into a smart contract, or unilaterally provide smart contract and the counterparty "click" to enter, smart contracts are not just cold computer code as shown on the surface, and expression of will and consensus behind them.

2   The Legal Effect of Smart Contract Codes

Since the smart contract can present the wills and consensus of both parties in the transaction, it can enter the field of contract law and become a legal contract or part of the legal contract. However, traditional contract forms are written or spoken, and do not include computer code forms which have very different appearances. Then whether the code appearance of a smart contract meets the legally required contract form and whether it has formal validity are the next questions we should answer immediately.

The parties agreed that writing on paper was a traditional written form recognized by most national contract laws. Are smart contracts essentially consistent with traditional written forms? Can it be recognized by contract law? The nature of smart contracts as modern digital high-tech products is loaded with information about transactions. Reviewing and exploring the inherent logic of the history of human information recording and transmission can help us answer the question.

2.1 The Nature of the Information Carrier in Traditional Written Form

Information is a reflection of the state and change of the subjective and objective world, and its expression forms vary with the progress of science and technology. In the ancient times, the rope was used for recording. The information of an event was recorded by a rope, but the content of expression was very limited and there were difficulties in transmission. Since human beings invented language, language has become an information carrier and an expression tool for expressing the subjective and objective world. Human natural language includes both oral and written forms. The oral form is based on phonemes and is fleeting without the recording of modern technology. Written records can be preserved and as proofs. Bamboo slips, silk crickets, and paper are such written forms with human constant exploration efforts for a long time. Paper has outstanding performance in terms of easy transmission, low cost, and popularity. As a historical selection, paper has become the main carrier of human languages. The information corresponding to human languages has also been recorded on paper in large quantities.

How to spread the text written on paper, and how to transmit the wills of trade parties when they do not meet each other? "Pigeon Mail" and "Messenger" are familiar methods to transmit information. In paper era, the only way to transfer text information is relying on the change of physical space of paper space. Although paper has a significant advantage in recording language information, it is difficult to transmit quickly, and the bread, deep and frequent commercial transactions need new ways to carry information.

2.2 Wired Transmission

How to increase the speed of information transfer between the parties who do not meet each other has become the direction for the further development of commercial transactions. Can information be expressed only in natural language? Can natural language transmission be transmitted only through paper writing? There are more than 6000 natural languages in the world expressing ideas and conveying information in different social groups. Observing from another perspective, we will find that natural language is actually a set of codes about subjective and objective world information, and the information encoding methods of different social groups are different. Since human natural language is just a way of encoding information, is there any other way of expressing information besides natural language? Inspired by the American doctor Jackson's discovery: "No matter how long the wire is, current can flow through quickly", Morse matched human language with electrical symbols (spark and spark length). When transmitting information, first step is to translate natural language into Electric symbols, the electric symbols are quickly transmitted to the destination with current, and then the electric symbols are translated into natural language, so that the information transfers much faster than in Paper Era.

While exploring the rapid transmission of text messages, people are constantly thinking about how to transmit sound over long distances. From the perspective of contract form, voice is oral. Transmission of sound over long distances allows people to make oral contracts over long distances. Although they are all converted into electrical symbols, the basic theory of conversion is different. The telegram converts text symbols into electrical symbols, while the telephone converts sound into electrical symbols. Specifically, it converts sound vibrations into electronic vibrations. Electronic vibration is reduced to sound. The long-distance transmission of human language in written and verbal through "wired" technology has greatly improved the efficiency of communication between the two parties of the transaction.

2.3   Wireless Transmission

Both the telegraph and the telephone were transmitted by wired at the early stage, relying on the laying of wired cables. Telegraphs or telephones cannot complete long distance transmissions without lines. The science and technology community began to think about how to get rid of the shackles of "wired" and to search for larger, faster, and more accurate information transmission. "Wireless" technology has entered the stage of history. Communication scientists use the characteristics of electromagnetic wave signals that can be transmitted in free space to convert text, sound, and images into electrical signals, which are then transmitted by a transmitter, transmitted in a transmission medium, and converted to text, sound or image by the receiver. The era of wireless transmission of information breaks through the constraints of wired connection, and further expands the number of transaction subjects represented by long-distance communication on the basis of wired transmission.

2.4   Digital Signal

Since the invention of computers by John von Neumann in 1946, people have begun to think about how information can be used and transmitted on computers due to their unparalleled superiority in computing, analysis, storage, and anti-interference. The computer uses binary. In order for the computer to identify, analyze, and process the information used in commercial transactions, the electrical symbols must be digitized. Then the logic of the generation and transmission of the information including the will representation has changed substantially: the text, sound, and image are converted into electrical signals, the aforementioned analog signals should be converted into digital forms of "0" and "1", and then the digital signals are modulated and transmitted. After receiving the digital signals, the receiver processes the reverse step to convert digital signals to text, sound, and images. With the advent of the digital age of the computer, electronic data, data messages, and electronic contracts have emerged as the means of communication between the parties. These methods can be almost synchronously transmitted remotely, which greatly improves the efficiency of negotiations.

2.5   Computer Codes Expressing Wills

Digital technology processes natural language into binary numbers, and traditional written contracts are converted into electronic contracts that can be transmitted by computers. However, digitization is much more than that. Numbers can not only be stored and transmitted, but can also be calculated under the premise of conforming to the rules of computer operation. The computing structure of "If ... then ..." can match the contract logic of "What kind of behavior should the parties do when some specific conditions are met?" The smart contract has discovered the mystery and pushed the electronic contract to be executable. If the electronic form expands the space for the parties to express and exchange wills, then smart contracts can automatically execute the terms of the transaction while converting the parties' intentions into digital codes. The "stationary" contract, which could only be contained on paper or electronic media, as the specific basis for the contractor to perform the contract, took a historic step and became an "active" contract.

From the perspective of information theory, computer codes are just like the "knots", languages, telegrams, telephones, data messages, etc. that appear in the long river of history, carrying the party's wills, and they have the logic basis of expression wills. The use of code calculations to execute transaction arrangements is also in line with the wills of the parties of the transaction (see Table. 1).

Human language, electronic symbols, and computer codes have enriched information sources about subjective and objective world. As long as the information sources is correct, the forms carrying information mentioned above is reliable, accurate, and reliable. Commercial contracts embody the parties' intentions, and the expressions of wills recorded on paper are mandatory acts that are bound by law. Since electronic symbols and computer numbers can also accurately express the wills of the parties, it is logical that electronic and digital forms can become legally recognized forms and have legal effect. Article 10 of Contract Law stipulates: "Forms of Contract: Writing Requirement A contract may be made in a writing, in an oral conversation, as well as in any other form." The basic type of form is intended to show the limits of freedom of contract [8]. Article 11 gives the definition of Writing: "A writing means a memorandum of

Table 1. History of information recording and transmission.

Information	Non-verbal era	Language era		Electronic form			Digital transmission and calculation
		Pre-paper period	Traditional writing	Wired transmission	Wireless transmission	Digital transmission	Digital code (smart contract)
Content	Things, viewpoints, ideas, etc. in the subjective and objective world						
Information expression	Behaviors, things, such as knotting notes	Languages	Languages	Electricity symbol	Electromagnetic wave	Digital (Electronic contract)	Digital code (smart contract)
Information record carrier	Other things	Bamboo slips, silk urns, walls	Paper	Cables, etc.	Radio waves, etc	Computer network	Computer, network, distributed ledger
Information dissemination	Physical transfer of things	Physical transfer of bamboo slips and silkworms	Physical transfer of paper	Language electric symbol	Languages Electricity symbol	Language electric symbol digital	Language electric symbol digital Code operation

contract, letter or electronic message (including telegram, telex, facsimile, electronic data exchange and electronic mail), etc. which is capable of expressing its contents in a tangible form." UCC 2-201 of the Uniform Commercial Code stipulates that "(1) except as otherwise provided in this section, a contract for the sale of goods for the price of $500 or more is not enforceable by way of action or defense unless there is some writing..." UCC expanded the written form as 1–201 states: "'writing' includes printing, typewriting, or any other intentional reduction to tangible form." US courts consider that tangible forms include electronic contracts because electronic contracts can be stored on a computer's hard drive and can be printed [9]. The electronics and data have been recognized by Chinese and American law.

A smart contract is computer code based on electronic data. Its direct expression is not electronic human natural language, but it records and conveys the meaning of the parties and can fulfill and realize the wills of the transaction parties. From the "principle of contract form freedom", the contract codes carry the parties' wills and can be "tangibly expressed", so smart contracts can become a contract form for parties to choose and computer codes can be considered as a kind of writing. The opinion of the US Commodity Futures Trading Commission is the proof of this logical reasoning, and they believe: "Existing law and regulation apply equally regardless what form a contract takes. Contracts or constituent parts of contracts that are written in code are subject to otherwise applicable law and regulation" [10].

3   The Legal Effect of Smart Contract Automatic Performance

Both Chinese law and American law have adopted expanded interpretation to recognize electronic contracts, which is actually giving legal effect to contract forms in a static state. The electronic contract only records the contents of the contract in the form of electronic and digital. In addition to the electronic form, the smart contract is more important in that it can automatically execute the terms of the contract. Traditional contract performance relies on both parties to "do it for themselves" or "commit other people to do it", and there must be "do" or "do not" by the transaction party. Once the smart contract is set up, no active or passive cooperation of the parties is required. As long as certain conditions are met, the smart contract performs automatically. Then whether the automatic performance without parties' behavior has legal effect should be carefully examined.

3.1   Structure of Smart Contracts

According to system theory, structure is the internal basis of function, and function is the external manifestation of elemental structure; certain structures always show certain functions, and certain functions are always generated by specific structures [11]. So how can smart contracts be fulfilled automatically?

It is bound to be inseparable from its special structure. The computer industry pays more attention to the smart contract structure. Experts generally think that smart contracts should generally have 5 basic elements: electronic contracts, blockchains, protocols, causality, and procedural steps [12]. The five-element structure of a smart contract facilitates its automatic performance. An electronic contract is an expression of the intention of the parties to the transaction. It is expressed in electronic data and can be stored in a computer and transmitted through the network. Electronic contracts are legal forms recognized by Chinese law and US law. This is just the electronicization of traditional written contracts. It is only the premise that smart contract performance has the most basic legal effect.

For smart contracts to achieve intelligence, they must be built in an environment where strangers can trust each other. The advent of the blockchain has broken through the problem of "zero information" but "trust" between strangers. Until the invention of blockchain, smart contracts can become a reality from a dream. Blockchain distributed storage, strangers witnessing transactions, and special consensus mechanisms enable smart contracts to facilitate strangers to establish trust relationships and enter trading contract relationships without the need for traditional third-party mechanisms.

The electronic contract written in natural language presents the wills of the parties, the parties can recognize, but the computer is not able to recognize, understand and process, because it just stores and transmits the electronic contract as a bunch of data. Therefore, for electronic contracts to be recognized and run by computers, human natural language needs to be translated into computer languages that computers can recognize. In the blockchain environment, Turing Complete programming language has completed this historical mission, and the protocols written by the programming language have completed the conversion from "human meaning" to "machine meaning".

Different from "cause leads effect", "cause leads no effect", "no cause but effect", and even completely random action modes in human world, computer behavior logic is more rigorous and strict. The logic structure of computer is "If … then …" the next behavioral requires a "trigger factor", and the computer can make the triggered behavior happen when the triggering factor is met. Computer protocols follow strict rule of causality and can form clear "cause must lead effect" behavior expectations. It is based on this point that the behavior path of computer operation can form symmetrical information between the parties and further increase the trust of transactions between strangers.

Computers operate on the logic of causal relation. It requires that the natural language contracts should be converted into causal relation, and then combed into a step-by-step process relationship. There are clear procedures and steps, one by one, and the goals of the protocol will be achieved step by step according to the sequence of events.

3.2 Institutional Significance of Smart Contracts Automatic Performance

Smart contracts have fundamentally changed the way contracts are performed. For thousands of years, the performance of a contract requires the parties to perform a specific action. Taking the most common commercial sales contract as an example, the seller needs to deliver the goods, the buyer needs to pay for the goods, and the performance of the contract requires the parties' positive/negative behavior. However, when facing the future that has not yet occurred and experienced, the information on whether the parties have performed the contract is distributed asymmetrically among the parties to the contract. After the contract is signed, because the specific behavior is future and uncertain, the parties' concerns can never be eliminated: the seller is worried that the buyer will not pay after delivery; the buyer is worried that the seller is not delivering or the quality of the goods is less than expected. In the performance of real-world contracts, the buyer and the seller fail to perform or perform the contract in accordance with various appearances, and the conflicts between the parties have never ceased. In the face of transaction disputes, how to settle disputes, civil and commercial law has developed a set of rules to deal with it after a long period of exploring: "the defense of Consecutive Performance", "the defense of Simultaneous Performance" and "the defense of Right to Suspend Performance". If the contract is not performed or not performed as agreed, the contract law has designed the liability for breach of contract. For disputes arising from the contract which is not performed or performed as agreed, the dispute settlement system arranges negotiation, mediation, litigation, and arbitration. Except for negotiation, which involves negotiation between the parties to solve the problem, the other methods need to pass through a third party, such as a mediation committee, court, arbitration commission, and relevant government departments. The third party also develops a set of substantive and procedural rules to solve contract disputes.

Smart contracts arrange specific performance actions into computer programs in advance. When specific "if" conditions are met, the computer automatically performs the specific actions of the parties to the transaction. In the context of asset digitization and the Internet of Things, transaction behavior will inevitably be performed as promised, which will change the asymmetric information between parties about whether the traditional contract is performed or not. The parties of the smart contract will form the expectations and confidence that the contract will perform. Smart contracts have the potential to create trust and facilitate transactions as promised, even without intermediaries. Once the contract can be performed accurately and realistically, the space of institutional arrangements such as contract non-performance and partial performance will be greatly reduced, and the cost of maintaining and operating these systems will also be reduced accordingly. Observing from the perspective of contract performance, transactions using smart contracts are improving in the direction of facilitating parties to enter transactions and reducing the cost of legal systems for contract performance.

3.3 Standards for Legal Effect of Smart Contracts Automatic Performance

Article 60 of Contract Law of the People's Republic of China stipulates: "Full Performance"; Performance in Good Faith The parties shall fully perform their respective obligations in accordance with the contract. The parties shall abide by the principle of good faith, and perform obligations such as notification, assistance, and confidentiality, etc. in light of the nature and purpose of the contract and in accordance with the relevant usage. In ordinary contract cases, it is not necessary to answer the question of whether the performance of the contract has legal effect. This is because the contract is legally valid after the parties with the capacity for civil conduct have expressed their unanimous agreement, meet the legal requirements, and obtain a positive evaluation of the law. A contract that is established and effective shall be performed in a comprehensive, complete, and honest manner, and the parties' performance in accordance with the contract is naturally legal and effective. It is worth noting that traditional contract performance is performed "by the parties" or other agents, and traditional contract law and judicial practice focus on whether the performance is in accordance with the provisions of a valid contract, and do not care about the law of contract performance The issue of validity is that the performance of "acting by yourself" in accordance with a legal and valid contract naturally has legal effect. However, the emergence of smart contracts has brought about different ways of contract performance, showing a parallel picture of "human performance" and "machine performance". Correspondingly, a question arises: is "computer performance" the method of legal effect?

From the structure of the smart contract, it can be seen that the electronic contract represented by electronic data written in natural language is the true meaning of the parties, and the computer actually runs the computer protocols (electronic protocols) written with the electronic contract as the script. The computer does the operation of computer protocols have the legal effect of performing the contract by parties? This may not be an easy question to answer. First of all, what is the standard of legal effect of contract performance? The legal effective contract is transferred to the computer to perform according to the rules of the computer, which involves the legal evaluation of the way the computer performs.

The basis for the legal effect of the "act for yourself" performance is that the contract itself is legally valid. Naturally performing acts such as "delivering the goods" and "paying the price" itself conform to the expression of intention of the parties to the effective contract and legal requirements. There is a change from the "electronic contract" to the "computer protocol". During the change, whether there is inconsistence between the "computer protocol" and the "electronic contract" should be decided. If there are inconsistencies, the legal effectiveness of smart contract performance may be questioned.

The meaning of natural human language does not match the meaning of machine word by word. First, electronic contracts are written by human natural language as electronic data storage and transmission, following the rules of natural language. Computer protocols follow the rules of the computer, and there is a big difference between the two. Human language is dominated by the human brain and inseparable from thinking, and the relationship between language symbols and things is arbitrary, so human natural language is very complicated. It is difficult for current science and technology to fully simulate the language function of the human brain, and it is not yet possible to achieve a complete translation between human language and computer language. Therefore, there are cases where the meaning of the machine does not match the meaning of natural human language. Second, the electronic contract is reached, drafted and signed by the parties through consensus. Computer protocols need to be written by professionals proficient in computer programming languages. In the context that machine language is not mastered by the vast majority of people, it is rare that both parties of the transaction are proficient in computer languages and can draft agreements. Computer agreements are often entrusted to smart contract producers to write. Different drafters of computer protocols and electronic contracts have different understandings of contract contents and legal points, which results the inconsistency of meaning conversion.

The meaning of the machine is inconsistent with legal requirements. The application of smart contracts in the financial field involves financial institutions, financial consumers, technology companies, and blockchain platform entities. The protection of financial security and financial consumers has always been an important legislative goal of multinational financial laws. Financial risks (including technical risks) control, financial consumer privacy, and protection of information rights are the focus of U.S. and Chinese laws. Computer protocols focus on the logical self-conformity of the computer world, lack of concern about technological risks, financial system risks, violation of consumer rights, and other violation of mandatory legal provisions. For example, in The DAO (Distributed Autonomous Organization Event) in 2016, participants holding DAO tokens jointly voted on the invested project. This project is legal under the premise of complying with relevant laws, but has been hacked because of its smart contract. The attack was "hard-forked" on the Ethereum platform, which meant that from the time the data was recovered to the forked transactions (both hacking and legitimate transactions), the legal effect of its performance was denied.

There is still room for judging the legal effect of smart contracts performance. The core that needs to pay attention to is whether the meaning written by the parties in the electronic contract are consistent with the meaning of the computer protocol written in machine language. When there are inconsistencies, are there rules for conflict resolution? Determine the effectiveness of smart contract performance according to the rules of dispute resolution. In addition, it is necessary to determine whether the computer protocol violates or has the possibility of violating the mandatory provisions of the law, and to determine the legal effect of performance according to whether the computer protocols are legal.

4   Legislative and Judicial Practice in U.S

The history of financial law in the United States is a history of continuously facing financial innovation and coping with financial risks. Under the fintech wave, how to stand at the forefront of the times, and how to allow technological innovation to further promote the in-depth development of the United States' finance and economy, have become an important issue pondered by the states and the United States. California, Vermont, Nevada, Delaware, and other states are actively developing blockchain technology legislation practices. 3 Arizona, Tennessee, and Wyoming are at the forefront of smart contract legislation. The state laws of the three states all involve smart contracts. Because of the different legislative contexts, the understanding and regulations of smart contracts have similarities as well as differences.

4.1   Legislative Practice in the United States

The United States is good at using legal activities to support and promote innovation. Many states have passed legislation to actively respond to innovative technologies such as blockchain and smart contracts. On March 29, 2017, Arizona commented on the revised regulations, Article 44-7061 stipulates signature and records protected by blockchain technology, smart contracts, and ownership of information; effective March 22, 2018 Note 47-10-201 of the Tennessee Code of Regulations provides for "distributed ledger technology"; the Wyoming Law Note, which came into effect on July 1, 2019, provides for "perfection of digital asset security interests". The three US legislative pioneers placed the provisions of smart contracts in different contexts (see Table 2). They both have the same legislative understanding and different emphasis.

Arizona is the earliest state to actively embrace smart contracts among the 50 states in the United States. It is after the advent of the blockchain environment that smart contracts have truly achieved their ideal vision, and before 2017 the blockchain was already at the technology Frontier positions, so the state's legislation is directly concerned with blockchain and smart contracts.

Tennessee passed an act the following year to recognize smart contracts, but the legislation adopted the term "distributed ledger technology" instead of blockchain. Legislators believe that the essence of blockchain is distributed ledger technology, and the latter is a superordinate concept of the former. In response to the open extension of technology, legal stability and forward-looking requirements, legislators adopted the more scientific concept "distributed ledger technology". The state further clarified the legal effect of executing transaction contracts through smart contracts. Tenn. Code Ann. §47-10-202 states that: "Smart contracts may exist in commerce. No contract relating to a transaction shall be denied legal effect, validity, or enforceability solely because that contract is executed through a smart contract." Arizona Annotated Revised Statutes § 44-7061: "C. Smart contracts may exist in commerce. A contract relating to a transaction may not be denied legal effect, validity or enforceability solely because that contract contains a smart contract term." The provisions of the two states

Table 2. Statutes about Smart Contract in Arizona, Tennessee and Wyoming.

State	Arizona	Tennessee	Wyoming
Code	44-7061	44-10-201; 44-10-202	34-29-103
Effective Dates	3/29/2017	3/22/2018	7/1/2019
Content	C. Smart contracts may exist in commerce. A contract relating to a transaction may not be denied legal effect, validity or enforceability solely because that contract contains a smart contract term 2. "Smart contract" means an event-driven program, with state, that runs on a distributed, decentralized, shared and replicated ledger and that can take custody over and instruct transfer of assets on that ledger	47-10-201: (2) "Smart contract" means an event-driven computer program, that executes on an electronic, distributed, decentralized, shared, and replicated ledger that is used to automate transactions, including, but not limited to, transactions that: (A) Take custody over and instruct transfer of assets on that ledger; (B) Create and distribute electronic assets; (C) Synchronize information; or (D) Manage identity and user access to software applications 44-10-202: (c) Smart contracts may exist in commerce. No contract relating to a transaction shall be denied legal effect, validity, or enforceability solely because that contract is executed through a smart contract	(B) A smart contract created by a secured party which has the exclusive legal authority to conduct a transaction relating to a digital asset. As used in this subparagraph, "smart contract" means an automated transaction, as defined in W.S. 40-21-102(a)(ii), or any substantially similar analogue, which is comprised of code, script or programming language that executes the terms of an agreement, and which may include taking custody of and transferring an asset, or issuing executable instructions for these actions, based on the occurrence or nonoccurrence of specified conditions

appear to be the same. However, in reality, there are substantial differences. On the basis of recognizing the effectiveness of smart contracts, Tennessean State legislation further recognizes the effectiveness of executing contracts through smart contracts.

Wyoming act regulates smart contracts in the "digital assets" section. Wyo. Statues believe that: "smart contract" means an automated transaction, which is comprised of code, script or programming language that executes the terms of an agreement. The state law also provides technical means such as private keys and multi-signature, but it is estimated that the purpose of applying technology is to trade, and the essence of the behavior is still digital asset trading, so it shows the legislative layout of smart contracts stipulated in Articles 34-29-103 in the "Performance of Digital Asset Rights" section. The state statute's understanding of smart contracts has actually gone a step further. Arizona and Tennessee consider smart contracts to be a program. Wyoming law considers smart contracts to be "negotiated in code, script, or programming language to execute an agreement". It is, affirming that computer languages such as the codes of a smart contract can be an expression of intention and form of consultation.

In spite of the above differences, the legislation of the three states also has the same understanding: First, there exist traditional transaction contracts and smart contracts. The former are contracts of electronic records and transactions, and they are in the electronic form written with human natural language, which is the legal form protected by law. Arizona and Tennessee consider the latter to be an event-driven process, and Wyoming considers smart contracts to compromise with code, script, or programming language to execute an agreement. Second, the signature in the blockchain and distributed ledger is an electronic signature, and the traditional transaction contract is an electronic record, which has legal effect. Third, smart contracts run on distributed, decentralized, shared, and replicable ledgers. Fourth, smart contracts can be used for automated transactions, including custody or transfer of digital assets. Fifth, smart contracts have legal effects. Wyoming and Tennessee recognize that the execution of transaction contracts through smart contracts is effective, that is to say they recognize the legal effect of smart contracts automatically performance. At the same time, it can be seen from the language used in the statute that there are also other negative factors to legal effect.

4.2   Judicial Practices of U.S. on Smart Contracts

On June 14, 2018, the Southern District Court of Florida in the United States ruled the first case involving smart contract, Rensel v. Centra Tech. Inc [13]. On the one hand, the case reflects that the U.S. courts have to face the new issue of smart contracts and consider them in judicial decisions; on the other hand, it can help us to see the judicial attitudes of the U.S. courts towards smart contracts. First, the court recognizes both sales agreements and smart contracts. The sales agreement is an electronic contract about Centra Tech.'S token CTR sales (referred to as the "sales agreement"). As the defendant did not have sufficient evidence to prove the plaintiff Rensel accepted the sales agreement, the defendant's claim that the parties signed the sales agreement was not recognized by the court. Smart contracts can automatically perform exchange transactions between CTR and other tokens, such as Ethereum and Bitcoin. The plaintiff conducts token transactions through smart contracts. Second, the sales agreement is inconsistent with the content of the smart contract. Entering a smart contract does not mean that the parties agree to the sales agreement. If the seller wants to claim that the buyer who is bound by the smart contract is also bound by the sales agreement, the seller bears the burden of proof. If the seller cannot prove it, two parties are only bound by the smart contract, and the content represented by the smart contract is recognized by the court, that is, the smart contract is no longer just a bunch of codes. Third, the court recognized the practical effects of automatic execution of smart contracts and transaction performance. The plaintiff used smart contracts to exchange 16.1 Ether for CTR and Bitcoin. The automatic transaction and transaction consequences were accepted by the court. The court held that the two parties have no objection to the transaction itself, but there was a dispute over the standard and amount of damages. Fourth, the defendant's obligation to compensate is not the defendant's use of smart contract as a contract method, but the defendant's public issuance of tokens CTR which were not registered with the securities registration department, violated the Securities Law.

5 Enlightenment from U.S. Legislation and Judicial Practice of Smart Contracts

5.1 Promoting Fintech by Legal Innovation

Finance is the core of a country's economy, and economic competition largely depends on financial competition. "A history of financial development is a history of scientific and technological progress. The financial industry and financial supervision have always followed the pace of scientific and technological innovation [14]." Although technology changes with each passing day and its changes are far greater than the corresponding legal changes [15], financial laws set boundaries and rules of conduct for fintech and the fate of emerging technologies will be directly determined. If FinTech with development prospects is denied by law, it may cause a country to miss a good opportunity to gain a competitive advantage in international financial competition. The history of the development of US financial law shows us its open and positive attitude to respond to high-tech. When mobile internet technology emerged in the United States and was widely used in the payment field, the effectiveness of mobile payment was solved in the subsequent Electronic Fund Transfer Act; when credit accounts such as credit cards were used for mobile payments, Regulation Z regulated the related cost and expense disclosure and dispute settlement procedures; electronic transactions such as mobile payments involve identification; the Fair and Accurate Credit Transaction Act solves the problem of identity theft in electronic transactions. Technology is neutral, and US law is good at playing its positive role and retaining its aggressiveness to improve technology innovation. Chinese law can learn experiences from the openness of American law in responding to science and technology. The legal system can be divided into legislative, judicial, and regulatory divisions, each with its own weight, promoting joint innovation and coordinating governance. The legislation recognizes the legal status of smart contracts and the legal effect of performing contracts, establishes a legal attitude that recognizes innovation and technology; the judiciary flexibly responds to various problems that arise in the practice of smart contracts; and the regulation departments focus on the prevention of risks in smart contracts.

It is worth mentioning that Arizona and Tennessee's legislation on smart contracts is worth learning. The states recognize the smart contract from the negative perspective: "No contract relating to a transaction shall be denied legal effect, validity, or enforceability solely because that contract is executed through a smart contract." Legislative techniques that affirm the legal validity of smart contracts from the opposite side can have two functions: one is that the adoption of smart contracts does not invalidate the transaction agreement; the other is that smart contracts themselves can not directly lead to the validity of transaction agreements, and they should also be examined under the legal framework of contract law.

5.2   Freedom of Contract Performance, and Recognition of Automatic Performance of Smart Contracts

Arizona, Tennessee, and Wyoming consider smart contracts as computer programs whose core functions include escrow and transfer of funds. All three states' legislation respects the principle of freedom of contract performance. China's legislation on the method of contract performance is in Article 12 of the Contract Law, which states: "The contents of a contract shall be agreed upon by the parties, and shall generally contain the following clauses: ... (6) Time limit, place and method of performance." It is also recognized that the contract performance mode reflects the parties' autonomy. Therefore, that the parties agreed to perform by smart contracts complies with the principle of contract freedom as well as the legal provisions.

While recognize the way smart contracts are performed, we should also be aware of the special nature of editing computer language. In legal system arrangements, the following two points must be noted: First, the fact parties using the smart contract formally indicates that the transaction parties choose the automatic execution. Once the smart contract is performed, one party proposes that it is not his/her genuine intent to transaction through smart contract. Under such a circumstance, the court should review, and may support the claim when the burden of proof is completed. Second, the application of smart contracts in the blockchain network environment violates the compulsory provisions of law, administrative regulations, and the results of the legal review about the performance by smart contract may be invalid.

5.3   Freedom of Contract Form, Leaving Space for Computer Codes

Although several states of the United States have reached a consensus on the issue of "smart contract as a form of contract performance", they have not answered from the front about whether "smart contracts are contracts" or "computer code is a form of contract". Arizona statutes state "contract contains smart contract terms". From a semantic perspective, this state recognizes that smart contract can be part of a contract. Since the smart contract is part of the contract content, to some extent the smart contract carries the wills of the transaction parties. Although Rensel v. Centra Tech. had a sales agreement and a smart contract, the plaintiff Rensel only traded with the defendant through the smart contract. The actual effect of the two transactions was recognized by the court. This shows that smart contracts can be obtained as an agreement between the parties.

With the widespread application of blockchain and smart contracts, the gradual popularization of computer programming languages, and the increasing degree of understanding and use of computer language, China responds more actively to blockchain and smart contracts. We have to face the tension between computer language and natural language. To resolve the conflict between the two, the following factors need to be considered: First, who drafts the smart contract? Does the counterparty understand the true meaning expressed by the smart contract computer code? Second, whether a party with an advantage of using computer language has abused its advantage to place the counterparty in an unfair position.

5.4 Hybrid Agreement

A hybrid agreement, involving electronic contracts and smart contracts, often occurs in real-world transactions. How should the relationship between the two be handled? The judicial practice in the United States has dealt with some conflicts between the two, and we can learn from them, but at the same time, when the legal system lays out rules for smart contract disputes in advance, it is necessary to clarify the relationship between the two and resolve conflicts in advance. From the real and possible situations that have appeared in the real world, they can be roughly divided into two categories: one is the coexistence of electronic contracts and smart contracts; the other is that the parties to the transaction only enter the electronic contracts or smart contracts.

As for the first type of situation, it can be divided into two cases where they are the same and they are not the same. If the two are consistent, the parties' intentions reflected in the smart contract can be confirmed in the electronic contract, and the electronic contract has been completely coded. If the two are inconsistent, it is necessary to carefully identify which of the following is inconsistent: First, the electronic contract has provisions, but the smart contract has no arrangements. This situation is equivalent to the smart contract only coding part of the content in the electronic contract, and the content in the electronic contract that is not automatically performed by the smart contract will continue to be performed according to the traditional performance method. Second, there are arrangements in smart contracts, but there is no agreement in electronic contracts, that is, the content of computer code programming in smart contracts is beyond the scope of electronic contracts. From the point of view of the transaction parties entering the smart contract, they have agreed to the content and method of automatic performance, and the performance of smart contract has legal effect. If the parties make claims that automatic performance was not genuine intentions, they shall take the burden of proof. Third, there are arrangements for electronic contracts and smart contracts, which conflict with each other. In response to such a situation, the core needs to determine which is the mutual assent reached by the parties. The judicial review should comprehensively consider the following factors: the process of electronic contract negotiation; the order in which electronic contracts and smart contracts are determined and made; the process of electronic contract negotiation; whether the smart contract producer has a computer language advantage and has abused that advantage; whether there was a test of smart contract completed by both parties; whether the transaction parties accept the result of the automatic execution of the smart contract.

The second type is that the transaction parties only enters into an electronic contract or a smart contract. If the two parties only have electronic contracts, this is actually a traditional electronic contract transaction, and the judicial rules are clear. If the two parties did not sign the electronic contract and only entered the smart contract, that is the case of Rensel, this situation is essentially the same as the content of the smart contract coded beyond the scope of the electronic contract. The party denied the result undertake the burden of proof (see Table 3).

Table 3. Relationship between electronic contracts and smart contracts.

Coexist	Only one form
The two are consistent: The expression of meaning embodied in smart contracts can be confirmed by electronic contracts	With electronic contract and no smart contract: Perform the contract as agreed in the electronic contract
The two are inconsistent: 1. There are regulations in electronic contracts, and there are no arrangements for smart contracts. Contents not fulfilled in smart contracts are performed in accordance with electronic contracts 2. Smart contracts have arrangements, and electronic contracts have no provisions. The execution of a smart contract has legal effect, unless there is evidence that the content and execution method of the smart contract do not meet the true intention of the transaction party 3. Conflicts between electronic contracts and smart contracts. The core needs to determine who belongs to the party's true meaning	With smart contracts, no electronic contracts: Perform according to smart contract, if the parties put forward the opposite claim, they must bear the burden of proof and proof

6   Conclusion

China's legislation has not changed for smart contracts. Although the judicial practice community has already engaged in smart contract disputes, the focus of the case disputes does not involve the smart contract itself, but other civil rights, such as Beijing Xinfubao Technology Co., Ltd. v. Qizhong Mu, Alibaba v. Shanghai Blockchain Net Technology Co. Both of these technology companies have adopted smart contracts, but disputes entering the judicial process are reputation rights and trademark rights. In China, smart contracts are increasingly used in practical scenarios especially in the financial field, and the possibility of judicial practice facing legal issues of smart contracts has become higher and higher. Chinese law has a legal basis for embracing smart contracts, and US state law has recognized the legal effect of smart contracts in legal form. The positive response of U.S. law to the form of smart contracts and contract performance methods has enlightened us to proactively lay out legal resources in the legislative and judicial fields to promote the safe and healthy development of high-tech technology in many fields such as finance.

References

1. Zetzsche, D., Buckley, R., Arner, D.: The distributed liability of distributed ledgers: legal risks of blockchain. Univ. Illinois Law Rev. **2**(5), 1361–1406 (2018)
2. China Blockchain Technology and Application Development White Paper (2016)
3. De Sevres, N.K., Chilton, B., Cohen, B.: The Blockchain Revolution, Smart Contracts and Financial Transactions. 21 Cyberspace Law, 3 (2016)
4. Alharby, M., Aad, V.M.: Blockchain based smart contracts: a systematic mapping study. In: 3rd International Conference on Artificial Intelligence and Soft Computing 2017, vol. 08, pp. 125–140. AIRCC Publishing Corporation (2017). https://doi.org/10.5121/csit.2017.71011
5. Baker, E.D.: Trustless Property Systems and Anarchy: How Trustless Transfer Technology Will Shape the Future of the Property Exchange, 45 Sw. L. Rev. 351, 360–361 (2015)
6. Kunz, C.L., Chomsky, C.L.: Contracts: A Contemporary Approach. West Academic Publishing, St. Paul (2010)
7. Mann, R.J.: Electronic Commerce. Wolters Kluwer, Alphen aan den Rijn (2011)
8. Zhu, G.: Written form and establishment of contract. Chin. J. Law **41**(5), 61–78 (2019)
9. Bazak Intern. Corp. v. Tarrant Apparel Group, 378 F. Supp. 2d 377, S.D.N.Y. (2005)
10. LNCS. https://www.cftc.gov/sites/default/files/2018-11/LabCFTC_PrimerSmart Contracts112718.pdf
11. Bertalanffy, L.V.: General System Theory: Foundations, Development, Applications. G. Braziller, New York (1987)
12. Möhring, M., Keller, B., Schmidt, R., Rippin, A.L., Schulz, J., Brückner, K.: Empirical insights in the current development of smart contracts. In: Twenty-Second Pacific Asia Conference on Information Systems 2018, Japan, p. 125 (2018). https://aisel.aisnet.org/pacis2018/146
13. Rensel, J.Z.T., et al., Plaintiffs v. Centra Tech, Inc., et al., Defendants: United States District Court, S.D. Florida. CASE NO. 17–24500-CIV-KING/SIMONTON
14. Zhou, Z.F., Li, J.W.: The transformation of the paradigm of financial supervision in the background of fintech. Chin. J. Law **40**(5), 3–19 (2018)
15. Taylor, K.C.: Fintech Law: A Guide to Technology Law in the Financial Services Industry. Bloomberg BNA, Arlington (2014)

Machine Learning

Convex Reconstruction of Structured Matrix Signals from Linear Measurements: Theoretical Results

Yuan Tian$^{(\boxtimes)}$

Software School, Dalian University of Technology, Dalian, People's Republic of China
tianyuan_ca@dlut.edu.cn

Abstract. The problem of reconstructing n-by-n structured matrix signal $X = (\mathbf{x}_1, \ldots, \mathbf{x}_n)$ via convex optimization is investigated, where each column $\mathbf{x}_j$ is a vector of s-sparsity and all columns have the same l_1-norm value. In this paper, the convex programming problem was solved with noise-free or noisy measurements. The uniform sufficient conditions were established which are very close to necessary conditions and non-uniform conditions were also discussed. In addition, stronger conditions were investigated to guarantee the reconstructed signal's support stability, sign stability and approximation-error robustness. Moreover, with the convex geometric approach in random measurement setting, one of the critical ingredients in this contribution is to estimate the related widths' bounds in case of Gaussian and non-Gaussian distributions. These bounds were explicitly controlled by signal's structural parameters r and s which determined matrix signal's column-wise sparsity and l_1-column-flatness respectively. This paper provides a relatively complete theory on column-wise sparse and l_1-column-flat matrix signal reconstruction, as well as a heuristic foundation for dealing with more complicated high-order tensor signals in, e.g., statistical big data analysis and related data-intensive applications.

Keywords: Compressive sensing · Structured matrix signal · Convex optimization · Column-wise sparsity · Flatness · Sign-stability · Support-stability · Robustness · Random measurement

1 Introduction

Compressive sensing develops effective methods to reconstruct signals accurately or approximately from accurate or noisy measurements by exploiting a priori knowledge about the signal, e.g., the signal's structural features [1,2]. So far in most works signals are only modeled as vectors of high ambient dimension. However, there are lots of applications in which the signals are essentially matrices or even tensors of high orders. For example, in modern MIMO radar systems [3], the measurements can be naturally represented as $y_{kl} = \sum_{ij} \Phi_{kl,ij} X_{ij} + e_{kl}$ where each y_{kl} is the echo sampled at specific time k and specific receiver element l in a linear or planar array; $\Phi_{kl,ij}$ is the coefficient of a linear processor

© Springer Nature Singapore Pte Ltd. 2020
J. Zeng et al. (Eds.): ICPCSEE 2020, CCIS 1257, pp. 189–221, 2020.
https://doi.org/10.1007/978-981-15-7981-3_13

determined by system transmitting/receiving and waveform features; e_{kl} is the intensity of noise and clutter; X_{ij}, if nonzero, is the scattering intensity of a target detected in specific state cell (i, j), e.g, a target at specific distance and radial speed, or at specific distance and direction, etc. In another class of applications related to signal sampling/reconstruction, multivariable functions (waveforms) in a linear space spanned by given basis e.g., $\{\psi_\mu(u)\phi_\nu(v)\}_{\mu,\nu}$, are sampled as $s(u_i, v_j) = \sum_{\mu\nu} \psi_\mu(u_i)\varphi_\nu(v_j)\chi_{\mu,\nu}$ where $\chi_{\mu,\nu}$ are the signal's Fourier coefficients to be recovered from the samples $\{s(u_i, v_j)\}$. These are typical examples to reconstruct matrix signals and many of them can be naturally extended to even more general tensor signal models.

In comparison with traditional compressive sensing methods which mainly deal with vector signals, the problem of reconstructing matrix, or more generally, tensor signals are more challenging. One difficulty is that such signals have richer and more complicated structures than vector signals. When solving the reconstruction problem via convex optimization, it is important to select the appropriate matrix norm (regularizer) for specific signal structure. For example, L_1-norm is only suitable for general sparsity, nuclear norm is suitable for singular-value-sparsity, and other regularizers are needed for more special or more fine-grained structures, e.g., column-wise sparsity, row-wise sparsity or some hybrid structure. Appropriate regularizer determines the reconstruction's performance. However, few theoretical results are developed beyond nuclear and L_1-norms.

In data science, machine learning and many related data-intensive processing fields, e.g., MIMO radar signal processing, matrix and high-order tensor signal reconstruction problems are emerging with more and more critical importance. However, so far the works on matrix or tensor signal reconstruction are relatively rare, among which typical works include low-rank matrix recovery [1,4], matrix completion, *Kronecker* compressive sensing [5,6,14], etc. Low-rank matrix recovery deals with how to reconstruct the matrix signal with sparse singular values from linear measurements using nuclear norm (sum of singular values) as the regularizer, *Kronecker* compressive sensing reconstructs the matrix signal from matrix measurements via matrix L_1-norm $\sum_{ij}|X_{ij}|$ as the regularizer, just a trivial generalization of vector model. Beyond these works, [15,16] investigated some general properties of arbitrary regularizers, among which lots of interesting problems are still open, and the related theory is far from complete.

Contributions and Paper Organization. In this paper we investigate the problem of reconstructing n-by-n matrix signal $X = (\mathbf{x}_1, \ldots, \mathbf{x}_n)$ by convex programming. Signal's structural features in concern are sparsity and flatness, i.e., each column x_j is a vector of s-sparsity and all columns have the same l_1-norm. Such signals naturally appear in some important applications, e.g., radar waveform space-time analysis, which will be investigated as an application in subsequent papers. The regularizer to be used is matrix norm $|||X|||_1 := max_j|\mathbf{x}_j|_1$ where $|.|_1$ is the l_1-norm on column vector space.

The contribution in this paper has two parts. The first part (Sect. 3 and Sect. 4) is about conditions for stability and robustness in signal reconstruction via solving the inf-$|||.|||_1$ convex programming from noise-free or noisy measurements.

In Sect. 3 we establish uniform sufficient conditions which are very close to necessary conditions and non-uniform conditions are also discussed. Similar as the $inf\text{-}l_1$ compressive sensing theory for reconstructing vector signals, a $|||.|||_1$-version RIP condition is investigated. In Sect. 4 stronger conditions are established to guarantee the reconstructed signal's support stability, sign stability and approximation-error robustness.

The second part in our contribution (Sect. 5 and Sect. 6) is to establish conditions on number of linear measurements for robust reconstruction in noise. We take the convex geometric approach [7–10] in random measurement setting and one of the critical ingredients in this approach is to estimate the related widths' bounds incase of Gaussian and non-Gaussian distributions. These bounds are explicitly controlled by signal's structural parameters r and s which determine matrix signal's column-wise sparsity and l_1-column-flatness respectively.

In comparison with other typical theoretical works on this topic, e.g., [5,14], our results are relatively more complete and our methods are more systematic and effective to be generalized for dealing with other regularizers. One important reason is that our methods are based upon matrix signal measurement operator's RIP-like properties and signal's intrinsic complexity, instead of measurement operator's coherence property. Our conclusions, together with their arguments and some auxiliary results, provide a relatively complete theory on column-wise sparse and l_1-column-flat matrix signal reconstruction, In addition, this work is a heuristic foundation for establishing further theory on more complicated high-order tensor signals, which demand is emerging in, e.g., statistical big data analysis and related data-intensive applications.

This paper is only focused on theoretical analysis. Algorithms, numerical investigations and applications will be the subjects in subsequent papers. Foundations for works in the first part is mainly general theory on convex optimization (e.g., first-order optimization conditions) in combination with the information on subdifferential $\partial|||X|||_1$, while foundations for the second part is mainly general theory on high-dimensional probability with some recent deep extensions.

2   Basic Problems, Related Concepts and Fundamental Facts

Conventions and Notations: In this paper we only deal with vectors and matrices in real number field and only deal with square matrix signals for notation simplification, but all results are also true for rectangle matrix signals in complex field. Any vector $\mathbf{x}$ is regarded as column vector, $\mathbf{x}^T$ denotes its transpose (row vector). For a pair of vectors $\mathbf{x}$ and $\mathbf{y}$, $<\mathbf{x},\mathbf{y}>$ denotes their scalar product. For a pair of matrices X and Y, $<X,Y>$ denotes the scalar product $tr(X^TY)$. In particular, the *Frobenius* norm $<X,X>^{1/2}$ is denoted as $|X|_F$.

For a positive integer s, $\sum_s^{n \times n}$ denotes the set of n-by-n matrices which column vectors are all of sparsity s, i.e., the number of non-zero components of each column vector is at most s. Let $S \equiv S_1 \cup \ldots \cup S_n$ be a subset of $\{(i,j): i,j = 1,\ldots,n\}$ where each S_j is a subset of $\{(i,j): i,j = 1,\ldots,n\}$

and its cardinality $|S_j| \leq s$, $\sum_s^{n \times n}(S)$ denotes the set of n-by-n matrices $\{M \colon M_{ij} = 0$ if (i,j) not in $S\}$. S is called the *matrix signal's s-sparsity pattern*. Obviously $\sum_s^{n \times n}(S)$ is a linear space for given S and $\sum_s^{n \times n} = \cup_S \sum_s^{n \times n}(S)$.

A matrix $M = (m_1, \ldots, m_n)$ is called l_1-*column-flat* if all its columns' l_1-norms $|m_j|_1$ have the same value.

If X_k is a group of random variables and $p(x)$ is some given probability distribution, then $X_k \sim^{iid} p(x)$ denotes that all these X_k's are identically and independently sampled under this distribution.

2.1 Basic Problems

In this paper we investigate the problem of reconstructing n-by-n matrix signal $X = (x_1, \ldots, x_n)$ with s-sparse and l_1-flat column vectors $x_1, \ldots, x_n$ (i.e., $|||X|||_1 = |x_j|_1$ for all j) by solving the following convex programming problems. The regularizer is matrix norm $|||X|||_1 := max_j |x_j|_1$.

$Problem MP^{(\alpha)}{}_{y,\Phi,\eta}$:
$$inf |||Z|||_1 \quad s.t. \ Z \in R^{n \times n}, |\mathbf{y} - \Phi(Z)|_\alpha \leq \eta \qquad (1a)$$

In this setting $\mathbf{y}$ is a measurement vector in R^m with some vector norm $|.|_\alpha$ defined on it, e.g., $|.|_\alpha$ being the l_2-norm. $\Phi \colon R^{n \times n} \to R^m$ is a linear operator and there is a matrix X (the real signal) satisfying $y = \Phi(X) + e$ where $|e|_\alpha \leq \eta$. In an equivalent component-wise formulation, $y_i = <\Phi_i, X> + e_i$ where $\Phi_i \in R^{n \times n}$ for each $i = 1, \ldots, m$.

$Problem MP^{(\alpha)}{}_{y,A,B,\eta}$:
$$inf |||Z|||_1 \quad s.t. Z \in R^{n \times n}, \left| \mathbf{Y} - AZB^T \right|_\alpha \leq \eta \qquad (1b)$$

In this setting Y is a matrix in space $R^{m \times m}$ with some matrix norm $|.|_\alpha$ defined on it, e.g., $|.|_\alpha$ being the *Frobenius*-norm. $\Phi_{A,B} \colon R^{n \times n} \to R^{m \times m}$: $Z \to AZB^T$ is a linear operator and there is a matrix signal X satisfying $Y = AXB^T + E$ and $|E|_\alpha \leq \eta$. In an equivalent component-wise formulation, $y_{kl} = <\Phi_{kl}, X> + e_{kl} = \sum_{ij} A_{ki} X_{ij} B_{lj} + e_{kl}$ where for each $1 \leq k, l \leq m$ Φ_{kl} is a n-by-n matrix with its (i,j)-entry as $A_{ki} B_{lj}$.

Remark: Throughout this paper we only consider the case $0 \leq \eta < |\mathbf{y}|_\alpha$ for problem $MP^{(\alpha)}_{y,\Phi,\eta}$ and $0 \leq \eta < |\mathbf{Y}|_\alpha$ for problem $MP^{(\alpha)}_{y,A,B,\eta}$ since otherwise the minimizers X^* of these problems are trivially O.

For the above problems, we will investigate the real matrix signal X's reconstructability and approximation error where the measurement operator Φ and $\Phi_{A,B}$ (actually matrix A and B) are deterministic or at random. In some cases problem $MP^{(\alpha)}_{y,\Phi,\eta}$ and $MP^{(\alpha)}_{y,A,B,\eta}$ are equivalent each other but in other cases some specific hypothesis is only suitable to one of them, so it's appropriate to deal with them respectively.

2.2   Related Concepts

Some related concepts are presented in this subsection which are necessary and important to our work. For brevity all definitions are only presented in the form of vectors, however the generalization to the form of matrices is straightforward.

A cone C is a subset in R^n such that tC is a subset of C for any $t > 0$. For a subset K in R^n, its polar dual $K^* := \{y : <\mathbf{x}, \mathbf{y}> \le 0 \text{ for all } \mathbf{x} \text{ in } K\}$. K^* is always a convex cone.

For a proper convex function $F(\mathbf{x})$, there are two important and related sets:

$$D(F, \mathbf{x}) := \mathbf{v} : F(\mathbf{x} + t\mathbf{v}) \le F(\mathbf{x}) \text{ for some } t > 0$$

$$\partial F(\mathbf{x}) = \{\mathbf{u} : F(\mathbf{y}) \ge F(\mathbf{x}) + <\mathbf{y} - \mathbf{x}, \mathbf{u}> \text{ for all } \mathbf{y}\}$$

and an important relation is $D(F, \mathbf{x})^* = \cup_{t>0} t \partial F(\mathbf{x})$.

Let $|.|$ be some vector norm and $|.|^*$ be its conjugate norm, i.e., $|\mathbf{u}|^* := max\{<\mathbf{x}, \mathbf{u}> : |\mathbf{x}| \le 1\}$ (e.g., $|||X|||_1^* = \sum_j |\mathbf{x}_j|_\infty$) then

$$\partial|\mathbf{x}| = \{\mathbf{u} : |\mathbf{x}| = <\mathbf{x}, \mathbf{u}> \text{ and } |\mathbf{u}|^* \le 1\} \tag{2}$$

Let K be a cone in a vector space L on which Φ is a linear operator, the minimum singular value of Φ with respect to K, norm $|.|_\beta$ on L and norm $|.|_\alpha$ on Φ's image space is defined as

$$\lambda_{min,\alpha,\beta}(\Phi; K) := inf\left\{|\Phi\mathbf{u}|_\alpha : \mathbf{u} \text{ in } K \text{ and } |\mathbf{u}|_\beta = 1\right\} \tag{3}$$

When both $|.|_\beta$ and $|.|_\alpha$ are l_2(or $Frobenius$) norms, $\lambda_{min,\alpha,\beta}(\Phi; K)$ is simply denoted as $\lambda_{min}(\Phi; K)$.

Let K be a cone (not necessarily convex) in normed space $(L, |.|_\beta)$, its conic *Gaussian width* is defined as

$$w_\beta(K) := E_g\left[sup\{<\mathbf{g}, \mathbf{u}> : \mathbf{u} \text{ in } K \text{ and } |\mathbf{u}|_\beta = 1\right] \tag{4}$$

where $\mathbf{g}$ is the random vector on L sampled under standard Gaussian distribution. When $|.|_\beta$ is l_2 or $Frobenius$ norm on L, $w_\beta(K)$ is simply denoted as $w(K)$.

2.3   Fundamental Facts

Our research in the second part (Sect. 5 and Sect. 6) follows the convex geometric approach built upon a sequence of important results, which are summarized in this section as the fundamental facts. Originally these facts were presented for vector rather than matrix signals [7–10]. We re-present them for matrix signals in consistency with the form of our problems. For brevity, all facts are only presented with respect to problem $\text{MP}_{y,\Phi,\eta}^{(\alpha)}$ except for FACT 6.

FACT 1.   (1) Let $X \in R^{n \times n}$ be any matrix signal and $\mathbf{y} = \Phi(X)$, X^* is the solution (minimizer) to the problem $\mathrm{MP}_{y,\Phi,\eta}^{(\alpha)}$ where $\eta = 0$, then $X^* = X$ iff $\ker \Phi \cap \mathrm{D}(|||\cdot|||_1, X) = \{O\}$.

(2) Let $X \in R^{n \times n}$ be any matrix signal and $\mathbf{y} = \Phi(X) + e$ where $|e|_\alpha \leq \eta$, X^* be the solution (minimizer) to the problem $\mathrm{MP}_{y,\Phi,\eta}^{(\alpha)}$ where $\eta > 0$, $|.|_\beta$ be a norm on signal space to measure the reconstruction error, then

$$|X^* - X|_\beta \leq 2\eta/\lambda_{min,\alpha,\beta}\left(\Phi; D\left(|||\cdot|||_1, X\right)\right)$$

FACT 2. K is a cone in $R^{n \times n}$ (not necessarily convex), $\Phi : R^{n \times n} \rightarrow R^m$ is a linear operator with entries $\Phi_{kij} \sim^{iid} N(0,1)$, then for any $t > 0$:

$$P[\lambda_{min}(\Phi; K) \geq (m-1)^{1/2} - w(K) - t] \geq 1 - exp(-t^2/2)$$

Combining these two facts, the following quite useful corollary can be obtained.

FACT 3. Let X and X^* be respectively the matrix signal and the solution to $\mathrm{MP}_{y,\Phi,\eta}^{(2)}$ as specified in FACT 1(2), $\Phi_{kij} \sim^{iid} N(0,1)$, then for any $t > 0$:

$$P[|X^* - X|_F \leq 2\eta/((m-1)^{1/2} - w(D(|||\cdot|||_1, X)) - t)_+] \geq 1 - exp(-t^2/2)$$

where $(u)_+ := max(u, 0)$. In particular, when the measurement vector's dimension $m \geq w^2(D(|||\cdot|||_1, X)) + Cw(D(|||\cdot|||_1, X))$ where C is some absolute constant, X can be reconstructed robustly with respect to the error norm $|X^* - X|_F$ with high probability by solving $\mathrm{MP}_{y,\Phi,\eta}^{(2)}$.

FACT 4. Let F be any proper convex function and zero matrix is not in $\partial F(X)$, then $w_\beta{}^2(D(F, X)) \leq E_G[inf\{|G - tV|_{\beta*}{}^2 : t > 0, V \text{ in } \partial F(X)\}]$ where $|.|_{\beta*}$ is the norm dual to $|.|_\beta$ and G is the random matrix with entries $G_{ij} \sim^{iid} N(0,1)$. In particular, when $|.|_\beta$ is $|.|_F$ then $w^2(D(F, X)) \leq E_G[inf\{|G - tV|_F{}^2 : t > 0, V \text{ in } \partial F(X)\}]$.

This fact is useful to estimate the squared Gaussian width $w_\beta{}^2(D(F, X))$'s upper bound.

FACT 5. Let X and X^* be respectively the matrix signal and the solution to $\mathrm{MP}_{y,\Phi,\eta}^{(2)}$ as specified in FACT 1(2), with the equivalent component-wise formulation $y_k = <\Phi_k, X> + e_k$, each $\Phi_k \sim^{iid} \Phi$ where Φ is a random matrix which satisfies the following conditions: (1) $E[\Phi] = 0$; (2) There exists a constant $\alpha > 0$ such that $\alpha \leq E[< \Phi, U >]$ for all $U : |U|_F = 1$; (3) There exists a constant $\sigma > 0$ such that $P[|< \Phi, U >| \geq t] \leq 2exp\left(-t^2/2\sigma^2\right)$. Let $\rho := \sigma/\alpha$, then for any $t > 0$:

$$P\left[|X^* - X|_F \leq 2\eta/\left(c_1 \alpha \rho^{-2} m^{1/2} - c_2 \sigma w\left(D\left(|||\cdot|||_1, X\right)\right) - \alpha t\right)_+\right] \geq 1 - exp\left(-c_3 t^2\right)$$

where c_1, c_2, c_3 are absolute constants.

FACT 6. Γ is a subset in n-by-n matrix space. Define the linear operator $\Phi_{A,B}$: $R^{n \times n} \to R^{m \times m} : Y = AXB^T$. In the equivalent component-wise formulation, $y_{kl} = <\Phi_{kl}, X> = \sum_{ij} A_{ki} X_{ij} B_{lj}$ for each $1 \leq k, l \leq m$, $\Phi_{kl} \sim^{iid} \Phi$ which is sampled under some given distribution. For any parameter $\xi > 0$, define

$$Q_\xi (\Gamma; \Phi) := inf \{ P \left[| < \Phi, U > | \geq \xi \right] : U \text{ in } \Gamma \text{ and } |U|_F = 1 \}$$

Furthermore, for each $1 \leq k, l \leq m$, let $\varepsilon_{kl} \sim^{iid}$ Rademacher random variable $\varepsilon (P[\varepsilon = \pm 1] = 1/2)$ which are also independent of Φ, and define

$$W(\Gamma; \Phi) := E_H [sup \{ < H, U > : U \text{ in } \Gamma \text{ and } |U|_F = 1 \}]$$

where $H := m^{-1} \sum_{kl} \varepsilon_{kl} \Phi_{kl} = m^{-1} A^T E B$, $E = [\varepsilon_{kl}]$;

$$\lambda_{min} (\Phi; \Gamma) := inf \{ \left(\sum_{kl} |U_{kl} \Phi_{kl}|^2 \right)^{1/2} : U \text{ in } \Gamma \text{ and } |U|_F = 1 \}$$

Then for any $\xi > 0$ and $t > 0$:

$$P[\lambda_{min} (\Phi; \Gamma) \geq \xi m Q_{2\xi} (\Gamma; \Phi) - 2W (\Gamma; \Phi) - \xi t] \geq 1 - exp \left(-t^2/2 \right)$$

Remark: In Fact 6 the definition of $\lambda_{min} (\Phi; \Gamma)$ is the matrix version of that in Subsect. 2.2 with respect to *Frobenius* norm. The proof of FACT 5 and 6 (with respect to vector signals) can be found in [8]'s Theorem 6.3 and Proposition 5.1.

3   Basic Conditions on Matrix Signal Reconstruction

In this and next section we investigate sufficient and necessary conditions on the measurement operator for accurate and approximate signal reconstruction via solving problems $MP^{(\alpha)}_{y,\Phi,\eta}$ and $MP^{(\alpha)}_{y,A,B,\eta}$. For notation simplicity, we only deal with problem $MP^{(\alpha)}_{y,\Phi,\eta}$ and the formulation can be straightforwardly transformed into problem $MP^{(\alpha)}_{y,A,B,\eta}$.

We present conditions for accurate, stable and robust reconstruction respectively. As will be seen, these conditions are similar as those related to the regularizers with so-called *decomposable* subdifferentials. The vector's l_1-norm and matrix's nuclear norm are such examples. However, $\partial |||X|||_1$ is not even weakly-decomposable (i.e., there is no W_0 in $\partial |||X|||_1$ such that $<W_0, W - W_0> = 0$ for all W in $\partial |||X|||_1$). At first we prove a technical lemma which describes $\partial |||X|||_1$'s structure.

Lemma 1. For n-by-n matrix $X = (\mathbf{x}_1, \ldots, \mathbf{x}_n)$ and matrix norm $|||X|||_1 := max_j |x_j|_1$, the subdifferential

$$\partial |||X|||_1 = \{ (\lambda_1 \xi_1, \ldots, \lambda_n \xi_n) : \xi_j \text{ in } \partial |x_j|_1 \text{ and } \lambda_j \geq 0$$

$$\text{for all } j, \lambda_1 + \ldots + \lambda_n = 1 \text{ and } \lambda_j = 0 \text{ for } j : |x_j|_1 < max_k |x_k|_1 \}$$

All proofs are presented in the appendices.

Remark: More explicitly, $\partial|||X|||_1 = \{(\lambda_1\xi_1,\ldots,\lambda_n\xi_n) : \lambda_j \geq 0 \text{ for all } j, \lambda_1 + \cdots + \lambda_n = 1 \text{ and } \lambda_j = 0 \text{ for } j : |x_j|_1 < max_k|x_k|_1; |\xi_j|_\infty \leq 1 \text{ for all } j \text{ and } \xi_j(i) = sgn(X_{ij}) \text{ for } X_{ij} \neq 0\}.$

3.1 Conditions on Φ for Accurate Reconstruction from Noise-Free Measurements

At first we investigate the conditions for matrix signal reconstruction via solving the following problem:

$Problem MP_{y,\Phi,0}$:

$$inf|||Z|||_1 \quad s.t. Z \in R^{n \times n}, \mathbf{y} = \Phi(Z) \tag{5}$$

Theorem 1. Given positive integer s and linear operator Φ, the signal $X \in \sum_S^{n \times n}$ is always the unique minimizer of problem $MP_{y,\Phi,0}$ where $\mathbf{y} = \Phi(X)$ if and only if

$$|||H_S|||_1 < |||H_{\sim S}|||_1 \tag{6}$$

for any $H = (\boldsymbol{h}_1,\ldots,\boldsymbol{h}_n) \in \ker \Phi \backslash \{O\}$ and any s-sparsity pattern S.

Remark: (6) provides the uniform condition for signal reconstruction which applies to all unknown column-wise sparse signals. By a similar proof, we can also obtain a (non-uniform) sufficient condition for individual signal reconstruction, namely, for given operator Φ and unknown signal X with unknown support $S = S_1 \cup \ldots \cup S_n$, if there exist $\lambda_1,\ldots,\lambda_n \geq 0: \lambda_1 + \ldots + \lambda_n = 1$ such that for any $H = (\boldsymbol{h}_1,\ldots,\boldsymbol{h}_n)$ in $\ker \Phi \backslash \{O\}$ there holds:

$$\left| \sum_{j=1}^{n} \lambda_j < \mathbf{h}_{j|Sj}, sgn(\mathbf{x}_j) > \right| < |||H_{\sim S}|||_1 \tag{7}$$

then X will be the unique minimizer of $MP_{y,\Phi,0}$ where $\mathbf{y} = \Phi(X)$.

On the other hand, from $MP_{y,\Phi,0}$'s first-order optimization condition, i.e., for its minimizer X there exist M in $\partial|||X|||_1$ and a multiplier vector $\mathbf{u}$ such that

$$M + \Phi^T(\mathbf{u}) = O \tag{8}$$

then for any $H = (\boldsymbol{h}_1,\ldots,\boldsymbol{h}_n)$ in $\ker \Phi$ we have

$$0 = \langle \Phi(H), \mathbf{u} \rangle = \langle H, \Phi^T(\mathbf{u}) \rangle = -\langle H, M \rangle = -\langle H, E + V \rangle$$

where $E = (\lambda_1 sgn(\mathbf{x}_1) \ldots, \lambda_n sgn(\mathbf{x}_n))$ and $V = (\lambda_1\xi_1,\ldots,\lambda_n\xi_n)$, $|\xi_j|_\infty \leq 1$, so $-\langle H,E \rangle = \langle H,V \rangle$. Note that for the left hand side $|\langle H,E \rangle| = |\sum_{j=1}^{n} \lambda_j \langle \mathbf{h}_{j|Sj}, sgn(\mathbf{x}_j) \rangle|$ and for the right hand side $|\langle H,V \rangle| \leq \sum_{j=1}^{n} |\lambda_j \langle \mathbf{h}_{j|\sim Sj}, \xi_j \rangle| \leq \sum_{j=1}^{n} \lambda_j |\mathbf{h}_{j|\sim Sj}|_1 |\xi_j|_\infty \leq max_j|\mathbf{h}_{j|\sim Sj}|_1 \sum_{j=1}^{n} \lambda_j|\xi_j|_\infty \leq max_j|\mathbf{h}_{j|\sim Sj}|_1 = |||H_{\sim S}|||_1,$

we obtain a (relatively weak) non-uniform necessary condition, namely, for given operator Φ and unknown signal X with unknown support $S = S_1 \cup \ldots \cup S_n$, if X is the minimizer of $MP_{y,\Phi,0}$ where $y = \Phi(X)$ then there exist $\lambda_j \geq 0$ for $j = 1, \ldots, n: \lambda_1 + \ldots + \lambda_n = 1$ such that for any $H = (h_1, \ldots, h_n)$ in ker Φ there holds the inequality

$$\left| \sum_{j=1}^{n} \lambda_j \langle h_{j|Sj}, sgn(\mathbf{x}_j) \rangle \right| \leq |||H_{\sim S}|||_1 \tag{9}$$

3.2 Conditions on Φ for Stable Reconstruction from Noise-Free Measurements

Now investigate the sufficient condition for reconstructing matrix signal via solving $MP_{y,\Phi,0}$ where $y = \Phi(X)$ for some signal X which is unnecessarily sparse but l_1-column-flat. The established condition guarantees the minimizer X^* to be a good approximation to the real signal X.

Theorem 2. Given positive integer s and linear operator Φ with the s-$|||.|||_1$ *Stable Null Space Property*, i.e., there exists a constant $0 < \rho < 1$ such that

$$|||H_S|||_1 \leq \rho|||H_{\sim S}|||_1 \tag{10}$$

for any H in ker Φ and sparsity pattern $S = S_1 \cup \ldots \cup S_n$ where $|S_j| \leq s$. Let $Z = (\mathbf{z}_1, \ldots, \mathbf{z}_n)$ be any feasible solution to problem $MP_{y,\Phi,0}$ where $y = \Phi(X)$ for some signal $X = (\mathbf{x}_1, \ldots, \mathbf{x}_n)$, then

$$|||Z - X|||_1 \leq (1 - \rho)^{-1}(1 + \rho)(2max_j \sigma_s(\mathbf{x}_j)_1 + max_j(|\mathbf{z}_j|_1 - |\mathbf{x}_j|_1)) \tag{11}$$

where $\sigma_s(\mathbf{v})_1 := |\mathbf{v}|_1 - (|v(i_1)| + \ldots + |v(i_s)|), v(i_1), \ldots, v(i_s)$ are $\mathbf{v}$'s s components with the largest absolute values.

In particular, for the minimizer X^* of $MP_{y,\Phi,0}$ where the real signal X is l_1-column-flat, there is the reconstruction-error bound:

$$|||X^* - X|||_1 \leq 2(1 - \rho)^{-1}(1 + \rho)max_j \sigma_s(\mathbf{x}_j)_1 \tag{12}$$

Remark: For any flat and sparse signal X, condition (10) guarantees X can be uniquely reconstructed by solving $MP_{y,\Phi,0}$ due to Theorem 1, while in this case the right hand side of (12) is zero, i.e., this theorem is consisted with the former one. In addition, (12) indicates that the error for the minimizer X^* to approximate the flat but non-sparse signal X is controlled column-wisely by X's non-sparsity (measured by $max_j \sigma_s(\mathbf{x}_j)_1$).

Remark: Given positive integer s, sparsity pattern $S = S_1 \cup \ldots \cup S_n$ where $|S_j| \leq s$ and linear operator Φ, Let Φ^T denote the adjoint operator and Φ_S denote the operator restricted on $\sum_s^{n \times n}(S)$ (so $\Phi_S(X) = \Phi(X_S)$). If $\Phi_S^T \Phi_S$ is a bijection and there exists a constant $0 < \rho < 1$ such that the operator norm satisfies the inequality

$$N((\Phi_S{}^T \Phi_S)^{-1} \Phi_S{}^T \Phi_{\sim S} : |||.|||_1 \to |||.|||_1) \leq \rho \tag{13a}$$

then for any H in ker Φ, from $\Phi_S(H_S) = \Phi(H_S) = \Phi(-H_{\sim S}) = \Phi_{\sim S}(-H_{\sim S})$ we can obtain $H_S = -(\Phi_S^T \Phi_S)^{-1} \Phi_S^T \Phi_{\sim S}(H_{\sim S})$ and then $|||H_{\sim S}|||_1 \leq N((\Phi_S^T \Phi_S)^{-1} \Phi_S^T \Phi_{\sim S} : |||.|||_1 \to |||.|||_1) |||H_{\sim S}|||_1 \leq \rho |||H_{\sim S}|||_1$, therefore (13a) is a uniformly sufficient condition stronger than s-$|||.|||_1$ Stable Null Space Property (10), similar as Tropp's exact recovery condition for vector signal's l_1-min reconstruction. By operator norm's duality $N(M : |.|_\alpha \to |.|_\beta) = N(M^T : |.|_\beta^* \to |.|_\alpha^*)$ an equivalent sufficient condition is:

$$N\left(\Phi_{\sim S}{}^T \Phi_S (\Phi_S{}^T \Phi_S)^{-1} : |||.|||_1^* \to |||.|||_1^*\right) \leq \rho \tag{13b}$$

The condition (13) can be enhanced to provide more powerful results for signal reconstruction (discussed in next section). Now we conclude this subsection with a simple condition for problem $MP_{y,\Phi,0}$ and $MP_{y,\Phi,\eta}$ to guarantee their minimizers' l_1-column-flatness.

Theorem 3 (Condition for Minimizer's l_1-Column-Flatness). Given positive integer s, sparsity pattern $S = S_1 \cup \ldots \cup S_n$ where $|S_j| \leq s$ and linear operator Φ, let Φ^T denote the adjoint operator and Φ_S denote the operator restricted on $\sum_s^{n \times n}(S)$ (so $\Phi_S(X) = \Phi(X_S)$). If $\Phi_S^T(z)$ doesn't have any zero-column for any $z \neq 0$, then any minimizer X^* of $MP_{y,\Phi,0}$ or $MP_{y,\Phi,\eta}$ with $supp(X^*)$ contained in S is l_1-column-flat.

For $MP_{y,\Phi,\eta}$ where $\Phi(Z)_k = tr(\Phi_k^T Z)$, the adjoint operator $\Phi^T(z) = \sum_{k=1}^m (z_k)\Phi_k : R^m \to R^{n \times n}$. For problem $MP_{Y,A,B}$, where $\Phi_{A,B}(Z) = AZB^T$, the adjoint operator $\Phi^T(\mathbf{Y}) = A^T Y B : R^{m \times m} \to R^{n \times n}$. In addition, $supp(\Phi_S^T(z))$ is always a subset in S.

Remark: For the unconstrained convex optimization problem

$$X^* = Arg \ inf |||Z|||_1 + (1/2)\gamma |\mathbf{y} - \Phi_S(Z)|_2{}^2$$

The sufficient condition for minimizer X^*'s l_1-column-flatness is the same: $\Phi_S^T(z)$ doesn't have any zero-column for $z \neq 0$. This fact will be used in Sect. 4.1.

In fact, the first-order optimization condition guarantees there is M^* in $\partial |||X^*|||_1$ such that $M^* + \gamma \Phi_S^T(\Phi_S(X^*) - \mathbf{y}) = O$, i.e., $M^* = \gamma \Phi_S^T(\mathbf{y} - \Phi_S(X^*))$ in $\partial |||X^*|||_1$. Under the above condition, M^* has no $\mathbf{0}$-column unless $\mathbf{y} - \Phi_S(X^*) = \mathbf{0}$. However, $\mathbf{y} - \Phi_S(X^*) = \mathbf{0}$ implies $M^* = O$ which cannot happen in $\partial |||X^*|||_1$ unless $X^* = O$. As a result, $|\mathbf{x}_j^*|_1 = max_k |\mathbf{x}_k^*|_1 = |||X^*|||_1$ for every j.

3.3   Conditions on Φ for Robust Reconstruction from Noisy Measurements

Now consider matrix signal reconstruction from noisy measurements by solving the convex optimization problem $MP_{y,\Phi,\eta} : inf|||Z|||_1$ s.t. $Z \in R^{n \times n}, |y - \Phi(Z)|_2 \leq \eta$ where $\eta > 0$.

Theorem 4. Given positive integer s and linear operator Φ with the s-$|||.|||_1$ *Robust Null Space Property*, i.e., there exist constant $0 < \rho < 1$ and $\beta > 0$ such that

$$|||H_S|||_1 \leq \rho|||H_{\sim S}|||_1 + \beta|\Phi(H)|_2 \tag{14}$$

for any n-by-n matrix H and sparsity pattern $S = S_1 \cup \ldots \cup S_n$ where $|S_j| \leq$ s. Let $Z = (z_1, \ldots, z_n)$ be any feasible solution to the problem $MP_{y,\Phi,\eta}$ where $y = \Phi(X) + e$ for some signal $X = (x_1, \ldots, x_n)$ and $|e|_2 \leq \eta$, then

$$|||Z - X|||_1 \leq (1-\rho)^{-1}(1+\rho)(2max_j\sigma_s(x_j)_1 + 2\beta\eta + max_j(|z_j|_1 - |x_j|_1)) \tag{15}$$

In particular, for the minimizer X^* of $MP_{y,\Phi,\eta}$ where the real signal X is l_1-column-flat, there is the error-control inequality:

$$|||X^* - X|||_1 \leq 2(1-\rho)^{-1}(1+\rho)(max_j\sigma_s(x_j)_1 + \beta\eta) \tag{16}$$

Remark: (16) indicates that the error for the minimizer X^* to approximate the flat but non-sparse signal X is up-bounded column-wisely by X's non-sparsity (measured by $max_j \sigma_s(x_j)_1$) and the noise strength η. If the signal X is both s-sparse and l_1-column-flat, the column-wise approximation error $|||X^* - X|||_1 \leq 2(1-\rho)^{-1}(1+\rho)\beta\eta = O(\eta)$, i.e., with linear convergence rate.

Remark: For any minimizer X^* of problem $MP_{y,\Phi,\eta}$, $M^* + \gamma^* \Phi_S^T(\Phi_S(X^*) - y) = O$ and $|y - \Phi_S(X^*)|_2 = \eta$ is the first-order optimization condition with positive multiplier $\gamma^* > 0$ and M^* in $\partial|||X^*|||_1$. Then for any $H = (h_1, \ldots, h_n)$ we have $<M^*, H> = \gamma^*<y - \Phi(X^*), \Phi(H)>$ where by Lemma 1 $M^* = E + V$, $E = (\lambda_1 sgn(x_1^*), \ldots, \lambda_n sgn(x_n^*))$, $V = (\lambda_1 \xi_1, \ldots, \lambda_n \xi_n), |\xi_j|_\infty \leq 1$ and note that $supp(E) = supp(X^*) = S = \sim supp(V)$, so $<E,H_S> = <E,H> = -<V,H> + \gamma^*<y - \Phi(X^*), \Phi(H)> = -<V,H_{\sim S}> + \gamma^* <y - \Phi(X^*), \Phi(H)>$. Since for the left hand side

$$|<E,H_S>| = |\sum\nolimits_{j=1}^{n} \lambda_j<h_{j|Sj},\ sgn(x_j^*)>|$$

and for the right hand side

$$|\langle V, H_{\sim S}\rangle| + |\gamma^*\langle y - \Phi(X^*), \Phi(H)\rangle|$$
$$\leq \sum\nolimits_{j=1}^{n} |\lambda_j\langle h_{j|\sim Sj},\ \xi_j\rangle| + \gamma^*|y - \Phi(X^*)|_2|\Phi H)|_2$$
$$\leq \sum\nolimits_{j=1}^{n} \lambda_j|h_{j|\sim Sj}|_1|\xi_j|_\infty + \gamma^*\eta|\Phi(H)|_2 \leq |||H_{\sim S}|||_1 + \gamma^*\eta|\Phi(H)|_2$$

we obtain a non-uniform necessary condition, namely, for given operator Φ and unknown signal X with unknown support $S = S_1 \cup \ldots \cup S_n$ and $\mathbf{y} = \Phi(X) + \mathbf{e}$, if X^* is the minimizer of $MP_{y,\Phi,\eta}$ with the correct support S and correct non-zero component signs as the real signal X, then there exist constants $\beta(= \gamma^*\eta) > 0$ and $\lambda_j \geq 0$ for all $j = 1, \ldots, n: \lambda_1 + \ldots + \lambda_n = 1$ such that for any $H = (\boldsymbol{h}_1, \ldots, \boldsymbol{h}_n)$ there holds the inequality

$$\left| \sum_{j=1}^{n} \lambda_j <\mathbf{h}_{j|S_j}, sgn(\mathbf{x}_j) > \right| \leq |||H_{\sim S}|||_1 + \beta|\Phi(H)|_2 \qquad (17)$$

3.4   M-Restricted Isometry Property

It's well known that *RIP* with appropriate parameters provide powerful sufficient conditions to guarantee l_1-*min* reconstruction for sparse vector signals. With our regularizer $|||X|||_1$ we propose a similar but slightly stronger condition to guarantee reconstruction robustness by solving the convex programming $MP_{y,\Phi,\eta}$.

Theorem 5. Given positive integer s and linear operator $\Phi: R^{n \times n} \to R^m$. Suppose there exist positive constants $0 < \delta_s < 1$ and $\Delta_s > 0$ such that:

(1) $(1 - \delta_s)|Z|_F^2 \leq |\Phi(Z)|_2^2 \leq (1 + \delta_s)|Z|_F^2$ for any $Z \in \sum_{s}^{n \times n}$; $\qquad (18)$

(2) $|<\Phi(Z), \Phi(W)>| \leq (\Delta_s/n) \sum_{j=1}^{n} |z_j|_2 |w_j|_2$ $\qquad (19)$

for any $Z = (\mathbf{z}_1, \ldots, \mathbf{z}_n) \in \sum_{s}^{n \times n}$ and $W = (\mathbf{w}_1, \ldots, \mathbf{w}_n) \in \sum_{s}^{n \times n}$ with $\text{supp}(Z) \cap \text{supp}(W) = \emptyset$. Under these two conditions, there are constants ρ and β such that

$$|||H_S|||_1 \leq \rho|||H_{\sim S}|||_1 + \beta|\Phi(H)|_2 \qquad (20)$$

for any n-by-n matrix H and any s-sparsity pattern S, where the constants can be selected as

$$\rho \leq \Delta_s/(1 - \delta_s - \Delta_s/4), \ \beta \leq s^{1/2}(1 + \delta_s)^{1/2}/(1 - \delta_s - \Delta_s/4) \qquad (21)$$

In particular, $\delta_s + 5\Delta_s/4 < 1$ implies the robust null space condition: $\rho < 1$.

Note: Condition (1) is the standard RIP which implies $|\langle\Phi(Z), \Phi(W)\rangle| \leq \delta_{2s}|Z_F|W|_F$ for s-sparse matrices Z and W with separated supports, slightly weaker than condition (2).

4 More Properties on Reconstruction from Noisy Measurements

In this section we establish stronger conditions on the measurement operator $\boldsymbol{\Phi}$ for some stronger results on sparse and flat matrix signal reconstruction from noisy measurements, e.g., conditions to guarantee uniqueness, support and sign stability as well as value-error robustness.

As in last sections, for notation simplification this section only deals with problem $MP_{y,\Phi,\eta}$ but all conclusions can be straightforwardly transformed into the formulation for problem $MP_{Y,A,B,\eta}$. At first we note the basic fact that $X^* = arginf |||Z|||_1$ s.t. $Z \in R^{n \times n}$, $|\mathbf{y} - \Phi(Z)|_2 \leq \eta$ if and only if their exists a multiplier $\gamma^* > 0$ (dependent on X^* in general) such that X^* is a minimizer of the unconstrained convex programming $inf |||Z|||_1 + (1/2)\gamma^*|\mathbf{y} - \Phi(Z)|_2^2$. In Sect. 4.1 we investigate some critical properties for the minimizer of unconstrained optimization (Sect. 4.1), then on basis of these results we establish conditions for robustness, support and sign stability in signal reconstruction via solving $MP_{y,\Phi,\eta}$ in Sect. 4.2.

In the following for given positive integer s, sparsity pattern $S = S_1 \cup \ldots \cup S_n$ where $|S_j| \leq s$ for all j and the linear operator $\Phi: R^{n \times n} \to R^m$, when $\Phi_S^T \Phi_S$ is a bijection for $\sum_s^{n \times n} (S) \to \sum_s^{n \times n} (S)$ we denote the pseudo- inverse $(\Phi_S^T \Phi_S)^{-1} \Phi_S^T$: $R^m \to \sum_s^{n \times n} (S)$ as Φ_S^{*-1}.

4.1 Conditions on Minimizer Uniqueness and Robustness for $MLP_{y,\Phi}(\gamma)$

Consider the convex programming (22) with given parameter $\gamma > 0$ (value of γ is independently set): *Problem* $MLP_{y,\Phi}(\gamma)$

$$inf|||Z|||_1 + (1/2)\gamma|y - \Phi(Z)|_2^2 \tag{22}$$

Lemma 2 indicates basic properties of its sparse minimizer under some sparsity-related conditions.

Lemma 2. Given $\mathbf{y}$, positive integer s and sparsity pattern $S = S_1 \cup \ldots \cup S_n$ where $|S_j| \leq s$ for all j, suppose the linear measurement operator $\Phi: R^{n \times n} \to R^m$ satisfies:

1. $\Phi_S^T(\mathbf{z})$ does not have any $\mathbf{0}$-column for $\mathbf{z} \neq \mathbf{0}$;
2. $\Phi_S^T \Phi_S$ is a bijection;
3. $\gamma\sup\{<\Phi_{\sim S}^T(\Phi_S \Phi_S^{*-1}(\mathbf{y})-\mathbf{y}), H> : |||H|||_1 = 1\}+\sup\{< \Phi_{\sim S}^T(\Phi_S^{*-1})^T M, H >: |||H|||_1 = 1, |||M|||_1^* \leq 1\} < 1$

Let $X_S^* = Arg\ inf_{supp(S)\ in\ S}|||Z|||_1 + (1/2)\gamma|\mathbf{y} - \Phi(Z)|_2^2$ be the minimizer of $MLP_{y,\Phi}(\gamma)$ with support in S, i.e.,

$$X_S^* = Arg\ inf|||Z|||_1 + (1/2)\gamma|y - \Phi_S(Z)|_2^2 \tag{23}$$

Then there are the following conclusions:

1. X_S^* is the unique minimizer of problem (23) and is l_1-column-flat;
2. X_S^* is also the unique minimizer of problem (22), i.e., the (global) minimizer of (4.1) is unique and is X_S^*.
3. Let $Y^* = \Phi_S^{*-1}(\mathbf{y}) \in \sum_s^{n \times n}(S)$, then $X_{S,ij}^* \neq 0$ for all (i,j): $|Y_{ij}^*| > \gamma^{-1} N((\Phi_S^T \Phi_S)^{-1} : |||\cdot|||_1^* \to |||\cdot|||_{max})$ where $N((\Phi_S^T \Phi_S)^{-1} : |||\cdot|||_1^* \to |||\cdot|||_{max})$ denotes $(\Phi_S^T \Phi_S)^{-1}$'s operator norm and matrix norm $|||M|||_{max} := max_{ij} |M_{ij}|$.
4. With the same notations as the above, if

$$min_{(i,j) \; in \; S} |Y_{ij}^*| > \gamma^{-1} N((\Phi_S^T \Phi_S)^{-1} : |||\cdot|||_1^* \to |||\cdot|||_{max}) \qquad (24)$$

then $sgn(X_{S,ij}^*) = sgn(Y_{ij}^*)$ for all (i,j) in S.

4.2   Conditions on Minimizer Uniqueness and Robustness for $MP_{y,\Phi,\eta}$

Now consider matrix signal reconstruction via solving the constrained convex programming $MP_{y,\Phi,\eta}$:

$$X^* = Arg \; inf|||Z|||_1 s.t. Z \in R^{n \times n}, |\mathbf{y} - \Phi(Z)|_2 \leq \eta \qquad (25)$$

Lemma 3. Given $\mathbf{y}$, positive integer s and sparsity pattern $S = S_1 \cup \ldots \cup S_n$ where $|S_j| \leq s$ for all j, suppose the linear measurement operator $\Phi: R^{n \times n} \to R^m$ satisfies:

1. $\Phi_S^T(\mathbf{z})$ does not have any $\mathbf{0}$-column for $\mathbf{z} \neq \mathbf{0}$;
2. $\Phi_S^T \Phi_S$ is a bijection;
3. $sup\{\langle \Phi_{\sim S}^T(\Phi_S \Phi_S^{*-1}(\mathbf{y}) - \mathbf{y}), H \rangle : |||H|||_1 = 1\} < \eta \Lambda_{min}(\Phi_S^T)(1 - N(\Phi_S^{*-1} \Phi_{\sim S} : |||\cdot|||_1^* \to |||\cdot|||_1^*))$
   where $\Lambda_{min}(\Phi_S^T) := inf\{|||\Phi_S^T(z)|||_1^* : |z|_2 = 1\}$.

Then there are the following conclusions:

1. As the minimizer of problem $MP_{y,\Phi,\eta}$, X^* is unique, S-sparse and l_1-column-flat;
2. Let $Y^* = \Phi_S^{*-1}(\mathbf{y}) \in \sum_s^{n \times n}(S)$, then for all (i,j) in S:
   $X_{ij}^* \neq 0$ and $sgn(X_{ij}^*) = sgn(Y_{ij}^*)$ for all (i,j): $|Y_{ij}^*| > \eta N(\Phi_S^T : l_2 \to |||\cdot|||_1^*) N((\Phi_S^T \Phi_S)^{-1}: |||\cdot|||_1^* \to |||\cdot|||_{max})$
3. For any given matrix norm $|.|_\alpha$ there holds:

$$|X^* - Y^*|_\alpha \leq \eta N(\Phi_S^{*-1} : l_2 \to |.|_\alpha) \qquad (26)$$

Remark: Under the conditions specified in this lemma, the minimizer X^* can have support stability, sign stability and component value-error robustness relative to any metric $|.|_\alpha$, e.g., we can take $|.|_\alpha$ as $|.|_F$, $|||.|||_1$, $|||.|||_{max}$, etc. The error is linearly upper-bounded by the measurement error η. When η decreases as small as possible, the error $|X^* - Y^*|_\alpha$ seems to be reduced also as small as possible. However, this is not true in general because condition (3) may require η not be two small.

In real world applications, the support S of the signal is of course unknown so Lemma 3 cannot be applied directly. However, on basis of Lemma 3 a stronger and uniform sufficient condition can be established to guarantee the uniqueness and robustness of the reconstructed signal from solving $MP_{y,\Phi,\eta}$.

Theorem 6. Given positive integer s and the linear measurement operator Φ : $R^{n\times n} \to R^m$, suppose Φ satisfies the following conditions for any s-sparsity pattern $S = S_1 \cup \ldots \cup S_n$ where $|S_j| \le s$ for all j:

1. $\Phi_S^T(\mathbf{z})$ does not have any **0**-column for $\mathbf{z} \ne \mathbf{0}$;
2. $\Phi_S^T\Phi_S$ is a bijection;
3. $N(\Phi_{\sim S}^T(\Phi_S\Phi_S^{*-1} - I_S) : l_2 \to |||.|||_1^*) < \Lambda_{min}(\Phi_S^T)(1 - N(\Phi_S^{*-1}\Phi_{\sim S} : |||.|||_1^* \to |||.|||_1^*))$ or equivalently $N(((\Phi_S^{*-1})^T\Phi_S^T - I_S^T)\Phi_{\sim S} : |||.|||_1 \to l_2) < \Lambda_{min}(\Phi_S^T)$ $(1 - N(\Phi_{\sim S}^T(\Phi_S^{*-1})^T : |||.|||_1 \to |||.|||_1))$ where $\Lambda_{min}(\Phi_S^T) := inf\{|||\Phi_S^T(z)|||_1^*$: $|z_2 = 1\}$ and $I_S :=$ the identical mapping on $\sum_s^{n\times n}(S)$.

Then for the minimizer X^* of problem $MP_{y,\Phi,\eta}$ (25) where $\mathbf{y} = \Phi(X) + e$ with noise $|e|_2 \le \eta$ and a real flat signal $X \in \sum_s^{n\times n}(R)$ of some s-sparsity pattern R, there are the following conclusions:

1. *Sparsity, flatness and support stability*:
   $X^* \in \sum_s^{n\times n}(R)$ and is l_1-column-flat and the unique minimizer of $MP_{y,\Phi,\eta}$;
2. *Robustness*: For any given matrix norm $|.|_\alpha$ there holds:

$$|X^* - X|_\alpha \le 2\eta N(\Phi_R^{*-1} : l_2 \to |.|_\alpha) \qquad (27)$$

3. *Sign Stability*: $sgn(X_{ij}^*) = sgn(X_{ij})$ for (i,j) in R such that:

$$|X_{ij}| > \eta(N(\Phi_R^{*-1} : l_2 \to |||.|||_{max})$$
$$+ N(\Phi_R^T : l_2 \to |||.|||_1^*)N((\Phi_R^T\Phi_R)^{-1} : |||.|||_1^* \to |||.|||_{max})) \qquad (28)$$

Remark: In this theorem condition (3) is independent with measurement error bound η, which implies $|X^* - X|_\alpha = O(\eta)$ can hold for any small value of η, and (28) indicates that with η small enough, all signal's non-zero components' signs can be correctly recovered. For given Φ, the associated value

$$Max_{s-sparse\,pattern\,S}N(\Phi_S^{*-1} : l_2 \to |||.|||_{max}) + N(\Phi_S^T : l_2 \to |||.|||_1^*)N((\Phi_S^T\Phi_S)^{-1} :$$
$$|||.|||_1^* \to |||.|||_{max})$$

or an enhancement $2Max_{s-sparse\,pattern\,S}N(\Phi_S^T : l_2 \to |||.|||_1^*)N((\Phi_S^T\Phi_S)^{-1}$: $|||.|||_1^* \to |||.|||_{max})$ can be regarded as a signal-to-noise ratio threshold for correct sign recovery.

Finally we mention that in condition (3) $\Lambda_{min}(\Phi_S^T) = inf\{\||\Phi_S^T(z)\||_1^* : z|_2 = 1\}$ for $MP_{y,\Phi,\eta}^{(2)}$ or $\Lambda_{min}(\Phi_S^T) = inf\{\||\Phi_S^T(Z)\||_1^* : |Z|_F = 1\}$ for $MP_{Y,A,B,\eta}^{(F)}$ is one of critical quantities which value should be as large as possible to satisfy this condition. Observe that there is a counterpart $\lambda_{min,\alpha\beta}(\Phi;K) = inf\{|\Phi(Z)|\alpha : Z$ in K and $|Z|_\beta = 1\}$ (e.g., see FACT 6) which is also critical for analysis in next sections in random measurement setting.

5   Number of Measurements for Robust Reconstruction via Solving $MP_{y,\Phi,\eta}^{(2)}$

After established the conditions for the measurement operator to guarantee desired properties (e.g., uniqueness, robustness, etc.) of the matrix signal reconstruction, next question should be about how to construct the measurement operator to satisfy such conditions with required number of measurements as few as possible. This and next sections deal with this problem in random approach. In this section we establish conditions on number of measurements m for robustly reconstructing the matrix signal X by solving the convex programming problem $MP_{y,\Phi,\eta}^{(2)}$:

$$inf\||Z\||_1 \text{ s.t } Z \in R^{n\times n}, |\mathbf{y} - \Phi(Z)|_2 \leq \eta$$

where $\mathbf{y} = \Phi(X) + e$ and $|e|_2 \leq \eta$, $\Phi : R^{n\times n} \to R^m$ is a linear operator. In the equivalent component-wise formulation, $y_i = <\Phi_i, X> + e_i$ where for $i = 1,\ldots,m, \Phi_i \in R^{n\times n}$ are random matrices independent each other and Φ_i's entries are independently sampled under standard Gaussian $N(0,1)$ or sub-Gaussian distribution. In Sect. 5.3, a simple necessary condition on m is established.

Instead of proving that M-RIP property (defined in Sect. 3.4) can be satisfied with high probability in context of the distributions in consideration, which are quite involved particularly for problem $MP_{Y,A,B,\eta}^{(F)}$, here we straightforwardly prove that the robustness in terms of *Frobenius*-norm error metric can be reached with high probability when number of measurements exceeds some specific bound. The investigation on how those conditions established in last sections can be satisfied by the random measurement operator will be subjects in subsequent papers.

5.1   Case 1: Gaussian Measurement Operator Φ

Based upon the fundamental facts presented in Sect. 2.2, one of the critical steps in this approach is to estimate the width $w(D(\||X\||_1, X))$'s upper bound for matrix signal $X = (\mathbf{x}_1,\ldots,\mathbf{x}_n)$ with s-sparse column vectors $\mathbf{x}_1,\ldots,\mathbf{x}_n$. This is done in Lemma 4.

Based on Lemma 1 and FACT 4, the upper bound of Gaussian width $D(\||.\||_1, X)$ with respect to *Frobenius* norm is estimated in the following lemma.

Lemma 4. Given n-by-n matrix $X = (x_1, \ldots, x_n)$ with s-sparse column vectors $x_1, \ldots, x_n$. Let r (called l_1-*column-flatness parameter* hereafter) be cardinality of the set $\{j : |x_j|_1 = max_k |x_k|_1\}$, i.e., the number of column vectors which have the maximum l_1-norm. Then

$$w^2(D(|||.|||_1, X)) \leq 1 + n^2 - r(n - slog(Cn^4 r^2)) \tag{29}$$

In particular, when $r = n$ then

$$w^2(D(|||.|||_1, X)) \leq 1 + nslog(Cn^6) \tag{30}$$

where C is an absolute constant.

Combing this lemma and FACT 3, we obtain the general result in the following:

Theorem 7. Suppose $\Phi_{kij} \sim^{iid} N(0,1)$, let $X \in \sum_s^{n \times n}$ be a columnwise s-sparse and l_1-column-flat matrix signal, $R^m \ni y = \Phi(X) + e$ where $|e|_2 \leq \eta$, X^* be the minimizer of the problem $MP_{y,\Phi,\eta}^{(2)}$. If the measurement vector y's dimension

$$m \geq (t + 2\eta/\delta + (nslog(Cn^6))^{1/2})^2$$

where C is an absolute constant, then $P[|X^* - X|_F \leq \delta] \geq 1 - exp(-t^2/2)$, i.e., such X can be reconstructed robustly with respect to the error norm $|X^* - X|_F$ with high probability by solving $MP_{y,\Phi,\eta}^{(2)}$.

5.2   Case 2: Sub-Gaussian Measurement Operator Φ

Combing Lemma 4, FACT 1(2) and FACT 5, the following result can be obtained straightforwardly:

Theorem 8. Let X and X^* be respectively the matrix signal and the minimizer of $MP_{y,\Phi,\eta}^{(2)}$ where the signal $X \in \sum_s^{n \times n}$ is columnwise s-sparse and l_1-column-flat, $y_k = < \Phi_k, X > + e_k$, each $\Phi_k \sim^{iid} \Phi$ where Φ is a random matrix satisfying the conditions (1) (2) (3) in FACT 5 with parameters α, ρ and σ. If the measurement vector y's dimension

$$m \geq (C_1 \rho^4/\alpha)(\alpha t + 2\eta/\delta + \sigma C_2(\rho^6 nslog(C_3 n^6))^{1/2})^2$$

where C_i's are absolute constants, then $P[|X^* - X|_F \leq \delta] \geq 1 - exp(-C_4 t^2)$, i.e., such X can be reconstructed robustly with respect to the error norm $|X^* - X|_F$ with high probability by solving $MP_{y,\Phi,\eta}^{(2)}$.

5.3   Necessary Condition on Number of Measurements

Theorem 9. Given measurement operator $\Phi: R^{n \times n} \rightarrow R^m$, if any $X = (x_1, \ldots, x_n)$ with sparse column vectors x_j in $\sum^{2S}$ for all j is always the unique solution to problem $MP_{y,\Phi,\eta}^{(\alpha)}$ where $\eta = 0$, then

$$m \geq C_1 nslog(C_2 n/s)) \tag{31}$$

where C_1 and C_2 are absolute constants.

Remark: For the measurement operator $\Phi_{A,B} : R^{n \times n} \to R^{m \times m} : Y = AXB^T$, the same result is true with $m^2 \geq C_1 n s log(C_2 n/s))$.

6   Number of Measurements for Robust Reconstruction via Solving $MP^{(F)}_{Y,A,B,\eta}$

Now we investigate the problem $MP^{(F)}_{Y,A,B,\eta}$ in which A and B are both Gaussian or sub-Gaussian m-by-n random matrices. Notice that in these cases the measurement operator $\Phi_{A,B}$ is neither Gaussian nor sub-Gaussian. The basis is FACT 6 and the critical step is also the width's upper bound estimation.

More explicitly, $\Phi_{A,B}: R^{n \times n} \to R^{m \times m}$: $Y = AZB^T$. In the equivalent component-wise formulation, $y_{kl} = <\Phi_{kl}, X> = \sum_{ij} A_{ki} X_{ij} B_{lj}$ for each $1 \leq k$, $l \leq m$, $A_{ki} \sim^{iid} B_{lj} \sim^{iid} N(0,1)$ or sub-Gaussian distribution, and A, B are independent each other. Let $\varepsilon_{kl} \sim^{iid}$ Rademacher random variable $\varepsilon(P[\varepsilon = \pm 1] = 1/2)$ which are also independent of A and B. The width is defined as (FACT 6)

$$W(\Gamma_X; \Phi_{A,B}) := E_H[sup\{\langle H, U \rangle : U \text{ in } \Gamma_X \text{ and } |U|_F = 1] \tag{32}$$

where

$$H := m^{-1} \sum_{kl} \varepsilon_{kl} \Phi_{kl} = m^{-1} A^T E B \text{ and } E = [\varepsilon_{kl}] \tag{33}$$

and $\Gamma_X = D(|||.|||_1, X)$ in our applications.

Let $\varepsilon_l : l = 1, \ldots, m$ be column vectors of Rademacher matrix $[\varepsilon_{kl}]$, then $H = (\boldsymbol{h}_1, \ldots, \boldsymbol{h}_n)$ has its column vectors $\boldsymbol{h}_j = \sum_{l=1}^m m^{-1} B_{lj} A^T \varepsilon_l \in R^n$. Notice that in problem $MP^{(F)}_{Y,A,B,\eta}$, m^2 is the measurement dimension.

6.1   Case 1: A and B are both Gaussian

Lemma 5. Given n-by-n matrix $X = (\mathbf{x}_1, \ldots, \mathbf{x}_n)$ with s-sparse column vectors $\mathbf{x}_1, \ldots, \mathbf{x}_n$, r is cardinality of $\{j : |\mathbf{x}_j|_1 = max_k |\mathbf{x}_k|_1\}$, $\Phi_{A,B}$, Γ_X, $W(\Gamma_X; \Phi_{A,B})$ are specified as the above and $A_{ki} \sim^{iid} B_{lj} \sim^{iid} N(0,1)$, then

$$W^2(\Gamma_X; \Phi_{A,B}) \leq 1 + n^2 - r(n - s log^2(cn^2 r))$$

where c is an absolute constant. Particularly, when $r = n$ (i.e., X is sparse and l_1-column-flat) then

$$W^2(\Gamma_X; \Phi_{A,B}) \leq 1 + n s log^2(cn^3) \tag{34}$$

To apply FACT 6 completely, now estimate the lower bound of $Q_\xi(\Gamma_X; \Phi_{A,B}) := inf\{P[|\langle \Phi_{kl}, U \rangle| \geq \xi] : U \text{ in } \Gamma_X \text{ and } |U|_F = 1\}$ for some $\xi > 0$ where $\langle \Phi_{kl}, U \rangle = \sum_{ij} A_{ki} U_{ij} B_{lj}$ for each $1 \leq k, l \leq m$, $A_{ki} \sim^{iid} B_{lj} \sim^{iid} N(0,1)$. The estimate is independent of the indices k and l, so we give a general and notational simplified statement on it.

Lemma 6. Let $M_{ij} = a_i b_j$ where $a_i \sim^{iid} b_j \sim^{iid} N(0,1)$ and all random variables are independent each other, there exists an positive absolute constant c such that $\inf\{P[|\langle M,U\rangle| \geq 1/\sqrt{2}] : U \text{ in } \Gamma_X \text{ and } |U|_F = 1\} \geq c$, as a result $Q_{1/\sqrt{2}}(\Gamma_X; \Phi_{A,B}) \geq c$.

Combing the lemmas and FACT 6, we obtain the general result in the following.

Theorem 10. Suppose $A_{ki} \sim^{iid} B_{lj} \sim^{iid} N(0,1)$ and independent each other, $X \in \sum_s^{n \times n}$ is a columnwise s-sparse and l_1-column-flat signal, $\mathbf{Y} = AXB^T + E \in R^{m \times m}$ with measurement errors bounded by $|E|_F^2 \leq \eta$, X^* is the minimizer of the problem $MP_{Y,A,B,\eta}^{(F)}$. If

$$m \geq t + 4\sqrt{2}\eta/\delta + C_1(ns)^{1/2} log(C_2 n^3)$$

where C_i's are absolute constants, then $P[|X^* - X|_F \leq \delta] \geq 1 - exp(-t^2/2)$, i.e., such X can be reconstructed robustly with respect to the error norm $|X^* - X|_F$ with high probability by solving $MP_{Y,A,B,\eta}^{(F)}$.

6.2 Case 2: A and B are both Sub-Gaussian

Lemma 7. Given n-by-n matrix $X = (\mathbf{x}_1, \ldots, \mathbf{x}_n)$ with s-sparse column vectors $\mathbf{x}_1, \ldots, \mathbf{x}_n$, r is cardinality of $\{j : |\mathbf{x}_j|_1 = max_k|\mathbf{x}_k|_1\}$, $\Phi_{A,B}$, Γ_X, $W(\Gamma_X; \Phi_{A,B})$ are specified as before, $A_{ki} \sim^{iid}$ Sub-Gaussian distribution and $B_{lj} \sim^{iid}$ Sub-Gaussian distribution with ψ_2-norms σ_A, σ_B respectively, then

$$W^2(\Gamma_X; \Phi_{A,B}) \leq \sigma_A^2 \sigma_B^2(1 + n^2 - r(n - slog^2(Cn^2r))) \quad (35)$$

where C is an absolute constant. Particularly, when $r = n$ then

$$W^2(\Gamma_X; \Phi_{A,B}) \leq \sigma_A^2 \sigma_B^2(1 + nslog^2(Cn^3)) \quad (36)$$

Theorem 11. Suppose random matrices A, B are independent each other, $A_{ki} \sim^{iid}$ Sub-Gaussian distribution, $B_{lj} \sim^{iid}$ Sub-Gaussian distribution, each with ψ_2-norm σ_A and σ_B. Let $X \in \sum_s^{n \times n}$ be a columnwise s-sparse and l_1-column-flat signal, $\mathbf{Y} = AXB^T + E \in R^{m \times m}$ with $|E|_F^2 \leq \eta$, X^* be the minimizer of $MP_{Y,A,B,\eta}^{(F)}$. If

$$m \geq t + 4\sqrt{2}\eta/\delta + C_1\sigma_A\sigma_B(ns)^{1/2} log(C_2 n^3)$$

where C_i's are absolute constants, then $P[|X^* - X|_F \leq \delta] \geq 1 - exp(-t^2/2)$, i.e., such X can be reconstructed robustly with respect to the error norm $|X^* - X|_F$ with high probability by solving $MP_{Y,A,B,\eta}^{(F)}$.

7   Conclusions, Some Extensions and Future Works

In this paper we investigated the problem of reconstructing n-by-n column-wise sparse and l_1-column-flat matrix signal X $= (\mathbf{x}_1, \ldots, \mathbf{x}_n)$ via convex programming with the regularizer $|||X|||_1 := max_j |\mathbf{x}_j|_1$ where $|.|_1$ is the l_1-norm in vector space. In the first part (Sect. 3 and Sect. 4), the most important conclusions are about the general conditions to guarantee uniqueness, value-robustness, support stability and sign stability in signal reconstruction. In the second part (Sect. 5 and Sect. 6) we took the convex geometric approach in random measurement setting and established sufficient conditions on dimensions of measurement spaces for robust reconstruction in noise. In particular, when $r = n$ (i.e., sparse and flat signal) the condition reduces to $m \geq t + 4\sqrt{2}\eta/\delta + C_1\sigma_A\sigma_B(ns)^{1/2}log(C_2n^3)$. The results established in this paper provide a relatively complete theory on column-wise sparse and l1-column-flat matrix signal reconstruction.

This paper is only focused on basic theoretical analysis. In future work, the algorithms to solve the $|||.|||_1$-optimization problems (e.g., generalized inverse scale space algorithms, etc.), related numeric investigations and applications (e.g., in radar space-time waveform analysis) will be further explored.

Appendix A: Proofs of Theorems in Section 3

Proof of Lemma 1. It's easy to verify the set $\{(\lambda_1\xi_1, \ldots, \lambda_n\xi_n) : \xi_j$ in $\partial|\mathbf{x}_j|_1$ and $\lambda_j \geq 0$ for all $j, \lambda_1 + \ldots + \lambda_n = 1$ and $\lambda_j = 0$ for $j : |\mathbf{x}_j|_1 < max_k|\mathbf{x}_k|_1\}$ is contained in $\partial|||X|||_1$: since for any M $\equiv (\lambda_1\xi_1, \ldots, \lambda_n\xi_n)$ in this set, we have

$$<\text{M,X}> = \sum_j \lambda_j <\xi_j, \mathbf{x}_j> = \sum_j \lambda_j |\mathbf{x}_j|_1 = |||X|||_1 \sum_j \lambda_j = |||X|||_1$$

and $|||.|||_1$'s conjugate norm $|||M|||_1^* = \sum_j \lambda_i |\xi_i|_\infty \leq \sum_j \lambda_i = 1$, as a result M is in $\partial|||X|||_1$.

Now prove that any M in $\partial|||X|||_1$ has the form specified as a member in the above set. Let M $\equiv (\boldsymbol{\eta}_1, \ldots, \boldsymbol{\eta}_n)$, $|||Y|||_1 \geq |||X|||_1 + <Y - X, M>$ for all Y $\equiv (\mathbf{y}_1, \ldots, \mathbf{y}_n)$ implies:

$$max_j |\mathbf{y}|_1 \geq max_j |\mathbf{x}_j|_1 + \sum_j <\mathbf{y}_j - \mathbf{x}_j, \boldsymbol{\eta}_j> \tag{37}$$

Let $\boldsymbol{\eta}_j = |\boldsymbol{\eta}_j|_\infty \xi_j$ (so $|\xi_j|_\infty = 1$ if $\boldsymbol{\eta}_j \neq 0$), then $max_j |\mathbf{y}_j|_1 \geq max_j |\mathbf{x}_j|_1 + \sum_j |\boldsymbol{\eta}_j|_\infty <\mathbf{y}_j - \mathbf{x}_j, \xi_j>$. For each $j : \boldsymbol{\eta}_i \neq 0$ we can select a i_j such that $|\xi_j(i_j)| = 1$ and let e_j^* be such a vector with component $e_j^*(i_j) = sgn\,\xi_j(i_j)$ and $e_j^*(i) = 0$ for all $i \neq i_j$, then for $\mathbf{y}_j = \mathbf{x}_j + e_j^*, j = 1, \ldots, n$, (37) implies

$$1 + max_j |\mathbf{x}_j|_1 \geq max_j |\mathbf{y}_j|_1 \geq max_j |\mathbf{x}_j|_1 + \sum_j |\boldsymbol{\eta}_j|_\infty <e_j^*, \xi_j>$$
$$= max_j |\mathbf{x}_j|_1 + \sum_j |\boldsymbol{\eta}_j|_\infty|\xi_j|_\infty = max_j |\mathbf{x}_j|_1 + \sum_j |\boldsymbol{\eta}_j|_\infty$$

As a result $1 \geq \sum_j |\boldsymbol{\eta}_j|_\infty$.

Furthermore for any given i, let $\mathbf{y}_j = \mathbf{x}_j$ for all $j \neq i$ and $\mathbf{y}_i$ be any vector satisfying $|\mathbf{y}_i|_1 \leq |\mathbf{x}_i|_1$, then substitute these $\mathbf{y}_1, \ldots, \mathbf{y}_n$ into (37) we obtain

$$max_j\,|\mathbf{x}_j|_1 \geq max_j\,|\mathbf{y}_j|_1 \geq max_j\,|\mathbf{x}_j|_1 + \sum_j <\mathbf{y}_j - \mathbf{x}_j, \boldsymbol{\eta}_j>$$
$$= max_j\,|\mathbf{x}_j|_1 + \left|\boldsymbol{\eta}_j\right|_\infty <\mathbf{y}_i - \mathbf{x}_i, \xi_j>$$

i.e., $<\mathbf{y}_i - \mathbf{x}_i, \xi_j> \leq 0$. As a result $<\mathbf{x}_i, \xi_j> \geq <\mathbf{y}_i, \xi_j>$ for all $\mathbf{y}_i : |\mathbf{y}_i|_1 \leq |\mathbf{x}_i|_1$ so $<\mathbf{x}_i, \xi_j> \geq |\mathbf{x}_i|_1 |\xi_i|_\infty = |\mathbf{x}_i|_1$, hence finally we get $<\mathbf{x}_i, \xi_j> = |\mathbf{x}_i|_1$. This (together with $|\xi_i|_\infty = 1$) implies ξ_i in $\partial |\mathbf{x}_i|_1$ if $\boldsymbol{\eta}_i \neq 0$, for any $i = 1, \ldots, n$.

In summary, we have so far proved that for any M in $\partial|||X|||_1$, M always has the form $(\lambda_1 \xi_1, \ldots, \lambda_n \xi_n)$ where ξ_j in $\partial |\mathbf{x}_j|_1$, $\lambda_j \geq 0$ for all j and $\lambda_1 + \ldots + \lambda_n \leq 1$ since $|||X|||_1 = <M, X> = \sum_j \lambda_j <\xi_j, \mathbf{x}_j> = \sum_j \lambda_j |\mathbf{x}_j|_1 \leq max_j\,|\mathbf{x}_j|_1 \sum_j \lambda_j \leq |||X|||_1$, as a result $\lambda_1 + \ldots + \lambda_n = 1$ and $\lambda_j = 0$ for $j : |\mathbf{x}_j|_1 < max_k\,|\mathbf{x}_k|_1$.

$\square$

Proof of Theorem 1. To prove the necessity, let S be a s-sparsity pattern and $H \in ker\,\Phi \backslash \{O\}$ Set $\mathbf{y} \equiv \Phi(H_S) = \Phi(-H_{\sim S})$ and $H_S \in \sum_s^{n \times n}$, Hs should be the unique minimizer of $MP_{y,\Phi,0}$ with $-H_{\sim s}$ as its feasible solution, hence $|||H_S|||_1 < |||H_{\sim S}|||_1$.

Now prove the sufficiency. Let $X = (\mathbf{x}_1, \ldots, \mathbf{x}_n)$ be a matrix signal with its support $S = S_1 \cup \ldots \cup S_n$ as a s-sparsity pattern (where $S_j = supp(\mathbf{x}_j)$) and let $\mathbf{y} = \Phi(X)$ For any feasible solution $Z(\neq X)$ of $MP_{y,\Phi,0}$, obviously there exists $H = (\boldsymbol{h}_1, \ldots, \boldsymbol{h}_n)$ in $ker\,\Phi \backslash \{O\}$ such that $Z = X + H$. since $\partial|||Z|||_1 \geq \partial|||X|||_1 + <H, M>$ for any M in $\partial|||X|||_1$, we have

$\partial|||Z|||_1 - \partial|||X|||_1 \geq \sup\{<H,M>: \text{for any M in } \partial|||X|||_1\}$

$= \sup\{<H, M>: M = E + V \text{ where } E = (\lambda_1 sgn(\mathbf{x}_1) \ldots, \lambda_n sgn(\mathbf{x}_n)) \text{ and } V$
$= (\lambda_1 \xi_1, \ldots, \lambda_n \xi_n), |\xi_j|_\infty \leq 1, \lambda_j \geq 0 \text{ for all } j, \lambda_1 + \ldots + \lambda_n = 1\}$ (by Lemma 1 and notice supp $(sgn(\mathbf{x}_j)) = S_j = \sim supp(\xi_j)$)

$\geq \sup\{-|<H, E>|+ <H, V>: \text{E and V specified as the above}\}$

$= \sup\left\{-|\sum_{j=1}^n \lambda_j <\boldsymbol{h}_{j|S_j}, sgn(\mathbf{x}_j)>| + \sum_{j=1}^n \lambda_j <\boldsymbol{h}_{j|\sim S_j}, \xi_j> : \lambda_j \text{ and } \xi_j \text{ specified as the above }\right\}$

$\geq -\sup\left|\sum_{j=1}^n \lambda_j <\boldsymbol{h}_{j|S_j}, sgn(\mathbf{x}_j)>\right|: \lambda_j \geq 0 \text{ for all } j, \lambda_1 + \ldots + \lambda_n = 1\} + \sup\{<H_{\sim S}, V>: |||V|||_1^* \leq 1\}$

(note that $|||V|||_1^* = \sum_j |\lambda_j \xi_j|_\infty \leq \sum_j |\xi_j|_\infty = 1$ where $||| \cdot |||_1^*$ is $||| \cdot |||_1$'s conjugate norm)

$= -\sup\left|\sum_{j=1}^n \lambda_j <\boldsymbol{h}_{j|S_j}, sgn(\mathbf{x}_j)>\right|: \lambda_j \geq 0 \text{ for all } j, \lambda_1 + \ldots + \lambda_n = 1\} + |||H_{\sim S}|||_1$

$= -max_j\,|<\boldsymbol{h}_{j|S_j}, sgn(\mathbf{x}_j)>| + |||H_{\sim S}|||_1$

$\geq -max_j\,|\boldsymbol{h}_{j|S_j}|_1 + |||H_{\sim S}|||_1 = -|||H_S|||_1 + |||H_{\sim S}|||_1 > 0$

under the condition (3.3). As a result, X is the unique minimizer of $MP_{y,\Phi,0}$ $\square$

The proof of Theorem 2 follows almost the same logic of proving l_1-min reconstruction's stability for vector signals under the l_1 Null Space Property assumption (e.g., see sec. 4.2 in [1]). For presentation completeness we provide the simple proof here. The basic tool is an auxiliary inequality (which unfortunately does not hold for matrix norm $|||.|||_1$): given index subset Δ and any vector $\mathbf{x}$, $\mathbf{z}$, then[1]

$$|(\mathbf{x} - \mathbf{z})_{\sim \Delta}|_1 \leq |\mathbf{z}|_1 - |\mathbf{x}|_1 + |(\mathbf{x} - \mathbf{z})_\Delta|_1 + 2 |\mathbf{x}_{\sim \Delta}|_1 \quad (38)$$

Proof of Theorem 2. For any feasible solution $Z = (\mathbf{z}_1, \ldots, \mathbf{z}_n)$ to problem $MP_{y,\Phi,0}$ where $\mathbf{y} = \Phi(X)$, there is $H = (\mathbf{h}_1, \ldots, \mathbf{h}_n)$ in ker Φ such that $Z = H + X$. Apply (38) to each column vector $\mathbf{z}_j$ and $\mathbf{x}_j$ we get

$$\left|\mathbf{h}_{j|\sim Sj}\right|_1 \leq |\mathbf{z}_j|_1 - |\mathbf{x}_j|_1 + \left|\mathbf{h}_{j|Sj}\right|_1 + 2 \left|\mathbf{x}_{j|\sim Sj}\right|_1$$

Hence $|||H_{\sim S}|||_1 \equiv max_j \left|\mathbf{h}_{j|\sim Sj}\right|_1 \leq max_j \left(|\mathbf{z}_j|_1 - |\mathbf{x}_j|_1\right) + |||H_S|||_1 + 2max_j \left|\mathbf{x}_{j|\sim Sj}\right| \leq max_j \left(|\mathbf{z}_j|_1 - |\mathbf{x}_j|_1\right) + \rho|||H_{\sim S}|||_1 + 2 max_j |\mathbf{x}_{j|\sim Sj}|_1$ (by (10)), namely:

$$|||H_{\sim S}|||_1 \leq (1 - \rho)^{-1} \left(2 \max_j \left|\mathbf{x}_{j|\sim Sj}\right|_1 + \max_j \left(|\mathbf{z}_j|_1 - |\mathbf{x}_j|_1\right)\right)$$

As a result $|||H|||_1 = |||H_S|||_1 + |||H_{\sim S}|||_1 \leq (1 + \rho)|||H_{\sim S}|||_1 \leq (1 - \rho)^{-1}(1 + \rho) \left(2max_j \left|\mathbf{x}_{j|\sim Sj}\right|_1 + \max_j \left(|\mathbf{z}_j|_1 - |\mathbf{x}_j|_1\right)\right)$ for any s-sparsity pattern S, which implies (11) since $min_S max_j \left|\mathbf{x}_{j|\sim Sj}\right|_1 = max_j \sigma_s (\mathbf{x}_j)_1$.

In particular, if Z is minimizer X^* and X is l_1-column-flat then $|\mathbf{x}_j|_1 = |||X|||_1$ for any j so $max_j(|\mathbf{x}_j^*|_1 - |\mathbf{x}_j|_1) = |||X^*|||_1 - |||X|||_1 \leq 0$ for minimizer X^*, which implies the conclusion. □

Remark: For any flat and sparse signal X, condition (10) guarantees X can be uniquely reconstructed by solving $MP_{y,\Phi,0}$ due to Theorem 1, while in this case the right hand side of (12) is zero, i.e., this theorem is consisted with the former one. In addition, (12) indicates that the error for the minimizer X^* to approximate the flat but non-sparse signal X is controlled column-wisely by X's non-sparsity (measured by $max_j \sigma_s(\mathbf{x}_j)_1$).

Proof of Theorem 3. Consider the problem $MP_{y,\Phi,\eta}$: inf $|||Z|||_1$ s.t. $Z \in R^{n \times n}$, $|\mathbf{y} - \Phi_S(Z)|_2 \leq \eta$ at first where $\eta > 0$. For any minimizer X^* of this problem with both its objective $|||.|||_1$ and constraint function $|\mathbf{y} - \Phi_S(.)|_2$ convex, according to the general convex optimization theory, there exist a positive multiplier $\gamma^* > 0$ and M^* in $\partial |||X^*|||_1$ such that

$$M^* + \gamma^* \Phi_S^T (\Phi_S (X^*) - \mathbf{y}) = O \text{ and } |\mathbf{y} - \Phi_S (X^*)|_2 = \eta \quad (39)$$

then $M^* = \gamma^* \Phi_S^T (\mathbf{y} - \Phi_S(X^*))$ can not have any zero column since $\mathbf{y} - \Phi_S(X^*) \neq \mathbf{0}$, which implies $|\mathbf{x}_j^*|_1 = max_k |\mathbf{x}_k^*|_1$ for every j according to Lemma 1.

Now consider the problem $MP_{y,\Phi}$: inf $|||Z|||_1$ s.t. $Z \in R^{n \times n}$, $\mathbf{y} = \Phi_S(Z)$. For its minimizer X^* there is a multiplier vector $\mathbf{u}$ such that $M^* + \Phi_S^T(\mathbf{u}) = O$.

If $\mathbf{u} \neq \mathbf{0}$ then M^* doesn't have any zero column which implies $|\mathbf{x}_j|_1 = max_k|\mathbf{x}_k|_1$ for every j according to Lemma 1. On the other hand, $\mathbf{u} = \mathbf{0}$ implies $M^* = O$ which cannot happen according to Lemma 1 unless $X^* = O$. □

Proof of Theorem 4. For any feasible solution $Z = (\mathbf{z}_1, \ldots, \mathbf{z}_n)$ to problem $MP_{y,\Phi,\eta}$ where $\mathbf{y} = \Phi(X) + e$, Let $Z - X = H = (\mathbf{h}_1, \ldots, \mathbf{h}_n)$, apply (38) to each column vector $\mathbf{z}_j$ and $\mathbf{x}_j$ we get $\left|\mathbf{h}_{j|\sim Sj}\right|_1 \leq |\mathbf{z}_j|_1 - |\mathbf{x}_j|_1 + \left|\mathbf{h}_{j|Sj}\right|_1 + 2\left|\mathbf{x}_{j|\sim Sj}\right|_1$
Hence $|||H_{\sim S}|||_1 \equiv max_j\left|\mathbf{h}_{j|\sim Sj}\right|_1 \leq max_j\left(|\mathbf{z}_j|_1 - |\mathbf{x}_j|_1\right) + |||H_S|||_1 + 2max_j\left|\mathbf{x}_{j|\sim Sj}\right| \leq max_j(|\mathbf{z}_j|_1 - |\mathbf{x}_j|_1) + \rho|||H_{\sim S}|||_1 + 2max_j\left|\mathbf{x}_{j|\sim Sj}\right|_1 + \beta|\Phi(H)|_2$
(by (14)), namely:

$$|||H_{\sim S}|||_1 \leq (1-\rho)^{-1}\left(2max_j\left|\mathbf{x}_{j|\sim Sj}\right|_1 + max_j\left(\mathbf{z}_j|_1 - |\mathbf{x}_j|_1\right) + \beta|\Phi(H)|_2\right)$$

As a result $|||H|||_1 = |||H_S|||_1 + |||H_{\sim S}|||_1 \leq (1+\rho)|||H_{\sim S}|||_1 + \beta|\Phi(H)|_2 \leq (1-\rho)^{-1}(1+\rho)((2max_j\left|\mathbf{x}_{j|\sim Sj}\right|_1 + max_j\left(|\mathbf{z}_j|_1 - |\mathbf{x}_j|_1\right)) + 2(1-\rho)^{-1}\beta|\Phi(X)|_2)$ for any s-sparsity pattern S, which implies (15) since $min_S max_j|\mathbf{x}_{j|\sim Sj}|_1 = max_j\sigma_s(\mathbf{x}_j)_1$.

In particular, if Z is a minimizer X^* and X is l_1-column-flat then $|\mathbf{x}_j|_1 = |||X|||_1$ for any j so $max_j(|\mathbf{x}_j^*|_1 - |\mathbf{x}_j|_1) = |||X^*|||_1 - |||X|||_1 \leq 0$ for minimizer X^*, which implies (16). □

Proof of Theorem 5. Let $H = (\mathbf{h}_1, \ldots, \mathbf{h}_n)$ be any n-by-n matrix. For each j suppose $|h_j(i_1)| \geq |h_j(i_2)| \geq \cdots \geq |h_j(i_n)|$, let $S_0(j) = \{(i_1,j), \ldots, (i_s,j)\}$, i.e., the set of indices of s components in column $\mathbf{h}_j$ with the largest absolute values, $S_1(j) = \{(i_{1+s},j), \ldots, (i_{2s},j)\}$ be the set of indices of s components in $\mathbf{h}_j$ with the secondary largest absolute values, etc., and for any $k = 0, 1, 2, \ldots$ let $S_k = \cup_{j=1}^n S_k(j)$, obviously $H = \sum_{k\geq 0} H_{Sk}$. At first we note that (20) holds for S as long as it holds for S_0, so we try to prove this in the following. Start from condition (1):
$(1 - \delta_s)|H_{S0}|_F^2 \leq |\Phi(H_{S0})|_2^2 = <\Phi(H_{S0}), \Phi(H) - \sum_{k\geq 1}\Phi(H_{Sk})>$
$= <\Phi(H_{S0}), \Phi(H)> - \sum_{k\geq 1} <\Phi(H_{S0}), \Phi(H_{Sk})>$
$\leq |\Phi(H_{S0})|_2 |\Phi(H)|_2 + (\Delta_s/n)\sum_{n\geq j\geq 1}\sum_{k\geq 1}\left|\mathbf{h}_{j|S0(j)}\right|_2\left|\mathbf{h}_{j|Sk(j)}\right|_2$ (by condition (2))
$\leq (1+\delta_s)^{1/2}|H_{S0}|_F|\Phi(H)|_2 + (\Delta_s/n)|H_{S0}|_F\sum_{n\geq j\geq 1}\sum_{k\geq 1}\left|\mathbf{h}_{j|Sk(j)}\right|_2$ (by condition (1) and $\left|\mathbf{h}_{j|S0(j)}\right|_2 \leq |H_{S0}|_F$)
$\leq (1+\delta_s)^{1/2}|H_{S0}|_F|\Phi(H)|_2 + (\Delta_s/n)|H_{S0}|_F\sum_{n\geq j\geq 1}\left(s^{-1/2}\left|\mathbf{h}_{j|\sim S0(j)}\right|_1 + \right.$
$(1/4)\left|\mathbf{h}_{j|S0(j)|2}\right|)$ (by the inequality $\left(\sum_{s\geq k\geq 1}a_k^2\right)^{1/2} \leq s^{-1/2}\sum_{s\geq k\geq 1}a_k + \left(s^{1/2}/4\right)(a_1 - a_s)$ for $a_1 \geq a_2 \geq \cdots \geq a_s \geq 0$ and the fact $min_{s\geq i\geq 1}\left|h_{j|Sk(j)}(i)\right| \geq max_{s\geq i\geq 1}\left|h_{j|Sk+1(j)}(i)\right|$ for any j)

$$\leq |H_{S0}|_F \left((1+\delta_s)^{1/2} |\Phi(H)|_2 + \left(s^{-1/2}\Delta_s/n \right) \sum_{n \geq j \geq 1} |h_{j|\sim S0(j)}|_1 \right.$$

$$\left. + (\Delta_s/4n) \sum_{n \geq j \geq 1} |h_{j|S0(j)}|_2 \right)$$

$$\leq |H_{S0}|_F \left((1+\delta_s)^{1/2} |\Phi(H)|_2 + s^{-1/2}\Delta_s \max_j |h_{j|\sim S0(j)}|_1 \right.$$

$$\left. + (\Delta_s/4n) n^{1/2} \left(\sum_{n \geq j \geq 1} |h_{j|S0(j)}|_2^2 \right)^{1/2} \right)$$

$$= |H_{S0}|_F \left((1+\delta_s)^{1/2} |\Phi(H)|_2 + s^{-1/2}\Delta_s |||H_{\sim S0}|||_1 + \left(\Delta_s/4n^{1/2} \right) |H_{S0}|_F \right)$$

Cancel $|H_{S0}|_F$ on both sides we get $(1-\delta_s)|H_{S0}|_F \leq (1+\delta_s)^{1/2}|\Phi(H)|_2 + s^{-1/2}\Delta_s|||H_{\sim S0}|||_1 + (\Delta_s/4n^{1/2})|H_{S0}|_F$ hence
$|H_{S0}|_F \leq (1-\delta_s - \Delta_s/4n^{1/2})^{-1}((1+\delta_s)^{1/2}|\Phi(H)|_2 + s^{-1/2}\Delta_s|||H_{\sim S0}|||_1)$
Note that $|||H_{S0}|||_1 = max_j|h_{j|S0(j)}|_1 \leq s^{1/2}max_j|h_{j|S0(j)}|_2 \leq s^{1/2}|H_{S0}|_F$
and combine this with the above inequality, we obtain (20) and (21) for S_0, which implies they hold for any S. □

Appendix B: Proofs of Theorems in Section 4

Proof of Lemma 2. (1) Observe that when $\Phi_S^T \Phi_S$ is a bijection, (23)'s objective function $L_S(Z) = |||Z|||_1 + (1/2)\gamma|y - \Phi_S(Z)|_2^2$ is strictly convex for variable $Z \in \sum_s^{n \times n}(S)$. According to general convex programming theory, its minimizer X_S^* is unique.

(2) Let $L(Z) := |||Z|||_1 + (1/2)\gamma|y - \Phi(Z)|_2^2$. To prove X_S^* is also the global minimizer of (22), we prove its perturbation by H will always increase the objective's value, i.e., $L(X_S^* + H) > L(X_S^*)$ under the conditions specified by (1) (2) (3). Since conclusion (1) implies $L(X_S^*+H) > L(X_S^*)$ for any $H \neq O$ with support in S and $L(Z)$ is convex, we only need to consider the perturbation $X_S^* + H$ with $H_S = O$.

Since X_S^* is the minimizer of (23), by first-order optimization condition there exists M^* in $\partial|||X_S^*|||_1$ such that

$$M^* + \gamma\Phi_S^T \left(\Phi_S \left(X_S^* \right) - y \right) = O \tag{40}$$

then $M^* = \gamma^*\Phi_S^T(y - \Phi_S(X_S^*))$ and in particular $M_{\sim S}^* = O$. Equivalently:

$$X_S^* = \Phi_S^{*-1}(y) - \gamma^{-1} \left(\Phi_S^T \Phi_S \right)^{-1} (M^*) \tag{41}$$

Now we compute $L(X_S^* + H) - L(X_S^*)$

$= |||X_S^* + H|||_1 - |||X_S^*|||_1 + (1/2)\gamma(|\Phi(X_S^*) - \mathbf{y}|_2^2 + 2 < \Phi(X_S^*) - \mathbf{y}, \Phi(H) >$
$+ |\Phi(H)|_2^2 - |\Phi(X_S^*) - \mathbf{y}|_2^2)$

$= |||X_S^* + H|||_1 - |||X_S^*|||_1 + \gamma < \Phi(X_S^*) - \mathbf{y}, \Phi(H) > + (1/2)\gamma|\Phi(H)|_2^2$

$= |||X_S^* + H|||_1 - |||X_S^*|||_1 + \gamma < \Phi(X_S^*) - \mathbf{y}, \Phi_{\sim S}(H) > + (1/2)\gamma|\Phi_{\sim S}(H)|_2^2$

$\geq |||X_S^* + H|||_1 - |||X_S^*|||_1 + \gamma < \Phi(X_S^*) - \mathbf{y}, \Phi_{\sim S}(H) >$

The first term $|||X_S^* + H|||_1 - |||X_S^*|||_1$

$= max_j(|\mathbf{x}_j^*|_1 + |\mathbf{h}_j|_1) - max_j|\mathbf{x}_j^*|_1(\text{supp}(X_S^*) \cap \text{supp}(H) = \varnothing)$

$= |||X_S^*|||_1 + |||H|||_1 - |||X_S^*|||_1$ (condition (1) implies X_S^*'s l_1-column-flatness: remark after Theorem 3)

$= ||| H |||_1$

By replacing X_S^* with (41), note $\text{supp}(\Phi_{\sim S}^T) = \sim S$ and $|||M|||_1^* \leq 1$, the second term

$\gamma < \Phi_S(X^*) - \mathbf{y}, \Phi_{\sim S}(H) >$

$= \gamma < \Phi_{\sim S}^T(\Phi_S\Phi_S^{*-1}(\mathbf{y}) - \mathbf{y}), H > - < M^*, \Phi_S^{*-1}\Phi_{\sim S}(H) >$

$\geq (-\gamma \sup\{< \Phi_{\sim S}^T(\Phi_S\Phi_S^{*-1}(\mathbf{y}) - \mathbf{y}), H >: |||H|||_1 = 1\} - \sup\{< \Phi_{\sim S}^T(\Phi_S^{*-1})^T$
$M, H >: |||H|||_1 = 1, |||M|||_1^* \leq 1\})|||H|||_1 \tag{42}$

Therefore

$L(X_S^* + H) - L(X_S^*)$
$\geq |||H|||_1(1 - \gamma \sup\{\langle \Phi_{\sim S}^T(\Phi_S\Phi_S^{*-1}(\mathbf{y}) - \mathbf{y}), H \rangle : |||H|||_1 = 1\}$
$- \sup\{\langle \Phi_{\sim S}^T(\Phi_S^{*-1})^T M, H \rangle : |||H|||_1 = 1 \text{ and } |||M|||_1^* \leq 1\}) \tag{43}$

and condition (3) implies the right hand side > 0. This proves X_S^* is the minimizer of (22) and the minimizer is unique.

(3) For $Y^* = \Phi_S^{*-1}(\mathbf{y}) \in \sum_s^{n \times n}(S)$(then supp$(Y^*)$ in S) and by (41) we have

$|X_{S,ij}^*|$

$= |Y_{ij}^* - \gamma^{-1}(\Phi_S^T\Phi_S)^{-1}(M^*)_{ij}|$

$\geq |Y_{ij}^*| - \gamma^{-1}|(\Phi_S^T\Phi_S)^{-1}(M^*)_{ij}|$

$\geq |Y_{ij}^*| - \gamma^{-1}max_{ij}|(\Phi_S^T\Phi_S)^{-1}(M^*)_{ij}|$

$= |Y_{ij}^*| - \gamma^{-1}|(\Phi_S^T\Phi_S)^{-1}(M^*)|_{max}$

$\geq |Y_{ij}^*| - N((\Phi_S^T\Phi_S)^{-1} : |||\cdot|||_1^* \to |||\cdot|||_{max})|||M|||_1^*$

$\geq |Y_{ij}^*| - N((\Phi_S^T\Phi_S)^{-1} : |||\cdot|||_1^* \to |||\cdot|||_{max})(|||M|||_1^* \leq 1)$

$> 0 \quad \text{for those}(i,j) : |Y_{ij}^*| > N((\Phi_S^T\Phi_S)^{-1} : |||\cdot|||_1^* \to |||\cdot|||_{max})$

(4) Note that for any non-zero scalars u and v, $sgn(u) = sgn(v)$ iff $|u| > |u-v|$. Therefore

$$sgn(X_{S,ij}^*) = sgn(Y_{ij}^*) \text{ iff } |Y_{ij}^*| > |Y_{ij}^* - X_{ij}^*| = \gamma^{-1}|(\Phi_S^T\Phi_S)^{-1}(M^*)_{ij}| \tag{44}$$

In particular, if $min_{(i,j)in\,S}|Y_{ij}^*| > \gamma^{-1}N((\Phi_S^T\Phi_S)^{-1} : |||.|||_1^* \to |||.|||_{max})$ then $|Y_{ij}^*| > \gamma^{-1}max_{(i,j)}|(\Phi_S^T\Phi_S)^{-1}(M^*)_{ij}|$ so $sgn(X_{S,ij}^*) = sgn(Y_{ij}^*)$ for all (i,j) in S.   □

Proof of Lemma 3. (1) Let $X_S^* \in Arg\,inf\,|||Z|||_1$ s.t. $Z \in R^{n\times n}, |y - \Phi_S(Z)|_2 \le \eta$. i.e., a minimizer with its support restricted on S. We first prove X_S^* is the only minimizer of this support-restricted problem, then we prove X^* is also the minimizer of problem $MP_{y,\Phi,\eta}$ (25), i.e., X_S^* is the global minimizer and (25)'s minimizer is unique.

According to general convex optimization theory, there exist a positive multiplier $\gamma^* > 0$ and M^* in $\partial|||X_S^*|||_1$ such that

$$M^* + \gamma^*\Phi_S^T(\Phi_S(X_S^*) - y) = O \text{ and } |y - \Phi_S(X_S^*)|_2 = \eta \qquad (45)$$

then equivalently

$$X_S^* = \Phi_S^{*-1}(y) - \gamma^{*-1}(\Phi_S^T\Phi_S)^{-1}(M^*) \qquad (46)$$

Suppose X^0 is another minimizer of $inf\,|||Z|||_1$ s.t. $Z \in R^{n\times n}, |y - \Phi_S(Z)|_2 \le \eta$, then there exist a positive multiplier $\gamma^0 > 0$ and M^0 in $\partial|||X^0|||_1$ such that

$$M^0 + \gamma^0\Phi_S^T(\Phi_S(X^0) - y) = O \text{ and } |y - \Phi_S(X^0)|_2 = \eta \qquad (47)$$

Equivalently, (45) shows that X_S^* is also a minimizer of $L_S(Z) = |||Z|||_1 + (1/2)\gamma^*|y - \Phi_S(Z)|_2^2$ which is a strictly convex function on $\sum_s^{n\times n}(S)$ since $\Phi_S^T\Phi_S$ is a bijection (condition(2)), as a result $L_S(Z)$'s minimizer is unique. However, since $|||X_S^*|||_1 = |||X^0|||_1$ we have $L_S(X_S^*) = |||X_S^*|||_1 + (1/2)\gamma^*|y - \Phi_S(X_S^*)|_2^2 = |||X_S^*|||_1 + \gamma^*\eta^2/2 = |||X^0|||_1 + (\gamma^*/2)|y - \Phi_S(X^0)|_2^2 = L_S(X^0)$, which implies $X_S^* = X^0$, i.e., X_S^* is the unique minimizer of the support-restricted problem $inf\,|||Z|||_1$ s.t. $Z \in R^{n\times n}, |y - \Phi_S(Z)|_2 \le \eta$.

X_S^*'s l_1-column-flatness is implied by condition (1) and Theorem 3.

Now prove X_S^* (which is S-sparse and l_1-column-flat) is also a minimizer of problem $MP_{y,\Phi,\eta}$ (25). Again we start with the fact that $X_S^* = ArginfL_S(Z) = Arginf|||Z|||_1 + (1/2)\gamma^*|y - \Phi_S(Z)|_2^2$ with some multiplier $\gamma^* > 0$ (which value depends on X_S^*) and by Lemma 2, X_S^* is the unique minimizer of the convex problem (without any restriction on solution's support)

$$inf|||Z|||_1 + (1/2)\gamma^*|y - \Phi(Z)|_2^2 \qquad (48)$$

under the condition

$$\gamma^* sup\{< \Phi_{\sim S}^T(\Phi_S\Phi_S^{*-1}(y) - y), H >: |||H|||_1 = 1\}$$
$$+ sup\{< \Phi_{\sim S}^T(\Phi_S^{*-1})^T(M), H >: |||H|||_1 = 1 \text{ and } |||M|||_1^* \le 1\} < 1 \qquad (49)$$

According to convex optimization theory, X_S^* (under condition (49)) being the unique minimizer of problem (48) means X_S^* is also a minimizer of $MP_{y,\Phi,\eta}$ (25), which furthermore implies that $MP_{y,\Phi,\eta}$'s minimizer is unique, S-sparse and l_1-column-flat.

In order to make condition (49) more meaningful, we need to replace the minimizer-dependent parameter γ^* with explicit information. From (48)'s first-order optimization condition (45) we obtain

$1 \geq |||M^*|||_1^* = \gamma^*|||\Phi_S^T(\Phi_S(X_S^*) - y)|||_1^* \geq \gamma^* \, min\{|||\Phi_S^T(z)|||_1^* \; : \; |z|_2 = 1\}|\Phi_S(X_S^*) - y|_2 = \gamma^*\eta\Lambda_{min}(\Phi_S^T)$

i.e.,

$$\gamma^* \leq \left(\eta\Lambda_{\min}\left(\Phi_S^T\right)\right)^{-1} \tag{50}$$

with this upper-bound of γ^*, (49) can be derived from a uniform condition

$$\left(\eta\Lambda_{\min}\left(\Phi_S^T\right)\right)^{-1} \sup\left\{< \Phi_{\sim S}^T\left(\Phi_S\Phi_S^{*-1}(y) - y\right), H\right) >: |||H|||_1 = 1\right\}$$
$$+ \sup\left\{< \Phi_{\sim S}^T\left(\Phi_S^{*-1}\right)^T (M), H >: |||H|||_1 = 1 \text{ and } |||M|||_1^* \leq 1\right\} < 1 \tag{51}$$

which is equivalent to condition (3).

From now on we denote X_S^* as X^*.

(2) For $Y^* = \Phi_S^{*-1}(y) \in \sum_s^{n \times n}(S)$ and by Lemma 2's conclusion (4), if $min_{(i,j) in S} |Y_{ij}^*| > \gamma^{*-1}N((\Phi_S^T\Phi_S)^{-1}: |||\cdot|||_1^* \to |||\cdot|||_{max})$ then $sgn(X_{S,ij}^*) = sgn(Y_{ij}^*)$ for all (i,j) in S. To replace multiplier γ^* with more explicit information in this condition, we need some lower bound of γ^* which can be derived from the first-order optimization condition $M^* = \gamma^*(y - \Phi_S^T(\Phi_S(X^*))$ again. Note that X^* is l_1-column-flat implies every column of X^* is not $\mathbf{0}$, further more M^* has no $\mathbf{0}$-column so $M^* = (\lambda_1\mathbf{u}_1, \ldots, \lambda_n\mathbf{u}_n)$ with $\lambda_j > 0$ for all $j, \lambda_1 + \ldots + \lambda_n = 1$ and $|\mathbf{u}_j|_\infty = 1$, as a result $|||M^*|||_1^* = \sum_j \lambda_j|\mathbf{u}_j|_\infty = 1$. Hence

$1 = |||M^*|||_1^* \leq \gamma^*|||\Phi_S^T(\Phi_S(X^*) - y)|||_1^* \leq \gamma^* N(\Phi_S^T : l_2 \to |||\cdot|||_1^*)|\Phi_S(X^*) - y|_2 = \gamma^*\eta N(\Phi_S^T : l_2 \to |||\cdot|||_1^*)$

i.e.,

$$\gamma^{*-1} \leq \eta N\left(\Phi_S^T : l_2 \to |||\cdot|||_1^*\right) \tag{52}$$

Replace γ^{*-1} with its upper-bound in (52), we obtain if $min_{(i,j) in S} |Y_{ij}^*| > \eta N(\Phi_S^T : l_2 \to |||\cdot|||_1^*)N((\Phi_S^T\Phi_S)^{-1}: |||\cdot|||_1^* \to |||\cdot|||_{max})$ then $sgn(X_{S,ij}^*) = sgn(Y_{ij}^*)$ for all (i,j) in S.

(3) $Y^* = \Phi_S^{*-1}(y) \in \sum_s^{n \times n}(S)$ implies $\Phi_S^T(\Phi_S(Y^*) - y) = O$ and then condition (1) leads to $\Phi_S(Y^*) = y$. Furthermore, $\Phi_S^T\Phi_S$ is a bijection for $\sum_s^{n \times n}(S) \to \sum_s^{n \times n}(S)$ and notice $X^* - Y^* \in \sum_s^{n \times n}(S)$, so for any matrix norm $|.|_\alpha$:

$|X^* - Y^*|_\alpha = |(\Phi_S^T\Phi_S)^{-1}(\Phi_S^T\Phi_S)(X^* - Y^*)|_\alpha = |\Phi_S^{*-1}\Phi_S(X^* - Y^*)|_\alpha = |(\Phi_S^{*-1}(\Phi_S(X^*) - y))|_\alpha$
$\leq N(\Phi_S^{*-1} : l_2 \to |.|_\alpha)|\Phi_S(X^*) - y|_2 = \eta N(\Phi_S^{*-1} : l_2 \to |.|_\alpha)$ ☐

Proof of Theorem 6. (1) Note that in case of $X \in \sum_s^{n \times n}(R)$ and $y = \Phi(X) + e = \Phi_R(X) + e, |e|_2 \leq \eta$, we have

$$\Phi_R\Phi_R^{*-1}(y) - y = (\Phi_R\Phi_R^{*-1} - I_R)e$$

It's straightforward to verify that in this situation condition (3) in this theorem leads to condition (3) in Lemma 3: $\sup\{< \Phi_{\sim R}^T(\Phi_R\Phi_R^{*-1}(y) - y), H >: |||H|||_1 = 1\} < \eta\Lambda_{min}(\Phi_R^T)(1 - N(\Phi_R^{*-1}\Phi_{\sim R} : |||\cdot|||_1^* \to |||\cdot|||_1^*))$

for any η. As a result, $X^* \in \sum_s^{n \times n}(R)$ and is l_1-column-flat and the unique minimizer of $MP_{y,\Phi,\eta}$.

(2) For $Y^* = \Phi_R^{*-1}(y) \in \sum_s^{n \times n}(R)$ and by Lemma 3(4), we obtain $|X^* - Y^*|_\alpha \leq \eta N(\Phi_R^{*-1}: l_2 \to \cdot_\alpha)$ for any given matrix norm $|.|_\alpha$. On the other hand, $Y^* = \Phi_R^{*-1}(y)$ implies $\Phi_R^T(\Phi_R(Y^*)-y) = O$ then condition (1) leads to $\Phi_R(Y^*) = y$, hence $\Phi_R(Y^*) = y = \Phi(X)+e = \Phi_R(X)+e$, namely $\Phi_R^T\Phi_R(Y^*) = \Phi_R^T\Phi_R(X)+\Phi_R^T(e)$, as a result:

$$Y^* - X = \left(\Phi_R^T\Phi_R\right)^{-1}\Phi_R^T(e) \equiv \Phi_R^{*-1}(e) \tag{53}$$

Since $|e|_2 \leq \eta$, we get $|Y^* - X|_\alpha \leq \eta N(\Phi_R^{*-1}: l_2 \to |.|_\alpha)$ for any given matrix norm $|.|_\alpha$. Combining with $|X^* - Y^*|_\alpha \leq \eta N(\Phi_R^{*-1}: l_2 \to |.|_\alpha)$ we get the reconstruction error bound $|X^* - X|_\alpha \leq 2\eta N(\Phi_R^{*-1}: l_2 \to |.|_\alpha)$.

(3) By the first-order optimization condition on minimizer X^* with the fact $supp(X^*) = R$, we have the equation $X^* = \Phi_R^{*-1}(y) - \gamma^{*-1}(\Phi_R^T\Phi_R)^{-1}(M^*) = Y^* - \gamma^{*-1}(\Phi_R^T\Phi_R)^{-1}(M^*)$ where M^* is in $\partial|||X^*|||_1$, namely:

$$X^* - Y^* = -\gamma^{*-1}\left(\Phi_R^T\Phi_R\right)^{-1}(M^*) \tag{54}$$

Combining with (53), we get

$$X^* - X = \Phi_R^{*-1}(e) - \gamma^{*-1}\left(\Phi_R^T\Phi_R\right)^{-1}(M^*) \tag{55}$$

Since $sgn(X_{ij}^*) = sgn(X_{ij})$ iff $|X_{ij}| > |X_{ij}-X_{ij}^*| = |\Phi_R^{*-1}(e)_{ij}-\gamma^{*-1}(\Phi_R^T\Phi_R)^{-1}(M^*)_{ij}|$, in particular, if X_{ij} can satisfy $|X_{ij}| > max_{ij}|\Phi_R^{*-1}(e)_{ij}| + \gamma^{*-1}max_{ij}|(\Phi_R^T\Phi_R)^{-1}(M^*)_{ij}|$ then the former inequality is true and as a result $sgn(X_{ij}^*) = sgn(X_{ij})$. It's straightforward to verify (by using (52)) that the condition (3) just provides a guarantee for this. $\square$

Appendix C: Proofs of Theorems in Section 5

Proof of Lemma 4. We start with (FACT 4) $w^2(D(|||.|||_1, X)) \leq E_G[inf\{|G-tV|_F^2: t > 0, V \text{ in } \partial|||X|||_1\}]$ where G is a random matrix with entries $G_{ij} \sim^{iid} N(0,1)$.

Set $G = (g_1,\ldots,g_n)$ where $g_j \sim^{iid} N(0,I_n)$. By Lemma 1, $V = (\lambda_1\xi_1,\ldots,\lambda_n\xi_n)$ where w.l.o.g. $\lambda_j \geq 0$ for $j = 1,\ldots,r, \lambda_1 + \ldots + \lambda_r = 1, \lambda_j = 0$ for $j \geq r + 1$; $|x_j|_1 = max_k|x_k|_1$ for $j = 1,\ldots,r$ and $|x_j|_1 < max_k|x_k|_1$ for $j \geq 1+r; \xi_j(i) = sgn(X_{ij})$ for $X_{ij} \neq 0$ and $|\xi_j(i)| \leq 1$ for all i and j. Then

$w^2(D(||| \cdot |||_1, X))$

$\leq E_G\left[inf_{t>0,\lambda_j,\xi_j \text{ specified as the above}} \sum_{j=1}^r |g_j - t\lambda_j\xi_j|_2^2 + \sum_{j=r+1}^n |g_j|_2^2\right]$

$\leq inf_{t>0, \text{ all } \lambda_j \text{ specified as the above}} E_G\left[inf_{\text{all } \xi_j \text{ specified as the above}} \sum_{j=1}^r\right.$

$\left.|g_j - t\lambda_j\xi_j|_2^2 + \sum_{j=r+1}^n |g_j|_2^2\right]$

$= inf_{t>0, \text{ all } \lambda_j \text{ specified as the above}} E_G\left[inf_{\text{all } \xi_j \text{ specified as the above}} \sum_{j=1}^r\right.$

$$|\mathbf{g}_j - t\lambda_j\xi_j|_2^2] + \sum_{j=r+1}^n E_G\left[|\mathbf{g}_j|_2^2\right]$$

$$= \inf{}_{t>0, \text{ all } \lambda_j \text{ specified as the above }} E_G\left[\sum_{j=1}^r \inf{}_{\xi_j \text{ specified as the above }}\right.$$

$$|\mathbf{g}_j - t\lambda_j\xi_j|_2^2] + (n-r)n$$

(since ξ_j is unrelated each other and $E_G\left[|\mathbf{g}_j|_2^2\right] = n$)

$$= \inf{}_{t>0, \text{ all } \lambda_j \text{ specified as the above }} \sum_{j=1}^r E_{gj}\left[\inf{}_{\xi_j \text{ specified as the above }} |\mathbf{g}_j - t\lambda_j\xi_j|_2^2\right] +$$

$(n-r)n$

For each $j = 1, \ldots, r$ let $S(S)$ be the support of $\mathbf{x}_j$(so $|S(j)| \leq s$) and $\sim S(j)$ be its complimentary set, then $|\mathbf{g}_j - t\lambda_j\xi_j|_2^2 = |\mathbf{g}_{j|S(j)}| - t\lambda_j\xi_{j|S(j)}|_2^2 + |\mathbf{g}_{j|\sim S(j)} - t\lambda_j\xi_{j|\sim S(j)}|_2^2$. Notice that all components of $\xi_{j|S(j)}$ are ± 1 and all components of $\xi_{j|\sim S(j)}$ can be any value in the interval $[-1, +1]$. Select $\lambda_1 = \ldots = \lambda_r = 1/r$, let $\varepsilon > 0$ be arbitrarily small positive number and select $t = t(\varepsilon)$ such that $P[|g| > t(\varepsilon)/r] \leq \varepsilon$ where g is a standard scalar Gaussian random variable (i.e., $\mathrm{g} \sim N(0,1)$ and ε can be $\exp\left(-t(\varepsilon)^2/2r^2\right)$) For each j and each i outside $S(j)$, set ξ_j's component $\xi_j(i) = rg_j(i)/t(\varepsilon)$ if $|g_j(i)| \leq t(\varepsilon)/r$ (in this case $\left|g_j(i) - t\lambda_j\xi_{j|}(i)\right| = 0$) and otherwise $\xi_j(i) = \mathrm{sgn}(g_j(i))$ (in this case $\left|g_j(i) - t\lambda_j\xi_{j|}(i)\right| = |g_j(i)| - t(\varepsilon)/r$), then $\left|\mathbf{g}_{j|\sim S(j)} - t\lambda_j\xi_{j|\sim S(j)}\right|_2^2 = 0$ when $|\mathbf{g}_{j|\sim S(j)}|_\infty < t(\varepsilon)/r$, hence:

$$E\left[\left|\mathbf{g}_{j|\sim S(j)} - t\lambda_j\xi_{j|\sim S(j)}\right|_2^2\right] = \int_0^\infty du P\left[\left|\mathbf{g}_{j|\sim S(j)} - t\lambda_j\xi_{j|\sim S(j)}\right|_2^2 > u\right]$$

$$= 2\int_0^\infty du u P\left[\left|\mathbf{g}_{j|\sim S(j)} - t\lambda_j\xi_{i|\sim S(j)}\right|_2 > u\right]$$

$$\leq 2\int_0^\infty du u P\left[\text{ There exists } (\mathbf{g}_{j|\sim S(j)} - t\lambda_j\xi_{i|\sim S(j)})\text{'s component with mag-}\right.$$
nitude $> (n-s)^{-1/2}u]$

$$\leq 2(n-s)\int_0^\infty du u P\left[|g| - t(\varepsilon)/r > (n-s)^{-1/2}u\right]$$

$$\leq 2(n-s)\int_0^\infty du u \exp\left(-\left((t(\varepsilon)/r) + (n-s)^{-1/2}u\right)^2/2\right)$$

$$\leq C_0(n-s)^2 \exp\left(-t(\varepsilon)^2/2r^2\right) \leq C_0(n-s)^2\varepsilon$$

where C_0 is an absolute constant. On the other hand:

$E_{gj}[|\mathbf{g}_{j|S(j)} - t\lambda_j\xi_{j|S(j)}|_2^2] = E_{gj}[|\mathbf{g}_{j|S(j)}|^2] + (t(\varepsilon)^2/r^2)|\xi_{j|S(j)}|_2^2 = (1 + t(\varepsilon)^2/r^2)s = (1 + 2log(1/\varepsilon))s$

Hence $w^2(D(\|\|.\|\|_1, X)) \leq (1 + 2log(1/\varepsilon))rs + (n-r)n + r(n-s)^2\varepsilon \leq n^2 - r(n - slog(e/\varepsilon^2)) + C_0n^2r\varepsilon$

In particular, let $\varepsilon = 1/C_0n^2r$ then we get $w^2(D(\|\|.\|\|_1, X)) \leq n^2 - r(n - slog(Cn^4r^2)) + 1$.   □

Proof of Theorem 9. For any $s < n$, there exist $k \geq (n/4s)^{ns/2}$ subsets $S^{(\alpha\beta\ldots\omega)} = S_1^{(\alpha)} \cup S_2^{(\beta)} \cup \ldots \cup S_n^{(\omega)}$ in $\{(i,j) : 1 \leq i,j \leq n\}$ where each $S_j^{(\mu)} = \{(i_1,j), \ldots, (i_s,j) : 1 \leq i_1 < i_2 < \ldots < i_s \leq n\}$ and $|S_j^{(\mu)} \cap S_j^{(\nu)}| < s/2$ for $\mu \neq \nu$. This fact is based on a combinatorial theorem [11] that for any $s < n$ there exist $l \geq (n/4s)^{s/2}$ subsets $R^{(\mu)}$ in $\{1, 2, \ldots, n\}$ where $|R^{(\mu)} \cap R^{(\nu)}| < s/2$ for any $\mu \neq \nu$. For the n-by-n square $\{(i,j) : 1 \leq i,j \leq n\}$, assign a $R^{(\mu)}$ to each column, i.e., set $S_j^{(\mu)} := \{(i,j) : i \in R^{(\mu)}\}$. As a result $|S_j^{(\mu)} \cap S_j^{(\nu)}| < s/2$ for $\mu \neq \nu$ since $|R^{(\mu)} \cap R^{(\nu)}| < s/2$ for $\mu \neq \nu$ and totally there can be $k = l^n$ such assignments $S^{(\alpha\beta\ldots\omega)} = S_1^{(\alpha)} \cup S_2^{(\beta)} \cup \ldots \cup S_n^{(\omega)}$ on the square.

Now we call the above $S_1^{(\alpha)} \cup S_2^{(\beta)} \cup \ldots \cup S_n^{(\omega)}$ a *configuration* on the n-by-n square. Let m be the rank of linear operator Φ. Consider the quotient space $L :=R^{n\times n}/\ker\Phi$, then $\dim L = n^2 - \dim\ker\Phi = m$. For any $[X]$ in L define the norm $|[X]| := inf\{|||X-V|||_1 : V \text{ in } \ker\Phi\}$. For any $X = (\mathbf{x}_1, \ldots, \mathbf{x}_n)$ with $\mathbf{x}_j$ in $\sum^{2S}$ for all j, the assumption about Φ implies $|[X]| = |||X|||_1$. Now for any configuration $\Delta = S_1 \cup S_2 \cup \ldots \cup S_n$ on the n-by-n square, define $X_{ij}(\Delta) := 1/s$ if $(i,j) \in S_j$ and 0 otherwise, then $|||X(\Delta)|||_1 = 1$, each $X(\Delta)$'s column $\mathbf{x}_j(\Delta) \in \sum^S$ and each column of $X(\Delta')-X(\Delta'')$ is in $\sum^{2S}$, furthermore $|[X(\Delta')]-[X(\Delta'')]| = |||X(\Delta')-X(\Delta'')|||_1 > 1$ because of the property $|S_j'\cap S_j''| < s/2$ for $S_j' \neq S_j''$. These facts imply that the set $\Theta := \{[X(\Delta)] : \Delta \text{ runs over all configurations}\}$ is a subset on normed quotient space L's unit sphere with distances between any pair of its members >1, i.e., a d-net on the sphere where $d > 1$. The cardinality of Θ = number of configurations $k \geq (n/4s)^{ns/2}$ and an elementary estimate derives $k \leq 3^{\dim L} = 3^m$, hence $m \geq C_1 n\log(C_2 n/s))$ where $C_1 = 1/2\log 3$ and $C_2 = 1/4$. $\square$

Appendix D: Proofs of Theorems in Section 6

Proof of Lemma 5. We start with a similar inequality as that in FACT 4 (the proof is also similar) $W^2(\Gamma_X; \Phi_{A,B}) \leq E_H[inf\{|H - tV|_F^2 : t > 0, V \text{ in } \partial|||X|||_1\}]$. With the same specifications for $V = (\lambda_1\xi_1, \ldots, \lambda_n\xi_n)$ as those in Lemma 4, i.e.(w.l.o.g.) $\lambda_j \geq 0$ for $j = 1, \ldots, r, \lambda_1 + \ldots + \lambda_r = 1$, $\lambda_j = 0$ for $j \geq r + 1$; $|\mathbf{x}_j|_1 = max_k|\mathbf{x}_k|_1$ for $j = 1, \ldots, r$ and $|\mathbf{x}_j|_1 < max_k|\mathbf{x}_k|_1$ for $j \geq 1 + r$, $\xi_j(i) = sgn(X_{ij})$ for $X_{ij} \neq 0$ and $|\xi_j(i)| \leq 1$ for all i and j. Let $\mathbf{h}_j \equiv \sum_{l=1}^m m^{-1}B_{lj}A^T\varepsilon_l$, we have

$$W^2(\Gamma_X; \Phi_{A,B})$$
$$\leq E_{A,B,E}\left[inf_{t>0,\lambda_j,\xi_j \text{ specified as the above}} \sum_{j=1}^n |\sum_{l=1}^m m^{-1}B_{lj}A^T\varepsilon_l - t\lambda_j\xi_j|_2^2\right]$$
$$= \sum_{j=r+1}^n E_{A,B,E}\left[|\mathbf{h}_j|_2^2\right] + E_{A,B,E}\left[inf_{t>0,\lambda_j,\xi_j \text{ specified as the above}} \sum_{j=1}^r |\mathbf{h}_j - t\lambda_j\xi_j|_2^2\right]$$
$$= I + II$$

The first and second terms are estimated respectively. The first term
$$I = \sum_{j=r+1}^n m^{-2}\sum_{l,k=1}^m E_B[B_{lj}B_{kj}]E_{A,E}[\varepsilon_l^T AA^T\varepsilon_k] = m^{-2}(n-r)\sum_{l,k=1}^m \delta_{lk}E_{A,E}[\varepsilon_l^T AA^T\varepsilon_l] = (n-r)n$$

To estimate II, for each $j = 1, \ldots, r$ let $S(j)$ be the support of $\mathbf{x}_j$ (so $|S(j)| \leq s$) and $\sim S(j)$ be its complimentary set, then
$$\sum_{j=1}^r |\mathbf{h}_j - t\lambda_j\xi_j|_2^2 = \sum_{j=1}^r |\mathbf{h}_{j|S(j)} - t\lambda_j\xi_{j|S(j)}|_2^2 + \sum_{j=1}^r |\mathbf{h}_{j|\sim S(j)} - t\lambda_j\xi_{j|\sim S(j)}|_2^2]$$

Notice that all components of $\xi_{j|S(j)}$ are ± 1 and all components of $\xi_{j|\sim S(j)}$ can be any value in the interval $[-1, +1]$. Select $\lambda_1 = \ldots = \lambda_r = 1/r$, let $\delta > 0$ be arbitrarily small positive number and select $t = t(\delta)$ such that $P_{A,B,E}[|h| > t(\delta)/r] \leq \delta$ where h is a random scalar such that $h_j(i) \sim h$ and i indicates the vector $\mathbf{h}_j$'s i-th component. For each j and i outside $S(j)$, set ξ_j's component $\xi_j(i) = rh_j(i)/t(\varepsilon)$ if $|h_j(i)| \leq t(\delta)/r$ and otherwise $\xi_j(i) = sgn(h_j(i))$, then

$\left|\boldsymbol{h}_{j|\sim S(j)} - t\lambda_j \xi_{j|\sim S(j)}\right|_2^2 = 0$ when $\left|\boldsymbol{h}_{j|\sim S(j)}\right|_\infty < t(\delta)/r$ and notice the fact that for independent standard scalar Gaussian variables a_l, b_l and Rademacher variables $\varepsilon_l, l = 1, \ldots, m$, there exists absolute constant c such that for any $\eta > 0$:

$$\mathrm{P}\left[\left|m^{-1}\sum_{l,k=1}^{m} b_l a_k \varepsilon_k\right| > \eta\right] < c\exp(-\eta) \qquad (56)$$

as a result, in the above expression δ can be $cexp(-t(\delta)/r)$ and:

$\mathrm{E}\left[\left|\boldsymbol{h}_{j|\sim S(j)} - t\lambda_j \xi_{j|\sim S(j)}\right|_2^2\right] = \int_0^\infty du \mathrm{P}\left[\left|\boldsymbol{h}_{j|\sim S(j)} - t\lambda_j \xi_{j|\sim S(j)}\right|_2 > u\right]$

$= 2\int_0^\infty du\, u \mathrm{P}\left[\left|\boldsymbol{h}_{j|\sim S(j)} - t\lambda_j \xi_{j|\sim S(j)}\right|_2 > u\right]$

$\leq 2\int_0^\infty du\, u \mathrm{P}\left[\text{ There exists } \left(\boldsymbol{h}_{j|\sim S(j)} - t\lambda_j \xi_{i\sim S(j)}\right)'\text{s component with mag-}\right.$

nitude $> (n-s)^{-1/2}u]$

$\leq 2(n-s)\int_0^\infty du\, u \mathrm{P}\left[|h| - t(\delta)/r > (n-s)^{-1/2}u\right]$

$\leq 2(n-s)\int_0^\infty du\, u \exp\left(-\left((t(\delta)/r) + (n-s)^{-1/2}u\right)\right)$

$\leq C_0(n-s)^2 \exp(-(t(\delta)/r)) \leq C_0(n-s)^2\delta$

where C_0 is an absolute constant. On the other hand $\left|\xi_{j|S(j)}\right|_2^2 \leq s$ for $j \geq 1+r$

so:

$\mathrm{E}_{A,B,E}\left[\inf_{t>0,\lambda_j,\xi_j}\sum_{j=1}^{r}\left|\boldsymbol{h}_{j|S(j)} - t\lambda_j \xi_{j|S(j)}\right|_2^2\right]$

$\leq \mathrm{E}_{A,B,E}\left[\sum_{j=1}^{r}\left|\boldsymbol{h}_{j|S(j)} - t(\delta)\xi_{j|S(j)}/r\right|_2^2\right]$

$\leq \sum_{j=1}^{r} \mathrm{E}_{A,B,E}\left[m^{-2}\left|\sum_{l=1}^{m}B_{lj}\left(\mathrm{A}^T\varepsilon_l\right)_{|S(j)}\right|_2^2\right] + rst(\delta)^2/r^2$

$= rs\left(1 + t(\delta)^2/r^2\right)$

hence $\mathrm{II} \leq rs(1 + t(\delta)^2/r^2) + nr\delta$. Combine all the above estimates we have:

$W^2(\Gamma_X; \Phi_{A,B}) \leq \mathrm{I} + \mathrm{II} \leq (n-r)n + rs(1 + t(\delta)^2/r^2) + C_0 n^2 r\delta = n^2 - r(n - s(1 + t(\delta)^2/r^2)) + C_0 n^2 r\delta$

Substitute $t(\delta)/r$ with $log(c/\delta)$ we get, for any $\delta > 0$:

$$W^2(\Gamma_X; \Phi_{A,B}) \leq n^2 - r(n - s(1 + log^2(c/\delta))) + C_0 n^2 r\delta$$

In particular, let $\delta = 1/C_0 n^2 r$ then $W^2(\Gamma_X; \Phi_{A,B}) \leq n^2 - r(n - s(1 + log^2(cn^2 r))) + 1$. $\qquad \square$

Proof of Lemma 6. By the second moment inequality $\mathrm{P}[Z \geq \xi] \geq (\mathrm{E}[Z] - \xi)_+^2/\mathrm{E}[Z^2]$ for any non-negative r.v. Z and any $\xi > 0$. Set $Z = |<\mathrm{M}, \mathrm{U}>|^2$ and $\xi = \mathrm{E}[|<\mathrm{M}, \mathrm{U}>|^2]/2$, we get:

$$\mathrm{P}\left[|<\mathrm{M},\mathrm{U}>|^2 \geq \mathrm{E}\left[|<\mathrm{M},\mathrm{U}>|^2\right]/2\right] \geq \mathrm{E}\left[|<\mathrm{M},\mathrm{U}>|^2\right]^2/4\mathrm{E}\left[|<\mathrm{M},\mathrm{U}>|^4\right] \qquad (57)$$

To estimate the upper bound of $\mathrm{E}[|<\mathrm{M},\mathrm{U}>|^2]$, let $\mathrm{U} = \sum_j \lambda_j \boldsymbol{u}_j \boldsymbol{v}_j$ be U's singular value decomposition, $\boldsymbol{u}_i^T \boldsymbol{u}_j = \boldsymbol{v}_i^T \boldsymbol{v}_j = \delta_{ij}$, $\lambda_j > 0$ for each j. Notice that $\mathrm{M} = \boldsymbol{a}\boldsymbol{b}^T$ where $\boldsymbol{a} \sim \boldsymbol{b} \sim N(0, I_n)$ and independent each other, then $<\mathrm{M},\mathrm{U}> = \boldsymbol{a}^T\mathrm{U}\boldsymbol{b} = \sum_j \lambda_j \boldsymbol{a}^T \boldsymbol{u}_j \boldsymbol{v}_j^T \boldsymbol{b}$ where $\boldsymbol{a}^T \boldsymbol{u}_i \sim \boldsymbol{v}_j^T \boldsymbol{b} \sim N(0,1)$ and independent each other, hence $\mathrm{E}[|<\mathrm{M},\mathrm{U}>|^2] = \sum_j \lambda_j^2 \mathrm{E}[|\boldsymbol{a}^T \boldsymbol{u}_j|^2]\mathrm{E}[|\boldsymbol{v}_j^T \boldsymbol{b}|^2] = \sum_j \lambda_j^2 = |\mathrm{U}|_F^2 = 1$ for U in the assumption.

On the other hand by Gaussian hypercontractivity we have

$$(E[|<M,U>|^4])^{1/4} \le C_0(E[|<M,U>|^2])^{1/2} = C_0$$

In conclusion $P[|<M,U>|^2 \ge 1/2] = P[|<M,U>|^2 \ge E[|<M,U>|^2]/2] \ge c$ for U: $|U|_F^2 = 1$. $\qquad\square$

The proof of Lemma 7 is logically the same as the proof of Lemma 5, the only difference is about the distribution tail of the components of vectors $h_j \equiv \sum_{l=1}^m m^{-1} B_{lj} A^T \varepsilon_l$ which $\sim^{iid} h \equiv m^{-1} \sum_{l,k=1}^m b_l a_k \varepsilon_k$ with independent scalar sub-Gaussian variables a_l, b_l and Rademacher variables ε_l, $l=1,\dots,m$. This auxiliary result is presented in the following lemma:

Lemma 8. For independent scalar zero-mean *sub-Gaussian* variables a_l, b_l and Rademacher variables $\varepsilon_l, l = 1, \dots, m$, let $\sigma_A \equiv max_l|a_l|_{\psi 2}$, $\sigma_B \equiv max_l|b_l|_{\psi 2}$ ($|.|_{\psi 2}$ denotes a *sub-Gaussian* variable's ψ_2-norm), then there exists absolute constant c such that for any $\eta > 0$:

$$P[|h| > \eta] < 2\exp\left(-c\eta/\sigma_A\sigma_B\right) \tag{58}$$

Proof. Notice that $a_\ell \varepsilon_k$ is zero-mean sub-Gaussian variable with $|a_k \varepsilon_k|_{\psi 2} = |a_k|_{\psi 2}$, for $b = m^{-1/2} \sum_{1 \le l \le m} b_l$ and $a = m^{-1/2} \sum_{1 \le k \le m} a_k \varepsilon_k$ we have $|b|_{\psi 2} \le Cm^{-1/2}\left(\sum_l |b_l|_{\psi 2}^2\right)^{1/2} C\sigma_B$ and $|a|_{\psi 2} \le Cm^{-1/2}\left(\sum_l |a_k|_{\psi 2}^2\right)^{1/2} \le C\sigma_A$ where C is an absolute constant. Furthermore, because the product of two sub-Gaussian variables a and b is sub-Exponential and its ψ_1 -norm $|ba|_{w1} \le |b|_{\psi 2}|a|_{\psi 2} \le C^2\sigma_A\sigma_B, h \equiv m^{-1}\sum_{l,k=1}^m b_l a_k \varepsilon_k = ab$ has its distribution tail $P[|h| > \eta] < 2\exp\left(-c\eta/\sigma_A\sigma_B\right)$ where c is an absolute constant. $\qquad\square$

Proof of Lemma 7. With the same logic as in the proof of Lemma 5 and based-upon Lemma 8, the auxiliary parameter δ in the argument can be $2exp(-ct(\delta)/r\sigma_A\sigma_B)$ and equivalently $t(\delta)/r = \sigma_A\sigma_B log(2/\delta)$ which derives the final result. $\qquad\square$

References

1. Foucart, S., Rauhut, H.: A Mathematical Introduction to Compressive Sensing. Birkhaeusser (2013)
2. Eldar, Y.C., Kutynoik, G. (eds.): Compressed Sensing: Theory and Applications. Cambridge University Press (2012)
3. Cohen, D., Eldar, Y.C.: Sub-Nyquist radar systems: temporal, spectral and spatial compression. IEEE Signal Process. Mag. **35**(6), 35–57 (2018)
4. Davenport, M.A., Romberg, J.: An overview of low-rank matrix recovery from incomplete observations. arXiv:1601.06422 (2016)
5. Duarte, M.F., Baraniuk, R.G.: Kronecker compressive sensing. IEEE Trans. Image Process. **21**(2), 494–504 (2012)
6. Dasarathy, G., Shah, P., Bhaskar, B.N., Nowak, R.: Sketching sparse matrices. arXiv:1303.6544 (2013)

7. Chandrasekaran, V., Recht, B., Parrio, P.A., Wilsky, A.S.: The convex geometry of linear inverse problems. Found. Comput. Math. **12**, 805–849 (2012)
8. Tropp, J.A.: Convex recovery of a structured signal from independent random linear measurements. In: Pfander, G. (ed.) Sampling Theory: A Renaissance: Compressive Sampling and Other Developments. Birkhaeusser (2015)
9. Mendelson, S.: Learning without concentration. J. ACM **62**, 3 (2014)
10. Mendelson, S., Pajor, A., Tomczak-Jaegermann, N.: Reconstruction and subgaussian operators in asymptotic geometric analysis. Geom. Func. Anal. **17**(4), 1248–1282 (2007)
11. Van Lint, J.H., Wilson, R.M.: A Course in Combinatorics. Springer-Verlag (1995)
12. Vershynin, R.: High-Dimensional Probability - with Applications to Data Science. Oxford University Press (2015)
13. Ledoux, M., Talagrand, M.: Probability in Banach Space: Isopermetry and Processes. Springer, Heidelberg (1991). https://doi.org/10.1007/978-3-642-20212-4
14. Dai, W., Li, Y., Zou, J., Xiong, H., Zheng, Y.: Fully decomposable compressive sampling with optimization for multidimensional sparse representation. IEEE Trans. Signal Process. **66**(3), 603–616 (2018)
15. Vatier, S., Peyre, G., Dadili, J.: Model consistency of partly smooth regularizers. IEEE Trans. Inform. Theory **64**(3), 1725–1747 (2018)
16. Oymak, S., Tropp, J.A.: Universality laws for randomized dimension reduction with applications. Inf. Infer. **7**, 337–386 (2017)

Design of Integrated Circuit Chip Fault Diagnosis System Based on Neural Network

Xinsheng Wang[✉], Xiaoyao Qi, and Bin Sun

Harbin Institute of Technology, Heilongjiang, China
xswang@hit.edu.cn

Abstract. This paper focused on the application of neural network in fault diagnosis and its implementation on FPGA. The function of the feature parameter processing module is to process the feature parameters into a form suitable for the input of the neural network model. The feature parameter processing module includes a receiving algorithm, a digital signal processing algorithm, a Kalman filtering algorithm, and a dispersion normalization algorithm, all of which are designed using Verilog language and implemented on an FPGA. The function of the neural network diagnosis module is to analyze the feature parameters and predict the failure state of the system to be tested; the neural network diagnosis module includes a neural network training platform and a feedforward neural network model, wherein the neural network training platform is designed using Python language and implemented by software; The feedforward neural network model is designed using Verilog language and implemented on an FPGA. The test results show that when the number of training exceeds 2000 times, the failure state diagnosis is more than 97% stable for the high-temperature failure diagnosis accuracy of the CMOS static memory cell circuit and the JFM4VSX55RT FPGA.

Keywords: Neural network · FPGA · Fault diagnosis

1 Introduction

In recent decades years, people have become more and more demanding on the reliability of integrated circuits, and the process and structure of large-scale integrated circuits have become increasingly complex. The amount of test data required for fault detection has increased dramatically, and the difficulty of testing and protection has increased. Changes have prompted scholars in related fields to break the tradition and seek new solutions [1]. With the advent of the era of big data in recent years, the maturity of neural network theory in deep learning and the improvement of high-speed parallel computing capabilities, the change of parameters is analyzed by machine learning to identify the failure state of the device under test (whether it fails and fails). The type and incentives have been explored by domestic and foreign scholars and carried out a series of practical explorations. Artificial Neural Network (ANN) has amazing performance in many fields such as computer vision, pattern recognition and biological natural science. It uses deep learning to detect high-dimensional data such as images, audio, video and natural science. The data is excellent in classification [2].

© Springer Nature Singapore Pte Ltd. 2020
J. Zeng et al. (Eds.): ICPCSEE 2020, CCIS 1257, pp. 222–229, 2020.
https://doi.org/10.1007/978-981-15-7981-3_14

At present, hundreds of application models have been developed based on neural network theory. Many neural computing models have developed into classic methods in various fields such as signal processing, computer vision and optimal design, which have promoted progress and interconnection in various fields. The entire scientific field has a milestone significance [3]. The powerful pattern recognition capability of neural networks is the key to achieving classification goals in various fields. The essence of failure diagnosis of integrated circuits is a pattern recognition (that is, classification of failure states, and normal working conditions are also one type). It is theoretically feasible to apply neural network theory to failure analysis in the field of integrated circuit reliability. And clever.

Applying neural network theory to the practice of failure analysis in the field of integrated circuit reliability has a milestone in the development of integrated circuit reliability. This will provide practical examples and data samples for theoretical research in the field of integrated circuit reliability to promote its improvement and development. At the same time, it is of great significance to carry out targeted early warning and protection planning for the failure of integrated circuit devices, which provides new ideas and new ways for the reliability maintenance of integrated circuit products. It marks that the Prognostics and Health Management (PHM) transitions from a physical model to a mathematical model based on an experience-driven phase into a data-driven phase [4].

2 Fault Diagnosis System Design

2.1 Overview of Neural Network Principles

ANN is a model that simulates human brain regions, neurons, and their ability to learn. It simulates biological neurons with a single neuron node structure, simulating the biological neural network with the topology of a single neuron node. At present, the research and application of ANN has achieved a lot of results, and has carried out many applications in pattern recognition, prediction and early warning functions, and provided new problem-solving problems in engineering, biomedicine, humanities and social sciences, and economics.

The development of ANN began with the construction of a single neuron (McCulloch-Pitts, M-P) model to simulate a single biological cell source. The structure of a single neural network node model is shown in Fig. 1.

The output calculation formula of a single neuron node model is shown in Eq. (1).

$$y = f(\sum_{i=1}^{n} w_i * x_i + b) \tag{1}$$

where y denotes the output of the neuron node; xi denotes the input of the neuron node; f denotes the activation function of the neuron node; wi denotes the weight of the neuron node and all nodes of the previous layer; b denotes the offset of the neuron node.

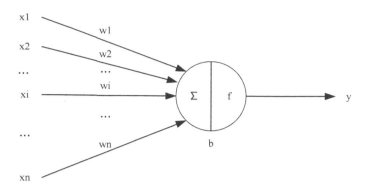

Fig. 1. Block diagram of a single neural network node model

With the development of science and technology in other fields, based on the M-P model, a variety of neural network failure prediction models are proposed and developed according to specific application requirements. Among them, the neural network failure prediction model with important significance and wide application is shown below [5].

(1) The next step in the M-P model is the development of a perceptron model. A single-layer neural network formed by multiple M-P model topologies has only one layer of computation between the input layer and the output layer. It is equivalent to a feedforward neural network without a hidden layer.

(2) Feedforward neural network is a multi-layer perceptron model. There are one or more hidden layers between the input layer and the output layer. The multi-layer depth structure makes the calculation accuracy higher. The feedforward neural network is a one-way non-feedback network model. The network structure and network parameters (weights and offsets) are fixed, or there is no independent learning function. Generally, it is not portable, only in the original code. Modifications or migrations to the model through other means of transmission.

(3) BP algorithm the neural network is a feedforward neural network with learning function. It adds training function to the feedforward neural network. The input samples are divided into training samples (with expected output) and test samples (unknown output). The training samples are input into the network model for one-way calculation. The network parameters are updated by comparing the expected output with the calculated output difference (feedback). When the expected output and the calculated output error rate reach the standard or the number of feedbacks reaches the standard, the network parameters are considered to be mature, and the test sample is calculated as a feedforward neural network.

(4) Recurrent Neural Network (RNN) is better at analyzing time series input sequences than feedforward neural networks. The time series input sequence means that the input characteristic parameters are variables that have a certain relationship with time, rather than only the reaction value. Constant data. RNN is mostly used for handwriting and speech recognition.

(5) Convolutional Neural Networks (CNN) is a feedforward neural network with convolutional computational power and depth structure that simulates biological vision, sharing network parameters (weights and biases) within its convolution kernel. And the characteristics of the pooled sparse representation are particularly suitable for processing samples with very large data volumes. CNN is mostly used for image recognition.

In addition to the above several neural network failure prediction models, a large number of other models have been proposed in different development periods. The characteristic parameter of this subject is the chip-level power supply current constant value. There are only 5 parameters in one sample setting, so BP neural network is used. However, the back- propagation algorithm of BP neural network is difficult to implement with FPGA, so FPGA implements the same feedforward neural network, software platform (training platform) realizes the training process, and trains the mature network parameters to transmit to the FPGA through serial protocol. Neural network module.

2.2 Feedforward Neural Network Implementation

The number of hidden layer neuron nodes is related to the computational requirements of the actual problem and the number of input and output nodes. If the number of hidden layer nodes is too small, the calculation accuracy will be affected; if the number of hidden layer nodes is large, the training process will be over-fitting. The empirical formulas for the selection of the number of hidden layer nodes can be found in Eqs. (2) and (3) [6].

$$n_1 = \sqrt{n+m} + a \tag{2}$$

$$n_1 = log_2 n \tag{3}$$

Where n denotes the number of input-layer nodes denotes the number of output-layer nodes; 1 denotes the number of nodes in the hidden layer; a denotes a constant in the interval [1, 10].

Due to the constraints of the look-up table (LUT) resources of the system carrier Xilinx XC7A100T FPGA, after the final design is completed, the number of input nodes is compressed to 5; the high-temperature fault injection test is performed for the system to be tested. It detects normal status and high temperature faults, and the number of output nodes is 2 (two classifications) [7-10]. The single hidden layer can solve most of the current pattern recognition problems, and the number of hidden layers is set to 2 layers, which provides a certain guarantee for calculation accuracy. The structure of the feedforward neural network is shown in Fig. 2.

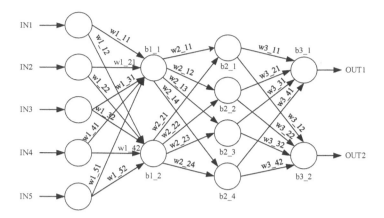

Fig. 2. Structure diagram of feedforward neural network

The functional simulation results of the feedforward neural network are shown in Fig. 3. Datain1–datain5 are the five parameters IN1–IN5 of the input sample, and OUT1_1–OUT1_2, OUT2_1–OUT2_4, OUT3_1–OUT3_2 are the output of each neuron node of the first hidden layer, the second hidden layer and the output layer respectively. Done1, Done2, and Done3 are the end flags of the feedforward operation, respectively.

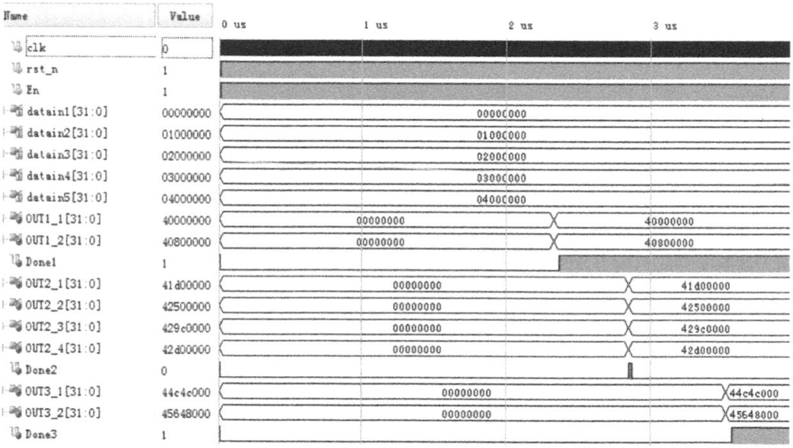

Name	Value	0 us	1 us	2 us	3 us
clk	0				
rst_n	1				
En	1				
datain1[31:0]	00000000		0000c000		
datain2[31:0]	01000000		0100c000		
datain3[31:0]	02000000		0200c000		
datain4[31:0]	03000000		0300c000		
datain5[31:0]	04000000		0400c000		
OUT1_1[31:0]	40000000	00000000		40000000	
OUT1_2[31:0]	40800000	00000000		40800000	
Done1	1				
OUT2_1[31:0]	41d00000	00000000		41d00000	
OUT2_2[31:0]	42500000	00000000		42500000	
OUT2_3[31:0]	429c0000	00000000		429c0000	
OUT2_4[31:0]	42d00000	00000000		42d00000	
Done2	0				
OUT3_1[31:0]	44c4c000	00000000			44c4c000
OUT3_2[31:0]	45648000	00000000			45648000
Done3	1				

Fig. 3. Functional simulation results of feedforward neural networks

2.3   Implementation of Output Classification Function

In the previous section, the neural network failure prediction model structure block diagram 3–16 input sample parameters IN1–IN5 are the processed chip-level power supply current parameters, the output values are OUT1 and OUT2, and the output values are the same fixed-point decimals as the input value format. As the last loop of the failure analysis system, the feedforward neural network module needs to convert the output value of the output node of the neural network failure prediction model into the classification result (normal state or high temperature fault) [11].

The output value of the output layer node is classified by a SoftMax function, also called a normalized exponential function. The essence of the SoftMax function is to normalize the array, highlighting the largest value and suppressing other components far below the maximum [12]. The meaning of the SoftMax function is: use the output value of all output nodes as an array, use the SoftMax function to process the array, get the SoftMax function value of each element in the array, they represent the contribution rate of each element to the array classification. For a neural network, the number of output nodes represents the number of classification categories, and the SoftMax function is: use the output value of all output nodes as an array, use the SoftMax function to process the array, get the SoftMax function value of each element in the array, they represent the contribution rate of each element to the array classification. For a neural network, the number of output nodes represents the number of classification categories, and the output node with the largest contribution represents the classification result. If SoftMax (OUT1) > SoftMax (OUT2), the classification result is 1; otherwise, the classification result is 2. The formula for calculating the SoftMax function is given by Eq. (4).

$$softmax(x_i) = \frac{\exp(x_i)}{\sum\limits_{i=1}^{n} \exp(x_i)} \tag{4}$$

Where xi is the value of each parameter in the sample. Figure 4 shows the simulation results of the SoftMax function module.

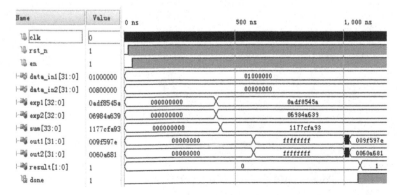

Fig. 4. Simulation result of SoftMax function module

As shown in Fig. 3, data_in1 and data_in2 are the output values OUT1 and OUT2 of the neural network failure prediction model. Exp1 and exp2 are their corresponding exponents. The RTL implementation of the exponential function is completed by instantiating the CORDIC core. The principle of the CORDIC core. See formula (5). Sum is the sum of the indices, which is the denominator of Eq. (3–11). Here out1 and out2 refer to SoftMax (OUT1) and SoftMax (OUT2). It can be seen that SoftMax (OUT1) > SoftMax (OUT2), so the result of result classification is 1.

$$exp(x) = sinh\,x + cosh\,x \qquad (5)$$

3   The Simulation Results and Analysis

Figure 5 and Fig. 6 show the resource application and power consumption report of the feedforward neural network (including the serial port receiving module) based on FPGA.

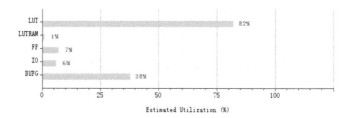

Fig. 5. Resource application of feedforward neural network module

This module occupies 82% of the lookup table (LUT) resources of the Xilinx XC7A100T FPGA, 1% of distributed RAM (LUTRAM) resources, 7% of flip-flops (FF) resources, 6% of I/O resources, and 38% Global Buffer Unit (BUFG) resource.

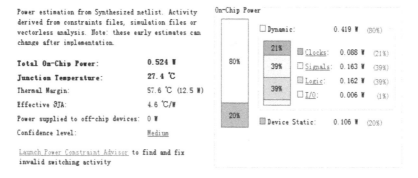

Fig. 6. Power consumption of the feedforward neural network module

The total power consumption of the module is 0.524 W, of which dynamic power consumption is 80% and static power consumption is 20%. The power consumed by clocks, signals, logic cells, and I/Os accounted for 21%, 39%, 39%, and 1% of total power consumption.

4 Conclusions

This paper implement the failure analysis system on the FPGA carrier. The failure analysis system includes a feature parameter processing module and a feedforward neural network module. The feature parameter processing module includes an SPI protocol algorithm, a digital signal recovery fixed point fractional algorithm, a Kalman filter algorithm, and a dispersion normalization algorithm; the feedforward neural network includes the establishment of a network structure. And output classification algorithms are proposed. The Verilog code is used to design each part of the algorithm, and some Xilinx auxiliary IP cores are called. The function simulation, synthesis, gate level simulation, place and route and post-simulation process are sequentially implemented on the vivado platform.

References

1. Fengshuang, L., Hengjing, Z.: Current status and development prospects of astronaut large scale integrated circuit assurance technology. Chin. Aerosp. **06**, 29–32 (2017)
2. Patel, H., Thakkar, A., Pandya, M., et al.: Neural network with deep learning architectures. J. Inf. Optim. Sci. **39**(1), 31–38 (2018)
3. Licheng, J., Shuyuan, Y., Fang, L., Shigang, W., Zhiwei, F.: Neural network for seventy years: retrospect and prospect. Chin. J. Comput. **39**(08), 1697–1716 (2016)
4. Du, T., Zhai, A., Wang, P., Li, Y., Li, P.: An integrated circuit PHM model based on BP neural network. Comput. Eng. Sci. **39**(01), 55–60 (2017)
5. Shen, H., Wang, Z., Gao, C., Qin, J., Yao, F., Xu, W.: Determination of the number of hidden layer elements in BP neural network. J. Tianjin Univ. Technol. **05**, 13–15 (2008)
6. Tian, X.C., Li, J., Fan, Y.B., Yu, X.N., Liu, J.: Design and implementation of SPI communication based-on FPGA. Adv. Mater. Res. **291–294**, 2658–2661 (2011)
7. Khamankar, R.B., McPherson, J.W.: Molecular model for intrinsic time- dependent dielectric breakdown in SiO2 dielectrics and the reliability implications for hyper-thin gate oxide. Semicond. Sci. Technol. **15**(5), 462–470 (2000)
8. Mahapatra, S., Kumar, P.B., Alam, M.A.: Investigation and modeling of interface and bulk trap generation during negative bias temperature instability of p-MOSFETs. IEEE Trans. Electron Devices **51**(9), 1371–1379 (2004)
9. Jian, S.L., Jiang, J.F., Lu, K., Zhang, Y.P.: SEU-tolerant restricted Boltzmann machine learning on DSP-based fault detection. In: International Conference on Signal Processing. IEEE (2014)
10. Chen, B., Li, J.: Research on fault diagnosis in wireless sensor network based on improved wavelet neural network. Acta Technica (2016)
11. He, F.J., Wang, Y.H., Ding, G.Q., Fei, Y.Z.: Radar circuit fault diagnosis based on improved wavelet neural network. Appl. Mech. Mater. **651–653**, 1084–1087 (2014)
12. Rimer, M., Martinez, T.: Softprop: softmax neural network backpropagation learning. In: IEEE International Joint Conference on Neural Networks. IEEE (2004)

Improving Approximate Bayesian Computation with Pre-judgment Rule

Yanbo Wang, Xiaoqing Yu, Pinle Qin, Rui Chai, and Gangzhu Qiao(✉)

School of Data Science and Technology, North University of China,
Taiyuan 030051, Shanxi, China
qiaogangzhu@sohu.com

Abstract. Approximate Bayesian Computation (ABC) is a popular approach for Bayesian modeling, when these models exhibit an intractable likelihood. However, during each proposal of ABC, a great number of simulators are required and each simulation is always time-consuming. The overall goal of this work is to avoid inefficient computational cost of ABC. A pre-judgment rule (PJR) is proposed, which mainly aims to judge the acceptance condition using a small fraction of simulators instead of the whole simulators, thus achieving less computational complexity. In addition, it provided a theoretical study of the error bounded caused by PJR Strategy. Finally, the methodology was illustrated with various examples. The empirical results show both the effectiveness and efficiency of PJR compared with the previous methods.

Keywords: Approximate Bayesian Computation · Bayesian inference · Markov Chain Monte Carlo · Pre-judgment rule

1 Introduction

The crucial component of Bayesian statistics is to estimate the posterior distribution of parameter θ with given observations y. The posterior distribution, denoted as $p(\theta|y)$, satisfies that

$$p(\theta|y) = \frac{p(y|\theta)p(\theta)}{p(y)} \propto p(y|\theta)p(\theta), \tag{1}$$

where $p(y) = \int p(y|\theta)p(\theta)d\theta$ is the normalizing constant and computationally inefficient in general. $p(y|\theta)$ and $p(\theta)$ represent likelihood function and the prior distribution, respectively. However, the likelihood $p(y|\theta)$ is not always intractable due to the lager sample size and high dimension of parameters. Approximate Bayesian Computation (ABC) methods provide likelihood-free approach for performing statistical inferences with Bayesian models [5, 17, 26]. The ABC method replaces the calculation of the likelihood function $p(y|\theta)$ in Eq. (1) with a simulation of the model that produces an artificial data set $\{x_i\}$. The most influential part of ABC is to construct some metric (or distance) and compare the simulated data $\{x_i\}$ to the observed data $\{y_i\}$ [6, 15]. Recently, ABC has gained popularity particularly for the analysis of complex problems arising out of biological sciences (e.g. in population genetics, ecology, epidemiology, and systems biology) [5, 24, 27].

© Springer Nature Singapore Pte Ltd. 2020
J. Zeng et al. (Eds.): ICPCSEE 2020, CCIS 1257, pp. 230–246, 2020.
https://doi.org/10.1007/978-981-15-7981-3_15

There are at least three leaps in the development of ABC, we denote as algorithms $\mathbb{A}$, $\mathbb{B}$ and $\mathbb{C}$. Algorithms of type $\mathbb{A}$, the simplest algorithm of ABC proposed in [25], is listed as follows:

- $\mathbb{A}$1. Sample θ from the prior distribution $p(\theta)$.
- $\mathbb{A}$2. Accept the proposed θ with probability h proportional to $p(y|\theta)$. Return to $\mathbb{A}$1.

Concretely, if θ^* denotes the maximum-likelihood estimator of θ, the acceptance probability h can be directly set as:

$$h = \frac{p(y|\theta)}{c}, \tag{2}$$

where c can be any constant greater than $p(y|\theta^*)$. Unfortunately, the likelihood function $p(y|\theta)$ is computationally expensive or even intractable. Hence Algorithm $\mathbb{A}$1 is not practical.

Many variants are proposed, among which one common approach is algorithms of type $\mathbb{B}$ [19]:

- $\mathbb{B}$1. Sample θ from the prior distribution $p(\theta)$.
- $\mathbb{B}$2. Generate x given the parameter θ via the simulator, i.e., $x \sim p(\cdot|\theta)$.
- $\mathbb{B}$3. Accept the proposed θ if $x = y$. Return to $\mathbb{B}$1.

The success of algorithm $\mathbb{B}$ depends on the fact that simulating from $p(\cdot|\theta)$ is easy for any θ, a basic assumption of ABC. To discriminate simulated data x from the observation y, we call x pseudo-observation here. Moreover, in Step $\mathbb{B}$3, $\mathbb{S}(x) = \mathbb{S}(y)$ is employed instead of $x = y$ in practice, where $\mathbb{S}(x)$ represents the summary statistics of x. It has been shown that if the statistics used in likelihood function are sufficient, then Algorithm $\mathbb{B}$ sample correctly from the true posterior distribution. Here, for ease of exposition, we use $x = y$ instead of $\mathbb{S}(x) = \mathbb{S}(y)$. Whereas the acceptance criteria $x = y$ is too restrictive here, leading the acceptance rate intolerably small. One might resort to relaxing the criteria as algorithm $\mathbb{C}$ [21]:

- $\mathbb{C}$1. Sample θ from the prior distribution $p(\theta)$.
- $\mathbb{C}$2. Generate x given the parameter θ via the simulator, i.e., $x \sim p(\cdot|\theta)$.
- $\mathbb{C}$3. Calculate the similarity between observations y and simulated data x, denoted $\rho(x, y)$[1].
- $\mathbb{C}$4. Accept the proposed θ if $\rho(x, y) \geq \xi$ (ξ is a prespecified threshold). Return to $\mathbb{C}$1.

Notice that in Step $\mathbb{C}$2, a quantity of pseudo-observations x are simulated from $p(\cdot|\theta)$ independently, i.e., $x = \{x_1, ..., x_S\}, x_i \sim p(\cdot|\theta)$ $i.i.d.$, where S is the number of simulators in each proposal and always fixed, independent of θ. The similarity $\rho(x, y)$ can be represented in terms of the average similarity between x_i and y such that

$$\rho(x, y) = \frac{1}{S} \sum_{i=1}^{S} \pi_\zeta(x_i|y),$$ where $\pi_\zeta(\cdot|y)$ is an ζ-kernel around observation y[2].

It is apparent that the choice of S plays a critical role in the efficiency of the algorithm. Obviously a large S will degrade the efficiency of ABC. In contrast, if S is small,

---

[1] $\rho(\mathbb{S}(x), \mathbb{S}(y))$ is replaced by $\rho(x, y)$, similar with Step $\mathbb{C}$3.
[2] E.g., ζ-kernel can be chosen as $\pi_\zeta(x_1|x_2) = (1/\sqrt{2\pi}\zeta) \exp(-\|x_1 - x_2\|^2/2\zeta^2)$.

though leading a significant reduction for each θ in computation, the samples may fail to converge to the target distribution [4]. Moreover, it is awful to spend amounts of computation (S simulations) for just 1 bit information, namely accept or reject the proposal. A natural question is proposed: can we simulate a small number of pseudo-observations in Step $\mathbb{C}2$ and maintain the convergence to the target distribution simultaneously? Or can we find a tradeoff between efficiency and accuracy? Here, we claim it is feasible.

In this paper, we devise Pre-judgment (PJR) rule, adjusting number of simulators dynamically, instead of using a constant S. In short, we firstly generate small amount of data and estimate a rough similarity. If the similarity is far away from the prespecified threshold (say, in Step $\mathbb{C}4$, ξ), then we judge (accept/reject) the proposal ahead. Otherwise, we draw more data from the simulator and repeat the evaluation until we have enough evidence to make the decision. Empirical results show that majority of these decision can be made based on a small amount of simulators with high confidence, thus lots of computations are saved.

The remainder of the paper is organized as follows. Section 2 describes our algorithm and Sect. 3 provides theoretical analysis. A toy model is shown in Sect. 4.1 to show some properties of PJR based method. Furthermore, the empirical evaluations are given in Sect. 4.2. Finally, the last section is devoted to conclude the paper.

2   Methodology

In this section, we will review the relative works and then present our method. Firstly, we introduce how pre-judgment rule (PJR) accelerate ABC rejection method. Then we adapt PJR strategy to ABC-MCMC framework [20].

2.1   Related Works

In this section, we briefly review the related studies. Firstly, we focus on recent developments in ABC community. Though allowing parallel computation, ABC is still in its infancy owing to the large computational cost. Many approaches are proposed to scale up ABC in machine learning community. Concretely, [22,29] introduced Gaussian process to accelerate ABC. [23] made use of the random seed in sampling procedure and transform ABC sampler into an deterministic optimization procedure. [21] adapted Hamiltonian Monte Carlo to ABC scenario, allowing noise in estimated gradient of log-likelihood by borrowing the idea from stochastic gradient MCMC framework [1,2,11,12,18,28] and pseudo-marginal MCMC methods [3,14].

In addition, theoretical works has become popular recently [4,7,8,30]. Some works focus on the selection of summary statistics [9,13]. Different from these methods, PJR strategy essentially alleviates the computational burden in ABC rejection step, which can be extended to any ABC scenario, e.g., ABC rejection approach and ABC-MCMC proposed in this paper.

2.2 PJR Based ABC: (PJR-ABC)

In the Algorithm A, the likelihood is not available explicitly. Thus we resort to approximate methods by introducing the simulated data x, as follows:

$$p(y|\theta) = \int \delta_D(x - y)p(x|\theta)dx \approx \int \pi_\zeta(x|y)p(x|\theta)dx \approx \frac{1}{S}\sum_{i=1}^{S} \pi_\zeta(x_i|y), \qquad (3)$$

where $\delta_D(\cdot)$ is the Dirac delta function. Then a relaxation is employed by introducing an ζ-kernel around the observation y. The last approximate equality use a Monte Carlo estimate of the likelihood via S draws of x from simulator $p(\cdot|\theta)$.

On the other hand, for Algorithm $\mathbb{C}$, the similarity between pseudo-observations x and raw observations y can be expressed as the mean similarity between each simulator output x_i and y

$$\rho(x, y) = \frac{1}{S}\sum_{i=1}^{S} \pi_\zeta(x_i|y). \qquad (4)$$

From Eq. (3) and (4), it is validated that Algorithm $\mathbb{A}$ is equivalent to Algorithm $\mathbb{C}$ in essence. Then acceptance conditions in both Step $\mathbb{A}2$ and Step $\mathbb{C}4$ are equivalent to performing a comparison (between z and z_0, defined later). Specifically, firstly we compute $z = \frac{1}{S}\sum_{i=1}^{S} \pi_\zeta(x_i|y)$, where $x_i \sim p(\cdot|\theta)$ $i.i.d.$, and then compare it with z_0, a constant. If $z > z_0$, accept the proposed θ. If $z \leq z_0$, reject it, where z_0 is a prespecified threshold, say, in Step $\mathbb{C}4$, z_0 corresponds to ξ^3.

To guarantee the convergence to the true posterior, S should be a large number, which means each proposal needs S simulations [4]. However, spending quantities of computation (i.e., simulating S pseudo-data $x_1, \ldots, x_S$) to get just one bit of information, namely whether to accept or reject a proposal, is likely not the best use of computational resources.

To address this issue, PJR is devised to speedup the ABC procedure. We are willing to tolerate small error in this step to achieve faster judgement. In particular, we firstly draw a small number of pseudo-observations x and estimate a rough z. If the difference between z and z_0 is significantly larger than the standard deviation of z, we claim that z is far away enough from z_0 confidently and make the decision by comparing the rough z with z_0. Otherwise, we draw more pseudo-observations to increase the precision of z until we have enough evidence to make the decision.

More formally, checking the acceptance condition can be reformulated to the following statistical hypothesis test.

$$H_1 : z > z_0, \quad H_2 : z \leq z_0.$$

In order to test the hypothesis, we are able to generate infinitely many pseudo-observations from $p(\cdot|\theta)$. On the other hand, we expect to simulate less pseudo-observations owing to computational cost.

---

[3] In Step A2, z_0 is more complex. Checking the acceptance condition is equivalent to judging $\frac{z}{c} > u$, where c is defined in Eq. 2 and $u \sim \text{Uniform}(0, 1)$.

To do this, we proceed as follows. We compute the sample mean $\bar{z}$ and sample standard deviation s_z as

$$z_i = \pi_\zeta(x_i|y), \quad \bar{z} = \tfrac{1}{n}z_i, \quad s_z = \sqrt{\tfrac{\overline{z^2}-(\bar{z})^2}{n-1}}, \tag{5}$$

where $\overline{z^2}$ represents the mean of z^2. Then we compute the test statistics t via

$$t = \tfrac{\bar{z}-z_0}{s_z}. \tag{6}$$

It is assumed that n is large enough here. Under this situation central limit theorem (CLT) kicks in and the test statistic t follows the standard Student-t distribution with $n-1$ degrees of freedom. Note that when n is large enough, Student-t distribution with $n-1$ degrees of freedom is close to the standard normal distribution. Then we compute η defined as:

$$\eta = 1 - \psi_{n-1}(|t|), \tag{7}$$

where $\psi_{n-1}(\cdot)$ is the cdf of the standard Student-t distribution with $n-1$ degrees of freedom.

Then we provide a threshold ϵ, e.g., $\epsilon = 0.1$. If $\eta < \epsilon$, we make a decision that z is significantly different from z_0. Then we accept/reject θ via comparing $\bar{z}$ and z_0. If $\eta \geq \epsilon$, it means that we do not have enough evidence to decide. Thus more pseudo-observations are drawn to reduce the uncertainty of z. Note that when S pseudo-observations are drawn, the procedure would be terminated and it reduces to previous ABC algorithm. The resulting algorithm can be seen in Algorithm 1.

The advantage of PJR-ABC is that we can often make confident decisions with s_i ($s_i \ll S$) pseudo-observations and reduce computation significantly. Though PJR-ABC brings error in judgement, we can use the computational time we save to draw more samples to offset the small bias. Worth to note that ϵ can be regarded as a knob. When ϵ approaches to 0, we make almost the same decision with the ABC rejection method but requires masses of simulators. On the other hand, when ϵ is high, we make decisions without sufficient evidence and the error would be high. This accuracy-efficiency trade-off will be empirically verified in Sect. 4.1.

2.3   PJR Based Markov Chain Monte Carlo Version of ABC: PJR-ABC-MCMC

The ABC rejection methods are easy to implement and compatible with embarrassingly parallel computation. However, when the prior distribution is long way from posterior distribution, most of the samples from prior distribution would be rejected, leading acceptance rate too small, especially in high-dimensional problem. To address this issue, a Markov Chain Monte Carlo version of ABC (ABC-MCMC) algorithm is proposed [20]. It is well-known that MCMC has been the main workhorse of Bayesian computation since 1990s and many state-of-the-art samplers in MCMC framework can be extended into ABC scenario, e.g., Hamiltonian Monte Carlo can be extended to Hamiltonian ABC [21]. Hence ABC-MCMC [20] is a benchmark in ABC community. Now we show that our PJR rule can be adapted to the ABC-MCMC framework. First, ABC-MCMC is briefly introduced:

Algorithm 1. PJR-ABC

Require: θ drawn from prior $p(\theta)$, $\{s_i\}_{i=0}^k$: a strictly increasing sequence satisfying that $s_i \in$ $\mathbb{N}_+$, $s_0 = 0$ and $s_k = S$. knob ϵ.

Ensure: accept/reject θ

1: **for** $i = 1 : k$ **do**

2:     draw $|s_i - s_{i-1}|$ pseudo-observations $x_{s_{i-1}+1}, x_{s_{i-1}+2}, \ldots, x_{s_i}$ from simulator $p(\cdot|\theta)$, compute the corresponding $z_{s_{i-1}+1}, \ldots, z_{s_i}$ and store, where $z_i = \pi_\zeta(x_i|y)$.

3:     Set $n = s_i$.

4:     Update the mean $\bar{z}$ and std s_z using Equation (5).

5:     Compute the test statistics t via Equation (6).

6:     Compute η via Equation (7).

7:     **if** $\eta < \epsilon$ **then**

8:         **if** $\bar{z} > z_0$ **then**

9:             accept the proposed θ and break.

10:         **else**

11:             reject the proposed θ and break.

12:         **end if**

13:     **end if**

14: **end for**

15: **if** $\bar{z} > z_0$ **then**

16:     accept the proposed θ.

17: **else**

18:     reject the proposed θ.

19: **end if**

- $\mathbb{D}1$. Given the current point θ, θ' is proposed according to a transition kernel $q(\theta'|\theta)$.
- $\mathbb{D}2$. Generate x' from the simulator $p(\cdot|\theta')$.
- $\mathbb{D}3$. Compute the acceptance probability α defined in Eq. 8.
- $\mathbb{D}4$. Accept θ' with probability α. Otherwise, stay at θ. Return to $\mathbb{D}1$.

In MCMC sampler, MH acceptance probability α is defined as

$$\alpha = \min \left\{ 1, \frac{p(\theta')p(y|\theta')q(\theta|\theta')}{p(\theta)p(y|\theta)q(\theta'|\theta)} \right\}. \tag{8}$$

In likelihood-free scenario, the acceptance probability of ABC-MCMC is

$$\alpha = \min \left\{ 1, \frac{p(\theta') \sum\limits_{s=1}^{S} \pi_\zeta(x'_s|y)q(\theta|\theta')}{p(\theta) \sum\limits_{s=1}^{S} \pi_\zeta(x_s|y)q(\theta'|\theta)} \right\},$$

where $x_s \sim p(\cdot|\theta)$ *i.i.d.* and $x'_s \sim p(\cdot|\theta')$ *i.i.d.* The acceptance of proposal is determined by following form:

$$u < \alpha = \min \left\{ 1, \frac{p(\theta') \sum\limits_{s=1}^{S} \pi_\zeta(x'_s|y)q(\theta|\theta')}{p(\theta) \sum\limits_{s=1}^{S} \pi_\zeta(x_s|y)q(\theta'|\theta)} \right\},$$

where $u \sim$ Uniform$(0, 1)$. This is equivalent to the following expression:

$$u < \frac{p(\theta')\frac{1}{S}\sum\limits_{s=1}^{S} \pi_\varsigma(x'_s|y)q(\theta|\theta')}{p(\theta)\frac{1}{S}\sum\limits_{s=1}^{S} \pi_\varsigma(x_s|y)q(\theta'|\theta)}.$$

Note that $\{x_1, ..., x_S\}$ is given in ABC-MCMC, then define the fixed part z_0 and test variable z, we obtain that

$$z_0 = \frac{p(\theta)}{p(\theta')}\frac{1}{S}\sum\limits_{s=1}^{S} \pi_\varsigma(x_s|y)\frac{q(\theta'|\theta)}{q(\theta|\theta')}u, \quad z = \frac{1}{S}\sum\limits_{s=1}^{S} \pi_\varsigma(x'_s|y),$$

where z can be further simplified into the following form, similar to PJR-ABC: $z = \frac{1}{S}\sum\limits_{i=1}^{S} z_i$, where $z_i = \pi_\varsigma(x'_i|y)$.

Following PJR-ABC, we test the following hypothesis $H_1 : z_0 > z$ vs $H_2 : z_0 < z$. Then the sample mean $\bar{z}$, the sample standard deviation s_z and the test statistics t can be calculated as shown in Eq. (5) and (6), same with PJR-ABC. The resulting algorithm is similar and not listed.

3   Theoretical Analysis

In this section, we study the theoretical properties for PJR strategy. Specifically, we provide the error analysis for both PJR-ABC and PJR-ABC-MCMC. Since every time we accept/reject a proposal in PJR-ABC/PJR-ABC-MCMC, we deal with a hypothesis testing problem. We are attempting to bound the error caused by such a testing problem first. Then we build the relationship between such a single test error and total error for both PJR-ABC and PJR-ABC-MCMC. Now we focus on the error caused by a single testing problem. In hypothesis testing problem, two types of error are distinguished. A type I error is the incorrect rejection of a true hypothesis while the type II error is the failure to reject a false hypothesis. Now we discuss the probabilities of these two errors in a single decision problem.

Theorem 1. *The probability of both the error I and II decreases approximately exponentially w.r.t. the sample size of z (sample size of z corresponds to $s_1, \ldots, s_k$ in Algorithm 1).*

Proof. We assume that $\psi_{n-1}(\cdot)$ is the cdf of standard Student-t distribution with degree $n - 1$. For simplicity, we first discuss the probability of type I error, i.e., the incorrect rejection of a true hypothesis. It would be easy to extend the conclusion into the type II error owing to the symmetry.

In this case, $z > z_0$. Suppose the number of sampled z is n. The test statistics t satisfies that $t = \frac{\bar{z}-z_0}{s_z}$, following the standard Student-t distribution with degree $n - 1$. The standard Student-t distribution is approaching to the standard normal distribution when the degree $n - 1$ is large enough. Hence, many properties of normal distribution can be shared.

Given the knob parameter ϵ, according to the monotonicity of the function $\psi_{n-1}(\cdot)$ on $\mathbb{R}$, we know that there exists a unique s such that $\psi_{n-1}(s) = \epsilon$. Moreover, since $\bar{z} = \frac{z_1+z_2+\dots+z_n}{n}$ and $t = \frac{\bar{z}-z_0}{s_z} \sim \psi_{n-1}(\cdot) \approx \mathcal{N}(0,1)$, we have that z_i can be seen as sampled independent identically distributed from $\mathcal{N}(z_0, ns_z)$, i.e., $z_i \sim \mathcal{N}(z_0, ns_z)$ $i.i.d.$

The type I error only occurs when $\frac{\bar{z}-z_0}{s_z} < s$. That is, $\sum_{i=1}^{n} z_i < n(s_z s + z_0)$. Thus, we can have the probability of type I error via integrating over the space $(z_1, z_2, \dots, z_n)$ and $\sum_{i=1}^{n} z_i < n(s_z s + z_0)$.

$$
\begin{aligned}
\Pr(\text{Type I error}) &= \Pr(\textstyle\sum_{i=1}^{n} z_i < n(s_z s + z_0)) \\
&= \int \cdots \int \int_{-\infty}^{(z_1,z_2,\cdots,z_n),\sum_{i=1}^{n} z_i < n(s_z s+z_0)} \psi'(z_1)\psi'(z_2)\dots\psi'(z_n)dz_1 dz_2 \dots dz_n \\
&= \int_{-\infty}^{n(s_z s+z_0)-z_1-\dots-z_{n-1}} \cdots \int \int_{-\infty}^{z_1} \psi'(z_1)\psi'(z_2)\dots\psi'(z_n)dz_1 dz_2 \cdots dz_n \\
&= \psi_{n-1}(z_1)\psi_{n-1}(z_2)\dots\psi_{n-1}(n(s_z s + z_0) - z_1 - \dots - z_{n-1})
\end{aligned}
$$

where $\psi'(\cdot)$ and $\psi_{n-1}(\cdot)$ represent the pdf and cdf of the standard Student-t distribution with $n-1$ degree of freedom.

This completes the proof.

The above theorem demonstrates that the error during a single judge can be negligible as long as the number of sampled z is large enough. Based on this theorem, the following assumption are reasonable.

Assumption 1. The probability of error produced by a single hypothesis testing problem in both PJR-ABC and PJR-ABC-MCMC can be upper-bounded, denoted by $\delta_1, \delta_2 \to 0_+$, for PJR-ABC and PJR-ABC-MCMC, respectively.

In Bayesian inference, we are interested in the posterior average, defined as $\bar{\phi} \triangleq \int_{\theta} \phi(\theta)p(\theta|y)d\theta$ for some test function $\phi(\theta)$ of interest. For a given numerical method (say, PJE-ABC or PJR-ABC-MCMC) with generated samples $\{\theta_1, \dots, \theta_M\}$, we use the sample average $\hat{\phi}$ defined as $\hat{\phi} = 1/M \sum_{l=1}^{M} \phi(\theta_l)$ to approximate $\bar{\phi}$. Before providing a bound for the bias of a PJR-ABC algorithm, we make a mild assumption first.

Assumption 2. The prior average of $\phi(\cdot)$ is bounded away from infinity, i.e.,

$$
\int_{\theta} \phi(\theta)p(\theta)d\theta < +\infty.
$$

Theorem 2. *Under Assumption (1) and (2), the bias of PJR-ABC can be upper-bounded as: $|\mathbb{E}\hat{\phi} - \bar{\phi}| \leq C_1 \delta_1$, where $C_1 = \frac{\int_{\theta} \phi(\theta)p(\theta)d\theta}{p(y)}$ is a constant, $p(y)$ denotes the normalizing constant.*

Proof. In ABC rejection method, each θ drawn from $p(\theta)$ is independent. The error at θ caused by PJR is denoted by $\xi(\theta)$, which is assumed to be a perturbation on the true likelihood. Thus the estimated likelihood function can be represented as $\hat{p}(y|\theta) = p(y|\theta) + \xi(\theta)$, where $|\xi(\theta)| \leq \delta_1$ owing to the boundedness of single error, described in Assumption 3.

$$
\begin{aligned}
\mathbb{E}(\hat{\phi}) &= \tfrac{1}{p(y)} \int \phi(\theta)\hat{p}(y|\theta)p(\theta)d\theta \\
&= \tfrac{1}{p(y)} \int \phi(\theta)p(y|\theta)p(\theta)d\theta + \tfrac{1}{p(y)} \int \phi(\theta)\xi(\theta)p(\theta)d\theta
\end{aligned} \tag{9}
$$

The first term in RHS of Eq. (9) is the expectation of the true posterior distribution. While the second term is the error. We can observe that the error is upper bounded.

$$\frac{1}{p(y)} \int \phi(\theta)\xi(\theta)p(\theta)d\theta \le \frac{1}{p(y)}|\delta_1| \int \phi(\theta)p(\theta)d\theta = C_1|\delta_1|,$$

where $C_1 = \frac{\int \phi(\theta)p(\theta)d\theta}{p(y)}$ is bounded followed from the fact that both $\frac{1}{p(y)}$ and $\int \phi(\theta)p(\theta)d\theta$ are bounded away from $+\infty$.

This completes the proof.

In PJR-ABC, each sample is independent with each other. However, in PJR-ABC-MCMC, all the samples are in a single chain, leading the analysis more complicated. Here, the distance between probability distributions is measured by the total variational distance (TVD),[4] described as follows.

Theorem 3. *Under Assumption 3, for any posterior distribution, there exists a constant C_2 such that the discrepancies between the true posterior distribution S_0 and the stationary distribution of our PJR-ABC-MCMC algorithm S_ϵ can be upper bounded as:* $d_v(S_0, S_\epsilon) \le C_2\delta_2.$

Proof. We firstly focus on the error for a single step. Based on this, the error about the stationary distribution is derived. The transition kernel of the ABC-MCMC algorithm can be written as

$$\mathcal{T}_0(\theta, \theta') = P_a(\theta, \theta')q(\theta'|\theta) + (1 - P_a(\theta, \theta'))\delta_D(\theta' - \theta),$$

where $\delta_D(\cdot)$ is the Dirac delta function, $P_a(\theta, \theta')$ is the acceptance probability. Similar definition of transition kernel of PJR-ABC-MCMC hold for $\mathcal{T}_\epsilon(\theta, \theta')$ and acceptance probability $P_{a,\epsilon}(\theta, \theta')$.

The discrepancies between $P_a(\theta, \theta')$ and $P_{a,\epsilon}(\theta, \theta')$ is defined as: $\delta P_a(\theta, \theta') \triangleq P_{a,\epsilon}(\theta, \theta') - P_a(\theta, \theta')$. For every (θ, θ'), according to the error for a single test, there exists an upper bound for $\delta P_a(\theta, \theta')$, i.e., $|\delta P(\theta, \theta')| \le \delta_{max}$ for $\forall (\theta, \theta')$.

Then the total variational distance for a single step can be upper bounded for any distribution P as:

$$
\begin{aligned}
\int_{\theta'} |(\mathcal{P}\mathcal{T}_\epsilon)(\theta') - (\mathcal{P}\mathcal{T}_0)(\theta')|d\Omega(\theta') &= \int_{\theta'} |\int_\theta (\mathcal{T}_0(\theta,\theta') - \mathcal{T}_\epsilon(\theta,\theta'))dP(\theta)|d\Omega(\theta') \\
&= \int_{\theta'} |\int_\theta (\mathcal{T}_0(\theta,\theta') - \mathcal{T}_\epsilon(\theta,\theta'))dP(\theta)|d\Omega(\theta') \\
&= \int_{\theta'} |\int_\theta (q(\theta'|\theta) - \delta_D(\theta' - \theta))(\delta P(\theta,\theta'))dP(\theta)|d\Omega(\theta') \\
&\le \int_{\theta'} \int_\theta |(q(\theta'|\theta) - \delta_D(\theta' - \theta)| \cdot |\delta_{max}| \cdot dP(\theta)d\Omega(\theta') \\
&\le \delta_{max} \int_{\theta'} |\int_\theta q(\theta'|\theta)dP(\theta')|d\Omega(\theta') \\
&\quad + \delta_{max} \int_{\theta'} |\int_\theta \delta_D(\theta' - \theta)dP(\theta')|d\Omega(\theta') = 2\delta_{max}
\end{aligned}
$$

Then apply Lemma 1, substitute $2\delta_{max}$ into δ in Eq. 10 we prove Theorem 3. This completes the proof.

---

[4] The total variation distance between two distribution P and Q, absolutely continuous w.r.t. measure Ω, is defined as $d_v(P, Q) \triangleq 1/2 \int_\theta |f_P(\theta) - f_Q(\theta)|d\Omega(\theta)$, where $f_P(\cdot)$ and $f_Q(\cdot)$ are their respective densities.

Lemma 1 [16]. *Given two transition kernels, T_0 and T_ϵ, whose stationary distributions are denoted by S_0 and S_ϵ, if T_0 satisfies the following contraction condition with a constant $\eta \in [0, 1)$ for all probability distribution $\mathcal{P}$:*

$$d_v(\mathcal{P}T_0, S_0) \leq \eta d_v(\mathcal{P}, S_0)$$

and the one step error between T_0 and T_ϵ is upper bounded uniformly with a constant $\delta > 0$ as:

$$d_v(\mathcal{P}T_0, \mathcal{P}T_\epsilon) \leq \delta, \forall \mathcal{P} \tag{10}$$

then the distance between S_0 and S_ϵ can be bounded as: $d_v(S_0, S_\epsilon) \leq \frac{\delta}{1-\eta}$

Theorem 2 and 3 indicate that the error is proportional to the single testing error. Combining this result with Theorem 1, we know that the bias of both PJR-ABC and PJR-ABC-MCMC can be bounded.

4 Numerical Validation

In this section, we use a toy model to demonstrate both PJR-ABC and PJR-ABC-MCMC.

4.1 Synthetic Data

We adopt the gamma prior with shape α and rate β, i.e., $p(\theta) = \mathrm{Gamma}(\alpha, \beta)$. The likelihood function is exponential distribution, i.e., $x \sim \exp(1/\theta)$. Let observations are generated via $y = \frac{1}{N} \sum_{i=1}^{N} e_i$, where $e_i \sim \exp(1/\theta^*)$, N is the number of observations. Regarding the selection of the sequence $\{s_i\}_{i=1}^k$ ($s_0 = 0$), we find geometric sequence is the usually the best choice, thus is used in both Sect. 4.1 and 4.2. The common ratio of the geometric sequence is usually set to 1.5–2. The true posterior is a gamma distribution with shape $\alpha + N$ and rate $\beta + Ny$, i.e., $p(\theta|y) = \mathrm{Gamma}(\alpha + N, \beta + Ny)$. In particular, we set $S = 1000$, $N = 20$, $y = 7.74$, $\alpha = \beta = 1$, $\theta^* = 0.15$ in this scenario. We run chains of length 50K for ABC-MCMC and PJR-ABC-MCMC and 100K for ABC and PJR-ABC. For each method, we conduct 5 independent trials and report the average value. In this paper, the choice of proposal distribution in both ABC-MCMC and PJR-ABC-MCMC is a Gaussian distribution centered at current θ.

First, we investigate how the performance (both efficiency and accuracy) changes as a function of the knob ϵ empirically. For each $\epsilon \in \{0, 0.01, 0.03, 0.07, 0.1, 0.2, 0.3\}$, we record both efficiency[5] and accuracy[6]. $\epsilon = 0$ means the PJR-ABC/PJR-ABC-MCMC reduce to ABC/ABC-MCMC approach. The results are reported in Fig. 1. We find that smaller ϵ usually leads to higher accuracy and less efficiency, validating the statement about ϵ mentioned in Sect. 2. Hence, the empirical trade-off between efficiency and accuracy can be controlled by adjusting ϵ. In the following, we set $\epsilon = 0.1$. In Fig. 3, we show the trace plots of the last 1K samples from a single chain for ABC-MCMC

---

[5] Measured in term of number of simulator.

[6] Measured in term of TVD with the true posterior distribution.

and PJR-ABC-MCMC. It is a positive result, indicating PJR-ABC-MCMC preserve the ability of exploration to the parameter space compared with ABC-MCMC. The empirical histograms of θ for all the methods are presented in Fig. 2. We find that all of them are close to the desired posterior. In Table 1 we show

- the average Total Variational Distance[7] (between the true posterior and the ABC posteriors) and the corresponding standard deviation using the first 10K samples and whole chain;
- the average number of simulators.

We can observe that our PJR based ABC rejection and ABC-MCMC achieve similar result with original algorithm in convergence to the target posterior distribution. Furthermore, PJR strategy can accelerate both ABC and ABC-MCMC in terms of number of simulators.

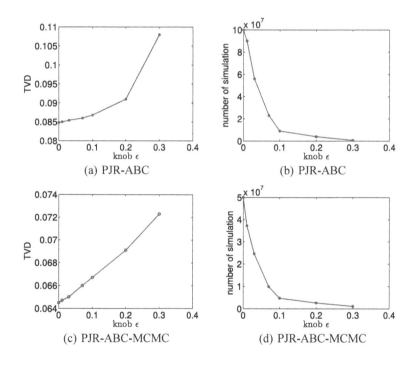

Fig. 1. Demonstration problem. TVD and number of simulations as a function of the knob ϵ.

4.2   Real Applications

The Popular Ricker Model. In this section, we show the application of our method on the popular Ricker model [31]. The Ricker model, a classic discrete population model

---

[7] Note that in experiment the total variational distance is estimated empirically owing to the absence of explicit formulae.

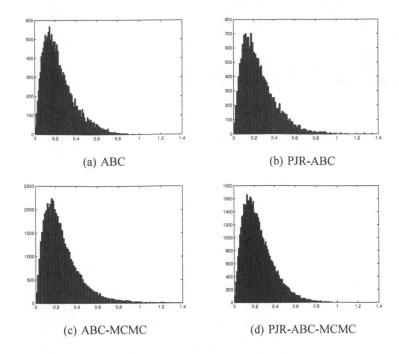

(a) ABC

(b) PJR-ABC

(c) ABC-MCMC

(d) PJR-ABC-MCMC

Fig. 2. Demonstration problem. The empirical histograms of θ for all the methods.

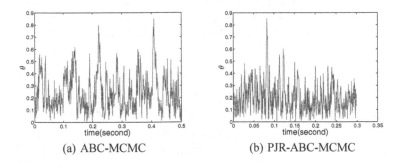

(a) ABC-MCMC

(b) PJR-ABC-MCMC

Fig. 3. Demonstration problem. Trace plot of last 1K samples, where $\epsilon = 0.1$.

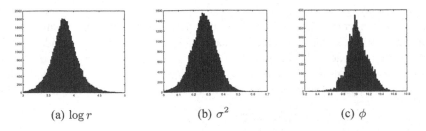

(a) $\log r$

(b) σ^2

(c) ϕ

Fig. 4. Ricker model. Empirical histogram of parameter $\theta = (\log r, \sigma, \phi)$ generated by ABC-MCMC.

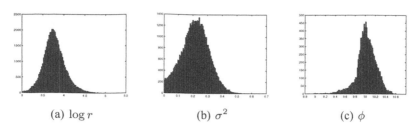

(a) $\log r$ (b) σ^2 (c) ϕ

Fig. 5. Ricker model. Empirical histogram of parameter $\theta = (\log r, \sigma, \phi)$ generated by PJR-ABC-MCMC.

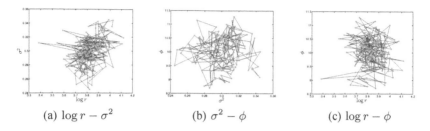

(a) $\log r - \sigma^2$ (b) $\sigma^2 - \phi$ (c) $\log r - \phi$

Fig. 6. Ricker model. Trajectories of each pair of two parameters over the last 200 time-steps generated by our PJR-ABC-MCMC.

Table 1. Results for the demonstration problem in terms of TVD (Total Variational Distance) and number of simulators. Note that for TVD the value below is the actual value times 100 (mean ± std). Simulators represent the total number of pseudo-observations from the simulator. For the first two approaches, we draw 100K samples while for the last two approaches, 50K samples are drawn.

Algorithm	10K	Whole chain	Simulators
ABC	8.48 ± 0.89	6.15 ± 0.03	100M
PJR-ABC	8.68 ± 0.45	6.11 ± 0.02	**9M**
ABC-MCMC	6.43 ± 0.03	5.96 ± 0.04	50M
PJR-ABC-MCMC	6.48 ± 0.07	6.03 ± 0.04	**4.7M**

used in ecology, gives the expected number of individuals in current generation as a function of number of individuals in previous generation. This model is commonly used as an exampler of complex model [29] because it cause the collapse of standard statistical methods due to near-chaotic dynamics [31]. In particular, N_t denote the unobserved number of individuals in the population at time t while the number of observed individuals is denoted by Y_t. The Ricker model is defined via the following relationships [31]

$$N_{t+1} = rN_t\exp(-N_t + e_t), \quad Y_t \sim \text{Poisson}(\phi N_t),$$
$$e_t \sim \mathcal{N}(0, \sigma^2),$$

where each e_t $(t = 1, 2, ...,)$ is independent and Y_t only depends on N_t. In this model, the parameter vector is $\theta = \{\log r, \sigma^2, \phi\}$. $y_{1:T} = \{y_1, ..., y_T\} \in \mathbb{R}^T$ is the time-series of observations. For each parameter, we adopt the uniform prior as

$$\log r \sim \text{Uniform}(3, 6), \quad \sigma \sim \text{Uniform}(0, 0.8),$$
$$\phi \sim \quad \text{Uniform}(4, 20).$$

The target distribution is the posterior of θ given observations $y_{1:T}$, i.e., $p(\theta|y_{1:T})$. Artificial dataset is generated using $\theta^* = (3.8, 0.3, 10.0)$. We compare PJR-ABC-MCMC method with ABC-MCMC. For ABC-MCMC, we run the simulator $S = 2000$ times at each θ to approximate the likelihood value. The knob ϵ is set to be 0.1. For summary statistics, we follow the methods described in [29], which contain a collection of phase-invariant measures, such as coefficients of polynomial autogressive models.

Effectiveness: Figure 4 and 5 show the empirical histogram of parameter of interest $\theta = (\log r, \sigma, \phi)$ generated by ABC-MCMC and PJR-ABC-MCMC, respectively. Furthermore, we present the scatter plots of trajectories for every two parameters in Fig. 6. We can observe that the mode of the empirical posterior is close to the θ^* and the posteriors produced by the two algorithms are similar, showing the success of PJR-ABC-MCMC in Ricker model.

Efficiency: The simulation procedure is complex and dominate in computational time. Therefore, the running time of samplers is almost proportional to the number of simulators. Specifically, sampling 1K parameters, ABC-MCMC requires 2M simulators ($S = 2000$) while PJR-ABC-MCMC only requires about 371K simulators. We conclude that majority of the decision can be made based on a small amount of simulators with high confidence. Hence, our PJR strategy accelerates ABC-MCMC algorithm greatly in Ricker model.

4.3 Apply to HABC-SGLD

In this part, we apply our method to SGLD (Stochastic Gradient Langevin Dynamics, [28]) version of HABC (Hamiltonian ABC) proposed in [21].

In each iteration of SGLD, a mini-batch $\mathcal{X}_n$ of size n is drawn to estimate the gradient of log-posterior. The proposal is

$$\theta' \sim q(\cdot|\theta, \mathcal{X}_n) = \mathcal{N}(\theta + \tfrac{\alpha}{2}\nabla_\theta\{\tfrac{N}{n}\sum_{i\in\mathcal{X}_n}\log p(x_i|\theta) + \log p(\theta)\}, \alpha)$$

It can be shown that when the stepsize α approaches to zero, the acceptance probability approaches to 1 [28]. Based on this, the MH correction step is ignored. However, the assumption that $\alpha \to 0$ is too restrictive. In practice, to keep the mixing rate high, we always choose a reasonably large α. Under this situation, SGLD can not converge to target distribution in some cases. The detailed reasons can be found in [16].

In ABC scenarios, conventional MH rejection step is time-consuming. So our method fit to this problem naturally. Specifically, we consider an L1-regularized linear regression model. This model has been used in [16] to explain the necessity of MH rejection in SGLD. We explore its effectiveness in ABC scenario.

Given a dataset $\{u_i, v_i\}_{i=1}^N$, where u_i are the predictors and v_i are the targets. Gaussian error model and Laplacian prior for parameter $\theta \in \mathbb{R}^D$ are adopted, i.e., $p(v|u, \theta) \propto \exp(-\frac{\lambda}{2}(v - \theta^T u)^2)$ and $p(\theta) \propto \exp(-\lambda_0 \|\theta\|_1)$. We generate a synthetic dataset of size $N = 10000$ via $v_i = \theta_0^T u_i + \xi$, where $\xi \sim \mathcal{N}(0, 1/3)$ and $\theta_0 = 0.5$, following [16]. For pedagogical reason, we set $D = 1$. Furthermore, we choose $\lambda = 1$ and $\lambda_0 = 4700$ so that the prior is not washed out by the likelihood.

Here, standard MCMC sampler is employed as the baseline method. And we run the HABC-SGLD without rejection and HABC-SGLD with rejection (PJR-HABC-SGLD). The empirical histograms of samples obtained by running different samplers are shown in Fig. 7. We observe that the empirical histogram of samples obtained from PJR-HABC-SGLD is much closer to the standard MCMC sampler than that of HABC-SGLD, thus verifying the effectiveness of PJR-HABC-SGLD.

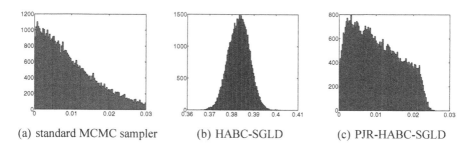

(a) standard MCMC sampler   (b) HABC-SGLD   (c) PJR-HABC-SGLD

Fig. 7. Application to HABC-SGLD. Empirical histogram of samples obtained by different samplers. We can observe that HABC-SGLD fails to converge to the posterior distribution in this situation. But PJR correction version of HABC-SGLD converges to the posterior.

5 Conclusion

In this paper, we have proposed pre-judgment Rule to accelerate ABC method. Computational methods adaptive to ABC rejection method and ABC-MCMC are provided as PJR-ABC and PJR-ABC-MCMC respectively. We analyze the error bound produced by PJR strategy. Our methodology establishes its practical value with desirable accuracy and efficiency. Finally, as a future direction, we plan to integrate PJR strategy with neural network as [24].

Acknowledgement. This study was funded by Scientific research fund of North University of China (No. XJJ201803).

References

1. Ahn, S., Korattikara, A., Welling, M.: Bayesian posterior sampling via stochastic gradient fisher scoring. In: Proceedings of the 29th International Conference on International Conference on Machine Learning, pp. 1771–1778 (2012)

2. Ahn, S., Shahbaba, B., Welling, M.: Distributed stochastic gradient MCMC. In: Proceedings of the 31st International Conference on Machine Learning (ICML-14), pp. 1044–1052 (2014)
3. Andrieu, C., Roberts, G.O.: The pseudo-marginal approach for efficient Monte Carlo computations. Ann. Stat. **37**, 697–725 (2009)
4. Barber, S., Voss, J., Webster, M., et al.: The rate of convergence for approximate Bayesian computation. Electron. J. Stat. **9**(1), 80–105 (2015)
5. Beaumont, M.A.: Approximate Bayesian computation in evolution and ecology. Annu. Rev. Ecol. Evol. Syst. **41**, 379–406 (2010)
6. Bernton, E., Jacob, P.E., Gerber, M., Robert, C.P.: Approximate Bayesian computation with the Wasserstein distance. J. Roy. Stat. Soc.: Ser. B (Stat. Methodol.) **81**(2), 235–269 (2019)
7. Biau, G., Cérou, F., Guyader, A., et al.: New insights into approximate Bayesian computation. In: Annales de l'Institut Henri Poincaré, Probabilités et Statistiques, vol. 51, pp. 376–403. Institut Henri Poincaré (2015)
8. Blum, M.G., François, O.: Non-linear regression models for approximate Bayesian computation. Stat. Comput. **20**(1), 63–73 (2010)
9. Blum, M.G., Nunes, M.A., Prangle, D., Sisson, S.A., et al.: A comparative review of dimension reduction methods in approximate Bayesian computation. Stat. Sci. **28**(2), 189–208 (2013)
10. Cabras, S., Nueda, M.E.C., Ruli, E., et al.: Approximate Bayesian computation by modelling summary statistics in a quasi-likelihood framework. Bayesian Anal. **10**(2), 411–439 (2015)
11. Chen, T., Fox, E., Guestrin, C.: Stochastic gradient Hamiltonian Monte Carlo. In: International Conference on Machine Learning, pp. 1683–1691 (2014)
12. Ding, N., Fang, Y., Babbush, R., Chen, C., Skeel, R.D., Neven, H.: Bayesian sampling using stochastic gradient thermostats. In: Advances in Neural Information Processing Systems, pp. 3203–3211 (2014)
13. Fearnhead, P., Prangle, D.: Constructing summary statistics for approximate Bayesian computation: semi-automatic approximate Bayesian computation. J. Roy. Stat. Soc.: Ser. B (Stat. Methodol.) **74**(3), 419–474 (2012)
14. Fu, T., Luo, L., Zhang, Z.: Quasi-Newton Hamiltonian Monte Carlo. In: Proceedings of the Thirty-Second Conference on Uncertainty in Artificial Intelligence, pp. 212–221 (2016)
15. Jiang, B., Wu, T.Y., Zheng, C., Wong, W.H.: Learning summary statistic for approximate Bayesian computation via deep neural network. Stat. Sin. **27**, 1595–1618 (2017)
16. Korattikara, A., Chen, Y., Welling, M.: Austerity in MCMC land: cutting the metropolis-hastings budget. In: International Conference on Machine Learning, pp. 181–189 (2014)
17. Lintusaari, J., Gutmann, M.U., Dutta, R., Kaski, S., Corander, J.: Fundamentals and recent developments in approximate Bayesian computation. Syst. Biol. **66**(1), e66–e82 (2017)
18. Ma, Y.A., Chen, T., Fox, E.: A complete recipe for stochastic gradient MCMC. In: Advances in Neural Information Processing Systems (2015)
19. Marin, J.M., Pudlo, P., Robert, C.P., Ryder, R.J.: Approximate Bayesian computational methods. Stat. Comput. **22**(6), 1167–1180 (2012)
20. Marjoram, P., Molitor, J., Plagnol, V., Tavaré, S.: Markov chain Monte Carlo without likelihoods. Proc. Natl. Acad. Sci. **100**(26), 15324–15328 (2003)
21. Meeds, E., Leenders, R., Welling, M.: Hamiltonian ABC. In: Proceedings of the Thirty-First Conference on Uncertainty in Artificial Intelligence, pp. 582–591 (2015)
22. Meeds, E., Welling, M.: GPS-ABC: Gaussian process surrogate approximate Bayesian computation. In: Proceedings of the Thirtieth Conference on Uncertainty in Artificial Intelligence, pp. 593–602 (2014)
23. Meeds, T., Welling, M.: Optimization Monte Carlo: efficient and embarrassingly parallel likelihood-free inference. In: Advances in Neural Information Processing Systems, pp. 2071–2079 (2015)

24. Mondal, M., Bertranpetit, J., Lao, O.: Approximate Bayesian computation with deep learning supports a third archaic introgression in Asia and Oceania. Nat. Commun. **10**(1), 246 (2019)

25. Pritchard, J.K., Seielstad, M.T., Perez-Lezaun, A., Feldman, M.W.: Population growth of human Y chromosomes: a study of Y chromosome microsatellites. Mol. Biol. Evol. **16**(12), 1791–1798 (1999)

26. Sisson, S.A., Fan, Y., Beaumont, M.: Handbook of Approximate Bayesian Computation. Chapman and Hall/CRC, New York (2018)

27. Sunnåker, M., Busetto, A.G., Numminen, E., Corander, J., Foll, M., Dessimoz, C.: Approximate Bayesian computation. PLoS Comput. Biol. **9**(1), e1002803 (2013)

28. Welling, M., Teh, Y.W.: Bayesian learning via stochastic gradient Langevin dynamics. In: Proceedings of the 28th International Conference on Machine Learning (ICML-11), pp. 681–688 (2011)

29. Wilkinson, R.: Accelerating ABC methods using Gaussian processes. In: Artificial Intelligence and Statistics, pp. 1015–1023 (2014)

30. Wilkinson, R.D.: Approximate Bayesian computation (ABC) gives exact results under the assumption of model error. Stat. Appl. Genet. Mol. Biol. **12**(2), 129–141 (2013)

31. Wood, S.N.: Statistical inference for noisy nonlinear ecological dynamic systems. Nature **466**(7310), 1102–1104 (2010)

Convolutional Neural Network Visualization in Adversarial Example Attack

Chenshuo Yu, Xiuli Wang[✉], and Yang Li

School of Information, Central University of Finance and Economics,
Beijing 100081, China
wangxiuli@cufe.edu.cn

Abstract. In deep learning, repeated convolution and pooling processes help to learn image features, but complex nonlinear operations make deep learning models difficult for users to understand. Adversarial example attack is a unique form of attack in deep learning. The attacker attacks the model by applying invisible changes to the picture, affecting the results of the model judgment. In this paper, a research is implemented on the adversarial example attack and neural network interpretability. The neural network interpretability research is believed to have considerable potential in resisting adversarial examples. It helped understand how the adversarial examples induce the neural network to make a wrong judgment and identify adversarial examples in the test set. The corresponding algorithm was designed and the image recognition model was built based on the ImageNet training set. And then the adversarial-example generation algorithm and the neural network visualization algorithm were designed to determine the model learning heat map of the original example and the adversarial-example. The results show that it develops the application of neural network interpretability in the field of resisting adversarial-example attacks.

Keywords: Adversarial-example · CNN · Visualization · Interpretability

1 Introduction

As an important content of the current computer industry, deep learning has gradually penetrated into the operation and development of various industries with the promotion of Internet plus. While deep learning is continuously advancing industrial form innovation and improving production efficiency, the internal ignorance of neural network models has also restricted the application of this technology, so it is imminent to promote research on the interpretability of neural networks. In addition, adversarial example attacks, as an important means of attack in image recognition, have a significant impact on deep learning. This paper applies the neural network visualization method to defend against adversarial example attacks, which not only enhances the credibility of the neural network visualization method, but also helps people to identify adversarial examples and make it clearer how the adversarial examples mislead the model [1, 2].

This work is supported by National Defense Science and Technology Innovation Special Zone Project (No. 18-163-11-ZT-002-045-04).

J. Zeng et al. (Eds.): ICPCSEE 2020, CCIS 1257, pp. 247–258, 2020.
https://doi.org/10.1007/978-981-15-7981-3_16

This article is different from the existing ideas of defending adversarial example: optimizing the network structure, training the adversarial examples, taking the optimized training examples as the main goal, it studies the possibility of defending adversarial examples from a new perspective, and strive to reduce the threat of adversarial example attack from the source.

2 Related Research

With the strong application ability of deep learning in various fields, the weak interpretability of deep neural networks has limited its application in the pillar field [3]. Therefore, in recent years, deep learning researchers, continuously improving deep learning capabilities, have begun to turn their attention to improving the interpretability of models. Well-known Internet companies such as Google and Alibaba are working hard to promote the development of AI interpretability and enhance the performance of human-machine dialogue. This move aims to clearly explain the decision-making or prediction behavior of Deep Neural Networks (DNN), and significantly improve model users' trust in the network, thereby reducing the potential hidden dangers caused by the use of deep learning technology in sensitive industries [4, 5].

Summarizing so many related work, it can be seen that there are three main types of ideas for indirect interpretation of CNN [6]: First, the hidden layer neuron feature analysis method. This method analyzes the hidden layer neurons in CNN to verify the behavior of the neural network, and collects the characteristics which can stimulate neurons to generate some brand-new images that can activate the entire network, as well as analyzing the features of the model. However, the obtained activation image is pretty abstract and not interpretable enough. The second is to imitate the model method. With the strong readability of the linear. An interpretable model is trained to simulate the input and output of the CNN. However, the model is heavily influenced by subjective factors; Third, the local linear replacement method. It selects a set of neighboring examples from the training set, and train a linear model to fit the original model according to the imitator model method. Local replacement reduces the huge error of the overall substitution, and can better grasp the characteristics of neurons inside the neural network. Correspondingly, this method also has the disadvantages of the above two methods.

The adversarial example attack, as a unique attack form in deep learning, is the key target of the robustness of deep learning. To defend adversarial example attacks, there are three main ideas. First, we can modify the training network, adjust the parameters of each convolutional layer and pooling layer, improve the redundancy of the model, and enhance the model's resistance to the adversarial examples. Correspondingly the recognition accuracy drops. And the structural reconstruction of the existing network model requires a lot of manpower and material resources, which is unrealistic in many cases; the second is to conduct adversarial example training [7], using the existing model to train the adversarial examples, so that the model can identify the characteristics of the adversarial examples, that is, we use the adversarial examples as the original training examples to train an adversarial example classification model. But this method also has its drawbacks. Obviously sorting out a large number of homogeneous adversarial examples requires a lot of effort, and the error interference of adversarial examples can be irregular. Thus, the accuracy of the obtained model needs to be verified. Third, we can optimize the training examples, such as strengthening the selection of training examples and searching the adversarial examples in the example set. This method also requires a lot of energy to distinguish the correct examples from the adversarial examples, and essentially avoid the occurrence of adversarial example attacks [8].

Different from mainstream adversarial example defense schemes, this paper proposes an adversarial example identification scheme based on a convolutional neural network visualization algorithm [9–11], which applies heat maps to example recognition to improve interpretability and effective resistance to adversarial example attacks.

3   Algorithm Principle

3.1   Implementation Principle of Adversarial Examples

As shown in Fig. 1, in an animal classification task, the example X is to be identified as the correct animal classification, and there are two classification processes. f_1 is a classifier based on deep learning. After selecting the digital features of the input image and discriminating them, the classification result is $Y = cat$. f_2 is a discrimination mechanism based on naked eye recognition. Participants read features of the image through the naked eye to identify salient features such as hair, eyes, and ears of cats. The final classification result is $Y = cat$. When the model recognition is correct, the recognition results of the machine and the user are the same, that is, $f_1(X) = f_2(X)$.

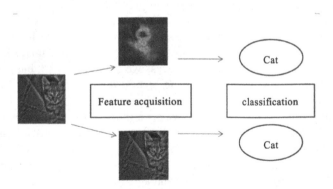

Fig. 1. Image identification

What the adversarial sample attack does is to increase the numerical interference limited to a threshold based on the original image X to obtain the adversarial example X', which makes the naked eye recognize the original image, and the machine distinguishes it into a completely new misclassification. Its mathematical meaning is:

$$Confirm\ X\prime\quad s.t\ \Delta(X, X\prime) < \varepsilon$$
$$f_2(X) = f_2(X\prime)$$
$$f_1(X) \neq f_1(X\prime)$$

This paper uses the gradient descent method to generate the adversarial examples. Under the conditions of minimizing example disturbance and maximizing the model's misleading effect, the optimization is based on the direction of the maximum gradient, which guarantees the basic work of the subsequent visualization CNN processing. The ImageNet project is a large-scale image database for computer vision research. Over ten million image samples have been sorted, classified, and annotated. It is one of the most comprehensive image recognition resource pools. In order to expand the content range of the adversarial samples, not limited to the common training sets like cats and dogs, this article selects ImageNet as the sample object of the algorithm.

Models based on the ImageNet have been pre-trained in the Slim image classification library in the deep learning framework Tensorflow. You can use ready-made models with simple function calls and obtain model parameters and weights [12]. The code is shown below:

$$saver = tf.train.Saver(restore\_vars)$$
$$saver.restore(sess, os.path.join(data\_dir,' inception\_v3.ckpt'))$$

In order to make the model universal, the algorithm supports processing all JPG format images. After reading the images according to the URL link, the data is processed to the image size required by the program with the help of the resize function. After obtaining the standardized image data, the Inception V3 model is used to perform initial output of the image, display the original image, and output the correct classification result for comparison with subsequent generated adversarial examples.

For a given image sample, the model will give the confidence of the Top-N term classification, that is, the corresponding probability $P(Y|x = X)$. In the process of generating adversarial examples, the algorithm aims to get a new image sample X', so that during the visual feature extraction process we get $f(X) = f(X')$, but it is judged as the confidence of the new category Y'. The highest degree, that is, the value of $P(Y'|x = X)$ is the largest.

In order to get adversarial examples, we set a threshold value and optimize sample constraints with back propagation and gradient descent. Iterate according to the following formula to maximize the confidence of the target classification of the adversarial examples and ensure that the visual characteristics of adversarial examples and the original images are within a certain range, and an adversarial example can be obtained.

$$f(X\prime) = X + \alpha \times \nabla \log P(Y\prime|x = X\prime)\, X\prime \in (X - \varepsilon, X + \varepsilon)$$

3.2   Convolutional Neural Network Visualization

For the complex structure of the neural network model, the most obvious step is to write a model structure diagram, analyze each convolutional layer, pooling layer, fully connected layer and other structures in the network, and record the parameters of each layer in detail. In the deep learning architecture keras, the constructed model architecture can be easily obtained through the function instructions of model. summary (). A part of network structure of the VGG16 model is shown in Fig. 2.

Layer (type)	Output Shape	Param #
input_3 (InputLayer)	(None, 224, 224, 3)	0
block1_conv1 (Conv2D)	(None, 224, 224, 64)	1792
block1_conv2 (Conv2D)	(None, 224, 224, 64)	36928
block1_pool (MaxPooling2D)	(None, 112, 112, 64)	0
block2_conv1 (Conv2D)	(None, 112, 112, 128)	73856
block2_conv2 (Conv2D)	(None, 112, 112, 128)	147584
block2_pool (MaxPooling2D)	(None, 56, 56, 128)	0
block3_conv1 (Conv2D)	(None, 56, 56, 256)	295168
block3_conv2 (Conv2D)	(None, 56, 56, 256)	590080
block3_conv3 (Conv2D)	(None, 56, 56, 256)	590080
block3_pool (MaxPooling2D)	(None, 28, 28, 256)	0
block4_conv1 (Conv2D)	(None, 28, 28, 512)	1180160
block4_conv2 (Conv2D)	(None, 28, 28, 512)	2359808
block4_conv3 (Conv2D)	(None, 28, 28, 512)	2359808
block4_pool (MaxPooling2D)	(None, 14, 14, 512)	0
block5_conv1 (Conv2D)	(None, 14, 14, 512)	2359808
block5_conv2 (Conv2D)	(None, 14, 14, 512)	2359808
block5_conv3 (Conv2D)	(None, 14, 14, 512)	2359808
block5_pool (MaxPooling2D)	(None, 7, 7, 512)	0

Fig. 2. VGG16 model structure diagram

Class Activation Mapping Select the output of the last convolutional layer block_conv3 before the Softmax function [13]. This is because the data loses the spatial information after passing the Softmax function, and it is impossible to obtain the identification situation of each part. Based on the traditional CNN model, CAM uses the Global Average Pooling layer instead of the original fully connected layer. While retaining the ability of the fully connected layer to reduce the data dimension, it also significantly reduces the parameters of CNN models (one-to-one correspondence in the fully connected layer makes the parameters in the network extremely huge), preventing the original model from overfitting.

The value of each feature map processed by the global average pooling layer is multiplied with the weight of each feature map and summed to obtain the heat map. Based on the CAM algorithm, it can be optimized by combining the ideas of back

252 C. Yu et al.

propagation and deconvolution to obtain a weighted class activation mapping method (Grad-CAM) [14, 15, 17], which avoids the reconstruction and training of the existing network. Using the global average of the gradient instead of the weight of each feature map, we can obtain heat maps in the same way, as shown in Fig. 3. Whether the weight got from each feature map directly is equivalent to the weight calculated with the global average gradient has been demonstrated in detail in the paper [16].

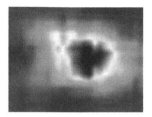

Fig. 3. Heat map

Based on the existing ordinary model, we get the output loss of the image label classification, and get the output of the last convolutional layer through the get_layer () function. The program can automatically calculate the average gradient by using the mathematical method of reverse automatic differentiation and taking the model output loss and the convolution layer output as parameters. Because it is complicated to manually derive the derivative expression of each parameter and then implement the code, you can use the gradients function encapsulated in Tensorflow to automatically calculate the gradient tensor and calculate the average value of the global gradient. Then we multiply the average gradient of all feature maps with the output of the global average pooling layer to get the gradient weight map. In order to enhance the visualization effect, we improve the color contrast effect, and a heat map is obtained to realize the visualization of the picture example classification model.

4 Algorithm Operation Effect Analysis

4.1 Visualization of Correct Sample Recognition

To ensure the effectiveness and universality of the algorithm, pictures are randomly selected from the Internet. Taking a common cat as an example, as shown in Fig. 4 (a). Needless to say, this is a test picture of a Egyptian cat (available from ImageNet tags). After the model discriminates, it can be known that the system has obtained the top three results of confidence: Egyptian cat, tabby cat and tiger cat, and the results predicted by the model are accurate. At the same time, several samples were selected for repeated experiments, as shown in Figs. 4 (b), 4 (c), and 4 (d). The principle is the same as that of cat samples.

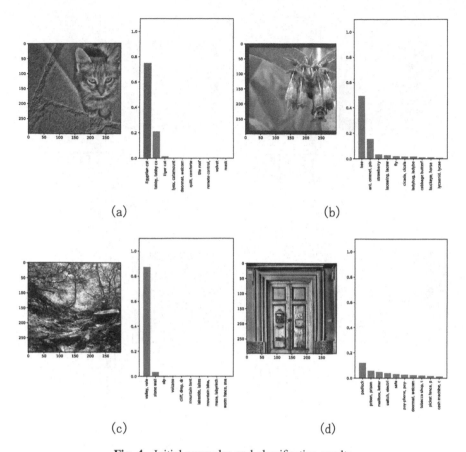

Fig. 4. Initial examples and classification results

With the implementation of the adversarial example generation algorithm, the target of the adversarial example is set as the 620th type in the ImageNet dataset—the PDA. After continuous optimization of projection and gradient descent, an adversarial example generated based on Fig. 4 can be obtained, as shown in Fig. 5.

Comparing the picture samples in Fig. 4 and Fig. 5, it can be clearly seen that the newly generated adversarial examples are consistent with the original pictures in visual effects and cannot be distinguished. But using the trained classification model to identify the adversarial examples, we can get that the images are recognized as handheld computers with a confidence level of almost 100%. Such a comparison result means that the two pictures that cannot be discerned by the naked eye are very different for CNN, which also proves the realization of the adversarial example attack. While the attack effect is good, the model further analyzes how the model recognizes the two picture samples as different objects. Therefore, the visualization algorithm is used to help the user understand which parts of the picture are recognized by the model. First,

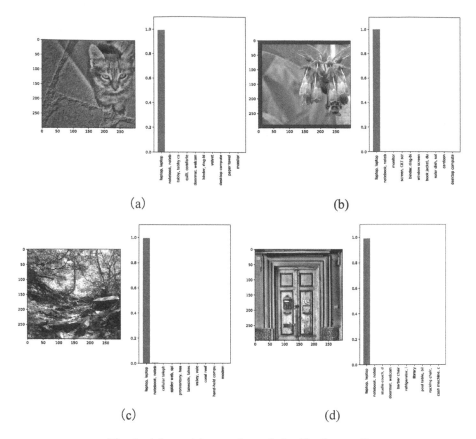

(a)   (b)

(c)   (d)

Fig. 5. Adversarial examples and classification results

the original picture is identified, and an initial heat map is obtained, as shown in Fig. 6 (a). The naked eye can clearly see that the identified areas are mainly the cat's head and body parts, roughly showing the outline of a cat, confirming the rationality of the original picture sample being identified as a tabby cat.

As a comparison, a model visualization is performed on a sample classified as a handheld computer, and a heat map is obtained as shown in Fig. 7. Comparing Fig. 6 and Fig. 7, we can see that there are obvious differences in the outlines of the two heat maps. The brightly colored parts of the heat maps of the adversarial samples are square and do not meet the visual characteristics of cats. The conclusion of the handheld computer in the picture.

The comparison of the examples above proves that the algorithm implemented in this paper has significant effects in adversarial example generation, CNN visualization, and adversarial example discrimination. However, the simple selection of correct prediction examples cannot guarantee the universality of the algorithm, so this article continues to verify the reliability of the algorithm from multiple perspectives.

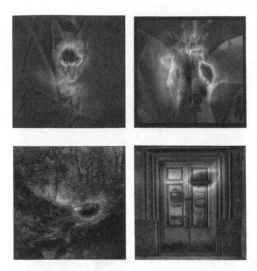

Fig. 6. Heat maps of the original examples

Fig. 7. Heat maps of the adversarial examples

4.2 Visualization of Wrong Sample Recognition

Compared with the first group of examples, a slightly more complex example is selected, as shown in Fig. 8. According to the label of the example, the original image should be a horse cart, but it is identified as an oxcart by the classification model. In addition, the target classification of the adversarial example is also set to a laptop. After inspection, the example with the same visual characteristics is successfully identified as a laptop by the model.

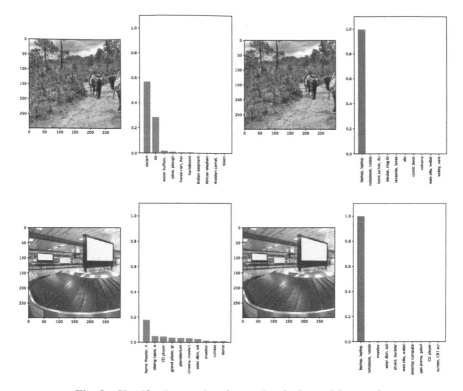

Fig. 8. Classification results of normal and adversarial examples

Unlike the first set of examples, the recognition result of the original sample is wrong. The user can find the cause of the identification error based on the heat map. As shown in Fig. 9, the model mainly recognizes the goods on horseback instead of the horse's head. Since the sample is not a frontal picture of the horse, it is easy to be misled, which makes the model mistake it for the oxcart, and then derives incorrect classification. Therefore, the heat map improves the interpretation ability of the model. Observing the heat map of the adversarial samples, it can be found that the features obtained by the model also do not match the visual characteristics of the horse or ox, which proves that the model can identify the anomalies of the adversarial examples and help distinguish between normal and adversarial examples.

After obtaining the CNN visualization algorithm based on the original neural network model, this paper repeatedly measures the accuracy of the algorithm to ensure the reliability of the algorithm. After 200 sample experiments, the model's prediction rate for image classification results is about 95%, but for the training set mixed with adversarial samples, the algorithm's success rate of identifying adversarial samples is close to 100%. This accuracy rate shows that the algorithm has a very high discrimination rate for the adversarial examples.

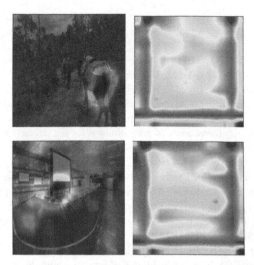

Fig. 9. Heat maps of normal and adversarial examples

5   Conclusions

This paper focuses on the application of visualized CNNs in adversarial example attacks. It studies the existing defense technologies and adversarial sample generation technologies, and considers that optimizing training examples is a more effective way to defend against adversarial example attacks. This paper implements the adversarial sample generation algorithm and the example recognition category activation mapping algorithm, which can accurately outline the recognition area of the original training examples and distinguish normal examples from adversarial examples according to the heat map. The algorithm can randomly select a picture for category determination and set target categories to generate adversarial samples. The visual neural network is used to delineate the image recognition area, determine the effective information of the picture, and find the anomaly in heat maps to distinguish the examples.

In addition, the visual CNN generated in this paper can accurately describe the recognition range of normal examples, and confirm anomalies in adversarial examples. And it is impossible to determine the recognition range of them. Follow-up studies will be conducted on how to improve the visualization performance and enhance the ability to explain the classification of adversarial examples.

References

1. Biggio, B., et al.: Evasion attacks against machine learning at test time. In: Blockeel, H., Kersting, K., Nijssen, S., Železný, F. (eds.) ECML PKDD 2013. LNCS (LNAI), vol. 8190, pp. 387–402. Springer, Heidelberg (2013). https://doi.org/10.1007/978-3-642-40994-3_25

2. Papernot, N., Mcdaniel, P., Wu, X., et al.: Distillation as a defense to adversarial perturbations against deep neural networks. In: Proceedings of the IEEE Symposium on Security and Privacy, pp. 582–598 (2016)
3. Li, P., Zhao, W., Liu, Q., et al.: Review of machine learning security and its defense technology. Comput. Sci. Explor. **2**, 171–184 (2018)
4. Marco, T.R., Sameer, S., Carlos, G.: Why should i trust you?: explaining the predictions of any classifier. In: Proceedings of the 22nd ACM SIGKDD International Conference on Knowledge Discovery and Data Mining, pp. 1135–1144 (2016)
5. Qiu, Y., Li, S.: Security threat analysis and solutions for the development and application of artificial intelligence. Netinfo Secur. **9**, 35–41 (2018)
6. Chu, L., Hu, X., Hu, J., Wang, L.J., et al.: Exact and consistent interpretation for piecewise linear neural networks: a closed form solution. In: Proceedings of the 24th ACM SIGKDD International Conference on Knowledge Discovery & Data Mining, pp. 1244–1253 (2018)
7. LeCun, Y., Bengio, Y., Hinton, G.: Deep learning. Nature **521**(7553), 436–444 (2015)
8. Yu, Y., Ding, L., Chen, Z.: Research on attacks and defenses towards machine learning systems. Netinfo Secur. **9**, 10–18 (2018)
9. Chris, O., Arvind, S.: The Building Blocks of Interpretability [OL]. https://opensource. googleblog.com/2018/03/the-building-blocks-of-interpretability.html. 3 June 2018
10. Zhang, Q., Zhu, S.: Visual interpretability for deep learning: a survey. Front. Inf. Technol. Electron. Eng. **19**(1), 27–39 (2018)
11. Li, Y., Yan, Z., Yan, G.: A edge-based 2-channel convolutional neural and its visualization. Comput. Eng. Sci. **41**(10), 1837–1845 (2019)
12. Zhang, S., Zuo, X., Liu, J.: The Problem of the Adversarial Examples in Deep Learning [OL]. http://kns.cnki.net/kcms/detail/11.1826.2018-1-20
13. Zhou, B., Khosla, A., Lapedriza, A., et al.: Learning deep features for discriminative localization. In: Proceedings of the IEEE Conference on Computer Vision and Pattern Recognition, pp. 2921–2929 (2016)
14. Amit, D., Karthikeyan, S., Ronny, L., Peder, O.: Improving Simple Models with Confidence Profiles[OL]. https://arxiv.org/pdf/1807.07506.pdf. 19 June 2018
15. Zeiler, M.D., Fergus, R.: Visualizing and understanding convolutional networks. In: Fleet, D., Pajdla, T., Schiele, B., Tuytelaars, T. (eds.) ECCV 2014. LNCS, vol. 8689, pp. 818–833. Springer, Cham (2014). https://doi.org/10.1007/978-3-319-10590-1_53
16. Zeiler, M.D., Taylor, G.W., Fergus, R.: Adaptive deconvolutional networks for mid and high level feature learning. In: Proceedings of 2011 IEEE International Conference on Computer Vision, pp. 2018–2025 (2011)
17. Ramprasaath, R.S., Michael, C., Abhishek, D., Devi, P., Ramakrishna, V., Dhruv, B.: Grad-CAM: visual explanations from deep networks via gradient-based localization. In: Proceedings of the IEEE International Conference on Computer Vision, pp. 618–626 (2017)

Deeper Attention-Based Network for Structured Data

Xiaohua Wu, Youping Fan, Wanwan Peng, Hong Pang, and Yu Luo[✉]

School of Information and Software Engineering,
University of Electronic Science and Technology of China, Chengdu 610054, China
luoyu@uestc.edu.cn

Abstract. Deep learning methods are applied into structured data and in typical methods, low-order features are discarded after combining with high-order featuresfor prediction tasks. However, in structured data, ignorance of low-order features may cause the low prediction rate. To address this issue, in this paper, deeper attention-based network (DAN) is proposed. With DAN method, to keep both low- and high-order features, attention average pooling layer was utilized to aggregate features of each order. Furthermore, by shortcut connections from each layer to attention average pooling layer, DAN can be built extremely deep to obtain enough capacity. Experimental results show DAN has good performance and works effectively.

Keywords: Structured data · DeepLearning · Feature aggregation

1 Introduction

Deep learning [12] has been receiving widespread attention since Krizhevsky et al. proposed AlexNet [11], which won the championship on imagenet dataset [3] in 2012. Numerous experiments have established that deep learning has led to many breakthroughs across various areas including natural language processing [4], computer vision [6], speech and audio processing and many other areas [19]. However, less research has focused on processing structured data using deep learning methods.

Data in many areas is structured, such as recommender systems [18]. In traditional methods, linear models are applied to structured inputs, such as LR, FTRL [13]. However, these methods lack the ability to combined features. In contrast, due to the ability of automatically combine features, in recent years, researchers have gradually started using deep learning methods to process structured data. Typically efforts fall into two groups. One is multilayer perceptron (MLP) [16], which is used as backbone to learn interactions among features, such as Wide&Deep [2], Deep&Cross [17], and DIN [20]. And for the other, common methods combine Factorization Machine [15] and MLP. Factorization Machine is used to model the pairwise interactions between features, and then high-order feature interactions are learned by MLP such as PNN [14], DeepFM [5]

Supported by Sichuan Science and Technology Program 2018GZDZX0042 and 2018HH0061.

J. Zeng et al. (Eds.): ICPCSEE 2020, CCIS 1257, pp. 259–267, 2020.
https://doi.org/10.1007/978-981-15-7981-3_17

and NFM [8]. These methods aim at combining low-order features to obtain higher-order features. Nevertheless, only higher-order features are used for predictions. Unlike meaningless single pixel which needs to be combined into higher-level abstract features in an image, features of all orders in structured data are meaningful for predictions.

Recently, the number of samples and feature sizes in structured data are getting larger and larger. It is necessary to build a network with large capacity. In other areas of deep learning, some methods have made many breakthroughs. Typically works to build large capacity networks include ResNet [7], DenseNet [10] and so on. The key point of these methods to make networks deeper is to address the gradient vanishing problem by shortcut connections.

In order to address the above problems, we propose Deeper Attention-Based Network (DAN), which is based on MLP. Considering the need for feature information of all orders, DAN use attention average pooling layer to convert the outputs of all MLP hidden layers into a fixed-length vector, which contains information of all order features. Furthermore, in DAN, only one dense layer is stacked after attention average pooling layer. Hence, all parameters of DAN are very closed to the output. DAN will not encounter gradient vanishing. In this paper, our main contributions are summarized as follows:

- We propose a novel network architecture DAN, which can aggregate the features of each order through attention average pooling layer. By automatically adjusting attention weight parameters in pooling layer, DAN can focus on higher-order or lower-order features to adapt different datasets.
- Also, we design shortcut connections through attention average pooling layer and only one stacked dense layer after it. Despite the depth is significantly increased, DAN does not suffer from gradient vanishing.

2 Preliminary

Before introducing our proposed method, we give a brief review of the feature representation and multilayer perceptron (MLP).

2.1 Feature Representation

Structured data includes numerical and categorical values. Numerical values refer to continuous and discrete values such as [height = 1.75, age = 23]. In contrast, categorical values are in a limited set, for example, [gender = Female]. The normalized numerical values can be used directly as the input of deep neural networks. In general, categorical value is represented by a one-hot vector such as [gender = Female] and [1, 0], [gender = male] and [0, 1].

However, due to sparsity and high dimensionality, one-hot vectors are not suitable as deep neural network inputs. In order to better extract categorical feature to improve performance, high dimensional sparse one-hot vectors are embedded into low dimensional dense vector spaces [1]. For i-th categorical feature t_i, which is a one-hot vector, let $E^i = [e_i^1, e_i^2, \cdots, e_i^k] \in \mathbb{R}^{d \times k}$ represents the i-th embedding matrix, where d is the dimension of low dimensional dense vector, k is the size of i-th feature value set.

Then, all embedding vectors of categorical features and normalized numerical features are concatenated into the input vector $x = [e_1, e_2, \cdots, e_m, v_1, \cdots, v_n]$, where e is the embeddings of categorical features with the number m. And v is the embeddings or values of numerical features with the number n.

2.2 MLP

Multilayer perceptron provides a universal approximation framework by stacking hidden layers [9]. Hence, it is fully capable of learning higher-order interactions among features in structured data. Firstly, categorical features are transformed into low-dimensional dense vectors through embedding matrix. Then low-dimensional dense vectors and continuous features are as the input of MLP. MLP extracts higher-order features by stacking more hidden layers. Formally, the definition of l-th hidden layers is as the follow:

$$A_l = \sigma(W_l A_{l-1} + b_l), \tag{1}$$

where W_l, b_l denotes weight matrix and bias, σ is non-linear activation function, A_l, A_{l-1} denotes the output of this and last layer, specially, A_0 is the input matrix X consisting of all input vectors x.

3   Deeper Attention-Based Network

Deeper Attention-Based Network (DAN) is based on MLP. There are two motivations behind it. Firstly, without discarding any information, DAN keeps the features of each order for predictions. In this paper, we use attention average pooling layer to aggregate the features of all orders. To build large capacity networks is our second motivation. Thus, there are shortcut connections in DAN. Making all learned parameters closed to the output, after aggregation, DAN only stack one dense layer. At last, we also theoretically prove why DAN can be built very deep. The architecture of DAN is shown in Fig. 1.

Notation: Let A_i denote the output of i-th layer, W_i, b_i denote the weight matrix and bias, A_0 is the input matrix X consisting of all input vectors x.

3.1   Feature Aggregation

In order to obtain the abstract features of each layer, we must extract the output of all hidden layers of the MLP. Next, we aggregate all outputs as the input of a shallow feedforward neural network with a fix-length input. Nevertheless, the number of the outputs changes as the hidden layers are stacked. It is a common practice to transform the list of vectors to get a fixed-length vector via a pooling layer:

$$C = pooling(A_0, A_1, \cdots, A_L), \tag{2}$$

where A_i is the output of i-th layer, L is the number of MLP hidden layers, C is a fix-length vector. The most common pooling operations are max pooling and average

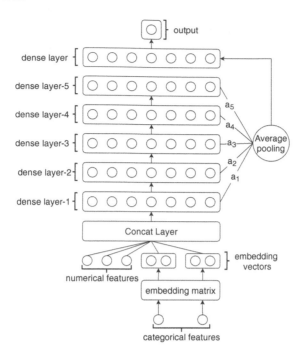

Fig. 1. The architecture of DAN based on MLP

pooling. Max pooling selects the maximum value and average pooling calculates the average value of each position among all vectors. In this paper, our purpose is to take into account the abstract features of each level, we use average pooling without dropping any values.

In addition, on various datasets and tasks, the output decision focuses on different levels of the abstract features. Attention average pooling with learnable weight parameters is used to address it in DAN:

$$pooling(A_0, A_1, \cdots, A_L) = \sum_{i=1}^{L} \alpha_i A_i, \tag{3}$$

where α_i denotes learnable weight parameter. Through backpropagation algorithm, DAN can automatically find one group of adaptive weight parameters for the specific datasets and tasks. For a DAN with l hidden layers, if $[\alpha_1, \alpha_2, \cdots, \alpha_{l-1}, \alpha_l] = [0, 0, \cdots, 0, 1]$, DAN is the same as a MLP. Hence, our proposed DAN contains MLP and has larger capacity than MLP. The optimal solution of DAN must be at least equal to the MLP. Intuitively, DAN has a better performance than MLP.

3.2   Deeper Network

Gradient vanishing is the problem that must be considered when building deeper networks. To tackle these problems, in this paper, we make the trainable parameters closer to the output by shortcut connections. In DAN, all layers containing trainable parameters have a path directly connected to the input of the pooling layer. After the pooling layer, we just stack a single dense layer to avoid the trainable parameters being too far from the output:

$$\hat{y} = \sigma(WC + b), \tag{4}$$

where C is the output of pooling layer, W, b is weight matrix and bias of last layer, $\hat{y}$ is the output of DAN. In last layer, two most commonly used non-linear activation function σ are $sigmoid$ and $softmax$.

3.3 Theoretical Proof

We study the back-propagated gradient of the $Loss$ function on the inputs weights at each layer in DAN. There are $L - l + 1$ path to the output from l-th layer in DAN. For weight matrix W_l of l-th layer, the gradient of the $Loss$ function is as follows:

$$\frac{\partial Loss}{\partial W_l} = \sum_{i=l}^{L} \frac{\partial A_l}{\partial W_l} \cdot \frac{\partial A_i}{\partial A_l} \cdot \frac{\partial C}{\partial A_i} \cdot \frac{\partial Loss}{\partial C}, \tag{5}$$

$$\frac{\partial Loss}{\partial W_l} = \sum_{i=l}^{L} \alpha_i \frac{\partial A_l}{\partial W_l} \prod_{j=l+1}^{i} \frac{\partial A_j}{\partial A_{j-1}} \cdot \frac{\partial Loss}{\partial C}, \tag{6}$$

where α_i denotes learnable weight parameter of i-th layer, A_j denotes the output of i-th layer. In the formula (6), there is always one term $\alpha_l \cdot \frac{\partial A_l}{\partial W_l} \cdot \frac{\partial Loss}{\partial C}$ where gradients are only propagated three times. Thus, no matter how deep the network layer is, DAN will not encounter the problem of gradient vanishing.

Since the gradient of all parameters will not be too small, during the training process, the convergence speed of DAN will be relatively fast. Fast convergence speed is very important for large data sets.

4   Experiments

4.1   Experimental Settings

Dataset. We evaluate the effectiveness of DAN on two public datasets. **Criteo** dataset was used for the Display Advertising Challenge hosted by Kaggle and includes 45 million users'click records. **Porto Seguro** dataset was used to predict the probability that an auto insurance policy holder files a claim.

Evaluation Metrics. To evaluate the performance, we adopt AUC (Area Under ROC) and Logloss. AUC is a widely used metric in CTR prediction field. It sets different thresholds to evaluate model performance. And Logloss is the value of $Loss$ function where a lower score indicates a better performance.

Models for Comparisons. We compare DAN with two models: basic model (MLP) and DeepFM which includes MLP and Factorization Machine (FM).

Hyper-parameter Settings. To be fair, all models use the same setting of hyper-parameters shown in Table 1.

Table 1. Hyper-parameters setting

	Criteo	Porto Seguro
Embedding size	32	8
Hidden layer width	256	64
Hidden layer depth	5, 10, 20, 50, 100	5, 10, 15, 20, 25
Batch size	256	256
Learning rate	3e−5 with Adam	3e−5 with Adam
Dropout rate	0.5	0.5
Training epochs	1 to 10	1 to 20

Table 2. The best result on two datasets

	Criteo		Porto Seguro	
	AUC	Logloss	AUC	Logloss
MLP	0.8028	0.4656	0.6290	0.1529
DeepFM	0.8015	0.4546	0.6260	0.1531
DAN	**0.8037**	**0.4478**	**0.6330**	**0.1524**

4.2 Performance Evaluation

In this section, we evaluate model performance on **Criteo** and **Porto Seguro** datasets based on the hyper-parameters listed in last section. While keeping other hyper-parameters constant, we gradually make the network deeper. Then the best results are chosen for different models on both dataset and shown as Table 2.

Overall, DAN beats other competitors in AUC and Logloss. In fact, compared with other models, our proposed model improves performance as the number of layers increases. In next section, we show the details through experiments of Comparison.

4.3 Layer Depth Study

In this section, first we list the details of different layers. Next, we observe the trend of the evaluation score with epochs. Furthermore, we analyze the reasons for the experimental results.

Details. In our experiments, we mainly explore the effect of the layer depth which is closely related to model capacity. Let layer depth be [5, 10, 20, 50, 100] or **Criteo** while keeping other hyper-parameters constant.

Table 3. AUC of different layers on Criteo

	5	10	20	50	100
MLP	0.8028	0.8017	0.7931	0.500	0.500
DeepFM	0.8015	0.8015	0.7987	0.7862	0.7866
DAN	**0.8032**	**0.8033**	**0.8035**	**0.8035**	**0.8037**

Table 4. Logloss of different layers on Criteo

	5	10	20	50	100
MLP	0.4656	0.4672	0.4673	0.5692	0.5692
DeepFM	0.4546	0.4633	0.4730	0.4638	0.4634
DAN	**0.4482**	**0.4481**	**0.4479**	**0.4479**	**0.4478**

Table 3 to Table 4 show AUC and Logloss with different layers. We have the following observations:

- Our proposed model DAN outperforms MLP and DeepFM with any layer depth. This verifies the usefulness of combining features of each order.
- The performance of MLP and DeepFM drops sharply when the network is extremely deep. However, due to no gradient vanishing, the deeper network has higher performance for DAN.
- DeepFM has better performance than MLP with network very deep. The reason is consistent with the intuition behind DAN. There is a short path from input to output through FM. The trainable parameters in FM component are closed to the output of DeepFM.

Trends. Firstly, we observe the trends of the AUC and Logloss curves under the condition that the layer depth is 5 without gradient vanishing. Then, to highlight the key advantage of DAN, we adjust the layer depth to 100 for **Criteo**.

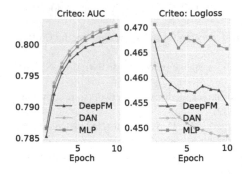

Fig. 2. Trends of layer depth 5 on Criteo

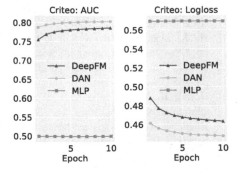

Fig. 3. Trends of layer depth 100 on Criteo

Figure 2 compares the AUC and Logloss of DeepFM, MLP and DAN of each epoch with the layer depth 5. We observe that DAN converges faster and obtains a better

Logloss and AUC score than MLP and DeepFM, which indicates that DAN can fit the data better because of greater capability. Faster convergence speed of DAN helps us get better performance in less time that is important for actual productions.

Figure 3 shows the key advantage of DAN, which can be built extremely deep and obtain better performance. The deeper networks mean that the greater capability. However, with the network depth significantly increasing, MLP completely degrades with AUC 0.5 which means random prediction. Although there is a short path to the output in Factorization Machine, DeepFM's performance still drops significantly. Only DAN gets benefit from the very deep architecture.

5   The Conclusions

In this paper, we propose DAN based on MLP. DAN gains performance improvement with any layer depth by extracting the features of each order and being built more deep. The results show that DAN has faster convergence speed and better performance when architecture is very deep.

References

1. Bengio, Y., Ducharme, R., Vincent, P., Janvin, C.: A neural probabilistic language model. J. Mach. Learn. Res. **3**, 1137–1155 (2003)
2. Cheng, H., et al.: Wide & deep learning for recommender systems. In: DLRS@RecSys, pp. 7–10. ACM (2016)
3. Deng, J., Dong, W., Socher, R., Li, L., Li, K., Li, F.: Imagenet: a large-scale hierarchical image database. In: CVPR, pp. 248–255. IEEE Computer Society (2009)
4. Devlin, J., Chang, M., Lee, K., Toutanova, K.: BERT: pre-training of deep bidirectional transformers for language understanding. In: NAACL-HLT (1), pp. 4171–4186. Association for Computational Linguistics (2019)
5. Guo, H., Tang, R., Ye, Y., Li, Z., He, X.: DeepFM: a factorization-machine based neural network for CTR prediction. In: IJCAI, pp. 1725–1731. ijcai.org (2017)
6. He, K., Gkioxari, G., Dollár, P., Girshick, R.B.: Mask R-CNN. In: ICCV. pp. 2980–2988. IEEE Computer Society (2017)
7. He, K., Zhang, X., Ren, S., Sun, J.: Deep residual learning for image recognition. In: CVPR, pp. 770–778. IEEE Computer Society (2016)
8. He, X., Chua, T.: Neural factorization machines for sparse predictive analytics. In: SIGIR, pp. 355–364. ACM (2017)
9. Hornik, K., Stinchcombe, M.B., White, H.: Multilayer feedforward networks are universal approximators. Neural Netw. **2**(5), 359–366 (1989)
10. Huang, G., Liu, Z., van der Maaten, L., Weinberger, K.Q.: Densely connected convolutional networks. In: CVPR, pp. 2261–2269. IEEE Computer Society (2017)
11. Krizhevsky, A., Sutskever, I., Hinton, G.E.: Imagenet classification with deep convolutional neural networks. In: NIPS, pp. 1106–1114 (2012)
12. LeCun, Y., Bengio, Y., Hinton, G.: Deep learning. Nature **521**(7553), 436–444 (2015)
13. McMahan, H.B., et al.: Ad click prediction: a view from the trenches. In: KDD, pp. 1222–1230. ACM (2013)
14. Qu, Y., et al.: Product-based neural networks for user response prediction over multi-field categorical data. ACM Trans. Inf. Syst. **37**(1), 5:1–5:35 (2019)

15. Rendle, S.: Factorization machines. In: ICDM, pp. 995–1000. IEEE Computer Society (2010)
16. Rosenblatt, F.: Principles of neurodynamics. Perceptrons and the theory of brain mechanisms. Technical report, Cornell Aeronautical Lab Inc., Buffalo, NY (1961)
17. Wang, R., Fu, B., Fu, G., Wang, M.: Deep & cross network for ad click predictions. In: ADKDD@KDD, pp. 12:1–12:7. ACM (2017)
18. Xi, W., Huang, L., Wang, C., Zheng, Y., Lai, J.: BPAM: recommendation based on BP neural network with attention mechanism. In: IJCAI, pp. 3905–3911. ijcai.org (2019)
19. Xinyi, Z., Chen, L.: Capsule graph neural network. In: ICLR (Poster). OpenReview.net (2019)
20. Zhou, G., et al.: Deep interest network for click-through rate prediction. In: KDD, pp. 1059–1068. ACM (2018)

Roofline Model-Guided Compilation Optimization Parameter Selection Method

Qi Du[1(✉)] [ID], Hui Huang[2], and Chun Huang[1]

[1] College of Computer Science, National University of Defense Technology, Changsha, China
nudtcsduqi@foxmail.com
[2] HuNan Institute of Scientific and Technical Information, Changsha, China

Abstract. In this paper, the method of roofline model-guided compilation optimization parameter selection (RMOPS) is proposed based on Roofline model to maximize the performance of targets. Through the orthogonal test design compiler, the problem of optimization parameter selection in complex dependencies was solved. The performance data generated by empirical roofline tool (ERT) were used to implement the optimization parameter selection decision. RMOPS method was evaluated on ARMv8 platform, and the feasibility of RMOPS method was verified by using SPEC CPU2017 and NPB. Experimental results show that the program performance obtained by using the optimal optimization parameters of RMOPS search is generally improved by 5%–33% compared with that achieved by -O3 optimization parameter setting.

Keywords: Roofline · RMOPS · ERT · NPB · SPEC CPU2017

1 Introduction

With the continuous development of computer architecture, applications are more and more dependent on program performance optimization technology to obtain higher runtime performance. The same code through a different compiler optimization can be run on different computer architecture, when programmers manual when optimizing a program for a system, the optimization Settings for another system usually is not the most optimal [1], the programmer is very difficult to full and accurate understanding of each machine on the architecture and implementation of difference, implemented by means of experience to optimize the application performance of portability difficult [2], the compiler optimization is to realize the important means to improve the performance of the program, in the compiler, there are many optional compiler optimizations can achieve different transformation of program, Using the right tuning parameters can lead to a substantial improvement in program performance, while using the wrong tuning parameters can lead to a small improvement in program performance, even less than before the optimization method was used. Therefore, optimization parameter selection is one of the core problems to be solved in program performance optimization. Aiming at this problem, a kind of technology called adaptive optimization or experience optimization is used to program optimization [3, 4], the iterative compilation techniques [5, 6] is adaptive for general program optimization techniques, it can be for

© Springer Nature Singapore Pte Ltd. 2020
J. Zeng et al. (Eds.): ICPCSEE 2020, CCIS 1257, pp. 268–282, 2020.
https://doi.org/10.1007/978-981-15-7981-3_18

different computer architecture, different programs have different characteristics, effective collection various optimization techniques, to choose a more good than the default compiler options optimization combination.

As an adaptive programming optimization technique, iterative compilation can achieve good optimization results and portable code performance for any compiler implementation and system environment. However, the long iteration time of iterative compilation is widely criticized, which makes iterative compilation not directly applicable to the optimization of large programs. For example, the GCC-8.3.0 compiler [7] has more than 200 optional compilation optimization options, some of which are switch options, while the other part can set specific parameters, so there are many combinations of possible optimization options. In high performance computing and practical engineering applications, many programs require a long time to compile and run, generally measured in hours and days. These large programs take an enormous amount of time to compile and run iteratively. At the same time, iterative compilation is a mechanical search that lacks the use of previous experience and sometimes requires human intervention to reduce search blindness. As the data scale grows, the developer's task becomes onerous.

Aiming at the problems above, this paper proposes a select compiler Optimization parameters to maximize the performance of target: based on the Optimization of the Parameter Selection method RMOPS Roofline Model (Roofline Model -based Optimization Parameter Selection), the method through the orthogonal test design compiler Optimization Parameter Selection problem in complex dependencies, to ERT (Empirical Roofline Tool) the performance of the generated data to carry out Optimization Parameter Selection decisions. The main contributions of this paper are as follows:

(1) the orthogonal experimental design and genetic algorithm are proposed to design the complex dependencies in the compilation optimization parameter selection problem, so as to reduce the optimization spatial search scale;
(2) the performance data generated by ERT(Empirical Roofline Tool) are used to implement the optimization parameter selection decision, so as to obtain better compilation optimization parameters with less time consumption;
(3) the obtained compilation optimization parameters were applied to NPB and SPEC CPU2006 test set, and a good performance improvement was achieved.

The structure of this paper is as follows: Sect. 2 introduces Roofline model and ERT; Sect. 3 introduces the orthogonal test method to optimize the search space of compilation optimization parameters. Section 4 is the experiment and analysis; Finally summarize the full text.

2   Roofline Model and ERT

2.1   Roofline Model

The diversity of processor architectures makes for a complex hardware environment. In 2009, Samuel Williams et al. proposed the Roofline model [8], which is a performance

270   Q. Du et al.

analysis model for modern computer architectures, paying special attention to floating-point computing performance and off-chip memory bandwidth. Although the model has been around for more than a decade, its usefulness has been widely recognized and it has been widely used in the industry.

Off-line memory bandwidth will remain a constrained resource for the foreseeable future [9]. The original intention of establishing Roofline Model by Samuel Williams et al. was to visually show the relationship between processor performance and out-of-chip memory bandwidth utilization, as shown in the figure below.

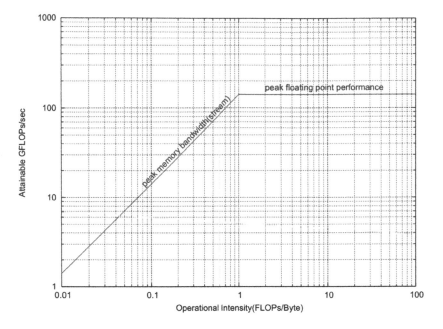

Fig. 1. Roofline model

In order to represent the relationship between them, the concept of Operational Intensity is introduced into Roofline model, which represents the number of floating point operations performed during the access of each byte DRAM. The formula is as follows:

$$OI = Flops/Bytes \tag{1}$$

Where *OI* is the Operational Intensity, *Flops* (Total Float Point Operations) and *Bytes* respectively represent the Total number of times of floating Point operation and the Total number of Bytes of DRAM access.

In the figure above, the Y-axis is GFlops/sec, whose relationship with the operation intensity is as follows:

$$\begin{matrix} Attainable \\ GFlops/sec \end{matrix} = min \left\{ \begin{matrix} Peak\ Floating\ Point \\ Performance \end{matrix}, \begin{matrix} Peak\ Memory \\ Bandwidth \end{matrix} \times \begin{matrix} Operational \\ Intensity \end{matrix} \right\}$$

As can be seen from the intersection of the slash and the horizontal line in the figure, if the point is very close to the right, it means that only the floating-point kernel with high operation density can achieve the maximum performance of the operation device. If it is very close to the left, more floating-point cores with low operation density can achieve the maximum performance of the operation device.

As we know, there are generally three ways to improve program performance: one is to optimize data locality to reduce memory access requests; the other is to improve instruction-level parallelism; the third is to improve data-level parallelism. All of these methods require the programmer and compiler to work together. The purpose of this paper is to improve the program performance by using Roofline model to select the optimal sequence of compiler optimization parameters.

2.2   ERT

At present, many tools using Roofline model have been widely used, such as ERT (Empirical Roofline Tool) [10], Intel Advisor, NVIDIA NVProf/NSight, LIKWID and SDE/VTune. In this paper, ERT is used to describe the target machine with Roofline.

ERT generates the required characterization of a machine empirically. It does this by running the kernel codes on the machine so the results are, by definition, Attainable by some code (s) and the options, informs the to compile and run the code are known. The ERT generates the bandwidth and gflop/sec data needed by the Roofline Performance Model, the input to the ERTIs a configuration file and the final output is a roofline graph in the PostScript format and the roofline parameters in JSON format. Currently, the ERT can utilize the MPI, OpenMP, and Cuda/GPU for parallelization [10].

The core of the ERT code is to add and FMA(multiply multiply-add) in a loop, as follows:

```
#define KERNEL1(a, b, c) ((a) = (b) + (c))
#define KERNEL2(a, b, c) ((a) = (a) * (b) + (c))
for (j = 0; j < ntrials; ++j) {
  for (i = 0; i < nsize; ++i) {
    ...
    KERNEL1(beta, A[i], alpha);
    ...
    KERNEL2(beta, A[i], alpha);
    ...
  }
}
```

FMA operations are included in IEEE 754-2008, and all ARMv8 processors support FMA. It can significantly improve the performance and accuracy of operations like these:

- The dot product;
- Matrix multiplication;
- Polynomial equation solution;
- Newton's method solves the zero of the function.

3 Roofline Model Based Optimization Parameter Selection

To achieve the desired performance of compiled target machine code, the compiler provides a number of compilation optimization options. Normally, compilers predefined criteria optimization sequence (-O0, -O1, -O2 - and -O3, -Os, etc.) will optimize the target code some performance, but these options in the general standard compiler optimization sequence according to the Settings and experience analysis, the optimization of these default options are hard to guarantee in any architecture compiler optimization any program can make the program to achieve optimal performance, or conform to the requirements of the users on the performance of the program. Literature [11] shows that for different programs to be compiled, personalized compilation optimization sequences can achieve better results than standard compilation optimization sequences. Of the large number of compiler optimizations the compiler to provide, such as the compiler GCC has more than 200 tuning options, from the large number of tuning options to choose the appropriate option to compile optimization sequence leads to the combination explosion problem, to compile the program as a way of using artificial manually choose efficient compiler optimization sequence is a time-consuming and laborious task.

In gcc-8.3.0, there are 239 compilation optimization options, among which '-O0' is the default optimization sequence, which reduces compilation time and enables debugging to produce expected results, and contains 80 optimization options. If you choose the appropriate option from the remaining 159 optimization options to form the compilation optimization sequence, there will be 159! combinations. In order to solve this problem, this paper designs the complex dependence relation of compiler optimization parameter selection problem by orthogonal experiment, selects the optimal combination as the initial value of genetic algorithm, and reduces the scale of optimization spatial search.

Orthogonal experimental design is a method for multi-factor experiments. According to the orthogonality, some representative points are selected from the whole experiment for the test. Genetic algorithm (GA) is an adaptive global optimization probabilistic search algorithm, which provides a general framework for solving non-linear, multi-objective, multi-model and other complex system optimization problems, and is a useful, efficient and robust optimization technology. However, genetic algorithms tend to fall into premature convergence and obtain local optimal solutions. The orthogonal experimental design method can solve the minimum deception problem in general genetic algorithms. In the combination of orthogonal test method and genetic algorithm, the orthogonal test method can determine the initial value range of the genetic algorithm and greatly reduce the computational amount of the genetic algorithm.

The algorithm steps of RMOPS, the optimization parameter selection method based on Roofline model combined with orthogonal test method and genetic algorithm, are as follows.

Input: test program P0, the number of compilation and optimization options is n, the initial value range selects the maximum iteration number of the phase N_0, and the maximum iteration number of the search phase N_1;

Output: initial value range OET, best optimization parameter list L^* and corresponding test program P0 running result $P(L^*)$.

Step 1: divide all compiler optimization options into $i = 0, 1, \cdots, k$ compiler optimization sequence L_i at different levels, test the given compiler optimization sequence through test program P0, and generate test result $P(L_i)$;

Step 2: put the corresponding compiler optimization options in the given compiler optimization sequence L_i into the optimization parameter list LT;

Step 3: generate orthogonal table T_i by orthogonal test for the given compilation optimization sequence L_i;

Step 4: generate test case $j = 0, 1, \cdots, l$ based on LT and orthogonal table T_i, calculate the program execution result of each test case $R(t_{ij})$;

Step 5: if the group with the best results in $R(t_{ij})$ is better than $P(L_i)$, append the compilation optimization options corresponding to test case j to OET, append the compilation optimization options corresponding to the best results to LT, and remove these optimization options from L_i.

Step 6: return to step 2 if L_i is not empty or the number of iterations is no more than N.

Step 7: if $i < k$, $i = i + 1$; Otherwise, output initial value range OET;

Step 8: generate the initial population according to the initial value range OET. The genetic algorithm will start the iteration with this initial group as the initial point.

Step 9: construct fitness function: define fitness function as the maximum value of running result of test program P0;

Step 10: application of selection operator: selection operation according to the fitness of the individual decided to the next generation of individuals can be selected according to the proportion of the fitness value;

Step 11: apply the crossover operator: take the crossover probability as 0.5;

Step 12: apply mutation operator: the value of mutation operator is 0.5;

Step 13: when N_1 generation is calculated, if the group convergence stops, the optimal parameter list L^* and the corresponding P0 running result $P(L^*)$ will be output. Return to step 9 if the termination condition is not met.

4 Experimental Analysis

4.1 Experimental Environment

We evaluated the RMOPS method on the ARMv8 platform, whose configuration is shown in Table 1. ERT test program was used as a performance evaluation tool, and each group of test cases was run twice. The average results of all test cases were compared, and the corresponding sequence of compilation optimization options with the best results was selected. The SPEC CPU2006 and NPB test sets were used to verify the feasibility of the RMOPS approach.

Table 1. ARMv8 experimental platform configuration

Configuration items	Value
Computing core frequency	2.6 GHz
Number of CPU cores	64
L1 cache	64 KB instruction cache
	64 KB data cache
L2 cache	512 KB
L3 cache	65536 KB
Memory	256 GB
Operating system	CentOS 8.2.0
Kernel version	4.19.23
Compiler	GCC-8.3.0

The SPEC CPU2017 [12] benchmark is a set of benchmark programs developed by the NASA standard performance evaluation organization to evaluate the performance of General-purpose CPU. It contains 10 integer benchmarks and 13 floating point benchmarks. The input sizes of the benchmark programs are divided into test, train and reference sizes. In this paper, the reference size is used for the test. Table 2 shows the application area and source code lines for each benchmark in the SPEC CPU2017.

Table 2. SPEC CPU2017 benchmarks [12]

Benchmark	Application area	KLOC
500.perlbench_r	Perl interpreter	362
502.gcc_r	GNU C compiler	1304
505.mcf_r	Route planning	3
520.omnetpp_r	Discrete event simulation - computer network	134
523.xalancbmk_r	XML to HTML conversion via XSLT	520
525.x264_r	Video compression	96
531.deepsjeng_r	Artificial Intelligence: alpha-beta tree search (Chess)	10
541.leela_r	Artificial Intelligence: Monte Carlo tree search (Go)	21
548.exchange2_r	Artificial Intelligence: recursive solution generator (Sudoku)	1
557.xz_r	General data compression	33
503.bwaves_r	Explosion modeling	1
507.cactuBSSN_r	Physics: relativity	257
508.namd_r	Molecular dynamics	8

(continued)

Table 2. (*continued*)

Benchmark	Application area	KLOC
510.parest_r	Biomedical imaging: optical tomography with finite elements	427
511.povray_r	Ray tracing	170
519.lbm_r	Fluid dynamics	1
521.wrf_r	Weather forecasting	991
526.blender_r	3D rendering and animation	1577
527.cam4_r	Atmosphere modeling	407
538.imagick_r	Image manipulation	338
544.nab_r	Molecular dynamics	259
549.fotonik3d_r	Computational electromagnetics	24
554.roms_r	Regional ocean modeling	14

The NAS Parallel Benchmarks (NPB) are a small set of programs designed to help evaluate the performance of parallel supercomputers [13]. NPB is an assembly developed by NASA and has been widely used in parallel computer test and comparison. NPB consists of 10 benchmarks, and the input size can be divided into S, W, A, B, C, D and E altogether. Table 3 shows the description of each benchmark in NPB and the number of source lines.

Table 3. The NAS parallel benchmarks [13]

Benchmark	Description	KLOC
BT	Block tri-diagonal solver	5.2
CG	Conjugate gradient, irregular memory access and communication	1.2
DC	Data cube	2.8
EP	Embarrassingly parallel	0.3
FT	Discrete 3D fast Fourier Transform, all-to-all communication	1.1
IS	Integer sort, random memory access	1.1
LU	Lower-upper gauss-seidel solver	5.4
MG	Multi-grid on a sequence of meshes, long- and short-distance communication, memory intensive	1.4
SP	Scalar penta-diagonal solver	3.2
UA	Unstructured adaptive mesh, dynamic and irregular memory access	8.2

4.2 Experiment and Result Analysis

The compiler provides a large number of compilation optimization options. The optimization options included in the predefined standard compilation optimization sequence do not completely cover all compilation optimization options. The statistics of compilation optimization options in GCC-8.3.0 are shown in Table 4, where compile optimization sequence O1 contains all the options in O0, O2 contains all the options in O1, and O3 contains all the options in O2.

Table 4. GCC-8.3.0 compilation optimization options statistics

	Compile optimization sequence O0	Compile optimization sequence O1	Compile optimization sequence O2	Compile optimization sequence O3	Other compilation options
Number of switch options	62	97	134	149	72
Number of specify parameter options	18	18	18	18	0

For the standard compilation optimization sequence, the best performance is not achieved in a specific platform, so the optimization options in the standard compilation sequence need to be tested. Based on Table 4, this article divides the compilation optimization options into the search hierarchy shown in Table 5.

Table 5. Compile optimization options search hierarchy

Hierarchy	Compile optimization options	Number of compilation optimization options
1st-level	Switch options in the compile optimization sequence O0	62
2nd-level	Remove the switch option in O0 from the compilation optimization sequence O1	35
3rd-level	Remove the switch option in O1 from the compilation optimization sequence O2	37
4th-level	Remove the switch option in O2 from the compilation optimization sequence O3	15
5th-level	Specify parameter options	18
6th-level	Other compilation options	72

Based on RMOPS algorithm, this paper USES the first to fifth levels to determine the initial value of the genetic algorithm, and the sixth level as the search space of the genetic algorithm. For the on/off option in the compilation optimization option, this option has both on and off states, so an orthogonal table with level 2 can be designed. For the specified parameter options in the compilation optimization options, this article limits the specified parameter options to a maximum of 6 sets of values, so an orthogonal table with a level of 6 can be designed. For example, without considering the interaction, the orthogonal table $L_{64}(2^{64})$ can be used for the first iteration of the first level, and the orthogonal table $L_{216}(6^{18})$ can be used for the first iteration of the fifth level.

The test program in RMOPS algorithm is ERT. Figure 2 shows the performance of the single-core implementation of GCC-8.3.0 -O3 optimization parameter setting for ERT in ARMv8 experimental platform with different data scale strength. As can be seen from Fig. 1, when the value of ERT_FLOPS is 4, the ERT test has the highest single-core performance, so the ERT_FLOPS value of the test program ERT is set to 4 in RMOPS algorithm. Other major configuration parameters of ERT such as ERT_ALIGN are set to 32 and ERT_PRECISION to FP64.

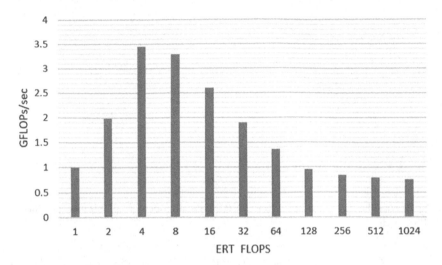

Fig. 2. The performance of the single-core implementation for ERT in ARMv8 platform

Figure 3 shows the Roofline graph generated by the result of a single-core run with the -O3 compilation optimization option for ERT. Figure 4 shows the Roofline graph generated by the results of a single-core run of ERT with the optimal optimization parameters of RMOPS search. By comparing Fig. 3 and Fig. 4, we can see that the performance obtained by using the optimal optimization parameter of RMOPS search is 34.28% better than that obtained by using only the -O3 compilation optimization option.

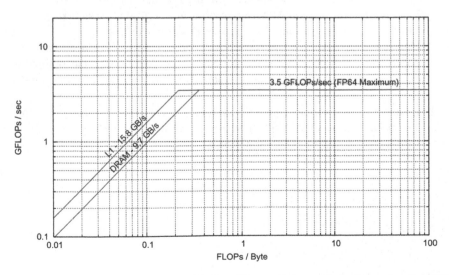

Fig. 3. The result of a single-core run with the -O3 compilation optimization option for ERT.

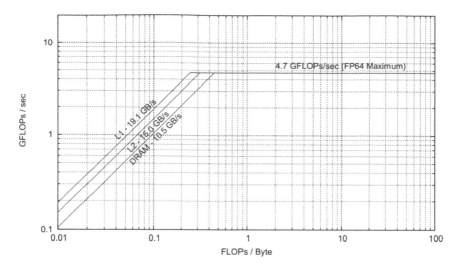

Fig. 4. The results of a single-core run of ERT with the optimal optimization parameters of RMOPS search.

Add the -fopenmp option to both sets of compilation options to compare the multi-core performance acceleration ratio. Figure 5 shows the Roofline graph generated by 64 threads running with the -O3 compilation optimization option for ERT. Figure 6 shows the Roofline graph generated by 64 threads running results for ERT using the optimal optimization parameter of RMOPS search. By comparing Fig. 5 and Fig. 6, it can be seen that the 64 core performance obtained by using the optimal optimization parameter of RMOPS search is 36.12% better than the 64 core performance obtained by using only the -O3 compilation optimization option.

Table 6 shows the percentage of program performance improvement relative to -O3 optimization parameter settings obtained from the optimal optimization parameters of RMOPS search for the NPB on ARMv8 platform. In this group of tests, except DC, the size of B, the rest of the test questions choose D size for testing.

Table 7 and Table 8 show the percentage of program performance improvement achieved on the ARMv8 platform for the SPEC CPU2017 using the optimal optimization parameters of RMOPS search relative to the -O3 optimization parameter settings.

Experiments show that the program performance obtained by using the optimal optimization parameters of RMOPS search is generally improved by 5%–33% compared with that achieved by -O3 optimization parameter setting.

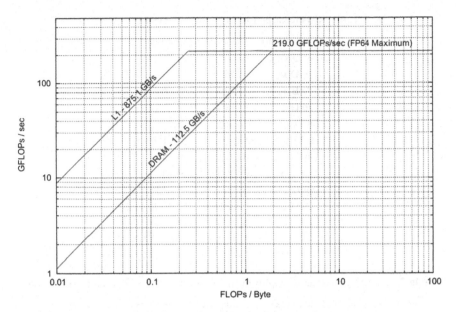

Fig. 5. The result of 64 cores run with the -O3 compilation optimization option for ERT.

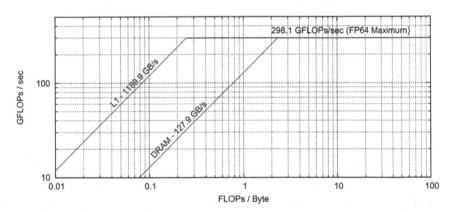

Fig. 6. The results of 64 cores run of ERT with the optimal optimization parameters of RMOPS search.

Table 6. Percent performance improvement for NPB

Benchmark	Single-core	64 cores
bt.D.x	10.46%	10.15%
cg.D.x	10.56%	7.75%
dc.B.x	7.32%	6.16%
ep.D.x	9.18%	7.87%
ft.D.x	7.21%	5.11%
is.D.x	20.56%	19.31%
lu.D.x	10.07%	9.15%
mg.D.x	13.35%	9.13%
sp.D.x	21.76%	20.13%
ua.D.x	34.08%	33.71%
Average	14.46%	12.85%

Table 7. Percent performance improvement for SPEC CPU2017 integer benchmarks.

Benchmark	Single-core	64 cores
500.perlbench_r	6.66%	5.59%
502.gcc_r	8.17%	6.78%
505.mcf_r	7.18%	6.04%
520.omnetpp_r	9.75%	7.25%
523.xalancbmk_r	10.15%	9.38%
525.x264_r	10.73%	8.12%
531.deepsjeng_r	13.74%	9.24%
541.leela_r	12.29%	9.03%
548.exchange2_r	11.92%	10.97%
557.xz_r	9.20%	8.75%
Average	9.98%	8.12%

Table 8. Percent performance improvement for SPEC CPU2017 floating point benchmarks.

Benchmark	Single-core	64 cores
503.bwaves_r	9.66%	8.86%
507.cactuBSSN_r	11.17%	9.57%
508.namd_r	7.18%	6.08%
510.parest_r	9.75%	7.35%
511.povray_r	10.15%	8.25%
519.lbm_r	8.73%	7.77%
521.wrf_r	18.74%	15.94%

(continued)

Table 8. (*continued*)

Benchmark	Single-core	64 cores
526.blender_r	12.29%	11.19%
527.cam4_r	11.92%	11.92%
538.imagick_r	23.20%	18.84%
544.nab_r	18.40%	16.00%
549.fotonik3d_r	16.20%	14.74%
554.roms_r	17.20%	14.69%
Average	13.43%	11.63%

5 Conclusion

This paper proposes a method for selecting compiler optimization parameters to maximize the performance of the target program – RMOPS, an optimization parameter selection method based on the Roofline model. This method designs the complex dependencies in the optimization parameter selection problem through the orthogonal test method, and implements the optimization parameter selection decision with the performance data generated by ERT. The evaluation of RMOPS method on ARMv8 platform achieves good results.

Acknowledgments. This work is supported by National Key Research and Development Program of China (No. 2017YFB0202003).

References

1. Ashouri, A.H., Palermo, G., Silvano, C.: Auto-tuning techniques for compiler optimization (2016)
2. Muchnick, S.S.: Advanced Compiler Design Implementation. Morgan Kaufmann, Burlington (1997)
3. Whaley, R.C.: Automated empirical optimization of high performance floating point kernels. Ph.D. thesis, Florida State University, December 2004
4. Diniz, P.C., Lee, Y.-J., Hall, M.W., et al.: A case study using empirical optimization for a large, engineering application. In: 2004 Proceedings of the 18th International Parallel and Distributed Processing Symposium, p. 200. IEEE (2004)
5. Kisuki, T., Knijnenburg, P.M.W., O'Boyle, M.F.P., et al.: Iterative compilation in program optimization. In: Proceedings of the CPC 2010 (Compilers for Parallel Computers), pp. 35–44 (2000)
6. Chen, Y., Fang, S., Huang, Y., et al.: Deconstructing iterative optimization. ACM Trans. Archit. Code Opt. (TACO) **9**(3), 21 (2012)
7. GCC Homepage. http://gcc.gnu.org. Accessed 21 Mar 2020
8. Williams, S., Waterman, A., Patterson, D.: Roofline: an insightful visual performance model for multicore architectures. Commun. ACM **52**(4), 65–76 (2009)
9. Patterson, D.A.: Latency lags bandwidth. Commun. ACM **47**(10), 71–75 (2004)
10. ERT Homepage. https://crd.lbl.gov/departments/computer-science/PAR/research/roofline. Accessed 21 Mar 2020

11. Hoste, K., Eeckhout, L.: COLE: compiler optimization level exploration. In: Proceedings of the 6th Annual IEEE/ACM International Symposium on Code Generation and Optimization, pp. 165–174 (2008)
12. SPEC CPU2017 Homepage. http://www.spec.org. Accessed 21 Mar 2020
13. NPB Homepage. https://www.nas.nasa.gov. Accessed 21 Mar 2020

Weighted Aggregator for the Open-World Knowledge Graph Completion

Yueyang Zhou, Shumin Shi$^{(\boxtimes)}$, and Heyan Huang

School of Computer Science and Technology, Beijing Institute of Technology,
Beijing, China
{3220180778,bjssm,hhy63}@bit.edu.cn

Abstract. Open-world knowledge graph completion aims to find a set of missing triples through entity description, where entities can be either in or out of the graph. However, when aggregating entity description's word embedding matrix to a single embedding, most existing models either use CNN and LSTM to make the model complex and ineffective, or use simple semantic averaging which neglects the unequal nature of the different words of an entity description. In this paper, an aggregator is proposed, adopting an attention network to get the weights of words in the entity description. This does not upset information in the word embedding, and make the single embedding of aggregation more efficient. Compared with state-of-the-art systems, experiments show that the model proposed performs well in the open-world KGC task.

1 Introduction

Knowledge graphs are multi-relationship graphs consisting of entities (nodes) and relations (edges of different types). Each edge is represented as a triple *(head, rel, tail)* which indicates that the two entities are connected by a specific relation. For example, a statement like *"Beijing is the capital of China."* can be represented as <Beijing, capitalOf, China>. Recently, KGs have been applied to question answering [1], information extraction [2] and so on. Although KGs can effectively represent structured information, there are some problems in application, such as data quality issues and data sparseness. Knowledge graph completion (KGC) is used to solve these problems. It aims to enrich incomplete graphs by assessing the possibility of missing triples. One common method is to learn low-dimensional representations for entities and relations using vector embeddings. These typical embedding-based KGC algorithms can be divided into fact-based models and additional information models. Fact-based models learn embeddings using only facts extract from triples, which can be further divided into translational distance models and semantic matching models. Translation distance models treat the relationship as a translation measured by distance between the two entities, such as TransE [3], TransH [4], TransR [5], TranSpare [6] and so on. Semantic matching models which are also called multiplicative models match latent semantics of entities and relations in their representations,

© Springer Nature Singapore Pte Ltd. 2020
J. Zeng et al. (Eds.): ICPCSEE 2020, CCIS 1257, pp. 283–291, 2020.
https://doi.org/10.1007/978-981-15-7981-3_19

such as RESCAL [7], DistMult [8], HolE [9], ComplEx [10], ANALOGY [11] and so on. Additional information models incorporate other information besides facts, e.g., entity types, relation paths and textual descriptions. These typical models are TKRL [12], PTransE [13], and DKRL [14]. The above models have a key problem that the existence of missing triples can be predicted by known entities only. But most existing KGs add new entities and relations very quickly. If we use the above models, we have to re-training the model repeatedly whenever joining a new entity or relation. It is impractical obviously. We call the above models as closed-world KGC.

Open-world KGC can predict relationships involving entities not present in the graph or those entities that have a few connections. It has a key step that it needs to aggregate the word embedding matrix of entity descriptions, entity names and relation names into a single embedding. ConMask [15] first processes the word embedding matrix of entity description with relation-dependent content masking and then aggregated with CNN. For entity and relation names, it directly uses the simple semantic averaging. OWE [16] finds that semantic averaging is better by comparing LSTM, CNN and semantic averaging when aggregating word embedding matrix. We argue that LSTM and CNN upset the information in the word embedding. In contrast, semantic averaging preserves the information as much as possible.

In this paper, we aim at extending OWE to model the weight of word embedding for open-world KGC. OWE is the latest open-world KGC and achieves the best experimental results in this field. In our model, instead of using simple semantic averaging when aggregating word embedding matrix, we propose a weighted aggregator, which aggregates words by a weighted combination of their embeddings. To estimate the weights in the weighted aggregator, we adopt an attention network [17] to model the information between the entity and the words in the entity's description.

To summarize, the present work makes the following contributions:

(1) We propose a novel aggregator to aggregate the embedding matrix with dynamic weights in the open-world KGC field and we describe the whole model as WOWE.
(2) We conduct extensive comparisons with conventional open-world KGC models on the link prediction task. The results show that the proposed model is better than them in some respects.

In the remainder of the paper, we describe our model in Sect. 2 and compare the experimental results of our model and other baselines to prove the validity of our proposed model in Sect. 3. We finally conclude the model and give some future work directions in Sect. 4.

2   Our Model

In this section, we introduce WOWE that extends the OWE to training the weights of word embeddings when aggregating them. Given an incomplete knowledge graph represented by a set of triples $T = (h, r, t), h, t \in E, r \in R$, where

h, r and t denote the head entity, relation, and tail entity, E and R denote entity set and relation set. Our model is expected to score the missing triples whose head entity or tail entity can be absent from the original entity set E with the text information of entity name and description. The higher the score, the more likely the missing triple is to exist.

2.1 OWE and WOWE

OWE has two parts, one is closed-world KGC and the other is an open-world extension. For closed-world KGC, a regular KGC model scores triples (h, r, t):

$$score(h, r, t) = \phi(u_h, u_r, u_t), \quad u_x \in R^d \tag{1}$$

where u_x is the embedding of entity/relation x and d is the dimension of u_x. ϕ is a scoring function that is expected to get a high score when (h, r, t) holds, and low otherwise. For open-world extension, its goal is to extend the closed-world KGC model to perform open-world KGC. Because the task of tail prediction and head prediction are the same concepts, we will only discuss tail prediction. In word embedding layer, given an absent head entity $head \notin E$ which is represented by its name and description, we need to concatenate it into a word sequence $W = (w_1, w_2, \ldots, w_n)$. The word sequence W is then transformed into a sequence of embeddings $(v_{w_1}, v_{w_2}, \ldots, v_{w_n})$. The sequence of embeddings can also be represented by a word embedding matrix $M \in R^{n*k}$, where n is the number of words and k is the dimension of word embedding. In aggregation layer, this word embedding matrix M is then aggregated to a single text-based embedding $v_h \in R^k$. OWE uses semantic averaging after comparing the performance of CNN, LSTM and semantic averaging. In close-world layer, the graph-based embedding $u_h \in R^d$ is obtained through close-world KGC model such as TransE, DisMult, CompleX and so on. In transformation layer, the text-based embedding v_h is transformed to the graph-based embedding u_h such that $\Psi^{map}(v_h) \approx u_h$. If we need to do the task of open-world KGC, we can score triples with an absent head entity by applying the graph-based model with the mapped text-based head description:

$$score(h, r, t) = \phi(\Psi^{map}(v_h), u_r, u_t) \tag{2}$$

As discussed above, OWE ignores the unequal nature of the different words of an entity description in aggregation layer. WOWE aggregates the word embeddings with attention weights instead of directly taking the weights as $1/n$. The other layers use the settings of OWE to achieve optimal performance, such as map functions using affine, close-world KGC model using ComplEx. The overall structure of WOWE is illustrated in Fig. 1. We also provide details in training the weighted aggregator.

2.2 Incorporating Attention Mechanism

As the OWE approach above, after the word embedding layer, we get a sequence of embeddings $(v_{w_1}, v_{w_2}, \ldots, v_{w_n})$. For the absent head entity $head$, the words

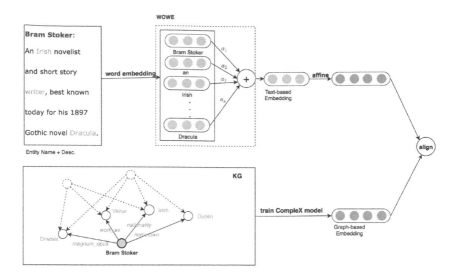

Fig. 1. General architecture of WOWE. First, the absent entity name and description is transformed into a sequence of embeddings. Then, the embeddings are aggregated to a text-based embedding with attention weights. Next, graph-based embedding is obtained through close-world KGC model called ComplEx. Finally, the text-based embedding is transformed to the graph-based embedding using affine. Our improvements are shown in the dotted boxes.

of its name and description should contribute differently to v_h according to its importance in representing *head*. To consider the different contribution, we employ an attention-based aggregating approach on the sequence of embeddings $(v_{w_1}, v_{w_2}, \ldots, v_{w_n})$. The additional information in the words W and *head* is then exploited when estimating the attention weights. Formally, the single text-based embedding v_h is computed as a weighted sum of these words:

$$v_h = \sum_{i=1}^{n} \alpha_i v_{w_i} \tag{3}$$

Here α_i is the attention weight for each word w_i.

2.3   A Weighted Aggregator

The goal of aggregator is to get a text-based embedding. The word embedding matrix M is aggregated into a k-dimension embedding. The difference between OWE and our model in aggregation layer is that OWE uses semantic averaging, which is equivalent to taking the weight of each word as $1/n$ and our model regards the weight of each word as a parameter for training. This not only does not upset information in the word embedding, but also make the single embedding of aggregation more efficient.

ConMask [15] uses the simple function called Maximal Word-Relationship Weights (MWRW) to get the weight of each word in entity description before aggregating the word embedding matrix. MWRW assigns higher scores to the words that appear in the relationship or are semantically similar to the relationship. Inspired by ConMask, we present a novel aggregator to OWE which enables it to satisfy the unequal nature of the different words of an entity description. The aggregator gets the single text-based embedding v_h by Eq. 3. To take the attention mechanism into consideration, we adopt an attention network [17].

Specifically, given an absent head entity *head*, the importance of a word in an entity's description is measured by

$$\alpha_i = softmax(\alpha_i{}') = \frac{exp(\alpha_i{}')}{\sum_{i=1}^{n} exp(\alpha_i{}')} \tag{4}$$

Here the unnormalized attention weight $\alpha_i{}'$ is given by an attention neural network as

$$\alpha_i{}' = v_a^\top tanh(W_a z_h + U_a v_{w_i}) \tag{5}$$

where $v_a \in R^k$, $W_a \in R^{k*k}$ and $U_a \in R^{k*k}$ are global attention parameters, while z_h is the semantic averaging of the absent head entity *head*'s name. All those attention parameters are regarded as parameters of the weighted aggregator, and learned directly from the data.

3 Experiments

3.1 Data Sets

In order to better compare the performance of WOWE and OWE, the data sets in this paper are same as OWE. For closed-world KGC, we used FB15k/FB20k, DBPedia50k and FB15k-237-OWE. For open-world KGC, we used FB20k, DBPedia50k and FB15k-237-OWE. The statistics of the datasets are listed in Table 1 and Table 2.

Table 1. Statistics of data sets for closed-world KGC

Dataset	$|E|$	$|R|$	Train	Valid	Test
FB15k/FB20k	14,904	1,341	472,860	48,991	57,803
DBPedia50k	24,624	351	32,388	123	2,095
FB15k-237-OWE	12,324	235	242,489	12,806	–

Table 2. Statistics of data sets for open-world KGC

Dataset	$|E^{open}|$	Train	Head pred.		Tail pred.	
			Valid	Test	Valid	Test
FB20k	5,019	472,860	–	18,753	–	11,586
DBPedia50k	3,636	32,388	55	2,139	164	4,320
FB15k-237-OWE	2,081	242,489	1,539	13,857	9,424	22,393

3.2   Evaluation Protocol

We follow the same protocol in OWE: For each testing triplet (h, r, t) with open-world head $h \notin E$, we replace the tail t by every entity e in the knowledge graph and calculate a different score on the corrupted triplet (h, r, e). After ranking the scores, we get the rank of the correct triplet. Aggregated over all the testing triplets, five metrics are reported: mean rank (MR), mean reciprocal rank (MRR), Hits@1, Hits@3, and Hits@10. This is called the "raw" setting. Note that if a corrupted triplet exists in the knowledge graph, as it is also correct, ranking it before the original triplet is not wrong. To eliminate this factor, we remove those corrupted triplets which exist in either training, valid, or testing set before getting the rank of each testing triplet. This setting is called "filtered".

Table 3. Comparison with other open-world KGC models on tail prediction.

Data sets	Model	Hits@1	Hits@3	Hits@10	MRR
DBPedia50k	DKRL	–	–	40.0	23.0
	ConMask	47.1	64.5	**81.0**	58.4
	OWE	51.9	65.2	76.0	60.3
	WOWE	**52.7**	**66.5**	76.9	**61.2**
FB20k	DKRL	–	–	–	–
	ConMask	42.3	57.3	**71.7**	53.3
	OWE	44.8	57.1	69.1	53.1
	WOWE	**45.2**	**58.3**	70.0	**54.1**
FB15k-237-OWE	DKRL	–	–	–	–
	ConMask	21.5	39.9	45.8	29.9
	OWE	31.6	43.9	56.0	40.1
	WOWE	**31.9**	**44.1**	**56.4**	**40.4**

3.3   Implementation

As the data sets are the same, we directly copy the experimental results of several baselines from OWE. In training WOWE, we use the OpenKE framework when training ComplEx. For training the transformation Ψ^{map}, we used the Adam optimizer with a learning rate of 10^{-3} and batch size of 128. For DBPedia50k, FB20k and FB15k-237-OWE, the value of dropout is 0.5, 0 and 0. We used the pretrained 300 dimensional Wikipedia2Vec embedding as the word embedding and used affine as the transformation.

3.4 Comparison with State of the Art

The results are reported in Table 3. We can see that our model WOWE performs competitively when compared to OWE. On FB15k-237-OWE, our model performs a little better than OWE. On DBPedia50k and FB20k, the results are better on FB15k-237-OWE. We argue that this is due to FB15k-237-OWE having very short descriptions. OWE used simple semantic averaging whose weights with fixed value 1/n. Shorter descriptions make dynamic weights less advantageous, but increase the complexity. On the other two data sets with longer descriptions, our model is obviously better than OWE.

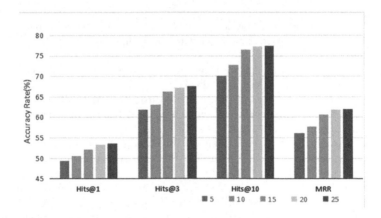

Fig. 2. Results on DBPedia50k with different lengths of entity descriptions. Different colors mean different lengths of entity descriptions

3.5 Influence of the Length of Entity Descriptions

According to the result of comparison with state of the art, it's reasonable to suppose that when the length of entity descriptions increases, WOWE's performance would be better. To figure out whether our WOWE's performance is related to the length of entity descriptions, we conduct link prediction on datasets with different lengths of entity descriptions. The results are displayed in Fig. 2. We observe that the increasing of length of entity descriptions certainly has a positive effect on the performance of WOWE. So we can consider WOWE more suitable for entities with longer descriptions.

4 Conclusion

In this paper, we have introduced WOWE, a new model to perform open-world knowledge graph completion with dynamic word embedding weights. WOWE extends OWE to aggregating words by a weighted combination of their embeddings while inheriting its advantages and adopt an attention network to get the weights. Experimental results show that WOWE achieves improvements in some respects as compared with ConMask, OWE and other baselines. The WOWE model uses a more complex aggregation function. For the weights of word embeddings, instead of estimating by attention mechanism, we will try other method in the future, such as logic rule.

Acknowledgements. This work was supported by the National Natural Science Foundation of China (Grant No. 61671064, No.61732005) and National Key Research & Development Program (Grant No. 2018YFC0831700).

References

1. Lukovnikov, D., Fischer, A., Lehmann, J., Auer, S.: Neural network-based question answering over knowledge graphs on word and character level. In: WWW, pp. 1211–1220 (2017)
2. Dong, X.L., et al.: Knowledge vault: a web-scale approach to probabilistic knowledge fusion. In: KDD, pp. 601–610 (2014)
3. Bordes, A., Usunier, N., Garcia-Duran, A., Weston, J., Yakhnenko, O.: Translating embeddings for modeling muti-relational data. In: Advances in Neural Information Processing Systems, pp. 2787–2795 (2013)
4. Wang, Z., Zhang, J., Feng, J., Chen, Z.: Knowledge graph embedding by translating on hyperplanes. In: Proceedings of the 28th AAAI Conference on Artificial Intelligence, pp. 1112–1119 (2014)
5. Lin, Y., Liu, Z., Sun, M., Liu, Y., Zhu, X.: Learning entity and relation embeddings for knowledge graph completion. In: AAAI 2015, pp. 2181–2187 (2015)
6. Ji, G., Liu, K., He, S., Zhao, J.: Knowledge graph completion with adaptive sparse transfer matrix. In: AAAI, pp. 985–991 (2016)
7. Maximilian, N., Tresp, V., Kriegel, H.P.: A three-way model for collective learning on multi-relational data. In: Proceedings of the 28th International Conference on Machine Learning, pp. 809–816 (2011)
8. Yang, B., Yih, W., He, X., Gao, J., Deng, L.: Embedding entities and relations for learning and inference in knowledge bases. In: ICLR (2015)
9. Nickel, M., Rosasco, L., Poggio, T.: Holographic embeddings of knowledge graphs. In: Proceedings of the 30th AAAI Conference on Artificial Intelligence, pp. 1955–1961 (2016)
10. Trouillon, T., Welbl, J., Riedel, S., Gaussier, E., Bouchard, G.: Complex embeddings for simple link prediction. In: Proceedings of the 33rd International Conference on Machine Learning, pp. 2071–2080 (2016)
11. Liu, H., Wu, Y., Yang, Y.: Analogical inference for multi-relational embeddings. In: ICML, pp. 2168–2178 (2017)
12. Xie, R., Liu, Z., Sun, M.: Representation learning of knowledge graphs with hierarchical types. In: IJCAI, pp. 2965–2971 (2016)

13. Lin, Y., Liu Z., Luan H., Sun, M., Rao, S., Liu, S.: Modeling relation paths for representation learning of knowledge bases. In: Proceedings of the Conference on Empirical Methods Natural Language Process, pp. 705–714 (2015)
14. Xie, R., Liu, Z., Jia, J., Luan, H., Sun, M.: Representation learning of knowledge graphs with entity descriptions. In: Proceedings of the 30th AAAI Conference on Artificial Intelligence, pp. 2659–2665 (2016)
15. Shi, B., Weninger, T.: Open-world knowledge graph completion. CoRR abs/1711.03438 (2017)
16. Shah, H., Villmow, J., Ulges, A., Schwanecke, U., Shafait, F.: An open-world extension to knowledge graph completion models. In: AAAI (2019)
17. Bahdanau, D., Cho, K., Bengio, Y.: Neural machine translation by jointly learning to align and translate. In: ICLR (2015)

Music Auto-Tagging with Capsule Network

Yongbin Yu[1(✉)], Yifan Tang[1], Minhui Qi[1], Feng Mai[1],
Quanxin Deng[1], and Zhaxi Nima[2]

[1] University of Electronic Science and Technology of China, Chengdu, China
ybyu@uestc.edu.cn
[2] Tibet University, Lasa, China

Abstract. In recent years, convolutional neural networks (CNNs) become popular approaches used in music information retrieval (MIR) tasks, such as mood recognition, music auto-tagging and so on. Since CNNs are able to extract the local features effectively, previous attempts show great performance on music auto-tagging. However, CNNs is not able to capture the spatial features and the relationship between low-level features are neglected. Motivated by this problem, a hybrid architecture is proposed based on Capsule Network, which is capable to extract spatial features with the routing-by-agreement mechanism. The proposed model was applied in music auto-tagging. The results show that it achieves promising results of the ROC-AUC score of 90.67%.

Keywords: Convolutional neural networks · Music Information Retrieval · Music auto-tagging · Capsule network

1 Introduction

Music Information Retrieval (MIR) is a growing field of research which contains many audio tasks such as genre classification [1], music auto-tagging [2] and melody extraction [3]. Over the last few years, much attention has been put on music auto-tagging with deep learning approaches [4–6]. Music auto-tagging focus on predicting descriptive tags of music from audio signals, such as genre (classical, rock, blues, pop), instrumentation (guitar, violin, piano), emotion (happy, sad, excited) and eras (60s–00s). Since tags provide high-level information from the listeners' perspectives, music auto-tagging plays an important role in music discovery and recommendation.

Since CNNs are one of the most popular approaches applied in image classification and other areas, CNN-based models have been widely adopted in MIR tasks especially in music auto-tagging [2, 4, 5]. Previous works have proved that CNNs are capable of extracting local features effectively and show great performance in music auto-tagging. However, one shortcoming of CNNs is that with the kernel-based convolution process, CNNs are not sensitive to spatial relationship or orientation information of input features [7]. To this end, Hinton et al. [8] proposed a novel network named Capsule Network, which is able to learn hierarchical relationship between different layers and it has achieved promising performance on image classification.

For music auto-tagging, most successful CNN-based model takes spectrogram as input [4, 5, 9], which is similar to images, containing 2D information in both time steps

© Springer Nature Singapore Pte Ltd. 2020
J. Zeng et al. (Eds.): ICPCSEE 2020, CCIS 1257, pp. 292–298, 2020.
https://doi.org/10.1007/978-981-15-7981-3_20

and frequency domain. Therefore, we assume that the latent spatial feature also exists in the spectrogram which is significant for this task. However, CNN are not good at capturing spatial relationship or position information of spectrogram. To this end, we proposed a hybrid architecture based on Capsule Network and applied it to music auto-tagging, using spectrogram as input. The experiment conducted on MagnaTagATune (MTAT) dataset [10] shows that our proposed model is effective in music auto-tagging.

In this paper, we introduced a hybrid Capsule Network for music auto-tagging and explored its architecture to reach a great performance. Compared with previous state-of-the-art models, our proposed model shows comparable performance in ROC-AUC score of 90.67% on MTAT dataset.

2 Models

2.1 Capsule Network

Hinton et al. [8] proposed Capsule Network with the routing-by-agreement mechanism. A capsule is defined as a group of neurons being represented by activity vectors and the length of the vector represents the probability that the feature exists. With the routing-by-agreement mechanism, the capsule network is able to learn the hierarchical relationship between capsules in adjacent layers. Assume there are two capsules i and j in the l capsule layer and $l+1$ capsule layer respectively. And b_{ij} represents the log priori probabilities that capsule i should be coupled to capsule j. The coupling coefficients c_{ij} is defined as:

$$c_{ij} = \frac{\exp(b_{ij})}{\sum_k \exp(b_{ik})} \tag{1}$$

With c_{ij}, we could calculate the s_j which means the total input of capsule j:

$$s_j = \sum_i c_{ij} W_{ij} u_i \tag{2}$$

Where W_{ij} is the weight matrix and u_i is the output of the capsule i. Then we can get the vector output of capsule j with a non-linear squashing function:

$$v_j = \frac{\|s_j\|^2}{1 + \|s_j\|^2} \frac{s_j}{\|s_j\|} \tag{3}$$

Which is used to ensure that short vectors get shrunk to almost zero length and long vector get shrunk to a length slightly below 1. Therefore, we could update b_{ij} with Eq. (4):

$$b_{ij} = b_{ij} + W_{ij} u_i v_j \tag{4}$$

The routing algorithm described with Eq. 1–4 and through iteration of routing-by-agreement, capsule output is only routed to the appropriate capsule in the next layer. Thus, capsules will get a cleaner input signal and will more accurately determine the pose of the object.

In addition, margin loss is selected for loss function in training:

$$L_k = T_k \max(0, m^+ - \|v_k\|)^2 + \lambda(1 - T_k)\max(0, \|v_k\| - m^-) \tag{5}$$

Where c is the number of output class and v_k is the final output capsule. $T_k = 1$ iff the target class is k, $m^+ = 0.9$, $m^- = 0.1$, $\lambda = 0.5$.

A basic Capsule Network architecture is also proposed in [8] with a convolution layer in the bottom, following by tow capsule layers called PrimaryCaps layer and DigitCaps layer. Furthermore, a decoder is built after the output capsules to reconstructs the input images, which consists of three fully connection layers with ReLU activation for the first two layers and sigmoid activation for the last layer. In addition, the loss function of decoder is mean square error (MSE).

2.2 Proposed Model

As illustrated in Fig. 1, we proposed a hybrid architecture based on the basic architecture of Capsule Network proposed in [8]. The hybrid architecture consists of 7 convolution layers with 128 filters and the kernel size is 3 × 3 which is used to extract features in low level from the input. Each convolution layer uses batch normalization [11] and ReLU activation. To provide a representation with appropriate size to the capsule, we also add max pooling layer behind ReLU activation layer. Furthermore, we use dropout [12] technique with each convolution layer to avoid overfitting. The output of the last convolution layer is fed to a parallel architecture which consists of two parts.

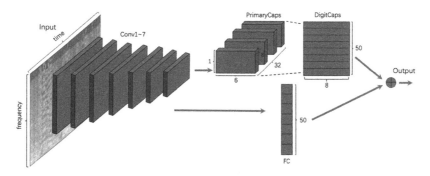

Fig. 1. A hybrid architecture based on Capsule Network.

The first part is Capsule Network which is used to capture spatial feature. The Capsule Network is composed of two capsule layers, PrimaryCaps layer and DigitCaps layer. In PrimaryCaps layer, the dimensionality of capsules is 8D which means that each capsule is comprised of 8 convolutional units and the kernel size is 3×3. Each capsule in the same channel of PrimaryCaps layer is sharing their weight with each other while the number of channels is set to 32. The DightCaps layer has one 16D capsule per digit class and each of these capsules receive input from capsules in the PrimaryCaps layer. Since capsules only exist in PrimaryCaps layer and DigitCaps layer, routing-by-agreement mechanism is only work between the two layers and the routing times is set to 3. As the final layer of Capsule Network, the classification result is output by DigitCaps layer while the number of capsules in this layer, which is the same as the number of target classes.

To retain the local feature extracted from the convolution layer, the second part is set as a fully connection layer with 50 units, which is corresponding to the 50 target classes. Finally, in order to integrate the prediction from fully connection layer and DigitCaps layer, the outputs of them is combined as the final output of the model.

3 Experiment

3.1 Dataset

The MagnaTagATune Dataset is one of the most widely used dataset for music auto-tagging which consists of 25,863 music clips with 188 tags including genres, instruments, emotion and other descriptive classes. Considering that we only select the most frequently used 50 tags, 21,108 music clips labeled with these tags are used for experiment. We set 15,247 music clips for training, 1,529 for validation and 4,332 for test.

3.2 Setup

Following previous work [4, 5, 9], all the music clips are trimmed to 29.1 s and resampled to 12,000 Hz by Librosa [13]. Audio signal from music clips are converted into log amplitude mel-spectrogam as the input of network, with the size of (96, 1366, 1), which represents frequency, time and channels respectively. Moreover, Adam [14] is chosen as the activation function.

3.3 Result

As shown in Fig. 2, we conducted experiments to explore the architecture in different aspects. Firstly, since [9] has proved that model with the number of convolution layers of 6 would have a promising performance on music auto-tagging, we tested the model with 5, 6, 7, 8 convolution layers at the bottom. The results shown in Fig. 2 (a) indicates that the model with CNN of 7 layers is most effective. We also tried different number of filters in convolution layers as shown in Fig. 2 (b). Secondly, Fig. 2 (c)(d)(e) illustrate the results of experiments with diverse parameters of the capsule layer including PrimaryCaps layer and DigitCaps layer. Moreover, we also tried different architecture with the fully connection layer followed by GRU or LSTM and attention mechanism such as SE-Block [15] and CBAM [16]. However, as shown in Fig. 2 (f), all of them did not improve the performance. Finally, with all the optimized parameters of our proposed model, we get a promising result of 90.67% ROC-AUC score.

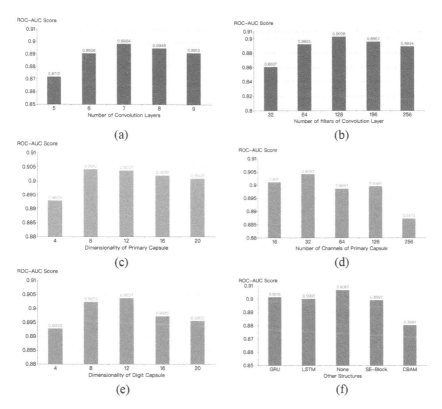

(a)　　　　　　　　　　　　　(b)

(c)　　　　　　　　　　　　　(d)

(e)　　　　　　　　　　　　　(f)

Fig. 2. Comparison of our proposed model with different architecture, (a) compared the number of convolution layers from 5 to 9, (b) compared the number of filters of convolution layers, (c) compared the dimensionality of Primary Capsule, (d) compared the number of channels of Primary Capsule, (e) compared the dimensionality of Digit Capsule, (f) compared diverse structures based on the proposed model.

Furthermore, to validate the effectiveness of Capsule Network and fully connection layer, we conduct experiments without these architectures respectively. The results shown in Table 1 imply that these structures are helpful for music auto-tagging and the combination of them shows better performance. We also tested the model without max-pooling layer since it may lead to the loss of valuable information. In the architecture without pooling layers, we increased the stride of convolution layer from 1 to 2, in order to reduce the size of the feature map and make it similar to that with pooling layer. However, the result indicates that the performance is worse without max-pooling layer. Based on this result, it can be assumed that not all the information in spectrogram is benefit for boosting the performance of music auto-tagging and max-pooling layer is able to eliminate the noise in spectrogram by discarding the redundant information and reduce the size of feature map in each convolution layer.

Table 1. Comparison of our proposed model without some structures

Model	AUC-ROC score
Proposed model without FC layer	0.8973
Proposed model without capsule layers	0.8961
Proposed model without max-pooling layer	0.8605
Proposed model	**0.9067**

Table 2 shows the comparison of our proposed model and previous state-of-the-arts model. It should be noted that our proposed model achieved a promising AUO-RUC score and outperforms all the state-of-the-art models. This indicates that the spatial feature learned by Capsule Network is useful for music auto-tagging.

Table 2. Comparison of our proposed model to prior state-of-the-art models

Model	AUC-ROC score
FCN [9]	0.894
Sample-CNN [2]	0.9055
Global model with both multi-features [4]	0.9021
Bag of multi-scaled features [17]	0.888
1-D CNN [18]	0.8815
Transferred learning [19]	0.882
Proposed model	**0.9067**

4   Conclusion

In this paper, we introduced a hybrid architecture base on Capsule Network, which is designed for extracting the spatial feature in spectrogram. Through the experiments on MTAT datasets with different architecture of our proposed model, it proved that the Capsule Network and max-pooling layer are capable of improving the performance in music auto-tagging. Finally, it measured the proposed model with ROC-AUC score, which showed a promising performance. In the future, the architecture of Capsule Network can be applied in music auto-tagging taking raw waveform as input will be investigated.

Acknowledgement. This work was supported by Research Fund for Sichuan Science and Technology Program (GrantNo. 2019YFG0190) and Research on Sino-Tibetan multisource information acquisition, fusion, data mining and its application (Grant No. H04W170186).

References

1. Nanni, L., Costa, Y.M.G., Lumini, A., et al.: Combining visual and acoustic features for music genre classification. Expert Syst. Appl. **45**, 108–117 (2016)
2. Lee, J., Park, J., Kim, K.L., et al.: Sample-level deep convolutional neural networks for music auto-tagging using raw waveforms. arXiv preprint arXiv:1703.01789 (2017)
3. Su, L.: Vocal melody extraction using patch-based CNN. In: 2018 IEEE International Conference on Acoustics, Speech and Signal Processing (ICASSP), pp. 371–375. IEEE (2018)
4. Lee, J., Nam, J.: Multi-level and multi-scale feature aggregation using pretrained convolutional neural networks for music auto-tagging. IEEE Signal Process. Lett. **24**(8), 1208–1212 (2017)
5. Choi, K., Fazekas, G., Sandler, M., et al.: Convolutional recurrent neural networks for music classification. In: 2017 IEEE International Conference on Acoustics, Speech and Signal Processing (ICASSP), pp. 2392–2396. IEEE (2017)
6. Pons, J., Nieto, O., Prockup, M., et al.: End-to-end learning for music audio tagging at scale. arXiv preprint arXiv:1711.02520 (2017)
7. Wu, X., Liu, S., Cao, Y., et al.: Speech emotion recognition using capsule networks. In: ICASSP 2019-2019 IEEE International Conference on Acoustics, Speech and Signal Processing (ICASSP), pp. 6695–6699. IEEE (2019)
8. Sabour, S., Frosst, N., Hinton, G.E.: Dynamic routing between capsules. In: Advances in Neural Information Processing Systems, pp. 3856–3866 (2017)
9. Choi, K., Fazekas, G., Sandler, M.: Automatic tagging using deep convolutional neural networks. arXiv preprint arXiv:1606.00298 (2016)
10. Law, E., West, K., Mandel, M.I., et al.: Evaluation of algorithms using games: the case of music tagging. In: ISMIR, pp. 387–392 (2009)
11. Ioffe, S., Szegedy, C.: Batch normalization: accelerating deep network training by reducing internal covariate shift. arXiv preprint arXiv:1502.03167 (2015)
12. Srivastava, N., Hinton, G., Krizhevsky, A., et al.: Dropout: a simple way to prevent neural networks from overfitting. J. Mach. Learn. Res. **15**(1), 1929–1958 (2014)
13. McFee, B., Raffel, C., Liang, D., et al.: librosa: audio and music signal analysis in python. In: Proceedings of the 14th Python in Science Conference, vol. 8 (2015)
14. Kingma, D.P., Ba, J.: Adam: a method for stochastic optimization. arXiv preprint arXiv: 1412.6980 (2014)
15. Hu, J., Shen, L., Sun, G.: Squeeze-and-excitation networks. In: Proceedings of the IEEE Conference on Computer Vision and Pattern Recognition, pp. 7132–7141 (2018)
16. Woo, S., Park, J., Lee, J.-Y., Kweon, I.S.: CBAM: convolutional block attention module. In: Ferrari, V., Hebert, M., Sminchisescu, C., Weiss, Y. (eds.) ECCV 2018. LNCS, vol. 11211, pp. 3–19. Springer, Cham (2018). https://doi.org/10.1007/978-3-030-01234-2_1
17. Nam, J., Herrera, J., Lee, K.: A deep bag-of-features model for music auto-tagging. arXiv preprint arXiv:1508.04999 (2015)
18. Dieleman, S., Schrauwen, B.: End-to-end learning for music audio. In: 2014 IEEE International Conference on Acoustics, Speech and Signal Processing (ICASSP), pp. 6964–6968. IEEE (2014)
19. Van Den Oord, A., Dieleman, S., Schrauwen, B.: Transfer learning by supervised pretraining for audio-based music classification. In: Conference of the International Society for Music Information Retrieval (ISMIR 2014) (2014)

Complex-Valued Densely Connected Convolutional Networks

Wenhan Li, Wenqing Xie, and Zhifang Wang$^{(\boxtimes)}$

Department of Electronic Engineering,
Heilongjiang University, Harbin 150080, China
wangzhifang@hlju.edu.cn

Abstract. In recent years, deep learning has made significant progress in computer vision. However, most studies focus on the algorithms in real field. Complex number can have richer characterization capabilities and use fewer training parameters. This paper proposes a complex-valued densely connected convolutional network, which is complex-valued DenseNets. It generalizes the network structure of real-valued DenseNets to complex field and constructs the basic architectures including complex dense block and complex transition layers. Experiments were performed on the CIFAR-10 database and CIFAR-100 datasets. Experimental results show the proposed algorithm has a lower error rate and fewer parameters than real-valued DenseNets and complex-valued ResNets.

Keywords: Deep learning · Complex field · Convolutional network

1 Introduction

In the past few years, significant progress has been made in deep neural networks. One of breakthroughs is Batch Normalization (BN) [1], which standardizes the input of each layer to ensure the relative stability of the input distribution during training, and weakens the dependence of the parameters between the layers. BN has been shown to regularize the network and accelerate network training. Shortcut connection is another breakthrough. This structure can construct a deep network and solve the problems of network degradation and gradient disappearance. Highway Networks [2], Residual Networks (ResNets) [3], and DenseNets [4] all use this method to solve gradient disappearance and network degradation. Other work has been done to improve performance. For example, attention module has been widely used in various types of deep learning tasks such as computer vision, language processing and speech recognition. However, as the number of network layers increases, the number of parameters, memory space, and computational complexity also increase. Researchers are turning to build deeper architectures with the lowest possible parameter cost.

Because of fewer training parameters [5] and better generalization ability [6], complex-valued neural networks have attracted researchers' attention. When training a neural network, complex-valued neurons have more complex representation than real-valued neurons, so complex-valued networks have better learning ability than real-valued networks. It has been proved that complex-valued recurrent neural networks

© Springer Nature Singapore Pte Ltd. 2020
J. Zeng et al. (Eds.): ICPCSEE 2020, CCIS 1257, pp. 299–309, 2020.
https://doi.org/10.1007/978-981-15-7981-3_21

have higher learning speed and stronger noise robustness [7–9]. Recently, the complex BN, complex convolution, and complex weight initialization are proposed and applied to ResNets [10]. DenseNets uses dense connectivity to improve information flow capabilities and reduce the number of parameters. This paper generalizes the real-valued DenseNets to complex field and proposes complex-valued DenseNets. The key characteristics are as follows: 1) Densely connected convolutional networks are generalized to complex field. 2) Complex dense block and complex transition layers are constructed. 3) The input formula of the layer also changes accordingly to accommodate complex-valued input data. 4) Experiments performed on CIFAR-10 and CIFAR-100 datasets show that our algorithm has better generalization capabilities, faster convergence speed and fewer training parameters than real-valued DenseNets and complex-valued ResNets.

2   Motivation

From biological and signal processing perspectives, there are many advantages to generalize neural networks to the complex field. Reichert and Serre realized a deep network with biological significance, which used complex-valued neuron units in order to construct a richer and more general expression [11]. The complex-valued representation can express the neuron's output in terms of its firing rate and the relative timing of its activity. The amplitude of the complex-valued neurons represents the former, and the phase represents the latter. Input neurons with large differences in phase information are called asynchronous neurons, otherwise they are called synchronous neurons. This characteristic of complex-valued neurons has some similarities with the gating mechanism in deep feedforward neural networks and recurrent neural networks. In a deep network with a gating mechanism, asynchronous means that its control gate keeps a low activation value as the input propagates, while synchronization means that its control gate keeps a high activation as the input propagates. Phase information is also very important in signal processing. Oppenheim and Lin have proved that the phase of a picture contains much more information than the amplitude [12]. That is, as long as the phase of the picture is known, it is enough to restore the original picture. Actually, phase provides a detailed description of objects as it encodes shapes, edges and orientations.

In fact, complex-valued neural networks [14–16] have been studied for a long time before deep learning was proposed [13]. Recently, the references [8] and [9] tried to establish the theoretical basis of complex-valued deep neural networks and their mathematical proofs. Its aim is to arouse people's attention and attention to complex-valued deep neural networks. Reference [17] has actually used complex-valued deep neural networks in visual tasks. Reference [10] extended ResNets to complex field and achieved better performance on visual tasks. Since DenseNets performs better than ResNets on vision tasks, our paper extends DenseNets to complex field and proposes complex-valued DenseNets.

3   Complex-Valued DenseNets

This section will introduce the network structure of complex-valued DenseNets. The definition of complex convolution is adopted [10]:

$$W * h = (A * x - B * y) + i(B * x + A * y) \tag{1}$$

Where $W = A + Bi$ is a complex-valued filter matrix, $h = x + yi$ is a complex-valued feature map matrix, A, B, x and y are all real matrices, $*$ represents convolution operation. The implementation process of complex convolution is shown in Fig. 1. The feature map is divided into two parts: the real part and the imaginary part. And the convolution kernel is also divided into the real part and the imaginary part.

In deep learning, BN is usually used to accelerate network training. The principle of BN is to keep the activation of each layer of the network at zero mean and unit variance. The original BN algorithm only applies to real numbers, not complex numbers. Therefore, this paper adopts the complex BN [10]. The flow is as follows:

$$\tilde{x} = V^{-\frac{1}{2}}(x - E(x)) \tag{2}$$

Where $x - E(x)$ represents 0 mean data, V is covariance matrix, and $\tilde{x}$ denotes standardized data.

The covariance matrix V is expressed as:

$$V = \begin{pmatrix} \text{Cov}\{R(x), R(x)\} & \text{Cov}\{R(x), I(x)\} \\ \text{Cov}\{I(x), R(x)\} & \text{Cov}\{I(x), I(x)\} \end{pmatrix} \tag{3}$$

The scaling parameters γ are expressed as:

$$\gamma = \begin{pmatrix} \gamma_{rr} & \gamma_{ri} \\ \gamma_{ir} & \gamma_{ii} \end{pmatrix} \tag{4}$$

$$\text{BN}(\tilde{x}) = \gamma\tilde{x} + \beta \tag{5}$$

In a general case, particularly when batch normalization is not performed, proper initialization is critical in reducing the risks of vanishing or exploding gradients. This paper adopts the complex weight initialization [10].

Complex weight can be expressed as:

$$W = |W|e^{i\theta} = R\{W\} + iI\{W\} \tag{6}$$

Where θ and $|W|$ are respectively the phase and magnitude of W.

Variance is defined as:

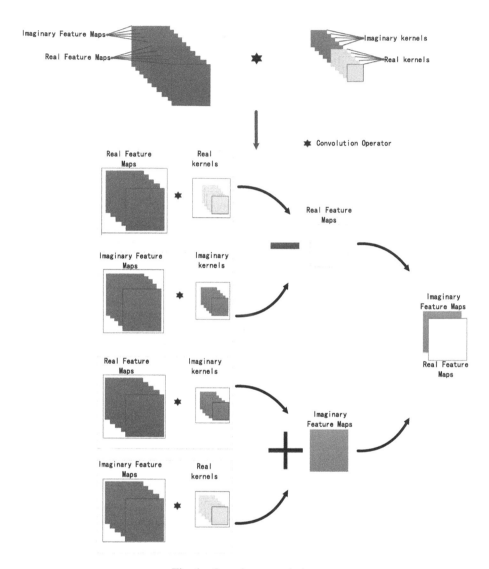

Fig. 1. Complex convolution.

$$\mathrm{Var}(W) = \mathrm{E}\left[|W|^2\right] - (\mathrm{E}[W])^2 \tag{7}$$

Where W is symmetric around 0, therefore, $(\mathrm{E}[W])^2$ is 0. Variance can be written as:

$$\mathrm{Var}(W) = \mathrm{E}\left[|W|^2\right] \tag{8}$$

According to [10], the Variance of W can be expressed as:

$$\begin{cases} \mathrm{Var}(W) = 2\sigma^2 \\ \sigma = 1/\sqrt{n_{in} + n_{out}} \quad or \quad \sigma = 1/\sqrt{n_{in}} \end{cases} \tag{9}$$

Where n_{in} and n_{out} are the number of input and output units respectively.

Referring to the spatial structure of real-valued DenseNets, this paper constructs 57-layer complex-valued DenseNets. It consists of input, preprocessing module (PB), complex Dense Block, complex transition layers (CTL), pooling layer (MP), softmax layer (SM), and output. The network structure is shown in Fig. 2.

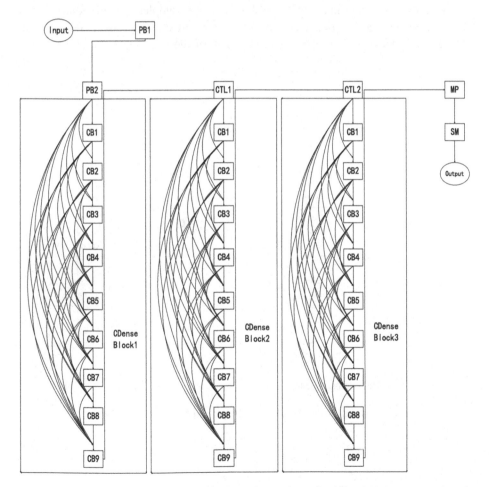

Fig. 2. The network structure of complex-valued DenseNets.

PB first uses a 3×3 convolution kernel to process the input picture so that the number of generated feature maps can be divided by two, which is convenient for constructing the real and imaginary parts of the complex feature maps. Then the 3×3 complex convolution kernel is used to perform the convolution operation with complex feature maps.

This paper extends the real dense block as Fig. 3(a) into complex field. Complex dense block shown as Fig. 3(b) is composed of several duplicate complex blocks stacked in a dense connectivity manner. Complex dense block establishes the connection relationship between different layers, and the input of each layer is added to the output of all the previous layers. Therefore, the input expression for l^{th} layer is:

$$x_l = H_l\left(\left[x_0^R, x_1^R, x_2^R, \cdots, x_{l-1}^R, x_0^I, x_1^I, x_2^I, \cdots, x_{l-1}^I\right]\right) \tag{10}$$

Where $H_l(\cdot)$ is expressed as composite functions with 6 consecutive operations: complex BN, a rectified linear unit (ReLU), an 1×1 complex convolution, complex BN, a ReLU and a 3×3 complex convolution. x_{l-1}^R and x_{l-1}^I respectively represent the real and imaginary parts of the input feature map of the $l-1$ layer.

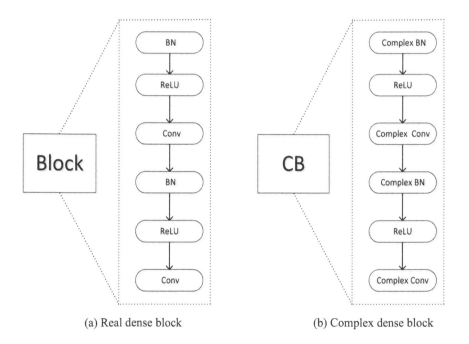

(a) Real dense block   (b) Complex dense block

Fig. 3. Comparison of real dense block and complex dense block.

The input of the lth layer can be represented by the magnitude and phase of the feature map:

$$\left\{ \begin{array}{c} x_l = H_l([|x_0|\cos\theta_0, \cdots, |x_{l-1}|\cos\theta_{l-1}, |x_0|\sin\theta_0, \cdots, |x_{l-1}|\sin\theta_{l-1}]) \\ |x_k| = \sqrt{\left(x_k^R\right)^2 + \left(x_k^I\right)^2}, \ 0 \le k \le l-1 \\ \theta_k = \arctan\frac{x_k^I}{x_k^R} \end{array} \right. \tag{11}$$

It can be seen from formula (11) that complex dense block makes full use of the amplitude and phase information of the complex feature map, while the real dense block does not consider this information. Complex dense block also enhances feature propagation and promotes feature reuse. The complex eature map output by the previous complex convolutional layer is used as the input of each subsequent complex convolutional layer. This method makes the network have a continuous memory function, and the state of each layer can be continuously transferred. It is more conducive to learning the nonlinear features of the network.

For complex dense block, each layer function H_l generates ck feature maps. Then the lth layer will have $ck_0 + ck \times (l-1)$ input feature maps, where ck_0 is the number of feature maps of the input layer, and hyperparameter ck is referred to as the growth rate of complex-valued DenseNets.

In Fig. 4(a), transition layers are composed by a BN layer, an 1×1 convolution layer and a 2×2 average pooling layer. As shown in Fig. 4 (b), complex transition layers include three parts: a complex BN layer, an 1×1 complex convolutional layer and a 2×2 average pooling layer. The main purpose of complex transition layers is to implement dimensionality reduction operations, which reduces training parameters and is conducive to network convergence. If complex-valued Dense block generates m

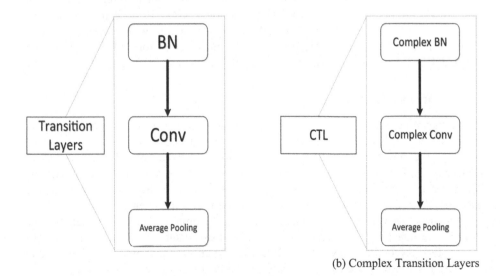

(b) Complex Transition Layers

Fig. 4. Comparison of transition layers and complex transition layers.

feature maps, the complex transition layers will generate $c\theta m$ feature maps, where $c\theta$ satisfies $0 < c\theta \leq 1$, which is referred to herein as the compression rate of complex-valued DenseNets.

4 Experiments and Analysis

This paper uses CIFAR-10 database and CIFAR-100 database for classification. The CIFAR-10 dataset consists of 60000 32 × 32 color images in 10 classes, with 6000 images per class. There are 50000 training images and 10000 test images. The CIFAR-100 dataset is just like the CIFAR-10, except it has 100 classes containing 600 images each. There are 500 training images and 100 testing images per class. The experimental hardware equipment in this paper is the graphics card Nvidia2080Ti, and the deep learning framework is Keras.

4.1 Experiment Design

In this paper, the back-propagation algorithm used a stochastic gradient descent method with Nesterov momentum set to 0.9. The norm of the gradient was clipped to 1, and a custom learning rate was used. It was set to 0.01 in the first 100 epochs, 0.001 in the 101–150 epochs, and 0.0001 in the last 50 epochs. The algorithm model of this paper would be compared with 57-layer real-valued DenseNets [4] and complex-valued ResNets [10].

This paper compares complex-valued DenseNets with real-valued DenseNets and complex-valued ResNets. To a certain extent, the deeper the network is, the better the performance is. In order to compare the algorithm performance under fair conditions, three algorithms have the same depth. Therefore, this paper compares the performance of the three algorithms on two databases and under different network widths. Four experiments are performed: (1) Experiment I adopts CIFAR-10 database with narrow network; (2) Experiment II adopts CIFAR-100 database with narrow network; (3) Experiment III adopts CIFAR-10 database with wide network; (4) Experiment IV adopts CIFAR-100 database with wide network.

4.2 Experiment Design

In this paper, error rate is chosen as the performance parameter. Of course, the lower the error rate, the better the algorithm performance. Under the condition of narrow network, the number of feature maps of complex-valued DenseNets, real-valued DenseNets and complex-valued ResNets is set as $ck = 14$, $k = 24$, $f = 8$. Under the condition of wide network, $ck = 28$, $k = 48$ and $f = 16$. Figure 5(a)–(d) respectively corresponds to the testing error rate curves of the four experiments.

As can be seen from the four pictures in Fig. 5, complex-value DenseNets has the lowest error rate curve in two network widths and two databases. Compared with real-valued DenseNets, complex-valued DenseNets that uses complex convolution and complex BN can improve the nonlinear expression ability of the network and improve the generalization ability of the network. Compared with complex-valued ResNets,

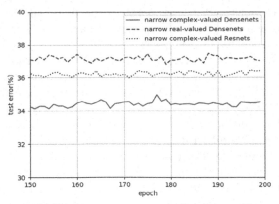

(b) Testing error rate curves of experiment II

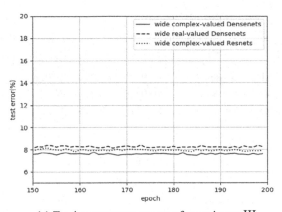

(c) Testing error rate curves of experiment III

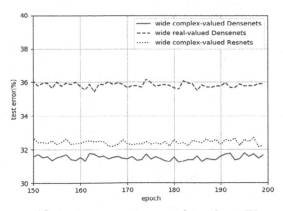

(d) Testing error rate curves of experiment IV

Fig. 5. Testing error rate curves of four experiments.

complex-valued DenseNets that uses dense connectivity alleviate the vanishing-gradient problem, strengthen feature propagation, encourage feature reuse, and substantially reduce the number of parameters.

To quantitatively compare algorithm performance, Table 1 shows the accuracy of three algorithms in four experiments. In experiment I and experiment III, the proposed algorithm has a weak advantage over real-valued DenseNets and complex-valued ResNets, because CIFAR-10 database contains a limited variety. In experiment II with CIFAR-100 database, the accuracy of our algorithm is 3% higher than that of real-valued DenseNets and 2% higher than that of complex-valued ResNets. In experiment IV, the accuracy improved by 4% compared with real-valued DenseNets. Figure 5 and Table 1 show the performance of our algorithm. Table 2 shows the number of parameters used by the three algorithms under the above experimental results. As can be seen from Table 2, under the condition of optimal performance of our algorithm, we also used the least parameters.

Table 1. Accuracy comparison of four experiments.

	I	II	III	IV
Real-valued Densenets	89.95%	63.78%	91.92%	65.14%
Complex-valued Resnets	90.07%	64.28%	92.27%	68.18%
Complex-valued Densenets	91.011%	66.31%	92.62%	69.12%

Table 2. The number of all parameters of four experiments.

	I	II	III	IV
Real-valued Densenets	253422	276912	1130094	1176624
Complex-valued Resnets	412278	418128	1627494	1639104
Complex-valued Densenets	227446	259936	854406	912816

5   Conclusion

This paper generalizes real-valued DenseNets to complex filed and proposes complex-valued DenseNets. The input formula of the layer also changes accordingly to accommodate complex-valued input data. The construction of complex dense blocks and complex transition layers increases the generalization ability and convergence speed of the proposed algorithm. We design four experiments on two network widths and two databases to compare our algorithm with real-valued DenseNets and complex-valued ResNets. It can be seen from the experimental results that our algorithm performs better than the other two algorithms and uses the least number of parameters.

References

1. Ioffe, S., Szegedy, C.: Batch normalization: accelerating deep network training by reducing internal covariate shift. In: Proceedings of the 32nd International Conference on Machine Learning, pp. 448–456 (2015)
2. Srivastava, R., Greff, K., Schmidhuber, J.: Training very deep networks. In: Proceedings of the 2015 Neural Information Processing Systems, pp. 2377–2385 (2015)
3. He, K.M., Zhang, X.Y., Ren, S.Q., et al.: Deep residual learning for image recognition. In: Proceedings of the 2016 IEEE Conference on Computer Vision and Pattern Recognition, pp. 770–778 (2016)
4. Huang, G., Liu, Z., Van Der Maaten, L., et al.: Densely connected convolutional networks. In: Proceedings of the 2017 IEEE Conference on Computer Vision and Pattern Recognition, pp. 2261–2268 (2017)
5. Gaudet, C.J., Maida, A.S.: Deep quaternion networks. In: Proceedings of the 2018 International Joint Conference on Neural Networks, pp. 1–8 (2018)
6. Hirose, A., Yoshida, S.: Generalization characteristics of complex-valued feedforward neural networks in relation to signal coherence. IEEE Trans. Neural Netw. Learn. Syst. **23**(4), 541–551 (2012)
7. Arjovsky, M., Shah, A., Bengio, Y.: Unitary evolution recurrent neural networks. In: Proceedings of the 33rd International Conference on Machine Learning, pp. 1120–1128 (2016)
8. Danihelka, I., Wayne, G., Uria, B., et al.: Associative long short-term memory. In: Proceedings of the 33rd International Conference on Machine Learning, pp. 1986–1994 (2016)
9. Wisdom, S., Powers, T., Hershey, J., et al.: Full-capacity unitary recurrent neural networks. In: Proceedings of the 2016 Neural Information Processing Systems, pp. 4880–4888 (2016)
10. Trabelsi, C., Bilaniuk, O., Zhang, Y., et al.: Deep complex networks. In: Proceedings of the 2018 International Conference on Learning Representations, pp. 94–102 (2018)
11. Reichert, D.P., Serre, T.: Neuronal synchrony in complex-valued deep networks. In: Proceedings of the 2014 International Conference on Learning Representations, pp. 68–76 (2014)
12. Oppenheim, A.V., Lim, J.S.: The importance of phase in signals. Proc. IEEE **69**(5), 529–541 (1981)
13. Hinton, G.E., Osindero, S., Teh, Y.W.: A fast learning algorithm for deep belief nets. Neural Comput. **18**(7), 1527–1554 (2006)
14. Georgiou, G.M., Koutsougeras, C.: Complex domain backpropagation. IEEE Trans. Circ. Syst. II: Analog Digit. Signal Process. **39**(5), 330–334 (1992)
15. Zemel, R.S., Williams, C.K., Mozer, M.C.: Lending direction to neural networks. Neural Netw. **8**(4), 503–512 (1995)
16. Kim, T., Adali, T.: Approximation by fully complex multilayer perceptrons. Neural Comput. **15**(7), 1641–1666 (2003)
17. Oyallon, E., Mallat, S.: Deep roto-translation scattering for object classification. In: Proceedings of the 2015 IEEE Conference on Computer Vision and Pattern Recognition, pp. 2865–2873 (2015)

Representation Learning with Deconvolution for Multivariate Time Series Classification and Visualization

Wei Song[1(✉)], Lu Liu[2], Minghao Liu[1,3], Wenxiang Wang[1,3], Xiao Wang[1], and Yu Song[4(✉)]

[1] Henan Academy of Big Data, Zhengzhou University,
Zhengzhou 450001, China
iewsong@zzu.edu.cn
[2] Department of Computational Linguistics, University of Washington, Seattle,
WA 98195, USA
[3] School of Information Engineering, Zhengzhou University,
Zhengzhou 450001, China
[4] Network Management Center, Zhengzhou University,
Zhengzhou 450001, China
ieysong@zzu.edu.cn

Abstract. We propose a new model based on the convolutional networks and SAX(Symbolic Aggregate Approximation) discretization to learn the representation for multivariate time series. The deep neural networks has excellent expressiveness, which is fully exploited by the convolutional networks with means of unsupervised learning. We design a network structure to obtain the cross-channel correlation with means of convolution and deconvolution, the pooling operation is utilized to perform the dimension reduction along each position of the channels. Discretization which based on the Symbolic Aggregate Approximation is applied on the feature vectors to extract the bag of features. We collect two different representations from the convolutional networks, the compression from bottle neck and the last convolutional layers. We show how these representations and bag of features can be useful for classification. We provide a full comparison with the sequence distance based approach on the standard datasets to demonstrate the effectiveness of our method. We further build the Markov matrix according to the discretized representation abstracted from the deconvolution, time series is visualized to complex networks through Markov matrix visualization, which show more class-specific statistical properties and clear structures with respect to different labels.

Keywords: Multivariate time-series · Deconvolution · Symbolic aggregate approximation · Deep learning · Markov matrix · Visualization

---

W. Song and L. Liu—These authors contribute equally to this work.

J. Zeng et al. (Eds.): ICPCSEE 2020, CCIS 1257, pp. 310–326, 2020.
https://doi.org/10.1007/978-981-15-7981-3_22

1   Introduction

1.1   A Subsection Sample

Sensors are now becoming easier to acquire and more ubiquitous in recent years to motivate the wide application of numerous time series data. For example, Non-invasive, continuous, high resolution vital signs data, such as Electrocardiography (ECG) and Photoplethysmography (PPG), are commonly used in hospital settings for better monitoring of patient outcomes to optimize early care. Industrial time series help the engineers to predict and get early preparation of the potential failure. We formulate these tasks as regular multivariate time series classification/learning problem. Compared to the univariate time series data, multivariate time series is more widespread, hence providing more patterns and insights of the underlying phenomena to help improve the classification performance. Therefore, multivariate time series classification is becoming more and more important in a broad range of applications, such as industrial inspection and clinical monitoring.

Multivariate time series data is not only characterized by individual attributes, but also by the relationships between the attributes [1]. The similarity among the individual sequences [2] is not sufficient to get the information. To solve the classification problem on multivariate time series, several similarity measurements including Edit distance with Real Penalty (ERP) and Time Warping Edit Distance (TWED) are summarized and tested on several benchmark dataset [3]. Recently, a symbolic representation for multivariate time series classification (SMTS) is proposed. Mining core feature for early classification (MCFEC) along the sequence is proposed to capture the shapelets in each channel independently [4]. SMTS builds a tree learner with two ensembles to learn the segmentations and a high-dimensional codebook [5]. While these methods provide new inspirations to analyse multivariate data, some are time consuming (e.g. SMTS), some are effective but cannot address the curse of dimensionality (distance on raw data).

Inspired by recent progress in feature learning on image classification, several feature-based approaches are proposed [5, 6]. Compared with those sequence-distance based approaches, the feature-based approaches can learn a hierarchical feature representation from raw data automatically thus skipping the tricky hand-crafted features. However, the feature learning approach are only limited on the scenario of supervised learning and few comparison towards distance-based learning approaches (like [7]). The method described in [6] is simple but not fully automated, instead they still need to design the weighting scheme manually.

Our work provides a new perspective to learn the underlying representations with convolutional networks accompanied by deconvolution (or convolution transpose) in a self-supervised learning manner, which can fully exploit the unlabeled data especially when the data size is large. We design the network structure to extract the cross-channel correlation with tied convolutions, forcing the pooling operation to perform the dimension reduction along each position of the individual channel. Inspired by the discretization approaches like Symbolic Aggregate Approximation (SAX) with its variations [6, 8, 9] and Markov matrix [10], we further show how this representation is applied on classification and visualization tasks. A full comparison with the sequence distance based approach is provided to demonstrate the effectiveness of our approach.

2 Background and Related Work

2.1 Deep Neural Networks

Since 2006, the techniques developed from deep neural networks (or, deep learning) have greatly influence on natural language processing, speech recognition and computer vision research [11, 12]. A successful deep learning architecture used in computer vision is convolutional networks (CNN) [13]. CNNs make use of the translational invariance by extracting features through receptive fields [14] and learning with weight sharing, CNN has become the state-of-the-art approach in various image recognition and computer vision tasks [15].

The exploration of unsupervised learning algorithms for generative models has got a lot of advancing results, such as Deep Belief Networks (DBN) and De-noised Auto-encoders (DA) [16, 17]. Numerous deep generative models are developed based on energy-based model or auto-encoders. Temporal autoencoding is integrated with Restrict Boltzmann Machines (RBMs) to improve generative models [18]. A training strategy inspired by recent work on optimization-based learning is proposed to train complex neural networks for imputation tasks [19]. A generalized Denoised Auto-encoder extends the theoretical framework and is applied to Deep Generative Stochastic Networks (DGSN) [20, 21].

However, since unsupervised pretraining has been proved to improve performance in both fully supervised tasks and weakly supervised tasks [22, 23], deconvolution and Topographic Independent Component Analysis (TICA) are integrated as unsupervised pretraining approaches to learn more diverse features with complex invariance [24–26]. We use convolution and deconvolution to obtain both the temporal and cross-channel correlation in the multivariate time series, rather than pretrain a supervised model.

2.2 Discretization and Visualization for Time Series

Time Series discretization is widely used in the research of symbolic approximation based approach. Aligned Cluster Analysis (ACA) is an unsupervised method to cluster the temporal patterns of human motion data [27], which is a development of kernel k-means clustering, however, this method requires a large computational ability. Persist is an unsupervised discretization method which can maximize the persistence measurement of each symbol [28]. Piece-wise Aggregate Approximation (PAA) method is proposed by Keogh [29] to reduce the dimensionality of time series, which is then upgraded to Symbolic Aggregate Approximation (SAX) [8]. In SAX method, every PAA process resulted aggregation values are mapped into the equiprobable intervals according to standard normal distribution, thus producing a sequence of symbolic representations. Among these symbolic approaches, SAX method has become one of the de facto standards to discretize time series and is using as the core of many effective classification algorithms.

The overall pipeline of SAX is to smooth the input time series with Piecewise Aggregation Approximation (PAA) and then assign symbols to the PAA bins. After these processes, the overall time series trend is extracted as a sequence of symbols. The algorithm involves three parameters: window length n, number of symbols w and alphabet size a. Different parameters would result in different representations of the time series. Given a normalized time series of length L, we first reduce the dimensionality by dividing it into [L/n] non-overlapping sliding windows with skip size 1. Each sliding window is partitioned into w subwindows. Then we need to compute the mean values to reduce volume and smooth the noise, the resulting is called PAA values. After that PAA values are mapped to a probability density function N(0, 1), which is divided into several equiprobable segments. Letters starting from A to Z are assigned to PAA values according to their corresponding positions on the segments (Fig. 1).

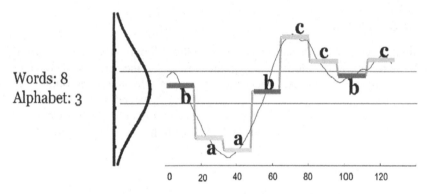

Fig. 1. PAA and SAX word for the ECG data. The time series is partitioned into 8 segments. In each segment we compute means to map them to the equiprobable interval. After discretization by PAA and symbolization by SAX, we convert the time series into SAX word sequence baabccbc.

Besides, reformulating time series to visual clues has become more important in computer science and physics in which the discretization method plays an important role. For example, Mel-frequency cepstral coefficients (MFCCs) or Perceptual Linear Prediction (PLP) are always used to represent the acoustic/speech data input to represent the temporal and frequency information. A lot of different network structures for time series have been come up to visually inspect the features or design the distance measurement. [30, 31] proposed the Recurrence Networks to analyze the structural properties of time series from complex systems. They build adjacency matrices from the predefined recurrence functions to represent the time series as complex networks. This recurrence plot paradigm for time series classification is extended by Silva et al. with compression distance [32]. There is also another effective way to build a weighted adjacency matrix, that is to extract transition dynamics from the first order Markov matrix [33]. These maps have demonstrated distinct topological differences among different time series, but the relationship between these topological properties and the original time series is still unknown because there are no exact inverse operations. [10] proposed an generalized Markovian encoding to map the complex correlations to image information and also preserve the temporal information at the same time. [34] stated a further exploration of the visual analytics method with Markov transition fields.

We build the Markov Matrix to visualize the topology of the complex networks generated by [33] to demonstrate the formation process of our learned feature.

3   Representation Learning with Convolution and Deconvolution

Deconvolution has the similar mathematical form with convolution. The difference is, convolutional networks contain the 'inverse' operation of convolution and pooling for reconstruction.

3.1   Deconvolution

The input of convolutional layers includes multiple input activations within a filter window, while they all connect to a single activation. On the contrary of convolutional layers, deconvolutional layers associate a single input activation with multiple outputs (Fig. 2). The output of a deconvolutional layer is an enlarged dense feature map. For the practical implementation, we crop the boundary of the enlarged feature map to make sure that the size of output map is the same with the one from the preceding layer.

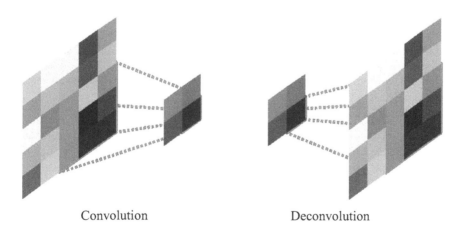

Convolution   Deconvolution

Fig. 2. Illustration of the deconvolution operations.

The learned filters in deconvolutional layers are matched to the bases to reconstruct the same shape of the input. Similar to the convolution network, we use a hierarchical structure of deconvolutional layers to capture different level of shape information. The low level filters tend to capture detailed features while the high level filters tend to capture more abstract features. According to all these reasons, the network can directly collect specific shape information for multi-scale feature capturing, which is often overlooked in other approaches that only based on convolutional layers.

For all the above reasons, we have designed two different feature extraction approaches to compare the quality of feature learned by compression and reconstruction, as shown in 3:

- Compression vector. The bottle nect layer can learn a set of compressed representations of MTS. The convolution and deconvolution enforce the compression to preserve the local temporal correlation across channels and the temporal order at the same time, which is different with PCA.
- Reconstruction vector. The last convolution layer contains the full information needed to reconstruct the raw signals. The kernel and pooling operation can determine the dimension of the layer, and the combination of the layer and the raw signal is an excellent baseline to compare the full performance of learned representations.

The pooling process in convolution network can abstract activations in a receptive field with a single representative value, it helps gain the robustness to noise and translation. Although it helps classification by retaining only robust activations in upper layers, spatial information within a receptive field is lost during pooling. This kind of information loss may be important for precise feature learning, however, the reconstruction and classification often require a highly precise feature learning.

Unpooling layers in deconvolution network is used to perform a reverse operation of pooling and reconstruct the original size of activations. During the implement process of unpooling, we first record the locations of the maximum activations selected during pooling operation in the transposed variables, which are employed to return each activation back to its original pooled location. This unpooling strategy is also very useful to reconstruct the structure of input object. The another important thing is that the output of an unpooling layer is an enlarged and sparse activation map, which might loss the delivery ability of the complex feature for reconstruction. In order to solve this problem, the deconvolution layers are used after the unpooling operation, in this way, the sparse activations can be densified through convolution-like operations with several learned filters.

We will show in the later experiments that the pooling and unpooling layers sometimes can help improve the quality of compression vectors.

3.2 Convolution for Multivariate Time Series

For the proposed algorithm, we need to pay special attention on the convolutional network to learn the feature precisely for multivariate time series data. This algorithm uses deep tied convolution along the channel and pooling across the channel to generate feature maps, which is different with the simple and common convolution and pooling methods that both performed with square kernels. We make full use of the successive operations of unpooling, deconvolution, and recti cation to get the dense element-wise deconvolutional map.

Different with the 1D convolution across temporal axis which is rational enough, the 2D convolution across channels are pointless in MTS data, because we can not make sure that the local neighbors of different channels have interaction and correlations among that temporal window. Through using tied 1D convolution kernel across different channels, the network can include the cross channel information into the learned representations.

Figure 3 illustrations an sample network structure which is layer by layer organized, this is a demo of the internal operations of our deconvolution network. In this structure, we use deconvolution with multiple 1×3 filters to obtain the correlation of temporal and implicit cross-channel information. The lower layers are easy to get an overall coarse configuration of the short term signals (e.g. location and frequency), higher layers can distinguish some more complex patterns. The pooling/unpooling layer and deconvolution process have different effects in the construction of learned features. Pooling/unpooling track each individual position with strong activations back to the signal space to abstract important information. The pooling/unpooling process can effectively reconstruct the details of multivariate signals in finer resolutions. On the contrary, the learned filters in deconvolutional layers tend to capture the overall generating shapes for the local temporal structure. With the processes of convolution, deconvolution and tied weights, the activations which are closely related to the generating distribution along each signal and cross the channels are amplified, the noisy activations from other regions are suppressed. Through the combination of both unpooling and deconvolution, our network is able to generate accurate information for the reconstruction of multivariate time series.

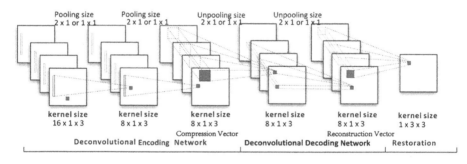

Fig. 3. Demo of the complete network architecture for deep deconvolutional net on the multivariate time series.

4   Visualization and Classification

We suppose to visualize the learned representation for further inspection and explanation, we discretize and convert the final encoding in the hidden layers of the deconvolutional networks to a Markov Matrix, thus visualizing the matrix as complex networks [33].

As in Fig. 4, a time series X splits into Q = 10 quantiles, each quantile qi is represented by a node $n_i \in N$ in the network G. Then, nodes n_i and n_j are connected by the arc of network, the weight ω_{ij} of the arc is the probability that a point in quantile q_i is followed by a point in quantile q_j. The transition between the quantiles are in large frequency, we use the thicker lines to represent large weights of the network. Although discretization is based on quantile bins at first, the SAX methods indicate that time series tend to follow the Gaussian distribution, we use Gaussian mapping to replace quantile bins for discretization.

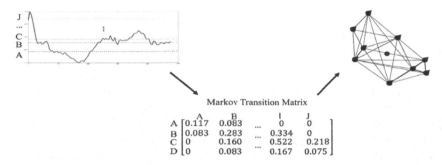

Fig. 4. Illustration of the conversion from time series to complex networks for visualization.

The deconvolution operation has a sliding window on time dimension, which means that the hidden representation should also has a significant temporal component, this would demands that the representation must within the application domain of SAX and bag-of-words approaches. The bag-of-words method is generated from the SAX words, its benefit comes from the invariance to locality. Compared with the original raw vector-based representation, these feature bags improve the classification performance, the feature bags can fit the temporal correlation and also reinforce the highlight of noise and outliers. In our experiments, we use both the raw hidden vector and the bag of SAX words for classification.

5 Experiments and Results

This section first states the settings and results of our representation learning with deconvolution. We analyze and evaluate the proposed representation according to the classification and visualization tasks.

We firstly test our algorithms on two standard datasets that are widely appeared in the literature about multivariate time series[1]. The ECG dataset contains 200 samples with two channels, among which 133 samples are normal and 67 samples are abnormal. The size of a MTS sample is between 39 and 153. The wafer datasets contain 1194 samples. 1067 samples are normal and 127 samples are abnormal. The length of a sample is between 104 and 198. We preprocess each dataset by standardization and

---

[1] http://www.cs.cmu.edu/~bobski/.

realigning all the signals along with the maximum of the length. All missing values are filled with 0. Table 1 gives the statistics summary of each dataset.

Table 1. Summary of the preprocessed datasets.

	Channel	Length	Class	Training	Test
Wafer	6	199	2	896	298
ECG	2	153	2	100	100

5.1 Representation Learning with Deconvolution

Figure 3 summarizes the detailed configuration of the proposed network. We use symmetrical configuration of convolution and deconvolution network, the center is around the output of the 2nd Convolutional layer. The input and output layers correspond to input signals and their corresponding reconstruction. We use SeLU [35] to be the activation function. The network is trained by Adadelta with learning rate 0.1 and $\rho = 0.95$.

Figure 5 and 6 show the reconstructions conducted by the convolutional networks. Because the deconvolution can train the filters to obtain both the temporal and cross-channel information, our network combines both the unpooling and deconvolution, the network can generate accurate reconstruction of the multivariate time series, the learned representations can be expressed explicitly. As shown in Fig. 7, the final encoding of each map learns different representation independently after the filter process by the deconvolution and pooling/unpooling. Diverse local patterns (shapes) of time series can be captured automatically. Through the deconvolution, the filters can distinguish different importance level of each feature through both the single channel and cross channel information.

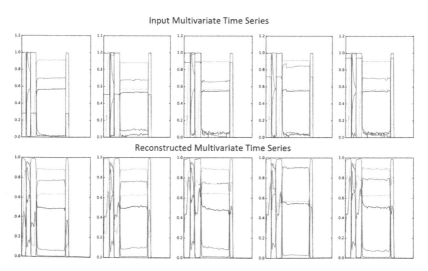

Fig. 5. Input and reconstruction of the convolutional network on the 'wafer' dataset with 6 channels.

Input Multivariate Time Series

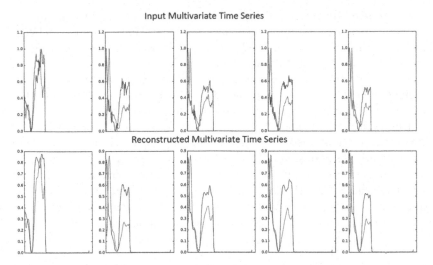

Reconstructed Multivariate Time Series

Fig. 6. Input and reconstruction of the convolutional network on the 'ECG' dataset with 2 channels

5.2 Classification

During the classification process, we feed both the learned representation vectors and the bag of SAX words into a logistic regression classifier. We only use training data to train the representation with convolutional networks, the test representation would be generated by the well trained model with a single forward pass on the test set. We use Leave-One-Out cross validation accompanied by Bayesian optimization [36] to choose the parameters of SAX from the training set, the parameters are window length n, number of symbols w and alphabet size a.

After discretization and symbolization, the sliding window of length n builds the bag of words dictionary, every subsequence is converted into w SAX words. Bag of words can capture the features that have same structure in even different instance and location. The discretized features are based on the bag of words histogram of word counts.

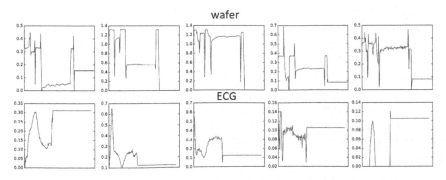

Fig. 7. 5 learned representation encoding in the convolutional networks.

We compared our model with several best methods for multivariate time series classification in recent literatures including Dynamic Time Warping (DTW), Edit Distance on Real sequence (EDR), Edit distance with Real Penalty (ERP) [3], STMS [5] and MCFEC [4] Early-Clf [4] and Pooling SAX [6].

Table 2 summarizes the classification results. Our model performs better than all other approaches. Our approach is able to capture the precise information through deconvolution to improve the classification performance even though the dataset has large channel size, for example, the wafer dataset which has 6 channels can be classified excellently. Unlike the supervised deep learning model which tends to overfit the label of datasets, our unsupervised feature learning framework takes advantage of the generic expressiveness of the neural networks, the framework builds precise feature set from the large number of weights. These features provide a larger possibility for precise classification. The compression vector shows better generalization capability compared with reconstruction vectors, thus performing the best in our experiments.

Table 2. Test error rate on the standard dataset.

	ECG	Wafer
Pooling SAX	0.16	0.02
SMTS	0.182	0.035
MCFEC	0.22	0.03
Euclidean	0.1778	0.0833
DTW (full)	0.1889	0.0909
DTW (window)	0.1722	0.0656
EDR	0.2	0.3131
ERP	0.1944	0.0556
Early-Clf (SI-clustering)	0.22	0.03
Early-Clf (Greedy method)	0.23	0.06
Ours (compression vector)	0.13	0.03
Ours (reconstruction vector)	0.13	0.0369
Ours (compression vector + SAX)	0.13	0.02
Ours (reconstruction vector + SAX)	0.13	0.02

Another comparison is performed between the vector and discretized bag-of-words representation from the convolutional networks (Table 3). Although discretization by SAX indicates more hyperparameters, both cross validation and test error rate are better than the compression and reconstruction feature vector alone. As the analyse above, the convolutional networks tend to preserve the high order abstract temporal information during the learning process. SAX and bag of words reinforce these information specifically for classification in the supervised way, and remove the less useful noise and outliers, the dependency on temporal locality is enfeebled by bag of words. To recapitulate, the bag of SAX words show advantage against the raw vector feature. The Bayesian optimization greatly boosts the searching process on the hyperparameter and also converges fast.

Table 3. Train and test error rates of the vector/discretized representation from the convolutional networks.

	Compression vector + SAX		Compression vector		Reconstruction vector	
	CV train	Test	cv train	Test	CV train	Test
Wafer	0.007	0.02	0.09	0.038	0.011	0.035
ECG	0.12	0.13	0.13	0.14	0.14	0.14

5.3   Visualization and Statistical Analysis

Figure 7 has shown how the compression vector is like as time series. In order to explicitly and intuitively understand the compression representation learned through convolution and the effect of discretization through SAX, we flatten and discretize each feature map (which is feed in the classifier as input) and visualize them as complex networks to inspect some more statistical properties.

For the problem of setting the number of discretization bins (or the alphabet size in our SAX settings), we set $Q = 60$ for the ECG dataset and $Q = 120$ for the wafer dataset. We use the hierarchical force-directed algorithm as the network layout [37]. As shown in Fig. 8 and 9, the demos on two datasets show different network structures. For ECG, The normal sample tends to have round-shaped layout while the abnormal sample always has a narrow and winded structure. For wafer data, normal sample is shown as a regular closed-form shape, but the structure of the abnormal sample is open-form while thicker edges piercing through the border.

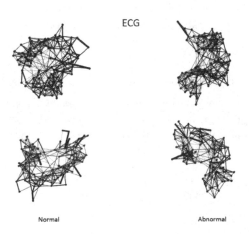

ECG

Normal   Abnormal

Fig. 8. Visualization of the complex network generated from the discretized convolutional features on the ECG dataset

Wafer

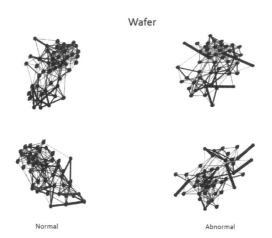

Normal   Abnormal

Fig. 9. Visualization of the complex network generated from the discretized convolutional features on the Wafer dataset.

Table 4 summarizes four statistics of all the complex networks generated from the convolutional representations: average degree, modularity class [38], Pagerank index [39] and average path length. The Pagerank index of the table denotes the largest value in its propagation distribution. For the ECG dataset, all the statistics are significantly different between the graphs with different labels under the rejection threshold of 0.05. However, for the Wafer dataset, only average path length shows significant difference between two labels. This might comes from the topological structure that there are thicker edges around the network structures, which indicates that the number of edge and their weights are highly skewed, this outlier would only effect the path degree, but not other statistics.

Table 4. Summary statistics of the complex networks generated by compression vector

	Avg. degree	Modularity	Pagerank	Avg. path length
ECG normal	7.0966	0.5718	16.8	2.7802
ECG abnormal	8.7202	0.5948	18.4	2.6384
P value	0.003	0.0124	0.0054	0.0024
Wafer normal	8.7866	0.5058	12	2.0878
Wafer abnormal	8.5734	0.5098	12.2	2.0684
P value	0.2989	0.1034	0.3786	0.0012

For the last step, we investigate visualization approaches proposed in [34] by building the Markov Transition Field from the discretized compression vectors. We apply two different encoding frameworks to encode the relevant network statistics (Table 5).

Table 5. Visual encoding framework

Flow	Vertex	Edge
Encoding		
Color	Time index	Markov matrix weight
Size	PageRank weight	Constant
Modularity	Vertex	Edge
Encoding		
Color	Module label	Module label of the target
Size	Clustering coefficient	Constant

In Fig. 10, we plot the MTF network structure of the Wafer dataset by modularity color encoding. Although the sample normal and abnormal signals have different community numbers, the blue community in the normal signal are almost shrink to the green community. As the statistics indicated before, only average path length are significant network statistics to distinguish the labels, the similar shape and color in the network plot are consistent with this conclusion. In the contrast, all network statistics are significant to tell different labels for ECG dataset, the ow encoding plot shows such big differences across different labels.

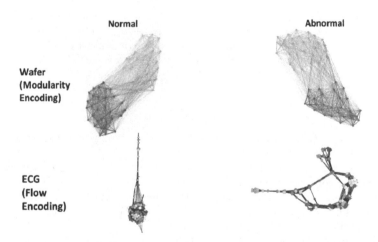

Fig. 10. Visualization of the complex network of MTF generated from the discretized compression features on Wafer and ECG dataset with different visualization encoding approaches.

6 Conclusion and Future Work

We propose a new model based on the convolutional networks and SAX discretization to learn the representation for multivariate time series. Convolutional networks can make full use of the advantage of powerful expressiveness of deep neural networks in an self-supervised manner. We design a network structure specifically to capture the cross-channel correlation with convolution and deconvolution, forcing the pooling operation to perform the dimension reduction along each position across the channels. We apply SAX discretization on the feature vectors to further extract the bag of features. We show how this representation and bag of features can be beneficial for classification. A full comparison with the sequence distance based approach is provided to demonstrate the effectiveness of our approach. Finally, we build the Markov matrix according to the discretized representation to visualize the time series as complex networks, which show more statistical properties and clearly different structures in terms of different labels.

As future work, we suppose to integrate grammar induction approach on the convolutional SAX words to further infer the semantics of multivariate time series. We would also further develop intelligent interfaces to enable human interaction, through which we can inspect, understand and even guide the feature learning and semantic inference for mechanical multivariate signals.

Acknowledgment. This work is supported by the International Cooperation Project of the Department of Science & Technology of Henan Province under Grant No. 172102410065; Basic Research Project of the Education Department of Henan Province under Grant No. 17A520057; Frontier Interdisciplinary Project of Zhengzhou University under Grant No. XKZDQY202010.

References

1. Bankó, Z., Abonyi, J.: Correlation based dynamic time warping of multivariate time series. Expert Syst. Appl. **39**(17), 12814–12823 (2012)
2. Weng, X., Shen, J.: Classification of multivariate time series using locality preserving projections. Knowl.-Based Syst. **21**(7), 581–587 (2008)
3. Lin, J., Williamson, S., Borne, K., DeBarr, D.: Pattern recognition in time series. In: Data Mining and Knowledge Discovery Series: Advances in Machine Learning and Data Mining for Astronomy, pp. 617–645. Chapman & Hall/CRC (2012)
4. He, G., Duan, Y., Peng, R., Jing, X., Qian, T., Wang, L.: Early classification on multivariate time series. Neurocomputing **149**, 777–787 (2015)
5. Baydogan, M.G., Runger, G.: Learning a symbolic representation for multivariate time series classification. Data Min. Knowl. Discov. **29**, 400–422 (2015)
6. Wang, Z., Oates, T.: Pooling sax-bop approaches with boosting to classify multivariate synchronous physiological time series data. In: FLAIRS Conference, pp. 335–341 (2015)
7. Zheng, Y., Liu, Q., Chen, E., Ge, Y., Zhao, J.L.: Time series classification using multi-channels deep convolutional neural networks. In: Li, F., Li, G., Hwang, S.-W., Yao, B., Zhang, Z. (eds.) WAIM 2014. LNCS, vol. 8485, pp. 298–310. Springer, Cham (2014). https://doi.org/10.1007/978-3-319-08010-9_33

8. Lin, J., Keogh, E., Lonardi, S., Chiu, B.: A symbolic representation of time series, with implications for streaming algorithms. In: Proceedings of the 8th ACM SIGMOD Workshop on Research Issues in Data Mining and Knowledge Discovery, pp. 2–11. ACM (2003)
9. Sun, Y., Li, J., Liu, J., Sun, B., Chow, C.: An improvement of symbolic aggregate approximation distance measure for time series. Neurocomputing 138, 189–198 (2014)
10. Wang, Z., Oates, T.: Imaging time-series to improve classification and imputation. In: 24th International Joint Conference on Artificial Intelligence, IJCAI 2015 (2015). arXiv:1506.00327
11. Bengio, Y.: Learning deep architectures for AI. Found. Trends® Mach. Learn. 2(1), 1–127 (2009)
12. Deng, L., Yu, D.: Deep learning: methods and applications, Technical report MSR-TR-2014-21, January 2014. http://research.microsoft.com/apps/pubs/default.aspx?id=209355
13. LeCun, Y., Bottou, L., Bengio, Y., Haffner, P.: Gradient-based learning applied to document recognition. Proc. IEEE 86(11), 2278–2324 (1998)
14. Hubel, D.H., Wiesel, T.N.: Receptive fields, binocular interaction and functional architecture in the cat's visual cortex. J. Physiol. 160(1), 106 (1962)
15. Krizhevsky, A., Sutskever, I., Hinton, G.E.: ImageNet classification with deep convolutional neural networks. In: Advances in Neural Information Processing Systems, pp. 1097–1105 (2012)
16. Hinton, G., Osindero, S., Teh, Y.-W.: A fast learning algorithm for deep belief nets. Neural Comput. 18(7), 1527–1554 (2006)
17. Vincent, P., Larochelle, H., Bengio, Y., Manzagol, P.-A.: Extracting and composing robust features with denoising autoencoders. In: Proceedings of the 25th International Conference on Machine Learning, pp. 1096–1103. ACM (2008)
18. Häusler, C., Susemihl, A., Nawrot, M.P., Opper, M.: Temporal autoencoding improves generative models of time series. arXiv preprint arXiv:1309.3103 (2013)
19. Brakel, P., Stroobandt, D., Schrauwen, B.: Training energy-based models for time-series imputation. J. Mach. Learn. Res. 14(1), 2771–2797 (2013)
20. Bengio, Y., Yao, L., Alain, G., Vincent, P.: Generalized denoising auto-encoders as generative models. In: Advances in Neural Information Processing Systems, pp. 899–907 (2013)
21. Bengio, Y., Thibodeau-Laufer, E.: Deep generative stochastic networks trainable by backprop. In: 31st International Conference on Machine Learning, ICML 2014, Beijing, China (2014)
22. Erhan, D., Bengio, Y., Courville, A., Manzagol, P.-A., Vincent, P., Bengio, S.: Why does unsupervised pre-training help deep learning? J. Mach. Learn. Res. 11, 625–660 (2010)
23. Grzegorczyk, K., Kurdziel, M., Wójcik, P.I.: Encouraging orthogonality between weight vectors in pretrained deep neural networks. Neurocomputing 202, 84–90 (2016)
24. Zeiler, M.D., Krishnan, D., Taylor, G.W., Fergus, R.: Deconvolutional networks. In: 2010 IEEE Conference on Computer Vision and Pattern Recognition (CVPR), pp. 2528–2535. IEEE (2010)
25. Ngiam, J., Chen, Z., Chia, D., Koh, P.W., Le, Q.V., Ng, A.Y.: Tiled convolutional neural networks. In: Advances in Neural Information Processing Systems, pp. 1279–1287 (2010)
26. Wang, Y., Xie, Z., Xu, K., Dou, Y., Lei, Y.: An efficient and effective convolutional auto-encoder extreme learning machine network for 3d feature learning. Neurocomputing 174, 988–998 (2016)
27. Zhou, F., Torre, F., Hodgins, J.K.: Aligned cluster analysis for temporal segmentation of human motion. In: 2008 8th IEEE International Conference on Automatic Face and Gesture Recognition, pp. 1–7. IEEE (2008)

28. Mörchen, F., Ultsch, A.: Finding persisting states for knowledge discovery in time series. In: Spiliopoulou, M., Kruse, R., Borgelt, C., Nürnberger, A., Gaul, W. (eds.) From Data and Information Analysis to Knowledge Engineering, pp. 278–285. Springer, Heidelberg (2006). https://doi.org/10.1007/3-540-31314-1_33
29. Keogh, E., Chakrabarti, K., Pazzani, M., Mehrotra, S.: Dimensionality reduction for fast similarity search in large time series databases. Knowl. Inf. Syst. 3(3), 263–286 (2001)
30. Donner, R.V., Zou, Y., Donges, J.F., Marwan, N., Kurths, J.: Recurrence networksa novel paradigm for nonlinear time series analysis. New J. Phys. 12(3), 033025 (2010)
31. Donner, R.V., et al.: Recurrence-based time series analysis by means of complex network methods. Int. J. Bifurcat. Chaos 21(04), 1019–1046 (2011)
32. Silva, D.F., De Souza, V.M., Batista, G.E.: Time series classification using compression distance of recurrence plots. In: 2013 IEEE 13th International Conference on Data Mining (ICDM), pp. 687–696. IEEE (2013)
33. Campanharo, A.S., Sirer, M.I., Malmgren, R.D., Ramos, F.M., Amaral, L.A.N.: Duality between time series and networks. PLoS ONE 6(8), e23378 (2011)
34. Liu, L., Wang, Z.: Encoding temporal markov dynamics in graph for time series visualization. arXiv preprint arXiv:1610.07273v2 (2016)
35. Klambauer, G., Unterthiner, T., Mayr, A., Hochreiter, S.: Self-normalizing neural networks. In: Advances in Neural Information Processing Systems. arXiv preprint arXiv:1706.02515 (2017)
36. Snoek, J., Larochelle, H., Adams, R.P.: Practical bayesian optimization of machine learning algorithms. In: Advances in Neural Information Processing Systems, pp. 2951–2959 (2012)
37. Hu, Y.: Efficient, high-quality force-directed graph drawing. Math. J. 10(1), 37–71 (2005)
38. Blondel, V.D., Guillaume, J.-L., Lambiotte, R., Lefebvre, E.: Fast unfolding of communities in large networks. J. Stat. Mech: Theory Exp. 2008(10), P10008 (2008)
39. Langville, A.N., Meyer, C.D.: Google's PageRank and Beyond: The Science of Search Engine Rankings. Princeton University Press, Princeton (2011)

Learning Single-Shot Detector with Mask Prediction and Gate Mechanism

Jingyi Chen, Haiwei Pan$^{(\boxtimes)}$, Qianna Cui, Yang Dong, and Shuning He

Harbin Engineering University, Harbin, People's Republic of China
panhaiwei2006@hotmail.com

Abstract. Detection efficiency plays an increasingly important role in object detection tasks. One-stage methods are widely adopted in real life because of their high efficiency especially in some real-time detection tasks such as face recognition and self-driving cars. RetinaMask achieves significant progress in the field of one-stage detectors by adding a semantic segmentation branch, but it has limitation in detecting multi-scale objects. To solve this problem, this paper proposes RetinaMask with Gate (RMG) model, consisting of four main modules. It develops RetinaMask with a gate mechanism, which extracts and combines features at different levels more effectively according to the size of objects. It firstly extracted multi-level features from input image by ResNet. Secondly, it constructed a fused feature pyramid through feature pyramid network, then gate mechanism was employed to adaptively enhance and integrate features at various scales with the respect to the size of object. Finally, three prediction heads were added for classification, localization and mask prediction, driving the model to learn with mask prediction. The predictions of all levels were integrated during the post-processing. The augment network shows better performance in object detection without the increase of computation cost and inference time, especially for small objects.

Keywords: Single-shot detector · Feature pyramid networks · Gate mechanism · Mask prediction

1 Introduction

One-stage detectors have been widely used in real life because of their high speed, especially in some real-time detection tasks, such as autonomous driving, face recognition in mobile phone and so on. Since object detection steps into deep learning based phrase, the accuracy of the two-stage detection models was much higher than that of the one-stage methods until the models, such as YOLO [22], SSD [20], RetinaNet [19], were successively proposed. They promote the accuracy of the one-stage methods to the comparative level as two-stage detectors. A number of excellent studies have been proposed to develop them, Cas-RetinaNet [30] develops RetinaNet [19] to a sequential-stage detector in a cascade manner. Adding deconvolutional layers, DSSD [6] improves the accuracy of SSD [20]

© Springer Nature Singapore Pte Ltd. 2020
J. Zeng et al. (Eds.): ICPCSEE 2020, CCIS 1257, pp. 327–338, 2020.
https://doi.org/10.1007/978-981-15-7981-3_23

because of introducing additional context into detection. How to promote the detection accuracy of single-shot methods while keeping the advantages of high speed and low computation cost is still a question worth studying.

So as to detect objects in various scales efficiently, feature pyramid construction has been popularly integrated into modern detection models. For example, SSD [20] directly applys the convolutional layers of backbone to generate pyramidal feature representation and adds predicting models on each layer. FPN [18] is a feature pyramid network which combines features at different levels following a top-down principle. In the past few years, FPN has been popularly adopted in the backbone of detectors due to its outstanding performance, and many excellent studies have been proposed based on FPN. NAS-FPN [8] proposes a novel feature pyramid network which fuses different-level features in a combination of top-down and bottom-up way, ReBiF [4] fuses features at different levels in a "residual" and bidirectional way. Based on the backbone composed of FPN [18] and ResNet [14], RetinaNet [19] employs a novel classification loss function Focal Loss to overcome the problem of class imbalance and promoted the classification accuracy of single-shot methods.

Moreover, two-stage detectors have advanced by learning with segmentation. Mask R-CNN [12] extends Faster R-CNN [26] with a branch for mask prediction, working in parallel with the branches for object classification and bounding boxes prediction. StuffNet [2] adds a segmentation network as a feature extractor to the framework to obtain additional features for object detection. Paper [32] describes that learning with segmentation would be helpful for category recognition and object localization. Inspired by Mask R-CNN [12], the single-shot detector RetinaMask [7] extends RetinaNet [19] by adding a branch for the purpose of instance mask prediction during training and achieves excellent performance,

Nevertheless, RetinaMask [7] still has its limitations. As other general FPN-based detectors, the feature fusion mechanism of RetinaMask [7] just focuses on how to combines the features in different levels, the contribution of each level to the final prediction is fixed. In other words, they didn't consider how to integrate the features in different channels and ignore the fact that small objects are easier to detect on features maps in low levels where the resolution is high, while large ones are easier to detect on in high-level features where the resolution is low. Inspired by GRF [28], this paper introduces a gate mechanism into the feature pyramid network of RetinaMask [7] to solve this issue and proposes RMG model. RMG is a one-stage detector learning object detection with the help of mask prediction and gate mechanism. First, it constructs a feature pyramid through ResNet. Second, the multi-level features are fused in a top-down principle through Feature Pyramid Network. Third, gate mechanism further extracts and combines the features in a more efficient way. Finally, the features are fed into three subnetworks respectively for three specific tasks: classification, bounding box regression and mask prediction, and the predictions of all levels are integrated in concatenate and select operations in the post-processing during the inference.

2   Related Works

On the basis of deep learning, object detection methods can be categorized into two fields: one is two-stage detectors, and the other is one-stage detectors (also called single-shot detectors).

Two-Stage Detectors. For the two-stage methods, such as R-CNN [10], SPP-Net [13], Fast R-CNN [9], Faster R-CNN [25], etc., a number of proposals are generated in the first step of the detectors, and the proposals are classified into specific classes and location regression in the second stage. A number of excellent works have developed those classical frameworks in various ways [3,12,21]. Generally, two-stage detectors are known for their high accuracy while single-shot detectors perform higher computational efficiency.

One-Stage Detectors. Different from two-stage detectors, one-stage methods are proposal-free, they just consists of a backbone, which works as a feature extractor, and predictors for anchors classification and bounding boxes regression. Compared to the two-stage methods, the accuracy of early single-shot detectors was much lower for a long time. Fortunately, the great performance of models such as SSD [20], YOLO [22] and YOLO v2 [24], YOLO v3 [23] attract researchers to pay attention to one-stage detectors again, those methods significantly improve the accuracy of single-shot detectors. Researchers found that the reason why one-stage detectors has a relatively low accuracy was the large class imbalance and they proposed RetinaNet [19], in which a new loss function named "Focal Loss" was proposed, it made the one-stage detectors achieve comparable accuracy for that of two-stage approaches while keep the advantages of high detection speed. RefinNet [18] filters out negative anchors before classification. Cas-RetinaNet [30] adopts a idea of cascade to the RetinaNet to improve its performance. ConerNet [17] explored a new method to detect object, it detects the bounding box of a object by predicting a series of keypoints. Detecting with keypoints is a intuitive idea, some new studies based on that keep achieving excellent performance [31].

Learning with Segmentation. Recent works suggest that learning with segmentation can promote the performance of object detection task. Mask R-CNN [12] extends Faster R-CNN [25] with a head for instance mask prediction, it detects objects while simultaneously generating a high-quality segmentation mask for them. RetinaMask [7] augments RetinaNet [19] by adding a branch for mask prediction during the process of training.

Feature Fusion. Fusing features in different resolutions leads the models to detect object at various scale more efficiently while shallow features have stronger invariance and deep features have higher equivariance. FPN [18] has been widely

used in feature fusion, which develops a top-down architecture with lateral connections to generate pyramidal hierarchy features at all scales. DSSD [5] adopts
deconvolutional layers to capture additional large-scale context. Recently, EfficientDet [29] shows remarkable performance by using weighted bi-directional
feature pyramid network and a new compound scaling method.

3   Methods

Figure 1 illustrates an overview of RMG, it's a single-shot network for object
detection which consists of a backbone, gate mechanism and three subnetworks.
The backbone is composed of ResNet and Feature Pyramid Network, it plays
a role of feature extractor. Working as a gate-controlled mechanism, the gate
mechanisms are added on each output layer of FPN to aggregate multi-view
features. Finally, the output of every gate mechanism is sent to three subnetworks
for different and specific tasks, including classification, localization and mask
prediction. Those building blocks will be elaborated in the following subsections.

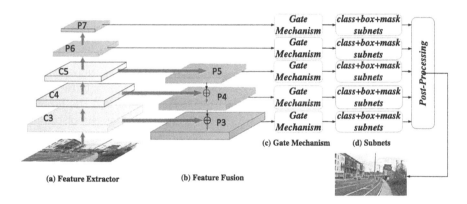

Fig. 1. The network framework of RMG. Our model is a single-shot object detector
learning with mask prediction and gate mechanism

3.1   Feature Extraction and Fusion

The backbone of our model composed of ResNet and Feature Pyramid Network
augments network by constructing feature pyramid from input image. It contains
rich feature information with a feature-fusing method. ResNet is a classical network working as a feature extractor with a residual learning principle, which
is popularly employed in models for object detection and classification. It can
be viewed as a five-stage network, the output feature maps of these stages are
denoted as $\{C3, C4, C5, C6, C7\}$ successively.

The feature maps $\{C3, C4, C5\}$ are fused following a top-down pathway in
Feature Pyramid Network. A convolution operation is performed on each feature

map in $\{C3, C4, C5\}$, in which kernel size is set to 1×1, the corresponding output of C5 is sent to a convolutional layer whose kernel size is 3×3 to generate P5, then a upsample operation is implemented on P5, and a element-wise addition followed by a convolution operation (conv3 $\times$ 3) is conduct between the result and the C4 after convolution to get the merged feature map P4. P3 is obtained in the same way as P4. Finally, a new pyramidal features $\{P3, P4, P5, P6, P7\}$ is generated.

3.2   Gate Mechanism

So as to drive the object detection network to automatically extract and combine significant information from features in different resolutions, a gate mechanism (also called attention mechanism) is introduced into our model, which consists of two levels attention components, one is channel-level, and the other is global-level. Gate mechanism can be seen as the attention of people, which distributes different weights for features in various resolutions according to the size of objects. SENet [15] firstly proposed a SElayer work as a attention mechanism, and adopted it into ResNet [14]. GFR [16] proposed a novel gate mechanism and added them before each prediction layer of DSOD [27] built in Caffe. Inspired by them, we introduce the latter mechanism, adding the gated layers on each output layer of the ResNet-FPN backbone as Fig. 2 shows.

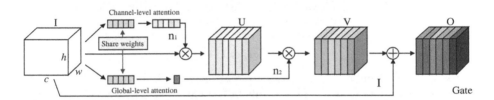

Fig. 2. The structure of gate mechanism which consists of a channel-level attention, a global-level attention and a mapping operation.

Gated layers can be viewed as a computational unit $\mathbf{F}_{tr}$ that can transforms the feature maps $\mathbf{I}$ to feature maps $\mathbf{O}$($\mathbf{F}_{tr}(\mathbf{I}) = \mathbf{O}$), where $\mathbf{I}, \mathbf{O} \in \mathbb{R}^{w \times h \times c}$. Let $\mathbf{I} = [i_1, i_2, i_3..i_c]$, i_c refers to the c_{th} feature map. Before elaborating it further, some symbols need be denoted. $\mathbf{U}, \mathbf{V} \in \mathbb{R}^{w \times h \times c}$ are intermediate states, $\mathbf{F}_{tr}$ can be seen as a series of transformation, denote $\mathbf{F}_{ca}: \mathbf{I} \rightarrow \mathbf{U}$, $\mathbf{F}_{ga}: \mathbf{U} \rightarrow \mathbf{V}$, $\mathbf{F}_{re}: \mathbf{V} \rightarrow \mathbf{O}$, Therefore:

$$\mathbf{F}_{tr}(\mathbf{I}) = \mathbf{F}_{re}(\mathbf{F}_{ga}(\mathbf{F}_{ca}(\mathbf{I}))) \tag{1}$$

Channel attention: The channel-level attention actually is a Squeeze-and-Excitation Blocks [15], it includes two stages: the first stage squeezes the global information of each channel by using a global pooling operation:

$$m_c = \frac{1}{w * h} \sum_{i=1}^{w} \sum_{j=1}^{h} i_c(i,j) \tag{2}$$

In Eq. (2), $\mathbf{m} \in \mathbb{R}^c$ and m_c is the c_{th} element of $\mathbf{m}$. The second stage is employed to model the dependencies between channels, so the second stage is defined as excitation stage. To achieve that goal, two fully connected layers followed by non-linearity function is employed:

$$\mathbf{n}_1 = \sigma(\mathbf{W}_2(\delta(\mathbf{W}_1 \mathbf{m}))) \tag{3}$$

Where δ is the ReLU activation and σ refers to the sigmoid activation, $\mathbf{W}_1 \in \mathbb{R}^{\frac{c}{16} \times c}$, $\mathbf{W}_2 \in \mathbb{R}^{c \times \frac{c}{16}}$. Then, the model resales the $\mathbf{n}$ and a channel-wise multiplication is adopted between the rescaled $\mathbf{n}$ and $\mathbf{I}$:

$$\mathbf{U} = [\text{rescale}(\mathbf{n}_1)] \otimes \mathbf{I} \tag{4}$$

Global attention: The output ($\mathbf{m}$) of the pooling operation (as Eq. (2) illustrates) is fed into this attention block: the feature maps are sent into linear layers followed by a sigmoid function:

$$\mathbf{n}_2 = \sigma(\mathbf{W}_4 \delta(\mathbf{W}_3 \mathbf{m})) \tag{5}$$

In Eq. (5), δ is the ReLU activation and σ refers to the sigmoid activation, $\mathbf{W}_3 \in \mathbb{R}^{\frac{c}{16} \times c}$, $\mathbf{W}_4 \in \mathbb{R}^{1 \times \frac{c}{16}}$, and $\mathbf{n}_2 \in \mathbb{R}^1$, then we reshape the $\mathbf{n}_2$, and implement a element-wise multiplication between $\mathbf{n}_2$ and $\mathbf{U}$.

$$\mathbf{V} = \mathbf{n}_2 \otimes \mathbf{U} \tag{6}$$

Finally, an element-wise addition operation is used between U and V to integrate them. The $\mathbf{O}$ is the final result of the gate mechanism:

$$\mathbf{O} = \mathbf{I} \oplus \mathbf{V} \tag{7}$$

3.3   Detecting with Mask Prediction

As is illustrated in Fig. 1, three subnetworks are implemented on the each output feature map of those gated layers for different and specific tasks: classification, bbox regression and mask prediction.

The classification subnetwork is responsible for predicting which class each anchor box belongs to, it is made up of 4 convolutional layers followed by a ReLU function, and 1 convolutional layer with sigmoid function, the kernel sizes of those convolution layers are set to 3×3. The box regression network aims to regress the offset between anchor boxes and the ground-truth boxes. Finally, it generates a series of bounding boxes, which denotes the predictions about objects' position, it also uses 4 convolution with a ReLU function, and a convolutional layer, the kernel size of each convolutional layer is 3×3.

The mask prediction subnetwork is designed for the purpose of instance mask prediction. The box regression network selects top-N bounding boxes with respect to score, and then the mask prediction subnetwork determines which feature map to operate according to the Eq. (8), where $k_0 = 4$, w, h denotes the weight and height of object. The mask prediction head is formed of 4 convolutional layers whose kernel size is 3×3, a transposed convolutional layer whose kernel size is 2×2 to upsample resolution and a convolutional layer for final prediction whose kernel size is 1×1.

$$k = \lfloor k_0 + log_2\sqrt{wh}/224 \rfloor \qquad (8)$$

As Fig. 1 shows, features in $\{P3, P4, P5\}$ of FPN are used for both bounding box predictions and mask prediction, and $\{P6, P7\}$ feature layers are just used for mask prediction.

It should be noticed that the mask branch is added to the performance of the object detection by using a multi-task loss function during training:

$$Loss = L_{cls} + L_{reg} + L_{mask} \qquad (9)$$

where L_{mask} denotes the mask prediction loss, L_{cls} is Focal Loss [19] used for computing the classification loss. L_{reg} represents the regression loss of bounding boxes. Following RetinaMask, this paper adopts Self-Adjusting Smooth L1 Loss as regression loss.

Self-adjusting Smooth L1 Loss: Current methods generally adopt smooth L1 Loss for bounding boxes regression, such as Faster R-CNN [25], SSD [20], RetinaNet [19]. Smooth L1 loss is showed in Eq. (10), where α is a fixed hyperparameter.

$$f(x) = \begin{cases} 0.5\frac{x^2}{\alpha}, & \text{if } x < \alpha \\ |x| - 0.5\alpha, & \text{otherwise} \end{cases} \qquad (10)$$

In the Self-Adjusting smooth L1 loss, α can be learned during training by operating the batch mean and variance as Eq. (11) illustrates, and C is a hyperparameter. And momentum(m) is assigned the value of 0.9 to update the means and variances. As is shown in Fig. 3, the α tends to be 0 in the process of training.

$$\mu_B = \frac{1}{n}\sum_{i=1}^{n} |x_i|, \ \sigma_B^2 = \frac{1}{n}\sum_{i=1}^{n} (|x_i| - \mu_B)^2$$

$$\mu_R = \mu_R * m + \mu_B * (1 - m) \qquad (11)$$

$$\sigma_R^2 = \sigma_R^2 + \sigma_B^2 * (1 - m)$$

$$\alpha = max(0, min(C, \mu_R - \sigma_R^2))$$

Based on the above-mentioned substructures, our model works as a single-shot detector by learning with gated feature pyramid network and mask prediction. During the inference, we set the score threshold at 0.05 to filter out low-confidence predictions, for each prediction layer, we select top-1000 bounding boxes with respect to scores. Then, Non-maximum suppression (NMS) is adopted for each class. In the end, the predictions whose scores are in top 100 are saved for the input image and others are discarded.

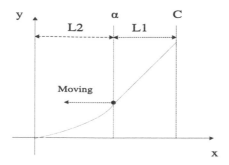

Fig. 3. Self-Adjusting Smooth L1 Loss. α splits it to two parts: L1 region and L2 region, α is limited in the range $[0, C]$ and it will get closer to 0 as the training processes

4   Experiments

We conduct our experiments on COCO datasets [1], which is a benchmark for object detection. It contains rich annotations for bounding boxes, classification and mask prediction. Following the general agreement, our model uses the `trainval135k` part for training, which consists of 80000 images chosen from the train set and 35000 images chosen from the val split, and evaluates on the `minival` set which consists of 5000 images chosen from the val set.

4.1   Implement Details

The code of our work is based on the Pytorch 1.0, and experiments are conducted on GPUs with 8 images in a batch. Following [7,19] and Linear Scaling policy [11], the base learning rate is set to 0.005, and the model adopts Stochastic Gradient Descent (SGD), in which the weight decay is set to 0.0001 and the momentum is 0.9 during training. For Self-Adjusting Smooth L1, the C is set to 0.11, which means the α will be adjust in the range $[0, 0.11]$. The model is trained on the server for 360k iterations, the learning rate is dropped to 0.0005 at 240k iteration and further dropped to 0.00005 at 320k iteration, and the model is trained with multi-scale $\{640, 800, 1200\}$. Those hyper-parameters will be adjusted appropriately when the backbone is ResNet-101-FPN.

4.2   Performance

This paper evaluates the object detection performance of our model by average precision at different IoU threshold, such as AP_{50}^{bb} (at threshold 0.5), AP_{75}^{bb} (at threshold 0.75) and so on, and also reports the average precision for objects of different size, including AP_{75}^{S}, AP_{75}^{M}, AP_{75}^{L}, AP_{75}^{S} means the average precision of small objects at IoU threshold 0.75.

The performance comparison between our model, the baseline and other related methods is displayed in Table 1. On the COCO minival dataset, the baseline RetinaMask with ResNet-50 achieves 39.4 mAP and the baseline with ResNet-101 achieves 41.4 mAP. The introduction of gate mechanism makes the model extract and use features in different channels and levels more efficiently, which distributes different weights to every level of feature pyramid according to the size of object. Compared to RetinaMask based on ResNet-50-FPN, our model shows promotion of 0.5 mAP. Also, RMG which uses ResNet-101-FPN as backbone is better than RetinaMask by 0.6 mAP. And the AP_{75}^{s} shows a improvement of 1.0 mAP, which proves that the introduction of gate mechanism helps the model achieve encouraging performance on detecting small objects.

Table 1. Detection accuracy comparison on COCO minival in terms of Average Precision

Method	Backbone	AP^{bb}	AP_{50}^{bb}	AP_{75}^{bb}	AP_{75}^{S}	AP_{75}^{M}	AP_{75}^{L}
Mask R-CNN [12]	ResNet-101-FPN	38.2	60.3	41.7	20.1	41.1	50.2
RetinaNet [19]	ResNet-101-FPN	39.1	59.1	42.3	21.8	42.7	50.2
RetinaMask [7]	ResNet-50-FPN	39.4	58.6	42.3	21.9	42.0	51.0
RetinaMask [7]	ResNet-101-FPN	41.4	60.8	44.6	23.0	44.5	**53.5**
RMG	ResNet-50-FPN	39.9	59.0	43.1	22.8	43.2	50.6
RMG	ResNet-101-FPN	**42.0**	**61.2**	**45.5**	**24.0**	**45.6**	53.1

In the Fig. 4, the objects with scores over 0.6 are visualized. As Fig. 4 shows, it proves that our model is more capable to detect small objects compared to RetinaMask, and performance suggests that a part of improvement of our model comes from a higher confidence prediction. For example, Fig. 4(a) is the detection result of RetinaMask while Fig. 4(b) is that of RMG, Our model successfully detects the person outside the window with score "0.75" in Fig. 4(a) that RetinaMask doesn't detect in Fig. 4(b), the scores for objects are normally higher. Similar situations can be seen in other pairs of images.

Fig. 4. Selective detection results on COCO minival using baseline or our model. The images in the left column (a, c, e, g) is the detection results of the baseline RetinaMask, the images in the right line (b, d, f, h) is the results of our model.

5 Conclusion

In this paper, we propose a single-shot object detection method based on Retina-Mask. Our model learns object detection with a mask prediction task during the training phrase and a gate mechanism to distribute various weights to different-level feature maps that the backbone generates. Experiments on COCO datasets demonstrate that our method achieves better performance comparing with the baseline, and obtain a competitive accuracy in the field of one-stage method.

In the future, we will try to introduce the adversarial learning into our framework, and focus on how to further promote the capability of our model to detect multi-size object.

Acknowledgements. The paper is supported by the National Natural Science Foundation of China under Grant No. 61672181

References

1. Lin, T.-Y., et al.: Microsoft COCO: common objects in context. In: Fleet, D., Pajdla, T., Schiele, B., Tuytelaars, T. (eds.) ECCV 2014. LNCS, vol. 8693, pp. 740–755. Springer, Cham (2014). https://doi.org/10.1007/978-3-319-10602-1_48
2. Brahmbhatt, S., Christensen, H.I., Hays, J.: StuffNet: using 'stuff' to improve object detection (2017)
3. Cai, Z., Vasconcelos, N.: Cascade R-CNN: delving into high quality object detection. In: The IEEE Conference on Computer Vision and Pattern Recognition (CVPR), June 2018
4. Chen, P.Y., Hsieh, J.W., Wang, C.Y., Liao, M.H.Y., Gochoo, M.: Residual bi-fusion feature pyramid network for accurate single-shot object detection (2019)
5. Fu, C.Y., Liu, W., Ranga, A., Tyagi, A., Berg, A.C.: DSSD: deconvolutional single shot detector (2017)
6. Fu, C., Liu, W., Ranga, A., Tyagi, A., Berg, A.C.: DSSD: deconvolutional single shot detector. CoRR abs/1701.06659 (2017). http://arxiv.org/abs/1701.06659
7. Fu, C., Shvets, M., Berg, A.C.: RetinaMask: learning to predict masks improves state-of-the-art single-shot detection for free. CoRR abs/1901.03353 (2019). http://arxiv.org/abs/1901.03353
8. Ghiasi, G., Lin, T.Y., Le, Q.V.: NAS-FPN: learning scalable feature pyramid architecture for object detection. In: The IEEE Conference on Computer Vision and Pattern Recognition (CVPR), June 2019
9. Girshick, R.: FAST R-CNN. Comput. Sci. (2015)
10. Girshick, R., Donahue, J., Darrell, T., Malik, J.: Rich feature hierarchies for accurate object detection and semantic segmentation. In: The IEEE Conference on Computer Vision and Pattern Recognition (CVPR), June 2014
11. Goyal, P., et al.: Accurate, large minibatch SGD: training ImageNet in 1 hour (2017)
12. He, K., Gkioxari, G., Dollár, P., Girshick, R.B.: Mask R-CNN. CoRR abs/1703.06870 (2017). http://arxiv.org/abs/1703.06870
13. He, K., Zhang, X., Ren, S., Sun, J.: Spatial pyramid pooling in deep convolutional networks for visual recognition. IEEE Trans. Pattern Anal. Mach. Intell. **37**(9), 1904–16 (2014)

14. He, K., Zhang, X., Ren, S., Sun, J.: Deep residual learning for image recognition. In: The IEEE Conference on Computer Vision and Pattern Recognition (CVPR), June 2016

15. Hu, J., Shen, L., Albanie, S., Sun, G., Wu, E.: Squeeze-and-excitation networks (2018)

16. Kollreider, K., Fronthaler, H., Bigun, J.: Evaluating liveness by face images and the structure tensor. In: 2005 Fourth IEEE Workshop on Automatic Identification Advanced Technologies (2005)

17. Law, H., Deng, J.: CornerNet: detecting objects as paired keypoints. In: The European Conference on Computer Vision (ECCV), September 2018

18. Lin, T.Y., Dollar, P., Girshick, R., He, K., Hariharan, B., Belongie, S.: Feature pyramid networks for object detection. In: The IEEE Conference on Computer Vision and Pattern Recognition (CVPR), July 2017

19. Lin, T., Goyal, P., Girshick, R.B., He, K., Dollár, P.: Focal loss for dense object detection. CoRR abs/1708.02002 (2017). http://arxiv.org/abs/1708.02002

20. Liu, W., et al.: SSD: single shot multibox detector. CoRR abs/1512.02325 (2015). http://arxiv.org/abs/1512.02325

21. Pang, J., Chen, K., Shi, J., Feng, H., Ouyang, W., Lin, D.: Libra R-CNN: towards balanced learning for object detection (2019)

22. Redmon, J., Divvala, S., Girshick, R., Farhadi, A.: You only look once: unified, real-time object detection. In: The IEEE Conference on Computer Vision and Pattern Recognition (CVPR), June 2016

23. Redmon, J., Farhadi, A.: YOLOv3: an incremental improvement (2018)

24. Redmon, J., Farhadi, A.: YOLO9000: better, faster, stronger. CoRR abs/1612.08242 (2016). http://arxiv.org/abs/1612.08242

25. Ren, S., He, K., Girshick, R., Sun, J.: Faster R-CNN: towards real-time object detection with region proposal networks. IEEE Trans. Pattern Anal. Mach. Intell. **39**(6), 1137–1149 (2015)

26. Ren, S., He, K., Girshick, R., Sun, J.: Faster R-CNN: towards real-time object detection with region proposal networks. In: Cortes, C., Lawrence, N.D., Lee, D.D., Sugiyama, M., Garnett, R. (eds.) Advances in Neural Information Processing Systems, vol. 28, pp. 91–99. Curran Associates, Inc. (2015). http://papers.nips.cc/paper/5638-faster-r-cnn-towards-real-time-object-detection-with-region-proposal-networks.pdf

27. Shen, Z., Liu, Z., Li, J., Jiang, Y.G., Chen, Y., Xue, X.: DSOD: learning deeply supervised object detectors from scratch. In: The IEEE International Conference on Computer Vision (ICCV), October 2017

28. Shen, Z., et al.: Learning object detectors from scratch with gated recurrent feature pyramids. CoRR abs/1712.00886 (2017). http://arxiv.org/abs/1712.00886

29. Tan, M., Pang, R., Le, Q.V.: EfficientDet: scalable and efficient object detection (2020)

30. Zhang, H., Chang, H., Ma, B., Shan, S., Chen, X.: Cascade RetinaNet: maintaining consistency for single-stage object detection. CoRR abs/1907.06881 (2019). http://arxiv.org/abs/1907.06881

31. Zhou, X., Zhuo, J., Krahenbuhl, P.: Bottom-up object detection by grouping extreme and center points. In: The IEEE Conference on Computer Vision and Pattern Recognition (CVPR), June 2019

32. Zou, Z., Shi, Z., Guo, Y., Ye, J.: Object detection in 20 years: a survey. CoRR abs/1905.05055 (2019). http://arxiv.org/abs/1905.05055

Network

Near-Optimal Transmission Optimization of Cache-Based SVC Vehicle Video

Xianlang Hu[1], He Dong[1], Xiangrui Kong[2], Ruibing Li[2],
Guangsheng Feng[2(✉)], and Hongwu Lv[2]

[1] Jiangsu Automation Research Institute, Lianyungang 222002, China
[2] College of Computer Science and Technology,
Harbin Engineering University, Harbin 150080, China
fengguangsheng@hrbeu.edu.cn

Abstract. A problem of video streaming of scalable video coding (SVC) is studied in vehicular networks. To improve the performance of the video streaming services and alleviate the pressure on backhaul links, the small cell base stations (SBS) is proposed with caching ability to assist the content delivery. In this paper, it introduced the problem of joint optimization of caching strategy and transmission path in SBS cache cluster. An integer programming problem was formulated to maximize the average quality of experience. In order to obtain the globally optimal solution, the primal problem was first relaxed, then an adaptive algorithm was used based on the joint KKT condition, and the branch definition algorithm was applied. Extensive simulations were performed to demonstrate the efficiency of our proposed caching strategy.

Keywords: SVC · Cache · Transmission method · Vehicle video

1 Introduction

Video streaming is the dominant contributor to the cellular network traffic. Currently, video content accounts for 50% of cellular network traffic and it is expected to account for around 75% of the mobile data traffic. This increase has forced service providers to strengthen their infrastructure to support high-quality video streaming. As the number of video stream users in digital media continues to increase, the locations of video transmission have become more diverse, gradually shifting from the traditional PC-side to mobile phones and vehicles. However, the difference between videos brings great challenge to video business, which requires video coding technology to have good video viewing effect. When the bandwidth resources available to the link are sufficient, the more sufficient bandwidth resources are used, the better the viewing effect of the user is. Obviously, the dynamic use of bandwidth resources requires the video streaming to have the function of adjusting the video bitrate, so that the video can make corresponding bitrate adjustments for different network environments. To address the challenges presented, the concept of scalable video coding has been proposed.

In Scalable Video Coding, videos are segmented into chunks and each chunk is encoded into ordered layers: a base layer (BL) and multiple enhancement layers. SVC has good scalability and strong robustness to channel changes. Compared with

© Springer Nature Singapore Pte Ltd. 2020
J. Zeng et al. (Eds.): ICPCSEE 2020, CCIS 1257, pp. 341–353, 2020.
https://doi.org/10.1007/978-981-15-7981-3_24

traditional dynamic adaptive video streaming, SVC occupies less cache space. When transmitting SVC streams over small cell networks, cache placement issues and the choice of transmission source become more challenging. Obviously, storing the same video in multiple SBS will bring higher channel diversity gain, and the total hit rate of all videos will decrease. Similarly, caching more video layers (for example, enhancing the scalability of the video) results in better Quality of Experience (QoE), but consumes more cache space. Therefore, how to correctly allocate cache resources and select transmission sources in small cell networks is particularly important.

Reference [1] reduced the buffer pressure of base stations in heterogeneous cellular networks through the joint deployment of SBS and macro cell base station (MBS). This optimizes the video transmission to users and maximizes the transmission rate of each video layer. The author of literature [2] introduced the concept of micro-millisecond caching, which gives BS high storage capacity to store popular files. He proved that the content placement of distributed femto-processors is NP-hard and proposed a greedy algorithm with a provable approximation ratio. A recent trend is to combine content caching with physical layer characteristics. In literature [3] the author proposed a cache placement algorithm to minimize the expected file download latency of a collaborative BSs cluster with cache facilities. BSs or relays that support caching can perform cooperative MIMO beamforming to support high-quality video streams. From an end-user perspective, QoE is related to the visual quality of the video sequence. In [4–6], they differed from the well-known quality of service because they considered the video sequence that is actually displayed, rather than a set of network-based parameters. QoE evaluation is based on subjective experiments. In the process, viewers are asked to give their opinions on the quality of their experience. In [7, 8], we can further extract the perceptual model from the experimental data and replace the viewer's point of view with objective quality measures. As mentioned earlier, the parameters used for SVC to encode each layer and the differences between layers will affect the quality of each layer and the quality of the entire stream. In order to evaluate the visual quality of scalable streams, the number of parameters to be considered is quite large, which is a challenge to be solved.

Therefore, for how to reasonably allocate cache resources and transmission paths for each video on the small cell network to improve user QoE, a series of interesting questions have arisen: 1) Which videos should be cached? 2) The bitrate (for example, video scalability) of each video should be cached, so as to optimize the average QoE. For streaming users, how to consider the limit of the cache capacity? 3) Because there are more than one transmission source in the scene, how to choose the transmission source is also a very important issue.

The goal of this paper is to maximize the average QoE of all vehicle users by jointly optimizing the transmission path and the cache strategy. The main contributions of this paper are summarized as follows:

- In the scenario of a small cell base station cache cluster with a cache function, this article optimizes the transmission path of video data to find the optimal point to maximize the average QoE of all users.

- Considering that the SBS cache capacity is limited, which part of the SBS cache video is one of the optimized variables in this article, the SBS cached video data will change as the user's request changes.
- In this paper, a rate adaptive selection algorithm based on joint KKT condition and branch definition is used to solve the problem.

2 System Model

In this paper, we consider a typical video streaming user in the range of the macro cell base station. In this scenario, vehicles can download video through the MBS network and surrounding SBS. The content of the video is considered to be of interest in the vehicle. For example, live of World Cup or other sporting events, advertising videos of surrounding institutions (shopping malls, shopping centers, travel agencies, etc.). As shown in Fig. 1, each vehicle is located at one cluster which is composed of several short-range SBS with caching capabilities. We assume that multiple SBS make up one cache cluster, using such cluster can ensure that one SBS cluster can serve multiple vehicles, the current vehicle will actively select the appropriate cluster for video transmission.

According to the prediction from content providers, some popular videos are preloaded into the local cache during off-peak hours. These SBS with caching capabilities will help MBS to provide content. In this scenario, Vehicles can communicate with MBS and SBS. However, communication between vehicles is not feasible by default.

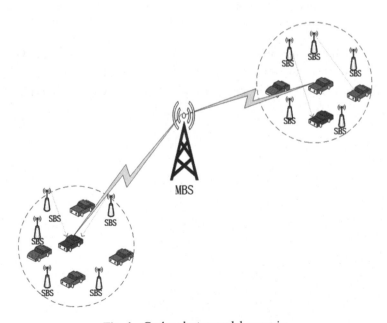

Fig. 1. Cache cluster model scenario.

In Fig. 1, it includes J cars, a collection of vehicle users denoted as $J = \{1, 2, 3 ..., J\}$. SBS can cache video and is distributed by cluster. Each SBS only communicates with vehicle users within the cluster. Assume that there are including I SBS in the scenario, so SBS is indexed as $I = \{1, 2, 3 ..., I\}$. The entire scenario contains a MBS, the base station can communicate with all vehicle users in the scene.

We assume that the upper limit of the storage video capacity of the I-th SBS is C_i. Owing to the intensive deployment of SBS, and divide into equal parts in the macro cell which has been cached in the small cell cluster. The cache capacity of each SBS is limited, so we assume that there are I SBS in the neighborhood of a typical user on average. SBS is indexed as $I = \{1, 2, 3 ..., I\}$. In order to adapt to different user needs for video bitrate, so we consider encoded by SVC to transmit video stream, the video requested by the user consists of N GOP, denoted as $N = \{1, 2 ... N\}$, Each GOP can be encoded into L video layers (one base layer, L-1 enhancement layer), and the size of the Nth GOP in the l layer is s_n^l (It has the dependence between video layers under SVC encoding. If the user stores the video of layer l, he must store the video of layer l-1, otherwise the video can't decode it). In order to improve the time efficiency, we propose that multiple SBS and MBS can simultaneously transmit one video layer data of one GOP and SBSs can cache video. The set of SBS that have cached video n is Ω_n.

After a user initiates a video request, determine whether the SBSs and MBS in the vehicle transmission range have the video data which requested by the original vehicle. If not, it is not eligible for transmission within the transmission range. We set that if SBS stores each layer of GOP, so it stores all layer video data of that GOP. If the i-th SBS has stored the n-th GOP, set $e_i^N = 0$.

If vehicle user j requests the nth GOP, set $r_j^n = 1$; If vehicle user j does not require the nth GOP, set $r_j^n = 0$.

In all transport sources (multiple SBS and MBS) that satisfy the transport conditions, by optimizing to choose the only transmission method to transmit the requested GOP.

If the nth GOP is transmitted from user i to user j, set $\chi_{ij}^n = 1$; If the nth GOP is not transmitted from user i to user j, set $q_{ij}^n = 0$.

If MBS transmits nth GOP to user j, set $\chi_{0j}^n = 1$; If MBS does not transmit nth GOP to user j, set $\chi_{0j}^n = 0$.

To improve resource utilization, each requested GOP is transmitted by only one video source. When SBS as the transmission source, the video bandwidth of the requested vehicle cannot exceed its own bandwidth limit:

$$\sum_{n=1}^{N}\sum_{l=1}^{L} e_i^n s_n^l \leq C_i \tag{1}$$

$$\chi_{ij}^n \leq r_j^n \cdot e_i^n \tag{2}$$

$$\sum_{i=1}^{I} \chi_{ij}^n + \chi_{0j}^n \leq 1 (\forall j, i \in \Omega_n) \tag{3}$$

V_{0j} represents the transmission rate of MBS to user j, V_{ij} represents the transmission rate of SBS i to user j. The transmission time of each requested GOP should be completed within the specified time. Otherwise the transmission will be considered failure, each video resource can only transmit one GOP at a time, so the transmission time of all GOPs requested by each vehicle user should be the maximum time for successful transmission of all provided sources. The time from the i-th SBS to the vehicle user j is t_{ij}, MBS time to vehicle user j is t_{0j}.

$$t_{ij} = \sum_{n=1}^{N}\sum_{l=1}^{L'} \frac{\chi_{ij}^n \cdot s_n^l}{v_{ij}} (i \in \Omega_i, \forall j, L' \in L) \tag{4}$$

$$t_{0j} = \sum_{n=1}^{N}\sum_{l=1}^{L'} \frac{\chi_{0j}^n \cdot s_n^l}{v_{0j}} (L' \in L) \tag{5}$$

$$t_{ij} \leq t_0 \tag{6}$$

$$t_{0j} \leq t_0 \tag{7}$$

According to the literature [1], the relationship between the bit rate of a video and the user experience can be written as:

$$QoE_n = 4 \cdot e^{(-c_1 \cdot (\frac{b_n}{b_{max}})^{-c2} + c_1)} \tag{8}$$

$$b_n = \sum_{1}^{L'} b_1 (L' \leq L) \tag{9}$$

QoE$_n$ represents the vehicle user gets the Nth GOP' QoE, and b_n represents bitrate of the nth GOP, bmax represents maximum bit rate provided by the video sources. c1 and c2 are attenuation coefficient in this formula. So, the average QoE of the n GOPs obtained by the j-th vehicle user can be obtained is:

$$
\overline{QoE_j} = \frac{\sum_{n=1}^{N} QoE_n \cdot r_j^n}{\sum_{n=1}^{N} r_j^n}
\tag{10}
$$

Therefore, the average QoE of the user is $\sum_{j=1}^{J} \overline{QoE_j}$.

The objective of this paper is to maximize the average user QoE, The optimization problem P0 can be given by

$$
P0: \quad \max \sum_{j=1}^{J} \overline{QoE_j}
\tag{11}
$$

$$
\text{s.t. } (1), (2), (3), (6), (7)
$$

Because in (3)–(15) r_j^n is constant, So problem P0 can be transformed into problem P1.

$$
P1: \quad \max \sum_{n=1}^{N} QoE_n
\tag{12}
$$

$$
\text{s.t. } (5), (6), (7), (10), (11)
$$

We can know from (8) and (9), QoE_n is related to the number of SBS and the number of layers for the required video, the number of SBS and QoE are positively correlated, So the problem can be transformed into maximizing the number of users receiving β.

$$
\beta = \sum_{j=1}^{J} \left(\sum_{i=1}^{I} \sum_{n=1}^{N} t_0 \cdot v_{ij} \cdot \chi_{ij}^n + \sum_{n=1}^{N} t_0 \cdot v_{0j} \cdot \chi_{0j}^n \right)
\tag{13}
$$

Then problem P1 is transformed into problem P2.

$$P2: \quad \max \beta$$
$$\text{s.t.} \quad (1), (2), (3) \tag{14}$$

3   Problem Transformation and Solution

According to this paper, problem P2 is obviously a nonlinear integer programming. By analyzing the properties of problem P2, we found that if the solution variable of problem P2 is relaxed to the continuous variable, the objective function and constraint conditions of the problem are both convex functions that are easy to solve. So in this paper, we relax problem P1, and transform problem P2 from a nonlinear integer programming problem to a nonlinear programming problem, and then obtain the solution of the nonlinear programming problem by applying KKT condition. Then the branch definition method is used to solve the 0-1 variable solution of the original problem P1.

First, the two group solutions in problem P1 $((\chi_{ij}^n, \chi_{0j}^n), e_i^n)$ are relaxed, so the problem P1 is transformed into a nonlinear programming problem, and we can use KKT condition to solve it.

Relevant KKT conditions can be obtained:

$$\frac{\partial L(\chi, \mathbf{e}, \lambda, \mu)}{\partial \chi} = -\frac{\overline{\partial QoE_j}}{\partial \chi} + \lambda + \mu_1 \frac{\partial g(\mathbf{e})}{\partial \chi} + \mu_2 \frac{\partial g(\chi, \mathbf{e})}{\partial \chi} = 0 \tag{15}$$

$$\frac{\partial L(\chi, \mathbf{e}, \lambda, \mu)}{\partial \mathbf{e}} = -\frac{\overline{\partial QoE_j}}{\partial \mathbf{e}} + \lambda \frac{\partial h(\chi)}{\partial \mathbf{e}} + \mu_1 \frac{\partial g(\mathbf{e})}{\partial \mathbf{e}} + \mu_2 \frac{\partial g(\chi, \mathbf{e})}{\partial \mathbf{e}} = 0 \tag{16}$$

$$g(\mathbf{e}) \leq 0, \quad g(\chi, \mathbf{e}) \leq 0 \tag{17}$$

$$\lambda \neq 0, \quad \mu_1, \mu_2, \mu_3 \geq 0 \tag{18}$$

$$\mu_1 g(\mathbf{e}) = 0, \quad \mu_2 g(\chi, \mathbf{e}) = 0 \tag{19}$$

Next, the branch definition algorithm is designed to solve the problem P1.

Algorithm: adaptive selection algorithm based on branch definition

---

Require:

Solution of the corresponding relaxation problem solved by KKT conditions;

Optimal objective function value for relaxation problems Z_{relax};

Ensure {

Optimal solution of problem P1 with 0-1 constraints X_{0-1};

Optimal solution of problem P1 with 0-1 constraints Z_{0-1} ;}

Initialization:

Set $K=0, L=0, U=Z_{relax}$;

Arbitrarily choose a solution from X_{relax} that does not meet the 0-1 constraint

Randomly generate a random number ε in the range (0,1)

If $0 \le X_j < \varepsilon$, **then**

add constraints X_j to problem *P1* to form sub-problem *Rmnum1*

else

add constraints $X_j = 1$ to problem *P1* to form sub-problem *Rmnum2*

end if

Continue to find the slack problem solutions for sub-problems *Rmnum1* and *Rmnum2*, denoted as X_K

Record the corresponding optimal objective function value as Z_K

$U=max\ Z_K, X_K \in [0,1]\ L=max\ Z_K, q_K \in [0,1]$

$U=max\ Z_K, X_K \in [0,1]\ L=max\ Z_K, X_K \in [0,1]$

If $Z_K < L$ **then**

Cut this branch

else

$Z_K > L$ Go back to the branch step and repeat

else

$Z_K = L$, The optimal solution for problem *P1* has been found, $Z_{0-1} = Z_K$, $X_{0-1} = X_K$

end if

---

4   Simulations

This section will perform numerical simulations to evaluate the cache placement and transfer selection strategies proposed in this paper. In the simulation, we first generate a test scenario with 100 users, which are randomly distributed in the test site, and then deploy a certain number of SBS in the test site. Each video is encoded into 4 layers. In our algorithm, the number of video layers does not have to be the same. Here we assume that the 4-layer SVC encoding mechanism is just for simplicity.

4.1   Selection of Different Transmission Sources

The transmission method proposed in this paper is the cooperative transmission of SBS and MBS, but at the same time a user vehicle can only have one transmission source. Next, we compare the transmission path optimized in this article with the baseline transmission path.

Figure 2 shows the comparison results of the common transmission scheme of MBS and SBS proposed by this paper with other baseline schemes. It can be seen from the figure that the results obtained by the cooperative transmission method of SBS and MBS are better. And the growth of QoE is also related to the caching strategy of this paper. The larger the number of SBS, the larger the cache capacity, and the QoE will increase accordingly. However, as the number of SBS continues to increase, QoE cannot be increased indefinitely, and it is also subject to other conditions.

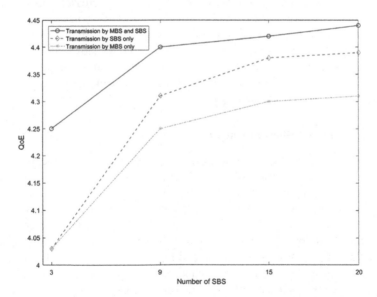

Fig. 2. Comparison of QoE values of protect different transmission sources under.

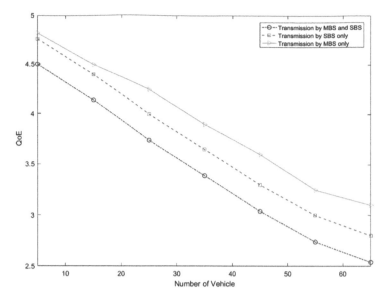

Fig. 3. Data comparison of three transmission forms under different vehicle numbers.

4.2 Different Numbers of SBS

Figure 3 shows that when the number of users gradually increases, the average QoE trend of users is gradually decreasing. Just because the increase in the number of users will inevitably lead to an increase in transmission pressure. Without increasing bandwidth and caching, the larger the number of requests and the slower the response time, the QoE will gradually decrease. However, it can be seen from the figure that using the transmission path proposed in this paper can slow down the trend of increasing the number of vehicles and decreasing QoE.

4.3 Comparison of Different Caching Strategies

The baseline caching scheme only caches one copy of each video in the SBS cluster in order to maximize the hit rate (MHR). We consider three bit rates for the baseline strategy, namely 4.8 mbps, 7.2 mbps, and 10.4 mbps. We consider a representative situation, when there are three SBSs near the user, the cache size of each SBS is 2 TB.

Figure 4 (N = 3, C = 2 TB) shows the average QoE of different cache strategies under different SNRs of SBS. We observe that the proposed cache strategy is superior to the baseline strategy both at low SNR and high SNR. The reason lies in the cache strategy optimized in this article, which chooses which SBS cached video and which part is one of the optimized variables. As the user's request changes, the video data of each SBS cache is changed accordingly, thereby achieving the effect of increasing the user's QoE.

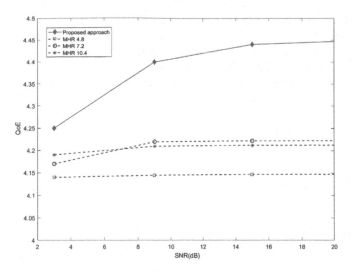

Fig. 4. Comparison of Baseline Cache protect Strategy and QoE of Proposed Strategy.

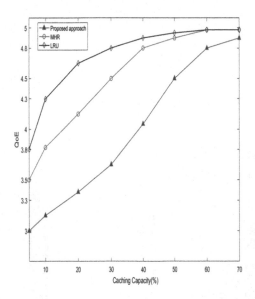

Fig. 5. Comparison of different cache protect strategies under different cache capacities.

Figure 5 is a comparison between the cache strategy obtained through optimization and the baseline strategy MHR and LRU cache strategy. In this comparison, we assume that the same transmission path, same bit rate, and transmit power are used, but the cache capacity of the SBS is adjusted to evaluate the effectiveness of these cache strategies. More specifically, we adjusted the SBS cache capacity from 5% of the total video size to 70%.

In Fig. 6, we assume that the transmission paths used for transmission are the same, that is, SBS and MBS are used for coordinated transmission, and have been optimized in Chapter 3. It can be seen from Fig. 6 that QoE decreases with the increase in the number of vehicles, and the cache strategy optimized in this paper is higher than the other two cache strategies under different numbers of vehicles.

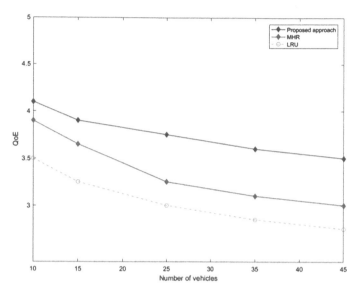

Fig. 6. Comparison of different cache protect strategies under different vehicle numbers.

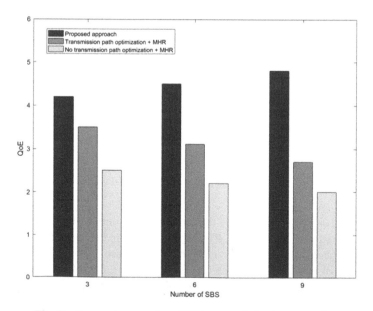

Fig. 7. Comparison results of different optimization strategies.

4.4 Comprehensive Experiment of Transmission Path Optimization and Cache Strategy Optimization

Figure 7 shows the optimization strategy proposed in this paper, compared with the other two cases. In the first case, we only optimize the transmission path and use the baseline cache scheme; in the second case, we do not optimize the path and use the baseline cache strategy. From the figure we can see that the optimization method proposed in this paper is superior to other strategies in the case of different numbers of SBS.

5 Conclusion

This paper mainly studies the SVC vehicle video transmission method based on the SBS cache cluster, and considers the scenario of a small cell base station cache cluster with a cache function to optimize the video data transmission path and finds the optimal point to maximize the number of video layers of the video block that all users can receive. Considering that the SBS cache capacity is limited, SBS cached video data and the parts of the optimization are variables optimized in this paper. And SBS cached video data will change as user requests change.

Acknowledgement. This work is supported by the System Architecture Project (No. 614000-40503), the Natural Science Foundation of China (No. 61872104), the Natural Science Foundation of Heilongjiang Province in China (No. F2016028), the Fundamental Research Fund for the Central Universities in China, and Tianjin Key Laboratory of Advanced Networking (TANK) in College of Intelligence and Computing of Tianjin University.

References

1. Moshref, M., Yu, M., Govindan, R., Vahdat, A.: Dream: dynamic resource allocation for software-defined measurement. In: ACM SIGCOMM Computer Communication Review, vol. 44, pp. 419–430. ACM (2014)
2. Agre, J., Clare, L.: An integrated architecture for cooperative sensing networks. Computer **33**(5), 106–108 (2000)
3. Cui, Y., et al.: Software defined cooperative offloading for mobile cloudlets. IEEE/ACM Trans. Netw. (TON) **25**(3), 1746–1760 (2017)
4. Wang, Y., Sheng, M., Wang, X., Wang, L., Li, J.: Mobile-edge computing: Partial computation offloading using dynamic voltage scaling. IEEE Trans. Commun. **64**(10), 4268–4282 (2016)
5. Munoz, O., Pascual-Iserte, A., Vidal, J.: Optimization of radio and computational resources for energy efficiency in latency-constrained application offloading. IEEE Trans. Veh. Technol. **64**(10), 4738–4755 (2014)
6. Zhang, W., Wen, Y., Wu, D.O.: Collaborative task execution in mobile cloud computing under a stochastic wireless channel. IEEE Trans. Wireless Commun. **14**(1), 81–93 (2014)
7. Barbera, M.V., Kosta, S., Mei, A., Stefa, J.: To offload or not to offload the bandwidth and energy costs of mobile cloud computing. In: 2013 Proceedings IEEE INFOCOM, pp. 1285–1293. IEEE (2013)
8. Yang, L., Cao, J., Yuan, Y., Li, T., Han, A., Chan, A.: A framework for partitioning and execution of data stream applications in mobile cloud computing. ACM SIGMETRICS Performance Eval. Rev. **40**(4), 23–32 (2013)

Improved Design of Energy Supply Unit of WSN Node Based on Boost and BUCK Structure

Shaojun Yu[1], Li Lin[2], Yujian Wang[1], Kaiguo Qian[1(✉)], and Shikai Shen[1]

[1] School of Information Engineering, Kunming University, Kunming 650214, Yunnan, China
qiankaiguo@qq.com
[2] College of Agricultural and Life Sciences, Kunming University, Kunming, Yunnan, China

Abstract. The energy supply unit of WSN node based on solar cell-powered outdoor environment monitoring needs storage battery for energy buffering, thus resulting in the distribution balance problem among solar battery, storage battery charge and discharge, and load energy consumption power. Based on the analysis of sensor network characteristics and node composition, the overall design of energy supply unit of wireless sensor network node on the basis of self-harvesting solar energy is carried out in this paper. At the same time, the improved circuit structure based on Boost and BUCK structure was designed, the working mode and energy distribution mode of the circuit were analyzed, and the circuit and control flow were comprehensively analyzed. In order to verify the energy distribution, experiments were implemented for test. The results of compensating the conversion efficiency of the power converter show that the energy distribution efficiency is significantly improved.

Keywords: Boost and BUCK circuits · Wireless sensor network · Node · Power supply unit · Design

1 Introduction

A wireless sensor network node usually consists of sensing module, data processing module, radio communication module and power supply module, and its basic structure is shown in Fig. 1. Nodes form a wireless network in a self-organizing and multi-hop manner to cooperatively sense, collect, process and transmit monitoring information of sensing objects in the geographical area covered by the network and report to users [3]. Sensor node is both the initiator and the forwarder of an information packet. At the same time, when nodes enter or exit the network, the network structure will be reorganized automatically. As shown in Fig. 1.

Based on the characteristics of wireless sensor network and the limitation of power supply energy of nodes, not just disposable batteries are generally used for power supply. In order to prolong the life cycle of nodes, researchers have conducted in-depth research on Energy Harvesting of nodes in recent years, integrating self-powered

© Springer Nature Singapore Pte Ltd. 2020
J. Zeng et al. (Eds.): ICPCSEE 2020, CCIS 1257, pp. 354–367, 2020.
https://doi.org/10.1007/978-981-15-7981-3_25

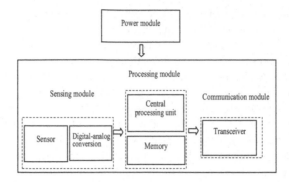

Fig. 1. WSN Node Structure.

technology into wireless sensor network nodes to prolong the network life cycle. These self-powered energy sources include electromagnetic energy, vibration energy, solar energy, wind energy, thermal energy and acoustic energy, etc. Various energy supply methods that may be self-harvested by wireless sensor network nodes have their own advantages and disadvantages. It is necessary to decide which method to adopt according to the functional configuration and operational environment of wireless sensor network nodes [1]. Light energy is the most abundant energy source in the environment, and is also the most easily obtained energy source. Its energy density is higher than that of other energy sources. The popularization and application of solar batteries make the cost of light energy for power supply much lower than other methods.

Although solar power supply has the advantage of high energy density, nodes need to be placed in a light environment that is as unobstructed as possible, and the appropriate maximum power point tracking algorithm needs to be studied. At present, the solar power supply control algorithm used by wireless sensor network nodes is relatively simple, the power control effect is not very good, and the overall performance of the energy storage unit is not considered. Based on the characteristics of outdoor environment monitoring wireless sensor network nodes and Boost structure improvement, the energy supply unit of nodes is designed and discussed in this paper.

2 Overall Design of Energy Supply Unit of WSN Node Based on Self-harvesting Solar Energy

Solar energy is the best choice of energy sources for outdoor WSN nodes at pre-sent, with the advantages of high energy density and low cost. However, the output of the solar battery has strong uncertainty and randomness, and cannot continuously supply electric energy at night and in rainy days. Therefore, it needs to be used together with other energy storage units. When the solar energy is sufficient, surplus energy is temporarily stored in the energy storage unit; when the solar energy is temporarily insufficient, the energy storage unit will supply energy for nodes. Energy storage units are generally rechargeable chemical batteries. Under fixed external illumination and temperature conditions, the actual output power of the solar battery is closely related to its output voltage or current value, and the actual output power is different under different output voltage and current states. Therefore, it is necessary to design a fast and efficient maximum power point tracking method for solar batteries to ensure that the solar battery works at the power close to the maximum power point in real time. However, the maximum power point tracking algorithm with higher precision for solar batteries at present generally requires a large amount of computation and the performance of the algorithm with a small amount of computation is relatively poor. The efficiency of the maximum power point tracking algorithm for solar energy of some WSN nodes is not more than 80%. Wireless sensor network nodes generally have small computing power and algorithms with too high computational complexity are not applicable. Therefore, it is necessary to study a maximum power point tracking method for solar batteries with high efficiency and low computation. Although solar batteries with higher power can be used to make up for the defect of low maximum power point tracking efficiency, it will bring the disadvantages of volume increase and cost increase. Therefore, it is necessary to study the maximum power point tracking algorithm for solar energy with small computation and high efficiency, and make full use of the photoelectric conversion capability of solar batteries.

At the same time, a large number of previous studies have proved that chemical batteries are constrained by charging strategies and their own cycle times, and improper charging methods will affect the charge efficiency and service life of batteries. Rechargeable chemical batteries are generally limited by the number of charges, overcharge or overdischarge will affect the service life of battery, and chemical batteries are greatly affected by temperature, and effective capacity and number of charge cycles are significantly reduced in a low temperature environment. Wireless sensor network nodes for outdoor environmental monitoring often suffer various environmental conditions, such as high and low temperature, lightning and static electricity. Therefore, it is necessary to study efficient charging methods of chemical batteries and try to reduce the number of charging and discharging cycles of chemical batteries, so as to prolong the service life of chemical batteries and ensure the continuous working time of nodes.

In addition, the circuit power conversion efficiency needs to be further improved. In the outdoor environment, the energy obtained by nodes from the environment is very limited, so it should be used as effectively as possible, rather than wasted in the circuit power conversion link. At present, the power of commonly used DC-DC con-version circuit is generally below 90%, so it is necessary to design and study efficient energy conversion methods to maximize the rational and effective use of energy produced by solar batteries.

Based on the above factors, the overall structure of the energy supply unit of nodes is shown in Fig. 2.

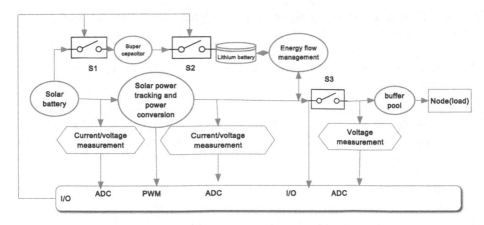

Fig. 2. Overall Structure of Energy Supply Unit of WSN Node.

3   Improved Structural Design of Solar Energy Supply Unit of Nodes Based on BOOST and BUCK Structure

3.1   Structure Analysis of Basic System of Solar Energy Supply Unit

The output power of solar batteries has strong randomness and uncertainty due to the influence of solar light intensity and ambient temperature, with a wide range of changes. In order to ensure the stability and continuity of the load system, the solar power supply system must be equipped with energy storage elements such as storage batteries or super capacitors to store and regulate electric energy.

In the traditional independent solar power supply system, the storage battery is directly connected to the load [1–3], as shown in Fig. 3.

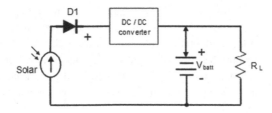

Fig. 3. Common solar power supply system structure.

R_L in Fig. 3 is equivalent load, and its current consumption changes dynamically. Obviously, this topology cannot effectively control the charge and discharge current of the storage battery. When the load changes suddenly, it may cause the charge and discharge current of the storage battery to change greatly, thus damaging the storage battery. The storage battery is directly connected in parallel with the load, and the battery voltage is not the same at different residual electric quantities, so the working voltage of the load is changing dynamically, which has certain influence on the stable operation of the load.

3.2   Structure Analysis of Improved Basic System of Solar Energy Supply Unit

In order to overcome the disadvantages of common solar energy power supply systems and accurately control and adjust the charge and discharge current of the storage battery, a DC power converter is inserted between the load power access terminal and the storage battery to control the current flow direction and magnitude of the storage battery.

In Fig. 4, the battery supplies power to the rear power supply main circuit through the unidirectional DC-DC power converter U1, and the storage battery is connected to the power supply main circuit through the bidirectional DC-DC power converter U2.

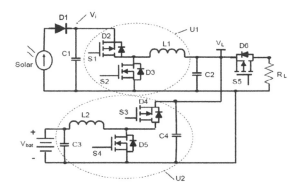

Fig. 4. Power Distribution Management Topology

This circuit has the following characteristics: (1) independent DC-DC power converter completes MPPT of solar batteries; (2) the charge and discharge control of the storage battery is independently carried out, and the U2 unit regulates the charge and discharge operation and the current magnitude of the storage battery; (3) according to the output power and load power of the solar battery, the working modes of U1 and U2 power converters are selected to realize the dynamic management of system energy.

4   Working Mode and Energy Distribution Mode

4.1   A Subsection Sample

In Fig. 4, the unidirectional DC power converter U1 is a unidirectional BUCK-structured converter, which converts a wide range of voltage output by the solar battery into a relatively stable voltage. The bidirectional DC power converter U2 can divide the circuit structure into BUCK and BOOST according to different situations. When the load of the solar battery energy supply node is surplus, U2 is a BUCK structure, and energy flows from the solar battery to the storage battery. When the energy of the solar battery is insufficient to support the node load, U2 is a BOOST structure, and energy flows from the storage battery to the post load.

In the power supply system combined with solar battery and storage battery, due to the dynamic changes of output power and load power of the solar battery, the energy flow direction and working mode between them also change dynamically. During these changes, the values of solar battery voltage (V_{solar}), storage battery voltage (V_{bat}), storage battery charge and discharge and storage battery charge and discharge current (I_{bat}) also change, and the circuit working mode is in a dynamic change process.

The node is always in a working state and consumes energy P_L, while the solar battery generates electric energy P_S, which can be used to charge the battery or supply power to the node for use. Namely,

$$P_S = P_{BAT} + P_L \tag{1}$$

In Eq. (1), P_{BAT} is the in-and-out energy of the battery. When P_{BAT} is larger than 0, the battery is charging and when P_{BAT} is smaller than 0, the battery is discharging.

In the working process, V_{solar} and V_{bat} have the corresponding normal ranges. For example, the voltage range of lithium iron phosphate battery is 2.7 V–3.6 V, and if exceeding the normal range, it means that state switching is required.

Table 1 summarizes the different working modes, where V_{bat_min} is the minimum voltage that the storage battery can safely discharge, and V_{bat_max} is the voltage when the storage battery is full.

Table 1. Working Mode

	$P_S = 0$	$P_S < P_L$	$P_S > P_L$
$V_{bat} < V_{bat\_max}$	M1: U1: Shutoff U2: Shutoff	M2: U1: Output constant current, non-MPPT state U2: BUCK charge (load not working to protect battery)	M2: U1: Output constant current, non-MPPT state U2: BUCK charge (load not working to protect battery)
$V_{bat\_min} < V_{bat} < V_{bat\_max}$	M3: U1: Shutoff U2: BOOST discharge	M4: U1: Output constant current, MPPT state U2: BOOST discharge	M5: U1: Output constant current, MPPT state U2: BUCK charge
$V_{bat} > V_{bat-max}$	M3: U1: Shutoff U2: BOOST discharge	M4: U1: Output constant current, MPPT state U2: BOOST discharge	M6: U1: Output constant voltage U2: BOOST discharge to maintain load access line voltage

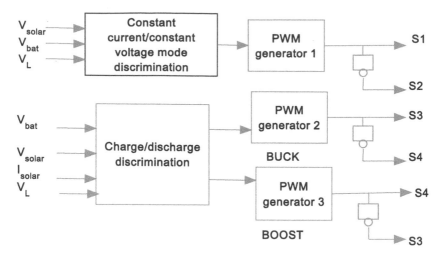

Fig. 5. Control Circuit Structure.

In Fig. 5, different energy distribution states can be switched by controlling the PWM switch signals of S1–S4. According to the working mode described in Table 1, a schematic diagram of the control circuit structure can be obtained. As shown in Fig. 5, three PWM signals are generated, corresponding to BUCK and BOOST modes of U1 and U2 respectively.

5 Circuit Structure Analysis

The U1 unit shown in Fig. 4 is a BUCK-structured circuit, energy flows from the solar battery to the battery and the subsequent power consumption unit, S1 and S2 are complementary switches, and the equivalent circuits are shown in Fig. 6(a) and (b) when they operate, wherein the post circuit is equivalent to resistor R_B.

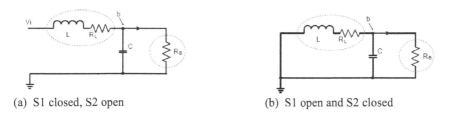

(a) S1 closed, S2 open   (b) S1 open and S2 closed

Fig. 6. Equivalent Structure of Control Circuit.

In Fig. 4, the current I_L flowing from the inductor L1 is divided into two parts, part flows to the capacitor C2 to charge it, and part flows to RB, namely

$$I_C = C\frac{dV_b}{dt} = I_L - \frac{V_b}{R_B} \qquad (2)$$

In the above equation, V_b is the voltage at point b, and I_C is the current flowing to the output filter capacitor C2.

In Fig. 6, the voltage across the pure inductor L is

$$U_L = L\frac{dI_L}{dt} = DV_i - V_B - I_L R_L \qquad (3)$$

In the above equation, R_L is the series equivalent resistance inside the inductor, and D is the duty cycle of PWM signal controlled by S1 and S2.

From Eqs. (2) and (3), the state observation equation of the circuit can be obtained as shown in Eq. (4).

$$\begin{bmatrix} \dot{V} \\ \dot{I_L} \end{bmatrix} = \begin{bmatrix} -\frac{1}{CR_B} & \frac{1}{C} \\ -\frac{1}{L} & -\frac{R_L}{L} \end{bmatrix} \cdot \begin{bmatrix} V_b \\ I_L \end{bmatrix} + \begin{bmatrix} 0 \\ \frac{1}{L} \end{bmatrix} \cdot V_i D \qquad (4)$$

The input-output transfer function of the circuit is

$$G(s) = D \cdot \frac{1/LC}{s^2 + s(1/R_B C) + 1/LC} \qquad (5)$$

From Eq. (5), it can be seen that the transfer function of this system has no zero point, the poles are negative, far from the actual control frequency, and the structure is a stable system.

When U2 is charging and discharging, it is of BUCK and BOOST structures respectively. BUCK circuit is the same as U1 unit, so the analysis results are the same.

In BOOST structure, similar to analysis method of BUCK circuit, we can obtain the input-output transfer function as follows

$$G(s) = \frac{1}{1-D} \cdot \frac{1/LC}{s^2 + s(1/RC) + 1/LC} \qquad (6)$$

In the equation, D is the duty cycle of PWM signal controlled by S3 and S4, R is the equivalent resistance of the post circuit, and L is the inductance value of inductor L2. The circuit is also a stable structure.

Due to the subsection design of the circuit, the division of labor of each part is clear and relatively independent, and the working state of the circuit is also relatively independent. There is a critical state when the output power of the solar cell is the same as or close to the load energy consumption power. At this time, the flowing power in the bidirectional DC power converter U2 is about 0. U2 works in an alternating state, changing back and forth between discharging and charging, and the circuit structure

362   S. Yu et al.

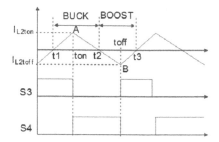

Fig. 7. Switching Tube State and Inductor Current under Equal Power State.

also switches back and forth between BUCK and BOOST. At this time, the relationship between the states of the switching tubes S3 and S4 in the circuit and the current in the inductor L2 is shown in Fig. 7.

(1) From 0 to t1, S3 is closed, S4 is open, and the circuit works in BOOST state

$$V_{bat} = L2 \cdot \frac{d(-I_{L2})}{dt} + V_L \tag{7}$$

The current of inductor L2 is

$$I_{L2}(t) = \frac{V_L - V_{bat}}{L2} - I_{L2toff} \tag{8}$$

(2) From t1 to ton, S3 is closed, S4 is open, and the circuit works in BUCK state

$$V_L = L2 \cdot \frac{d(I_{L2})}{dt} + V_{bat} \tag{9}$$

The current of inductor L2 is

$$I_{L2}(t) = \frac{V_L - V_{bat}}{L2} - I_{L2toff} \tag{10}$$

(3) From ton to t2, S3 is open, S4 is closed, and the circuit works in BUCK state

$$L2 \cdot \frac{dI_{L2}}{dt} + V_{bat} = 0 \tag{11}$$

Inductor current is

$$I_{L2}(t) = \frac{V_{bat}}{L2} - I_{L2ton} \tag{12}$$

(4) From t2 to toff, S3 is open, S4 is closed, and the circuit works in BOOST state

$$L2 \cdot \frac{d(-I_{L2})}{dt} - V_{bat} = 0 \qquad (13)$$

Inductor current is

$$I_{L2}(t) = \frac{-V_{bat}}{L2} + I_{L2ton} \qquad (14)$$

In Fig. 7, t1A and Bt3 are parallel, and the areas of the two triangular areas enclosed by the two lines are equal. $S_{\Delta t1 At2} = S_{\Delta t2Bt3}$, t1 is the midpoint of 0ton segment and t2 is the midpoint of t1t3 segment, then

$$|I_{L2ton}| = |I_{L2toff}| \qquad (15)$$

When t = t1,

$$|I_{L2toff}| = \frac{V_L - V_{bat}}{L2} t1 \qquad (16)$$

When t = t2-ton = t2-2t1,

$$|I_{L2ton}| = \frac{V_{bat}}{L2} (t2 - 2t1) \qquad (17)$$

The duty cycle D is

$$D = \frac{ton - t1}{t2 - t1} = \frac{t1}{t2 - t1} \qquad (18)$$

According to Eq. (15), it can deduce that

$$t2 = (\frac{V_L}{V_{bat}} + 1) \qquad (19)$$

According to Eq. (18), it can deduce that

$$D = \frac{V_{bat}}{V_L} \qquad (20)$$

From the above analysis, it can be seen that the duty cycle D of the control signal depends only on the storage battery voltage and the load access line voltage when operating at the equal power point, which is a fixed value. The fluctuation amplitude of inductor current is only related to inductance value L2 of inductor. Therefore, the control method is relatively simple.

6   Introduction

Figure 8 shows the flow of system working mode switching, where V_{solar_min} is the minimum voltage for solar batteries to have outward discharge capability, and V_{L_min} is the minimum value of VL voltage. During operation, the controller collects V_{solar}, V_{bat}, V_L, I_{solar}, I_{bat} and other data, and according to these data, determines the power

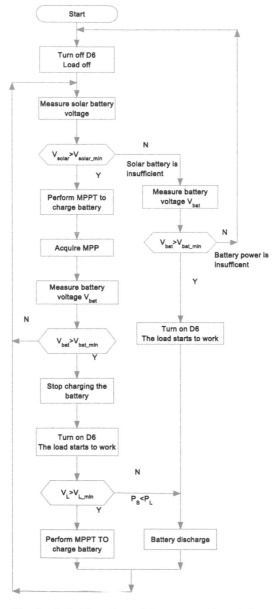

Fig. 8. Switching Flow of System Working Mode.

generation capacity of the solar battery, the energy storage state of the storage battery and the relative relationship between the output power of the solar battery and the load energy consumption, thus setting the working mode of the system and completing the distribution and flow of energy.

7 Introduction

In order to verify the energy distribution efficiency of solar battery, storage battery and node load designed above, the following experimental tests were carried out:

The solar battery, the battery (3.6 V lithium iron phosphate battery) and the specific WSN node were connected through the above circuit, the WSN node operated periodically, and the power consumption was different in different operation stages. The output power of the solar battery changes gradually, experiencing a process of increasing and then decreasing. The above circuit controls the charging and discharging of the storage battery. The experiment was carried out for 2 h. The relationship between the output power of the solar battery, the node consumed power and the charge and discharge power of the storage battery during the experiment is shown in Fig. 9.

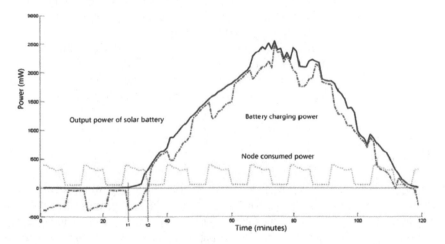

Fig. 9. Node Power Distribution.

During the period from 0 to t1, the output power of the solar battery is almost 0. The battery independently supplies power to nodes, and its discharge power is equivalent to the node consumed power. The energy flow structure of the system is shown in Fig. 10(a) (where Solar represents the solar battery, Bat represents the storage battery, and Load represents the load node).

During the period from t1 to t2, the output power of the solar battery gradually increases, but the output value is smaller than the node consumed power. Therefore, the

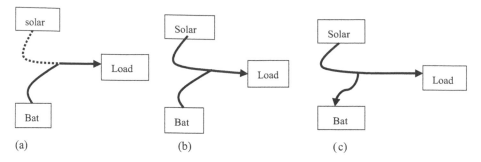

Fig. 10. Energy flow pattern

storage battery discharges and the solar battery and the storage battery support the node to work together. The energy flow structure is shown in Fig. 10(b).

During the period from t2 to later stage of the experiment, the output power of the solar battery is large enough to support the node to work, and there is surplus. Therefore, the surplus electric quantity is charged into the storage battery, and the storage battery is in a charged state at this time. The energy flow structure is shown in Fig. 10(c).

(a) The storage battery supplies power separately (b) The solar energy and the storage battery supply power simultaneously (c) The solar energy supplies power to the nodes and charges the storage battery simultaneously.

After numerical calculation, the power conversion efficiency of the DC power conversion circuit is compensated. When $P_S > 0$, $(P_B + P_L)/P_S = 0.991$; when $P_S = 0$, $P_B = -1.01P_L$ (P_B, P_L and P_S are respectively storage battery charge and discharge power, load consumed power and solar battery output power; $P_B < 0$ means battery discharge).

8   Conclusion

The energy supply unit of WSN node based on solar battery-powered outdoor environment monitoring needs storage battery for energy buffering, thus resulting in the distribution balance problem among solar battery, storage battery charge and discharge, and load energy consumption power. In this paper, the possible working states in such structural circuits are analyzed. Based on these working states, the corresponding control circuit structures and strategies are designed, and the circuits and control processes are comprehensively analyzed. In order to verify the energy distribution, corresponding experiments are for test. The results of compensating the conversion efficiency of the power converter show that the energy distribution efficiency is significantly improved.

Acknowledgment. This work was supported by the Yunnan Local Colleges Applied Basic Research Projects (2018FH001-061, 2018FH001-010, 2017FH001-042, 2017FH001-059), National Natural Science Foundation of China (61962033).

References

1. Brunelli, D., Moser, C., Thiele, L.: Desing of a solar-harvesting circuit for batteryless embedded systems. IEEE Trans. Circ. Syst. **56**(11), 2519–2528 (2009)
2. Dondi, D., Bertacchini, A., Brunelli, D.: Modeling and optimization of a solar energy harvester system for self-powered wireless sensor networks. IEEE Trans. Ind. Electron. **55** (7), 2729–2766 (2008)
3. Porcarelli, D., Brunelli, D., Magno, M.: A multi-harvester architecture with hybrid storage devices and smart capabilities for low power systems. In: International Symposium on Power Electronics, Electrical Drives, pp. 946–951(2012)
4. Porcarelli, D., Brunelli, D., Magno, M.: A multi-harvester architecture with hybrid storage devices and smart capabilities for low power systems. In: International Symposium on Power Electronics, Electrical Drives, pp. 2946–2951(2012)
5. Yu, S., Lin, L., Li, X.: Dynamic energy consumption analysis and test of node in wireless sensor networks. In: Proceedings of 13th IEEE International Conference on Electronic Measurement & Instruments (ICEMI) (2017)
6. Xiaoping, Z., Wenhao, S., Hongping, Z.: Multi-objective Optimization of the main circuit and control parameters of the buck-boost matrix converter. J. Syst. Simul. **30**(08), 3042–3049 (2018)
7. Zheng, M., Zhao, X., Hu, E., Zheng, J.: Research on MPPT control method for solar energy. J. Power Supply Collect. Circ. Boost Struct. **15**(6), 36–41 (2018)
8. Zilong, Y., Yibo, W.: Research on measurement and control technology for PV power generation system. Acta Energiae Solaris Sinica **36**(4), 1023–1028 (2015)
9. Sun, L.: Wireless Sensor Network. Tsinghua University Press (TUP), Beijing (2005)
10. Xiao, Y.: Research and design of intelligent charger for lead-acid battery. Tianjin University, Tianjin (2010)
11. Zhu Zhou, Yu., Bo, S.Y., Weidong, Y.: Designing solar lithium battery management system in wireless sensor nodes. J. Electron. Measur. Instrum. **29**(12), 1798–1805 (2015)
12. Zhu, Z., Yi, W., Yu, S.: Designing high-efficiency solar charging management system. Electron. Measur. Technol. **38**(9), 58–65 (2015)
13. Eguchi, K., Kuwahara, K., Ishibashi, T.: Analysis of an LED lighting circuit using a hybrid buck–boost converter with high gain. Energy Rep. **6**(2(Supl.2)), 250–256 (2020)
14. Jia, L.: Micro-energy harvesting system and its application in wireless sensors. Nanjing University of Posts and Telecommunications (2019)

Disjoint-Path Routing Mechanism in Mobile Opportunistic Networks

Peiyan Yuan[1,2,3]([⊠]), Hao Zhang[1,2,3], and Xiaoyan Huang[1,2,3]

[1] School of Computer and Information Engineering, Henan Normal University,
Xinxiang 453007, Henan, China
peiyan@htu.cn, zh.haohao@foxmail.com,
hxiaoyan96@foxmail.com
[2] Engineering Laboratory of Intellectual Business and Internet of Things
Technologies, Xinxiang 453007, Henan, China
[3] Big Data Engineering Laboratory for Teaching Resources & Assessment
of Education Quality, Xinxiang 453007, Henan, China

Abstract. The prevalent multi-copy routing algorithms in mobile opportunistic networks (MONs) easily cause network congestion. This paper introduces a disjoint-path (DP) routing algorithm, where each node can only transmit packets once except the source node, to effectively control the number of packet copies in the network. The discrete continuous time Markov chain (CTMC) was utilized to analyze the state transition between nodes, and the copy numbers of packets with the DP routing algorithm were calculated. Simulation results indicate that DP has a great improvement in terms of packet delivery ratio, average delivery delay, average network overhead, energy and average hop count.

Keywords: Multi-copy · Disjoint-path · Mobile opportunistic networks · Routing · Continuous time Markov chain

1 Introduction

MONs is one of the intermittently connected networks that do not require the support of infrastructures such as base stations, so an end-to-end connected path rarely exist. The data transmission is based upon the store-carry-and-forward style for the sake of carrying data packets. With this unique feature, MONs is widely used in the sparse environment including vehicle communication, military field etc. To accomplish data communications between the disconnected source-destination pair, packets transmission needs to rely on the node mobility, thus, forming opportunistic contacts. When two nodes enter into the communication range of each other, a contact happens and packets can be swapped and carried between them [1]. Using the mobility of nodes, the researcher community proposes some multiple copies routing strategies, such as the Epidemic [2], Spray and Wait [3], Spray and Focus [4], etc.

All these strategies have one thing in common: each node can transmit a packet several times, which means a node can participate in multiple paths. We call these strategies joint-path routing (JPR). In JPR, nodes appearing in multiple paths play an important role in the network performance. Once they are attacked or energy depletion,

© Springer Nature Singapore Pte Ltd. 2020
J. Zeng et al. (Eds.): ICPCSEE 2020, CCIS 1257, pp. 368–379, 2020.
https://doi.org/10.1007/978-981-15-7981-3_26

the path collapses. Considering these facts, a Disjoint-Path (DP) routing mechanism is proposed. This routing mechanism can strongly improve the event of node collapse due to energy consumption. Since this mechanism selectively sends messages, to effectively controlling the number of message copies. For the sake of observation, the transition between node states is shown by the CTMC model. In the last part, the thesis verifies the effectiveness of this system by experiments and data analysis. By DP mechanism, the system burden is alleviated and the energy consumption is reduced, so as to prolong the lifetime of MONs.

The rest of the paper is organized as follows. Related works are surveyed in Sect. 2. In Sect. 3, we present, model and analyze the performance of DP routing algorithm. In Sect. 4, a comparison between DP and other classical routing algorithms is presented. Finally, the conclusion and future work are discussed in Sect. 5.

2 Related Work

Based on the number of copies of a single message that may exist in the network, current opportunistic routing algorithms can be classified into two categories: single-copy and multiple-copy. Single-copy routing schemes use only one copy of messages, For example, the Direct Transmission [5] scheme sends the message if and only if the source encounters the destination. The authors of [6] proposed several single-copy routing algorithms. And [7] utilizes an optimized single-copy data transmission to achieve high energy efficiency. However, those approaches are vulnerable in mobile opportunistic networks given the nondeterministic and intermittent connectivity setting. Besides the single-copy routing scheme, a large number of multi-copy multi-path strategies have also been proposed. Most of them come from the flooding mechanism [2], which was proposed by Vahdat and Becker in 2000. It forwards the message to all neighboring nodes so that the message is constantly flooded and eventually reaches the destination node. In recent years, using social factors is a hot topic in MONs. The literature [8] combines social importance to design a social similarity-based routing algorithm:SSR. SimBet Routing [9] is proposed which exploits the centrality metrics and social similarity to the destination node. Inspired by PageRank [10], PeopleRank [11] gives higher weight to nodes if they are socially connected to other important nodes of the networks. Rank nodes based on their social property are used as a guide for forwarding decisions. Considering the social and location information of mobile users, the authors of [12] utilized the social characteristics of mobile nodes to make forwarding decisions. Employing the location history as a social feature, a new Geo-Social indicator is proposed to forward the message to the node with high utility value. The [13] proposed a multi-copy routing algorithm based on machine learning named iPRoPHET, which has a bigger cost than other similar multi-copy routing algorithms. [14] describes a query message routing algorithm based on a time-varying interest community, which improves data transmission efficiency but does not consider energy consumption. In a short conclusion, the multi-copy scheme has a high delivery probability; it however has increased the energy consumption. Therefore, the forwarding decisions are typically guided by the desire to reduce the number of replicas of packets in the network. We address this challenge by proposing DP strategy, and make effective decision for each transmission opportunity.

3   The Proposed Scheme

In intermittently connected mobile networks, there generally do not exist a fully connected end-to-end path from the source to the destination during data transmission. Both pedestrians with smart mobile devices and vehicles can be seen as mobile nodes in such scenarios. Because of their intermittent connectivity and unpredictable mobility, there is no guarantee that the full connection path between two communicating at any time. Therefore, collecting and exchanging data in opportunistic networks is cumbersome. In general, it is difficult to use global knowledge of network topology to make efficient routing and forwarding decisions. In MONs, nodes use Bluetooth, Wi-Fi, or any other wireless technology to exchange and forward data in an opportunistic hop by hop manner depending on the mobility of the node and the opportunistic contact. Normally, data will be forwarded from the source node to the destination node after multiple hops.

Vahdat and Becker presented a routing protocol for intermittently connected networks [2], called Epidemic Routing, which is primarily depending on node mobility and opportunistic contact to complete data transmission. More specifically, when two nodes meet, they first exchange the data ID in the buffer queue of the other party, and then judge and exchange the data packets carried by the other party but not by themselves. And in this way, messages will spread like an epidemic of some disease and "infect" each other. Since there is no restriction on the diffusion of packet copies, the opportunistic contact between nodes will result in a large amount of packet copies in the network. The Epidemic algorithm has optimal delay performance, but at the same time, the network has the heaviest load and the worst scalability. And the existence of a large number of copies can easily lead to network congestion. In this paper, in order to effectively control the number of packet copies in the network, we develop a systematic approach to guide forwarding decisions in an opportunistic network. The key idea of the policy is that every node except the source node can only transmit data once when they encounter the non-destination node. More specifically, it is defined as a mechanism for only one data transmission of any node except the source node, and no intermediate nodes will appear on both paths until the destination node is encountered. The intuition behind this idea is that forwarding with our routing strategy can reduce the number of cost (message copies), while the possibility of messages arriving at the destination increases (success rate) due to multiple transmissions from the source node. Note that since the source node forwards the data unrestricted, the probability of the data reaching its destination increases to some extent.

We apply the DP mechanism to the flooding mechanism, restrict the forwarding decisions of the nodes that carry the messages, and effectively reduce the copies of packets. In the process of changing the network topology, if the non-source node carrying the data has forwarded one packet to a contacted node, the packet is not forwarded to later contacted nodes unless the destination node. In the next section, we discuss the presentation and implementation of DP routing mechanism in detail and give the relevant description of the strategy.

3.1 Forwarded Symbol Setting

We set a binary flag, "Forwarded", to indicate whether a packet is forwarded or not. For each packet, the initial value of the flag is false. When an intermediate node forwards a packet, the flag of the packet becomes true, which means that the packet cannot be forwarded again unless the destination node is encountered. In this way, the joint-path situation is effectively avoided. The following Algorithm 1 describes the DP mechanism within Epidemic.

Algorithm 1 EpidemicDP Routing

When node A with a message M encounters node B

1: **if** node B is a non-destination node **then**

2:   **if** node A is the source node **then**

3:   node A forwards the message M to B

4:   **else if** node A is the intermediate node **then**

5:   // judge M's Forwarded flag

6:   **if** M_F=false **then**

7:   node A forwards the message M to B then

8: M_F=true

9:   **else if**

10:   no message transmission between A and B

11: **if** node B is destination node **then**

12:   node A forwards the message M to B

3.2 Modeling

In this subsection, we use the Epidemic model to analyze the number of copies in DP scheme. Since the Epidemic algorithm is similar to the spread of infectious diseases in the population, for convenience, this paper presents the data transmission as a virus transmission process; and we adopt the terminology from Epidemiology to denote the model. The source of infection can spread the virus to each uninfected individual in contact with it. According to the process of virus transmission, we call the node (source node) that generates the message as the "infection source"; a node that does not store copies of the message is called a "susceptible"; if a node stores the packet, it has been "infected". Note that in DP scheme, an "infected" node can only spread the virus once. Once the "infected" node spreads the virus, it enters into the state called "waiting for treatment" and the destination node is called "hospital". The rule is that all infected individuals can only recover in a designated hospital. Therefore, the "waiting for treatment" node can only transmit the data again when it encounters the destination node. In summary, the susceptible node can eventually be transformed into two states: infected or waiting for treatment state, which are all carriers of the virus.

Through the above analysis, the change in the number of infected nodes can be modeled as Markov chains [15]. And according to the mathematical model in [16], we present a three-state Markov chain model as shown in Fig. 1, where the three-state is represented as the susceptible state (S), the infectious state (A), and the state of waiting for treatment (W). The susceptible state (S) changes to the infected state (A) after the first infection, in which the infected state has only one chance to spread the virus. And the state of loss of activity after a virus spread is called waiting for treatment (W), in which case it can only wait for recovery without reinfection susceptible nodes.

Fig. 1. Markov chain model of an infectious disease with Susceptible, Active, and Wait state.

3.3   Analyzing the Model

One interesting phenomenon is that the change in the number of infected nodes does not depend on the spread of the virus from the infected node; it depends only on the spread of the source infection. For example, we assume that "infection source" does not infect the susceptible node at time T, and the infection node transmits the virus to the susceptible node when it encounters the susceptible node. After spread of the virus, the infection node turns into a node waiting for treatment, and the susceptible node becomes an infection node. As we expected, there is no change in the number of infectious nodes, and the number of nodes waiting for treatment increases. Therefore, it is understood that the change in the number of infectious nodes depends on the spread of "infectious sources", and the number of nodes waiting to be treated is dependent on the infection of the infected node.

Since the goal of this paper is to reduce the number of duplicate packets in the network so as to reduce the pressure on the network, it is necessary to focus on analyzing the change of the number of duplicate packets in the network. The number of duplicate packets is here understood as the number of carriers carrying the virus, including the infected node and the node waiting to be treated. According to the above model and related analysis, we use the CTMC model to track the changes in the number of infected nodes of each type. As shown in Fig. 1, the susceptible state $S(t)$ represents the number of "susceptible", $I_A(t)$ represents the number of "active", $I_W(t)$ represents the number of "waiting for treatment", $I(t)$ represents the total number of "carriers of the virus", and β represents the contact rate of the nodes. Before discussing the changes in the number of data copies in the system, we first need to explain the state changes of the nodes accordingly. And then through the correlation transition between the state nodes, it is easy to know the number of changes. We make the following assumptions. Suppose there are N nodes in the system, one is the source of infection, S susceptible nodes. Then a node contacts $\beta(N-1)$ other nodes per unit time, of which $S/(N-1)$ do not yet have the disease. And the total number of infected nodes in the system is $I = I_A + I_W$. Therefore, the transition rate from state S to state I_A becomes:

$$
\begin{aligned}
& Infected\ infection\ rate \\
& = (\#infection\ source)(contact\ rate)(\#susceptible\ nodes) \\
& = [\beta(N-1)][S/(N-1)] \\
& = \beta S
\end{aligned}
\tag{1}
$$

Because the wait state I_W is the infection state after the virus spreads, the wait state is indicated as the next change following the active infection state. The transition rate from state I_A to state I_W becomes:

$$
\begin{aligned}
& Waiting\ for\ treatment\ infection\ rate \\
& = (\#infected)(contact\ rate)(\#susceptible\ nodes) \\
& = [I_A - 1][\beta(N-1)][S/(N-1)] \\
& = \beta S(I_A - 1)
\end{aligned}
\tag{2}
$$

Through the above analysis and the following related calculation, we construct functions of I_A and I_W over time. And we assume that different diseases in the system propagate independently. At the initial moment, only the source node carries the virus in the system, so the initial conditions for the system are:

$$
S = N - 1, \quad I_A(0) = 1, \quad I_W(0) = 0
\tag{3}
$$

What we are interested in is the number of carriers in the system, through the analysis of the above state transition rate, the transient solution of Markov chain is given:

$$
\begin{aligned}
\frac{dI_A}{dt} &= \beta S \\
\frac{dI_W}{dt} &= \beta S(I_A - 1)
\end{aligned}
\tag{4}
$$

Note that the sum of virus carriers and susceptible nodes in the system is N, so it is inferred that $S = N - I_A - I_W$. We substitute S into the above transient solution to obtain the following two differential equations for the variables I_A and I_W.

$$
\frac{dI_A}{dt} = \beta(N - I_A - I_W)
\tag{5}
$$

$$
\frac{dI_W}{dt} = \beta(N - I_A - I_W)(I_A - 1)
\tag{6}
$$

After some algebra, we can get the expression about I_W. By deriving both sides of equation I_W and substituting into Eq. (6), we can get the Second-order Differential Equation expression about I_A. To solve this complex differential equation, we adopt the method of substitution to simplify the complex problem. And using the method of variables separation and integrals, we can get the differential equation only for I_A, and then substituting the obtained equations into the expression of I_W. According to the initial value, $I_W(0) = 0$, the relation expression of I_W and I_A is calculated. And then finally, using the same method, the I_A equation is calculated based on the initial value $I_A(0) = 1$.

$$I_A(t) = \frac{\sqrt{2N-1} + \dfrac{N-\sqrt{2N-1}}{N-1}e^{-\beta\sqrt{2N-1}t}}{1 - \dfrac{N-\sqrt{2N-1}}{N-1}e^{-\beta\sqrt{2N-1}t}} \tag{7}$$

$$I_W(t) = \frac{I_A^2}{2} + \frac{1}{2} - I_A$$

3.4   PageRank DP

Section A describes the DP mechanism within Epidemic. In this section, we discuss how to implement the DP into other opportunistic routing algorithms. We here take the social based mechanism as an example since the researcher community show that utilizing the social property of nodes can substantially improve the performance of opportunistic routing. Among them, PageRank is a representative of the social routing algorithm. In the next section, we illustrate the fact that the PageRank with DP mechanism can still have an excellent performance in most of the routing metrics.

The main idea of PageRank is that nodes with a higher centrality value will generally be more important in the network. More specifically, PageRank models the opportunity network as a social network graph. Firstly, the social property of nodes is obtained through the social relations between nodes, where social property is expressed as the weight of nodes. If a node is an important node, then the node has a higher weight and a higher PageRank value. Next, rank nodes based on their social property, i.e., the centrality value in PageRank algorithm. Finally, the forwarding decisions are made according to the ranking, which greatly improves the probability of data transmission.

We now apply the DP routing mechanism to PageRank. Similarly, as nodes in the DP routing mechanism are permitted to transmit data packets only once except the source, the binary flags of each packet storing in nodes have the same initial value: Forwarded = false. When the source node generates data packets, it can transmit the data to all nodes with higher PageRank value than itself. When an intermediate node encounters a non-destination node, the node forwards the data packet if and only if the forwarded flag is false and the relay node has a larger PageRank value than the intermediate node. After that, the forwarded flag of the packet is set to true, which means the intermediate node cannot forward the packet to other later relay nodes even the later has a higher centrality value, unless it encounters the destination node. The following Algorithm 2 describes the DP mechanism within PageRank.

Algorithm 2 PageRankDP Routing

When node A with a message M encounters node B

1: **if** node B is a non-destination node **then**

2:   **if** node A is the source node **then**

3:     // judge $PageRank_B$'s value

4:     **if** $PageRank_B > PageRank_A$ **then**

5:       node A forwards the message M to B

6:   **if** node A is the intermediate node **then**

7:     // judge M's Forwarded flag and $PageRank_B$'s value

8:     **if** M_F=true and $PageRank_B > PageRank_A$ **then**

9:       node A forwards the message M to B

10:       M_F=true

11: **if** node B is destination node **then**

12:   node A forwards the message M to B

---

4 Simulation Results

4.1 Validating the Model

In this section, we discuss the contact rate β which is a key parameter to evaluate the Markov model in our proposed approach. The contact rate β is calculated by recording the number of contact between nodes, and it is mainly reflected in two aspects: S and I_A, I_A and I_W. Let Δt record the time in the system from the state $I_A = i$ to the state $I_A = i + 1$, and from $I_W = j$ to $I_W = j + 1$. Note that the change in the number of nodes from state S to A is instantaneously solved as βS, then $\beta = 1/S\Delta t$, and the rate from state A to state W is accordingly $\beta S(I_A - 1)$, then $\beta = 1/(S(I_A - 1)\Delta t)$. That is, we can get the value of β taking the average.

We used the real dataset KAIST [17] (Korea Advanced Institute of Science and Technology) to emulate the movement of nodes. The simulation time is 15000 s and the number of nodes in the network is 90. Figure 2 shows the number of infected nodes over time.

As we expected, the theoretical results are in good agreement with the real situation, so we can use this model to represent the change in the number of infected nodes, and the establishment of this model is reasonable.

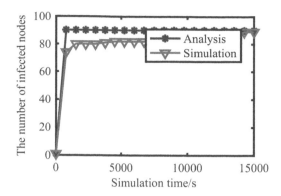

Fig. 2. Markov chain model of an infectious disease with Susceptible, Active, and Wait state.

4.2 Simulation Analysis

We use the Visual C++ platform to integrate the four routing algorithms: PageRank, EpidemicDP, PageRankDP and GeoSocial. The mobility of nodes is based on the real dataset KAIST. The maximum communication distance of each node is 250 m, the simulation time is set to 15000 s, and we randomly selected 1000 pairs of nodes as source-destination pairs. The initial energy of each node is 10000 units, it is assumed that data needs to consume one energy to receive or forward a message. The detailed simulator information can be found in [18]. Simulation results are reported in Fig. 3.

And the following indicators are used to analyze the performance of the routing algorithm.

(1) Average delivery delay: the ratio of the total delay of all successfully submitted messages to the number of successfully submitted messages.
(2) Total energy consumption: energy consumed by data transmission.
(3) Average network overhead: the ratio of the number of all message copies in the network to the total number of messages generated by the system.
(4) Average hop count: hop count indicates the number of nodes through which the message generated by the source node reaches the destination.
(5) Delivery ratio: the ratio of the total number of successfully received nodes to the total number of messages generated by the source node.

From Fig. 3(a), we can observe that the DP scheme effectively reduces the delivery delay. Compared with PageRank and GeoSocial, the PageRankDP algorithm improves the delay from 3400 s, 2700 s to 1000 s at the simulation moment 15,000 s, respectively. And in Fig. 3(b), we can see that the energy consumption of PageRankDP is between PageRank and GeoSocial. At 12000 s, the curve of PageRankDP gradually coincides with PageRank. Calculate the residual energy of each of the 90 nodes at intervals of 750 s, and obtain the standard deviation of the number of node energy consumption is shown in Fig. 3(c). From this figure, the uneven distribution of messages in the initial network results in a large fluctuation in energy consumption, and over time, the curve flattens out. This indicates that it has reached a state of equilibrium

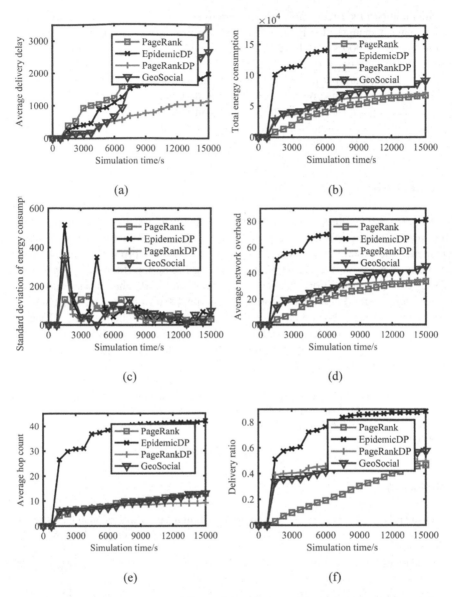

Fig. 3. Simulation Results: Average delivery delay, Energy consumption, Standard deviation of energy consumption, Average network overhead, Average hop count and Delivery ratio.

in energy consumption. In this way, some nodes can be prevented from "dying" due to energy exhaustion. The network overhead curve of PageRankDP in Fig. 3(d) is relatively stable and very close to that of PageRank. At 15000 s, the two curves almost overlap, and it is expected to be less than PageRank after 15000 s. From Fig. 3(e), the hops of the other three curves except EpidemicDP changed roughly the same in 0–3000 s; the PageRankDP curve was lower than PageRank at 6000 s and lower than

GeoSocial at 6750 s. Figure 3(f) shows the delivery rate. EpidemicDP's delivery rate is the best, and it changes slowly after 7500 s; PageRankDP is higher than PageRank and GeoSocial at 0–7000 s and lower than GeoSocial algorithm after 7000 s, gradually close to PageRank delivery rate.

Through comparative tests, DP routing mechanism has a practical research value. At the condition of consuming less energy, the delivery rate is higher. Over time, the energy consumption of nodes in the network can maintain a balance, which is one of the focuses of our current research. And it is pleased that delivery delay, hop and network overhead have all improved.

5 Conclusion

MONs relies on the multiple copies and multiple paths to transmit data packets quickly. However, resources in opportunity network are limited, and too many copies will degenerate the system performance. Aiming to solve the problems, the DP routing scheme is proposed to reduce copies and prolong the system lifetime. After using the CTMC to analyze transitions between node states, the DP scheme was implemented into Epidemic and PageRank algorithm and the performance was evaluated. The experimental results show that the DP scheme achieves a better tradeoff between delivery ratio, delay and the cost and number of hops. It also provides a reference for studying disjoint path routing strategies in MONs. Displayed equations are centered and set on a separate line.

Acknowledgements. This work was supported in part by the National Natural Science Foundation of China under Grants U1804164, 61902112 and U1404602, in part by the Science and Technology Foundation of Henan Educational Committee under Grants 19A510015, 20A520019 and 20A520020.

References

1. Yuan, P., Fan, L., Liu, P.: Recent progress in routing protocols of mobile opportunistic networks: a clear taxonomy, analysis and evaluation. J. Netw. Comput. Appl. **62**, 163–170 (2016)
2. Vahdat, A., Becker, D.: Epidemic routing for partially connected ad hoc networks. Durham North Carolina: Duke University CS2200006 (2016)
3. Spyropoulos, T., Psounis, K., Raghavendra, C.S.: Spray and wait: an efficient routing scheme for intermittently connected mobile networks. In: Proceedings of the 2005 ACM SIGCOMM Workshop on Delay-Tolerant Networking, pp. 252–259 (2005)
4. Spyropoulos, T., Psounis, K., Raghavendra, C.S.: Spray and focus: efficient mobility-assisted routing for heterogeneous and correlated mobility. In: Fifth Annual IEEE International Conference on Pervasive Computing and Communications Workshops (PerComW 2007), March, pp. 79–85 (2007)
5. Shah, C., Roy, S., Jain, S.: Modeling a three-tier architecture for sparse sensor networks. Ad Hoc Netw. **1**(2), 215–233 (2003)

6. Spyropoulos, T., Psounis, K., Raghavendra, C.S.: Single-copy routing in intermittently connected mobile networks. In: 2004 First Annual IEEE Communications Society Conference on Sensor and Ad Hoc Communications and Networks, IEEE SECON 2004, pp. 235–244 (2004)
7. Ayele, E., Meratnia, N., Havinga, P.J.M.: An asynchronous dual radio opportunistic beacon network protocol for wildlife monitoring system. In: IFIP International Conference on New Technologies (2019)
8. Yuan, P., Pang, X., Song, M.: SSR: using the social similarity to improve the data forwarding performance in mobile opportunistic networks. IEEE Access 7, 44840–44850 (2019)
9. Daly, E.M., Haahr, M.: Social network analysis for routing in disconnected delay-tolerant MANETs. In: Proceedings of the 8th ACM International Symposium on Mobile Ad Hoc Networking and Computing, pp. 32–40. ACM (2007)
10. Brin, S., Page, L.: The anatomy of a large-scale hypertextual web search engine. Comput. Netw. ISDN Syst. 30(1), 107–117 (1998)
11. Mtibaa, A., May, M., Diot, C., Ammar, M.: PeopleRank: social opportunistic forwarding. In: Proceedings of the 29th IEEE International Conference on Computer Communications, Joint Conference of the IEEE Computer and Communications Societies, pp. 111–115. IEEE, San Diego (2010)
12. Ying, Z., Zhang, C., Li, F., Wang, Y.: Geo-social: routing with location and social metrics in mobile opportunistic networks. In: IEEE International Conference on Communications, pp. 3405–3410. IEEE (2015)
13. Srinidhi, N., Sagar, C., Shreyas, J.: An improved PRoPHET - Random forest based optimized multi-copy routing for opportunistic IoT networks. Elsevier B.V. 11 (2020)
14. Bi, J., Li, Z., Li, F.: Time-variant interest community based query message routing algorithm in opportunity social network. J. Commun. 40(09), 86–94 (2019)
15. Ahn, J., Sathiamoorthy, M., Krishnamachari, B., Bai, F., Zhang, L.: Optimizing content dissemination in vehicular networks with radio heterogeneity. IEEE Trans. Mob. Comput. 13(6), 1312–1325 (2013)
16. Small, T., Haas, Z.J.: The shared wireless infestation model - a new ad hoc networking paradigm. In: Proceedings of the Fourth ACM International Symposium on Mobile Ad Hoc Networking and Computing, MobiHoc 2003, pp. 233–244 (2003)
17. Rhee, I., Shin, M., Hong, S., Lee, K., Kim, S.J., Chong, S.: On the levy-walk nature of human mobility. IEEE/ACM Trans. Netw. (TON) 19(3), 630–643 (2011)
18. Yuan, P., Song, M.: MONICA: one simulator for mobile opportunistic networks. In: Proceedings of the 11th EAI International Conference on Mobile Multimedia Communications. ICST (Institute for Computer Sciences, Social-Informatics and Telecommunications Engineering), pp. 21–32 (2018)

Double LSTM Structure for Network Traffic Flow Prediction

Lin Huang[1], Diangang Wang[1], Xiao Liu[1], Yongning Zhuo[2(✉)], and Yong Zeng[3]

[1] State Grid Sichuan Information and Communication Company, Chengdu 610041, China

[2] University of Electronic Science and Technology of China, Chengdu 611731, China
zyning@uestc.edu.cn

[3] The 10th Research Institute of China Electronics Technology Group Corporation, Chengdu 610036, China

Abstract. The network traffic prediction is important for service quality control in computer network. The performance of the traditional prediction method significantly degrades for the burst short-term flow. In view of the problem, this paper proposes a double LSTMs structure, one of which acts as the main flow predictor, another as the detector of the time the burst flow starts at. The two LSTM units can exchange information about their internal states, and the predictor uses the detector's information to improve the accuracy of the prediction. A training algorithm is developed specially to train the structure offline. To obtain the prediction online, a pulse series is used as a simulant of the burst event. A simulation experiment is designed to test performance of the predictor. The results of the experiment show that the prediction accuracy of the double LSTM structure is significantly improved, compared with the traditional single LSTM structure.

Keywords: Time sequence · Long-short term memory neural network · Traffic prediction · Service quality control

1 Introduction

The network traffic prediction is an important step in service quality control and also a long-term hot research topic in network technology. In recent years, the deep neural network for the time series prediction has become an important research direction. The long and short memory neural network (LSTM), performs well by learning the short and long term information in time series and finding the changing pattern in the series. LSTM contains many time memory units and multiple hidden layers, and acts well in dealing with interval and delay events in time series. It so far has been applied to fields

This work was supported by the research plan of State Grid Sichuan Electric Power Company, China,and supported by the research plan of the 10th Research Institute of China Electronics Technology Group Corporation (KTYT-XY-002).

J. Zeng et al. (Eds.): ICPCSEE 2020, CCIS 1257, pp. 380–388, 2020.
https://doi.org/10.1007/978-981-15-7981-3_27

such as the natural language processing, weather prediction, transportation flow prediction [1, 3, 4].

However, though LSTM has achieved good results in some fields, it is still based on the condition that there exists some pattern of variation in the flow [2], and the fewer the flow pattern switches, the better the LSTM predictor works. In computer network, the amount of the information traffic flow can change in large range. There are many variation pattern in flow and the pattern can change rapidly. Due to the random generation of burst traffic, it is impossible to know the time when the burst flow occur. The traditional LSTM can't get any information about the burst traffic in advance during the training. It is then poorly trained when the business flows contain many different variation pattern and switches rapidly, as the LSTM predictor will still treat burst traffic as an ordinary business fluctuation, which will lead to large error in the prediction [3, 8–12].

For the above analysis, this paper proposed a new predictor structure and a new training mode. The parallel LSTM predictor structure contains two LSTMs trained with the same method, one of which acts as the main predictor, another as the detector of the time when the burst flow starts. In the training stage, firstly the moments at the flow burst starts will be detected, then a indication signal of the burst traffic will be produced, which will participate in subsequent training process as an extra input parameter. The two neural networks can exchange their internal states, and the main predictor can use the information obtained from the detector to perform a multivariable prediction, then makes itself easy to find and learn the new flow pattern caused by burst flow, thus makes it possible to improve the prediction.

2 Characteristics Analysis of LSTM

LSTM is a special kind of recurrent neural network(RNN), which is formed by adding long and short term memory units in recurrent neural network(RNN). The structure of LSTM includes a set of memory units and three gate structures(in- gate, forgetting gate, out-gate) to control the use of historical information. The logical structure of LSTM neural network can be shown in Fig. 1 [2, 5–7]:

In above structure, $f(t)$ is the output of forgetting gate and the forgetting gate determines what information needs to be discarded from the previous step. The in-gate consists of two parts. The first part uses sigmoid activation function(namely σ function) and its output is $i(t)$, and the second part uses *tanh* activation function and its output is $g(t)$. The output of the in-gate and the forgetting gate combine to form the long-term state information s(t) of LSTM. The out-gate determines the output of current neuron $o(t)$ and the short-term hidden status information $h(t)$ which will be passed to the next step. The output will be obtained by using the short-term state of the previous moment, long-term state value and the reserved part of present input value.

Assuming that the weight coefficient and bias term of the hidden layer neurons of the forgetting gate are W_f, U_f, b_f, the weight coefficient and bias term of the hidden layer neurons of the input gate are W_i, U_i, b_i, W_g, U_g, b_g, and the weight coefficient and bias term of hidden layer neurons of the output-gate are W_o, U_o, b_o. $\odot$ represents Hadarmad product, and then the process of forward algorithm used by LSTM to predict is as following [2]:

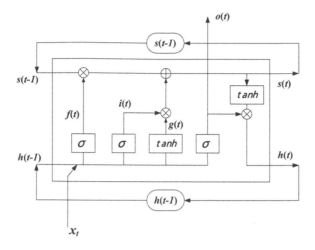

Fig. 1. Traditional LSTM architecture

1) Update output of forgetting gate:

$$f(t) = \sigma(W_f h(t-1) + U_f x_t + b_f)$$

2) Update two parts of in-gate's output:

$$i(t) = \sigma(W_i h(t-1) + U_i x_t + b_i)$$

$$g(t) = \tanh(W_g h(t-1) + U_g x_t + b_g)$$

3) Update the long-term state of neurons:

$$s(t) = s(t-1) \odot f(t) + i(t) \odot g(t)$$

4) Update output of out-gate:

$$o(t) = \sigma(W_o h(t-1) + U_o x_t + b_o)$$

$$s(t) = s(t-1) \odot f(t) + i(t) \odot g(t)$$

From the above structure and process we can see that once the LSTM is well trained, the weight coefficients in each layer of neurons are fixed and the predicted output just depends on current input variable x_t (namely characteristic), the long-term state $s(t-1)$ and short-term state $h(t-1)$ passed down from the previous step. Therefore, if the coefficients in each layer of two LSTM predictors are the same, the output will remain the same as long as the internal state and input criteria remain the same.

3   Training for Burst Traffic

One important characteristic of network information flow is suddenness, which can be determined by a certain threshold value and is usually caused by a burst event. The burst traffic may cause changes of network traffic patterns. For this purpose, our scheme is to detect the burst traffic in training data and form a indication signal of simulated burst events, which can be a pulse signal. This signal is regarded as an input signal p(t) other than the flow signal $\times$(t), and they jointly form the input signal x_t of LSTM at time t for training.

In the off-line training stage, we use the following Depth-Backstep algorithm to detect the burst flow's starting point:

1) *Set the threshold value of flow changes TH and TL and find the depth range value Depth, back step value Backstep.*
2) *Find the low flow point in Depth region. If the low point is current low point, move to next step and record it as a low point.*
3) *If the difference between the current low point and the previous low point is less than the threshold TL, no processing is performed. Otherwise, it is considered that a new low point is found and the previous Backstep time points above the current low point will be emptied.*
4) *Find the high point in Depth region. If the high point is current high point, move to next step and record it as a high point.*
5) *If the difference between the current high point and the previous high point is less than the threshold TH, no processing is performed. Otherwise, it is considered that a new high point is found and the previous Backstep time points below the current high point will be emptied.*

The traffic has different patterns in different periods, and the transition point in the pattern means that there is a burst of traffic. For different patterns, learning with the same LSTM network will not work well. Therefore, we give a signal to LSTM at transition point in the pattern so as to make the neural network recognize this pattern will shift and make subsequent training more purposefully. The above process is to detect the start point of the burst flow. Figure 2 shows a detection result for a sequence containing burst flow, in which the rod pulse indicates that a burst traffic starts. We can also regard the pulse as a sudden network event, which not only produce the pulse flow at that time, but also affects the flow's variation pattern in a following short period of time.

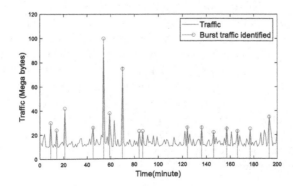

Fig. 2. Flow burst points detected through burst flow detection algorithm

4 Double LSTM Structure

Based on the above analysis, we can realize the predictor for the burst traffic based on LSTM. We use two parallel LSTMs to construct the predictor structure as shown in Fig. 3:

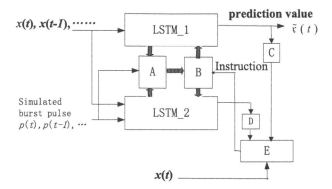

Fig. 3. parallel LSTM predictor structure A: register of the internal states and input data of two LSTMs; B: allocator of new states and input data for two LSTMs; C, D: register of output of LSTM; E: comparator

In above structure, the LSTM_1 is the main predictor and it uses actual traffic sequence data $X(t) = [x(t), x(t-1),...]$ as its input and then output the predicted traffic flow at next time step $\tilde{x}(t+1)$. LSTM_2 is the burst detector which uses $X(t)$ and burst pulse series $P(t) = [p(t), p(t-1),...]$ as input data. $P(t)$ is used to simulate burst events and it's a pulse series with a fixed period (the period is equal to the minimum interval of the known burst flow sequence). The prediction results of LSTM_2 will not be directly exported, but compared with the results of LSTM_1 and present input $x(t)$ at comparator to determine when the actual burst event in real flow occurs. The two LSTMs are trained with same mode in advance, so their coefficients of each layer are the same, and the internal state information of the two LSTM neural networks could be exchanged. Its working process is as follows:

(1) At time $t-1$, for LSTM_1, the current internal state ($c1$ $(t-1)$ and $h1$ $(t-1)$) is temporarily stored in the state register A; then x $(t-1)$ is inputted, forming a input vector signal $x_{t-1} = [x$ $(t-1)$ $0]$. The forward algorithm is carried out to compute and produce the predicted flow value $\tilde{x}_1(t)$ at the next time, The value is outputted and also stored in the result register C. For LSTM_2, it firstly copy LSTM_1's state ($c1$ $(t-1)$ and $h1($ $(t-1))$) as its internal state, and then take x $(t-1)$ and current burst pulse p $(t-1)$ as input, namely $x_{t-1} = [x$ $(t-1)$ p $(t-1)]$. The forward algorithm is carried out to obtain and output the predicted flow value $\tilde{x}_2(t)$ at the next time, and then it is stored in the prediction result register D. Meanwhile the state after forward operation ($c2(t)$ and $h2(t)$) will also be temporarily stored in the state and input registers A.

(2) At time t, for LSTM_1, the $x(t)$ is arrived. Since the true value of $x(t)$ is obtained, it is compared with the stored temporary $\tilde{x}_1(t)$ and $\tilde{x}_2(t)$ in comparator E. If meet:

$$|x(t) - \tilde{x}_1(t)| > Th1 \text{ and } |x(t) - \tilde{x}_1(t)| \leq Th2$$

where *Th1* and *Th2* is the threshold value of an experience, then it is considered that there is the burst flow occurred in the real flow, and the updating process (step (3)) of the internal state and new prediction process of LSTM_1 is started. Otherwise, return to step (1).

(3) Updating process of predicted value of burst flow:

The allocator B update the internal status register of LSTM_1 by using the state of the LSTM_2 *(c2(t)* and *h2(t))* that was temporarily stored previously in A, and copy the input value pulse *p(t)* to the input end of LSTM_1 as one of the inputs. The new predicted value $\widetilde{\tilde{x}}_1(t+1)$ at time t + 1 will be obtained by using $x(t)$, pulse signal *p(t)* and internal state *(c2(t)* and *h2(t))* on the LSTM_1, and will be used to update original predicted value $\tilde{x}_1(t+1)$ obtained at step(2) at time $t + 1$, then be outputted. After the updating process, the LSTM's internal state(after forward operation) *(c1(t)* and *h1(t))* will be stored to the state register A.

In above process, it is assumed that an sudden event of network flow occurs at time $(t-1)$ and it leads to a wide variation of the flow data at time t, therefore a predicted value at time t can be obtained at time $(t-1)$ by LSTM_2. Later the comparator will compare the predicted flow and the real flow at time t. If the error between the real flow value and LSTM_1's predicted flow is large and the error between the real flow value and LSTM_2's predicted one is within the threshold, it is considered that a burst of traffic has occurred and LSTM_2 will copy its internal states to the main predictor LSTM_1, then the main predictor will predict the new flow value at time $t + 1$ by using the new state and the input $x(t)$ at time t (actually it is to update the predicted value at time $t + 1$ made by LSTM_1 with its state at time t). If the error between the actual flow $x(t)$ at time t and LSTM_1 is within the threshold while LSTM_2's predicted result is outside the threshold, it is considered that there is no sudden event occured and no burst flow, and LSTM_1 will still output a flow predicted value at time t + 1 obtained by its present internal states. In the above prediction process, *p(t)* is actually used to simulate an burst event, making LSTM change from using single variable to multi-variable. Although the predicted value at time t when the burst event happens has not changed, because the time in which burst flow happens is detected and the internal state is updated, the flow predicted value after the time t can still be revised.

5 Experiment and Result Analysis

To verifying the estimated performance of parallel double LSTM predictor for burst flow, the publicly available data set on the Internet [13] will be selected for the validation. We select different burst flow intensity and patterns respectively for the experiment (Figs. 4, 5 and 6).

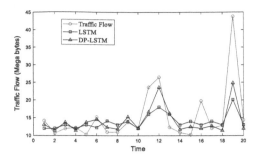

Fig. 4. The example 1 of flow prediction based on conventional LSTM and double parallel LSTMs (DP-LSTM)

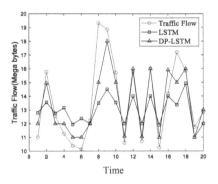

Fig. 5. The example 2 of flow prediction based on conventional LSTM and double parallel LSTMs (DP-LSTM)

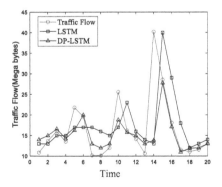

Fig. 6. The example 3 of flow prediction based on conventional LSTM and double parallel LSTMs(DP-LSTM)

As seen from the above experimental results, though at the moment the burst flow occured the double parallel LSTM predictor DP-LSTM didn't work well, in the subsequent time it did made a good prediction of the flow, showing that the predicted flow value did not lag behind. On the other hand, the traditional LSTM (with single variable) had a relatively large error in the flow predicted value when an burst event occurs, and also in the following period of time. Table 1 shows the statistic of the average error in the three predictions, from which we can see that compared with traditional single LSTM structure, the estimated performance of parallel double predictor DP-LSTM has improved by about 30% ~ 45%. This proves that the parallel double LSTM has better performance in the application in detecting burst events and predicting the network flow with burst part.

Table 1. The average error of flow prediction (Mbytes)

	Example 1	Example 2	Example 3
Conventional LSTM	18.23	5.23	19.57
Parallel LSTM	12.35	3.86	10.93

6 Conclusion

This paper proposed a parallel double LSTM predictor structure to settle the problem that the sudden change of flow leading to the decline of performance in the prediction of the network information flow. The main characteristics of its training and prediction process are:

Firstly detect the burst flow time of training data and establish the pulse train to simulate the burst events.

Perform the same training for two LSTM predictors in the double-parallel LSTM by using the training sequence and the simulated burst pulse sequence, making the training results of predictors consistent.

In the real prediction process, one LSTM of double-parallel LSTM works as the main predictor and the other one as the detector of burst events. The detector and main predictor exchange their internal states. When the burst flow is detected, the burst signal is simulated to change the internal state and the input signal of the main LSTM predictor, the main predictor can correct the predicted value and make itself adapt to the change of the flow pattern.

The simulation experiments show that double-parallel LSTM structure can adapt to different intensity of flow changes and its prediction accuracy is improved compared with the traditional LSTM.

References

1. Lu, H., Yang, F.: A network traffic prediction model based on wavelet transformation and LSTM Network. In: The Proceeding of 2018 IEEE 9th International Conference on Software Engineering and Service Science (ICSESS), Beijing, China, pp. 1131–1134. IEEE (2018)
2. Hochreiter, S., Schmidhuber, J.: Long short-term memory. Neural Comput. 9(8), 735–1780 (1997)
3. Liang, S., Nguyen, L. Jin, F.: A multi-variable stacked long-short term memory network for wind speed forecasting. In: Proceeding of 2018 IEEE International Conference on Big Data, Seattle, WA, USA, pp 4561–4564. IEEE (2018)
4. Lv, Y., Duan, Y., Kang, W., Li, Z., Wang, F.Y.: Traffic flow prediction with big data: a deep learning approach. IEEE Trans. Intell. Transp. Syst. 16(2), 865–873(2015)
5. Zhang, J., Zheng, Y., Li, D.: DNN-based prediction model for spatiotemporal data. In: Proceeding of ACM SIGSPATIAL International Conference on Advances in Geographic Information Systems, Burlingame, California, pp. 92–96. ACM (2016)
6. Zhu, Z., Peng, B, Xiong, C., Zhang, L.: Short-term traffic flow prediction with linear conditional Gaussian Bayesian network. J. Adv. Transp. 50(5), 1111–1123 (2016)
7. Jeong, Y., Byon, Y., Castro-Neto, M., Easa, S.: Supervised weighting online learning algorithm for short-term traffic flow prediction. IEEE Trans. Intell. Transp. Syst. 149(4), 1700–1707 (2013)
8. Chang, H., Lee, Y., Yoon, B., Baek, S.: Dynamic near-term traffic flow prediction: system oriented approach based on past experiences. IET Intell. Transp. Syst. 6(3), 292–305 (2012)
9. Smith, B., Williams, B., Oswald, R.: Comparison of parametric and nonparametric models for traffic flow forecasting. Transp. Res. Part C, 10(4), 303–321(2002)
10. Fan, D., Zhang, X.: Short-term traffic flow prediction method based on balanced binary tree and K-nearest neighbor nonparametric regression. In: proceeding of 2017 2nd International Conference on Modelling, Simulation and Applied Mathematics, Bangkok, Thailand, pp. 118–121. Atlantis press (2017)
11. Jun, M., Xiao, L., Meng, Y.: Research of urban traffic flow forecasting based on neural network. Acta Electronica Sinica, 37(5), 1092–1094 (2009)
12. Huang, W., Song, G., Hong, H., Xie, K.: Deep architecture for traffic flow prediction: deep belief networks with multitask learning. IEEE Trans. Intell. Transp. Syst. 15(5), 2191–2201 (2014)
13. Network Traffic Dataset. https://sites.google.com/site/dspham/downloads/network-traffic-datasets. Accessed 7 Mar 2020

Super Peer-Based P2P VoD Architecture for Supporting Multiple Terminals

Pingshan Liu[1,2], Yaqing Fan[2(✉)], Kai Huang[2], and Guimin Huang[2]

[1] Guangxi Key Laboratory of Trusted Software, Guilin University of Electronic Technology, Guilin, China
ps.liu@foxmail.com
[2] Business School, Guilin University of Electronic Technology, Guilin, China
fan.yaq@foxmail.com

Abstract. With the development of wireless network and the popularity of mobile terminal devices, users can watch videos through both the traditional fixed terminals and smart mobile terminals. However, network resources cannot be effectively shared between fixed terminals and mobile terminals, which causes a huge waste of resources. To solve this problem, this paper proposes a super peer-based P2P VoD architecture for supporting multiple terminals. In the architecture, resources of various terminals can be shared and server load can be reduced. An architecture was first built based on super peers. And then, a super peer selection algorithm was proposed to select super peers to manage other terminals. Considering the characteristics of the different types of terminals, a corresponding caching mechanism was designed for different terminals to achieve efficient resource sharing. Final, the maintenance of the architecture was discussed. The experiment demonstrates that the architecture is feasible. The architecture can effectively improve the quality and fluency of video playback, and reduce the server load.

Keywords: Peer-to-peer · Video-on-demand · Super peer · Architecture

1 Introduction

Due to the development of network technology and the popularity of mobile terminal devices, more and more people use smartphones and tablet computers. People are particularly fond of watching videos on their smartphones and tablet computers. At the same time, existing research [1] shows that the status of smart TV or computer in people's family life has not decreased. When watching video, a smart TV or computer is as important as a smartphone or tablet computer. Therefore, users want to share the resources on fixed terminals (computers and smart TVs) and mobile terminals (smartphones and tablet computers). However, the existing P2P VoD systems often construct different systems for different types of terminals. This systems result in the resources between the fixed terminals and mobile terminals cannot be shared. The completely separate systems result in a great waste of resources.

The research content of this paper is to build a super peer-based P2P VoD architecture for supporting multiple terminals. In the architecture, resources of various

© Springer Nature Singapore Pte Ltd. 2020
J. Zeng et al. (Eds.): ICPCSEE 2020, CCIS 1257, pp. 389–404, 2020.
https://doi.org/10.1007/978-981-15-7981-3_28

terminals can be shared, so the architecture can effectively improve the resource utilization in network. First we built a P2P VoD architecture based on super peers. Second, in order to deal with the dynamic of mobile terminals, we add super peers to the architecture. And we use super peers to manage mobile terminals and fixed terminals. The super peers can effectively reduce the server load in the architecture. Then, considering the characteristics of different types of terminals, we design the corresponding caching mechanism for different terminals to realize efficient resource sharing. Finally, we consider the VCR operations of various terminals in the architecture and discuss the maintenance of the architecture. The P2P VoD architecture proposed in this paper can achieve resource sharing among multiple types of terminals.

The organization of this paper is as follows. In the first section, we introduce the research background and significance. In Sect. 2, we introduced related work. In Sect. 3, we introduced the P2P VoD architecture model and the super peer selection algorithm. Then, we introduced the cache mechanism for different terminals and discuss the maintenance of the architecture in Sect. 4. In Sect. 5, we verify the performance of the P2P VoD architecture through simulation experiments. Final, we summarize the paper and draw conclusions.

2 Related Work

The application of P2P VoD has been very extensive. The existing researches on P2P VoD are mainly aimed at fixed terminals. The researches of papers [2–4] on P2P VoD are aimed at fixed terminals. Mo et al. proposed a clustering mechanism of peers for fixed terminals in paper [2]. The clustering mechanism include a merging algorithm of peer cluster and a segmentation algorithm of peer cluster. Then they proposed the maintenance algorithm of peer cluster. Final, they shown that the clustering mechanism has low time complexity and low communication complexity. Xin et al. analyzed the problems and difficulties in providing VoD services for fixed terminals [3]. And then, they proposed a load balancing (RQLB) strategy based on request queue and a cache replacement (EICR) strategy based on elimination index to solve the problems they analyzed. The authors of [4] studied the buffer-map exchange problem in a pull-based P2P VoD streaming system. They proposed an adaptive mechanism to reduce overhead. The mechanism sent the buffer-maps based on playing position of peers. Experiments shown that the mechanism effectively reduced the bandwidth overhead of buffer-map exchange in P2P VoD streaming systems. However, these researches are considered on fixed terminals, and mobile terminals are not considered.

There are also some P2P VoD researches aimed at mobile terminals to build mobile peer-to-peer networks (MP2P). Studies on MP2P for mobile terminals are papers [5–7]. The authors of [5] analyzed the characteristics of mobile terminals. Then, they pointed out the difficulties and challenges in building a mobile peer-to-peer networks. Final, they proposed solutions to these difficulties and challenges. The authors of [6] proposed a method for transmitting data in P2P VoD for mobile Adhoc networks (MANET). This method distributed the load of a mobile terminal to multiple terminals and multiple paths. So the method can achieve load balancing of each terminal. The authors of [7] proposed a scheme for video sharing in P2P VoD. Experiments shown that the scheme can effectively reduce the data transmission delay and packet loss rate, and effectively improve the network throughput. However, these studies were carried out on mobile terminals, without considering fixed terminals.

In this paper, super peers are added to manage other peers when constructing the P2P VoD architecture. So, we involve a super peer selection algorithm. The research work related to the super peer selection is [8, 9]. The authors of [8] proposed that the selection of super peers is an important factor affecting the smoothness and quality of video playback in P2P VoD. Then, they proposed an algorithm for selecting super peers based on peers' reputation and service capabilities. The authors of [9] built a super-peer system based on the small-world model. In the system, they considered the geographic distance and relational proximity of the peers. Experiments shown that the system can effectively reduce user search delay in P2P-SIP networks. Unlike the research in our paper, these studies are also performed on fixed terminals.

In the past, research work considered both fixed terminals and mobile terminals focused on multi-screen interactions, such as papers [10, 11]. The authors of [10] implemented a technology for multi-screen interaction in a wide area network. Then they used this technology for face recognition. The authors of [11] proposed a multi-screen aggregation system. In the system, users can use multiple screens to serve them at the same time. But unlike the research in our paper, the system they proposed is to replicate the same content in multiple screens.

Different from previous studies, the research of this paper is to build a P2P VoD architecture for supporting multi-terminals. First, we build a super peer-based P2P VoD architecture for supporting multi-terminals. Then, we propose a super peer selection algorithm. In this architecture, the super peers manage other peers. Final, we devise the cache mechanism for different terminals and discuss the maintenance of the architecture.

3   P2P VoD Architecture Model and Super Peer Selection

In our P2P VoD architecture for supporting multi-terminal, terminal types include computer, smart TV, smartphone, and tablet computer. In these terminals, computers and smart TV are relatively stable and have large storage capacity and bandwidth. Smartphone and tablet computer have strong mobility and small storage capacity. In addition, smartphone and tablet computer are greatly affected by battery power. Therefore, this paper fully considers the characteristics of various types of terminals. Then, we build a P2P VoD architecture that organizes various types of terminals together. In this paper, computers and smart TVs are called fixed terminals. Smart phones and tablet computers are called mobile terminals.

The mobile terminals in the P2P VoD architecture makes the architecture very dynamic. And it is difficult for a server to manage mobile terminals. Therefore, we add super peers when constructing the architecture. Super peers are used to manage other terminals. The super peers can effectively reduce the management pressure of the server and increase the stability of the architecture. Because the performance of fixed terminal is better than that of the mobile terminal, we choose fixed terminals with better performance as super peers.

3.1   P2P VoD Architecture Model

In the architecture, the performance of computer and smart TV is similar. So we call them ordinary peers and handle them uniformly. The performance of mobile terminals such as smartphones and tablets computer is similar. We call them mobile peers and handle them uniformly. Then we select super peers to manage ordinary peers and mobile peers. Therefore, our P2P VoD architecture structure includes a streaming source server, a tracker server, super peers, ordinary peers, and mobile peers. Figure 1 shows the model of our P2P VoD architecture structure.

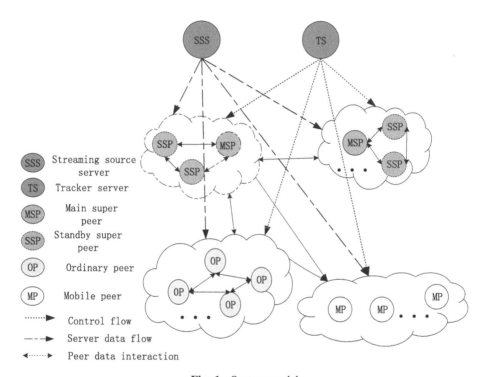

Fig. 1. System model

A streaming source server stores complete video content. It divides the entire video content into video segments and sending the video segments to certain peers. When a new video is received in the architecture, the streaming source server sends the new video to some seed peers. When a video segment requested by a peer cannot be queried in other peers, the streaming source server sends the required video segment to the peer, thereby ensuring smooth playback of video.

The tracker server stores a peer information in the architecture structure, such as the peer ID, the peer's bandwidth, the cached video resources and the peer's IP address. It also organizes the peers to form peer groups. When a super peer cannot query a certain video segment, it will query the tracker server. The tracker server returns information of other super peers with the video segment. Then, tracker server indicates the requesting peer which peer should request.

A super peer is a peer with better performance selected from ordinary peers. In our architecture, super peers manage ordinary peers and mobile peers. In order to manage other peers, the super peer constructs and maintains an index list. The list includes video information peers cached and peer ID. Then, super peers manage the joining and leaving of ordinary peers. Final, the super peer processes information exchange between peers. The super peers effectively reduces the load pressure on the streaming source server and the tracker server. In our architecture, super peers are included main super peers and standby super peers. In this paper, a super peer generally mean a main super peer.

The standby super peer is the standby of main super peer. The performance of standby super peer is worse than that of the main super peer, but better than that of the ordinary peer. The standby super peer obtains the information of other peers from the main super. Then, standby super peers saves the information. When the main super peer leaves, the standby super peer manages other peers instead of the main super peer.

Ordinary peers include computer and smart TV. Ordinary peers provide cached video content and free bandwidth for other peers. At the same time, the ordinary peers manage the mobile peers and provide its video content to mobile peers.

Mobile peers include smartphones, and tablet computers. The storage capacity and battery power of mobile peers are very limited, so mobile peers cannot provide their own resources for other peers. Therefore, in our architecture, the mobile peer only serves as the acquirer of the resource. That is, the mobile peer only obtains video from other peers, and does not provide video to other peers.

3.2   Metrics Considered in Super Peer Selection

When selecting a super peer, we consider the peer's own ability and the peer's reliability. Because the performance of mobile peers is far inferior to ordinary peers, we only consider ordinary peers when selecting super peers. In order to evaluate ordinary peers and select super peers, we construct a super peer evaluation function. The super peer evaluation function considers two indices included the peer's own ability and the peer's reliability. We choose a peer with a large super peer evaluation function value as a super peer.

1) Peer's own ability

In peer's own ability index, we mainly consider the peer's memory, storage capacity, and bandwidth. Memory is the memory that the CPU can directly access. The performance indicators of memory mainly include memory capacity and access speed. Storage capacity mainly refers to external storage, including storage other than computer memory and CPU cache. The bandwidth includes the upload bandwidth and download bandwidth of peers. Our P2P VoD architecture does not consider the computing power of a peer, because the architecture does not consider the calculation of video transcoding on super peers.

The calculation formula of peer's own ability is as follows:

$$Ability = \mu_1 Memory + \mu_2 Storage + \mu_3 Bandwidth \tag{1}$$

In Eq. (1), *Ability* is a peer's own ability. *Memory* is the memory of a peer, μ_1 is the weight coefficient of *Memory*. *Storage* is the storage capacity of a peer, μ_2 is the weight coefficient of *Storage*. *Bandwidth* is the upload bandwidth and download bandwidth of a peer, μ_3 is the weight coefficient of *Bandwidth*. According to previous papers [11], when selecting a super peer, the *Storage, Memory* and *Bandwidth* are equal importance. So $\mu_1 = \mu_2 = \mu_3 = 1/3$.

2) Peer's reliability

Peer's reliability includes the success interaction rate between peers and the online time of peers. The reliability calculation formula of a peer is:

$$Reliability = \gamma SI + (1 - \gamma)OT \tag{2}$$

In Eq. (2), *Reliability* is the reliability of a peer. *SI* is the success interaction rate between peers. *OT* is the online time of peers. γ is the weight coefficient of *SI*. In the process of data transmission by a peer, the success interaction rate is more important than its online time. Through experiments we get that the value of *Reliability* is more reasonable when $\gamma = 0.6$.

The formula for Success interaction rate *SI* in Eq. (2) is:

$$SI = \begin{cases} \alpha PSI + (1 - \alpha)SI(t) \\ 0.5 \end{cases} \tag{3}$$

In formula (3), *PSI* is the past success interaction rate of a peer. *SI(t)* is the success interaction rate in period t. If there is no information interaction between peers, *SI* is taken as 0.5. If there is information interaction between peers, it is calculated according to *PSI* and *SI(t)*. According to the conclusions of previous papers [12], the value of α is taken to be 0.6.

The formula for past success interaction rate *PSI* is:

$$PSI = \frac{S}{T} \tag{4}$$

In Eq. (4), where S is the number of past success interaction of a peer. T is the total number of past information interactions of a peer.

The formula for Success interaction rate in period t *SI(t)* is:

$$SI(t) = \frac{S(t)}{T(t)} \tag{5}$$

In Eq. (5), where *S(t)* is the number of success interaction of a peer in period t. *T(t)* is the total number of information interactions of a peer in period t.

The online time OT of a peer is obtained by weighting the POT and $OT(t)$. The calculation formula of OT is:

$$OT = \beta POT + (1 - \beta)OT(t) \tag{6}$$

In Eq. (6), POT is the total past online time of the peer. $OT(t)$ is the online time of the peer in the period t. According to the conclusions of peer's success interaction rate, it is concluded that the total past online time of peers is more important. Therefore, the value of the weighting factor β is 0.6.

3) Super peer evaluation function

Our super peer evaluation function evaluates the peer's own ability and peer reliability. We choose the peer with large evaluation function value as the super peer. The formula for the super peer evaluation function is:

$$W_{sp} = \omega_a \, Ability_i + \omega_r \, Reliability_i \tag{7}$$

In Eq. (7), where ω_a and ω_r are weighting factors. In the study of this paper, the ability of a peer and the reliability of a peer are equally important in the selection of the super peer, so the weight coefficient value is the same $\omega_a = \omega_r = 0.5$.

3.3   Super Peer Selection Algorithm

According to the super peer evaluation function, we give the pseudo code of the super peer selection algorithm in Table 1. First, the tracker server obtains the peer IP address of each peer j and determines the geographical area A_j where the peer j is located. Second, the algorithm obtain the information (storage, memory and bandwidth) of each peer i in the A_j. Then, the algorithm calculate the ability of peer i ($Ability_i$) according to the storage, memory and bandwidth of peer i. Third, the algorithm calculate the success interaction rate (SI_i) of peer i. If there is information interaction between peer i and other peers, SI_i is calculated according to the peer's past success interaction rate (PSI_i) and peer's success interaction rate in period t ($SI_i(t)$). If there is no information interaction between peer i and other peers, SI_i is set to 0.5. Fourth, the algorithm calculate the online time (OT_i) of peer i. The value of OT_i is obtained according to the peer's past online time (POT_i) and peer's online time in period t ($OT_i(t)$). Fifth, the algorithm calculate the reliability of peer i ($Reliability_i$) according to the SI_i and OT_i of the peer i. Sixth, according to the peer's $Ability_i$ and $Reliability_i$, the super peer evaluation function value (W_{sp}^i) is obtained. Final, a peer with the larger evaluation function value is selected as a super peer. We assume that m is the total number of peers in the area A_i, and n is the number of peers managed by a super peer. Then the number of super peers we choose is m/n. After the super peer selection is completed, two standby super peers are selected for each super peer according to the super peer selection algorithm.

In our architecture, a peer group includes a super peer, two standby super peers, ordinary peers and mobile peers. To avoid super peer overload, the number of peers in a peer group should not be too large. However, if the number of peers in a peer group is too small, the architecture will need more super peers. Too many super peers will increase the complexity of super peer management. Therefore, the number of peers in a

Table 1. The pseudo code of the super peer selection algorithm.

Algorithm 1
1   **For** each peer j **do**
2   Obtain the IP address of peer j and determine the geographical area A_j where peer j is located
3   **For** peer i in area A_j **do**
4   Get the information (storage, memory and bandwidth) of peer i
5   Compute the ability of peer i $Ability_i = \mu_1 Storage + \mu_2 Memory + \mu_3 Bandwidth$
6   **If** peer i and peer n have information interaction **then**
7   Compute the past success interaction rate $PSI_i = S_i / T_i$
8   Compute the success interaction rate in period t $SI_i(t) = S_i(t)/T_i(t)$
9   Compute the success interaction rate of peer i $SI_i = \alpha PSI_i + (1-\alpha) SI_i(t)$
10   **Else** $SI_i = 0.5$
11   **End if**
12   Compute the online time of peer i $OT_i = \beta POT_i + (1-\beta) OT_i(t)$
13   Compute the reliability of peer i $Reliability_i = \gamma SI_i + (1-\gamma) OT_i$
14   Compute the super peer evaluation function $W_{sp}^i = \omega_a Ability_i + \omega_r Reliability_i$
15   Arrange $W^i{}_{sp}$ in descending order, choose the top m/n as super peers
16   Select two peers with the lower $W^i{}_{sp}$ value than the super peer in each peer group as the standby super peers
17   **End for**
18   **End for**

peer group must be limited. Referring to previous studies [12], when the number of peers in a peer group is set to 20, the performance of the peer group is the best. We have confirmed the conclusion by experiments.

4   Cache Mechanism for Multi-terminal and Maintenance of Architecture

In this section, we introduce the cache mechanism of different terminals and the maintenance of P2P VoD architecture. First, we introduce the cache mechanism for various types of terminals in detail. Then, we discuss the joining and leaving of different types of peers in the architecture. Final, we discuss how our architecture responds to the VCR operations of peers in VoD.

4.1 Cache Mechanism for Various Types of Terminals

According to the characteristics of terminals, we know that ordinary peers have larger bandwidth and storage space. So ordinary peers can contribute more resources. Mobile peers have limited bandwidth and storage space, besides they are affected by battery power. So Mobile peers can contribute fewer resources. According to the characteristics of the peers, the cache mechanism of them is designed. The cache mechanism is shown in Fig. 2 (a) and Fig. 2 (b). From Fig. 2 (a) we can see that an ordinary peer has three buffer areas, including a play buffer, a supply buffer and a hard disk buffer. The play buffer is used to buffer the videos which will play. Videos played by ordinary peers can be copied to the supply buffer and provided to other peers. The hard disk buffer is used to cache the videos owned by peer. The videos in hard disk buffer can also provide to other peers. From Fig. 2 (b) we can see that a mobile peer only has a play buffer. Mobile peers do not provide video resources to other peers due to the limited battery power and the limited storage resources. That is, the mobile peers only gets resources from other peers. Therefore, each mobile peer only has a playback buffer.

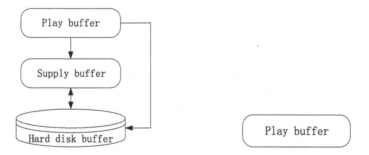

Fig. 2. (a) Cache structure of ordinary peers (b) Cache structure of mobile peers

4.2 Peer Join

When a new peer i requests to join the architecture, it first informs the tracker server. The tracker server obtains the IP address of peer i. Then tracker server determines which geographic area peer i should join. The geographic area that peer i joins is called A_i. Second, peer i looks for super peers in A_i. Third, peer i sends a detection packet to each super peer and measures the network delay (Round Trip time RTP) between it and each super peer. The smaller the RTP, the smaller the distance between peer i and a super peer. Final, peer i connect to the nearest super peer and joins the peer group.

After the new peer i is added, the tracker server calculates the super peer evaluation function value of the new peer. Then tracker server judges the performance of the new peer. If the performance of the new peer far exceeds the performance of the super peer, the new peer instead of the super peer (Table 2).

Table 2. The pseudo code of a new peer joins

Algorithm 2
1   The new peer i requests to join the architecture
2   Obtains the IP address and playback position of peer i
3   Determine the geographic area A_i to which this peer i belongs
4   Find the super peers in A_i
5   **For** each super peer SP_j **do**
6   Peer i sends a detection packet to the super peer SP_j
7   Measure the network delay RTP_j from peer i to super peer SP_j
8   **End for**
9   Sort network delay RTP_j in ascending order
10   Peer i connect to super peer SP_j with the smallest RTP_j
11   Peer i joins the group that the super peer SP_j belongs to

4.3   Peer Leave

In this section, we discuss the leave of peers. The peer leave includes normal leave and abnormal leave. The normal leave is the active leave of a peer. The abnormal leave is the passive leave of a peer, such as the peer leave caused by device abnormality or network abnormality.

1) The super peer leave

Super peer normal leave: In our P2P VoD architecture, when a super peer leaves, a standby super peer immediately replaces the super peer to manage other peers. There are two standby super peers in each peer group. The standby super peers obtain the information of other peers from the super peer. The standby super peers store the same information as the super peer. When the super peer works normally, the standby super peers do the same thing as the ordinary peers. When the super peer leaves normally, it first informs its standby super peers. Then, one of the standby super peers becomes a super peer. The new super peer manages other peers in the peer group. Then the new super peer selects a standby super peer for itself to ensure that it has two standby super peers. Since the possibility that one super peer and two standby super peers are leave at the same time is very small, our architecture is stable.

Super peer abnormal leave: In our architecture, the standby super peer always sends information to the super peer to confirm whether the super peer is online. When the super peer abnormal leaves, the standby super peer will immediately know that the super peer is offline. Then the standby super peer immediately replaces the super peer and manages other peers in the peer group. Afterwards, the new super peer selects a standby super peer for itself. This standby super peer candidate mechanism can ensure the architecture works well after the super peer leave. Besides the candidate mechanism effectively improve the stability of the architecture.

In our architecture, the tracker server saves the ID and the performance of super peer SP_i. When super peer SP_i rejoins the architecture, the tracker server compares the performance of the SP_i with the existing super peer SP_j. If the performance of SP_i is better than SP_j, SP_i become to a super peer again. Otherwise, SP_i exists as an ordinary peer in the architecture.

2) The ordinary peers leave

Ordinary peer normal leave: When ordinary peer normal leaves, it first informs the super peer. Then the super peer deletes the index information of the peer. The ordinary peer no longer provides video for other peers.

Ordinary peer abnormal leave: In our architecture, super peers send detection messages to ordinary peers to check whether ordinary peers are online. When an ordinary peer leaves abnormally, the super peer will detect that the ordinary peer is offline in time. Then the super peer immediately deletes the index information of the peer. The ordinary peer no longer provides video for other peers.

3) The standby super peer leave

The standby super peer leave is the same as the ordinary peer leave. The only difference is that when the standby super peer leaves, the main super peer will choose another standby super peer. Therefore, the super peer can guarantee that it has two standby super peers at the same time.

4) The mobile peer leave

Because the mobile peers do not provide their own cached video for any other peers, the mobile peers have no effect on the architecture after leaving. Therefore, after the mobile peers leave, we do not need to do any more processing.

4.4 Peer Drag and Drop

Peers will frequently generate VCR operations when playing videos in P2P VoD architecture. So peers will frequently make requests to request new video segments. The process when a peer requests a video segment is shown in Table 3. When a peer requests a video segment, it first sends a request to the super peer that messages it. Second, the super peer searches the peers with the video segment in the group. If only one peer owns the requested segment, the peer sends the segment to the requesting peer. Third, if more than one peer has the request segment, we first calculate the upload bandwidth of the peer. Then select a peer with largest available uplink bandwidth as a sender. Fourth, if no peer has the requested segment, the super peer sends the request to other super peers. Then the other super peers will search the peers with the segment in their group. The peers with the segment in other peer group as a sender. The sender send the segment to the requesting peer. Fifth, if the request segment is not found, the streaming source server sends the request segment to the request peer. Therefore, the speed of segment transmission and the load balance of peers can be guaranteed.

Table 3. The pseudo code of peer's VCR operations

Algorithm 3
1   Peer i requests segment S
2   Super peer SP_i receives request from peer i
3   SP_i searches peers with segment S
4   **If** SP_i finds peers with segment S in its peer group **then**
5   **If** only one peer j with segment S **then**
6   peer j sends segment S to peer i
7   **Else** compares the available upload bandwidth of peers with segment S
8   Select the peer k with the largest available upload bandwidth
9   Peer k sends segment S to peer i
10   **End if**
11   **Else** super peer SP_i sends the request of peer i to super peer SP_j
12   **If** SP_j finds peer l with segment S in its peer group **then**
13   Peer l sends segment S to peer i
14   **Else** streaming server sends segment S to peer i
15   **End if**
16   **End if**

5 Experimental Evaluation

In order to evaluate the performance of the P2P VoD architecture for supporting multi-terminal, we simulated the architecture on the PeerSim simulation platform.

5.1 Simulation Steps and Evaluation Indicators

In the experiments, we randomly set the ability of ordinary peers. And we randomly initialize the online time and success interaction rate of each peer. In the simulation, we randomly select 1% of peers to join or leave the architecture. We set the ability of mobile peers to 0.1 times that of ordinary peers. In the experiment, the number of peers in a peer group is set to 20. And we set the upload bandwidth and download bandwidth of the tracking server to 10 Mb. We compare the performance of P2P VoD architecture with that of the P2P-DASH VoD scheme proposed in Paper [13]. It can be seen from the experiments that the performance of P2P VoD architecture is better than that of the P2P-DASH VoD scheme.

We use perceived video quality, continuity index, and load balancing factor [14, 15] to evaluate the performance of P2P VoD architecture. We can see from the improvement of perceived video quality and continuity index that the architecture can realize the resource sharing among various terminals. And it can be seen from the decrease of load balancing factor that the architecture can effectively reduce the sever load.

5.2   Simulation Results

1) Perceived video quality

We use the PSNR value of a segment to represent the quality of the segment. Perceived video quality is defined as the ratio of the total PSNR value of the segments on-time received to the number of total segments. The higher the ratio, the higher the perceived video quality of a peer. From Fig. 3 we can see that the PSNR value of our P2P VoD architecture is significantly higher than that of the P2P-DASH VoD scheme. In our architecture, 90% of peers with PSNR over 36 dB. In the P2P-DASH VoD scheme, 90% of peers with PSNR over 33 dB. This shows that the perceived video quality of our architecture is better than that of the P2P-DASH VoD scheme.

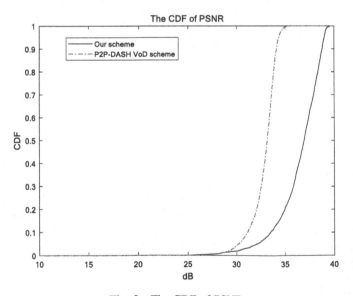

Fig. 3. The CDF of PSNR

2) Continuity index

Continuity index is the ratio of the number of video segments on-time received by a peer to the total number of video segments requested by the peer. The higher the continuity index, the smoother the video play. As can be seen from Fig. 4, the continuity index of our architecture is significantly better than that of the P2P-DASH VoD scheme. In our architecture, the continuity index value of 90% of peers reached 0.95 while the P2P-DASH VoD scheme only reached 0.8. This shows that, users watch videos more smoothly in our architecture.

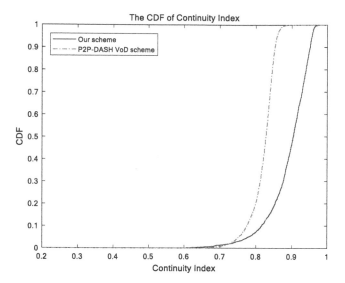

Fig. 4. The CDF of continuity index

3) Load balancing factor

We use the standard deviation of upload bandwidth utilization of a peer to represent the load balancing factor. The smaller the value, the more balanced the load of peers in the architecture. From Fig. 5, we can see that the load balancing factor of our architecture is significantly lower than that of the P2P-DASH VoD scheme. In our architecture, 85% of peers with load balancing factor lower than 0.15. The load balancing factor of 85% of peers reached 0.2 in the P2P-DASH VoD scheme. From Fig. 5, we can see that our architecture can effective reduce server load.

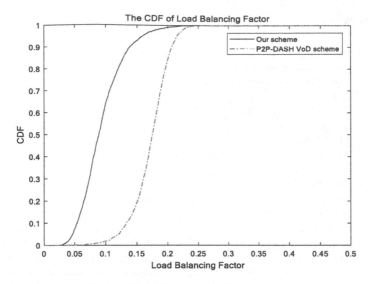

Fig. 5. The CDF of load balancing factor

6 Conclusion

This paper mainly implements a super peer-based P2P VoD architecture for supporting multiple terminals. The architecture organizes fixed terminals and mobile terminals together to achieve resource sharing among terminals. We use super peers to manage other peers when constructing the architecture. This super peer management strategy can effectively improve the stability of the architecture and the efficiency of reducing the server load. First, we built an architecture based on super peers. Second, we proposed a super peer selection algorithm to select super peers to manage other terminals. And then, we design the caching mechanism for different terminals to realize efficient resource sharing. Final, we discussed the maintenance of the architecture. Experiments show that the architecture can effectively improve the quality and fluency of video playback, and effectively reduce the server load. It can be seen that our proposed P2P VoD architecture is effective.

Acknowledgement. The research was supported by the National Natural Science Foundation (No. 61762029, No. U1811264, No. 61662012), Guangxi Natural Science Foundation (No. 2016GXNSFAA380193), Guangxi Key Laboratory of Trusted Software (No. kx201726), and the Foundation of Key Laboratory of Cognitive Radio and Information Processing, Ministry of Education (No. CRKL150105).

References

1. Widdicks, K., Hazas, M., Bates, O.: Streaming, multi-screens and YouTube: the new (unsustainable) ways of watching in the home. In: CHI 2019 Proceedings of the 2019 CHI Conference on Human Factors in Computing Systems, Glasgow, Scotland UK, vol. 466, pp. 1–13 (2019)
2. Cao, L., Zhong, J., Feng, Y.: A distributed node clustering mechanism in P2P networks. In: Cao, L., Zhong, J., Feng, Y. (eds.) ADMA 2010. LNCS (LNAI), vol. 6441, pp. 553–560. Springer, Heidelberg (2010). https://doi.org/10.1007/978-3-642-17313-4_57
3. Wei, X., Ding, P., Zhou, L.: QoE oriented chunk scheduling in P2P-VoD streaming system. IEEE Trans. Veh. Technol. **68**(8), 8012–8025 (2019)
4. Sheshjavani, A.G., Akbari, B.: An adaptive buffer-map exchange mechanism for pull-based peer-to-peer video-on-demand streaming systems. Multimed. Tools Appl. **76**(5), 7535–7561 (2016). https://doi.org/10.1007/s11042-016-3425-z
5. Lederer, C., Altstadt, S., Andriamonje, S.: A holistic approach in developing an ensemble mobile peer-to-peer computing. In: 2013 5th International Conference on Computational Intelligence, Communication Systems and Networks, Madrid, Spain, pp. 311–315 (2013)
6. Muhammad, S.R., Saeid, I., Raad, R.: A novel energy-efficient video streaming method for decentralized mobile ad-hoc networks. Pervasive Mob. Comput. **40**, 301–323 (2017)
7. Jia, S., Xu, C., Guan, J.: A MP2P-based VoD solution for supporting VCR-like operations in MANETS. In: 2012 IEEE 2nd International Conference on Cloud Computing and Intelligence Systems, Hangzhou, China, pp. 991–995 (2012)
8. Rongfei, M.A.: Super node selection algorithm combining reputation and capability model in P2P streaming media network. Pers. Ubiquit. Comput. **23**(3–4), 435–442 (2019). https://doi.org/10.1007/s00779-019-01219-y
9. Jun, L., Zhang, S., Wang, H.: A construction of SIP based Peer-to-Peer network and performance analysis. Chin. J. Electron. **18**(2), 374–378 (2009)
10. Hu, H., Jin, Y., Wen, Y.: Toward a biometric-aware cloud service engine for multi-screen video applications. Computer communication review: a quarterly publication of the special interest group on data communication. ACM SIGCOMM Comput. Commun. Rev. **44**(4), 581–582 (2014)
11. Kim, G.-H.: Multi-screen patterns and multi-device experiences in a multi-screen ecosystem. Adv. Comput. Sci. Ubiquit. Comput. **474**, 1051–1056 (2018)
12. Chen, Z., Liu, J., Li, D.: SOBIE: a novel super-node P2P overlay based on information exchange. J. Comput. **4**(9), 853–861 (2009)
13. Liu, P., Fan, Y., Huang, K., Huang, G.: A DASH-based peer-to-peer VoD streaming scheme. In: Gao, H., Feng, Z., Yu, J., Wu, J. (eds.) ChinaCom 2019. LNICSSITE, vol. 312, pp. 402–416. Springer, Cham (2020). https://doi.org/10.1007/978-3-030-41114-5_30
14. Zhang, M., Xiong, Y., Zhang, Q.: Optimizing the throughput of data-driven peer-to-peer streaming. IEEE Trans. Parallel Distrib. Syst. **20**(1), 97–110 (2009)
15. Shen, Y., Hsu, C.H., Hefeeda, M.: Efficient algorithms for multi-sender data transmission in swarm-based peer-to-peer streaming systems. IEEE Trans. Multimed. **13**(4), 762–775 (2011)
16. Deltouzos, K., Denazis, S.: Tackling energy and battery issues in mobile P2P VoD systems. Comput. Netw. **113**, 58–71 (2017)

Localization Algorithm of Wireless Sensor Network Based on Concentric Circle Distance Calculation

KaiGuo Qian[1], Chunfen Pu[2], Yujian Wang[1], ShaoJun Yu[1(✉)], and Shikai Shen[1]

[1] School of Information Engineering, Kunming University, Kunming, China
qiankaiguo@qq.com, 454918730@qq.com,
mftsy544756770@qq.com, ysjll@163.com
[2] School of Teacher Education, Kunming University, Kunming, China
49248139@qq.com

Abstract. Node Localization is one of the key technology in the field of wireless sensor network (WSN) that has become a challenging research topic under the lack of distance measurement. In order to solve this problem, a localization algorithm based on concentric circle distance calculation (LA-CCDC) is proposed. The LA-CCDC takes the beacon as the center of the concentric circle, then divides the task area into concentric circles with the k communication radius of sensor, which forms concentric rings. The node located in the k hops ring intersects the concentric circle with $(k-1)$ r radius that forms an intersection area. This area is used to calculate the distance from the beacon to the unknown node, hyperbola is then adopted to locate the unknown node. In the application scenario with node random distribution, the simulation results show that the LA-CCDC algorithm gets the node location with low error under different node number, different beacons and different communication radius of sensor.

Keywords: Wireless sensor network (WSN) · Localization algorithm · Concentric circles · Average relative positioning error (ARPE)

1 Introduction

A multi-hop self-organizing wireless sensor network [1] (Wireless Sensor Network: WSN) has become an important technical form of the underlying network of the Internet of things, which consists of a large number of sensor nodes deployed in the monitoring area, these nodes send and receive data each other. The owner can collect the required information at any time, any place and under any environmental conditions through the wireless sensor network. Location information is crucial to the application of sensor network [2], for the one is the context representing the perceived data, the other is the necessary information for the design of key technologies of wireless sensor network, such as coverage control and routing technology. Sensor nodes cannot be fully equipped hardware to obtain position due to requirements of the large-scale and low-cost deployment for the application, therefore, the design of effective node

© Springer Nature Singapore Pte Ltd. 2020
J. Zeng et al. (Eds.): ICPCSEE 2020, CCIS 1257, pp. 405–415, 2020.
https://doi.org/10.1007/978-981-15-7981-3_29

self-positioning algorithm becomes more popular research topic for application of wireless sensor networks and Internet of things [3]. The existing research results are classified into range-based and range-free positioning algorithms, the range-based require hardware module to measure distance, such as RSSI (Received Signal Strength Indicator) [4], TOA (Time of Arrival) [5], TDOA (Time Difference of Arrival) [6], and AOA (Angle of Arrival) [7]. The hardware module consume the limited power of sensor which shorten the life cycle and increase deployment costs for sensor networks. The range-free positioning algorithms estimate the distance between the beacon carried location and the normal node unknown position by the means of communication relationship of sensors such as DV-Hop (Distance Vector Hop) algorithm [8], Amorphous algorithm [9], Centroid [10], APIT (Approximate Perfect point-in-triangulation Test) algorithm [11], etc., which do not need hardware support resulted in low cost and power consumption. So the latter are more suitable for applications of resource limited sensor network. However, it increases error by the hops between nodes to estimate distance in range-free localization algorithm and so as to lead low positioning accuracy.

A localization algorithm based on concentric circle distance calculation is proposed according to the principle of range-free algorithm based on concentric ring division. By ring division, the distance calculation is completed without the introduction of measurement module. The organization of this paper is as follows. Related work is presented in Sect. 2. Section 3 discusses the LA-CCDC algorithm details. This is followed by the simulation analysis in Sect. 4. Finally, Sect. 5 gives the conclusion of this paper.

2 Related Works

The DV-Hop firstly obtains the hops between sensor nodes by broadcasting message each other, secondly calculates the average hop size of the network using the whole distance divided by total hops of beacons, thirdly, hop size and the hops are multiplied to calculate the distance between normal node and beacon, at last, least square method (LSM) is adopted to localize the normal node. It is similar to dv-hop algorithm, but the difference for Amorphous algorithm is that it uses the number of neighbor nodes to calculate the average hop size. Centroid algorithm takes the centroid of the beacons that receive the message as the position of the normal node which requires a high density of beacons to ensure positioning accuracy. APIT is similar to the centroid algorithm, which uses the centroid of the triangle set composed of beacons as the normal node position which node transmits a message to the beacon node. The range-free algorithms do not require hardware to measure distances, but the hops of nodes are used to estimate the distance, which leads to large localization error. Centering on improving the positioning accuracy, Literature [12] uses the average hop estimation error between beacons to correct the mean hop distance of the whole network, which way enhances distance precision from unknown node to beacons. Advanced DV-Hop [13] uses the hop-size of the beacon node, and weighted least square algorithm is used to reduce the influence of beacon nodes with large hops to unknown nodes and to reduce the positioning error by means of Hop weighting. A fuzzy logic method was introduced to calculate the distance between beacon nodes and unknown nodes, and then the centroid algorithm was used to determine the normal node in Literature [14], which improved the positioning accuracy of Centroid algorithm, but it required higher beacon node density. WDV-Hop [15] uses the hop size error to correct estimation distance so as to make the distance estimation more accurate, and then weighted hyperbola method is used to calculate coordinates for normal node. A series of means including construction of backbone network, iterative positioning and 2-hop neighbor distance estimation were used to joint positioning for normal node to obtain higher positioning accuracy in Literature [16]. In recent years, researchers have applied the intelligent algorithm to the improvement of range-free positioning algorithm, Literature [17] uses particle swarm optimization to correct the distance estimation between sensor nodes and to reduce the distance error between nodes.

3   A Localization Algorithm of Wireless Sensor Network Based on Concentric Circle Distance Calculation

The LA-CCDC algorithm divides the area into concentric rings taking the beacon as the rings center, then calculates the distance from beacon to unknown node through the intersection area that is constituted of beacon concentric circle and the circle of unknown node.

3.1   Localization Problem

Undirected graph $G = \{V, E\}$ represents a wireless sensor network, where, $V = \{v_i | i = 1, 2, \cdots, m, \cdots, n\}$ represents n nodes form network. There are m beacons which know its location and n-m normal nodes that are waited be located.

$E = \{e_{ij} | i, j = 1, 2, \cdots n\}$, where, e_{ij} is 1 if the distance between node i and node j is less than the r which is communication radius of sensor, otherwise e_{ij} is 0. Localization Problem is described that calculate coordinates of the n-m normal nodes according to the m beacons in wireless sensor networks. The distance estimation model of the range-free algorithm is shown in Fig. 1, which introduces range ambiguity that the distances are estimated to be equal if the two normal nodes have same hops to the one beacon, but this distances are different, for example, the error between the estimated distance and the actual distance to beacon i is close to r for that node j and node k are located in the 2-hop range of the beacon node i. At the same time, this mean brings curved path problem that the Euclidean path deviates significantly from the shortest hop path which is shown as $i \rightarrow j \rightarrow k \rightarrow u \rightarrow v$ shortest hop path, this path introduces nearly 2r distance estimation errors.

3.2   Concentric Ring Division

When estimating the distance between the unknown node and the beacon node, the LA-CCDC algorithm firstly forms concentric circle that takes the beacon node as the center and takes the k * r (k = 1, 2...) as radius, in this way, the monitoring area is divided into k concentric circle ring, and the k * r concentric circle and (k − 1) * r concentric circle constitute the k ring. It is shown as Fig. 2. The unknown node j, located in the k ring, whose communication circle will intersect the k − 1 concentric circle taking the

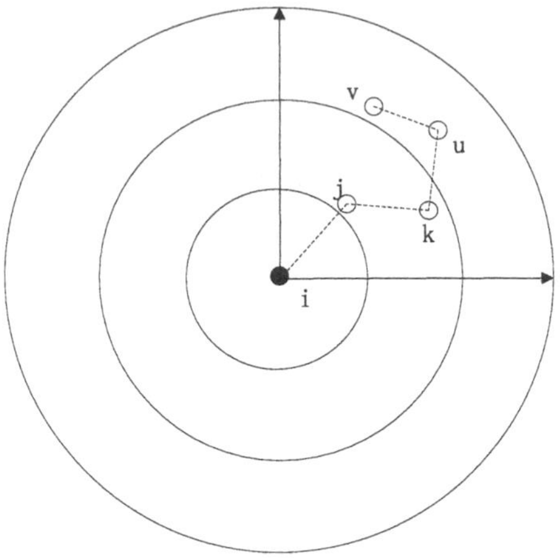

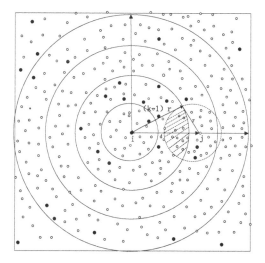

Fig. 1. Distance estimation model of range-free positioning algorithm

Fig. 2. Distance calculation model for concentric circle division

beacon i as center forms a intersection area shown as Fig. 2. The intersecting area can be used to calculate the distance between beacon node i and unknown node j.

3.3   Distance Calculation

The distance from unknown node j with k hop from beacon i, whose communication circle domain intersects the k − 1 hop circle domain, to i is set as d which is shown in Fig. 2 according to the concentric circle division model.

When k = 1, unknown node j and beacon node j are neighbor nodes to each other, and the distance is calculated according to Eq. 1:

$$d = \frac{1}{2}(1 - \frac{2|N_i \cap N_j|}{|N_i| + |N_j| + 2})\pi r \tag{1}$$

Where, N_i is the number of neighbor nodes of node i, and N_j is the number of neighbor nodes of node j.

When k > 1, it is shown in the shaded section in Fig. 2. The area of the intersection area is calculated as follow Eq. 2.

$$S = r^2 \arccos(\frac{d^2 + r^2 - (k-1)^2 r^2}{2dr})$$
$$+ (k-1)^2 r^2 \arccos(\frac{d^2 + (k-1)^2 r^2 - r^2}{2d(k-1)r}) \tag{2}$$
$$- \frac{1}{4} sqrt[(k^2 r^2 - d^2)(d^2 - (k-2)^2 r^2)]$$

Because of the n nodes are randomly distributed in the monitoring area and the principle of random distribution, the estimated area of the intersection region is calculated as Eq. 3.

$$\widehat{S} = \frac{n_{ij} + 1}{n} A \tag{3}$$

In the above equation, n_{ij} is the number of sensor nodes in the intersection area between the (k − 1) r circle domain of beacon node i and the communication circle domain of unknown node j, A is the monitoring area. We get a nonlinear function which respect to the independent variable d for that the area of the intersection area is replaced by the estimated area of the intersection area.

$$d = f(A) \tag{4}$$

The distance d can be obtained by solving the above nonlinear function with the zero point method or the secant method.

3.4 Node Localization

After obtaining the distance to all beacon nodes, we adopt two-dimensional hyperbola to determine the unknown node location. Let the position of unknown node j is (x, y), and the position of beacon node i is (x_i, y_i), the distance of j to i is d_i according to Eq. 3, and the positioning process is shown as follows:

$$(x_i - x)^2 + (y_i - y)^2 = d_i^2 \qquad (5)$$

There are m beacons, then:

$$\begin{cases} -2x_1x - 2y_1y + x^2 + y^2 = d_1^2 - x_1^2 - y_1^2 \\ -2x_2x - 2y_2y + x^2 + y^2 = d_2^2 - x_2^2 - y_2^2 \\ \cdots\cdots\cdots\cdots\cdots\cdots\cdots\cdots\cdots \\ -2x_mx - 2y_my + x^2 + y^2 = d_m^2 - x_m^2 - y_m^2 \end{cases} \qquad (6)$$

Let

$$A_i = x_i^2 + y_i^2, B = x^2 + y^2$$

$$Z = [x,\ y,\ B]^T$$

$$G = \begin{pmatrix} -2x_1 & -2y_1 & 1 \\ -2x_2 & -2y_2 & 1 \\ \vdots & \vdots & \vdots \\ -2y_m & -2y_m & 1 \end{pmatrix}$$

$$H = \begin{pmatrix} d_1^2 - A_1 \\ d_1^2 - A_1 \\ \vdots \\ d_m^2 - A_m \end{pmatrix}$$

The matrix expression of Eq. 6 is shown as

$$GZ = H \qquad (7)$$

Then,

$$Z = (G^T G)^{-1} GH \qquad (8)$$

The coordinates of the unknown node j are:

$$(x, y) = (Z(1), Z(2)) \qquad (9)$$

4   Experimental Results and Analysis

Extensive simulation experiments are conducted to test the performance of the proposed algorithm in MATLAB R2014a.300 sensor nodes were randomly and uniformly placed in the 1000 m * 1000 m task area. Network topology is shown as Fig. 3.

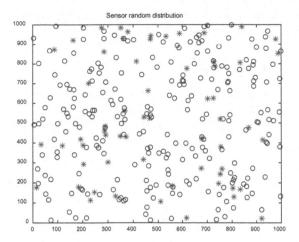

Fig. 3. Network topology

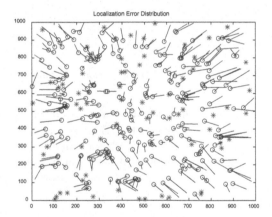

Fig. 4. Location error diagram of nodes, ARPE = 24.002%

We test the Positioning performance of the LA-CDCC algorithm in terms of the Average Relative Positioning Error (ARPE) of the overall network. ARPE is defined as Eq. 12.

$$ARPE = \frac{1}{n-m} \sum_{i=n-m}^{n} \frac{\sqrt{(\widehat{x}_i - x_i)^2 + (\widehat{y}_i - y_i)}}{r} \tag{12}$$

Where, $(\widehat{x}_i, \widehat{y}_i)$ is the estimated position coordinates of wireless sensor network nodes, (x_i, y_i) is the actual coordinates of nodes, and r is the wireless communication radius of sensor nodes.

The communication radius is set as 200, the radio of beacon nodes is set as 20%, and the positioning error is shown in Fig. 4.

4.1   Node Density Influence on ARPE

The number of sensor nodes randomly deployed is 220, 250, 280, 310, 340 and 370. The communication radius of the nodes is set as 200 m, and the proportion of beacon nodes is set as 20%. The positioning error of LA-CCDC algorithm is shown in Fig. 5.

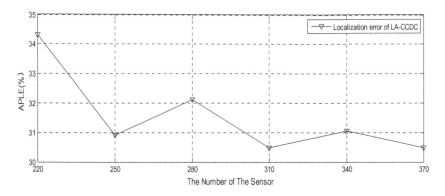

Fig. 5. ARPE with different sensor node number

It can be seen from the figure that the ARPE almost remains at about 31%, and the node density has a small impact on the ARPE. This distance calculation method only has an error in the distance of the k-hop, and has nothing to do with the k − 1 hop distance. Therefore, no matter how the node density changes, the distance calculation error is always within the same range, so the positioning error caused by the range error is not large.

4.2   The Influence of Beacon Node Ratio on ARPE

The number of randomly and uniformly deployed sensor nodes is 300, the communication radius is set to 200 m, the proportion of beacon nodes is set to 10%–50%, and the step length is increased by 5%. The ARPE of LA-CDCC algorithm is shown in Fig. 6, which shows the localization result, it is clearly observed that the ARPE tends to decrease as the number of beacon nodes increase. This is because the hops will reduce and the error of distance calculation is also reduced, so is the ARPE.

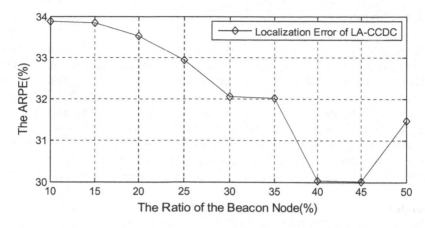

Fig. 6. The ARPE with different beacon proportions

4.3 Influence of Node Communication Radius on ARPE

We randomly deploy 300 sensor nodes in the 1000 m * 1000 m monitoring area, set the proportion of beacon node to 20%. Communication radius is varied from 170 m, to 250 m. The positioning performance is shown in Fig. 7, we observe that the ARPE tends to decrease varying the communication radius.

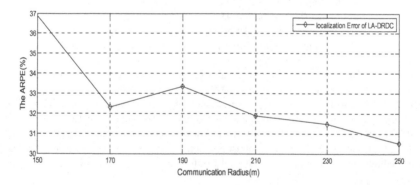

Fig. 7. The ARPE varied with Communication radius

The reason is that as the radius increases, the k of the whole network decreases, the distance calculation error decreases, and the positioning error also decreases.

5 Conclusion

Localization technology is one of the basic technologies for the application of wireless sensor networks. We adopt the division based on the equal radius hop ring, and then calculate the distance from the normal node to beacon by using intersection area which is constituted of the unknown node circle and the concentric circle. This intersection area is replaced by the node distribution ratio. Thus, the unknown node of wireless sensor network can be located by two-dimensional hyperbolic calculation method. In the future work, the accuracy of intersecting area will be replaced by node distribution ratio, and the accuracy quantization method will be found to make the distance calculation more accurate, so as to improve the accuracy of this positioning method.

Acknowledgment. This work was supported by the Yunnan Local Colleges Applied Basic Research Projects (2017FH001-059, 2018FH001-010, 2018FH001-061), National Natural Science Foundation of China (61962033).

References

1. Gupta, S., Singh, M., Srivastava, S.: Wireless sensor network: a survey. Int. J. Innov. Adv. Comput. Sci. **2**(4), 78–85 (2015)
2. Gumaida, B.F.: ELPMA: efficient localization algorithm based path planning for mobile anchor in wireless sensor network. Wirel. Pers. Commun. Int. J. **100**, 721–744 (2018). https://doi.org/10.1007/s11277-018-5343-z
3. Zhu, X., Dong, W., Renfa, L.I., et al.: Localization and nodes location-aware in Internet of Things. Scientia Sinica (Informationis) **43**(10), 1265–1287 (2013)
4. Nieoleseu, D., Nath, B.: Ad-hoc positioning systems (APS). In: Proceedings of the 2001 IEEE Global Telecommunications Conference, vol. 5, pp. 2926–2931. IEEE Communications Society, San Antonio (2001)
5. Patwari, N., Hero, A.O., Perkins, M., et al.: Relative location estimation in wireless sensor networks. IEEE Trans. Signal Process. **51**(8), 2137–2148 (2003)
6. Girod, L., Estrin, D.: Robust range estimation using acoustic and multimodal sensing. In: Proceedings of the IEEE/RSJ International Conference on Intelligent Robots and Systems (IROS 2001), vol. 3, pp. 1312–1320. IEEE Robotics and Automation society, Maui (2001)
7. Lazos, L., Poovendran, R.: POPE: robust position estimation in wireless sensor networks. In: Proceedings of the 4th IEEE International Conference on Information Processing in Sensor Networks, pp. 324–331 (2005)
8. Niculescu, D., Nath, B.: DV based positioning in ad hoc networks. Telecommun. Syst. **22**(1/4), 267–280 (2003)
9. Shen, S., Yang, B., Qian, K., et al.: An improved amorphous localization algorithm for wireless sensor networks. In: International Conference on Networking & Network Applications. IEEE (2016)
10. Bulusu, N., Heidemann, J., Estrin, D.: GPS-less low-cost outdoor localization for very small devices. IEEE Pers. Commun. **7**(5), 28–34 (2000)
11. He, T., Huang, C., Blum, B.M., et al.: Range-free localization and its impact on large scale sensor networks. ACM Trans. Embed. Comput. Syst. **4**(4), 877–906 (2005)
12. Goyat, R., Rai, M., Kumar, G., et al.: Improved DV-Hop localization scheme for randomly deployed WSNs. Int. J. Sensors Wirel. Commun. Control **10**(1), 94–109 (2020)

13. Qiao, X., Chang, F., Ling, J.: Improvement of localization algorithm for wireless sensor networks based on DV-Hop. Int. J. Online Biomed. Eng. **15**(06), 53–65 (2019)
14. Amri, S., Khelifi, F., Bradai, A., et al.: A new fuzzy logic based node localization mechanism for Wireless Sensor Networks. Future Gener. Comput. Syst. (2017). S0167739X17303886
15. Mass-Sanchez, J., Ruiz-Ibarra, E., Cortez-González, J., et al.: Weighted hyperbolic DV-Hop positioning node localization algorithm in WSNs. Wirel. Pers. Commun. **96**, 5011–5033 (2016)
16. Stanoev, A., Filiposka, S., In, V., et al.: Cooperative method for wireless sensor network localization. Ad Hoc Netw. **40**, 61–72 (2016)
17. Li, G., Zeng, J.: Ranging distance modified by particle swarm algorithm for WSN node localization. J. Jilin Univ. **56**(231(03)), 188–194 (2018)

Data Transmission Using HDD as Microphone

Yongyu Liang, Jinghua Zheng[(⊠)], and Guozheng Yang

National University of Defense Technology, Hefei 230037, China
zhengjinghua@nudt.edu.cn

Abstract. The technology of Covert Channel is often used for communications between computers and the Internet with high sensitivity or security levels. Presently, some research were carried out on covert channel using computer screen light radiation, speakers, electromagnetic leakage, etc. In this paper, the technology of SoundHammer is studied. It is a technical bridgeware that use acoustic waves to transmit data from sound device into an air-gapped network. Firstly, an idea was proposed for data transmission by covert channel through acoustic waves. Then, this method was validated by experiment and the risks of the air-gapped network were confirmed. Finally, some countermeasures for detecting and eliminating such covert channels were listed.

Keywords: Covert channel · Air-gapped network · Data exfiltration · Acoustic waves

1 Introduction

In order to counter against the threats of network covert channels, physical isolation is conducted in almost every top-secret organization to keep the networks with high level separated from the less secure and Internet. This type of isolation is known as air-gapped. Some important or sensitive websites are physically isolated, such as military network, financial network, critical infrastructure network and so on. But there are still many researches about data exfiltration from air-gapped network.

Presently, some researchers have proposed different types of covert channels to explore the feasibility of data exfiltration through air-gapped. Electromagnetic radiation methods using different components of computers are proposed [1–4]. N. Matyunin et al. [1] build a covert channel through the magnetic field sensor in mobile. M. Guri et al. [5] achieved data exfiltration by USB cable. The optical covert channel [6], thermal covert channels [7], and acoustic covert channels were proposed successively [8–10]. M. Guri studied the data leak method which was called LED-it-GO via hard drive LED indicators [11]. M. Guri et al. [12] proposed a new method to send acoustic signals without speakers.

Lopes et al. presented a malicious device [13] to leak data using an infrared LEDs. But Lopes' prototype still needs a malicious internal person who holds a receiver to obtain the infrared signal. It is very difficult to be realized in the place with a high secret level. M. Guri et al. [14] achieved that the malware called Fansmitter, could be acoustically penetrated from air-gapped computers when no audio hardware and speakers in 2016. And they demonstrated the effective transmission of encryption keys

© Springer Nature Singapore Pte Ltd. 2020
J. Zeng et al. (Eds.): ICPCSEE 2020, CCIS 1257, pp. 416–427, 2020.
https://doi.org/10.1007/978-981-15-7981-3_30

and passwords from a distance of zero to eight meters, with the speed of 900 bits/hour. Zheng Zhou et al. [6] achieved the method called IREXF, an infrared optical covert channel from a well-protected air-gapped network via a malicious infrared module implanted previously into a keyboard. They validated the method by experiment in which the speed of the covert channel was up to 2.62 bps. M. Guri et al. [7] proposed a unique method of supporting bidirectional communication based on thermal covert channel. They used their heat emissions and built-in thermal sensors. But this paper only exchanged information between two adjacent personal computers (PCs) that are part of separate, physically unconnected networks.

Many researchers work on various infiltration channels, including electro-magnetic, acoustic, thermal and optical covert channels. And it is necessary for a variety of equipment such as speaker, audio hardware or infrared device. But most researches only get data from the target network system. But it is very hard to inject the data into HDD in the air-gapped target system.

In this paper, we validate a new method called Soundhammer by which the data can be injected into the air-gapped computer. That is to say, there are some risks in the air-gapped network. This paper makes three important contributions:

1) We propose a new method of covert channel through acoustic waves that can realize covert communication of data in the place with a high secret level.
2) Secondly, we validate our method by experiment in which the speed of the covert channel was up to 1.24 bps. In the experiment, we discuss issues such as signal modulation, data transmission, reception, and demodulation algorithms. On the other hand, we confirm the risks of the air-gapped network.
3) Thirdly and last, some countermeasures for detecting and eliminating such covert channels are listed.

The rest of this paper is organized as follows: In Sect. 2, we will introduce the related work. In Sect. 3, based on our method, we propose the attack model, such as scenario, transmitter, receiver. We will explicate our dataset and experiment, and give the experiment results followed by a discussion about the results in Sect. 4. Finally, we will make conclusion in Sect. 5.

2 Preliminary

The study of covert channels, whether electromagnetic, optical, thermal or acoustic, is mostly used for information leakage, which is not capable of acquiring outside information, Luke Deshotels confirmed the feasibility of transmitting information by shock waves [15] by experiment on information transmission through shock and accelerometers of mobile phones. Accelerometers are ubiquitous in mobile devices, but are not available in computers. Our method solves this problem.

Our method can send data to receiver programs that have infected an air-gapped computer. The data is transmitted to the HDD of the infected computer by the acoustic waves, and the receiver program reads the status of HDD and gets the data. It is the first time to use HDD as an external signal receiver, as shown in Fig. 1. In our method, the receiver does not require a microphone, and the transmitter just needs speaker or other audio related hardware.

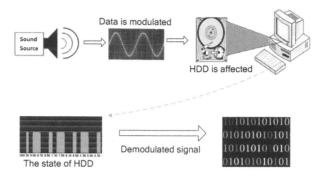

Fig. 1. Our method.

The key technology of our method is the resonance of HDD by acoustic waves. Then the speed of reading and writing data of HDD will be interfered by the resonance.

HDD Structure
A typical HDD consists of two basic components: platters, flat circular disks that are covered by a thin film of a ferromagnetic material, and read-write heads, which are positioned very closed to platters. The data is stored on the platters. The task of read-write head is reading and writing data, and the speed is very fast, up to 5,400 or 7,200 rotations per minute.

How the read operation works? Firstly, the disc rotates, and the head moves itself to the appropriate position above the platter which contains the requested data. Then, in magnetization induces a current in the head that is decoded by the HDD internal circuitry and converted to a binary value.

And how the write operation works? The HDD's internal circuits control the movement of the head and the rotation of the disk, and perform writes on demand from the disk controller. For a write operation, the corresponding head (one of multiple heads that has access to the requested location of the platter) first moves to a designated area on the platter, where it can modify the magnetization of the data cells as they pass under it.

The innovation of our method has the following three points:

1) It is the first time to use HDD as an external signal receiver.
2) Our experiment shows that data can be transferred between the HDD and the speaker. So the air-gapped network is also dangerous because of the strong concealment.
3) Some countermeasures for detecting and eliminating such covert channels are listed based on our research.

3 Our Method

It is so complicated to transmit data for air-gapped network. Our method proposes a new channel called SoundHammer, a bridge that use acoustic waves to transmit data from air-gapped computer.

Our design and implementation of data injection form sound equipment to HDD will be presented in this section. The flow chart is shown in Fig. 2. The transmitter is any object that can generate acoustic waves, such as a speaker. The receiver is HDD. The data is modulated onto the sound wave. The software in the HDD monitors the HDD's read/write state. The HDD interrupts and the data is demodulated. Then the data is successfully injected.

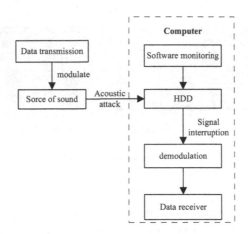

Fig. 2. The method's flow char.

This method also confirms the risk of air-gapped network. When the receiver code has entered the isolated computer network, it can realize the real-time control by the author. That is to say the attacks on air-gapped networks can be implemented. So our experiment proposes a warning for the security of air-gapped network.

3.1 Scenario

Our method uses the acoustic resonance to realize data transfer for covert channel. The sound source generates a periodic signal at a certain frequency close to the object, causing mechanical vibration. Our goal is to use acoustic resonance to cause the vibrations of HDD.

Assume that the data have been injected into the target HDD. The command will be sent to the codes. The command data can be modulated to acoustic waves and sent to the objected HDD. Our experiment is based on the research of Mohammad Shahrad et al. [8]. And some of the symptoms should disappear after stopping the sound. Thus our code can monitor the data transmission rate of the HDD and get the data from outside, as shown in Fig. 3. The normal state of HDD is shown in Fig. 3(a), and the state of being affected in Fig. 3(b). We can see there are some intermittent in the read and write of HDD. The command will be gotten by our code. That is to say it is to read the intermittent of HDD to receive the data from outside, and the intermittent is modulated to 0 and 1.

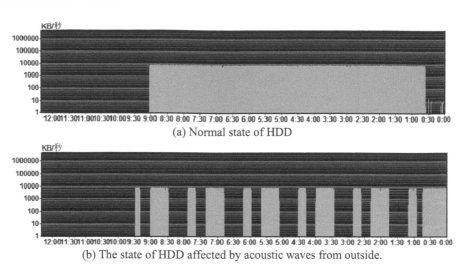

(a) Normal state of HDD

(b) The state of HDD affected by acoustic waves from outside.

Fig. 3. The state of HDD.

3.2 Modulate of Data

In order to achieve data transmission through acoustic waves, the data must be modulated. Based on our laboratory conditions, On-Off Keying (OOK) modulation is used. OOK modulation is a special case of ASK modulation (Amplitude Shift Keying). In OOK modulation, one amplitude is 0 and the other amplitude is non-zero, as shown in Table 1. Two amplitude levels of the carrier (acoustic waves) represent the values of binary data.

We define the minimum time that the acoustic waves can successfully interfere with HDD as dT. And dT is determined to be about 0.81 s by experiments. That is to say, as long as the vocal time of speaker is maintained for 0.81 s, the HDD will be affected. And the HDD reading and writing will produce the effect shown in Fig. 3(b). In order to realize the communication experiment, we define the value T, which must be greater than dT. Thus we can modulate the command data onto the acoustic waves, similar to the Morse code. And '0' is replaced by T, and '1' is replaced by 2T. So the receiver code in the HDD can accept external commands by monitoring the read and write status of the HDD.

Table 1. The way of modulate.

Acoustic amplitude	Duration	Value
1	T	"0"
1	2T	"1"

In our experiments, the data transfer rate is inversely proportional to the size of T. The longer the T is, the lower the data transfer rate is, and the more precise the reliability.

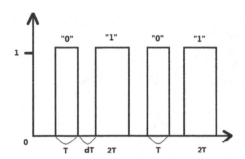

Fig. 4. The way of modulate.

In order to maintain synchronization between the receiver and the transmitter, the small data frame is used, as shown in Fig. 4. The data frame is consisted of preamble (4 bits) and payload (12 bits). 4-bits preamble is '1010', and 12-bits payload which is the data we want to transmit, as shown in Table 2.

Table 2. The structure of data frame.

Preamble (4 bits)	Payload (12 bits)	CRC
1010	111010101110	1110010

The preamble is transmitted at beginning of every packet. It consists of sequence of 4 alternating bits ('1010'). The preamble helps the receiver detect the beginning of the transmission of the each packet. It is very important to set the preamble, because the transmission might be interrupted.

We arbitrarily choose 12 bits as the payload which contains the actual packet. That is to say the data is divided into groups by 12 bits. And we use CRC as the error detection for each group. 8-bit CRC code is added to the end of the data frame. The receiver calculates the CRC for the received payload. And if it differs from received CRC, the current data packet is error. This package will be discarded.

3.3 Our Method

For the program has injected in the air-gapped target object, it is still able to receive external data by our method based on SoundHammer method. The SoundHammer method can cause a temporary interruption of the HDD data transmission. After the SoundHammer is terminated, the HDD will immediately resume the data transmission. Therefore, it is only necessary to monitor the state of the HDD data transmission to sense the state of the external modulated data on acoustic waves.

In order to receive externally transmitted data efficiently and accurately, the receiving software in the target host needs to construct continuous data read and write operations and simultaneously monitor the data transmission rate. The attack state of the external acoustic waves is discriminated according to the data transmission rate.

Algorithm HDD read and write speed monitoring and data extraction method
1.   While (state != end)
2.   T1 = GetCurrentTime(); //Get Current time
3.   Buff = ReadFile(FILE_1); //Read file FILE_1
4.   WriteFile(File_2,Buff); //Write the content to the file FILE_2
5.   FlushFileBuffers(FILE_2); //The data is written directly to the HDD immediately
6.   T2 = GetCurrentTime(); //Get the time after reading and writing
7.   t = T2 – T1; //Calculate time difference
8.   if(t == T)
9.   Data = '0';
10.   else if (t == 2T)
11.   Data = '1' //Modulated data
12. end

The function FlushFileBuffers(*) is to data directly to the disk platter instead of the read and write buffer. Thus our task will be executed immediately.

In general, the data is read and written to the hard disk cache instead of directly to the hard disk. Thus the receiver code can't monitor the status of the HDD in real time. The fifth line of code makes it possible to read and write data directly to the hard disk.

The IO data of HDD is obtained through software which is designed in advance in order to get precise experiment result. In practical applications, it is possible to perform monitoring through normal IO data and extract data to be transmitted externally.

4   Result and Discussion

4.1   Result

In our experiment, SoundHammer is used to air-gapped HDD based on external speaker. A scenario including signal transmission and reception is designed to describe our method of covert channel. In this scenario, the transmitter is any object that can generate acoustic waves, such as a speaker. The receiver is HDD, as is shown in Fig. 5. We choose Seagate Momentus 5400 (160 GB) as the sound receiver. Lenovo laptop X250 is used as the experiment platform, whose CPU is Intel Core i7-5200U, chipset is Intel Q87. In order to verify our experimental results, a program is implanted in the Lenovo X250 in advance. This computer is isolated from the outside world. Remote control of the program will be implemented by our method. Commands are passed to the Lenovo X250 through the Seagate HDD. In addition, the experimental consists of a dynamic signal analyzer for oscilloscope.

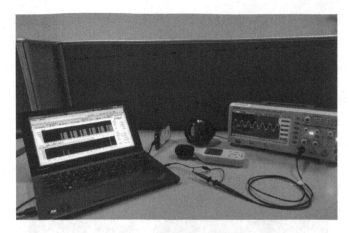

Fig. 5. An exfiltration scenario.

In our experiments, we also selected several other HDDs as experimental objects, including Westdata, Toshiba, HGST. The experimental parameters of these HDDs are shown in the Table 3.

Table 3. The parameters of HDDs.

Bland	Rotating speed (rpm)	Capacity (GB)	Frequence (Hz)	dT (s)	Sound intensity (dB)
Seagate	5400	160	[1,305–1,505]	0.81	95
Westdata	5400	1024	[1,230–1,465]	0.92	93
Toshiba	5400	1024	[1,280–1,560]	0.85	95
HGST	7200	1024	[1,170–1,490]	0.89	96

From the Table 3, we can see that the HDD is most affected when the acoustic waves attack is at the 1.305–1.505 kHz frequency for Seagate bland, 1.230–1.465 kHz frequency for Western digital bland, 1.280–1.560 kHz for Toshiba, and 1.170–1.490 kHz for HGST.

The minimum value of dT when it was not affected was obtained from the experiments is 0.81 s for Seagate, 0.92 s for Westdata, 0.85 s for Toshiba, and 0.89 s for HGST. And the average speed of the covert channel was up to 1.24 bps through several experiments. Although the experimental results are not particularly perfect, we validate the effectiveness of our method, and provide the foundation for the further work.

In our experiment, the sound intensity of the speaker must about 95 dB, and the data can be effectively transmission from external device to air-gapped computer. Thus the background noise is very weak relatively, and the impact on the hard disk is minimal. There is high decibel in our experiment, so if this happens near the air-gapped network, we should pay attention to it. Our experiment only confirms this technology of covert channel. With the development of technology, this covert channel may become less and less noticeable. It cannot be ignored for protection against air-gapped network.

Our computer lab has normal background noise. The modulation frequency of the speaker is 1.37501 kHz, as shown in Fig. 6. When the range of the abscissa of Fig. 5 is increased, that is Fig. 7. From Fig. 6, we can see the transmission status of the data "1010101".

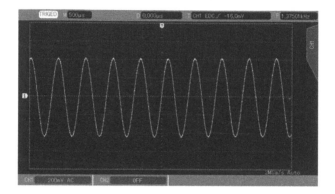

Fig. 6. The modulation frequency.

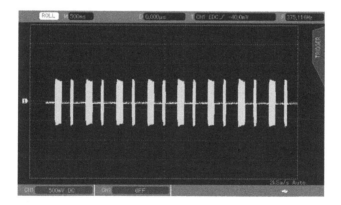

Fig. 7. ASK modulation of "1010101" in 60 s.

The data transmission status map after HDD is attacked can be seen by combined with Fig. 6 and Fig. 3(b). Our experiment shows that HDD can be used as a microphone to receive an external sound signal. If this covert channel is used in field of network confrontation, then the security of the physical isolation network will be threatened.

4.2   Discussion

Our method provides a new channel of the transmission from external device to a physically isolated computer, provides a new research idea for security transmission and defense of air-gapped network.

Table 4 lists the comparisons between the methods proposed by other researchers and the method in this paper.

Table 4. The detailed lists of different methods.

References	Method	Device	Function
[1–4]	Electro-magnetic	Internal computer components	Data leakage
[6]	Optical	Infrared remote control signals	Data leakage
[7]	Thermal	Thermal sensors in computers	Bidirectional communication
[8]	Acoustic	Acoustic resonance	DoS attack
[9, 10]	Acoustic	Speaker	Data leakage
	This paper	Speaker	Data injection

The countermeasures for the information security risks caused by covert communication technologies based on acoustic waves are threefold: the management-based strategy, the software-based strategy, and the physical or hardware-based strategy.

In the management-based strategy, the method of dividing regions is used. The electronic devices, such as speaker, are prohibited in places where sensitive computers or devices are used. This method is more effective, but the sounding device of the speaker is widely used in various types of devices, so the strategy may not completely eliminate the hidden danger. And it doesn't sound realistic that such device as speaker is forbidden. In the software-based strategy, the monitoring software is used. It monitors the working status of the HDD. And if an abnormality occurs, it will promptly issue a warning. And then the communication link is cut off artificially. In the physical or hardware-based strategy, the HDD will be soundproofed. Protected computer case covers a special soundproof cover to reduce the possibility of acoustic waves. Another way is that the HDD is replaced with SSD storage, and this may increase the cost of the device and the burden on the user.

5   Conclusions

In this paper, we propose a new method, called SoundHammer, for injecting data into a physically isolated computer by acoustic waves. External data will be injected into a computer that is pre-implanted into the receiving program. And only an audio device, such as a microphone, can do it. That is to say, acoustic waves can interfere with the

HDD efficiently, and can be monitored by the program. Our experiment provides a warning for the security of air-gapped network.

Our work makes three important contributions. Firstly, we propose a new idea for data injection by covert channel through acoustic waves. Secondly, we validate our method by experiment. Thirdly, some countermeasures for detecting and eliminating such covert channels are listed.

It shows that even if the microphone device is removed, there is still a risk of acoustic waves attack. But it is necessary to continuously monitor the HDD read and write status in our method. And the intensity of the sound reaches 95 dB, so it can be heard by the human ear. Our future work will focus on the security of air-gapped network and making the security strategy.

Acknowledgements. This work was partly financially supported through grants from the National Natural Science Foundation of China (No. 61602491), the University fund of National University of Defense Technology (No. KY19A013). We also would like to thank the Institute of Psychology, Chinese Academy of Sciences for generously supporting our research.

References

1. Matyunin, N., Szefer, J., Biedermann, S., Katzenbeisser, S.: Covert channels using mobile device's magnetic field sensors. In: the 21st Asia and South Pacific Design Automation Conference (ASP-DAC), Place, Macao, pp. 525–532, January 2016
2. Guri, M., Kachlon, A., Hasson, O., Kedma, G., Mirsky, Y., Elovici, Y.: GSMem: data exfiltration from air-gapped computers over GSM frequencies. In: 24th USENIX Security Symposium, Washington, D.C., August 2015
3. Robert, C., Alenka, Z., Milos, P.: A practical methodology for measuring the side channel signal available to the attacker for instruction-level events. In: 47th Annual IEEE/ACM International Symposium on Microarchitecture (MICRO), Cambridge, UK, pp. 242–254, December 2014
4. Guri, M., Gabi, K., Assaf, K., Yuval, E.: AirHopper: bridging the air-gap between isolated networks and mobile phones using radio frequencies. In: IEEE Malicious and Unwanted Software: in the 9th International Conference on MALWARE, Fajardo, PR, USA, pp. 58–67, October 2014
5. Guri, M., Monitz, M., Elovici, Y.: USBee: air-gap covert-channel via electromagnetic emission from USB. In: The 14th Annual Conference on Privacy, Security and Trust (PST), Auckland, New Zealand, pp. 264-268, December 2016
6. Zhou, Z., Zhang, W., Yu, N.: IREXF: data exfiltration from air-gapped networks by infrared remote control signals. http://home.ustc.edu.cn/~zhou7905/IREXF. Accessed January 2018
7. Guri, M., Monitz, M., Mirski, Y., Elovici, Y.: BitWhisper: covert signaling channel between air-gapped computers using thermal manipulations. In: IEEE 28th Computer Security Foundations Symposium (CSF), Verona, Italy, pp. 276–289, July 2015
8. Shahrad, M., Mosenia, A., Song, L., Chiang, M., Wentzlaff, D., Mittal, P.: Acoustic denial of service attacks on HDDs. http://arxiv.org/abs/1712.07816. Accessed December 2017
9. O'Malley, S.J., Choo, K.K.R.: Bridging the air gap: inaudible data exfiltration by insiders. In: Proceedings of the 20th Americas Conference on Information Systems, Savannah, Georgia, USA, pp. 1–11, August 2014

10. Guri, M., Solwicz, Y., Daidakulov, A., Elovici, Y.: MOSQUITO: covert ultrasonic transmissions between two air-gapped computers using speaker-to-speaker communication. In: IEEE Conference on Dependable and Secure Computing, Kaohsiung, Taiwan, December 2018
11. Guri, M., Zadov, B., Elovici, Y.: LED-it-GO: leaking (a lot of) data from air-gapped computers via the (small) hard drive LED. In: Polychronakis, M., Meier, M. (eds.) DIMVA 2017. LNCS, vol. 10327, pp. 161–184. Springer, Cham (2017). https://doi.org/10.1007/978-3-319-60876-1_8
12. Guri, M., Solewicz, Y., Daidakulov, A., Elovici, Y.: Acoustic data exfiltration from speakerless air-gapped computers via covert hard-drive noise ('DiskFiltration'). In: Foley, S.N., Gollmann, D., Snekkenes, E. (eds.) ESORICS 2017. LNCS, vol. 10493, pp. 98–115. Springer, Cham (2017). https://doi.org/10.1007/978-3-319-66399-9_6
13. Lopes, A.C., Aranha, D.F.: Platform-agnostic low-intrusion optical data exfiltration. In: International Conference on Information Systems Security and Privacy (ICISSP), Porto, Portugal, pp. 474–480, December 2017
14. Guri, M., Solewicz, Y., Daidakulov, A., Elovici, Y.: Fansmitter: acoustic data exfiltration from (speakerless) air-gapped computer. https://www.wired.com/wp-content/uploads/2016/06/Fansmitter-1.pdf. Accessed January 2016
15. Deshotels, L.: Inaudible sound as a covert channel in mobile devices. In: Proceedings of the 23rd Annual International Conference on Mobile Computing and Networking, San Diego, CA, p. 16, October 2017

Android Vault Application Behavior Analysis and Detection

Nannan Xie, Hongpeng Bai$^{(\boxtimes)}$, Rui Sun, and Xiaoqiang Di

Changchun University of Science and Technology, Changchun 130022, China
2018100597@mails.cust.edu.cn

Abstract. With the widespread application of Android smartphones, privacy protection plays a crucial role. Android vault application provides content hiding on personal terminals to protect user privacy. However, some vault applications do not achieve real privacy protection, and its camouflage ability can be maliciously used to hide illegal information to avoid forensics. In order to solve these two issues, behavior analysis is conducted to compare three aspects of typical vaults in the third-party market. The conclusions and recommendations were given. Support Vector Machine (SVM) was used to distinguish vault from normal applications. Extensive experiments show that SVM can achieve 93.33% classification accuracy rate.

Keywords: Android vault · Behavior analysis · Malware detection · Privacy protection

1 Introduction

In 2019, Android smartphones have accounted for 86.1% of the global smartphone sales market [1]. The rapid growth of various types of applications have brought more and more malware, which is becoming an important threat to privacy protection and network security. Public safety and social safety incidents caused by the mobile terminals' security occur frequently.

Android vault applications provide content hiding on the phone, which is a privacy protection that has gradually emerged in recent years. Vaults protect private information through encryption, camouflage, and hiding.

The hidden information capability of vaults has two aspects. On the one hand, from the user's perspective, it can be used to hide sensitive information and improve security. At this time, it is necessary to consider whether the vault can truly hide information. For example, the passwords set by certain vault can be obtained in files, and even encrypted pictures can be obtained in other files. On the other hand, from the perspective of forensics, some functions such as disguised icons in some vault are difficult to find once they are maliciously used, which may lead to store and transmit illegal information by criminals.

The contributions of this work are according to the above two problems. We mainly studies vault by behavior analysis and malware detection. First, the encryption and protection capabilities of commonly used vault on the market are compared, such as login password setting, password saving, and storage of the encrypted information.

© Springer Nature Singapore Pte Ltd. 2020
J. Zeng et al. (Eds.): ICPCSEE 2020, CCIS 1257, pp. 428–439, 2020.
https://doi.org/10.1007/978-981-15-7981-3_31

Second, vault applications are decompiled to extracted features and classified by SVM. The extensive experiments show that we can achieve a classification accuracy of 93.3%.

2   Related Work

2.1   Android File Structure

The design of vault is related to the file structure of Android system, the encrypted information is usually stored in some specific file structure of Android system. Android operation system adopts a layered structure design, which are application layer, application framework layer, libraries & Android runtime layer, Linux kernel layer from high to low. Based on this architecture, the security mechanism of Android is defined. In addition to the Linux file access control mechanism, it also includes unique signature mechanism, permission mechanism, process sandbox isolation mechanism, and process communication mechanism. The installation package of Android application software is apk. After decompiling the file, it mainly contains the following 7 directions.

- AndroidManifest.xml. It is the system control file and global description file of the application. It introduces configuration information, with features which can be used as classification features for distinguish applications.
- META-INF. The file stores signature information, including the developer's private key and application digital signature, the file's Sha1 value, etc., used to ensure the integrity and security of the application.
- Res. It is the resource directory, which contains many subdirectories, including pictures resource files, interface layout XML files, animation resource files, and other types of XML files.
- Lib. The local library files are stored in this directory, which stores the underlying library of the system.
- Class.dex. It is a bytecode file generated after Java compilation, since the Android system is implemented based on Java.
- Assets. Some extra resources and configuration files are stored in this directory, aiming to categorize and manage files.
- Resources.arsc. It contains different information corresponding to types, names and IDs of all resources in the res directory in different language environments, and records the mapping of resource files to resource file IDs.

2.2   Android Vault

Android vault is a privacy protection application which has gradually emerged in recent years. After installing vault, users can hide personal information such as pictures, audios, videos, and SMS messages in their mobile phones, thus bypass the detection of forensic software. however, there is few research efforts on vault. Android forensics technology is mainly classified as online forensics and offline forensics. Online forensics, also known as "global forensics technology", uses technologies such as the abd debugger to connect to the file system, such as "/data/data" and "/sdcard", to obtain the files and parse the data. Offline forensics, also known as "physical forensics technology", forensics of system backup files or physical images, which can obtain the original data of the phones.

From the perspective of computer forensics, the research related with vault gradually emerged around 2010, is initially based on data acquisition. Adb is a way to obtain physical information by gaining root authority [2], and Hoog's research obtain memory information through the chip [3]. The research by Zhang X et al. [4] is an earlier study on vault, which analyzes 18 typical vault software and compares their differences.

2.3   Android Malware Detection

Android malware detection has been developing for many years. It has some similarity with vault detection. Malware detection protect user privacy and terminal security by detect existing malware and potential threats quickly, accurately, and efficiently. Malware detection on traditional personal computers are represented by viruses, trojans, worms, and Android malware detection represented by remote connections, information leaks, malicious deductions, which most of them employed machine learning algorithms to detection with extracted features. The research focuses on two aspects of different machine leaning methods and different detection features. For different detection features, there are combined research using permission and API [5], behavior analysis with opcode [6, 18], complex-flows [7], and classification with Dalvik code [8]. Among them, permissions are the most adopted features [9, 10]. For the choice of machine learning algorithms, SVM [11, 12], deep learning algorithms [13–15], probabilistic generative models [16], and a cluster-ensemble classifier [17] are widely employed.

The difference between our work and existing work is that the vault is studied separately from the perspectives of users and forensics. It can not only compare vault's ability to hide information, but also prevent the illegal application of vault due to disguise and concealment, which is crucial for application computer forensics.

3   Vault Behavior Analysis and Detection Framework

This article focuses on the behavior analysis and detection of Android vault, including the steps of data set construction, decompilation, simulation and classification. We build data sets of Android vault and normal applications, and analyze the encryption behavior of specific typical vaults. SVM method is used to distinguish vault from normal applications. The framework is shown in Fig. 1.

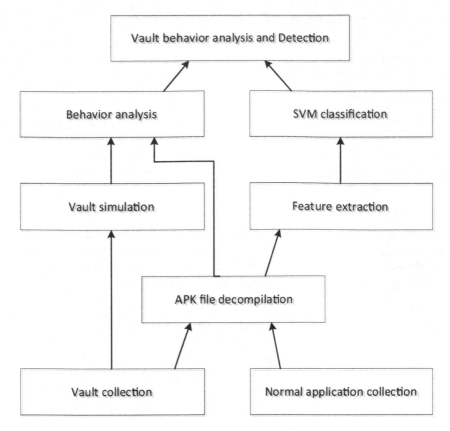

Fig. 1. Framework of vault behavior analysis and classification

As shown in Fig. 1, the vault behavior analysis and detection framework is divided into four modules.

(1) Data set construction

At present, there are few vaults in the application markets, and no standard data set has been formed. Therefore, we first build the dataset in this study. We have collected a total of 38 vaults through the markets. At the same time, as a comparison, we also collected 64 normal applications and saved their apk files.

(2) Apk file decompilation

We use the open source tool APKTool to decompile the above files for the following purposes: First, we study the encryption mode of vaults through decompilation and check the storage location of files encrypted by them; Second, we extract the permission features of Android application's permission features, which is the basic of the data set classified by SVM.

(3) Simulation and feature processing

We use an emulator to run the vaults and analyze the directories to hidden files, then track the encryption method and storage location of these files. Meanwhile, the features extracted in the previous step are pre-processed for further classification.

(4) Behavior analysis

In the analysis of vault storage passwords and encrypted information, we focus on three directories: /data/app/package name, /data/data/package name, and /Scard. Three following aspects are analyzed: login password setting, password saving, and storage of encrypted information, in order to analyze whether their encrypted information can be obtained through other ways.

(5) Vault classification

We use SVM to train and predict the constructed vault and normal application data sets. The features extracted from the apk samples are used to train the parameters of SVM algorithm. We distinguish vault from the normal applications, in order to solve the illegal application caused by camouflage and hiding of vault.

4   Behavior Analysis of Typical Vault

We collect 40 Android vaults, and 14 typical applications are selected as examples according to their functional similarity. In the simulated environment, we take 50 pictures from camera, 20 audio files, 10 text files, and 10 short messages as the source information to be hidden. The name of the vaults and the installation files are shown in Table 1.

Table 1. Vault name and installation package

Vault name	Installation package name
Private photo	com.enchantedcloud.photoVault.apk
Hide	com.morrismobi.secret.apk
GalleryVault	com.smartgallery.safelock.apk
Vaulty	com.theronrogers.Vaultyfree.apk
FileVault	in.zendroid.fileVault.apk
Calculator	com.yptech.privategalleryioc.apk
Photo safe	com.thinkyeah.galleryVault.apk
Privacy space	com.netqin.ps.apk
Yo vault	com.yo.media.Vaultioc.apk
Vaultlock	com.lock.screen.apk
MyphotoVault	com.rvappstudios.myphotoVault.apk
ApplockandgalleryVault	com.newsoftwares.ApplockandgalleryVault.apk
SmsVault	com.biztech.smssecurity.apk
Sms vault lite	com.macrosoft.android.smsVault.light.apk

4.1   Private Photo Vault

(1) Login password setting

Logging into the application need to set a four-digit password as "1234". After entering the application, users can choose to set a password lock or graphic lock. If pictures are added to it directly, the album will be automatically named without password. Password has to be set by users, which could be letters, symbols, or numbers.

(2) Password storage

The 4-digit password of this application is saved in the directory: shared_prefs/com. enchantedcloud.photoVault_preferences.xml. Particularly, it is stored in the form of SHA1 using cipher text, which has a good effect of encryption.

(3) Encrypted information storage

Unencrypted photos are stored in /storage/emulated/legacy/DCIM/camera, and encrypted photos are stored in /data/data/com.enchantedcloud.photoVault/files/media/ orig. The contents of the photos can be accessed in this directory after logging in. Without logging in, the encrypted photos and videos cannot be decoded. After deleting the encrypted photos, the photos will be deleted directly rather than being moved to the original camera's album.

4.2   Hide

(1) Login password setting

Login passwords are combined arbitrarily between numbers, *, and #, and the password questions need to be set as well. When the login password is lost, users can log in and change the password by answering the questions.

(2) Password storage

The password is saved in the form of a hidden file, which is stored in the directory / data/data/.com.morrismobi.secret/shared_prefs/com.morrismobi.secret_preferences. xml. The password cannot be explicitly found in this file, which is a more secure storage.

(3) Encrypted information storage

This vault has a good hidden function for photos, videos, files, and contacts. The photos are mainly hidden in the dictionary /data/media/0/.com.morrismobi.secret/1,/ mnt/shell/emulated/0/.com.morrismobi.secret/1,/storage/emulated/0/.com.morrismobi. secret/1. The photos are hidden in an encrypted form and the content cannot be seen directly. The hidden files can be seen in the dictionary /storage/emulated/legacy/com. morrismobi.secret/files, but the content of the file is displayed in encrypted form which cannot be seen. The hidden content is in the dictionary /storage/emulated/legacy/com.

morrismobi.secret/backup/databases, which is also encrypted. This application hides different types of information with high secure effect.

4.3 Gallery Vault

(1) Login password setting

The default password is "0000", which can be changed after login. The new user's password can be digit combination password lock or gesture lock. If password is lost, users can either retrieve it by answering a preset question or reset it to "0000". This encryption method is lack of security.

(2) Password storage

The password is stored in /com.com.smartgallery.safelock_preferdences.xml via Hash mode. This vault provides another password setting, which set CrackedPass to "963258758741", and the values can be seen in the above file. In other words, even if we do not know the CrackedPass in advance, we can login to this vault through this xml file, as long as you are familiar with the operation principle of the application. Although this setting helps to retrieve the lost password, it significantly reduces the security of this vault.

(3) Encrypted information storage

After the photos are locked, the original album no longer contains them, and the photos unlocked from the application will be transferred to the original album. The encrypted photos are displayed in hexadecimal in the dictionary: storage/emulated/0/GalleryLock/storage/emulated/0/DCIM/Camera. This method protects the original photos from being lost, but in a cost of lowing its hiding effect.

4.4 Vaulty

(1) Login password setting

When entering the application, users can choose to set a password. The password can be set to any combination of numbers in 0–9, or a mixed password with questions and answers. If the password is lost, users can answer the question and relogin this vault.

(2) Password storage

In the directory data/data/com.theronrogers.Vaultyfree/shared_prefs/com.theron-rogers.Vaultyfree_preferences.xml, all the contents are displayed in hexadecimal numbers, thus no information can be exposed directly. The password cannot be found out manually. Thus this vault has achieve a better hiding ability.

(3) Encrypted information storage

When this vault locks the photo, the locked photo can be shared but the shared photo's content is invisible. If the locked photo is deleted, it will be moved to the

original camera album. In the dictionary /storage/emulated/0/documents/Vaulty/.sdata,/
storage/emulated/legacy/documents/Vaulty/.sdata, we can see the contents of encrypted
photos, i.e., the photos are not encrypted. This indicates that this vault is lack of hiding
effect.

4.5   Photo Safe

(1) Login password setting

It sets an initial password of six digits. After logging in, users need to add a mailbox
to avoid password lost. The special feature of this vault is that a camouflaged password
can be set. In particular, when logging in with a false password, it will display
camouflaged content. In addition, it can disguise the icon as a "calculator". It has a
"break-in alarm" function, which will automatically take pictures if the login fails.

(2) Password storage

In the directory /data/data/com.thinkyeah.galeryVault/shared_prefs/com.thinkyeah.
galleryVault_preferences.xml, the user ID and password are set to a 40-bit hexadecimal
number, which ensures high security.

(3) Encrypted information storage

This vault can encrypt photos, videos, cards, documents, and files. The encrypted
file is stored in the /storage/emulated/legacy/.galleryVault_DoNotDelete_1546837573,
and there is a.txt file that contains the encrypted file. Under the encrypted file path, the
files that have been processed can be found, but we cannot see the contents of these
source files. The information hiding security of this software is high.

4.6   Summary

We analyze the above 5 vaults as typical cases indicating different vaults have different
privacy protection degrees. Some vault applications, such as Vaulty, although it
encrypts the photo, but the content of the photo still can be seen in another way (e.g.,
through by the storage path). What's more, when logging into SMS Vaultlite, we can
clearly see the password in the preservation file, and password can be easily cracked to
find hidden information. On the contrary, there are some vaults that can achieve a good
hiding effect, such as "photo safe", which can not only hide photos and files, but also
record violent login without password. We compare and analyze 14 typical vaults, of
which the results are shown in Table 2.

In Table 2, we mainly analyze the login password setting and storage, and the
strength of hidden information. The columns in the table are: "A: whether the login
password is visible (0 is visible, 1 is invisible)", "B: whether the hidden content is
visible (0 is visible, 1 is invisible)", "C: whether the vault has other security measures
(0 is No, 1 is Yes)", "D: application security degree (1 is high, 0 is low)". Among them,
"application security degree" is obtained according to the first three items, if the vault
satisfies two of them, its security degree is "high". Additionally, "whether the vault has
other security measures" refers to whether the application itself provides other functions, such as password camouflage, new fake privacy space, hidden icons, etc.

Table 2. Typical vaults comparative analysis

Vault name	A	B	C	D
Private photo	1	0	0	0
Hide	1	1	0	1
GaleryVault	0	0	1	0
Vaulty	1	0	0	0
FileVault	1	1	0	1
Calculator	0	0	1	0
Photo safe	1	1	1	1
Private space	1	1	1	1
Yo vault	0	1	0	0
Vaultlock	0	0	0	0
MyphotoVault	1	0	0	0
ApplockandgalleryVault	0	1	1	1
SMS vault	1	1	0	1
SMS vault lite	0	1	0	0

In the above analysis, there are 6 vaults perform better encryption or information hidding, which realize the function of vault, account for 43% of the analyzed samples. The rest vaults cannot achieve satisfied privacy protection.

5   Vault Detection

5.1   Detection Framework

From the above analysis, we can see that if the vault has better encrypt ability, it can hide information with more security, even camouflage its own icon, or create a special space in the system for camouflage. These vaults perform well when applied to protect personal privacy. However, when they are used to hide malicious or illegal information, such disguises make it difficult to collect evidence. Therefore, in this paper we exploit machine learning algorithms to propose a novel method to differentiate between vault applications and normal applications. The framework is shown in Fig. 2.

The classification framework based on machine learning mainly includes three parts: dataset collection, feature set extraction and classification.

(1) Dataset collection. It mainly includes the process of obtaining APK files from the Android market. The vaults are manually labelled, while the normal application is labelled by VirusTotal to detect malice, aiming to build a pure training dataset as much as possible.

(2) Feature set extraction. The open source tool APKTool is ultilized to extract the permission features. Permission feature is the most widely used feature in the current research, having a good classification effect with fewer feature dimensions. A total of 119 permission features are extracted as input data of the classification algorithm.

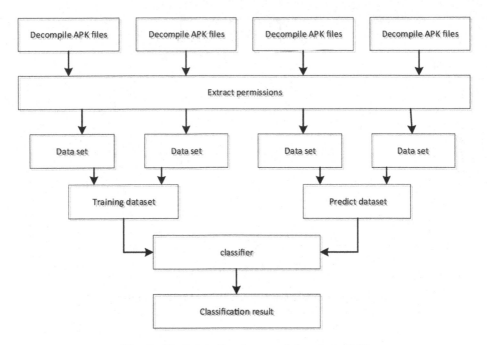

Fig. 2. Vault detection framework based on SVM

(3) Classification. The SVM algorithm is used to train and classify the above constructed data sets, whereas the classification accuracy is to evaluate the effectiveness and feasibility.

5.2 Experiment and Analysis

The apk file is decompiled by APKTool, and 119 permission features are extracted. Due to the small number of Vault samples, the training set contains 102 applications, including 38 vaults and 64 normal applications. Among them, 18 vaults and 24 normal samples are randomly selected as the test dataset, and the 5-fold cross-validation is used. One of the classification results is shown in Fig. 3.

From Fig. 3, we can see that the real category and the predicted category of samples 6 and 13 are different, and a total of 16 predictions are correct among the 18 samples. After a 5-fold cross-validation experiment, the average accuracy reaches 93.33%, which is a satisfied classification result. Experiment shows that using the SVM algorithm with the extracted permission features, the vaults can be effectively distinguished from the normal applications.

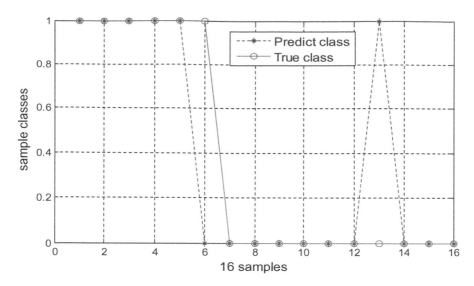

Fig. 3. Comparison of predicted and actual results

6   Conclusion

Considering the hidden information ability of Android vault, the work of this study comprises of two parts. On one hand, from the user's perspective, we focus whether the vaults can really hide information, so we conduct behavior analysis on 14 typical vaults and the evaluation to compare their information hiding ability. On the other hand, from the perspective of forensics, since the camouflage function of vault is difficult to be found in computer forensics, the SVM method is explored to distinguish vaults from normal software.

Through the analysis of 14 typical vaults, we find that 43% of them can achieve better information hiding and privacy protection. Photo safe and Private space can achieve better information hiding. Although other vaults also set passwords to hide the photos, they do not provide good information hiding. For example, we can either find the login password in another way, or find unencrypted information hidden in the file of the storage location. Furthermore, by using SVM to train the extracted permission features, the prediction result can reach 93.33% classification accuracy, which can better distinguish the vaults from the normal applications. By analyzing and classifying the behavior of vault, the situation of vault in the market is analyzed from two perspectives of user's safety and vault's detectability in order to give recommendations based on the above conclusions.

There are still some deficiencies in this work. Due to the small number of vault samples, the accuracy of the algorithm is slightly insufficient, and it is difficult to find a comparative analysis of similar work. Therefore, in further work, we will enrich the data set in order to improve the conclusions of the experiment.

Acknowledgement. This work was supported in part by the 13th Five-Year Science and Technology Research Project of the Education Department of Jilin Province under Grant No. JJKH20200794KJ, the Innovation Fund of Changchun University of Science and Technology under Grant No. XJJLG-2018-09, the fund of Key Laboratory of Symbolic Computation and Knowledge Engineering of Ministry of Education (Jilin University) under Grant No. 93K172018K05.

References

1. IDC Corporation: Smartphone OS Market Share [DB/OL]. IDC Corporation, USA (2020). https://www.idc.com/promo/smartphone-market-share/os
2. Lessard, J., Kessler, G.C.: Android forensic: simplifying cell phone examinations. Small Scale Digit. Device Forensics J. **4**(1), 1–12 (2010)
3. Hoog, A.: Android Forensics: Investigation, Analysis and Mobile Security for Google Android. ACM, New York (2011)
4. Zhang, X., Baggili, I., Breitinger, F.: Breaking into the vault: privacy, security and forensic analysis of Android vault applications. Comput. Secur. 1–14 (2017)
5. Hou, S., Ye, Y., Song, Y., et al.: HinDroid: an intelligent Android malware detection system based on structured heterogeneous information network. In: KDD 2017, pp. 1507–1515 (2017)
6. Ding, Y., Dai, W., Yan, S., Zhang, Y.: Control flow-based opcode behaviour analysis for malware detection. Comput. Secur. **44**(2), 65–74 (2014)
7. Shen, F., Del Vecchio, J., Mohaisen, A., et al.: Android malware detection using complex-flows. IEEE Trans. Mob. Comput. **18**(6), 1231–1245 (2019)
8. Backes, M., Künnemann, R., Mohammadi, E.: Computational soundness for Dalvik bytecode. In: CCS 2016, pp. 717–730 (2016)
9. Fang, Z., Han, W., Li, Y.: Permission based Android security: issues and countermeasures. Comput. Secur. **43**(6), 205–218 (2014)
10. Wang, W., Wang, X., Feng, D., et al.: Exploring permission-induced risk in Android applications for malicious application detection. IEEE Trans. Inf. Forensics Secur. **9**(11), 1869–1882 (2017)
11. Zhang, L., Thing, V.L.L., Cheng, Y.: A scalable and extensible framework for Android malware detection and family attribution. Comput. Secur. **80**, 120–133 (2019)
12. Li, J., Xue, D., Wu, W., et al.: Incremental learning for malware classification in small datasets. Secur. Commun. Netw. **2020** (2020)
13. McLaughlin, N., del Rincon, J.M., Kang, K., et al.: Deep Android malware detection. In: CODASPY 2017, pp. 301–310 (2017)
14. Yuan, Y., Yu, Y., Xue, Y.: DroidDetector: Android malware characterization and detection using deep learning. Tsinghua Sci. Technol. **21**(1), 114–123 (2016)
15. Kim, T.G., Kang, B.J., Rho, M., et al.: A multimodal deep learning method for Android malware detection using various features. IEEE Trans. Inf. Forensics Secur. **14**(3), 773–788 (2018)
16. Peng, H., et al.: Using probabilistic generative models for ranking risks of Android apps. In: CCS 2012, pp. 241–252 (2012)
17. Badhani, S., Muttoo, S.K.: CENDroid-a cluster-ensemble classifier for detecting malicious Android applications. Comput. Secur. **85**, 25–40 (2019)
18. Zhang, H., Xiao, X., Mercaldo, F., et al.: Classification of ransomware families with machine learning based on N-gram of opcodes. Future Gener. Comput. Syst. **90**, 211–221 (2019)

Research on Detection and Identification of Dense Rebar Based on Lightweight Network

Fang Qu, Caimao Li$^{(\boxtimes)}$, Kai Peng, Cong Qu, and Chengrong Lin

School of Computer and Cyberspace Security, Hainan University, Haikou
570228, Hainan, China
lcaim@126.com

Abstract. Target detection technology has been widely used, while it is less applied in portable equipment as it has certain requirements for devices. For instance, the inventory of rebar is still manually counted at present. In this paper, a lightweight network that adapts mobile devices is proposed to accomplish the task more intelligently and efficiently. Based on the existing method of detection and recognition of dense small objects, the research of rebar recognition was implemented. After designing the multi-resolution input model and training the data set of rebar, the efficiency of detection was improved significantly. Experiments prove that the method proposed has the advantages of higher detection degree, fewer model parameters, and shorter training time for rebar recognition.

Keywords: Target detection · Object recognition · Rebar recognition · Lightweight network

1 Introduction

At the construction site, inspectors have to count the rebars manually. The rebar cannot be unloaded until the quantity has been reviewed and confirmed. This process is tedious, energy-wasting and very slow. Because of these above problems, we hope to complete this task with a more efficient method.

Rebar detection requires simple and durable equipment. It should also have the ability to transmit data in real-time. Deep networks can often get excellent results in deep learning, but they also have the disadvantages of long training time, large weight files, and high equipment requirements. Therefore, it is necessary to design a lighter network algorithm to adapt mobile devices to match the actual usage scenario. Rebar detection technology faces the following key difficulties:

- High precision required

Rebar is relatively expensive and has a large number in actual use. Both false detection and missed detection need to be manually found out in a large number of marked points. Therefore, high accuracy is required to reduce the workload of the acceptance personnel. It is necessary to optimize the detection algorithm specifically for this dense target.

© Springer Nature Singapore Pte Ltd. 2020
J. Zeng et al. (Eds.): ICPCSEE 2020, CCIS 1257, pp. 440–446, 2020.
https://doi.org/10.1007/978-981-15-7981-3_32

- Photo quality varies

In actual scenes, the rebar is not neatly placed. Rebar may be blocked by other objects. The quality of rebar photos is also affected by light, shooting angle, and equipment.

- Rebar sizes vary

There are many types of rebars used in the construction site, and their cross-sectional sizes and shapes vary due to different uses. Sometimes there are multiple types of rebars on one car.

- Difficulty boundaries' separation

As can be seen from Fig. 1, the cross-section of the rebar in the picture is very small, and the objects are also dense. So rebar recognition requires not only distinguishing objects from backgrounds but also detecting the edges of rebar.

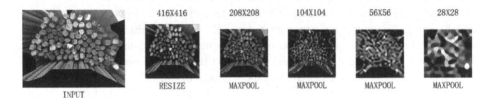

Fig. 1. Edge learning

Aiming at the above problems, this paper designs a fast and accurate rebar detection algorithm to achieve the detection of rebars. The accuracy of the algorithm for the detection of rebars is 0.9024. The weight file of this algorithm after training is only 4.7M, which can be adapted to mobile portable devices. Under the premise of ensuring accuracy, this method reduces the requirements of the algorithm network on the device and improves the convenience and practicability of the algorithm network.

2 Method of Target Detection

In recent years, due to the improvement of GPU performance, the development of distributed computing and the increase in data volume, the application of deep neural networks has been becoming more and more popular.

The SSD [1] algorithm is a method that uses a single deep neural network to detect objects in the image. By referring to the two-stage target detection algorithm, it adopts the prior boxes with different scales and aspect ratios, and uniformly distributes different sizes of default boxes to feature maps with different scales. It is excellent for large objects, but it has poor detection performance for small objects, because small objects may not leave information at the top level. The YOLO V2 [2] algorithm is a detection algorithm based on a convolutional neural network. It uses the convolutional neural network to predict the position, size, category, and confidence of multiple targets in the image. It uses a multi-scale training method and is suitable for inputs with different resolutions. It uses a 416 * 416 picture resolution input to obtain a 13 × 13 feature map. The 13 × 13 featured map has a central cell to predict the object located in the central of the picture and tends to detect larger objects. The YOLO V3 [3] algorithm is an improved algorithm of YOLO V2, which also uses a 416 * 416 picture input. On the coco dataset, it is the same as the MPA of the SSD, but the detection speed is increased by 3 times. It does improve the detection accuracy of small objects. However, there are some problems with long training time and many model parameters because the network's depth is deeper.

The FPN [4] algorithm mainly solves the multi-scale problem in object detection: Although there is less information in the underlying feature map, the target location is accurate. In the meantime, the target location is rough as high-level feature maps are rich in information. The difference between FPN and SSD is that FPN can make independent predictions on different feature maps through a simple network connection. Thus, the performance of small object detection is greatly improved without increasing the amount of calculation of the original model substantially.

The RFB [5] algorithm is inspired by the receptive field structure in the human visual system. The algorithm proposes a novel RF module, which enhances the discernibility of the features and the robustness of the model by simulating the relationship between the size of RF and the eccentricity. The RFB module is a module that increases the receptive field, which can strengthen the learning of object features by lightweight networks.

Over the years, many different algorithms have been proposed to improve the disadvantages that lightweight networks cannot learn more objects' features. From the SSD and YOLO V3 algorithms, it can be found that lightweight networks and high detection accuracy have not been able to achieve both. Lightlayer (LL) implements the detection of dense small objects based on the above algorithm and has achieved the goals of reducing the network depth, shortening the training time, and improving the detection accuracy.

3   Rebar Detection Algorithm and Structure

Lightlayer is inspired by the SSD algorithm. It can set multiple feature layers to extract target information. This article attempts to learn the SSD algorithm to set more feature layers to obtain target detection information. However, it is experimentally confirmed that the improvement is not that obvious. The reason is Lightlayer is a lightweight network. The high-level convolution does not lose much target position information. On the contrary, increasing the feature layer will aggrandize the number of model parameters. Concerning YOLO V3 with 3 feature maps, we found that the overall effect would be better when the number of feature layers is 3.

Through observation and research on the data set, it is showed that most pieces of rebar are in the center of pictures. So Lightlayer uses the series of YOLO algorithms for reference setting the picture input resolution as 416 * 416. Because the total step size of LightLayer sampling is 32, for 416 * 416 images, the final feature map size is 1313, and the dimension is odd, so the feature map just has one center position. For this data set, their center points often fall into the center of the image, and it is relatively easy to use a center point of the feature map to predict the boundary boxes of these objects.

The crucial key of detection technology is not only to distinguish rebar from the background as the cross-sectional area of the rebar is small and dense, but also to distinguish the boundaries between any pieces of rebar. This requires relatively steep learning about the edge features of rebar. It can be seen from Fig. 1 that the features extracted by the convolutional layer become more and more semantic as the number of layers increases and the outlines of the rebar become more and more blurred. Because different feature maps can extract different features in the image (Fig. 1), SSD algorithm can effectively detect and extract high-level features from maps that have a low resolution with high semantic information and low-level features from maps that have high resolution with low semantic information. However, the SSD treats them as the same for classification and regression. It can not make full use of local details and global semantic features. Lightlayer draws on the network feature fusion method of FPN, which makes reasonable use of semantic information at different scales in the feature pyramid. So it can combine low-level detail features with high-level semantic features effectively. Currently, good target detection is mainly based on deep networks (such as Resnet, VGG-16) with the disadvantages that a large amount of computational costs and slow training, while some lightweight networks are faster in training with the detection accuracy is relatively low. The RFB module was proposed to strengthen The deep features learned from the lightweight convolution model make the lightweight network detect objects quickly and accurately. However, since the calculated amount of the RFB module is also large, it is considered that the effect is best when a module is added through repeated experiments. Based on the above analysis, a lightlayer network structure diagram is designed (Fig. 2).

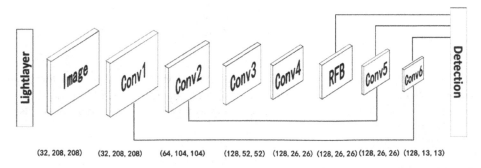

(32, 208, 208)   (32, 208, 208)   (64, 104, 104)   (128, 52, 52) (128, 26, 26) (128, 26, 26)(128, 26, 26) (128, 13, 13)

Fig. 2. Lightlayer network structure diagram

4   Results

4.1   Data Set Processing

There are 250 photos in the data set. Through data augmentation methods such as rotation, cropping, and occlusion, we can obtain more training and test samples (1500 here). It is processed into a VOC data set format and encoded to see the correctness of the data conversion. The data set is divided according to a ratio of 1:4. 1200 pieces as training data and 300 pieces as test data.

4.2 Training

Lightlayer is based on the RFBNet code modification. It is necessary for LL to set the border and aspect ratio in advance. The k-means clustering is used to analyze the annotation boxes given by the dataset to obtain the size and length-width ratio of the anchor box. The accuracy of the clustering is not enough, so we need to check the size and length-to-width ratio of the labeled box information on the training set and make some appropriate adjustments to the clustering results. Looking at the change of the loss function and the change of the coding design learning rate, the loss function will get a tendency to converge. Finally, the weight file obtained by training the rebar dataset is tested, the weight file with the highest accuracy is selected and this weight is used for detection whereafter.

4.3 Object Detection

Use the obtained weight file to test the test set, and evaluate the effect of the algorithm by the accuracy of the rebar recognition (Eq. 1) and the recall rate (Eq. 2) (Table 1).

$$\text{Accuracy} = TP / (TP + FP) \tag{1}$$

$$\text{Recall} = TP / (TP + FN) \tag{2}$$

Table 1. Test results of Lightlayer

Backbone	Precision	Recall	Weight
Lightlayer	0.9024	0.7760	4.7M

TP is the number of samples that the algorithm detects as positive and the true result is also true. FP is the number of samples that the algorithm detects as positive but the true result is false. FN is the number of samples that the algorithm detects as false and the true result is false.

5 Model Comparison

On the rebar data set, this paper compared the algorithms of YOLO V3, FPN, SSD, and Lightlayer from the aspects of accuracy, recall rate, and weight file size. To ensure the fairness of the test, we have not to change the parameters and processing functions but only replaced the network model accordingly.

It can be seen from Table 2 that the Lightlayer has the highest accuracy under similar parameters. Simultaneously, the Recall has a significant ascent relative to SSD. But it is still lower than FPN and YOLO V3 and the weight file is the smallest compared to other algorithms. At the same batch size, the training time of each epoch is also the shortest.

Table 2. Comparison results of various algorithms

Backbone	Precision	Recall	Weight
YOLO V3	0.8971	0.8844	230.4M
FPN	0.8483	0.9579	67.2M
SSD	0.8788	0.7287	49.1M
Lightlayer	0.9024	0.7760	4.7M

6 Conclusion

The rebar recognition algorithm uses a lightweight network to identify the rebar. It can achieve high accuracy on a smaller data-sets. The training time and weight files are greatly reduced compared to deep networks such as YOLO V3. What is remarkable about Lightlayer is that it tries to learn information on the features as much as possible from convolutional layers of the objects to avoid deepening the depth of the network. Nevertheless, according to the current recall rate, the loss of accuracy happens when identifying other objects as rebar. Hence, the accuracy of the algorithm could be further improved by data augmentation and network optimization in the later stage.

Acknowledgment. This work is partially supported by Hainan Science and Technology Project, which is Research and development of intelligent customer service system based on deep learning (No. ZDYF2018017). Thanks to Professor Caimao Li, the correspondent of this paper.

References

1. Liu, W., et al.: SSD: single shot multibox detector. In: Leibe, B., Matas, J., Sebe, N., Welling, M. (eds.) ECCV 2016. LNCS, vol. 9905, pp. 21–37. Springer, Cham (2016). https://doi.org/10.1007/978-3-319-46448-0_2
2. Redmon, J., Farhadi, A.: YOLO9000: better, faster, stronger. In: Proceedings of the IEEE Conference on Computer Vision and Pattern Recognition (2017)
3. Redmon, J., Farhadi, A.: Yolov3: an incremental improvement. arXiv preprint arXiv:1804.02767 (2018)
4. Lin, T.-Y., et al.: Feature pyramid networks for object detection. In: Proceedings of the IEEE Conference on Computer Vision and Pattern Recognition (2017)
5. Liu, S., Huang, D.: Receptive field block net for accurate and fast object detection. In: Proceedings of the European Conference on Computer Vision (ECCV) (2018)
6. Girshick, R.: Fast R-CNN. In: Proceedings of the IEEE International Conference on Computer Vision (2015)

7. Redmon, J., et al.: You only look once: unified, real-time object detection. In: Proceedings of the IEEE Conference on Computer Vision and Pattern Recognition (2016)
8. Tianshu, W., Zhijia, Z., Yunpeng, L., Wenhui, P., Hongye, C.: A lightweight small object detection algorithm based on improved SSD. Infrared Laser Eng. 47(7), 703005 (2018)
9. Wen-xu, S.H.I., Dai-lun, T.A.N., Sheng-li, B.A.O.: Feature enhancement SSD algorithm and its application in remote sensing images target detection. Acta Photonica Sinica 49(1), 128002 (2020)
10. Moran, J., Lu, H., Wang, Z., He, M., Chang, Z., Hui, B.: Improved YOLO V3 algorithm and its application in small target detection. Acta Opt. Sinica 39(7), 0715004 (2019)
11. Chen, Z.B., Ye, D.Y., Zhu, C.X., Liao, J.K.: Object recognition method based on improved YOLOv3. Comput. Syst. Appl. 29(1), 49–58 (2020)
12. Jiangrong, X., et al.: Enhancement of single shot multibox detector for aerial infrared target detection. Acta Opt. Sinica 39(6), 0615001 (2019)
13. Xiaoning, L.I., et al.: Detecting method of small vehicle targets based on improved SSD. J. Appl. Opt. 41(1), 150–155 (2020)
14. Shan, Y., Yang, J., Wu, S., et al.: Skip feature pyramid network with a global receptive field for small object detection. CAAI Trans. Intell. Syst. 14(06), 1144–1151 (2019)
15. Lei, J., Gao, C., Hu, J., Gao, C., Sang, N.: Orientation adaptive YOLOv3 for object detection in remote sensing images. In: Lin, Z., et al. (eds.) PRCV 2019. LNCS, vol. 11857, pp. 586–597. Springer, Cham (2019). https://doi.org/10.1007/978-3-030-31654-9_50

Automatic Detection of Solar Radio Spectrum Based on Codebook Model

Guoliang Li[1], Guowu Yuan[1,2(✉)], Hao Zhou[1], Hao Wu[1],
Chengming Tan[2], Liang Dong[3], Guannan Gao[3], and Ming Wang[3]

[1] School of Information Science and Engineering, Yunnan University,
Kunming 650091, China
yuanguowu@sina.com
[2] CAS Key Laboratory of Solar Activity, National Astronomical Observatories
of Chinese Academy of Sciences, Beijing 100012, China
[3] Yunnan Observatories, Chinese Academy of Sciences,
Kunming 650011, China

Abstract. Space weather can affect human production and life, and solar radio burst will seriously affect space weather. Automatic detection of solar radio bursts in real time has a positive effect on space weather warning and prediction. Codebook model is used to simulate solar background radio to achieve automatic detection of solar radio bursts in this paper. Firstly, channel normalization was used to eliminate channel difference of original radio data. Then, a new automatic detection method for solar radio bursts based on codebook model was proposed to detect radio bursts. Finally, morphological processing was implemented to obtain burst parameters by detecting binary burst area. The experimental results show that the proposed method is effective.

Keywords: Solar radio burst · Codebook model · Channel normalization

1 Introduction

The Sun is the only star that can be observed carefully. At the same time, the solar system is the only known star system with life and reproduction behavior. Therefore, solar physics is not only the most active field in astronomy, but also extremely important for researching the origin and evolution of life in the Universe. The violent burst of the Sun is the most intense energy release process in the solar system. It causes severe disturbance to the solar-terrestrial space environment and geomagnetic field, and directly affects some modern technology facilities such as spacecraft, communications, and electricity. It has a lot of adverse effects on human production and life. The energy release of the solar burst is manifested in the entire electromagnetic spectrum, and the radio band observation is an indispensable means.

In the radio band, solar radio radiation is divided into quiet solar radio and solar radio burst. Quiet solar radio is a background radio radiation that is always present, and is a radio component that basically does not change. Solar radio burst is a rapid and dramatic increase in the amount of radio current. The duration of solar radio burst ranges from less than 1 s to more than a few hours [1].

© Springer Nature Singapore Pte Ltd. 2020
J. Zeng et al. (Eds.): ICPCSEE 2020, CCIS 1257, pp. 447–460, 2020.
https://doi.org/10.1007/978-981-15-7981-3_33

Solar radio data is obtained by solar radio spectrometer. It is intensity spectrum changing with time at each frequency, which is called solar radio dynamic spectrum. According to radio signal changes on the dynamic spectrum, we can obtain the characteristic information of the radio burst and radio fine structure, which will greatly help the study of solar physics.

With the rapid development of solar observation instruments, a large amount of observation data is collected, but only 1% of the collected data is solar burst data [2]. Astronomers are usually more interested in the solar radio burst, but it is a tedious and boring task to find out the solar radio burst event in the massive observation data. Burst judgment is inefficient and error-prone. Therefore, it is very necessary to automatically detect the solar radio data using computers.

Some research achievements have been made on the automatic detection of solar radio spectrum. Lobzin used Hough transform and Radon transform to detect straight line segments in radio spectrum image, and identified burst events based on straight line matching [3, 4]. P.J. Zhang proposed an improved active contour model based on Hough transform to detect bursts, and they tracked the main part of radio bursts and computed burst frequency drift rate [5]. L. Xu and S. Chen used deep learning models to classify radio spectrum images [2, 6]. D. Singh achieved burst detection based on statistical methods [7]. N. Z. M. Afandi used an effortless and hassle-free tool for burst detection in order to monitor solar bursts during a complete cycle of 11 years [8]. Qing-Fu Du proposed a self-adaptive method for determining solar bursts by selecting 20 frequencies for radio flux profiles [9].

When the Sun bursts, the radio flux increases obviously compared with that in the quiet Sun. The radio flux in the quiet Sun can be regarded as background, and the flux of solar radio burst can be regarded as foreground. According to the idea of background subtraction, solar radio burst detection is realized.

In this paper, codebook model is used to detect solar radio burst. Firstly, we normalize the channel of solar radio data to reduce the difference between frequency channels. Then, the simplified codebook model is used to simulate the solar background radio and realize the automatic detection of solar radio burst. Finally, through the binary morphological operation, the post-processing of the binary image of the burst area is carried out. According to the binary area of the processed spectrum image, the burst time, frequency and other parameters can be obtained. Codebook model of each pixel can be renewed after a certain period of time to adapt to the slow change in the amount of solar radiation after a long time period. Codebook background model is less computationally intensive and takes up less memory.

2 Sources of Radio Spectrum Data

Yunnan astronomical observatories of the Chinese academy of sciences have two digital solar radio spectrometers. These two spectrometers have high spectral and time resolution, which can observe more elaborate and rapidly changing radio bursts and fine structures. Our solar radio spectrum data used in ours experiments in this paper comes from the 11-meter-diameter meter-wave solar radio spectrometer at the Fuxian

Lake solar observation and the 10-meter-diameter decimeter-wave solar radio spectrometer at Yunnan observatories headquarters.

The data file from the two spectrometers of Yunnan observatories is named with observation time. Each file only records 512 s, the metric spectrometer records 3150 frequency channels, and the decimetric spectrometer records 3500 frequency channels. Figure 1 is a more obvious image of solar radio burst dynamic spectrum data used in this paper. This data is a burst event collected by the metric spectrometer on March 9, 2012.

Considering the contradiction between time resolution and frequency resolution, the time resolution is set to 80 ms and the frequency resolution to 200 kHz during daily work (Table 1).

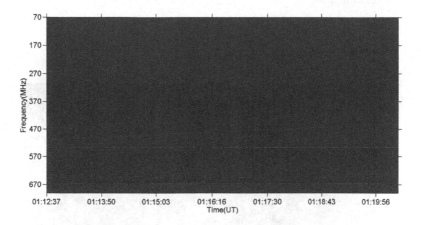

Fig. 1. A solar radio burst observed by the metric solar radio spectrometer on March 9, 2012

Table 1. Parameters of the two spectrometers in Yunnan Observatories [10]

Parameter	Metric spectrometer	Decimetric spectrometer
Frequency range	70–700 MHz	625–1500 MHz
Antenna	11 m meshed parabola	10 m solid parabola
Recording time	0:00–10:00 UT	0:00–10:00 UT
Highest time resolution	2 ms	2 ms
Highest spectral resolution	200 kHz	200 kHz

3   Channel Normalization of Solar Radio Spectrum

Due to the reasons of the observation equipment, even the adjacent frequency channels in the solar radio data will have great differences [6]. When the signal value of corresponding channel exceeds the instrument's sensitive range, instrument channel effect is caused by the channel oversaturation [11, 12]. We use channel normalization to eliminate channel effects.

In channel normalization algorithm, each channel's average value is subtracted firstly in each frequency channel to obtain the net solar radio flux, and then the net solar radio flux is divided by the channel's average value. Let $s(t,f)$ as the original channel flux in frequency y at time t, $\overline{s_f}$ as the average flux in frequency f, and $g(t,f)$ as the channel flux in frequency y at time t after channel normalization. The channel normalization is defined as:

$$g(t,f) = s(t,f) - \overline{s_f}/\overline{s_f} = s(t,f)/\overline{s_f} - 1 \tag{1}$$

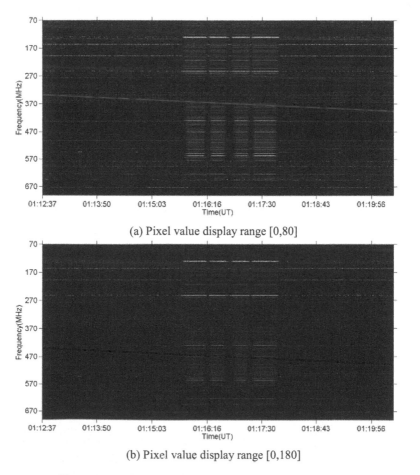

(a) Pixel value display range [0,80]

(b) Pixel value display range [0,180]

Fig. 2. Result of Fig. 1 after adjusting the pixel value range

The difference between each frequency channel has been reduced after channel normalization. However, the image contrast after channel normalization is too low. By adjusting the pixel value range of gray-scale spectrum image, we can see the spectrum details after channel normalization. Figure 2 is the result of Fig. 1 after channel normalization and adjusting the pixel value range.

4 Automatic Detection of Solar Radio Spectrum Based on Codebook Model

K. Kim et al. proposed codebook model to detect moving targets in 2005 [13]. The traditional codebook model achieves foreground segmentation by establishing a codebook for each pixel to record background information [14]. There are three channels for each pixel's color in the natural scene. But there is only one channel for each pixel in solar radio spectrum, and the change of background radio is not as complex as in the natural scene. Therefore, we can simplify codebook model and reduce computational complexity.

It is possible to establish a codebook for each pixel to describe background radio. When the spectral data after channel normalization arrives, each pixel's value is compared with the codeword in the corresponding codebook. If the value is between the minimum value and the maximum value, the pixel is considered as background pixel. If the value is greater than the maximum value, the pixel is determined to be burst pixel.

A codebook is established for each frequency channels after channel normalization. The codebook of each frequency channel contains multiple codewords, each codeword represents a possible background radio value on the channel. The codebook of each frequency channel is independent of each other. Let the codebook of a frequency channel as $CodeBook = \{c_1, c_2, \cdots c_L\}$, and the codebook has L codewords.

4.1 Background Construction and Update of the Codebook Model

We choose N-frame spectrum data as training sequence of codebook model to construct codebook background model. The number of codewords in each codebook is related to the change range of radio intensity value. We match the radio intensity values of all pixels in the spectrum data with the codewords in the corresponding codebook. If there is a matching codeword in the codebook, we will update the statistical data of the matched codeword. If there is no matching codeword, we will create a new codeword.

In this application, there is only one channel's radio intensity value, which can also be regarded as the gray value of the pixel. So we can omit the color distortion in the traditional codebook model, and only match the brightness distortion with the codewords. The construction and update process of background codebook model is as follows:

(1) Initialize the codebook. Apply storage space for the codebook, and the codebook is set to null;
(2) The training sequence images of solar radio spectrum are obtained to initialize codebook model. They are assumed to be $V = \{v_1, v_2, \cdots v_N\}$;

(3) Simplify the criterion of codeword matching. The criterion of codebook matching is defined.

$$Brightness(I, \langle I_{low}, I_{high} \rangle) = \begin{cases} Ture, (1 - \varepsilon)\hat{I}_i \leq v_i \leq \min\{(1 + \varepsilon)\hat{I}_i, \tilde{I}_i/(1 - \varepsilon)\} \\ False, otherwise \end{cases}$$

(2)

where, we simplify the parameters the traditional codebook model, and set the brightness threshold as ε. Therefore, the calculation formula of the brightness judgment thresholds I_{low} and I_{high} is simplified as follows:

$$I_{low} = \alpha\hat{I}_i = (1 - \varepsilon)\hat{I}_i$$

(3)

$$I_{high} = \min\{\beta\hat{I}_i, \tilde{I}_i/\alpha\} = \min\{(1 + \varepsilon)\hat{I}_i, \tilde{I}_i/(1 - \varepsilon)\}$$

(4)

(4) Construction and update of the codebook background model. Each spectrum image inputted will be matched with the codewords according to the criterion formula 2. If no match is successful, a new codeword c_l is created. The color vector of the new codeword is $w_l = (\bar{A}_t, 0, 0)$, and the new codeword is $auw_l = \langle I, I, 1, t - 1, t, t \rangle$. If the match is successful, the matched codeword is updated according to the formula 5 and 6:

$$w_m = (\frac{f_m\bar{A}_m + A_t}{f_m + 1}, 0, 0)$$

(5)

$$auw_m = \left\langle \min(\sqrt{A_t^2}, \tilde{I}_m), \max(\sqrt{A_t^2}, \hat{I}_m), f_m + 1, \max(\lambda_m, t - q_m), p_m, t \right\rangle$$

(6)

(5) Obtain a pure background codebook. Since the used training sequence may still have noise interference or include radio bursts, it is necessary to filter out the code words representing noise and bursts according to formula 7:

$$\mu = \{c_m| c_m \in \varphi \cap \lambda_m \leq \frac{N}{2}\}$$

(7)

(6) Repeating steps 4 and 5 until the training of the codebook background model is completed (Fig. 3).

4.2 Automatic Detection of Solar Radio Burst

After training the background model of codebook, the model can be used to detect the solar radio burst automatically. When the new time's spectrum data is obtained, the radio value is matched with the corresponding codebook background model in each frequency channel. If the matching is successful, it is determined as the background radio pixel; if the matching is not successful, it is determined as the solar radio burst pixel.

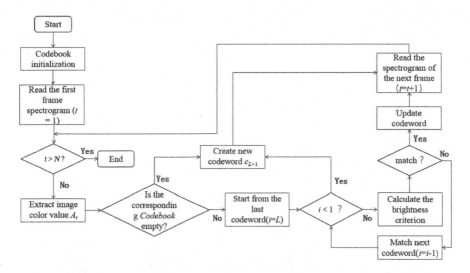

Fig. 3. Codebook background model training flowchart

The method steps are described as follows.

(1) Read the spectrum data $g(t,f)$ after channel normalization;
(2) According to the formula 2, $g(t,f)$ is determined whether it is a background pixel or a solar radio burst pixel.
(3) Obtain the detection result. Set the pixel determined as background to 0, and the pixel determined as solar radio burst to 1. The burst region binary image is obtained.

A more detailed flow chart is shown in Fig. 4.
The detection result of solar radio burst area in Fig. 2 is shown in Fig. 5.

5 Binary Morphological Processing

Binary morphology is used to process the binary image of the detected solar radio burst area, and it can remove the independent noise points and small structures [15].

We use the Open operation to deal with the solar radio burst area according to the formula 8.

$$A \circ B = (A \Theta B) \oplus B \tag{8}$$

The Open operation $\circ$ includes first corrosion Θ and subsequent expansion $\oplus$. The result is recorded as A, and the structural element as B (Fig. 6).

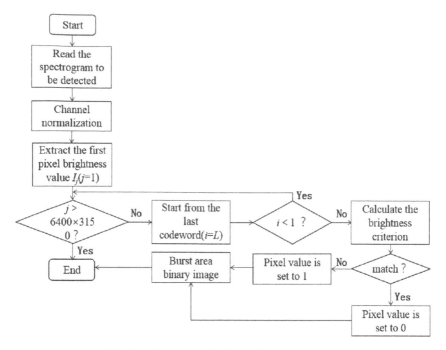

Fig. 4. Flow chart of solar radio burst detection

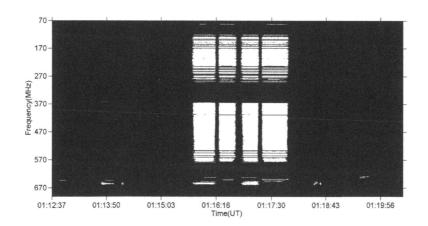

Fig. 5. Detection result of solar radio burst area in Fig. 2 using Codebook model

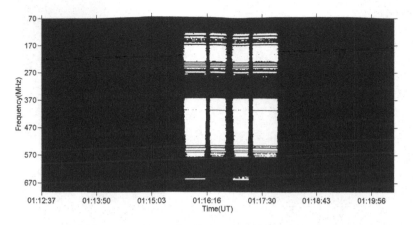

Fig. 6. Result of the Fig. 5 after binary morphological processing

6   Comparison and Analysis of Experimental Results

The official website of the Yunnan Observatories (http://www.ynao.ac.cn/solar/) can be used to query the records of metric solar radio dada after March 2012 and the decimetric solar radio data after September 2009. We use the method proposed in this paper to detect two solar radio bursts. The results are shown in Fig. 7 and Fig. 8.

In reference [16], a method based on the set threshold is proposed to detect solar radio bursts. Our results are compared with the results in the reference [16] and the records of Yunnan Observatories. The following is comparison data.

From the comparison results shown in the Table 2 and 3, our method proposed in this paper is effective. Compared with the reference 13 and the records of Yunnan Observatories, the method in this paper has a better agreement on the start or end time and start or end frequency.

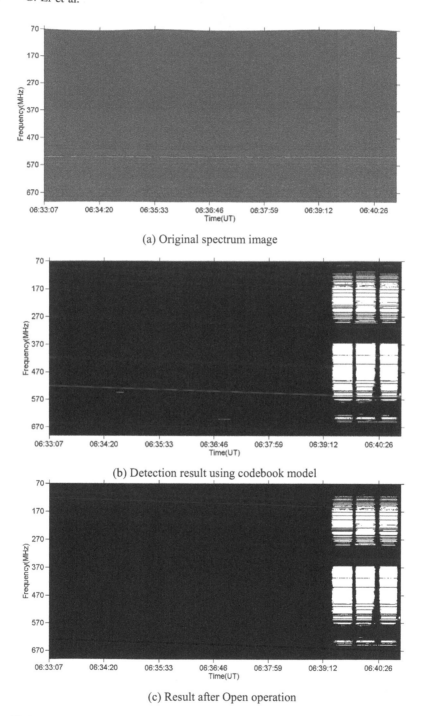

(a) Original spectrum image

(b) Detection result using codebook model

(c) Result after Open operation

Fig. 7. Detection results recorded at the Yunnan Observatories on March 7, 2012

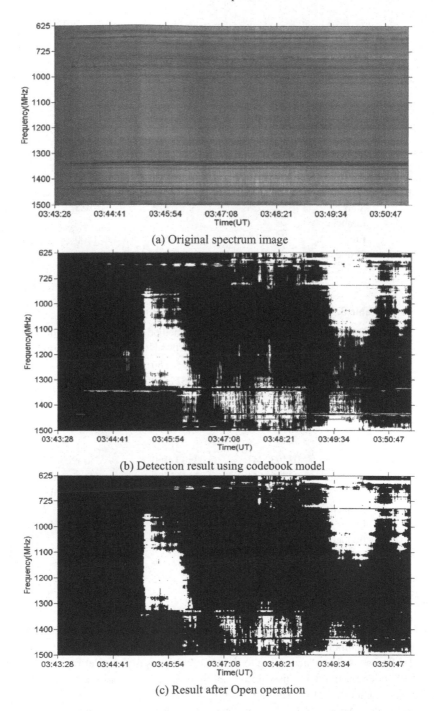

(a) Original spectrum image

(b) Detection result using codebook model

(c) Result after Open operation

Fig. 8. Detection results recorded at the Yunnan Observatories on March 5, 2012

458   G. Li et al.

Table 2. Detection results of solar radio burst on February 9, 2011

	Burst parameter	Our method in this paper	The method in reference [16]	Records of Yunnan Observatories
1st solar radio burst	Start time	012557157	012626128	0126
	End time	012713329	012711888	0127
	Start frequency	625	643	625
	End frequency	1493	1500	1500
2nd solar radio burst	Start time	012714935	012720001	0128
	End time	012936171	012921761	0129
	Start frequency	625	625	625
	End frequency	1500	1500	1500

Table 3. Detection results of solar radio burst on March 5, 2012

	Burst parameter	Our method in this paper	The method in reference [16]	Records of Yunnan Observatories
1st solar radio burst	Start time	034548855	034533360	0346
	End time	034600983	034622880	0347
	Start frequency	665.5	775.6	625
	End frequency	1310	1346.4	900
2nd solar radio burst	Start time	034933341	035007360	No record
	End time	035200641	035153680	No record
	Start frequency	625	625.4	No record
	End frequency	1124.5	1020.8	No record

7 Summary and Prospect

In this paper, image processing technology is used to automatically detect solar radio burst. Firstly, channel normalization is used to process original data to eliminate channel differences. Then, according to the characteristics of the solar radio spectrum, a simplified codebook model is used to model the background radio, and the solar radio burst area is detected. However, no further feature extraction of fine structure was explored in this paper. Perfect feature extraction and statistical analysis of feature parameters are hoped to be realized on the basis of this paper.

Acknowledgments. This work is supported by the Natural Science Foundation of China (Grant No. 11663007, 11703089, 41764007, 61802337, U1831201), Open Project of Key Laboratory of Celestial Structure and Evolution, Chinese Academy of Sciences (Grant No. OP201510), the Research Foundation of Yunnan Province (No. 2018FB100), the Scientific Research of Yunnan Provincial Education Department (No. 2018JS011) and the Action Plan of Yunnan University Serving for Yunnan.

References

1. Gao, G.N.: The Solar Radio Bursts and Fine Structuresin Metric and Decimetric Bands. University of Chinese Academy of Sciences (2015)
2. Ma, L., Chen, Z., Long, X., Yan, Y.: Multimodal deep learning for solar radio burst classification. Pattern Recogn. **61**(1), 573–582 (2017)
3. Lobzin, V.V., Cairns, I.H., Robinson, P.A., et al.: Automatic recognition of type III solar radio bursts: automated radio burst identification system method and first observations. Space Weather-Int. J. Res. Appl. **7**(4), S04002 (2009)
4. Lobzin, V.V., Cairns, I.H., Robinson, P.A., et al.: Automatic recognition of coronal type II radio bursts: the automated radio burst identification system method and first observations. Astrophys. J. Lett. **710**(1), L58 (2010)
5. Zhang, P.J., Wang, C.B., Ye, L.: A type III radio burst automatic analysis system and statistic results for a half solar cycle with Nançay Decameter Array data. Astron. Astrophys. **618**, A165 (2018)
6. Chen, S., Long, X., Lin, M., et al.: Convolutional neural network for classification of solar radio spectrum. In: IEEE International Conference on Multimedia & Expo Workshops (2017)
7. Singh, D., Raja, K.S., Subramanian, P., Ramesh, R., Monstein, C.: Automated detection of solar radio bursts using a statistical method. Sol. Phys. **294**(8), 1–14 (2019). https://doi.org/10.1007/s11207-019-1500-0
8. Afandi, N.Z.M., Sabri, N.H., Umar, R., Monstein, C.: Burst-finder: burst recognition for E-CALLISTO spectra. Indian J. Phys. **94**(7), 947–957 (2019). https://doi.org/10.1007/s12648-019-01551-2
9. Du, Q.-F., Chen, C.-S., Zhang, Q.-M., Li, X., Song, Y.: A self-adaptive method for the determination of solar bursts for high-resolution solar radio spectrometer. Astrophys. Space Sci. **364**(6), 1–9 (2019). https://doi.org/10.1007/s10509-019-3584-2
10. Gao, G.N., Lin, J., Wang, M., Xie, R.X.: Research progress of type II and type III radio storms and fine structure observation of solar metric and decimetric. Prog. Astron. **30**(1), 35–47 (2011)
11. Luo, G., Yuan, G., Li, G., Wu, H., Dong, L.: A noise reduction method for solar radio spectrum based on improved guided filter and morphological cascade. In: Liu, Y., Wang, L., Zhao, L., Yu, Z. (eds.) ICNC-FSKD 2019. AISC, vol. 1075, pp. 815–822. Springer, Cham (2020). https://doi.org/10.1007/978-3-030-32591-6_88
12. Zhao, R.Z., Hu, Z.Y.: Wavelet NeighShrink method for eliminating image moire in solar radio bursts. Spectrosc. Spectral Anal. **27**(1), 198–201 (2007)
13. Kim, K., Chalidabhongse, T.H., Harwood, D., Davis, L.: Real-time foreground-background segmentation using codebook model. Real-Time Imaging **11**(3), 172–185 (2005)
14. Qin, Z.Y.: Research on background modeling method based on codebook learning. University of Electronic Science and Technology of China (2015)

15. Gonzalez, R.C., Woods, R.E.: Digital Image Processing. Electronic Industry Press, Beijing (2007)
16. Yuan, G., et al.: Solar radio burst automatic detection method for decimetric and metric data of YNAO. In: Cheng, X., Jing, W., Song, X., Lu, Z. (eds.) ICPCSEE 2019. CCIS, vol. 1058, pp. 283–294. Springer, Singapore (2019). https://doi.org/10.1007/978-981-15-0118-0_22

Graphic Images

Self-service Behavior Recognition Algorithm Based on Improved Motion History Image Network

Liping Deng[✉], Qingji Gao, and Da Xu

Robotics Institute, Civil Aviation University of China, Tianjin 300300, China
2447749610@qq.com

Abstract. Aiming at the problem of automatic detection of normal operation behavior in self-service business management, with improved motion history image as input, a recognition method of convolutional neural network is proposed to timely judge the occurrence of anomie behavior. Firstly, the key frame sequence was extracted from the self-service operation video based on the method of uniform energy down-sampling. Secondly, combined with the timing information of key frames to adaptively estimate the decay parameters of the motion history image, adding information contrast to generating a logic matrix can improve the calculation speed of the improved motion history image. Finally, the formed motion history image was input into the established convolutional neural network to obtain the class of self-service behavior and distinguish anomie behavior. In real scenarios of self-service baggage check-in for civil aviation passengers, the typical check-in behavior data set is established and tested in actual self-service baggage check-in system of the airport. The results show that the method proposed can effectively identify typical anomie behaviors and has high practical value.

Keywords: Anomie behavior · Depth image · Improved motion history image · Behavior recognition

1 Introduction

Due to the popularity of self-service withdrawals, self-service shopping, self-service check-in [1] and other operational business, the normative monitoring of operational behavior has attracted attention. This kind of problem belongs to behavior recognition in fixed field of vision. In recent years, human behavior recognition is a research hotspot in the field of computer vision [2], which has been widely used in human-computer intelligent interaction [3], video monitoring, smart home [4], virtual reality and other fields. With the reduction of depth image acquisition cost and the improvement of research focus, there are behavior recognition methods based on human skeleton [5] and human segmentation figure [6], which improve the recognition accuracy, and more often use convolutional neural network (CNN) methods [7]. Its C3D [8] and other 3D-CNN model algorithms are an end-to-end recognition method, which is more suitable for the second level short-time video behavior recognition. P3D [9] (pseudo3d recurrent networks, P3D) algorithm uses '2 + 1' D convolution kernel to

© Springer Nature Singapore Pte Ltd. 2020
J. Zeng et al. (Eds.): ICPCSEE 2020, CCIS 1257, pp. 463–475, 2020.
https://doi.org/10.1007/978-981-15-7981-3_34

replace the 3D convolution kernel and combines with the existing residual neural network, which overcomes the shortcomings of C3D with many calculation parameters and low accuracy. MiCT-Net (Mixed 3D/2D Convolutional Tube Network, MiCT-Net) [10] improves the response speed by further reducing the complexity of 3D-CNN models. The above mentioned 3D-CNN method has effectively improved the recognition rate and generalization ability. However, these algorithms have many parameters, large amount of calculation and high performance requirements for hardware equipment.

Air passenger self-service baggage check-in (SSBC) is a self-service business that has attracted much attention in recent years. Due to the long time, many actions and high complexity of self-service operation, it is easy to have anomie behavior, which needs automatic monitoring. However, the above existing recognition methods can not meet the practical needs. Based on this background, this paper studies a fast and effective method to identify the anomie behavior of SSBC. By extracting the key frame of check-in behavior to reduce redundant information, and using the logic matrix to improve motion history image (MMHI) based on the key frame, it is input to the lightweight CNN with few layers, through learning and generalization application, the classification of check-in behavior is realized, and the anomie behavior is identified.

2   Definition of Check-in Behavior

Behavior B refers to all purposeful activities of people. It consists of N simple action sequences $a_1, \ldots, a_i, \ldots, a_N$. That is $B = \{a_1, \ldots, a_i, \ldots, a_N\}$.

The SSBC business can be generally divided into seven behaviors, i.e. identity authentication, seat selection and printing of boarding passes, notice for confirming SSBC, printing of baggage tags, attaching baggage tags, putting baggage, and obtaining baggage check-in vouchers. Among them, baggage put behavior B_p refers to that passengers put the labeled baggage on the transmission belt, which is easy to put multiple pieces of baggage at one time, so it needs to be monitored.

$$B_p = \{a_1, a_2, \cdots, a_n\} \tag{1}$$

Baggage put behavior B_p consists of n depth image frames. B_p consists of two subsets, both normative behavior B_n and anomie behavior B_a, B_p satisfies:

$$\begin{cases} B_p = B_n \cup B_a \\ B_a \cap B_n = \emptyset \end{cases} \tag{2}$$

In actual operation, B_p generally includes six classes: B_0: Put a hard bag, B_1: Stack two pieces of baggage, B_2: Put two lines side by side, B_3: Remove one piece after putting two pieces of baggage, B_4: Remove one piece of baggage after putting one piece of baggage, B_5: Put one piece of baggage to straighten. Among them, $\{B_0, B_3, B_5\} \subseteq B_n$, $\{B_1, B_2, B_4\} \subseteq B_a$.

3   Recognition of Check-in Behavior Based on MMHI and CNN

3.1   Self-service Check-in Recognition System

The SSBC is shown in Fig. 1(a). The system consists of three parts: weighing, tag identification and baggage detection. The Kinect (Fig. 1(c)) depth image acquisition device is optimally arranged on the device to collect the passenger's check-in behavior video, and the location layout is shown in Fig. 1(b). Figure 2 is part of the image of the check-in behavior.

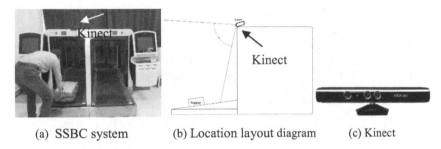

(a) SSBC system   (b) Location layout diagram   (c) Kinect

Fig. 1. SSBC environment

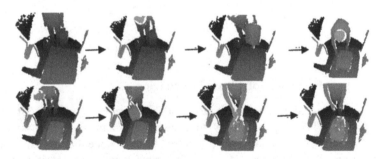

Fig. 2. The sequence of check-in behavior

3.2   Recognition Process of Check-in Behavior

The check-in behavior recognition is divided into extracting key frames, get MMHI behavior modeling and CNN calculation. The recognition process model is shown in Fig. 3.

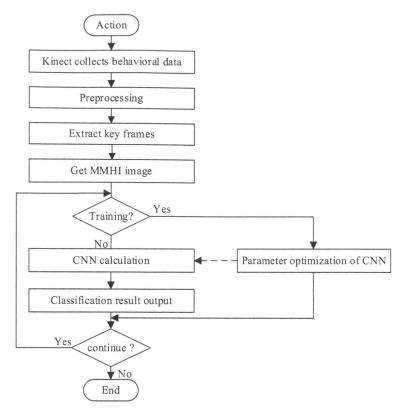

Fig. 3. Check-in behavior recognition process

3.3 Key Frame Extraction of Check-in Behavior

In order to remove redundant information and improve the efficiency of behavior feature extraction [11], it is necessary to extract key frame images from behavior sequence. The number of key frames is not necessarily the same, but does not affect the overall structure of the behavior [12]. A key frame extraction algorithm based on energy uniformity is proposed.

The pixel change of the human body between two adjacent frames on the image reflects the energy contained in the human body's motions in this period of time, when the pixel change of the human body between the first frame and the subsequent frame exceeds the set threshold value, it is considered that there is a large enough energy change between the two frames, that is, the motion amplitude is large enough, then the two frames are recorded, and the later frame is the new starting point until all frames of the entire behavior process are traversed. Set energy threshold to T. The calculation method of the key frame is shown in formula (3).

$$K(i) = \begin{cases} j & \|P(x,y,K(i-1)) - P(x,y,j)\| > T \\ 0 & others \end{cases} \qquad (3)$$

Where, j represents the sequence number of the behavior image sequence frame, i represents the sequence number of the key frame, $K(i)$ represents the sequence number of the i-th key frame in the behavior image sequence, $P(x,y,j)$ represents the image of the j-th frame, $\|P(x,y,K(i-1)) - P(x,y,j)\|$ represents the measurement result of the deviation between the image of the j-th frame and the image of the previous key frame, which can be the difference of the pixel gray degree or the sum of the pixel numbers with gray changes.

3.4   Get MMHI Image of Check-in Behavior

MMHI is used to describe the key frames of the above check-in behavior. MHI is calculating the change of pixels at the same position in a behavior sequence to obtain the recent motion situation, where the gray value of each pixel represents the pixel at that position. As shown in reference [13], the traditional MHI calculation method needs to estimate three parameters, namely duration τ, decay parameter δ and difference threshold ξ, so that the change of external environment has a great impact on MHI. When a new picture is entered, it needs to traverse all the corresponding position pixels of the two images before and after comparison to update the MHI, which is computationally expensive and time-consuming. However, MMHI only estimates the threshold value ξ of inter-frame difference, and calculates it in matrix, which reduces the complexity of calculation to a great extent, and improves the calculation speed. MMHI is shown in Fig. 4, the algorithm principle is as follows:

1) Firstly, the first key frame is binarized to get the matrix $[E(i)]$, the non-zero position pixel value is 255, initialize the motion history image $H(K(i))$, as shown in Eq. (4), where $i = 1$, $[\delta]$ is the attenuation coefficient matrix, which is calculated adaptively by the current frame number as shown in Eq. (5), the same as the key frame dimension, and Q is the total number of key frames. M' is the difference Boolean logic matrix obtained by comparing $[E(i)]$ and $[\delta]$ matrices, where the position greater than or equal to δ is set to 1, otherwise it is 0.

$$H(K(i)) = [[E(i)] - [\delta]]. * \mathbf{M'} \qquad (4)$$

$$\delta = 255 \times i/Q \qquad (5)$$

2) Then update the motion history image for the following key frames in turn, as shown in Eq. (6):

$$H(K(i+1)) = 255 \times \mathbf{N} + \mathbf{N'}. * \mathbf{M}. * [[H(K(i))] - [\delta]] \qquad (6)$$

Where, N is the difference Boolean logic matrix obtained by comparing $|[K(i+1)] - [K(i)]|$ matrix and the inter-frame difference threshold matrix $[\xi]$, where the position greater than or equal to ξ is set to 1, otherwise it is 0, which represents the part of motion between two frames, $[\xi]$ is the same as the key frame dimension, according to the experiment, ξ is 5, $255 \times N$ represents the pixel value of the motion change position is set to 255, which represents the brightest. N' is the reverse operation of N, representing the position to be attenuated where there is no change between two frames, M is the difference Boolean logic matrix obtained by comparing $[H(K(i))]$ and $[\delta]$, where the position greater than or equal to δ is set to 1, otherwise it is 0, so as to ensure that the minimum pixel value after attenuation is still 0.

Fig. 4. MMHI image

3.5 Check-in Behavior Recognition Based on Convolution Neural Network

The structure of CNN is shown in Fig. 5.

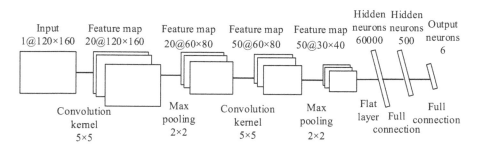

Fig. 5. Network structure

Take the MMHI image as input; Two convolution layers, the size of convolution kernel is 5×5, and the number is 20. ReLU activation function is adopted. Two pooling layers, maximum pooling, convolution kernel size is 2×2; Two full connection layers, Softmax function is used to calculate the probability of the class of checking behavior.

4 Experiment and Analysis

4.1 Experimental Environment and Data Set

The experiment is carried out under the Keras deep learning framework based on Tensorflow under Windows 10.

Data sets are established for the first time in the studies carried out. This data set consists of 800 groups of behavioral video sequences shot by 10 actors in the same scene, with a frame rate of 30 fps and a resolution of 320×240. The behavior of the experimental data set is shown in Table 1. Classes 0, 3, and 5 are normative behaviors, and the other classes are anomie behaviors. 80% of each class sample is the training set, a total of 640 samples; 20% is the test set, a total of 160 samples.

Table 1. Data set behavior table

Class	Check-in behavior
0	B_0
1	B_1
2	B_2
3	B_3
4	B_4
5	B_5

4.2 Convolution Neural Network Parameter Setting

When the input adopts 120×160 dimensions, the network structure parameters are shown in Table 2.

Table 2. Network structure parameter table

Number	Network layer	Output shape	Parameter amount
1	Convolution layer 1	(None, 160, 120, 20)	520
2	Activation layer 1	(None, 160, 120, 20)	0
3	Maximum pooling layer 1	(None, 80, 60, 20)	0
4	Convolution layer 2	(None, 80, 60, 50)	25050
5	Activation layer 2	(None, 80, 60, 50)	0
6	Maximum pooling layer 2	(None, 40, 30, 50)	0
7	Flat layer 1	(None, 60000)	0
8	Full connection layer 1	(None, 500)	30000500
9	Activation layer 3	(None, 500)	0
10	Full connection layer 2	(None, 6)	3006
11	Activation layer 4	(None, 6)	0

The total number of parameters is 30029076 and the training parameters are 30029076, batch-size set to 32; In view of the small sample size, Adam optimizer is used to achieve faster optimization speed; cross entropy loss function is used as the optimization objective function.

The key frames are extracted based on the method of uniform energy, according to the experiment, the thresholds T are 0, 230, 250 and 280 respectively. The energy distribution between frames before and after key frame extraction is shown in Fig. 6. The abscissa is the frame number and the ordinate is the number of inter-frame difference pixels.

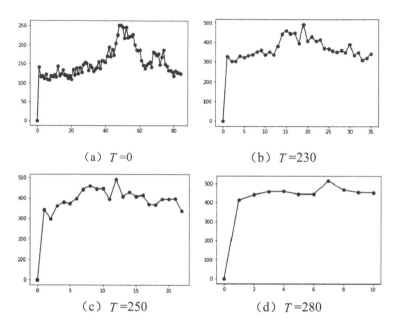

Fig. 6. Energy distribution of key frames at different thresholds

It can be seen from Fig. 6 that the algorithm can effectively uniformize the energy between frames, that is, the energy contained between two adjacent frames is approximately the same, and the larger the threshold value, the better the uniformity effect, and the fewer the number of key frames.

Table 3 shows the average number of frames for each class of behavior before and after key frame extraction. Then calculate to get the MMHI image and sent to the established neural network for training. The corresponding classification accuracy and loss function are shown in Table 4.

Table 3. Average frames of behavior before and after key frame extraction

number ╲ class T	0	1	2	3	4	5
0	92	131	137	172	150	122
230	46	72	82	104	67	64
250	32	45	51	62	39	39
280	19	19	21	24	17	16

Table 4. Accuracy and loss function of behavior test before and after key frame extraction

Threshold (T)	Training loss	Training accuracy	Test loss	Test accuracy
0	0.23	0.94	0.05	0.98
230	0.19	0.93	0.12	0.95
250	0.24	0.93	0.04	0.99
280	0.37	0.87	0.12	0.97

It can be seen from Table 4 that when the thresholds are 0 and 230, the test accuracy of check-in behavior recognition is slightly lower than that when the threshold is 250. Because the smaller the threshold is, the more the number of key frames is, the description of check-in behavior is more complete, and the recognition accuracy is relatively high, but its redundant information is also more, which has a small impact on the accuracy of check-in behavior recognition. When the threshold value is 280, the number of key frames is the least, and the relative information is less, which has a certain impact on the training and testing accuracy of check-in behavior. When the threshold value is 250, considering the recognition accuracy of training and testing, the

Fig. 7. Loss function and accuracy curve

check-in behavior recognition accuracy is the highest and the test effect is the best. The change of loss function and accuracy corresponding to testing and training is shown in Fig. 7, and the confusion matrix of behavior recognition accuracy is shown in Fig. 8.

It can be seen from Fig. 7 that through continuous iterations, the loss function of training and testing shows a downward trend, and the recognition accuracy rate shows an upward trend, and the two recognition accuracy curves gradually tend to be stable at 60 and 70 epochs, respectively, which shows that the network model has basically reached the optimal state, and the test accuracy rate can reach 98.5%.

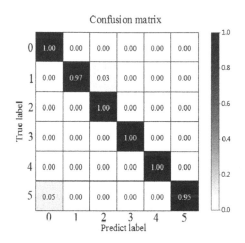

Fig. 8. Confusion matrix for behavior recognition

As can be seen from Fig. 8, the two types of normative behaviors of class 0 and class 5 are confused during recognition, because there is only one baggage on the belt conveyor as a result of these two types of behaviors, the key frames generated in the process of behavior are very similar. Because of the high similarity between these two types of check-in behaviors and the interference of different human bodies, the execution process of different types of check-in behaviors will be more similar, which eventually leads to confusion of recognition. There is also a confusion between the two types of anomie behaviors of class 1 and class 2 during the recognition, for similar reasons. The results of these two types of behaviors are two pieces of baggage on the belt conveyor, and the difference in human body interference leads to the execution process of the check-in behavior and the extracted behavior key frames are also very similar, so the final identification is confused. The differences between the check-in behavior of class 3 and class 4 is greater than the other four types, so their recognition accuracy is high and it is not easy to cause confusion. On the whole, there is no confusion in the algorithm of this paper in identifying normative behavior and anomie behavior, that is, the current network can well distinguish between normative behavior and anomie behavior. Experiments show that the trained CNN can meet the task of anomie behavior recognition in the SSBC system.

4.3 Experimental Analysis

Comparison Experiment Before and After the Improvement of Motion History Image. Compare the running speed of the improved MMHI and traditional MHI algorithms on the original data without key frame extraction, and compare the influence on training time and recognition accuracy, the speed is tested on a CPU model of I5-7500, the GPU is NVIDIA-1060, and the total training time is the sum of the calculated MHI time and the training CNN time. The results are shown in Table 5.

Table 5. Performance before and after MHI improvement

Algorithm name	Speed (fps)	Total training time (s)	Accuracy (%)
MHI	370	4070 + 41	98.5
MMHI	26430	58 + 41	98.5

From Table 5, it can be seen that the improved MMHI algorithm can increase the speed by 70 times with only CPU operation without affecting the recognition accuracy, greatly reducing the training time.

Comparison of Typical Behavior Recognition Algorithms. On the CPU system with the main frequency of 3.4 GHz, the accuracy and real-time speed of common algorithms such as SIFT-BOW-SVM [14] (scale invariant feature transform bag of visual words support vector machine), C3D [8], MiCT-Net [10] and the algorithm in this paper are compared. The experimental results are shown in Table 6.

Table 6. Average accuracy and speed

Algorithm name	Accuracy (%)	Speed (behavior/s)
MHI-SIFT-BOW-SVM	71.5	about 10
C3D	70.5	about 0.2
MiCT-Net	97.7	about 0.6
Article method	98.5	about 40

From Table 6, it can be seen that the accuracy of MHI-SIFT-BOW-SVM and C3D algorithm is low, and the processing speed is very slow. They process about 10 and 0.2 behaviors per second, respectively, which can not meet the real-time requirements at all. The MiCT-Net method, which is also a 3D-CNN framework, improves the processing of long time series problems, and the recognition accuracy reaches 97.7%, but the speed can only reach 0.6 behaviors per second, this is because the algorithm based on 3D convolution framework performs multiple convolution neural network calculations on the whole behavior sequence in units of 16 frames to get the final recognition results. However, the entire behavior of the algorithm in this paper only needs to perform convolutional neural network calculation once, and the improved MMHI calculation speed is 26430fps, which takes very little time, and the accuracy is the highest, so the algorithm in this paper has advantages in real-time.

In summary, the MMHI and CNN behavior recognition framework proposed in this paper is more suitable for the requirements of SSBC behavior recognition system.

5 Conclusion

An automatic detection method for self-service behavior is proposed to realize visual automatic detection of self-service operation behavior. Based on the lightweight CNN network, a uniform energy method is applied to extract key frame sequences from self-service videos, and the efficiency of generating motion history image is improve through a logical matrix comparison method. Taking the motion history image as the dimension reduction input of CNN, the learning and generalization application on the actual airport check-in system are carried out, which realizes the rapid classification of the self-service operation behavior, and the monitoring of the anomie behavior. The research method has practical significance for system application deployment.

References

1. Dong, H., Wang, W., Gao, Q., Luo, Q.: Transmission control strategy for self-service baggage checking system based on petri net. Control Eng. China **21**(01), 14–17 + 22 (2014)
2. Xu, G., Cao, Y.: Action recognition and activity understanding: a review. J. Image Graph. **14** (02), 189–195 (2009)
3. Ding, C.-Y., Liu, K., Li, G., Yan, L., Chen, B.-Y., Zhong, Y.-M.: Spatio-temporal weighted posture motion features for human skeleton action recognition research. Chin. J. Comput. **43** (1), 29–40 (2020)
4. Fan, Z., Zhao, X., Lin, T.: Attention-based multiview re-observation fusion network for skeletal action recognition. IEEE Trans. Multimedia **21**(2), 363–374 (2019)
5. Yang, Y., Deng, C., Gao, S., Liu, W., Tao, D., Gao, X.: Discriminative multi-instance multitask learning for 3D action recognition. IEEE Trans. Multimedia (S1520-9210) **19**(3), 519–529 (2017)
6. Kim, Y.-S., Yoon, J.-C., Lee, I.-K.: Real-time human segmentation from RGB-D video sequence based on adaptive geodesic distance computation. Multimedia Tools Appl. **78**(20), 28409–28421 (2017). https://doi.org/10.1007/s11042-017-5375-5
7. Qian, L., Li, P., Huang, L.: An adaptive algorithm for activity recognition with evolving data streams. Chin. J. Sens. Actuators **30**(6), 909–915 (2017)
8. Tran, D., Bourdev, L., Fergus, R.: Learning spatiotemporal features with 3D convolutional networks. In: Proceedings of the IEEE International Conference on Computer Vision, pp. 4489–4497 (2015)
9. Qiu, Z., Yao, T., Mei, T.: Learning spatio-temporal representation with pseudo-3D residual networks. In: Proceedings of the IEEE International Conference on Computer Vision, pp. 5533–5541 (2017)
10. Zhou, Y., Sun, X., Zha, Z.J.: Mict: Mixed 3D/2D convolutional tube for human action recognition. In: Proceedings of the IEEE Conference on Computer Vision and Pattern Recognition, pp. 449–458 (2018)
11. Wei, X., Le, Y., Han, J., Lu, Y.: Vehicle behavior dynamic recognition network based on long short-term memory. J. Comput. Appl. **39**(07), 1894–1898 (2019)

12. Shi, X., Liu, S., Zhang, D.: Human action recognition method based on key frames. J. Syst. Simul. **27**(10), 2401–2408 (2015)
13. Zhao, Q., Sun, Y.: 3D tracking algorithm for quickly locating image scales and regions. J. Image Graph. **21**(1), 114–121 (2016)
14. Xie, L., Yu, S., Zhou, G., Li, H.: Classification research of oil-tea pest images based on BOW model. J. Central South Univ. Forestry Technol. **35**(05), 70–73 (2015)

Supervised Deep Second-Order Covariance Hashing for Image Retrieval

Qian Wang[1], Yue Wu[1], Jianxin Zhang[2](✉), Hengbo Zhang[2], Chao Che[1],
and Lin Shan[3](✉)

[1] Key Lab of Advanced Design and Intelligent Computing (Ministry of Education),
Dalian University, Dalian, China
[2] School of Computer Science and Engineering, Dalian Minzu University,
Dalian, China
jxzhang0411@163.com
[3] School of Economic and Management,
Dalian University of Science and Technology, Dalian, China
linshan_dl@163.com

Abstract. Recently, deep hashing methods play a pivotal role in image retrieval tasks by combining advanced convolutional neural networks (CNNs) with efficient hashing. Meanwhile, second-order representations of deep convolutional activations have been established to effectively improve network performance in various computer vision applications. In this work, to obtain more compact hash codes, we propose a supervised deep second-order covariance hashing (SDSoCH) method by combining deep hashing with second-order statistic model. SDSoCH utilizes a powerful covariance pooling to model the second-order statistics of convolutional features, which is naturally integrated into the existing point-wise hashing network in an end-to-end manner. The embedded covariance pooling operation well captures the interaction of convolutional features and produces global feature representations with more discriminant capability, leading to the more informative hash codes. Extensive experiments conducted on two benchmarks demonstrate that the proposed SDSoCH outperforms its first-order counterparts and achieves superior retrieval performance.

Keywords: Deep hashing · Point-wise manner · Covariance pooling · Image retrieval

1 Introduction

Hashing methods have become increasingly popular in large-scale image retrieval for low storage and time demands [1–14]. With the prosperity of deep learning in various computer vision tasks, deep hashing, which integrates hash encoding and deep neural networks, has recently attracted more and more attention due to the significant performance improvement compared with traditional hashing. As a pioneer of supervised deep hashing, Xia et al. employ the convolutional

© Springer Nature Singapore Pte Ltd. 2020
J. Zeng et al. (Eds.): ICPCSEE 2020, CCIS 1257, pp. 476–487, 2020.
https://doi.org/10.1007/978-981-15-7981-3_35

neural network (CNN) as feature extractors and propose convolutional neural network hashing (CNNH) model [4]. CNNH includes two learning stages, in which the approximate hash codes learned in the first stage guide the learning of image representations in the second stage, whereas its limitation is that the learned representations from the second stage cannot help the learning of approximate hash codes. To overcome the shortcomings of non-end-to-end deep hashing, Lin et al. present deep learning of binary hash codes (DLBHC) [5] in an end-to-end manner. As a precedent for hashing methods supervised by label information, DLBHC learns hash functions using point-wise label information, and designs loss function based on semantics to preserve the similarity of the original and mapping spaces. Different with point-wise deep hashing, deep pairwise supervised hashing (DPSH) [6] simultaneously learns hash function and feature representation by using pairwise label information of two input images, while deep supervised hashing with triplet labels (DTSH) [7] performs feature representations and hash code learning through supervised label information of image triples. Besides, other methods, such as DNNH [8], deep supervised hashing (DSH) [9], deep hashing network (DHN) [10], asymmetric deep supervised hashing (ADSH) [12], deep quantization network (DQN) [11], and deep forest hashing (DFH) [13], are also put forward to promoting the development of deep hashing.

Meanwhile, recent research works have convinced the outstanding effect of modeling high-order/second-order statistics of deep convolutional features on computer vision tasks [15–21]. Representative high-order statistical models consist of deep second-order pooling (DeepO2P), bilinear CNN (B-CNN), matrix power normalized covariance pooling (MPN-COV). DeepO2P [15] introduces a learnable second-order layer into convolutional neural networks with a matrix backpropagation algorithm for end-to-end manner learning. As a counterpart, B-CNN [16] obtains second-order statistics by conducting an outer product of convolutional features extracted from two different convolutional networks. Particularly, MPN-COV proposed by Li et al. [21] well demonstrates the powerful representation ability of covariance statistics, which captures global feature statistics of images with preserving local features as well. Inspired by this, we focus on embedding global covariance statistics into deep hashing frameworks to improve the performance of deep hashing methods.

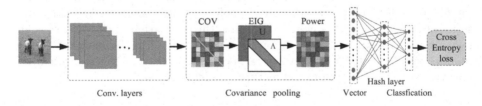

Fig. 1. The overall architecture of the proposed SDSoCH

In this work, we propose a novel supervised deep second-order covariance hashing (SDSoCH) method, whose overall architecture is shown in Fig. 1. SDSoCH utilizes matrix power normalized covariance pooling to model second-order statistics of convolutional features, which is naturally introduced into the existing point-wise hashing network in an end-to-end manner. The embedded covariance pooling operation well captures the interaction of convolutional features and produces global feature representations, which could lead to the more informative hash codes. We further perform compared experiments on the CIFAR-10 and NUS-WIDE datasets to evaluate its effectiveness. Experiment results demonstrate that SDSoCH outperforms its first-order counterparts and achieves superior retrieval performance.

2   Methods

2.1   Network Architecture

As illustrated in Fig. 1, our SDSoCH mainly includes four parts, i.e., the basic feature network for extracting first-order deep features, the high-order pooling module for estimating global covariance statistics, the hash layer to realize hash activation and encoding, and the final classification layer to implement label supervision and network optimization through cross-entropy constraints. Passing through these four parts, images can be converted into unified global and detailed hashing representations after the forward and backward propagation.

Given an image $x_i \in X = [x_1, x_2, ..., x_N], \{i = 1, 2, ..., N\}$ with its corresponding label $y_i \in Y = [y_1, y_2, ..., y_M]$ as the input of CNN-F [22] network, which is adopted to learn the first-order deep features and can be replaced by other networks such as [23,24], the input image is mapped layer by layer to obtain its deep feature representation. Assume that the final output of the basic feature network is recorded by Z which can be denoted as

$$Z = \Phi(X; \theta), \tag{1}$$

where Φ and θ denote the mapping relationship and parameters of network, respectively. Then, we set Z as the outputted first-order feature, which tends to ignore correlation among different activations.

Motivated by [21], we try to embed the matrix manifold structure into the existing hashing architectures using covariance pooling. Specifically, after obtaining the basic feature Z, the covariance pooling module is inserted to perform a series of matrix operations, which maintains the correlation of the feature channels, determining the feature distribution to enhance the performance of feature descriptors. We firstly construct a sample covariance matrix M to represent the pairwise interaction of the initial features, which is defined as:

$$M = Z\bar{I}Z^T. \tag{2}$$

Here, $\bar{I} = \frac{1}{n}\left(I - \frac{1}{n}JJ^T\right)$, where I represents the identity matrix of $n \times n$ and J is an n-dimension column vector whose elements are set to 1. The superscript T indicates the transpose matrix.

Although the sample covariance is obtained by the classic maximum likelihood estimation (MLE) estimation, its overall robustness is weak with high dimensions but few samples, making the retrieval performance unsatisfactory. However, the competitive MPN can simultaneously achieve covariance estimation and maintain manifold space when dealing with unfriendly situations. Due to the property of symmetric positive semi-definite of the sample covariance matrix, eigenvalue decomposition can be performed on M:

$$\mathrm{M} = U\Lambda U^T \tag{3}$$

$\Lambda = diag\,(\lambda_1, ..., \lambda_d)$ indicates a diagonal matrix composed of eigenvalues, and U represents an orthogonal matrix composed of eigenvectors.

Relying on the manifold structure of the covariance matrix M, we then establish robust global feature statistics as

$$P \triangleq M^m = U F\,(\Lambda)\,U^T. \tag{4}$$

In Eq. (4), $F\,(\Lambda) = diag\,(\lambda_1^m, ..., \lambda_d^m)$, and m is a positive real number ranging from 0 to 1 which generally sets to 0.5. Considering that Λ is an orthogonal matrix and λ_i denotes its eigenvalue, we can apply the eigenvalues to the global estimation to simplify matrix power depending on the eigenvalue decomposition [15] and diagonal matrix principles, and determine a robust covariance estimation by utilizing the Riemannian characteristic of the covariance structure. Different from the traditional first-order pooling method, this layer aims to obtain high-order features P as global feature representations, which can be written as

$$P \triangleq U\Lambda^m U^T. \tag{5}$$

In addition, we also perform norm $(M - \ell_2)$ or $(M - Fro)$ to normalize matrix elements after the power on the matrix as [21].

Then, the hash layer devotes to convert deep covariance features into a group of hash codes by adding a hidden hash layer and an activation layer to implement activation of codes [25], jointly transforming the local descriptor based deep covariance representations to codable features. This process can be expressed by

$$H = W_H^T P + \sigma, \tag{6}$$

where W_H^T represents the connection weights and σ is a parameter vector. Notably, we apply different activation approaches during training and testing. During the training, the optimization of loss function relies on continuous outputs of the hash hidden layer, thus the representations should be relaxed through the Sigmoid activation function. However, in the testing process, we perform constraint on the continuous outputs using a threshold function shown in Eq. (7) to get the final binary hash code.

$$b_i = sign\,(h_i) = \begin{cases} 1, & h_i > 0 \\ -1, & h_i \le 0 \end{cases} \tag{7}$$

2.2   Loss Function and Optimization

Given the hashing representation H of the network, we can utilize classification layer to predict the label as

$$C = W_f^T H + \tau, \tag{8}$$

where W_f^T indicates the weight matrix of the classification layer, and τ is the bias vector.

   To optimize hash codes, we adopt the Softmax and the Sigmoid based cross-entropy loss to perform the training of single-label datasets and multi-label datasets, respectively. The optimization is to determine the hash function of multi-classification tasks and multi-label tasks, maintaining the image retrieval performance. The loss function is written as

$$L = -\frac{1}{N} \sum_{i=1}^{N} y_i \log \tilde{y}_i, \tag{9}$$

where N denotes the quantity of images. y_i and $\tilde{y}_i$ represent the true label and the label predicted by the network, respectively. Notably, the input value of cross-entropy constraints depend on the image classification tasks, which should be written as Softmax function $\tilde{y}_i = e^{h_i} / \sum_{j=1}^{m} e^{h_j}$ for multi-classification tasks and Sigmoid function $\tilde{y}_i = 1/\left(1 + e^{-h_i}\right)$ for multi-label tasks.

   According to the loss function, for reducing the difference between the predicted label and the real label, we employ stochastic gradient descent (SGD) algorithm to realize the back-propagation, and solve partial derivatives for each parameter through the chain rule. Here, we can obtain the gradient relation between the final loss function and the approximate hash code through Eq. (8):

$$\frac{\partial L}{\partial H} = -\sum_{i=1}^{N} \left(\frac{\partial L}{\partial \tilde{y}_i} \cdot \frac{\partial \tilde{y}_i}{\partial c_i} \cdot \frac{\partial c_i}{\partial h_i} \right) = -\sum_{i=N}^{K} \frac{y_i}{\tilde{y}_i} \cdot \frac{\partial \tilde{y}_i}{\partial c_i} \cdot \frac{\partial c_i}{\partial h_i}, \tag{10}$$

   It is worth noting that the above formula can be rewritten to the following two forms as for corresponding value of $\tilde{y}_i$ used for different tasks:

$$\frac{\partial L}{\partial H} = \begin{cases} \sum_{i=1}^{N} (\tilde{y}_i - y_i)\frac{\partial c_i}{\partial h_i}, multi - classification \\ \sum_{i=1}^{N} y_i(1 - \tilde{y}_i)\frac{\partial c_i}{\partial h_i}, multi - label \end{cases} \tag{11}$$

   On the basis of introducing covariance estimation into hash codes illustrated in Eq. (5) and Eq. (6), the gradient representation of hash codes to covariance features can be obtained through matrix back-propagation and intermediate variables generated during the eigenvalue decomposition process, which is expressed as below:

$$\frac{\partial L}{\partial P} = tr\left(\left(\frac{\partial H}{\partial U}\right)^T \mathrm{d}U + \left(\frac{\partial H}{\partial \Lambda}\right)^T \mathrm{d}\Lambda\right) \tag{12}$$

Finally, depending on Eq. (1) and Eq. (2), the back-propagation from covariance representations to basic features Z can be obtained:

$$\frac{\partial P}{\partial Z} = \overline{I} \cdot Z\left(\frac{\partial P}{\partial M} + \left(\frac{\partial P}{\partial M}\right)^T\right) \tag{13}$$

In the entire training process, parameters of the network are iteratively updated according to the BP algorithm to realize the network optimization.

3 Experiments and Results

In this section, we firstly introduce the benchmarks adopted for performance evaluation. Then, a brief description of evaluation metrics is given. Finally, compared experiment results on the two datasets are reported and discussed. All experiments are carried out on a PC equipped with 3.30 GHz CPU, 64 GB RAM, and NVIDIA GTX 1080 GPU using MatLab R2014b.

3.1 Datasets

Two commonly used datasets, i.e., CIFAR-10 [26] and NUS-WIDE [27] are utilized in this work for the performance evaluation of SDSoCH. CIFAR-10 consists of 60,000 32×32 color images even from ten categories. For fair comparisons, the dataset is divided into training set containing 50,000 images and test set containing 10,000 images following [5]. NUS-WIDE is a multi-label dataset containing nearly 270k images belonging to one or more of the 81 categories. Following [6], we select images associated with the 21 most frequent concepts, each of which associates with at least 5,000 images, where 500 images are selected for training set and 100 images for test set. The training set is utilized to learn network parameters, whereas the test set aims at performance judgment at the test stage. Depending on the ground truth of each image, whether two sample points are true neighbors can be determined by checking whether they share at least one category label.

3.2 Evaluation Metrics

We employ mean average precision (mAP), top-k accuracy and precision-recall rate (PR) to evaluate the performance of various hashing methods. Top-k accuracy is a ranking based criterion for image retrieval evaluation, calculating the percentage of related images in the first k retrieved images as

$$P@k = I_r/K, \tag{14}$$

where I_r and K denote the number of ground truth relevance and all samples returned, respectively. PR rate can be reported as a curve measuring the proportion of true positive samples in the recall samples and true positive samples among all positive samples. Based on the average precision (AP), mAP can be regarded as the area below the PR curve, which is defined as follows.

$$AP(X_q) = \frac{1}{I_r} \sum_{k=1}^{N} P@k \cdot \psi(K), \tag{15}$$

$$mAP = \frac{1}{M} \sum_{q=1}^{M} AP(X_q), \tag{16}$$

where X_q indicates a query image, $\psi(K)$ is an indicator function, and M represents the number of query set. $\psi(K) = 1$ if the retrieval image has the same label with X_q; otherwise, $\psi(K) = 0$.

3.3   Experimental Results on CIFAR-10 Dataset

We firstly carry out experiments on CIFAR-10 dataset to evaluate the retrieval performance of SDSoCH. We also compare SDSoCH with six typical deep hashing methods including CNNH, DFH, DPSH, etc. The compared mAP results with four various hashing code bits are listed in Table 1. As shown in this table, SDSoCH gains its optimal mAP values of 0.741, 0.756, 0.771 and 0.779 on 12, 24, 32, and 48 encoding bits, respectively. Meanwhile, it outperforms the other six deep hashing models on all of the four bits. Compared with its first-order counterpart of DLBHC, SDSoCH obtains significant gains of 6.7%, 5.2%, 7.0%, and 7.2% on 12, 24, 32, and 48 encoding bits, respectively. Though DPSH utilizes the pair-wise label to optimize hash encoding, SDSoCH is still superior to DPSH by 2.65% gains averagely. The superiority illustrates the effectiveness of SDSoCH for image retrieval task, which can be mainly contributed to the robustness of feature interactions causing by the covariance pooling.

Table 1. Compared mAP results on CIFAR-10 dataset.

Methods	12-bit	24-bit	32-bit	48-bit
CNNH [4]	0.439	0.511	0.509	0.522
DFH [13]	0.457	0.513	0.524	0.559
DNNH [8]	0.552	0.566	0.558	0.581
DLBHC [5]	0.674	0.704	0.701	0.707
DHN [10]	0.681	0.721	0.723	0.733
DPSH [6]	0.713	0.727	0.744	0.757
SDHoCH (Ours)	**0.741**	**0.756**	**0.771**	**0.779**

Besides, to further demonstrate the validity of embedding second-order covariance statistics into deep hashing network, we exhibit compared Top-k accuracy curves and PR curves of five deep hashing methods using 48-bit hash code in Fig. 2 and Fig. 3. As shown in Fig. 2, the retrieval performance of hashing methods gets improved continuously with the increasing number of retrieved images. SDSoCH achieves obvious performance improvement over CNNH, DNNH and DLBHC. Although it is slightly worse than DPSH when the number of retrieved images is 100, it is superior to DPSH on all of the other cases. Observing the compared PR curves, it is clearly that SDSoCH shows consistent performance advantages over other counterparts. Therefore, compared Top-k accuracy curve and PR curve results also effectively demonstrate the superiority of embedding second-order covariance statistics into deep hashing for image retrieval.

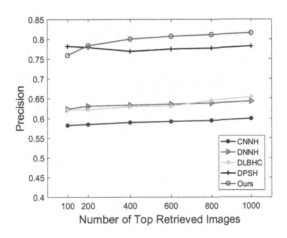

Fig. 2. Compared top-k accuracy results on CIFAR-10 (48-bit hashing code)

3.4 Experimental Results on NUS-WIDE Dataset

The comparison of retrieval mAPs of SDSoCH with other deep hashing methods on the multi-label NUS-WIDE dataset is reported in Table 2. It can be found that SDSoCH achieves the best mAP results of 0.761, 0.793, 0.796 and 0.807 on 12, 24, 32, and 48 encoding bits, respectively. Compared with the first-order baseline of DLBHC, SDSoCH shows a 6.0%, 5.4%, 4.9%, and 5.2% improvement on 12, 24, 32, and 48 encoding bits, respectively. It outperforms other deep hash methods in most cases except for a slightly lower performance than DPSH [6] on 48-bit code. It should be noted that the main reason for this inferiority is the dual-input architecture of DPSH using pair-wise labels for similarity preserving. In general, SDSoCH shows advantages in enhancing image retrieval performance over these first-order deep hashing methods.

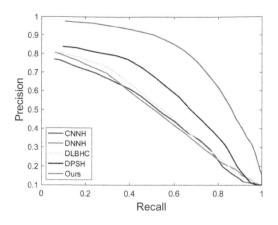

Fig. 3. Compared PR curves on CIFAR-10 (48-bit hashing code)

Table 2. Compared mAP results on NUS-WIDE dataset.

Methods	12-bit	24-bit	32-bit	48-bit
CNNH [4]	0.611	0.618	0.625	0.608
DFH [13]	0.622	0.659	0.674	0.695
DLBHC* [5]	0.701	0.739	0.747	0.755
DQN [11]	0.768	0.776	0.783	0.792
DPSH* [6]	0.752	0.790	0.794	**0.812**
SoPDH (Ours)	**0.761**	**0.793**	**0.796**	0.807

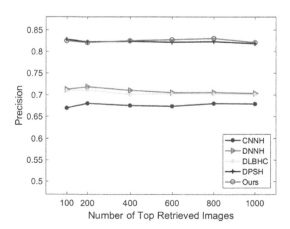

Fig. 4. Compared top-k accuracy results on NUS-WIDE (48-bit hashing code)

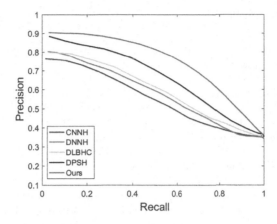

Fig. 5. Compared PR curves on NUS-WIDE (48-bit hashing code)

Similarly, for verifying the effectiveness of SDSoCH on the multi-label dataset, we compare and report the Top-k accuracy curve and Precision-Recall curve using 48-bit code. The fluctuation of Top-k accuracy curve is shown in Fig. 4, it can be concluded that our SDSoCH is comparable to DPSH for the stability advantage of pairwise similarity preservation. Meanwhile, it is much better than other deep hash methods, especially for the significant enforcement to the baseline DLBHC, illustrating the superior performance of SDSoCH in image retrieval tasks. Likewise, the Precision-Recall rates of different deep hashing methods shown in Fig. 5 are broadly consistent with the aforementioned comparisons, demonstrating the obvious advantage in SDSoCH over other methods. In summary, the point-wise deep hashing networks embedded into high-order representations of covariance pooling can achieve superior performance in image retrieval applications.

4 Conclusions

In this paper, we present a second-order deep supervised hashing method named SDSoCH for image retrieval tasks. SDSoCH generates more discriminant hashing codes by embedding the second-order covariance statistics into existing deep hashing frameworks based on point-wise label learning. The experimental results on two benchmarks demonstrate the effectiveness of SDSoCH as well. Our future work will evaluate SDSoCH on more image datasets or for other visual tasks. In addition, we will also explore more powerful high-order statistical information and perform them on various deep hashing frameworks.

Acknowledgements. This work was partially supported by the National Key R&D Program of China (2018YFC0910506), the National Natural Science Foundation of China (61972062), the Natural Science Foundation of Liaoning Province (2019-MS-011), the Key R&D Program of Liaoning Province (2019 JH2/10100030) and the Liaoning BaiQianWan Talents Program.

References

1. Gong, Y., Lazebnik, S., Gordo, A., et al.: Iterative quantization: a procrustean approach to learning binary codes. In: IEEE Conference on Computer Vision and Pattern Recognition (CVPR), pp. 817–824 (2011)
2. Weiss, Y., Torralba, A., Fergus, R.: Spectral hashing. In: Advances in Neural Information Processing Systems (NIPS), pp. 1753–1760 (2009)
3. Shen, F., Shen, C., Liu, W., et al.: Supervised discrete hashing. In: IEEE Conference on Computer Vision and Pattern Recognition (CVPR), pp. 37–45 (2015)
4. Xia, R., Pan, Y., Lai, H., et al.: Supervised hashing for image retrieval via image representation learning. In: Twenty-Eighth AAAI Conference on Artificial Intelligence (AAAI), pp. 2156–2162 (2014)
5. Lin, K., Yang, H.F., Hsiao, J.H., et al.: Deep learning of binary hash codes for fast image retrieval. In: IEEE Conference on Computer Vision and Pattern Recognition Workshops (CVPRW), pp. 27–35 (2015)
6. Li, W.J., Wang, S., Kang W.C.: Feature learning based deep supervised hashing with pairwise labels. In: The Twenty-Fifth International Joint Conference on Artificial Intelligence (IJCAI), pp. 1711–1717 (2016)
7. Wang, X.F., Shi, Y., Kitani, K.M.: Deep supervised hashing with triplet labels. In: Asian Conference on Computer Vision (ACCV-16), pp. 70–84 (2016)
8. Lai, H., Pan, Y., Liu, Y., et al.: Simultaneous feature learning and hash coding with deep neural networks. In: IEEE Conference on Computer Vision and Pattern Recognition (CVPR), pp. 3270–3278 (2015)
9. Liu, H.M., Wang, R.P., Shan, S.G., et al.: Deep supervised hashing for fast image retrieval. In: Proceedings of the IEEE Conference on Computer Vision and Pattern Recognition (CVPR), pp. 2064–2072 (2016)
10. Zhu, H., Long, M.S., Wang, J.M., et al.: Deep hashing network for efficient similarity retrieval. In: The Thirtieth AAAI Conference on Artificial Intelligence (AAAI), pp. 2415–2421 (2016)
11. Cao, Y., Long, M.S., Wang, J.M., Wen, Q., et al.: Deep Quantization Network for efficient image retrieval. In: Thirtieth AAAI Conference on Artificial Intelligence (AAAI), pp. 3457–3463 (2016)
12. Jiang, Q.Y., Li, W.J.: Asymmetric deep supervised hashing. In: Thirty-Second AAAI Conference on Artificial Intelligence (AAAI), pp. 3342–3349 (2018)
13. Zhou, M., Zeng, X.H., Chen, A.Z.: Deep forest hashing for image retrieval. Pattern Recogn. **95**, 114–127 (2019)
14. Cao, Y., Liu, B., Long, M., et al.: HashGAN: deep learning to hash with pair conditional Wasserstein GAN. In: IEEE Conference on Computer Vision and Pattern Recognition (CVPR), pp. 1287–1296 (2018)
15. Ionescu, C., Vantzos, O., Sminchisescu, C.: Matrix backpropagation for deep networks with structured layers. In: IEEE International Conference on Computer Vision (ICCV), pp. 2965–2973 (2015)
16. Lin, T.Y., RoyChowdhury, A., Maji, S.: Bilinear CNN models for fine-grained visual recognition. In: IEEE international Conference on Computer Vision (ICCV), pp. 1449–1457 (2015)
17. Valmadre, J., Bertinetto, L., Henriques, J., et al.: End-to-end representation learning for correlation filter based tracking. In: IEEE Conference on Computer Vision and Pattern Recognition (CVPR), pp. 2805–2813 (2017)

18. Arnab, A., Jayasumana, S., Zheng, S., Torr, P.H.S.: Higher order conditional random fields in deep neural networks. In: Leibe, B., Matas, J., Sebe, N., Welling, M. (eds.) ECCV 2016. LNCS, vol. 9906, pp. 524–540. Springer, Cham (2016). https:// doi.org/10.1007/978-3-319-46475-6_33

19. Sun, Q.L., Wang, Q.L., Zhang, J.X., et al.: Hyperlayer bilinear pooling with application to fine-grained categorization and image retrieval. Neurocomputing **282**, 174–183 (2018)

20. Wu, Y., Sun, Q.L., Hou, Y.Q., et al.: Deep covariance estimation hashing. IEEE Access **7**, 113223–113234 (2019)

21. Li, P.H., Xie, J.T., Wang, Q.L., et al.: Is second-order information helpful for large-scale visual recognition? In: IEEE International Conference on Computer Vision (ICCV), pp. 2089–2097 (2017)

22. Chatfield, K., Simonyan, K., Vedaldi, A., et al.: Return of the devil in the details: delving deep into convolutional net. In: British Machine Vision Conference (BMVC) (2014)

23. Krizhevsky, A., Sutskever, I., Hinton, G.E.: ImageNet classification with deep convolutional neural networks. In: Advances in Neural Information Processing Systems (NIPS), pp. 1097–1105 (2012)

24. He, K.M., Zhang, X.Y., Ren, Q.S., et al. Deep residual learning for image recognition. In: IEEE Conference on Computer Vision and Pattern Recognition (CVPR), pp. 770–778 (2016)

25. Yang, H.F., Lin, K., Chen, C.S.: Supervised learning of semantics-preserving hash via deep convolutional neural networks. IEEE Trans. Pattern Anal. Mach. Intell. **40**(2), 437–451 (2018)

26. Krizhevsky, A., Hinton, G.: Learning multiple layers of features from tiny images. Technical report, May 2009

27. Chua, T.-S., Tang, J., Hong, R., et al.: NUS-WIDE: a real-world web image database from National University of Singapore. In: International Conference on Image and Video Retrieval (CIVR), pp. 368–375 (2009)

Using GMOSTNet for Tree Detection Under Complex Illumination and Morphological Occlusion

Zheng Qian[1,2], Hailin Feng[1,3(✉)], Yinhui Yang[1,2], Xiaochen Du[1,3],
and Kai Xia[1,3]

[1] School of Information Engineering, Zhejiang A&F University,
Hangzhou, China
hlfeng@zafu.edu.cn
[2] Key Laboratory of Forestry Intelligent Monitoring and Information
Technology of Zhejiang Province, Hangzhou, China
[3] Key Laboratory of State Forestry and Grassland Administration on Forestry
Sensing Technology and Intelligent Equipment, Hangzhou, China

Abstract. Trees are an integral part of the forestry ecosystem. In forestry work, the precise acquisition of tree morphological parameters and attributes is affected by complex illumination and tree morphology. In order to minimize a series of inestimable problems, such as yield reduction, ecological damage, and destruction, caused by inaccurate acquisition of tree location information, this paper proposes a ground tree detection method GMOSTNet. Based on the four types of tree species in the GMOST dataset and Faster R-CNN, it extracted the features of the trees, generate candidate regions, classification, and other operations. By reducing the influence of illumination and occlusion factors during experimentation, more detailed information of the input image was obtained. Meanwhile, regarding false detections caused by inappropriate approximations, the deviation and proximity of the proposal were adjusted. The experimental results showed that the AP value of the four tree species is improved after using GMOSTNet, and the overall accuracy increases from the original 87.25% to 93.25%.

Keywords: Tree detection · Illumination · Occlusion · Deep learning

1 Introduction

Trees are invaluable to the global environment and human life. By mapping and characterization of individual trees to realize the work of collecting relevant tree composition and biomass information. It is necessary to achieve sustainable management of trees. However, the shape and state of trees are constantly changing. They are vulnerable to insects and other creatures during their growth process, which leads to abnormalities or even death. The collection of information of tree species and their locations is key to using a variety of methods to evaluate the health status of trees and protect trees [1, 2]. In the past, airborne laser scanners (ALS) and mobile laser scanning (MLS) data have been widely used in the calculation of morphological parameters of

J. Zeng et al. (Eds.): ICPCSEE 2020, CCIS 1257, pp. 488–505, 2020.
https://doi.org/10.1007/978-981-15-7981-3_36

trees [3–6]. With the widespread application of a neural network UAV images [7–9], remote sensing data, and other methods [10–12] are used to detect trees in combination with the neural network [13–15].

Under complex site environments, the illumination and the trees' morphology can significantly affect the detection of smaller trees [16]. In individual tree detection, high accuracy can be ensured for trees with visible tops, while the trees in the lower canopy cannot be detected reliably [17]. Under complex site environment, the accuracy of high-altitude detection primarily depends on the way trees are arranged in space [18, 19]. A lot of woodlands around the world, especially public lands, are subject to structural diversity management in terms of stand and landscape scale. To sum up, traditional tree detection methods are inefficient and limited by large equipment which is difficult to carry and expensive. With the widespread application of deep learning, the efficiency of forestry operations have been improved [20, 21]. However, with the influence of the characteristics of trees and the complex site environment, the accuracy and applicability of tree detection need to be improved. Considering that tree detection is affected by the morphological characteristics of trees and the complex site environment, this article proposes a GMOSTNet-based ground tree detection method to solve problems caused by poor illumination and tree morphology and achieve sustainable management of trees.

2 Material and Methods

2.1 Study Area and Data Collection

Currently, most datasets are leaf datasets and image datasets for plant recognition published online include PlantNet, Flavia, and Leafsnap. There are also overall image datasets for trees that haven't been published yet. Therefore, this article uses the combination of autonomous photographs and online crawling to construct the GMOST dataset (see Fig. 1), with four tree species: Ginkgo Biloba, Maple, Osmanthus Fragrans, and Sapindus Mukorossi, as the research objects. The study area of this article is located in Donghu Campus of Zhejiang Agriculture and Forestry University, Lin'an District, Hangzhou City of Zhejiang Province (see Fig. 2). A total of 1,533 pictures were collected, including 610 pictures that were taken autonomously and 923 pictures that were crawled online.

Fig. 1. GMOST dataset

Fig. 2. Autonomously photographed areas and tree species (Autonomously photographed areas: Donghu Campus of Zhejiang Agriculture and Forestry University, tree species: Ginkgo biloba, Maple, Osmanthus fragrans and Sapindus mukorossi)

In order to improve the classification accuracy of the network and prevent over-fitting, this article uses diagonal flip, affine transformation, color adjustment, partial cropping and fuzzy processing to augment the data, as shown in Table 1 (using Ginkgo Biloba as an example).

Table 1. The example of data augmentation

	Original	Diagonal Flip	Fuzzy Processing	Color Adjustment	Partial Cropping	Affine Transformation
Ginkgo biloba						
Maple						
Osmanthus fragrans						
Sapindus mukorossi						

2.2 Methods

As mentioned above, this paper proposes a method (as shown in Fig. 3). The main goal of the GMOSTNet is to reduce the effects of illumination and occlusion in tree detection.

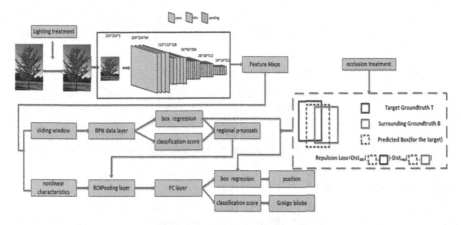

Fig. 3. GMOSTNet method

Treatment of Illumination in Tree Detection

There are obvious differences in the illumination of the tree images collected in the GMOST dataset. In order to reduce the impact of illumination conditions on the acquisition of specific location information of the trees, these images need to be processed. From the tree images in GMOST dataset, we find some pixels always have at least one color channel with a very low value in most of the non-sky local areas. In other words, the minimum value of the illumination intensity in this area is a small number, which can be expressed by:

$$J^{dark}(x) = min_{y \in \Omega(x)} \left(min_{c \in \{r,g,b\}} J^C(y) \right) \quad (1)$$

where J^C represents each channel of the color images, x represents pixeland $\Omega(x)$ represents a window centered on the pixel x. The minimum RGB component of each pixel is calculated for the input tree images, and the result is stored in a grayscale image of the same size as the original input tree image. And then the minimum filter is conducted on the grayscale image. The filter radius is determined by window size. In general, the window size equal to (2 * Radius + 1). With J^{dark} tending toward 0, the image after illumination treatment can be expressed using Eq. (2):

$$\frac{I^c(x)}{A^c} = t(x)\frac{J^c(x)}{A^c} + 1 - t(x) \quad (2)$$

Among them, I(x) is the image to be processed, J(x) is the image we want to recover, A is the global atmospheric illumination component, t(x) is the transmittance, and superscript C is the three channels R/G/B. The same object may have differences at different angles, so a factor between [0, 1] is introduced to t(x). As J^{dark} tends toward 0, we can obtain:

$$t(x) = 1 - \omega min_{y \in \Omega(x)} \left(min_c \frac{I^C(y)}{A^C} \right) \tag{3}$$

The final processing result is as follows:

$$J(x) = \frac{I(X) - A}{\max(t(x), t_0)} + A \tag{4}$$

By processing the tree image, an image suitable for subsequent processing is obtained.

Object Detection Network

The purpose of GMOSTNet is to frame the specific position of trees in the image of trees so as to facilitate the calculation of attribute information in subsequent forestry work. Considering the particularity of the detected object in the experiment, this paper adopts Faster R-CNN.

After illumination treatment, the tree images were sent to Faster R-CNN for feature extraction, candidate region generation, and tree detection classification. The input tree images were resized to 224 × 224 pixels before entering the feature extraction network. If the image was smaller than 224 × 224, the edge was complemented by 0. Then the convolution feature was extracted, and the process could be described as,

$$x_i = act(x_{i-1} \otimes k_i + b_i) \tag{5}$$

Where x_i is the characteristic map of the ith layer of the convolutional neural network, k_i is the ith layer convolution kernel, b_i is the offset vector of the ith layer, $\otimes$ is the convolution operator symbol, and act($\cdot$) is the activation function.

In this article, the VGG16 network and ReLU activation functions were adopted. In the whole Conv layers, after the tree image passed through four pooling layers, the feature map became smaller in space, yet deeper in depth. In order to obtain candidate region blocks, the tree feature maps were input in the region proposal network (RPN), and a 3 × 3 sliding window was used to convolve the feature maps to obtain 256-dimensional feature vectors. The proportion of candidate frames generated by the anchor of the pixel points on the feature map was adjusted. For the area score of the rectangular candidate frames, the non-maximal suppression algorithm was adopted to output the first N area proposals. It was input to the RoI pooling layer, through which regional recommendation features were obtained. Finally, it was input to the full connect layer, and then the classification scores and locations of areas were got. RoI Pooling is responsible for collecting proposal, calculating proposal feature maps, and sending them to the subsequent network. Through the ROI pooling layer, the boundary of the candidate frame was quantized into integer point coordinate values for segmentation and quantization.

During the classification part, the proposal feature map obtained ran through the full connect layer and softmax to calculate each proposal, classify them, and finally to perform bounding box regression on the proposals to obtain the rectangular box with a higher precision.

Treatment of Occlusion in Tree Detection

During the whole process of feature extraction, candidate region generation, and tree detection classification by Faster R-CNN, it can be found that there is an obvious occlusion problem due to the impact of the morphological particularity of trees, which also needs to be addressed. In the experiment, multi-task loss was followed in training and the objective function was minimized. The loss function is defined as:

$$L(\{p_i\}, \{t_i\}) = \frac{1}{N_{cls}} \sum_i L_{cls}(p_i, p_i^*) + \lambda \frac{1}{N_{reg}} \sum_i p_i^* L_{reg}(t_i, t_i^*) \tag{6}$$

Where N is the total amount of anchors; p_i is the probability that the anchor is predicted to be the target; GT label: when the anchor is a positive sample, $p_i^* = 1$, when the anchor is a negative sample, $p_i^* = 0$; $t_i = \{t_x, t_y, t_w, t_h\}$ is a vector representing the four parameterized coordinates of the bounding box bound by the prediction; t_i^* is the coordinate vector of the Ground Truth bounding box corresponding to the positive anchor; $L_{cls}(p_i, p_i^*)$ is the logarithmic loss of two categories (target vs non-target), $L_{reg}(t_i, t_i^*)$ is the regression loss.

$$L_{cls}(p_i, p_i^*) = -\log[p_i^* p_i + (1 - p_i^*)(1 - p_i)] \tag{7}$$

$$L_{reg}(t_i, t_i^*) = smooth_{L1}(t_i - t_i^*) \tag{8}$$

$$smooth_{L1}(x) = \begin{cases} 0.5x^2 & if \ |x| < 1 \\ |x| - 0.5 & otherwise \end{cases} \tag{9}$$

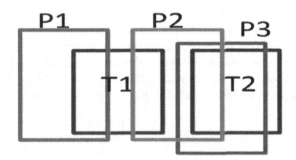

Fig. 4. Schematic diagram of P and T in case of false recall and missed detection (T1, T2 are Groundtruth boxes, P1, P2, P3 are anchors)

In order for the predicted object window to approach the Ground Truth window, the original proposal was mapped through regression or fine-tuning to obtain a regression window closer to the target Ground Truth. When the difference between the input proposal and Ground Truth is small, linear regression was used to fine-tune the window to remove redundant proposals, as well as to consider any false recalls and missed detections (Fig. 4). In order to suppress such false detections, non-maximum suppression (NMS) was used for prediction screening mechanism. When the IoU was less than 0.3, greater than 0.7, or when the overlap ratio exceeded its maximum, the anchor was set to be a negative sample, and IoU is defined as

$$IoU = \frac{A \cap B}{A \cup B} \tag{10}$$

In view of this problem, referring to [22], the inaccuracy of the tree occlusion and the leak recall were processed through Eq. 11. Considering the case where the proposal was shifted to other targets or approached to the proposals corresponding to other targets, misdetection caused by leak recall due to approximation and unsuitable approximation was suppressed. L_{Attr} used the SmoothL1 distance metric to specify the degree of approximation between the candidate frame and the target frame corresponding to the anchor. It needed to consider all anchors for each image. L_{RepGT} used the SmoothL1 distance to measure the ratio of the intersection of the candidate frame corresponding to the anchor and the target frame to the area of the candidate frame corresponding to the anchor. L_{RepBox} was used to process the candidate boxes corresponding to different targets that were too close to miss recall. The candidate boxes of different targets were separated. The smaller the IoU between the candidate boxes, the better, and the candidate boxes corresponding to different targets were separated. The detection results became more robust than the same gained through the NMS algorithm.

$$L = L_{Attr} + \alpha * L_{RepGT} + \beta * L_{RepBox} \tag{11}$$

Evaluation Indicator

In this article, the mAP (mean Average Precision) is the average value of each category accuracy, and was selected as the performance evaluation model. When an IoU threshold was set, all GT and DT were classified by category. Calculating the GTs and DTs of the same class resulted in a single performance (i.e., AP), and the performance of the four species of trees studied in this article was averaged (mAP), which was the performance under the IoU threshold. The formulas were as follows:

$$\text{Recall} = \frac{TP}{TP + TN} \tag{12}$$

$$\text{Precision} = \frac{TP}{TP + FP} \tag{13}$$

$$\text{AP} = \int_0^1 P dR \tag{14}$$

$$\text{mAP} = \frac{1}{|Q_R|} \sum_{q \in Q_R} AP(q) \tag{15}$$

3 Results

3.1 Evaluate the Training Process

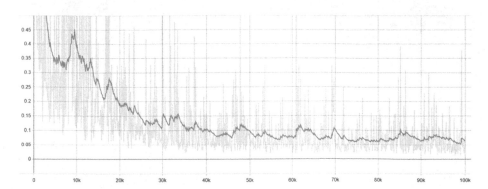

Fig. 5. The loss map diagram for 100,000 iterations

The neural network adopted an end-to-end training method (simultaneous training of RPN and the classification network). The GMOST dataset was sent to the neural network for training in batches. The number of training iterations was 100,000 and the sum of the loss values is shown in Fig. 5. From Fig. 5, there was a strong loss fluctuation at around the 10,000 times training because of the instability in the process of finding the anchor position. The network encountered some inaccurate bounding boxes, which made the effective features of the trees unable to be learned, resulting in strong fluctuation of the regression loss function values. Despite this phenomenon, as the number of iterations increased, the loss function value and the fluctuation intensity of the regression loss function gradually decreased, indicating that the training results converged better. The local fluctuation of the curve was caused by repeatedly confirming the size of the bounding box to accurately depict the tree from the depth image, yet the overall descending trend indicated the training still had better convergence results.

3.2 Summary of Test Results of Four Tree Species

Fig. 6. Effect diagram of single-tree

Fig. 7. Effect diagram of multiple-tree

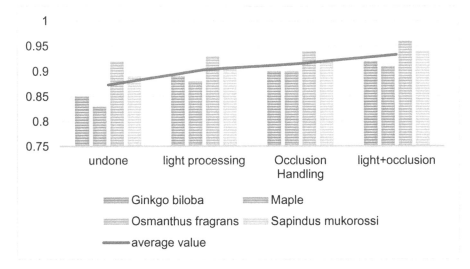

Fig. 8. Comparison of detection accuracy of four types of tree species

The accurate identification and framing of tree images is conducive to the acquisition of specific information about tree attributes and the advancement of subsequent forestry work. Illumination differences and morphological factors affect the work to varying degrees. In this article, four tree species were selected to evaluate the applicability of GMOSTNet. Based on the GMOST dataset, the overall results of single and multiple tree detection were good. They can accurately identify tree species and mark specific locations of trees after illumination and occlusion treatment (Fig. 6, Fig. 7). In order to reduce the influence of illumination on experiments, contrast was augmented for the over-dark and over-bright parts and image quality was improved to a certain degree. The mAP of the four tree species increased from 87.25% to 90.25% (Fig. 8). For tree occlusion problem, the focus was on controlling the interference of one side GT on the prediction box and constraining the predicted proposal, as a result the detection effect of the experiment was more robust. After occlusion treatment, the mAP of the four types of trees was increased from 87.25% to 91.5% (Fig. 8).

4   Discussion

4.1   Analysis of the Influence of Illumination on Tree Detection

Fig. 9. Original image detection map and comparison with prediction box (blue box is the actual label box, red box is the prediction box)

In real life (Fig. 9), due to various factors, the illumination in images is not the most suitable for experimentation. Under illumination from different angles, trees take on different shapes. A same tree may have multiple manifestations in images that are brighter or darker due to illumination. Images with partially blank space or incomplete framing were not conducive to the calculation of specific parameters of trees. Objects merged with the background in some pictures with different light and shade areas and uneven illumination, which made it difficult to find and distinguish target objects. By processing the images to a "suitable" state, the progress of the entire experiment was facilitated. In the process of image processing, filtering is a commonly used method in image processing. By replacing every pixel in the image with the calculated average value of each pixel and its surrounding pixels, which improve the image quality of trees to some extent. Illumination treatment was conducted for the tree images to find the weakest areas of illumination intensity in the input image. That then became the minimum value of the RGB components in each pixel. That value was then stored in the grayscale image and minimum filtering was performed to obtain the maximum point of the gray value. Then, the maximum channel image and the maximum gray value in the RGB image were obtained, and the gray values of the two points were averaged. In order to keep the characteristic of image edge, the difference of pixel space and intensity was considered in the experiment, and the ideal filtering effect is obtained by processing the content of the image. At the same time, considering the difference in illumination in different situations in real life, certain lights were retained according to weather conditions during the illumination treatment of images. There are edges with different intensity differences in the tree image, and the filtering effect can be adjusted by adjusting the weight accordingly.

Fig. 10. Image detection map after illumination treatment

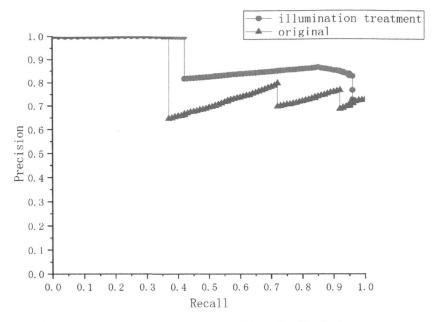

Fig. 11. Comparison diagram of Ginkgo biloba after illumination treatment

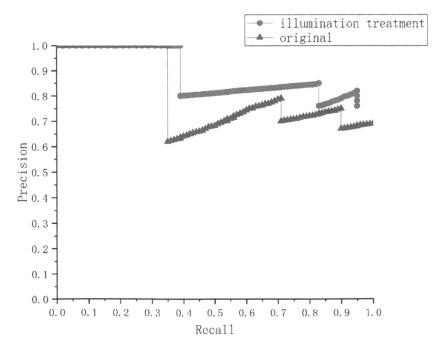

Fig. 12. Comparison diagram of Maple after illumination treatment

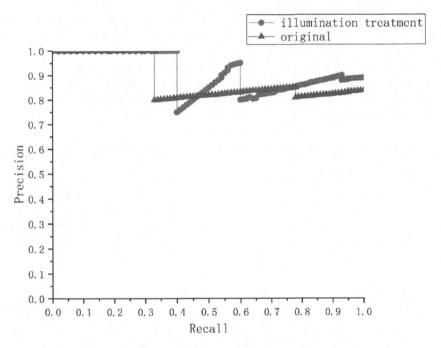

Fig. 13. Comparison diagram of Osmanthus fragrans after illumination treatment

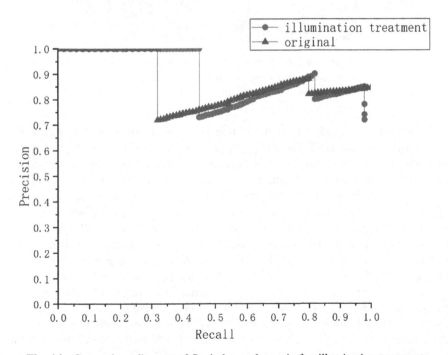

Fig. 14. Comparison diagram of Sapindus mukorossi after illumination treatment

By processing the difference area of pixel intensity, the gradient can be maintained at the edge of the tree image where the intensity gradient is large or small. Processing the darker or the brighter parts of an image can effectively reduce local shadow and illumination changes of the image, markedly improve the image quality, and make the combined image of the target and the background significantly improve the frame of the tree edge (Fig. 10). The mAP of the four tree species improved to varying degrees after illumination treatment (Fig. 11, Fig. 12, Fig. 13, Fig. 14).

4.2   Analysis of the Influence of Tree Occlusion on Tree Detection

Different from other detection objects, trees have unique morphological characteristics. Branches and occlusion between trees make information loss of tree species larger, as information cannot be captured and distinguished well. Due to the obvious "interference" of appearances and adjacent trees on detection, such problems as missed detection, large difference between detection frame and originally marked frame, and frame position offset may be caused.

Fig. 15. Diagram of difference between tree occlusion prediction box and actual box (the orange and blue boxes represent two Ground Truth, and the red represents the anchor or proposal on the GT assigned to the orange or blue) (Color figure online)

As shown in Fig. 15, although the tree types were correctly distinguished, the detection quality was poor. The red proposal was far from the real target due to the influence of orange and blue GT, resulting in crowded situation. In addition to eliminate redundancy in the experiment, the cases of false recall and missed inspections also need to be considered. The overall optimization goal is to realize that all proposals approximate the corresponding target box and multiple proposals belonging to the same target box as far as possible. Based on part3, RepGT and RepBox were mainly used to deal with occlusions. RepGT was used to punish the proposals shifting to other targets, so as to achieve repulsion. RepBox was used to punish the proposals approaching to the same corresponding to other targets, so as to achieve repulsion. The α and β are hyper parameters used to balance the weight of different effects. Throughout the process, the proposed proposal is constrained so that it is not only close to target T, but also away from other objects and their corresponding proposals. If the surrounding Ground Truth of T contains goals other than the object, the predictive proposal is required to stay away from all these goals. The task of tree detection is simplified, only foreground and background were considered to calculate the IoU of the anchor and Ground Truth boxes, and the largest IoU was regarded as the target box. The SmoothL1 distance was used to measure the degree of approximation of the two boxes. In the case of repulsion, the IoG (Intersection over Ground-Truth) indicator was used to measure the degree of approximation. The prediction boxes corresponding to different targets were separated - the smaller the IoU between the prediction boxes was better. This avoided a leak recall situation caused by the merge prediction boxes in NMS post-processing.

Fig. 16. Detection diagram after handling the occlusion

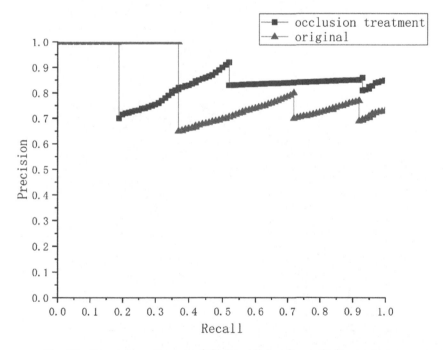

Fig. 17. Comparison diagram of Ginkgo biloba after occlusion treatment

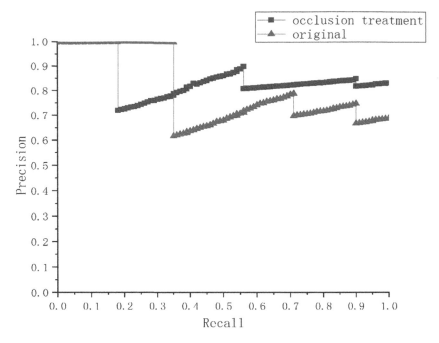

Fig. 18. Comparison diagram of Maple after occlusion treatment

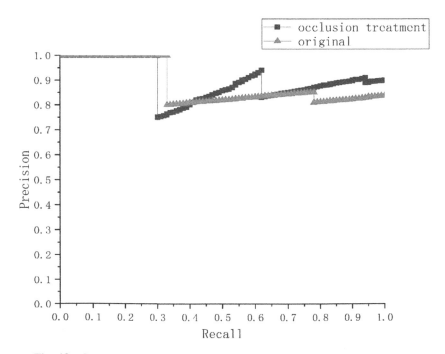

Fig. 19. Comparison diagram of Osmanthus fragrans after occlusion treatment

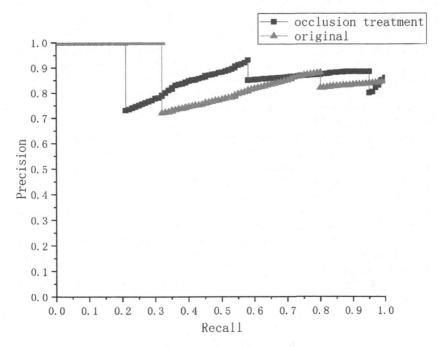

Fig. 20. Comparison diagram of Sapindus mukorossi after occlusion treatment

After dealing with the occlusion, the problems of missed detection and false detection had been well processed (Fig. 16). The mAPs of the four types of trees had been improved to some extent after handling the occlusion (Fig. 17, Fig. 18, Fig. 19, Fig. 20).

5   Conclusions

In this paper, a method is proposed based on the overall tree image GMOST dataset to solve the problem of low mAP caused by illumination and occlusion. Under the application of GMOSTNet, the accuracy of tree detection is improved from 87.25% in the original to 93.25%, which makes tree detection work significantly easier, more convenient, more adaptable to complex terrain, and free from the limit of conservation methods and areas. It also reduces a series of unpredictable problems, such as tree damage, ecological damage and even destruction, due to delayed or poorly timed acquisition of tree information. In the future work, if current work can be applied into new areas and a more detailed direction, it will be an interesting exploration to combine the fine-grained problems of trees.

Acknowledgement. This work is supported by National Natural Science Foundation of China (U1809208).

References

1. Piiroinen, R., Fassnacht, F.E., Heiskanen, J., Maeda, E., Mack, B., Pellikka, P.: Invasive tree species detection in the Eastern Arc Mountains biodiversity hotspot using one class classification. Remote Sens. Environ. **218**, 119–131 (2018)
2. Aval, J., et al.: Detection of individual trees in urban alignment from airborne data and contextual information: a marked point process approach. ISPRS J. Photogramm. Remote Sens. **146**, 197–210 (2018)
3. Zhang, J., He, L., Karkee, M., Zhang, Q., Zhang, X., Gao, Z.: Branch detection for apple trees trained in fruiting wall architecture using depth features and regions-convolutional neural network (R-CNN). Comput. Electron. Agric. **155**, 386–393 (2016)
4. Duncanson, L., Dubayah, R.: Monitoring individual tree-based change with airborne lidar. Ecol. Evol. **8**(10), 5079–5089 (2018)
5. Ferraz, A., Saatchi, S., Mallet, C., Meyer, V.: Lidar detection of individual tree size in tropical forests. Remote Sens. Environ. **183**, 318–333 (2016)
6. Jeronimo, S.M., Kane, V.R., Churchill, D.J., McGaughey, R.J., Franklin, J.F.: Applying LiDAR individual tree detection to management of structurally diverse forest landscapes. J. For. **116**(4), 336–346 (2018)
7. Kansanen, K., Vauhkonen, J., Lähivaara, T., Seppänen, A., Maltamo, M., Mehtätalo, L.: Estimating forest stand density and structure using Bayesian individual tree detection, stochastic geometry, and distribution matching. ISPRS J. Photogramm. Remote Sens. **152**, 66–78 (2019)
8. Dempewolf, J., Nagol, J., Hein, S., Thiel, C., Zimmermann, R.: Measurement of within-season tree height growth in a mixed forest stand using UAV imagery. Forests **8**(7), 231 (2017)
9. Goldbergs, G., Maier, S.W., Levick, S.R., Edwards, A.: Efficiency of individual tree detection approaches based on light-weight and low-cost UAS imagery in Australian Savannas. Remote Sens. **10**, 161 (2018)
10. Gebreslasie, M.T., Ahmed, F.B., Van Aardt, J.A., Blakeway, F.: Individual tree detection based on variable and fixed window size local maxima filtering applied to IKONOS imagery for even-aged Eucalyptus plantation forests. Int. J. Remote Sens. **32**(15), 4141–4154 (2011)
11. Gougeon, F.A., Moore, T.: Individual tree classification using MEIS-II imagery. In: 1988 Geoscience and Remote Sensing Symposium, IGARSS 1988. Remote Sensing: Moving Toward the 21st Century, International, vol. 2, p. 927 (1988)
12. Heinzel, J., Koch, B.: Investigating multiple data sources for tree species classification in temperate forest and use for single tree delineation. Int. J. Appl. Earth Obs. Geoinf. **18**, 101–110 (2012)
13. Nevalainen, O., et al.: Individual tree detection and classification with UAV-Based photogrammetric point clouds and hyperspectral imaging. Remote Sens. **9**, 185 (2017)
14. Ren, S., He, K., Girshick, R., Sun, J.: Faster R-CNN: towards real-time object detection with region proposal networks. NIPS **1**, 91–99 (2015)
15. Roth, S.I., Leiterer, R., Volpi, M., Celio, E., Schaepman, M.E., Joerg, P.C.: Automated detection of individual clove trees for yield quantification in northeastern Madagascar based on multi-spectral satellite data. Remote Sens. Environ. **221**, 144–156 (2019)
16. Binkley, D., Stape, J.L., Bauerle, W.L., Ryan, M.G.: Explaining growth of individual trees: light interception and efficiency of light use by Eucalyptus at four sites in Brazil. For. Ecol. Manag. **259**(9), 1704–1713 (2010)

17. Maltamo, M., Eerikäinen, K., Pitkänen, J., Hyyppä, J., Vehmas, M.: Estimation of timber volume and stem density based on scanning laser altimetry and expected tree size distribution functions. Remote Sens. Environ. **90**(3), 319–330 (2004)
18. Korpela, I., Anttila, P., Pitkänen, J.: The performance of a local maxima method for detecting individual tree tops in aerial photographs. Int. J. Remote Sens. **27**(6), 1159–1175 (2006)
19. Li, J., Hu, B., Noland, T.L.: Classification of tree species based on structural features derived from high density LiDAR data. Agric. For. Meteorol. **171–172**, 104–114 (2013)
20. Li, W., Dong, R., Fu, H., Yu, L.: Large-Scale oil palm tree detection from high-resolution satellite images using two-stage convolutional neural networks. Remote Sens. **11**, 11 (2019)
21. Vauhkonen, J., et al.: Comparative testing of single-tree detection algorithms under different types of forest. Forestry **85**(1), 27–40 (2011)
22. Wang, X., Xiao, T., Jiang, Y., Shao, S., Sun, J., Shen, C.: Repulsion loss: detecting pedestrians in a crowd. In: 2018 IEEE/CVF Conference on Computer Vision and Pattern Recognition, Salt Lake City, UT, pp. 7774–7783 (2018)

Improved YOLOv3 Infrared Image Pedestrian Detection Algorithm

Jianting Shi[1](✉), Guiqiang Zhang[1], Jie Yuan[2], and Yingtao Zhang[3]

[1] School of Computer and Information Engineering,
Heilongjiang University of Science and Technology, Harbin 150022, China
229468764@qq.com
[2] Shanghai Aerospace Electronics Technology Research Institute,
Shanghai 201109, China
[3] School of Computer Science and Technology, Harbin Institute of Technology,
Harbin 150001, China

Abstract. Security surveillance is widely used in daily life. For nighttime or complicated monitoring environments, this article proposes an infrared pedestrian monitoring based on YOLOv3. In the original YOLOv3 network structure, two aspects of optimization were made. One was to optimize the scale in the residual structure, and the rich features of the deconvolution layer were added to the original residual structure. The other was to use the DenseNet network to enhance the features. The optimization of fusion ability and delivery ability effectively improves the detection ability for small targets, and the pedestrian detection performance based on infrared images. After comparative testing, compared with YOLOv3, the overall mean average precision is improved by 4.39% to 78.86%.

Keywords: Infrared · YOLOv3 · DenseNet

1 Introduction

With the development of computer vision technology, the field of pedestrian detection is also becoming more and more popular. Pedestrian detection is to study and judge the given image or whether there is a pedestrian to be detected in each video sequence, and can accurately and quickly find the specific location of the target. In today's era, science and technology are constantly improving, and hardware technology is gradually improving. In response to the call of the times, deep learning is stepping into the global development track step by step. Therefore, in computer vision, the research of pedestrian detection is gradually culminating. In recent years, road safety issues have attracted more and more attention, and people are looking for ways to reduce the occurrence of traffic accidents, and pedestrian detection technology can effectively reduce them. In terms of future driverless technology, pedestrian detection technology is even more important. Therefore, the aspect of pedestrian detection technology is receiving more and more attention from the society.

Traditional visible light equipment cannot meet the requirements of nighttime or unmanned driving. Compared with the traditional situation, infrared thermal imaging is

© Springer Nature Singapore Pte Ltd. 2020
J. Zeng et al. (Eds.): ICPCSEE 2020, CCIS 1257, pp. 506–517, 2020.
https://doi.org/10.1007/978-981-15-7981-3_37

based on the information of the relative temperature of the object, and is less affected by various additional factors. In intelligent monitoring, vehicle assisted driving, human body Behavior analysis and other fields have broad application prospects [1, 2]. However, the infrared image has no color, and its accuracy is low when detecting pedestrians. Pedestrian detection algorithms can be divided into traditional algorithms and deep learning-based algorithms. Traditional pedestrian detection algorithms include Haar wavelet features [3], HOG-SVM [4], DPM [5], etc. Traditional pedestrian detection mainly uses artificially designed methods to extract image features, combined with machine learning related algorithms, to identify and classify image features, but traditional algorithms are complex to design, sometimes it is difficult to design reasonable methods in complex scenes, weight parameters are difficult to get more accurate values, and generalization ability is not strong.

Convolutional Neural Network (CNN) [6] has made a significant breakthrough in pedestrian detection. CNN can automatically learn the original representation of the target through a large amount of data. Compared with the features designed by hand, it has more advantages strong discrimination and generalization. Then, based on the deep learning algorithm, after the RCNN algorithm was proposed, a new boom was ushered in. The performance of deep learning methods in multiple image processing fields surpassed the traditional algorithms [7, 8], that is, a series of improved algorithms appeared, including Fast RCNN [9], Faster RCNN [10–12], SSD [13], YOLO [12] and other algorithms. There are two-stage and one-stage algorithms for deep learning. Compared with traditional methods, deep learning methods have both improved the efficiency and speed of detection. Among them, before the advent of YOLO, deep learning was not fast in detection speed and could not guarantee real-time performance, especially in the future in driverless technology. Redmon et al. [14] proposed the YOLO (You Only Look Once, Unified, Real-Time Object Detection) algorithm, and thus entered the field of one-stage target detection. In recent years, with the continuous development of deep learning, methods applied to target recognition and model prediction have been continuously introduced [15–17].

The one-stage concept solves the problem of speed in object detection, while ensuring a certain accuracy, greatly improving real-time performance. Although the speed is improved, compared with other algorithms, the accuracy is not very high. Then came YOLOv2, YOLO9000, YOLOv3. Among them, YOLOv3 has a simple and efficient network structure, which makes it easy to deploy and has a wide range of application scenarios. It is one of the preferred algorithms in many commercial fields. Combined with our actual application scenarios, it is applied to large outdoor surveillance to detect areas where pedestrians are prohibited. And for small object detection and in the case of pedestrian detection in infrared images, YOLOv3 has great application prospects. It not only uses better backbone networks, such as classifiers from backbone networks such as DarkNet or ResNet, but also can detect quickly. The main thing is that the setup environment is simple, the background detection error rate is low, and the versatility is strong. Although the YOLOv3 network uses multi-scale prediction and combines with better classifiers, it has great advantages. However, YOLOv3 still has the following disadvantages: compared with other RCNN series object detection algorithms, the accuracy of identifying objects is poor, and the recall rate is low.

In view of the above problems, this paper improves the YOLOv3 network framework and borrows the ideas of the DenseNet network. Through the improvement, compared to the original YOLOv3 detection effect on pedestrians in infrared images, the (accuracy) MAP is increased by 4.39%. Compared with the network before the improvement, the improved network also has an increase of 2.36% over the network intersection ratio (IOU) before improvement.

2   YOLOV3 Network Algorithm Structure

YOLOv3 is the beginning of one-stage detection. It is a single neural network-based object detection system proposed by Joseph Redmon and Ali Farhadi and others in 2015. In 2017, CVPR Joseph Redmon and Ali Farhadi published YOLOv2, which further improved the accuracy and speed. After further improvement appeared YOLOv3 algorithm. YOLOv3 network algorithm structure Fig. 1 is as follows: YOLOv3 is mainly divided into three aspects, namely: network input, structure, output where defined.

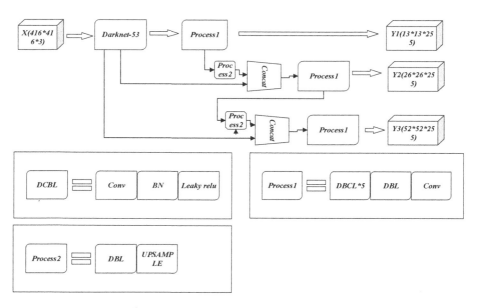

Fig. 1. YOLOv3 network algorithm structure

2.1   Network Input

The general input size of YOLOv3 network is 320 * 320, 416 * 416, 608 * 608, this article introduces 416 * 416, the size must be an integer multiple of 32, so as to facilitate the subsequent training test and analysis. The YOLOv3 network mainly uses 5 downsampling. The YOLOv3 network is based on the DarkNet-53 backbone network, and the step size of each sampling is 2, so the largest size of the backbone network is.

2.2   Network Structure

The general input size of YOLOv3 network is 320 * 320, 416 * 416, 608 * 608, this article introduces 416 * 416, the size must be an integer multiple of 32, so as to facilitate the subsequent training test and analysis. The YOLOv3 network mainly uses 5 downsampling. The YOLOv3 network is based on the DarkNet-53 backbone network, and the step size of each sampling is 2, so the largest size of the backbone network is.

The YOLOv3 network is a fully convolutional network that uses the first 52 layers of DarkNet-53, but does not use a fully connected layer and a pooling layer, and uses many residual structures for layer hopping connections. The use of the residual structure helps the network structure to stay in a deep situation and maintain convergence, so that the training continues, and the deeper the network, the better the training result, the better the results obtained by classification and detection, and the 1 * 1 convolution reduces the amount of calculation to a certain extent. YOLOv3 uses three types of downsampling, which are $32\times$ downsampling, $16\times$ downsampling, and $8\times$ downsampling. In order to ensure that the deeper the characteristics of the network, the better the effect is displayed. Sampling, this can be used for target detection of deep features. YOLOv3 performs shallow features generated by downsampling. YOLOv3 wants to make use of shallow features, so it has a route layer. Tensor stitching (concat) is then performed, and the upsampling of the DarkNet middle layer and a later layer is stitched.

2.3   Output

Taking the input 416 * 416 * 3 as a reference, the output in Fig. 1 above has three scales, which are 52 * 52, 26 * 26, and 13 * 13. This gives the grid a center position. Can detect targets of different sizes. The network output needs to predict the anchor box, and predict three bounding boxes for each cell of the feature map. For each bounding box, it will predict the bounding box. Three aspects, (1) the position of each box. (2) Detect objectness prediction (3) M categories. The bounding box coordinate prediction formula is as follows:

$$b_x = \sigma(t_x) + c_x \tag{1}$$

$$b_y = \sigma(t_y) + c_y \tag{2}$$

$$b_w = p_w e^{t_w} \tag{3}$$

$$b_h = p_h e^{t_h} \tag{4}$$

among them t_x, t_y, t_w, t_h Predicted output representing the bounding box, c_x, c_y Represents the coordinates of the grid cell, p_w, p_h Represents the size of the bounding box before prediction, that is, the width and height of the anchor box. b_x, b_y, b_w, b_h Represents the center coordinates and size of the obtained bounding box.

The confidence formula of the YOLOv3 algorithm is as follows:

$$C_i^j = P_r(Object) * IOU_{pred}^{truth} \tag{5}$$

among them $P_r(Object)$ The probability that the bounding box has an object, IOU_{pred}^{truth} When representing the boundary and the object, the value of the intersection of its predicted boundary and the true boundary of the detected object, C_i^j Confidence of the j-th bounding box representing i grids cell.

Suppose a picture is divided into S * S grids and B anchor boxes, that is, each grid has B bounding boxes, each bounding box has four position parameters, 1 confidence level, and one is set. The parameters are: when the confidence degree indicates that there is an object in the current bounding box, the probability of the class is, and there are classes of class probability, then the output dimension formula of the final output layer of the model is:

$$S * S * [B * (4 + 1 + classes)] \tag{6}$$

Where S represents the length and width of a grid, B represents the number of anchor boxes, and classes represents the class probability.

3   Improved YOLOV3 Infrared Pedestrian Detection Algorithm

3.1   Problems

Under the conditions of good lighting conditions and high imaging quality, the YOLO algorithm can detect pedestrians with an accuracy of more than 96.5%. However, when the lighting conditions at night are insufficient and pedestrians and the background are mixed, the detection accuracy of the YOLO algorithm is only 68.4%. There are more missed inspections. At present, a common solution is to use an infrared camera for shooting. Depending on the principle of thermal imaging, the infrared camera can effectively separate pedestrians from the background, so that pedestrians can be more intuitively identified in monitoring. Although the use of infrared camera can improve the detection efficiency, it still has the following four problems: The first point is that the brightness of the object in the infrared image is related to the surface temperature of the object. When there is more clothing wrapped in winter, the imaging result is poor. The second point is that the infrared image has no color information and the target does not have detailed features such as texture. At the same time, the infrared image has low resolution and many noises, which has a certain impact on the convolution operation. In addition, the subject of this experiment is infrared small target pedestrian detection. Compared with scene pedestrian detection, the third point is that the picture taken by the infrared camera only retains the boundary information of the heating source, but if the pedestrian is wearing thick clothes. Or when there is interference from other heat sources, the boundary between pedestrians and the surrounding environment will not be too obvious. At the same time, due to the characteristics of infrared imaging, additional noise will be introduced, and the effect of noise on the convolutional neural network increases with the number of network layers. The fourth point and the main problem is that the pedestrian target to be detected has a small pixel area in the image and is not easy to detect.

3.2   Improvement Plan

In order to solve the above problems, this paper optimizes and improves the network based on YOLOv3 algorithm. The improvement work mainly includes the following two measures.

The first one point is that this article has replaced the residual module of YOLOv3. At the same time, this article also optimized the structure in the residual block. The target of this test belongs to the category of small targets. We add a deconvolution layer in the residual module to expand the input feature map to twice the original size and fuse it with the output of the previous scale module. The optimized residual membrane block structure is shown in Fig. 2 as follows:

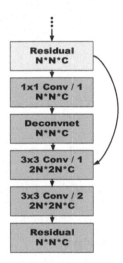

Fig. 2. Structure of the optimized residual module

Second point: The residual structure used in the YOLOv3 feature extraction network is designed with reference to the ResNet network. The network structure is simple and has good feature extraction effects, but for images with less feature information such as infrared images, its feature extraction capability obviously insufficient. Analyzing the data set samples used in this experiment, it can be found that in the image pedestrians currently only have appearance contours, and because of the characteristics of thermal imaging, features such as texture structure cannot be extracted. Convolutional neural networks can only be discriminated by the contours of pedestrians. The simple upper-lower network connection relationship of the ResNet network cannot be applied to this scenario. Therefore, we have borrowed the ideas of DenseNet and strengthened the connection between the shallow network and the deep network. For neural networks, the types of feature information extracted from the deep layer and the shallow layer are different. This improvement is that the features that can be extracted in the shallow network are mostly simple features, such as the edge structure and texture color of the object. With the deepening of the network, the stronger the learning ability of the network, the features contained in the deep network have richer semantic information. But for infrared images, simply deepening the number of network layers does not have a good effect, so this article borrowed the DenseNet network and input the shallow feature information into the deep network in turn. In the YOLOv3 network, there will be five downsamplings, and the pixel area occupied by pedestrians in the image is originally small. After five times of zooming, the feature discrimination ability of the target in the feature map will be greatly reduced. The feature map is expanded by deconvolution. This measure can improve the saliency of the target feature in the feature map, so that the extraction operation performed on the feature map residual block of the original 13 * 13 size is also 26 * 26-size feature map is performed, which improves the ability to extract small target features.

The optimized convolutional neural network structure of YOLOv3 is shown in Fig. 3.

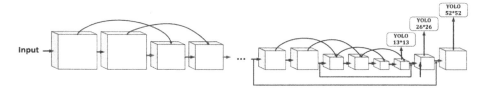

Fig. 3. Convolutional neural network structure of YOLOv3

Based on the above network optimization strategy, we named the optimized network YOLO-I (intimate).

3.3   Experimental Results and Analysis

The data set of the experiments in this paper is from the public data set. First, the data set is cleaned, and 1500 ordinary samples, 400 difficult samples, and 200 negative samples are selected to form a training set of 2100. And a 300 test set with 200 ordinary samples and 100 difficult samples. YOLOv3 introduced the anchor mechanism. The default anchor size is not suitable for this dataset. The K-means clustering method is utilize to recalculate the nine anchors suitable for small target pedestrian datasets.

The loss convergence curve during the network training process is as shown in Fig. 4. The abscissa represents the number of iterations, around 79000. When the network iterates around 60,000 times, it tends to be stable, the parameters change basically stable, and finally the loss value drops to 0.29. Judging from the convergence of this parameter, the network training results are ideal.

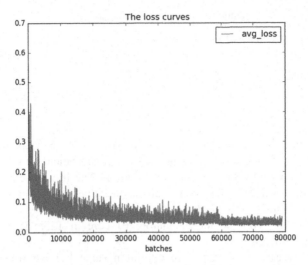

Fig. 4. Loss curve of YOLO-I

As shown in Fig. 5, the test set is tested to obtain the Precision-Recall curve corresponding to the improved model:

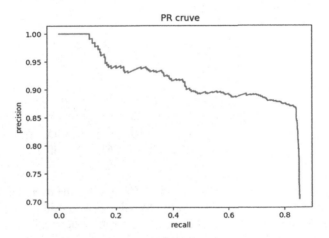

Fig. 5. Precision-recall curve of YOLO-I

It can be seen from Fig. 5 that after the recall rate is 0.8, precision shows a rapid downward trend.

After training the YOLOv3 network before improvement and YOLO-I after improvement, this article then conducted a horizontal comparison test of YOLO-I and YOLOv3. The test environment is the same. In order to reflect the robustness of the network after optimization in more detail, To improve, this article have added test indicators for a more comprehensive comparison. The test results are shown in Table 1.

514   J. Shi et al.

Table 1. Horizontal comparison table

Contrast situation	Comparison of YOLO-I and YOLOv3				
	Precision	Recall	IOU (%)	F1-score	mAP (%)
YOLOv3	0.86	0.68	60.75	0.76	74.47
YOLO-I	0.89	0.73	63.11	0.80	78.86

The full name of IOU is Intersection-over-Union. A concept used in object detection is the overlap ratio of the generated candidate bound and the ground truth bound, that is, the ratio of their intersection to union. The full name of mAP is Mean Average Precision, which is the average precision value. Is to average the accuracy value of multiple validation set individuals.

Among them: the formula for calculating F1score in Table 1: 2 * Precision * Recall/(Precision + Recall).

From the comparison of the F1-score column in Table 1 above, we can see that the overall robustness of YOLO-I is better than that of YOLOv3. The IOU crossover ratio also reflects the multi-level nesting and fusion of the positional characteristics of the shallow network into the deep Networks are of great help in improving the accuracy of predicting bounding boxes.

Finally, we compared the YOLO-I with the SSD-ResNet and Faster RCNN networks. The comparison results are shown in Table 2.

Table 2. Vertical comparison table

Contrast situation	Comparison of YOLO-I with SSD-ResNet and Faster RCNN			
	Precision	Recall	IOU	F-score
Faster RCNN	0.87	0.72	63.21%	0.79
SSD-ResNet	0.81	0.70	61.66%	0.75
YOLO-I	0.89	0.73	63.11%	0.80

A Subsection Sample It can be known from Table 2 that the overall data of YOLO-I is better than that of SSD-ResNet, and the SSD-ResNet algorithm adopted by SSD. For Faster RCNN network, due to its two-stage network structure, the candidate frame area is extracted beforehand for target classification detection, which reduces the influence of invalid background images. The accuracy of the network and the intersection ratio are better than SSD-ResNet Network, slightly lower than YOLO-I. The actual test comparison chart for the three networks are as shown in Fig. 6, 7 and 8.

Fig. 6. Faster RCNN

Fig. 7. SSD-VGG16

Fig. 8. YOLO-I

4 Conclusion

This paper proposes an improved infrared image pedestrian detection algorithm YOLO-I based on YOLOv3. The optimized YOLO-I has a significantly improved detection capability for grayscale images and small targets, which improves the practicality of infrared detection. This paper mainly aims at the detection environment of low pixels and small targets. Based on the actual detection situation, it is optimized on the basis of YOLOv3. The first is to increase the richness of the feature map size in the residual module. The deconvolution network and sliding step size are the convolution kernel of 2 performs upsampling and downsampling operations, and has two scale feature maps in the same residual module. Compared with the previous, the richness of feature information and the positioning ability of small targets were improved. It strengthened the utilization of shallow features and network-wide features, borrowed dense connections from DenseNet, enhanced the ability to transfer feature information, and effectively improved detection accuracy. After testing and optimization, YOLO-I is targeted at small infrared targets. The detection accuracy was significantly improved in the detection scene. The network in this paper is a reference value for pedestrians and vehicles driving at night. Our future work will explore a vehicle-mounted infrared camera equipped with the improved YOLO-I network, or equipped with the improved network in traffic. The camera is for pedestrians and drivers passing by at night, and hopes to improve safety.

Acknowledgements. This paper was supported by the Fundamental Research Funds for the Local Universities of Hei longjiang Province in 2018 (Grant No. 2018-KYYWF-1189) and Shanghai Aerospace Science and Technology Innovation Fund (Grand No. SAST2017-104).

References

1. Cui, M.: Application field and technical characteristics of infrared thermal imager. China Secur. Protect. **12**, 90–93 (2014)
2. Carlo, C., Salvetti, O.: Infrared: a key technology for security systems. Adv. Opt. Technol. **2012**, 838752 (2012)
3. ViolaI, P., Jones, J.M., Snow, D.: Detecting pedestrians using patterns of motion and appearance. Int. J. Comput. Vis. **63**(2), 153–161 (2005)
4. Dalal, N., Triggs, B.: Histograms of oriented gradients for human detection. In: International Conference on Computer Vision & Pattern Recognition (CVPR 2005), vol. 1, pp. 886–893. IEEE Computer Society (2005)
5. Felzenzwalb, P.F., Grishick, B.R., Mcallister, D., et al.: Object detection with discriminatively trained part-based models. IEEE Trans. Pattern Anal. Mach. Intell. **32**(9), 1627–1645 (2010)
6. Lecun, Y., Bottou, L., Bengio, Y., et al.: Gradient-based learning applied to document recognition. Proc. IEEE **86**(11), 2278–2324 (1998)
7. Ning, S., Liang, C., Guang, H., et al.: Research on deep classification network and its application in intelligent video surveillance system. Electro Opt. Control **22**(9), 77–82 (2015)
8. Jensen, M.B., Nasrollahi, K.T., Moeslund, B.: Evaluating state-of-the-art object detector on challenging traffic light data. In: Proceedings of the IEEE Conference on Computer Vision and Pattern Recognition Workshops, pp. 9–15 (2017)
9. Girshick, R.: Fast R-CNN. In: Proceedings of the IEEE International Conference on Computer Vision, pp. 1440–1448 (2015)
10. Ren, S., He, K., Girshick, R., et al.: Faster-R-CNN: towards real-time object detection with region proposal networks. IEEE Trans. Pattern Anal. Mach. Intell. **39**(6), 1137–1149 (2017). https://doi.org/10.1109/TPAMI.2016.2577031
11. Zhang, Z., Wang, H., Zhang, J., et al.: Aircraft detection algorithm based on faster-RCNN for remote sensing image. J. Nanjing Normal Univ. (Eng. Technol. Edn. **41**(4), 79 (2018). https://doi.org/10.3969/j.issn.1001-4616.2018.04.013
12. Yang, W., Wang, H., Zhang, J., Zhang, Z.: An improved algorithm for real-time vehicle detection based on faster-RCNN. J. Nanjing Univ. (Nat. Sci.) **55**(2), 231–237 (2019). https://doi.org/10.13232/j.cnki.jnju.2019.02.008
13. Liu, W., et al.: SSD: single shot multibox detector. In: Leibe, B., Matas, J., Sebe, N., Welling, M. (eds.) ECCV 2016. LNCS, vol. 9905, pp. 21–37. Springer, Cham (2016). https://doi.org/10.1007/978-3-319-46448-0_2
14. Redmon, J., Divvala, S., Girshick, R., et al.: You only look once: Unified, real-time object detection. In: Proceedings of the IEEE Conference on Computer Vision and Pattern Recognition, pp. 779–788 (2016)
15. Ang, D., Jiang, Y.: Face recognition system based on BP neural network. Software **36**(12), 76–79 (2015)
16. Zhang, X., Yi, H.: Scene classification based on convolutional neural network and semantic information. Software **39**(01), 29–34 (2018)
17. Gao, W., Li, Y., Zhang, J., et al.: Research on forecast model of high frequency section of urban traffic. Software **39**(2), 81–87 (2018)

Facial Expression Recognition Based on PCD-CNN with Pose and Expression

Hongbin Dong$^{(\boxtimes)}$, Jin Xu, and Qiang Fu

Computer Science and Technology College, Harbin Engineering University,
Harbin, China
donghongbin@hrbeu.edu.cn

Abstract. In order to achieve high recognition rate, most facial expression recognition (FER) methods generate sufficient labeled facial images based on generative adversarial networks (GAN) to train model. However, these methods do not estimate the facial pose before passing the images to the generator, which affects the quality of generated images. And mode collapse is prone to occur during the training process, leading to generate a single-style facial images. To solve these problems, a FER model is proposed based on pose conditioned dendritic convolution neural network (PCD-CNN) with pose and expression. Before passing the facial images to the generator, PCD-CNN was used to process facial images, effectively estimating the facial landmarks to detect face and disentangle the pose. In order to accelerate the training speed of the model, PCD-CNN was based on the ShuffleNet-v2 framework. Every landmark of facial image was modeled by a separate ShuffleNet-DeconvNet, maintaining better performance with fewer parameters. To solve the mode collapse during image generation, we theoretically analyzed the causes, and implemented mini-batch processing on the discriminator in the model and directly calculated the statistical characteristics of the mini-batch samples. Experiments were carried out on the Multi-PIE and BU-3DFE facial expression datasets. Compared with current advanced methods, our method achieves higher accuracy 93.08%, and the training process is more stable.

Keywords: Pose estimation · Mode collapse · Expression recognition

1 Introduction

FER is a biometric recognition technology that uses computers to obtain and analyze facial expressions so that computers can recognize and even understand human emotions, thereby achieving the purpose of human-computer interaction [1]. However, the existing FER is based on frontal facial images and videos. The recognition rate of FER in the case of non-frontal faces is not very ideal. Sariyanidi [2] pointed out that the current research on FER faces five major problems: head deflection, illumination changes, registration errors, face occlusion, and identity differences. In these problems, head deflection is an important cause of registration errors and face occlusion. In addition, the lack of sufficient training samples can cause overfitting during the learning process. To solve these problems, Zhang [3] proposed a FER model by jointing pose

© Springer Nature Singapore Pte Ltd. 2020
J. Zeng et al. (Eds.): ICPCSEE 2020, CCIS 1257, pp. 518–533, 2020.
https://doi.org/10.1007/978-981-15-7981-3_38

and expression. This model can simultaneously synthesize facial images and recognize facial expressions. The model of Zhang made an ideal recognition accuracy under the condition of non-frontal facial expressions. However, before conveying the facial image to the generator, Zhang used the lib face detection method with 68 facial landmarks for face detection [3], which could not effectively estimate the facial pose and affected the generation of facial images. And the model is based on GAN, the training process is unstable, mode collapse is prone to occur, resulting in only a certain style of facial images generated. Kumar [4] proposed a PCD-CNN model based on CNN, which can conditionally estimate the facial landmarks to disentangle the facial pose of the image. The PCD-CNN is a dendritic structure model, can effectively capture shape constraints in a deep learning framework, but due to the many model parameters, the running time is very long.

In response to the above problems, we improved method of Zhang and proposed a FER model based on PCD-CNN with pose and expression. It has three advantages: (1) we use the PCD-CNN to preprocess facial images before conveying the facial images to the generator. Compared with Zhang's method, ours can not only accurately detect the facial area, but also effectively estimate facial pose, (2) the PCD-CNN in our model is based on the ShuffleNet-v2 [5] architecture, which makes our model maintain better performance with fewer parameters, and (3) we perform mini-batch processing on the discriminator in the model and directly calculate the statistical characteristics of the mini-batch samples, making the model more stable during training, avoiding the occurrence of mode collapse as much as possible. Our model is trained and tested on the Multi-PIE [6], BU-3DFE [7] facial expression datasets. Compared with Zhang's and the latest methods, our improved model has higher recognition accuracy and achieves 93.08%.

2 Related Work

In non-frontal FER, disentangling the facial pose can improve the extraction of facial information features by the classifier, thereby improving the accuracy of expression recognition [4]. Common facial pose estimation methods include geometric method, tracking method and nonlinear regression method [8]. The geometric method focuses on the extraction of facial landmarks information and improves the correlation between facial landmarks and facial poses. The tracking method mainly focuses on tracking a person's head in a video. Literature [9] proposed a method of target tracking based on joint probability. Firstly, the target area was located by graph structure method, then the probability model of target tracking was built, and the particle filter was used to track the target of a single frame of image. Finally, the face pose was estimated. The nonlinear regression method focuses on establishing the mapping relationship between poses. Different from existing methods, Kumar [4] proposed PCD-CNN model to disentangle facial pose in unconstrained facial image alignment. PCD-CNN follows the Bayesian formula and effectively disentangles the facial pose of the image by adjusting the facial landmarks.

In addition, the public dataset of non-frontal FER lacks sufficient training samples. In this situation, it is difficult to effectively train non-frontal FER model based on deep neural networks, which may cause overfitting. To solve this problem, we generally generate enough labeled training samples through the model. Since the GAN proposed by Goodfellow [10] has wide applications in image generation, this inspired us to use GAN to generate labeled samples to enrich the training dataset. As a generative model with excellent performance, GAN has two outstanding advantages [11]: (1) does not rely on any prior assumptions; (2) the method of generating samples is very simple (Fig. 1).

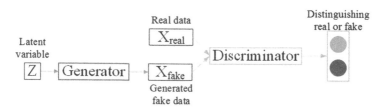

Fig. 1. Basic architecture of GAN. It is mainly composed of generator and discriminator. The task of generator is used to generate "fake data", and the task of the discriminator is to distinguish whether the data is "real data" or "fake data".

However, mode collapse is very common in general GAN model. In simple terms, mode collapse means that the generator produces a single or limited pattern. Common solutions are experience replay and using multiple GAN. Experience replay refers to minimizing the jumps between patterns by displaying previous generated samples to the discriminator at regular intervals. Using multiple GAN is to achieve the purpose of covering all patterns in the sample by merging multiple GAN, avoiding the occurrence of a single pattern. Zhang [3] proposed a model based on GAN for non-frontal FER. The generator in this method is based on the encoder-decoder structure, and can learn generative and discriminative representations in facial images. It can simultaneously synthesize facial images and recognize facial expressions in different poses and expressions. Due to it based on GAN, mode collapse may occur during the training, resulting in only a single style of facial images generated or the effect of generated images is not ideal.

In view of the superior performance and shortcomings of the model proposed by Zhang, we improved it and proposed a FER model based on PCD-CNN with pose and expression. The experimental results prove that the recognition accuracy of improved algorithm is higher than existing advanced algorithms, and the training process is more stable.

3   Our Method

We first introduce the improved PCD-CNN for face detection and facial pose estimation in this section. Then we describe the overall architecture of our model and how to synthesize labeled facial images for training. Finally, the improvement of the discriminator in the model is introduced to improve the stability of the model during training.

Facial Expression Recognition Based on PCD-CNN 521

3.1 Improved PCD-CNN

In order to accurately locate the facial landmarks and estimate the pose, Kumar [4] proposed PCD-CNN following the Bayes formula. The facial pose of image is accurately disentangled by conditionally adjusting the facial landmarks based on PCD-CNN. Inspired by this, we use PCD-CNN to perform face detection and pose estimation on the facial image before conveying the image to the generator. According to Kumar's settings [4], following the natural hierarchy between image I, head pose P and keypoint C variables, the joint and conditional probability are represented as Eq. (1) and Eq. (2) respectively.

$$p(C,P,I) = p(C\,|\,P,I)p(P\,|\,I)p(I) \tag{1}$$

$$p(C,P\,|\,I) = \frac{p(C,P,I)}{p(I)} = \underbrace{p(P\,|\,I)}_{CNN}\underbrace{p(C\,|\,P,I)}_{PCD-CNN} \tag{2}$$

In Eq. (2), in order to achieve $p(P\,|\,I)$, we train a CNN based on facial image to roughly predict the facial pose. In order to achieve $p(C\,|\,P,I)$, we combine convolutional neural network with multiple deconvolutional neural network arranged in a dendritic structure. The convolutional neural network maps the image to a lower dimension, and then the output of the deconvolutional neural network is used to form facial landmark heatmap. As shown in Fig. 2, PCD-CNN includes a PoseNet and a KeypointNet, which are used for the pose and keypoint estimation, respectively. The different keypoints coordinate position is determined according to the mutual relationship between the keypoints. In order to capture the relationship between different keypoints, the nose is used as the root node. The correlation between different keypoints is modeled by special functions $f_{i,j}$ [4], which can also be achieved by convolution. When adding responses corresponding to neighboring nodes, the low confidence for specific keypoints is strengthened [4]. The experimental results of Kumar prove that the model can effectively capture the shape constraints in the deep learning framework, conditionally estimate the facial landmarks to disentangle the facial pose of the image.

However, the PCD-CNN is based on the Squeezenet-11 architecture. Because more parameters are more likely to cause overfitting during training, and the computing performance of the machine is also higher, the processing time is longer. In order to maintain the performance with fewer parameters, we base the PCD-CNN model on the ShuffleNet-v2 [5] architecture. Due to the special structure of ShuffleNet-v2, the calculation cost is greatly reduced, not only the calculation complexity is very low, but also the accuracy is very high. ShuffleNet-v2 deprecated the 1×1 group convolution operation and directly used 1×1 ordinary convolution with the same number of input/output channels. A new type of channel separation operation is proposed. The input channel of the module is divided into two parts, one part is directly passed down and the other part is used for true backward calculation. At the end of the module, the output channels on the two branches are directly connected in series. Then perform random array operation on the output feature map of the final output, so that the information between the channels is communicated. It also provides a variant of the module that requires downsampling. In order to ensure that the total number of output channels is increased during downsampling, it cancels the random array operation at the beginning of the module, so that the channels are processed separately and then stitched together to double the final output channel number.

Our improved PCD-CNN model is based on the ShuffleNet-v2 structure and is implemented by combination of a convolutional neural network and multiple deconvolutional neural network. As shown in Fig. 3, we call it ShuffleNet-DeconvNet. We perform a convolution operation on the eighth pooling layer of PoseNet in the model, and then feed it back to the eighth pooling layer of KeypointNet. We perform ReLU nonlinearity and batch normalization after each convolution layer. Every landmark in the improved PCD-CNN is modeled by a separate ShuffleNet-DeconvNet, so that the parameters in the model can be effectively reduced. Before conveying the facial image to the generator, we use the improved PCD-CNN to preprocess it first. Compared with the lib face detection algorithm with 68 facial keypoints used by Zhang, our method can not only accurately detect human faces, but also effectively estimate facial poses.

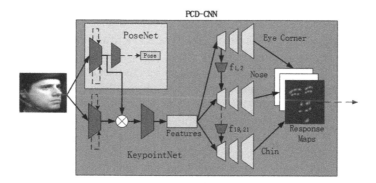

Fig. 2. The overall architecture of the PCD-CNN.

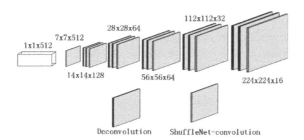

Fig. 3. The architecture of ShuffleNet-DeconvNet.

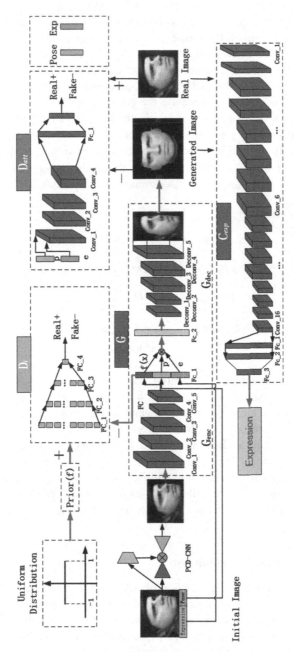

Fig. 4. The overall framework of our model. First, we use improved PCD-CNN model to detect human face and estimate pose. Image is then entered into the encoder-decoder structure generator G for facial image synthesis in different poses and expressions. At last, the generated images and the real images are used for training of the classifier C_{exp} to realize non-frontal FER.

3.2   The Overall Framework of the Model

After improved PCD-CNN preprocesses the facial images, we follow the work of Zhang to put the processed images into a generator based on the encoder-decoder structure, and learn the identity of the images at the same time, as shown in Fig. 4. The model includes a generator G, two discriminators D_{att}, D_i, and a classifier C_{exp}. The generator G is divided into an encoder G_{enc} and a decoder G_{dec}. G_{enc} learns the mapping $f(x)$ from the facial image to the identity, and then $f(x)$ connect the pose and expression into the decoder G_{dec} to generate the corresponding pose and expression facial image. The task of the discriminator D_{att} in the model is to distinguish whether the entered facial images are generated images or real images, which is a binary classification problem. In the training process, the distribution of training data is $P_{d(x,y)}$, the label of the training image is y, and following the image generation principle of GAN, we can train our generator G and discriminator D_{att} by Eq. (3).

$$\min_{G} \max_{D_{att}} E_{x,y \sim p_d(x,y)} [\log D_{att}(x,y)] \\ + E_{x,y \sim p_d(x,y)} [\log(1 - D_{att}(G(x,y),y))] \tag{3}$$

$$\min_{G} \max_{D_i} E_{f^* \sim Prior(f)} [\log D_i(f^*)] \\ + E_{x \sim p_d(x)} [\log(1 - D_i(G_{enc}(x)))] \tag{4}$$

$$L_c(G,C) = E_{x,y^e} [-y^e \log C(G(x),y^e) \\ -y^e \log C(x,y^e)] \tag{5}$$

The role of the discriminator D_i is mainly to be able to improve pose smooth and expression conversion. It can be seen from Fig. 4 that $f(x)$ is an identity mapping of facial image. It needs to be clear that when the generator generates a facial image, the pose and expression are changed, and the identity characteristics of the image remain unchanged. $Prior(f)$ is a prior distribution, and $f^* \sim Prior(f)$ represents a random sampling process from $Prior(f)$ [3]. The generator G and discriminator D_i are trained by the min-max objective function, as shown in Eq. (4). The role of the classifier C_{exp} is mainly to classify all the images in different poses to achieve expression recognition, include generated images and real images. The loss function of the classifier is softmax, as shown in Eq. (5).

3.3   Improvements to the Discriminator

In order to avoid the mode collapse as much as possible and improve the stability of the model during the image generation process, we improved the discriminator model of Zhang [3]. The model has two discriminators D_i and D_{att}. The role of the discriminator D_i is mainly to improve pose smooth and expression conversion. The task of the discriminator D_{att} is not only to distinguish whether the samples come from the real samples or the generated samples, but also to feedback the generator information to generate samples with dissimilar styles as much as possible. When mode collapse occurs during the image generation process, the generator G often maps different hidden variables Z to the same pattern X. After updating the discriminator D_{att}, the discriminator will soon find that the pattern X is abnormal. Therefore, the degree of trust in the pattern X is reduced, that is, the probability of the sample in the training dataset to generate pattern X, so the generator G will map many different hidden variables Z to another pattern Y. Similarly, the discriminator will find that the pattern Y is also abnormal, so the discriminator and generator enter a meaningless loop [22]. Regarding the feedback of the discriminator, the response of generator is excessive. Ideally, the generator should map some hidden variables Z to pattern Y and map some hidden variables Z to pattern X. The reason for the mode collapse is related to the way we train the generator. The objective function is shown in Eq. (6), The generator G generates m samples $\{x_1, x_2, \ldots, x_i, \ldots, x_m\}$, and then sends these x_i to the discriminator D_{att} separately to obtain gradient information. Because the discriminator can only process one sample independently at a time, the gradient information obtained by the generator on each sample lacks unified coordination and points in one direction [23]. And there is no mechanism that requires that the output results of the generator differ greatly from each other. For example, due to the discriminator does not trust the pattern X and trusts the pattern Y very much, the discriminator will guide generator to approach the pattern Y for any randomly generated sample: $G(z) \to Y$. That is, for any sample, the gradient direction passed by the discriminator D_{att} to the generator G is the same, the generator updates the parameters according to the gradient direction, and it is very easy to transfer all the hidden variables Z to the pattern Y, thereby generating a single-style facial images. To solve this problem, Goodfellow [12] proposed a mini-batch discriminator. For each sample x_i of a mini-batch, the result of an intermediate layer $g(x)$ of the discriminator is led out. It is an n-dimensional vector, which is multiplied by a learnable $n \times p \times q$-dimensional tensor T, the $p \times q$-dimensional feature matrix M_i of the sample x_i is obtained, which can be regarded as $p \times q$-dimensional features. Calculate the sum of the r-th feature difference between each sample x_i and other samples in the mini-batch, as shown in Eq. (7), where $M_{i,r}$ represents the r-th row of the matrix M_i, and the difference between the two vectors is represented by the $L1$ norm.

$$\min_{\theta_G} \frac{1}{m} \sum_{i=1}^{m} \log(1 - D(G(z^{(i)}))) \tag{6}$$

$$o(x_i)_r = \sum_j \exp\left(-\|M_{i,r} - M_{j,r}\|_{L1}\right) \tag{7}$$

Then each sample will calculate and get a corresponding vector, as shown in Eq. (8).

$$o(x_i) = \left[o(x_i)_1, o(x_i)_2, \ldots, o(x_i)_p\right]^T \tag{8}$$

$$o(x_i) = \frac{1}{n} \sum_{i=1}^{n} (\sigma_i) \tag{9}$$

$$\sigma_i = \sqrt{\frac{1}{m-1}\sum_{j=1}^{m}\left(g\left(x_j\right)_i - \hat{g}_i\right)^2} \tag{10}$$

Finally, $o(x_i)$ is connected to the next layer of the corresponding intermediate layer. We simplified the calculation method of the mini-batch discriminator to make the calculation process easier. For the input sample x_i of the discriminator, we extract an intermediate layer as the n-dimensional feature $\{g(x_1), g(x_2), \ldots, g(x_i), \ldots, g(x_m)\}$, calculate the standard deviation of each dimension and average the values, as shown in Eq. (9) and Eq. (10). Similarly, $o(x_i)$ is connected to the output of the corresponding intermediate layer as a feature map. The improved mini-batch discriminator does not contain the parameter T to be learned, and directly calculates the statistical characteristics of the batch samples, which is more concise. The idea of the two methods is basically the same, no longer let the discriminator process only one sample at a time, but process all samples of a mini-batch at the same time [13]. The specific implementation is based on the original discriminator intermediate layer add a mini-batch layer, whose input is $g(x_i)$ and the output is $o(x_i)$. The difference is that method of Goodfellow also includes a learning parameter T, and the calculation process involves norm, which is more complicated. We will extract an intermediate layer of D_{att} as the n-dimensional feature $\{g(x_1), g(x_2), \ldots, g(x_i), \ldots, g(x_m)\}$, calculate the standard deviation of each dimension and average it. When mode collapse occurs and the generator needs to be updated, generator G first generates mini-batch of samples $\{G_1, G_2, \ldots, G_i, \ldots, G_m\}$. Since these samples are in pattern X, The D_{att} will determine how close one sample is to other samples in mini-batch, so as to distinguish these samples that lack diversity. And the mini-batch discriminator will not simply give all samples x_i the same gradient direction. Thereby avoiding the occurrence of mode collapse, and improving the stability during model training.

4 Experimental Results and Analysis

In order to prove the effectiveness of improved model, we conducted experiments on the Multi-PIE and BU-3DFE standard datasets, respectively, then compared with Zhang and the latest algorithm.

4.1 Experimental Datasets

The datasets used in our experiments are Multi-PIE and BU-3DFE, as shown in Table 1. These are two standard facial expression datasets with various poses, and gradually become an important test set in the field of FER.

Table 1. Details of each dataset used in the experiment.

Datasets	Poses	Expressions	Samples
Multi-PIE [6]	5	6	7655
BU-3DFE [7]	35	6	21000

The Multi-PIE is developed on the basis of the CMU-PIE facial dataset, and contains more than 75,000 multi-pose, illumination and expression facial images of 337 volunteers. The pose and illumination change images were also collected under strict constraints. According to the work of Zhang [3], we selected the facial images of 270 volunteers and captured 1,531 images at five pan angles of $(\pm30°, \pm15°, 0°)$ respectively, so we have a total of $1531 \times 5 = 7,655$ facial images in our experiments. The expressions of the images were divided into six categories: disgust, neutral, scream, smile, squint and surprise. Similarly, we use 5-fold cross-validation. So we have 6,124 training samples and 1,531 testing samples respectively [3]. Due to our method could generate facial images with different style, the generated images and the real images together are $6124 \times 5 \times 6 + 6124 = 189,844$ images to train our classifier.

The BU-3DFE facial dataset is a sequence of 606 facial expressions obtained from 100 volunteers. It contains 6 expressions of anger, disgust, fear, happy, sad, and surprise, and is mostly used for 3D facial expression analysis [7]. Similarly, we follow the work of Zhang, the poses of the used facial images include 7 pan angles $(\pm45°, \pm30°, \pm15°, 0°)$ and 5 tilt angles $(\pm30°, \pm15°, 0°)$ [7]. We randomly divided 100 volunteers into 80 as the training dataset and 20 as the testing dataset. Therefore, in our experiments, there are $100 \times 6 \times 5 \times 7 = 21,000$ facial images, including 16,800 training samples and 4,200 test samples.

4.2 Experiment Introduction

The overall architecture of our method is shown in Fig. 4. Firstly, we detected faces and estimated facial poses based on the improved PCN-CNN model. Then according to the settings of Zhang [4], the facial image is cropped to size 224×224, and the image intensity is linearly scaled to the range of [1,1]. In the model, the generator is implemented based on the encoder-decoder structure. The encoder and decoder are connected through $f(x)$ that identifies the characteristics. $f(x)$ associates the pose p with the expression e and outputs it through the fully connected layer in the network. We use fractionally-strided convolution to transform the cascaded vector into a generated image of pose p and expression e with the same size as the real image. The main role of D_{att} is to be able to distinguish whether the samples are generated samples or real samples, discriminator performs batch normalization after each convolutional layer. The classifier network C_{exp} is implemented based on VGGNet-19. In our model, the classifier C_{exp} is trained by using both generated samples and real samples. Our model is implemented based on TensorFlow [14] and is trained with the ADAM optimizer [15]. It has a learning rate of 0.0002 and momentum 0.5. All weights are initialized from a zero-centered normal distribution with a standard deviation of 0.01.

4.3 Experimental Results

Experimental Results on the Multi-PIE Dataset. The red part represents the experimental results of our improved model, and the blue part represents the results of Zhang, as shown in Fig. 5. We can observe that except disgust (DI), the accuracy rate of the other five expressions are higher than the model of Zhang, and the accuracy of four expressions exceeds 93%. Figure 6 shows the each pose accuracy rate in two methods. It can be seen that our model has higher recognition accuracy in any pose than the model of Zhang, and the average recognition accuracy is 93.08%. Because before the facial image is passed to the generator, we use the improved PCD-CNN model to replace the lib face detection model. The improved model can capture facial landmarks well. Thus, the face area is accurately detected, and the facial pose is estimated, which improves the quality of synthetic facial image in different poses and expressions, and makes the classifier easier extract facial features in the image. In order to compare the convergence rate of the two methods, we selected N real images. Through the generative model, we can get $5 \times 6 \times N = 30N$ samples for training the classifier. From Fig. 7, we can see that our model has a faster convergence rate than model of Zhang, indicating that improved PCD-CNN has high computational efficiency. Because improved PCD-CNN is based on the ShuffleNet-v2 framework, it is able to maintain better performance with fewer parameters.

In addition, we also compare the improved model with the current state-of-the-art algorithms [16, 17]. Detailed results are shown in Table 2, we can clearly see the expression recognition accuracy in different poses and average accuracy of each algorithm. The highlighted results in the table are marked in bold. Obviously, our model has the best recognition accuracy at the pan angles +30°, +15°, 0°, and the average accuracy is 93.08%, which is better than all current methods. By careful observation, it can be found that other methods cannot achieve better recognition accuracy when the facial image is on the front. Because our method could synthesize different style facial images to make the training sample more sufficient, so that the classifier can get better performance.

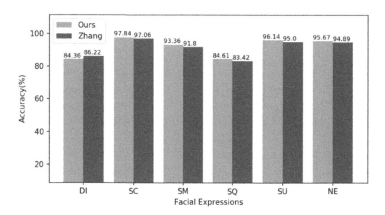

Fig. 5. Expression recognition rate in two methods on the Multi-PIE dataset.

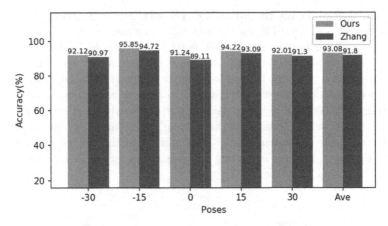

Fig. 6. Expression recognition rate of different poses in two methods on the Multi-PIE dataset.

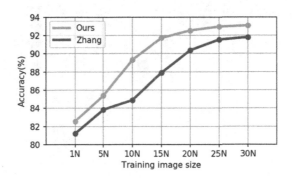

Fig. 7. Comparison of the convergence rate in two methods on the Multi-PIE dataset.

Table 2. Comparison with the latest methods on the Multi-PIE dataset.

Methods	+30	+15	0	−15	−30	Average
DS-GPLVM	90.11	89.97	82.42	**96.96**	**93.55**	90.60
D-GPLVM	86.04	85.96	78.70	93.51	91.65	87.17
GMLPP	87.36	87.22	78.16	94.13	91.86	87.74
GMLDA	85.72	86.64	76.60	94.18	90.47	86.72
GPLRF	86.01	85.66	77.59	93.77	91.65	86.93
MvDA	87.89	87.10	77.51	94.22	92.49	87.84
LDA	87.47	87.07	77.21	94.37	92.52	87.72
kNN	74.78	75.03	68.36	81.74	80.88	76.15
Zhang	91.30	93.09	89.11	94.72	90.97	91.80
Ours	**92.01**	**94.22**	**91.24**	95.85	92.12	**93.08**

Experimental Results on the BU-3DFE Dataset. Figure 8 shows the confusion matrix of our model for FER on the BU-3DFE dataset. It shows that recognition accuracy of surprise reaches the highest at 93.86%, followed by happy at 91.22%. Because when people are surprised or happy, the texture of the facial muscles is obvious. Generator is relatively easy to generate this kind of facial images, and it is easier to extract facial features for classifier to achieve expression recognition than other expressions. From Fig. 9, we can see our improved model convergence speed faster than the model of Zhang, and the training process is more stable. Because we perform mini-batch processing on the discriminator in the model and simplify the calculation method, thereby improving the rate of convergence and stability of the model.

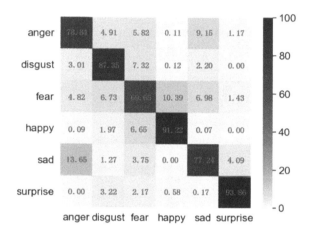

Fig. 8. The confusion matrix of our model for FER on the BU-3DFE dataset.

Similarly, we performed many experiments on the BU-3DFE dataset with improved model and compared the results with the latest methods. As shown in Table 3, our average expression recognition rate reached the highest, at 83.03%, which is higher than the method of Zhang (81.20%) and other methods. Because we have improved model of Zhang, not only generating facial images with various poses and expressions, but before passing the facial images to the generator, we use a more superior PCD-CNN to detect face and estimate facial pose. This method not only has fewer parameters, but also has higher calculation efficiency. In order to avoid the mode collapse as much as possible, we use a mini-batch discriminator to process and directly calculate the statistical characteristics of the mini-batch samples, which is more concise. These measures make generator generate facial images with different styles and greatly to improve the quality and quantity of training samples, facilitate the extraction of feature information for classifier, and improve the accuracy of non-frontal FER.

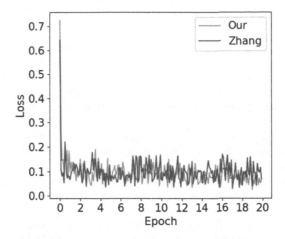

Fig. 9. Comparison of the convergence speed in two methods on the BU-3DFE dataset.

Table 3. Comparison with the current state-of-the-art methods on the BU-3DFE dataset.

Methods	Pan (°)	Tilt (°)	Total	Average
Ekweariri 2017 [8]	(−45, +45)	(−30, +30)	35	75.40
Zhou 2019 [7]	(−45, +45)	(−30, +30)	35	78.64
Jiao 2019 [1]	(−45, +45)	(−30, +30)	35	81.06
Zhang 2018[3]	(−45, +45)	(−30, +30)	35	81.20
Ours	(−45, +45)	(−30, +30)	35	83.03
Ozdemir 2019 [18]	(0, +90)	–	5	80.21
Gacav 2018 [19]	(0, +90)	–	5	78.90
Kim 2019 [20]	(0, +90)	–	5	80.46
Kaleekal 2019 [21]	(0, +90)	–	5	81.32

5 Conclusion

In this paper, we propose a FER model based on PCD-CNN with pose and expression. Firstly, before conveying facial images to the generator, we use the PCD-CNN model to detect faces and estimate facial poses; secondly, in order to accelerate the training speed of the model, the PCD-CNN is based on the ShuffleNet-v2 framework; finally, to avoid the mode collapse in training process, we carry out mini-batch processing on the discriminator in the model and simplify the calculation method. Compared with other methods, our method achieves better results and the training process is more stable. In future work, we will further improve the stability of model training, improve the quality of generated facial image, and consider the impact of other factors on FER, such as illumination changes, identity differences.

Acknowledgment. We would like to acknowledge the support from the National Science Foundation of China (61472095).

References

1. Jiao, Y., Niu, Y., Zhang, Y., Li, F., Zou, C., Shi, G.: Facial attention based convolutional neural network for 2D + 3D facial expression recognition. In: IEEE Visual Communications and Image Processing, pp. 1–4 (2019)
2. Sariyanidi, E., Gunes, H., Cavallaro, A.: Automatic analysis of facial affect: a survey of registration, representation, and recognition. IEEE Trans. Pattern Anal. Mach. Intell. **37**(6), 1113–1133 (2015)
3. Zhang, F., Zhang, T., Mao, Q., Xu, C.: Joint pose and expression modeling for facial expression recognition. In: Conference on Computer Vision and Pattern Recognition, pp. 3359–3368 (2018)
4. Kumar, A., Chellappa, R.: Disentangling 3D pose in a dendritic CNN for unconstrained 2D face alignment. In: Conference on Computer Vision and Pattern Recognition, pp. 430–439 (2018)
5. Dong, J., Yuan, J., Li, L., Zhong, X., Liu, W.: An efficient semantic segmentation method using pyramid ShuffleNet V2 with vortex pooling. In: IEEE 31st International Conference on Tools with Artificial Intelligence, pp. 1214–1220 (2019)
6. Gross, R., Matthews, I., Cohn, J., Kanade, T., Baker, S.: Multi-PIE. In: IEEE International Conference on Automatic Face & Gesture Recognition, pp. 1–8 (2008)
7. Zhou, W., Zhao, C., Lu, L., Zhao, Q.: Dense correspondence of 3D facial point clouds via neural network fitting. In: IEEE International Conference on Image Processing, pp. 3731–3735 (2019)
8. Ekweariri, A.N., Yurtkan, K.: Facial expression recognition using enhanced local binary patterns. In: 9th International Conference on Computational Intelligence and Communication Networks (CICN), pp. 43–47 (2017)
9. Yang, J., Zhang, F., Chen, B., Khan, S.U.: Facial expression recognition based on facial action unit. In: Tenth International Green and Sustainable Computing Conference (IGSC), pp. 1–6 (2019)
10. Goodfellow, I., et al.: Generative adversarial nets. In: NIPS, pp. 2672–2680 (2014)
11. Yin, R.: Multi-resolution generative adversarial networks for tiny-scale pedestrian detection. In: IEEE International Conference on Image Processing (ICIP), pp. 1665–1669 (2019)
12. Salimans, T., Goodfellow, I., Zaremba, W., Cheung, V., Radford, A., Chen, X.: Improved techniques for training GANs (2016)
13. Karras, T., Aila, T., Laine, S., Lehtinen, J.: Progressive growing of GANs for improved quality, stability, and variation (2017)
14. Zeng, Z., Gong, Q., Zhang, J.: CNN model design of gesture recognition based on Tensorflow framework. In: IEEE 3rd Information Technology, Networking, Electronic and Automation Control Conference (ITNEC), pp. 1062–1067 (2019)
15. Zhang, Z.: Improved adam optimizer for deep neural networks. In: IEEE/ACM 26th International Symposium on Quality of Service, Banff, pp. 1–2 (2018)
16. Kim, S., Kim, H.: Deep explanation model for facial expression recognition through facial action coding unit. In: IEEE International Conference on Big Data and Smart Computing (BigComp), pp. 1–4 (2019)
17. Divya, M.B.S., Prajwala, N.B.: Facial expression recognition by calculating euclidian distance for eigen faces using PCA. In: International Conference on Communication and Signal Processing (ICCSP), pp. 0244–0248 (2018)
18. Ozdemir, M.A., Elagoz, B., Alaybeyoglu, A., Sadighzadeh, R., Akan, A.: Real Time Emotion Recognition from Facial Expressions Using CNN Architecture. In: Medical Technologies Congress (TIPTEKNO), pp. 1–4 (2019)

19. Gacav, C., Benligiray, B., Özkan, K., Topal, C.: Facial expression recognition with FHOG features. In: 26th Signal Processing and Communications Applications Conference (SIU), pp. 1–4 (2018)
20. Kim, D.H., Baddar, W.J., Jang, J., Ro, Y.M.: Multi-objective based spatio-temporal feature representation learning robust to expression intensity variations for facial expression recognition. IEEE Trans. Affect. Comput. **10**(2), 223–236 (2019)
21. Kaleekal, T., Singh, J.: Facial Expression recognition using higher order moments on facial patches. In: 10th International Conference on Computing, Communication and Networking Technologies (ICCCNT), pp. 1–7 (2019)
22. Huang, R., Dong, H., Yin, G., Fu, Q.: Ensembling 3D CNN framework for video recognition. In: International Joint Conference on Neural Networks (IJCNN), pp. 1–7. Budapest, Hungary (2019)
23. Fu, Q., Wang, X., Dong, H., Huang, R.: Spiking neurons with differential evolution algorithm for pattern classification. In: IEEE International Conference on Systems, Man and Cybernetics (SMC), Bari, Italy, pp. 152–157 (2019)

Regression-Based Face Pose Estimation with Deep Multi-modal Feature Loss

Yanqiu Wu[1], Chaoqun Hong[1(✉)] [iD], Liang Chen[2], and Zhiqiang Zeng[1]

[1] School of Computer Science and Information Engineering,
Xiamen University of Technology, Xiamen 361024, Fujian, China
cqhong@xmut.edu.cn
[2] School of Data and Computer Science, Sun Yat-Sen University,
Guangzhou 510006, China

Abstract. Image-based face pose estimation tries to estimate the facial direction with 2D images. It provides important information for many face recognition applications. However, it is a difficult task due to complex conditions and appearances. Deep learning method used in this field has the disadvantage of ignoring the natural structures of human faces. To solve this problem, a framework is proposed in this paper to estimate face poses with regression, which is based on deep learning and multi-modal feature loss (M^2FL). Different from current loss functions using only a single type of features, the descriptive power was improved by combining multiple image features. To achieve it, hypergraph-based manifold regularization was applied. In this way, the loss of face pose estimation was reduced. Experimental results on commonly-used benchmark datasets demonstrate the performance of M^2FL.

Keywords: Face pose estimation · Deep learning · Multi-modal features

1 Introduction

Image-based Face pose estimation aims at computing the facial direction or head postures with 2D facial images. Recently, it has attracted plenty of attention since it provides some important information such as communicative gestures, saliency detection and so on [6,14,16]. Therefore, it is critical in human activity analysis, human-computer interface and some other applications [7,17,19,25]. It is also applied to improve the performance of face detection and recognition [5,11].

This work was partly supported in part by the National Natural Science Foundation of China (61871464 and 61836002), the Fujian Provincial Natural Science Foundation of China (2018J01573), the Foundation of Fujian Educational Committee (JAT160357), Distinguished Young Scientific Research Talents Plan in Universities of Fujian Province and the Program for New Century Excellent Talents in University of Fujian Province.

© Springer Nature Singapore Pte Ltd. 2020
J. Zeng et al. (Eds.): ICPCSEE 2020, CCIS 1257, pp. 534–549, 2020.
https://doi.org/10.1007/978-981-15-7981-3_39

Similar to the other applications of computer vision, feature representation of facial images is the key to the problem of face pose estimation [1, 26]. Therefore, a descriptive representation is critical. The pioneering work on face pose estimation was proposed by Robertson and Reid [22] which used a detector based on template training to classify face poses in eight directional bins. This approach is heavily reliant on skin colour model. Subsequently this template-based technique was extended to a color invariant technique by Benfold et al. [2]. Based on the template features, they proposed a randomized fern classifier for hair face segmentation for matching. This work was later improved by Siriteerakul et al. [26] using pair-wise local intensity and colour differences. However, in keeping with all template based techniques in head-pose estimation, these suffer from two major problems: first, it is non-trivial to localize the head in low resolution images; second, different poses of the same person may appear more similar compared to the same head-pose of different persons. Recently, some researchers propose to represent face images in different feature spaces that have more discriminatory property for face pose independent of people. Chen et al. [4] proposed unconstrained coupled face-pose and body-pose estimation in surveillance videos. They used multi-level Histogram of Oriented Gradients (HOG) to represent face and body poses for adaptive classification using high dimensional kernel space methods. A similar idea is applied by Flohr et al. to jointly estimate face poses and body poses [9]. However, face poses are complex and uncertain, which makes these methods unsuitable for fine-grained human interaction or saliency detection.

Recently, face pose estimation benefits from the development of neural networks. Most of existing methods depends facial landmark detection [3, 15]. However, it requires more computation and leads to bigger models. Besides, the performance is reduced with blurred or low-resolution images. Hopenet tried to achieve landmark-free face pose estimation [23]. FSA-Net learned a fine-grained structure mapping for spatially grouping features before aggregation [29]. In this way, different model variants can be generated and form a complementary ensemble. The above methods focus on improve the regression model. Some other methods still try to improve the descriptive power of feature representation. Inspired by previous multi-view ideas, Mukherjee and Robertson further proposed a multimodal method [18]. It uses both RGB images and depth images to improve the performance of pose estimation. Yan et al. defined classification in different views as different tasks. In this way, face pose classification based on multi-task learning is applied in an environment captured by multiple, large field-of-view surveillance cameras [28].

In a word, current trend of face pose estimation is improving the regression model and using multi-view images or multi-modal data. In traditional methods, multi-view integration is often over-simplified. However, in reality, the correspondence between face images and poses is complicated. It is often ambiguous or even arbitrary. Therefore, we try to combine the above idea and use multi-modal features to improve the regression modal. In this way, we develop a big data-driven strategy for face pose estimation in this paper. Specifically, we design a novel deep architecture with Multi-modal Feature Loss (M^2FL) to represent human face data. Different from existing face pose estimation methods, the proposed method applies multiple features are used as the losses of deep neural networks. It is flexible and compatible with different types of features. The contributions of this paper are summarized below:

1. First, we propose a new multi-modal learning framework based on Deep Neural Networks (DNN). In this framework, DNN-based regression is applied to face pose estimation, which provides an end-to-end solution of using multimodal features.
2. Second, in the proposed framework, multi-modal features are represented in a unified feature space by using hypergraph-based manifold learning. In this way, performance is improved by capturing the connectivity among images. It is based on a real-valued form of the combinatorial optimization problem.
3. Finally, we conduct comprehensive experiments to analyze our method on several challenging benchmark datasets. The experimental results validate the effectiveness.

The remainder of this paper is organized as follows. The proposed M^2FL is presented in Sect. 2. After that, we demonstrate the effectiveness of M^2FL by experimental comparisons with other state-of-the-art methods in Sect. 3. We conclude in Sect. 4.

2   Face Pose Estimation with Deep Multi-modal Feature Loss

2.1   Overview of the Proposed Method

The flowchart of the proposed framework is shown in Fig. 1. To get rid of the influences of background, we extract faces in images first. This process depends on the definitions of different datasets. In some datasets, the positions and sizes are provided and they can be used directly. However, in some other datasetes and application scenarios, we need face detection or face tracking to get the face area. Then, we use different types of feature descriptors to represent facial images. These features are integrated by the process of Hypergraph-based Multimodal Feature Fusion. Finally, the fusion feature is used in the loss function of DNN-based regression.

2.2   Regression-Based Deep Neural Networks

In the problem of image-based face pose estimation, the training set contains face images $X = x_n|n = 1, ..., N$ and the pose vector $Y = y_n|n = 1, ..., N$, where N is the number of samples and y_n is the 3D vector (Yaw, Pitch and Roll) corresponding to x_n. The goal of network training is to find a function F, so that it predicts $\bar{y}_n = \mathcal{F}(x_n)$ that provides the optimized error between the real

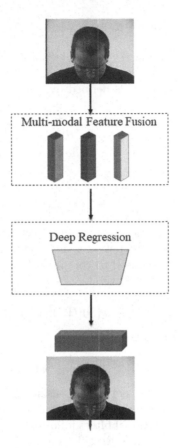

Fig. 1. The flowchart of the proposed framework is shown.

pose y_n and the estimated pose $\overline{y}_n$. Usually, mean absolute error (MAE) is used as the measurement of angular errors, which is defined as:

$$\mathcal{L}(X) = \frac{1}{N} \sum_{n=1}^{N} \| \overline{y}_n - y_n \| . \tag{1}$$

In the process of DNN-based regression, image features x is forwarded by:

$$(x)^l = f((W)^l(x)^{l-1} + (b)^l), \tag{2}$$

where $(W)^l$ and $(b)^l$ are the mapping parameters of the l-th layer. They are computed by the process of optimizing CNNs. $f(\cdot)$ is called the activation function. In computational networks, the activation function of a neuron defines how its output is activated. In the scenario of the deep neural network, activation functions project x to higher level representation gradually by learning a sequence of non-linear mappings. For a CNN with L layers, Eq. (1) will be concatenated L times. To reduce the size of tensors computed by convolutional layers, pooling layers are usually used. In this way, we can get the output δ^L with L convolutional layers.

To optimize the weighted matrix W which contains the mapping parameters, we use a back-propagation strategy. For each echo of this process, the weighted matrix is updated by ΔW, which is defined by:

$$\Delta W = -\eta \frac{\partial \mathcal{Z}}{\partial W}. \tag{3}$$

η is the learning rate and $\mathcal{Z}$ is the energy loss of neural networks. Equation (5) can be further defined as:

$$\frac{\partial \mathcal{Z}}{\partial W} = (y - \mathcal{R}(x))x^T. \tag{4}$$

where $\mathcal{R}$ is the mapping function from input to estimated output. In this way, we try to minimize the differences between the groundtruth y and the estimated output $\mathcal{R}(x)$. To achieve it, we need to compute the backward output, which is defined by:

$$(\delta)^{l-1} = (\delta)^l \cdot rot180((W)^l) \odot f((\mathcal{Z})^l). \tag{5}$$

In this way, W can be updated by:

$$(W)^l = (W)^l - \eta(\delta)^l \cdot rot180(x)^l. \tag{6}$$

Classic convolutional neuron networks consist of alternatively stacked convolutional layers and spatial pooling layers. The convolutional layers generate feature maps by linear convolutional filters followed by nonlinear activation functions (rectifier, sigmoid, tanh, etc.). Rectified Linear Units (ReLU) is defined by:

$$f_r(x) = \begin{cases} 0 \ , \ x < 0 \\ x \ , \ x \geq 0 \end{cases}. \tag{7}$$

With ReLU, the feature map can be calculated as follows:

$$\mathcal{F}(X;W) = \mathcal{R}(x_i) = max(0, wx_i) = \begin{cases} 0 \ , \ x_i < 0 \\ wx_i \ , \ x_i \geq 0 \end{cases}. \tag{8}$$

where w is the element in W corresponding to the output weights of x_i.

2.3 Hypergraph-Based Multi-modal Feature Fusion

Image features X in loss function defined by Eq. (1) play a key role for the above regression process. There are several features to represent facial images based on deep learning, such as HyperFace [21], KEPLER [15], FaceNet [24] and so on. Since single type of features could not completely describe one image, researchers have proposed learning methods by combining different types of features[27]. Inspired by previous works, we use multi-modal features in Eq. (1) and propose Multi-modal Feature Loss (M^2FL). However, traditional methods usually

Table 1. Definition of symbols in the hypergraph.

Symbol	Definition
u, v	Vertices in the hypergraph
e	Edges in the hypergraph
$\omega(e)$	The weight of an edge e
$\epsilon(e)$	The degree of an edge, e. It illustrates how many vertices are connected by e. In traditional graph representation, $\epsilon(e) = 2$
$d(v)$	The degree of a vertex, v. It is calculated by summing the weighting values of edges connected to this vertex.
D_v	The diagonal matrix containing the vertex degrees
D_e	The diagonal matrix containing the edge degrees
H	In this matrix, $H(v,e) = 1$ if $v \in e$
Ω	The diagonal matrix containing the weights of hyperedges
V	The set of vertices
E	The set of edges

applied simple concatenation, which were oversimplified and limited the descriptive power of multi-modal features.

The proposed M^2FL is implemented with Hypergraph-based Multi-modal Feature Fusion (HM^2F), which is based on Patch Alignment Framework [10, 30]. In the graph learning process, features are represented by vertices of this graph and their connectivity is represented by edges. The advantage of using hypergraph in graph learning is that it is able to capture complex connectivity, which can be seen in Fig. 2. Definitions of symbols are shown in Table 1 and the flowchart is shown in Fig. 3. The key to feature fusion in a unified feature space with manifold learning is the construction of Laplacian matrix, which is the third step in Fig. 3. The Laplacian matrix M for feature fusion can be computed with two stages.

1. Part Optimization: Features are represented as the vertices in hypergraph. Furthermore, we define one patch to be the vertices connected by one hyperedge. Thus, the patch in the feature fusion process is defined by:

$$\arg\min_{f \in R^{|V|}} \sum_{m,n \subset e} \frac{w(e)}{\epsilon(e)} (\frac{y_m}{\sqrt{d_m}} - \frac{y_n}{\sqrt{d_n}})^2 \quad (9)$$

For one patch, we should compute:

$$\sum_{m,n \subset e} \frac{w(e)}{\epsilon(e)} (\frac{y_m}{\sqrt{d_m}} - \frac{y_n}{\sqrt{d_n}})^2, \quad (10)$$

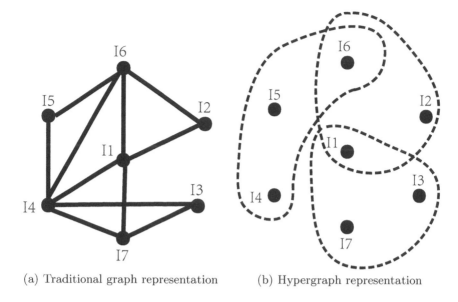

(a) Traditional graph representation   (b) Hypergraph representation

Fig. 2. Comparison of traditional graph and hypergraph

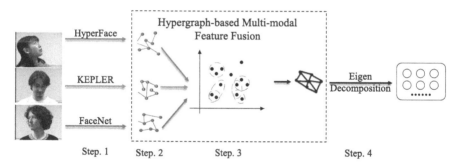

Fig. 3. The flowchart of the multi-modal feature fusion is shown. The process consists of four steps. First, different features are extracted for all the faces. These features become vertices in the graph. Second, affinity matrices are constructed with inner representations. In this way, different features are represented in a unified feature space and features are vertices of this space. Third, affinity matrices are integrated to form the hypergraph Laplacian matrix. Lastly, standard eigen decomposition is applied to get the fused features.

which means that we randomly choose two vertices in the subset of vertices contained by a hyperedge, e, and sum the value of

$$\frac{w(e)}{\epsilon(e)}(\frac{y_m}{\sqrt{d_m}} - \frac{y_n}{\sqrt{d_n}})^2. \tag{11}$$

Expanding (9) and combining items, we can get the patch optimization for each hyperedge:

$$\frac{1}{2}\sum_{v \subseteq e} \frac{F}{DV_v^{\frac{1}{2}}} EH_e' \frac{\Omega}{DE} H_e E' \frac{F}{DV_v^{\frac{1}{2}}}. \tag{12}$$

Matrix E is

$$\begin{bmatrix} -e^T \\ I \end{bmatrix} \tag{13}$$

where $e = [1, ..., 1]^T$, I is an $n \times n$ identity matrix.
This stage is computed for each feature of each modal.

2. Whole alignment: In the hypergraph, the weight of a hyperedge is computed by summing the similarity scores of all the pairs of vertices contained in this hyperedge. The similarity score of any pair of vertices is defined as the distance of image features:

$$S(u, v) = exp(-\frac{1}{\sigma} dist(feat(u), feat(v))), \tag{14}$$

where $feat(u)$ represents the image feature vector of vertex u, and $dist(x, y)$ is usually set to be the L2 distance. With the hyper edge weighting matrix, the multi-view hypergraph Laplacian can be computed by summing the patch optimization defined in Eq. (12) of all the hyperedges:

$$M = \frac{1}{2}\sum_{e \in E}\sum_{v \in e} \frac{F}{DV_v^{\frac{1}{2}}} EH_e' \frac{\Omega}{DE} H_e E' \frac{F}{DV_v^{\frac{1}{2}}}. \tag{15}$$

After this stage, different features can be represented in a unified feature space and the descriptive information is embedded in M.

Solve the standard eigen-decomposition of M to obtain the projection matrix, whose vectors are the eigenvectors corresponding to the k-smallest eigenvalues. Then we get the fused features with k dimensions.

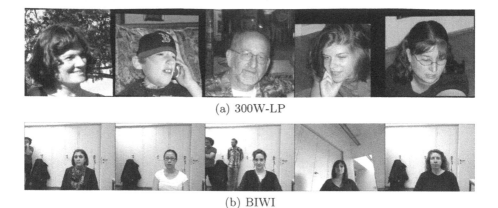

(a) 300W-LP

(b) BIWI

Fig. 4. Sample images of 300W-LP and BIWI

3 Experimental Results and Discussions

3.1 Benchmark Datasets

In this section, we demonstrate the effectiveness of the proposed approach by conducting experiments on two face pose benchmark datasets: 300W-LP [32] and Biwi Kinect Head Pose Database (BIWI) [8]. The 300W-LP dataset was derived from the 300 W dataset which unifies several datasets for face alignment with 68 landmarks. Zhu et al. used face profiling with 3D image meshing to generate 61225 samples across large poses and further expanded to 122450 samples with flipping. The synthesized dataset is named as the 300 W across Large Poses (300W-LP). The BIWI dataset contains 24 videos of 20 subjects in the controlled environment. There are a total of roughly 15000 frames in the dataset. Some sample images are shown in Fig. 4.

To evaluate the performance of different methods, we estimate yaw, pitch and roll degrees. MAE between the estimation and the ground truth are computed. In model training and testing, all the face images are resized to 64×64 and used as the inputs of training models. Then, degrees are used as the outputs. 50% samples are used for training and the rest are used as testing. Experiments are repeated 20 times to get the average performance and standard errors. Besides, we make use of SeetaFace [12] to locate the position of faces.

3.2 Comparison of Different Features

We use three types of features to represent facial images.

- HyperFace [21]. HyperFace learns common features by different intermediate layers in CNNs and obtains 3072 dimensional features. In this way, multi-task learning is achieved for simultaneously performing face detection, facial landmark localization, face pose estimation and gender recognition.

- KEPLER [15]. KEPLER learns global and local features by Heatmap-CNN to explore structural dependencies. It uses 21 key points and get 1029 dimensional features.
- FaceNet [20]. FaceNet provides a method to directly learn an embedding into an Euclidean space for face verification. In this way, 512 dimensional features are computed.

To emphasize the improvement of using multiple features, we compare the performance of face pose estimation of the above three feature and the proposed the proposed hypergraph-based multi-modal feature fusion HM^2F. The results are shown in Table 2 and 3. The items with the best performance are highlighted. Generally speaking, using multiple features provides better performance, except for Roll of 300W-LP. Although HyperFace obtains better average performance, the proposed method provides more stable performance.

Table 2. Performance comparison of different types of features and HM^2F (300W-LP)

Methods	Yaw	Pitch	Roll
HyperFace	7.61 ± 1.73	6.13 ± 2.56	$\mathbf{3.92 \pm 1.31}$
KEPLER	6.45 ± 2.01	5.85 ± 2.09	8.75 ± 2.09
FaceNet	8.12 ± 1.95	5.79 ± 2.13	7.38 ± 2.13
HM^2F	$\mathbf{5.98 \pm 1.66}$	$\mathbf{5.18 \pm 1.87}$	3.98 ± 0.91

Table 3. Performance comparison of different types of features and HM^2F (BIWI)

Methods	Yaw	Pitch	Roll
HyperFace	6.92 ± 1.29	5.78 ± 1.79	4.78 ± 1.19
KEPLER	5.89 ± 1.67	6.32 ± 1.69	6.88 ± 1.38
FaceNet	6.65 ± 1.33	5.39 ± 1.65	5.41 ± 1.45
HM^2F	$\mathbf{5.39 \pm 1.21}$	$\mathbf{4.89 \pm 1.15}$	$\mathbf{4.17 \pm 1.07}$

3.3   Comparison of Different Feature Fusion Methods

To demonstrate the effectiveness of the proposed HM^2F, we compare it with existing manifold learning methods, such as LDA, DLA, LPP, NPE, LSDA and ISOMAP [31]. The results with different settings of dimensionality k are shown in Fig. 5 and 6. We can figure out HM^2F achieves the best performance under 500d to 600d, although it is not always the best under each dimensionality. According to the results, we can see that HM^2F is capable of discovering the nonlinear degree of freedom that underlies complex natural observations.

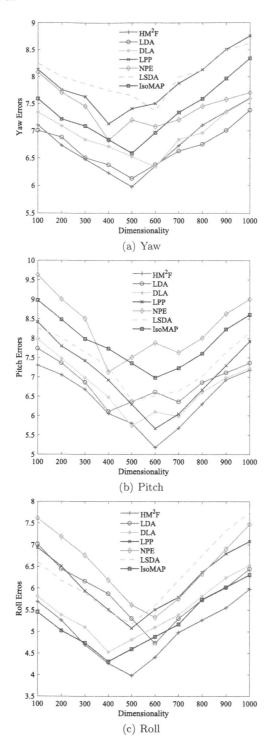

(a) Yaw

(b) Pitch

(c) Roll

Fig. 5. Comparison of different manifold learning methods (300W-LP)

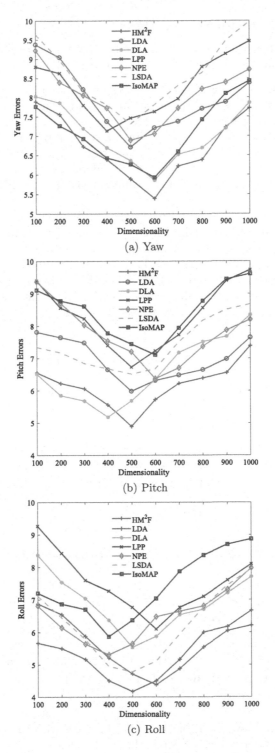

(a) Yaw

(b) Pitch

(c) Roll

Fig. 6. Comparison of different manifold learning methods (BIWI)

3.4   Comparison with State-of-the-Arts

For face pose estimation, we compare the proposed Face Pose Estimation with Deep Multi-modal Feature Loss (M^2FL) with the following existing methods.

- Dlib [13]. Dlib estimates facial landmarks on gradient boosting for learning an ensemble of regression trees that optimizes the sum of square error loss. With the landmarks, we can estimate facial poses.
- FAN [3]. We construct, for the first time, a very strong baseline by combining a state-of-the-art architecture for landmark localization with a state-of-the-art residual block, train it on a very large yet synthetically expanded 2D facial landmark dataset and finally evaluate it on all other 2D facial landmark datasets.
- HopeNet [23]. In this method, Euler angles (yaw, pitch and roll) directly from image intensities through joint binned pose classification and regression.
- FSA-Net [29]. FSA-Net learns a fine-grained structure mapping for spatially grouping features before aggregation. The fine-grained structure provides part-based information and pooled values. By utilizing learnable and non-learnable importance over the spatial location, different model variants can be generated and form a complementary ensemble.

Table 4. Performance comparison of face pose estimation (300W-LP)

Methods	Yaw	Pitch	Roll
dlib	7.13 ± 1.79	5.89 ± 1.92	6.39 ± 1.55
FAN	6.87 ± 1.87	5.91 ± 1.89	6.16 ± 1.19
HopeNet	6.28 ± 1.73	5.56 ± 1.91	5.22 ± 1.21
FSA-Net	6.22 ± 1.63	5.34 ± 2.01	4.19 ± 1.03
M^2FL	$\mathbf{5.98 \pm 1.66}$	$\mathbf{5.18 \pm 1.87}$	$\mathbf{3.98 \pm 0.91}$

Table 5. Performance comparison of face pose estimation (BIWI)

Methods	Yaw	Pitch	Roll
dlib	6.85 ± 1.65	5.71 ± 1.49	5.89 ± 1.87
FAN	7.13 ± 1.49	5.65 ± 1.58	5.18 ± 1.81
HopeNet	5.89 ± 1.51	5.26 ± 1.33	4.44 ± 1.71
FSA-Net	5.76 ± 1.34	5.01 ± 1.24	$\mathbf{3.93 \pm 1.56}$
M^2FL	$\mathbf{5.39 \pm 1.21}$	$\mathbf{4.89 \pm 1.15}$	4.17 ± 1.07

The results are shown in Table 4 and 5. The items with the best performance are highlighted. Based on it, we can make the following summarizations:

1. The methods without key points such as HopeNet, FSA-net and M^2FL achieve better performance.
2. The performance the propose M^2FL are better than state-of-the-arts except roll of BIWI.
3. In addition to better performance, M^2FL also provides more stable results.

4 Conclusion

In this paper, we propose a face pose estimation method. It improves previous methods by using multimodal features in DNN-based regression. Thanks to the application of hypergraph manifold learning, we obtain better representation of facial images and the representation is used in the loss function of training DNN. Compared with state-of-the-arts, estimating errors have been reduced by about 10%. In the future, we will consider different levels of neural networks to improve the performance, such as better definitions of neurons, improved network layers and more reasonable ways to fuse different networks.

References

1. BenAbdelkader, C.: Robust head pose estimation using supervised manifold learning. In: Daniilidis, K., Maragos, P., Paragios, N. (eds.) ECCV 2010. LNCS, vol. 6316, pp. 518–531. Springer, Heidelberg (2010). https://doi.org/10.1007/978-3-642-15567-3_38
2. Benfold, B., Reid, I.D.: Colour invariant head pose classification in low resolution video. In: British Machine Vision Conference, pp. 1–10 (2008)
3. Bulat, A., Tzimiropoulos, G.: How far are we from solving the 2D & 3D face alignment problem? (and a dataset of 230,000 3D facial landmarks). In: IEEE International Conference on Computer Vision, pp. 1021–1030 (2017)
4. Chen, C., Odobez, J.M.: We are not contortionists: coupled adaptive learning for head and body orientation estimation in surveillance video. In: IEEE Conference on Computer Vision and Pattern Recognition, pp. 1544–1551. IEEE (2012)
5. Ding, C., Xu, C., Tao, D.: Multi-task pose-invariant face recognition. IEEE Trans. Image Process. **24**(3), 980–993 (2015)
6. Drouard, V., Horaud, R., Deleforge, A., Ba, S., Evangelidis, G.: Robust head-pose estimation based on partially-latent mixture of linear regressions. IEEE Trans. Image Process. (2016)
7. Du, G., Zhang, P., Liu, X.: Markerless human manipulator interface using leap motion with interval Kalman filter and improved particle filter. IEEE Trans. Ind. Inf. **12**(2), 694–704 (2017)
8. Fanelli, G., Dantone, M., Gall, J., Fossati, A., Gool, L.V.: Random forests for real time 3D face analysis. Int. J. Comput. Vis. **101**(3), 437–458 (2013). https://doi.org/10.1007/s11263-012-0549-0
9. Flohr, F., Dumitru-Guzu, M., Kooij, J.F.P., Gavrila, D.M.: A probabilistic framework for joint pedestrian head and body orientation estimation. IEEE Trans. Intell. Transp. Syst. **16**(4), 1872–1882 (2015)
10. Hong, C., Yu, J., Li, J., Chen, X.: Multi-view hypergraph learning by patch alignment framework. Neurocomputing **118**, 79–86 (2013)

11. Huang, Q.Y., Jia, C.K., Zhang, X.F., Ye, Y.M.: Learning discriminative subspace models for weakly supervised face detection. IEEE Trans. Ind. Inf. **13**(6), 2956–2964 (2017)
12. Kan, M., Kan, M., Shan, S., Shan, S., Chen, X.: Funnel-structured cascade for multi-view face detection with alignment-awareness. Neurocomputing **221**, 138–145 (2016)
13. Kazemi, V., Sullivan, J.: One millisecond face alignment with an ensemble of regression trees. In: IEEE Conference on Computer Vision & Pattern Recognition, pp. 1867–1874 (2014)
14. Kong, S.G., Mbouna, R.O.: Head pose estimation from a 2D face image using 3D face morphing with depth parameters. IEEE Trans. Image Process. **24**(6), 1801–1808 (2015)
15. Kumar, A., Alavi, A., Chellappa, R.: KEPLER: keypoint and pose estimation of unconstrained faces by learning efficient H-CNN regressors. Computer Vision and Pattern Recognition (2017)
16. Lu, F., Sugano, Y., Okabe, T., Sato, Y.: Gaze estimation from eye appearance: a head pose-free method via eye image synthesis. IEEE Trans. Image Process. **24**(11), 3680–93 (2015)
17. Luo, R.C., Chen, S.Y.: Human pose estimation in 3-D space using adaptive control law with point-cloud-based limb regression approach. IEEE Trans. Ind. Inf. **12**(1), 51–58 (2016)
18. Mukherjee, S.S., Robertson, N.M.: Deep head pose: gaze-direction estimation in multimodal video. IEEE Trans. Multimedia **17**(11), 2094–2107 (2015)
19. Pawiak, P., Sonicki, T., Niedwiecki, M., Tabor, Z., Rzecki, K.: Hand body language gesture recognition based on signals from specialized glove and machine learning algorithms. IEEE Trans. Ind. Inf. **12**(3), 1104–1113 (2016)
20. Rajagopal, A.K., Subramanian, R., Ricci, E., Vieriu, R.L., Lanz, O., Sebe, N., et al.: Exploring transfer learning approaches for head pose classification from multi-view surveillance images. Int. J. Comput. Vis. **109**(1), 146–167 (2013). https://doi.org/10.1007/s11263-013-0692-2
21. Ranjan, R., Patel, V.M., Chellappa, R.: Hyperface: a deep multi-task learning framework for face detection, landmark localization, pose estimation, and gender recognition. IEEE Trans. Pattern Anal. Mach. Intell. **41**, 121–135 (2018)
22. Robertson, N., Reid, I.: Estimating gaze direction from low-resolution faces in video. In: Leonardis, A., Bischof, H., Pinz, A. (eds.) ECCV 2006. LNCS, vol. 3952, pp. 402–415. Springer, Heidelberg (2006). https://doi.org/10.1007/11744047_31
23. Ruiz, N., Chong, E., Rehg, J.M.: Fine-grained head pose estimation without keypoints. In: The IEEE Conference on Computer Vision and Pattern Recognition (CVPR) Workshops, June 2018
24. Schroff, F., Kalenichenko, D., Philbin, J.: FaceNet: a unified embedding for face recognition and clustering. In: Computer Vision and Pattern Recognition, pp. 815–823 (2015)
25. Simao, M.A., Neto, P., Gibaru, O.: Unsupervised gesture segmentation by motion detection of a real-time data stream. IEEE Trans. Ind. Inf. **13**(2), 473–481 (2016)
26. Siriteerakul, T., Sugimura, D., Sato, Y.: Head pose classification from low resolution images using pairwise non-local intensity and color differences. In: Pacific-Rim Symposium on Image and Video Technology, pp. 362–369. IEEE (2010)
27. Vadakkepat, P., Lim, P., Silva, L.C.D., Jing, L., Ling, L.L.: Multimodal approach to human-face detection and tracking. IEEE Trans. Ind. Electron. **55**(3), 1385–1393 (2008)

28. Yan, Y., Ricci, E., Subramanian, R., Liu, G., Lanz, O., Sebe, N.: A multi-task learning framework for head pose estimation under target motion. IEEE Trans. Pattern Anal. Mach. Intell. **38**(6), 1070–1083 (2016)
29. Yang, T.Y., Chen, Y.T., Lin, Y.Y., Chuang, Y.Y.: FSA-Net: learning fine-grained structure aggregation for head pose estimation from a single image. In: The IEEE Conference on Computer Vision and Pattern Recognition (2019)
30. Zhang, T., Tao, D., Li, X., Yang, J.: Patch alignment for dimensionality reduction. IEEE Trans. Knowl. Data Eng. **21**, 1299–1313 (2009)
31. Zhang, T., Tao, D., Li, X., Yang, J.: Patch alignment for dimensionality reduction. IEEE Trans. Knowl. Data Eng. **21**(9), 1299–1313 (2009)
32. Zhu, X., Zhen, L., Liu, X., Shi, H., Li, S.Z.: Face alignment across large poses: a 3D solution. In: IEEE Conference on Computer Vision & Pattern Recognition, pp. 146–155 (2016)

Breast Cancer Recognition Based on BP Neural Network Optimized by Improved Genetic Algorithm

Wenqing Xie, Wenhan Li, and Zhifang Wang$^{(\boxtimes)}$

Department of Electronic Engineering, Heilongjiang University,
Harbin 150080, China
wangzhifang@hlju.edu.cn

Abstract. In recent years, the incidence of breast cancer is increasing and becomes one of the main causes of female death. The BP neural network optimized by standard genetic algorithm has slow convergence speed and is prone to local optimization, which makes the diagnosis accuracy of breast cancer decrease. This paper uses the improved genetic algorithm to optimize BP neural network by improving the selection operator of the standard genetic algorithm. The population diversity was first increased, and the probability of crossover and mutation was adaptively adjusted. Then deep optimization was executed on the initial weight threshold of BP network to speed up the network's convergence, and the number of iterations was reduced. Finally breast cancer diagnose was performed. The experiment results show that both the fitness of the improved genetic algorithm and the recognition accuracy of breast cancer are improved. The shortcomings of the standard genetic algorithm optimized BP neural network algorithm in breast cancer diagnosis are well solved.

Keywords: Breast cancer · Genetic algorithm · BP neural network

1 Introduction

Breast cancer is one of the main malignant tumors in women, and its morbidity and mortality have been rising, causing serious harm to women's health [1]. In medicine, the pathological research of breast cancer has not made any breakthrough. The diagnosis of breast cancer mainly relies on prevention. The sooner it is found, the more effective the treatment will be. Because women do not pay enough attention to breast cancer, coupled with medical and economic constraints, they often lose opportunities for early diagnosis and cure. In recent years, based on the development of computer-aided diagnosis technology based on medical imaging, doctors can better diagnose breast cancer with the help of computer-assisted diagnosis and improve the diagnosis accuracy of breast cancer. Therefore, the automatic breast cancer diagnosis system has become a hot topic in current research.

The traditional breast cancer diagnosis method is to perform medical imaging on the lesion and then consult the doctor. Experts judge the images based on experience and obtain classification results. This method is mainly based on the subjective

J. Zeng et al. (Eds.): ICPCSEE 2020, CCIS 1257, pp. 550–562, 2020.
https://doi.org/10.1007/978-981-15-7981-3_40

judgment of experts, and requires experts to have a high degree of expertise in breast cancer diagnosis to effectively reduce the rate of misdiagnosis, otherwise the diagnosis results are extremely unreliable. In recent years, with the development of digital image processing technology and artificial intelligence technology, new approaches have been provided for breast cancer automatic diagnosis. Among them, artificial neural network is a relatively mature pattern classification technology, which has associative memory and self-learning ability, and has become an effective tool for computer-aided breast diagnosis. Among them, the BP (back propagation) neural network based on error back propagation is the most widely used model [2].

However, the BP neural network prediction model has two major shortcomings: first, the random initial assignment of the BP network connection weights before the model training, which tends to slow down the convergence rate and fall into local extreme points; second, the size of the BP neural network structure is difficult to determine. It is easy to cause the phenomenon of "overfitting" and insufficient learning ability in training. These two defects cause insufficient nonlinear learning and generalization ability. However, genetic algorithms (GA) often suffer from premature convergence and poor convergence performance [3]. In view of the above research deficiencies, this paper proposes a method of optimizing BP neural network based on improved genetic algorithm. The key characteristics of the proposed algorithm are as follows: (1) Improved selection operator of standard genetic algorithm; (2) Used adaptive adjustment of crossover probability; (3) Adaptively adjust mutation probability. The improved genetic algorithm is used to optimize the initial weights and thresholds of BP neural network to overcome the random defect of parameter initialization.

2   BP Neural Network

BP neural network is a kind of multilayer feedforward network, which consists of input layer, hidden layer, and output layer. Generally, only a single hidden layer of BP network can be used to approximate any continuous on a bounded region with arbitrary accuracy in regression prediction. Function, the number of neurons (nodes) in the input and output layers is generally equal to the input and output vector dimensions of the training sample [4], so the size of the BP network structure is ultimately determined by the number of nodes in the hidden layer.

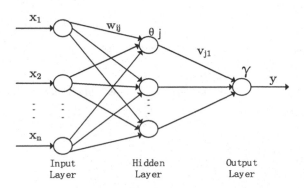

Fig. 1. Structure of the BP neural network with single hidden layer and single-output.

A typical single hidden layer, single output BP neural network topology is shown in Fig. 1: $x_i = (x_1, x_2, \cdots, x_n)$ is a set of input vectors, the expected value is y, the number of hidden layer nodes is m, and $w_{ij}(i = 1, 2, \cdots, n; j = 1, 2, \cdots, m)$ is the distance between the input layer and the hidden layer. Connection weight, $v_{j1}(= 1, 2, \cdots, m)$ is the connection weight between the hidden layer and the output layer, and $\theta_j(j = 1, 2, \cdots m)$ and γ are the hidden layer and output node thresholds, respectively. The BP neural network prediction model mainly implements its training process through the forward propagation of the input signal and the backward propagation of the error signal: the input vector x_i is propagated layer by layer from the input layer the hidden layer $\rightarrow$ the output layer, and the weights w_{ij}, The random initial assignment of v_{j1} and threshold values θ_j and γ and the calculation of the excitation function are used to calculate the predicted output $\hat{y}$ of the output layer; the error e between the output value $\hat{y}$ and the expected y is propagated layer by layer from the output layer $\rightarrow$ the hidden layer $\rightarrow$ the input layer, Modify the connection weights and thresholds of each layer along the direction of error reduction; repeat the above two processes until the algorithm converges. The specific calculation steps of BP neural network prediction algorithm are as follows [5]:

Forward transmission of information:

The input layer of the BP neural network is equivalent to a buffer, and its input is equal to the output, that is, the output of the input layer is

$$f(x) = x \tag{1}$$

The output of the j-th neuron node of the hidden layer can be expressed as:

$$a_{1j} = f_1\left(\sum_{i=1}^{P} w_{ij}P_i - \theta_j\right) \tag{2}$$

The output of the k-th neuron node in the output layer can be expressed as:

$$a_{2k} = f_2\left(\sum_{j=1}^{n} a_{1j}h_k - \varphi_k\right) \tag{3}$$

The error function of the network is:

$$E = \frac{1}{2}\sum_{k=1}^{m}(t_k - a_{2k})^2 \tag{4}$$

Among them, $f_1(x), f_2(x)$ all use sigmoid transfer function, t_k is the target output of the network.

Error back propagation based on gradient descent:
See the following formula for the weight change of the output layer:

$$\Delta h_k = -\eta \frac{\partial E}{\partial h_{jk}} = -\eta \frac{\partial E}{\partial h_{jk}} \frac{\partial a_{2k}}{\partial h_{jk}} \tag{5}$$

$$= -\eta(t_k - a_{2k})f_2' a_k = \delta_{jk} a_{1j}$$

In the formula, $e_k = t_k - a_{2k}$, $\delta_{jk} = e_k f_2'$, η represents the learning rate.

$$\Delta \varphi_k = -\eta \frac{\partial E}{\partial \varphi_k} = -\eta \frac{\partial E}{\partial a_{2k}} \frac{\partial a_{2k}}{\partial \varphi_k} = \eta \delta_{jk} \tag{6}$$

See the following formula for the weight change of the hidden layer:

$$\Delta w_{ij} = -\eta \frac{\partial E}{\partial w_{ij}} = -\eta \frac{\partial E}{\partial a_{2k}} \frac{\partial a_{2k}}{\partial a_{1j}} \frac{\partial a_{1j}}{\partial w_{1j}} = \delta_{ij} P_i \tag{7}$$

Among them, $\delta_{ij} = e_j f_1'$, $e_j = \sum_{k=1}^{m} \delta_{jk} h_{jk}$, then.

$$\Delta \theta_j = \eta \delta_{ij} \tag{8}$$

3 Genetic Algorithm

The genetic algorithm [6] is developed by Holland J. of University of Michigan. It was first proposed by the professor in 1975. It simulates the biological evolution process of genetic selection and natural elimination proposed by biologist Darwin. It is a kind of intelligent optimization algorithm with self-organizing, adaptive and self-learning capabilities. It introduces the biological evolution principle "natural selection, survival of the fittest" into the coding tandem population formed by parameter optimization. According to the fitness function, the genetic operations of selection, crossover and mutation are used to delete individuals and retain the fitness value. Individuals, eliminate individuals with poor fitness, and generate new groups that inherit the information of the parent and are superior to the parent. This cycle is repeated until the conditions are met.

3.1 Basic Concept

The basic concepts in genetic algorithm are as follows [7]:

(1) Chromosome: the individual in the population. A chromosome is actually a series of data or arrays. It is a string of symbols represented by one-dimensional string structure data. Each position of the string structure corresponds one-to-one to the value of the gene. Chromosomes can be represented by code strings formed by

different encoding methods, such as binary character strings or decimal character strings. The string of genes is the chromosome.

(2) Population: a certain number of individuals form a population, a population is a set, which contains several subsets of individuals, the number of individuals in the population is called the population size, because each individual represents a solution to the problem, so a population is equivalent to the solution set of the problem.

(3) Fitness: It indicates that a single individual shows the degree of fitness to the environment. Fitness is often used to distinguish the advantages and disadvantages of individuals or solutions in a population. Each individual is equivalent to a solution x of the problem. Each solution x corresponds to a function value $f(x)$, and each individual's function value $f(x)$ is regarded as its fitness to the environment. For different problems, the definition method of fitness function is also different.

(4) Selection: It selects several pairs of individuals from the group according to a certain probability. The choice reflects the principles of survival of the fittest and elimination of the unsuitable. It is to select good parents and let them reproduce the next generation of good individuals. The purpose of selection is to select individuals with strong adaptability.

There are two operation modes:

① Roulette selection

The higher the fitness value, the greater the probability that such an individual will be selected. Its fitness value is compared to the sum of individual fitness values in the entire population. This ratio is the probability that each individual enters the next generation.

② Optimal retention option

In the group, the individual with the highest fitness value does not perform crossover and mutation operations on it; the individuals with low fitness in the current generation group are still not ideal after the crossover or mutation operation, and the individual with the highest fitness value will come Make substitutions.

③ Random competition selection

Basically the same as ①, first select a pair of individuals each time according to the roulette method, then select the individuals with higher fitness values among the two individuals, and eliminate the individuals with lower fitness values.

(5) Crossover: Crossover is the operation of swapping two chromosomes. Through the cross operation, a new generation of individuals can be obtained, and the idea of information exchange is embodied in the cross. The main operation in the genetic algorithm is the cross operation.

① Single cross

Single-point crossover is also called simple crossover. It is a process of randomly setting an intersection in the individual coding string, and then exchanging the part before or after the intersection of the two parent individuals, thereby forming a new individual.

② Two-point cross, multi-point cross

Similar to single-point intersection, except that there are two or more intersections. As shown in Fig. 2.

(6) Mutation: Mutation produces mutation at a gene point on the chromosome, and the newly generated individual after the mutation operation is different from other individuals. The probability of mutation occurring in the genetic algorithm is very small, and the value is generally small. There are two ways to mutate:

① Basic bit variation

With a certain mutation probability, in an individual coding string, a certain gene or genes are randomly assigned to perform the mutation operation. The specific operation process is as follows:

First, a certain mutation probability specifies the mutation point. Specifically, each locus on the individual is designated as the mutation point. Next, operate on the gene value of each specified mutation point: either take the negation operation, or replace with other allele values to generate the next generation of new individuals.

② Uniform variation

For the original gene value at each locus in the individual coding string, a random number uniformly distributed within a certain range is selected and replaced with a certain small probability.

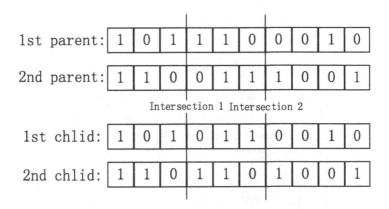

Fig. 2. Schematic diagram of two-point intersection

3.2 Coding and Fitness Function Selection

The operation object of the genetic algorithm is a population, and the population is represented by a binary string. Each chromosome corresponds to a solution to the problem. First, the initial population is coded, and then the selection strategy based on the W fitness ratio is used to select individuals from the current population, and the crossover and mutation operations are applied to the individuals to generate the next generation of the population. In this way, the inheritance continues from generation to generation. When the operation reaches the termination condition, the operation process stops. The generated offspring population continues to breed and evolve through selection, crossover, mutation, and other operations, and cycle from generation to generation, until it converges to a group of outstanding individuals who are most suitable for the environment, then the algorithm terminates, and the optimal solution to the problem is obtained at this time. For complex optimization problems, genetic algorithms do not need to be modeled, only simple operations and three operators using GA algorithm can find the optimal solution. It is especially suitable for dealing with complex and non-linear problems that traditional methods are difficult to solve, so it is mostly suitable for learning, optimization, adaptation and other problems [8]. The detailed workflow of genetic algorithm is shown in Fig. 3.

Step 1: First, a certain number of initial groups are randomly generated by coding.

Step 2: According to the set function, calculate the individual fitness value, and then judge whether the termination condition is met; if it is satisfied, the result is output, the result is the optimal solution, and the calculation is completed. If not satisfied, go to the third step.

The third step: generate new individuals through selection, crossover and mutation operations. The regenerative individuals are selected by evaluating the fitness value: selecting individuals with high fitness and eliminating individuals with low fitness. Return to the second step.

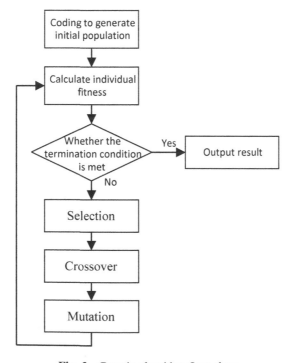

Fig. 3. Genetic algorithm flow chart

4   Improved Genetic Algorithm to Optimize BP Neural Network

Traditional BP neural network has complex mapping relationship between input and output, slow update of weights and thresholds, and fixed step size for each iteration. Local search may make the actual problem unable to find the global optimal solution, resulting in training failure. There are slow learning speeds and unsatisfactory local convergence, which affects its performance [9]. Because genetic algorithms have global optimization capabilities and BP neural networks have local optimization capabilities, this paper uses genetic algorithms to optimize BP neural networks to achieve global optimization the combination of local optimization can effectively improve the learning performance and convergence of the neural network. The flow of the optimization algorithm is shown in Fig. 4.

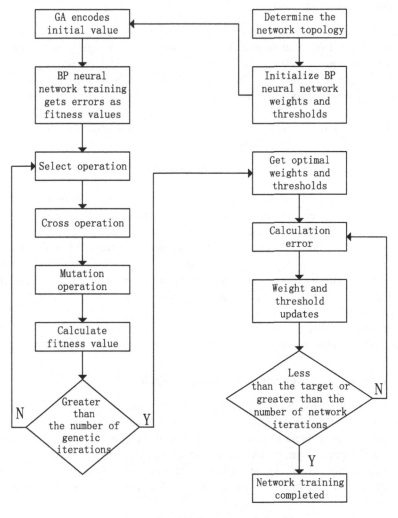

Fig. 4. The flow of the proposed algorithm

4.1 Coding and Fitness Function Selection

The weight threshold of BP network is a decimal number between -1 and 1, and there are a large number of them. It is not suitable to use binary coding. In this paper, the rules of real number coding are used. The value of each gene position represents a weight or threshold. Generally speaking, an individual with a larger fitness value indicates that the individual is better. In this paper, the squared and inverse of the output error of the BP network is used as the fitness function of the individual. The larger the fitness value of the individual, the better the individual is. The formula is:

$$fitness(x) = 1/E(x) \tag{9}$$

where is the sum of squared output errors of individual x.

4.2 Genetic Manipulation

Selection operators often adopt roulette or optimal individual preservation strategies. However, the roulette method may generate random errors [10], that is, individuals with high fitness may be eliminated, and individuals with low fitness may be selected, reflecting poor competitiveness. The optimal individual preservation strategy ignores the possibility that the remaining individuals destroy the diversity of the population and cause local convergence. This paper proposes a new selection operator that not only expands the diversity of the population, but also preserves the best individuals. The improved selection operator is as follows:

(1) Determine an initial population and calculate the fitness value of each individual in the population;
(2) Sort the individuals in the population in descending order of fitness;
(3) The first two bodies are selected to be inherited directly to the next generation, and the remaining individuals enter the crossover operation;
(4) For the population after cross-operation, randomly select 25% of the individuals to directly inherit to the next generation;
(5) All the populations after the crossover are subjected to mutation operation, and 25% of the individuals are randomly selected to directly inherit to the next generation;
(6) All mutated populations are ranked in descending order of fitness, and the first 50% of individuals are selected to be passed on to the next generation.

4.3 Adaptive Crossover and Mutation Probability

Standard genetic algorithms use fixed crossover and mutation probabilities, and parameters are difficult to adjust to the best. If a larger crossover rate is used, new individuals can be generated more quickly, but at the risk of destroying good individuals in the population; a smaller crossover rate can easily make the algorithm premature and stagnate. If a larger mutation rate is used, GA becomes a purely random algorithm; if a smaller mutation rate is used, it is difficult to generate a new pattern structure [11]. Aiming at the shortcomings of the standard genetic algorithm, Srinvas et al. [12] proposed adaptive genetic algorithm (adaptive GA, AGA). The formulas are Eqs. (10) and (11).

$$P_c = \begin{cases} \frac{k_1(f_{max}-f')}{f_{max}-f_{avg}}, & f' \geq f_{avg} \\ k_2, & f' < f_{avg} \end{cases} \tag{10}$$

$$P_m = \begin{cases} \frac{k_3(f_{max}-f)}{f_{max}-f_{avg}}, & f \geq f_{avg} \\ k_4, & f < f_{avg} \end{cases} \tag{11}$$

Where f' is the larger fitness value of the two individuals to be crossed; f_{max} is the maximum fitness of the population; f_{avg} is the average fitness of the population; f is the fitness of the mutated individual; $k_1 \sim k_4$ R is the adaptive control parameter. However, in actual use, it was found that AGA was not ideal in the early stage of evolution. This is because in the early stage of evolution, the better individuals in the group are almost in a non-occurring state, and the better individuals at this time may not be the global optimal solution, which easily makes the algorithm fall into the local optimal. An improved algorithm [13] (improved adaptive genetic algorithm (IAGA)) currently used is Eqs. (12) and (13), and this paper will also use this algorithm for optimization.

$$p_c = \begin{cases} p_{c1} - \frac{(p_{c1}-p_{c2})(f'-f_{avg})}{f_{max}-f_{avg}}, & f' \geq f_{avg} \\ p_{c1}, & f' < f_{avg} \end{cases} \tag{12}$$

$$p_m = \begin{cases} p_{m1} - \frac{(p_{m1}-p_{m2})(f_{max}-f)}{f_{max}-f_{avg}}, & f \geq f_{avg} \\ p_{m1}, & f < f_{avg} \end{cases} \tag{13}$$

$p_{c1} = 0.9$, $p_{c2} = 0.6$, $p_{m1} = 0.1$, $p_{m2} = 0.001$.

4.4   The Implementation Steps of the Proposed Algorithm

(1) Determine the experimental samples and normalize them;
(2) Set operating parameters such as population size N, maximum number of iterations G, and the number of nodes in the initial hidden layer;
(3) The initial population of N individuals is randomly generated and coded in real numbers;
(4) Decode each individual, obtain the initial connection weight threshold of the BP network, use the sample data to train the network, and calculate the individual fitness according to formula (5);
(5) Select, cross and mutate operations according to individual fitness values according to 4.2 and 4.3;
(6) Decode each individual of the offspring population, obtain new weights and thresholds, train the network again, and calculate the individual fitness value;

(7) Determine whether the maximum number of evolutions has been reached, if not, return to (5), if yes, proceed to (8);
(8) Decode the individual with the best fitness as the initial weight and threshold of the BP network, and build a model.

5 Experiments and Analysis

This experiment is a programming experiment on the Pycharm platform. The parameters of the BP neural network are set to: the maximum number of iterations is 3000, the target accuracy is 0.000001, and the learning rate is 0.5; The GA parameters are set as follows: the population size is 100, the maximum number of evolutions is 500, and the crossover probability p_c and mutation probability p_m use the IAGA adaptive algorithm in Sect. 3.3. A breast tumor data set of 569 cases was selected from the University of Wisconsin Medical School, of which 357 were benign and 212 were malignant. In breast tumor data, 500 sets of data were randomly selected as the training set, and the remaining 69 sets were used as the test set.

As shown in Fig. 3, the standard genetic algorithm finds the fitness value of the best individual when it converges and evolves to 163 generations 17.01; The improved genetic algorithm finds the fitness value of the best individual when it converges and evolves to 393 generations 37.14. The standard genetic algorithm has shrunk by 2.17 times. It can be seen that as the number of evolutionary generations increases, the fitness value of the optimal individual rises in steps, and the global optimization ability of the improved genetic algorithm is significantly better than that of the standard

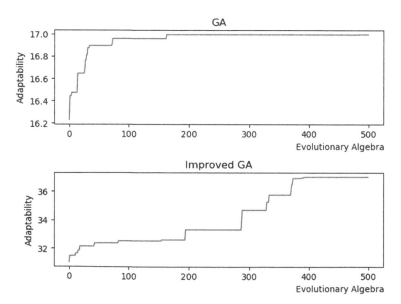

Fig. 5. Standard genetic algorithm and improved genetic algorithm optimal individual fitness value curve.

Table 1. Table captions should be placed above the tables.

	1	2	3	4	5	Average
BP	93.21%	90.16%	93.10%	92.14%	93.40%	92.40%
KNN						94.90%
GABP	95.80%	96.50%	95.10%	95.10%	95.10%	95.52%
GA-ELM-NN						97.28%
Improved GABP	98.27%	97.43%	97.43%	97.56%	96.83%	97.50%

genetic algorithm, and the optimization ability is significantly enhanced. It converges quickly, but it falls into the local extreme. From the figure, it can be seen that the improved algorithm proposed in this paper can find the optimal individual (Fig. 5 and Table 1).

By testing the recognition rate of the three algorithms five times and taking the average value, it can be seen that the improved genetic algorithm used in this paper optimizes the BP neural network algorithm. By improving the selection operator of the standard genetic algorithm, it increases the population diversity, adaptively adjusts the crossover and Probability of mutation, deep optimization of the initial weight threshold of the BP network, which makes the genetic algorithm's global optimization ability stronger and reduces the probability of falling into local extreme values. Compared with the BP neural network algorithm alone, the recognition rate is improved by about 5.1%. Compared with the standard genetic algorithm optimized BP neural network algorithm, the recognition rate is increased by about 1.98%, and the accuracy of recognition is effectively improved. Compared with the k-Nearest Neighbor (KNN) proposed in [14], the recognition rate is improved by about 2.6%, and compared with the genetic algorithm optimized extreme learning machine neural network (GA-ELM-NN) proposed in [15], it is improved by about 0.22% The recognition rate shows that the algorithm proposed in this paper is superior to most other algorithms.

6 Conclusions

The traditional BP neural network has the defects of slow learning convergence and being easy to fall into local minima. In order to solve this problem, this paper uses an improved genetic algorithm to optimize the weights and thresholds of the BP neural network, and conducts a classification study on the Wisconsin breast cancer dataset. The experimental results show that the proposed method has good generalization ability and stability. Compared with the standard genetic algorithm to optimize the BP neural network algorithm, it has higher adaptability and better recognition effect. It lays the foundation for further research and has good practical value.

References

1. Jemal, A., Siegel, R., Ward, E., Hao, Y.: Cancer statistics. CA. Cancer J. Clin. **58**(2), 71–96 (2008)
2. Xu, H., et al.: A combined parallel genetic algorithm and support vector machine model for breast cancer detection. J. Comput. Methods Sci. Eng. **16**(4), 773–785 (2016)
3. Guo, B., Cheng, L., Xu, J., et al.: Prediction of the heat load in central heating systems using GA-BP algorithm, pp. 441–445. IEEE Computer Society (2017)
4. Zhu, S.: Neural Network Application Foundation. Northeastern University Press, ShenYang (2000)
5. Li, S., Liu, L., Zhai, M.: Short-term traffic flow prediction with the improved particle swarm optimization BP neural network. Syst. Eng. Theory Pract. **32**(9), 2045–2049 (2012)
6. Holland, J.H.: Adaptation in Natural and Artificial Systems: An Introductory Analysis with Applications to Biology, Control, and Artificial Intelligence. MIT Press, Cambridge (1992)
7. Goldberg, D.E.: Genetic Algorithm in Search, Optimization, and Machine Learning, pp. 294–312. Addison-Wesley, Boston (1989)
8. Abdullah, A.H., Enayatifar, R., Lee, M.: A hybrid genetic algorithm and chaotic function model for image encryption. AEUE – Int. J. Electron. Commun. **66**(10), 806–816 (2012)
9. Wang, Z., Wang, B., Liu, C., et al.: Improved BP neural network algorithm to wind power forecast. J. Eng. **2017**(13), 940–943 (2017)
10. Chen, C., Chellali, R., Xing, Y.: Improved genetic algorithm optimize speech emotion recognition of BP neural network. Appl. Res. Comput. **36**(2), 344–346+361 (2019)
11. Zhang, D.: Research and Application of Improved Adaptive Genetic Algorithm. Kunming University of Science and Technology (2019)
12. Feng, J., Yin, Y., Huang, G.: A heterogenic improved adaptive genetic algorithm. J. Southwest Minzu Univ. (Nat. Sci. Ed.) **45**(05), 511–516 (2019)
13. Guo, C., Guo, X., Bai, L.: Research on improved BP algorithm for genetic simulated annealing algorithm. J. Chin. Comput. Syst. **40**(10), 2063–2067 (2019)
14. Sethi, A.: Analogizing of evolutionary and machine learning algorithms for prognosis of breast cancer. In: 7th International Conference on Reliability, Infocom Technologies and Optimization, Noida, India, pp. 252–255 (2018)
15. Nemissi, M., Salah, H., Seridi, H.: Breast cancer diagnosis using an enhanced extreme learning machine based-neural network. In: International Conference on Signal, Image, Vision and Their Applications, Guelma, Algeria, pp. 1–4 (2018)

Amplification Method of Lung Nodule Data Based on DCGAN Generation Algorithm

Minghao Yu[1], Lei Cai[1], Liwei Gao[2], and Jingyang Gao[1(✉)]

[1] College of Information Science and Technology, Beijing University
of Chemical Technology, Beijing 100029, China
gaojy@mail.buct.edu.cn
[2] Department of Radiation Oncology, China-Japan Friendship Hospital,
Beijing 100029, China

Abstract. Early diagnosis of lung cancer can effectively reduce the mortality of patients. Doctors use low-dose spiral CT to detect lung nodules, which is time-consuming and prone to omissions. Deep learning has achieved good results in the field of medical image sub-processing, which can reduce the pressure of doctors to a certain extent. However, in the actual lung CT images, the images containing lung nodules account for less than 1% of the total images. The lack of data increases the difficulty of detecting lung nodules by using deep learning methods. This paper proposes an amplification method using deep convolutional anti-generation network (DCGAN) to generate lung nodule data. Compared with different amplification methods, and the effectiveness of this method is confirmed. Experiments can prove that the use of DCGAN to generate data can better solve the problems of high false positive rate and low sensitivity of lung nodule classification than the graphical data amplification mode. Compared with the existing methods, this experimental method greatly improves the accuracy, sensitivity and F1 score of lung nodule detection, and achieves good results of 99.98%, 99.15% and 99.55%, respectively.

Keywords: Lung nodule · DCGAN · Data amplification

1 Introduction

The incidence and mortality of lung cancer ranked first in all cancer types, accounting for 11.6% and 18.4% of all cancer cases and all cancer deaths respectively [1]. The reason why lung cancer has the highest mortality rate is that lung cancer can not be diagnosed before the advanced stage, and the overall prognosis is very poor. If diagnosed in the early stage of lung cancer, the 5-year survival rate (18%) can be increased to 56%. Kostis et al. confirmed that 88% of patients with stage I lung cancer can survive for more than 10 years after diagnosis [2]. Therefore, the early diagnosis of lung cancer is very important to improve the treatment decision and prognosis.

National lung screening experiments in the United States have shown that the use of low-dose CT to screen lung cancer for high-risk lung cancer patients at an early stage can reduce lung cancer mortality by more than 20% [3]. The most important step in early screening is the detection of lung nodules in CT images. The radiologist screened

© Springer Nature Singapore Pte Ltd. 2020
J. Zeng et al. (Eds.): ICPCSEE 2020, CCIS 1257, pp. 563–576, 2020.
https://doi.org/10.1007/978-981-15-7981-3_41

the CT images to find the suspected nodules and further excluded the non-nodular parts. With the development of CT imaging, there are more and more CT imaging data. But radiologists have not increased much, resulting in a large workload for physicians. Along with this, the misjudgment rate of physicians is also increasing, and these misjudgments also easily affect the mentality of patients and increase the difficulty of treatment.

In recent years, deep learning has made quite good progress in medical image processing tasks. Deep learning has published many excellent papers in the analysis and application of lung cancer, breast cancer, prostate cancer and brain cancer [4]. At the same time, many CADE systems for lung nodule detection and classification have been proposed to achieve the effect of assisting physicians. However, a system with a high false-positive rate will cause doctorkack-propagation algorithm to train a deep convolutional neural network (DCNN) to extract valuable volume features from the input data [5]; Oseas et al. demonstrated that the use of texture features can improve lung nodules Classification effect [6]; Zhu et al. proposed a gradient accelerator (GBM) with three-dimensional dual path network (DPN) features to achieve classification of lung nodules [7]. Kim et al. proposed a new multiscale progressively integrated convolutional neural network (MGI-CNN) for learning end-to-end feature integration [8]. They all try to use different network models to learn the deep features of the image, and then improve the sensitivity of lung nodule classification.

As the depth of convolutional neural networks continues to increase, more and more medical data is required, but the required case data is usually less or difficult to obtain. In order to solve this problem, although the traditional graphics-based data augmentation method can expand the data, these amplified images are similar in features to the original images, and will not greatly improve the performance of the neural network. Generative adversarial network (GAN) is a new learning framework that can learn discriminative features from images and generate real samples [9]. Many people have used GAN for the generation of medical images and achieved good results. Wolterink et al. used GAN to generate CT images from MRI images and obtain images similar to the reference CT [10]; At the same time, they also proposed a depth generation model to synthesize the geometric structure of blood vessels and apply it to coronary heart CT Angiography [11]; Calimeri et al. used GAN to synthesize MRI images of human brain slices [12]. Zhao et al. proposed a data augmentation method based on generative adversarial network (GAN), called forward and backward GAN (F&BGAN), which can produce high-quality synthetic medical images [13]. Recently, more and more papers on generation adversarial networks used for medical image generation have also shown that they have better effects on medical image processing tasks.

Based on the contributions made by previous generations, this paper proposes a method for data amplification of lung nodules based on deep convolutional generation adversarial network (DCGAN). DCGAN is a new network that combines GAN and convolutional networks to solve the problem of GAN training instability after the groundbreaking GAN paper of Goodfellow in 2014 [14]. The main innovations of this paper are as follows: 1. We apply the generation adversarial network to medical images to generate higher-quality lung nodule images, and solve the problems of small sample data and imbalance of positive and negative samples in medical image processing.

2. The effects of GAN generated data and traditional data amplification on the classification of lung nodules were explored. 3. The method we use to generate data using the DCGAN ultimately improves the sensitivity of lung nodule classification and reduces the false positive rate compared with current methods.

2 Methods

2.1 Generative Adversarial Network

The full name of GAN is generative adversarial networks, as shown in Fig. 1. It consists of a generator and a discriminator. The function of the generation network is to generate an image, which generates the input noise into a desired image. The function of the discrimination network is to distinguish the images generated by the generation network from the real images. During the training process, the generating network tries to generate more realistic images to make the identification network difficult to distinguish. The identification network tries its best to separate the real and generated images. This is the zero-sum game idea of opposing the generating network.

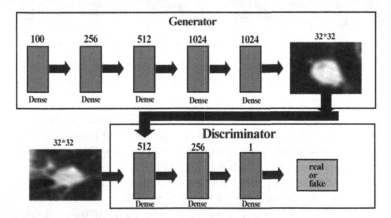

Fig. 1. The framework structure of GAN. It contains two networks. The generator is used to fit the noise into a false image with a similar distribution to the original image. The discriminator is used to identify the real image and the false image and extract more features of the real image.

The input of the GAN network generator is a 100-dimensional random noise, and a 32×32 size image is generated through 4 fully connected layers. The discriminator input is a 32×32 size real and generated image. After flattening, it passes through three fully connected layers and finally uses the sigmoid function to distinguish the real and generated images.

566   M. Yu et al.

2.2   Deep Convolution Generating Adversarial Network

The full name of DCGAN is deep convolution generative adversarial networks, as shown in Fig. 2. It is an improvement on the original GAN. It retains the ability of GAN to generate data and combines the advantages of convolutional neural network feature extraction. It replaces all the fully connected layers in GAN with different convolutional layers, which improves the analysis and processing capabilities of the image, thereby generating a more realistic image.

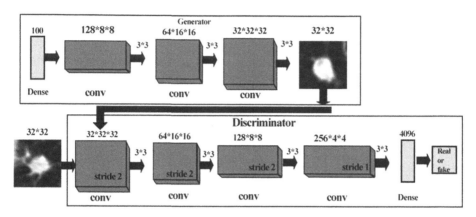

Fig. 2. DCGAN framework structure. It replaces the fully connected layer in the GAN with a convolutional layer to generate a more realistic image.

The structure of its identification network is: the input of the network is a $32 \times 32 \times 1$ size lung nodule image, which is transformed into a $256 \times 4 \times 4$ size tensor through 4 paddings of the same convolution and a 3×3 size convolution layer. The activation function uses LeakyReLU and adds the dropout layer, then flatten the data through the flatten layer, and finally a fully connected layer with an activation function of sigmoid to discriminate the image probability.

The structure of the generated network is: The input of the network is a 100-dimensional vector, which becomes a tensor of $128 \times 8 \times 8$ through a fully connected layer and reshape. After 3 layers of padding for the same convolution and a size of 3×3 convolutional layer, a $32 \times 32 \times 1$ size image is obtained. Corresponds to an upsampling layer. BatchNormolization is added before the activation layer, the activation function of the first two layers is ReLU, and the activation function of the output layer uses tanh. The reason for using tanh here is to normalize the data to $[-1,1]$.

2.3   Batch Normolization

In deep learning, in order to better learn more features, people will use deep neural networks for training. In this process, we need to constantly adjust parameters such as learning rate to make our network converge quickly. The slow network convergence rate is mainly due to the high correlation between the various layers in the neural network. This correlation will cause the input distribution of each layer to change when the underlying network parameters change during the process. This requires the upper layer network to adapt to these distribution changes, which makes our network training more difficult. This situation is called Internal Convariate Shift (internal covariance shift) [15]. Because the model has to constantly adapt to these changing input distributions, parameter learning is slow. If the distribution of the input layer can be fixed so that they all conform to the Gaussian distribution, then the model parameter learning will be accelerated.

The operation of normalizing the input distribution of each layer is called Batch Normolization. The first step of BN is to independently normalize each feature. Consider a batch of training, pass in m training samples, select the jth dimension of a certain layer, and normalize the dimension:

$$u_j = \frac{1}{m} \sum_{i=1}^{m} X_j^{(i)} \tag{1}$$

$$\sigma_j^2 = \frac{1}{m} \sum_{i=1}^{m} \left(X_j^{(i)} - u_j \right)^2 \tag{2}$$

$$\hat{X}_j = \frac{X_j - u_j}{\sqrt{\sigma_j^2 + \varepsilon}} \tag{3}$$

The batch normalization operation solves the problem of internal covariance translation, making the input distribution of each layer stable, but it causes the loss of parameter information learned from the underlying network. Therefore, BN also introduces two learnable parameters γ and β to restore the expressive ability of the data itself, and linearly transform the normalized data:

$$\tilde{X}_j = \gamma_j \hat{X}_j + \beta_j \tag{4}$$

By using BN, there are several benefits: BN can make the distribution of the input data of each layer relatively stable, and speed up the learning speed of the model; BN makes the model less sensitive to the parameters in the network, simplifies the parameter adjustment process, and makes the network learn More stable; BN can alleviate the problem of gradient disappearance to a certain extent; BN has a certain regularization effect.

2.4 LeakyReLU

LeakyReLU has the advantages of faster convergence speed and solving the problem of gradient disappearance compared with the saturation activation functions Sigmoid and tanh. At the same time, compared with ReLU, it can solve the Dead Neuron problem faced by ReLU [16].

LeakyReLU's activation function can be expressed as a mathematical formula:

$$LeakyReLU(x) = \begin{cases} x, x \geq 0 \\ ax, x < 0 \end{cases} \tag{5}$$

You can see the difference between ReLU and LeakyReLU as shown in Fig. 3: ReLU sets all negative values to 0, while LeakyReLU gives negative values to a non-zero slope. When the input is positive, both activation functions can learn gradient descent, and when the input is negative, the first derivative of the ReLU function at this time is zero, which will cause the weights to be unable to be updated, and a Dead Neuron problem will occur. When the Leaky ReLU input is negative, there will be a small first derivative, and the network will continue to train.

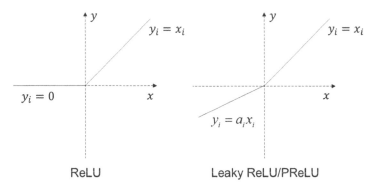

Fig. 3. ReLU and Leaky ReLU activation functions

3 Experiment

3.1 Data Preprocessing

The Lung Image Database Consortium (LIDC) and the Image Database Resource Project (IDRI) have completed a public reference database LIDC/IDRI for the medical imaging research community [17]. The LIDC/IDRI database is designed to facilitate computer-aided diagnostic (CAD) research for the detection, classification, and quantitative assessment of lung nodules. This dataset contains 1018 low-dose CT images of the lungs, each image containing a series of multiple axial slices of the thorax. The CT image is shown in Fig. 4. These CT images are divided into 3 categories. Those with a diameter greater than 3 mm are referred to as lung nodule, those with a diameter less than 3 mm are referred to as micronodules, and bronchial walls and blood vessels are similar to nodules as non-nodules. Each CT image slice size is 512 × 512, the image information is stored in a dicom file, and the xml file stores the position coordinates of the nodule. The LUNA16 dataset is a subset of the LIDC/IDRI dataset. It deletes CT images in the LIDC/IDRI dataset with slice thicknesses greater than 3 mm and lung nodules less than 3 mm and less than 3 physician-labeled CT images, leaving 888 low-dose lungs. CT image data [18]. The CT images of the LUNA16 dataset are stored in

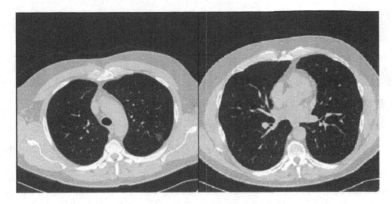

Fig. 4. Visualization of lung CT images

MetaImage (mhd/raw) format. Each mhd file stores a separate corresponding raw binary file for storing pixel data.

The first thing we need to do is convert mhd/raw to dicom format. Here we go back to get the dicom format file. The process of Dicom conversion to mhd/raw is: First, get the dimensions of the 3D picture composed of the dicom image, get the Spacing in the x direction and the y direction, and get the layer thickness in the z direction to get the origin of the image. Then iterate through all the dicom files, read the image data, store the data in the numpy array, and convert the current numpy array into mhd and raw files by the SimpleITK tool [19]. We use the method of dicom to mhd/raw to perform the inverse calculation. The coordinate transformation is used to calculate the x and y coordinates of the lung nodule through the thickness of the Spacing and z directions. The data of the coordinate position is then found using numpy. image. Monkam et al. used 16 × 16, 32 × 32 and 64 × 64 three different sizes of nodule images for testing, and found that 32 × 32 size nodule images achieved the best results in the detection [20]. Most of the lung nodules selected in this experiment are less than 32 in diameter, so the size of 32 × 32 is also used for the experiment.

The size of each dicom image is 512 × 512. We take the image without nodule label and cut it into 256 images of 32 × 32 size as non-nodular data. Repeat this operation on the middle 200 slices.

The final sample of the data we obtained is shown in Fig. 5, which includes a total of 1187 nodule data, 20,000 nodule data generated using DCGAN, 3000 nodule data generated using GAN, 3000 sample data amplified using traditional data amplification methods and 30,000 non-nodular data.

3.2   Experimental Process and Results

In this experiment, nodule data and non-nodule data are divided into three parts according to a ratio of 7: 2: 1, of which 70% is used as the training set, 20% as the test set, and the remaining 10% as the validation set. The learning rate used by the DCGAN is 0.0002, the optimizer uses Adam, and the batch_size is 32. The learning rate used by the CNN network is 0.0001, the SGD used by the optimizer, batch_size is 64, and

a,Non-nodular image   b,Nodule image   c,Generated nodule image

Fig. 5. Non-nodular images contain many structures such as blood vessels and calcification points, which are similar to nodular images and interfere with classification results. By comparing the original nodule image with the generated image, it can be seen that the nodule image generated using the DCGAN can generate many features not found in the original nodule image, thereby enhancing the ability of the classification network.

epochs is 50. The GPU used in the experiment was an Nvidia Tesla P100 graphics card with 12 GB of video memory. The deep learning framework used was Keras. The main evaluation indicators of this experiment are accuracy, sensitivity and F1 score. Accuracy is the ratio of the number of correct predictions to the total number of predictions. Sensitivity is the ratio of the number of nodule images that are correctly predicted as nodules in all nodule images. The F1 score is the harmonic mean of precision and sensitivity.

Exploring the Impact of Amplified Data on Lung Nodule Classification. In experiment 1, the ratio of nodule to non-nodule was 1:25 using the original data. Experiment 2 using nodules with the proportion of nodules 1:50. Experiment 3 using nodal data 1:100 (Table 1).

Table 1. Pulmonary nodule classification results under different proportions of positive and negative samples.

Positive and negative sample ratio	Precision (%)	Sensitivity (%)	F1 (%)
1:25	64.19	61.87	62.73
1:50	55.26	52.11	53.64
1:100	*51.17*	50.59	50.88

It can be seen that when the difference between the positive and negative samples is too large, the classification network used is almost useless and cannot be distinguished.

In experiment 4, DCGAN was used to generate 1000 pieces of data, and the generated data and real data were sent into the classification network together. In experiment 5, DCGAN was used to generate 3000 data. In experiment 6, DCGAN was used to generate 20,000 pieces of data. The results from raw data in experiment 1 were compared with those generated by GAN.

When using the original data, when the ratio of positive and negative samples is 1:25, the obtained indicators are the worst, only f1 scores of 62.73%, 61.87% sensitivity, and 64.19% accuracy are obtained. With the increase of nodule data, the proportion of positive and negative samples has continued to shrink, and the accuracy has been significantly improved. It achieved the best result of 99.99%, the sensitivity increased to 99.15%, and the F1 score also achieved the highest score of 99.55% (Table 2).

Table 2. Classification results of lung nodules under different sample proportions

	Positive and negative sample ratio	Precision (%)	Sensitivity (%)	F1 (%)
Raw data	1:25	64.19	61.87	62.73
DCGAN-1000	1:14	86.49	84.32	85.17
DCGAN-3000	1:7	97.27	93.21	94.87
DCGAN-20000	1:1.4	99.98	99.15	99.55

Explore the Impact of Different Amplification Methods on Lung Nodule Classification. In experiment 7, 1000 images of nodules were amplified by turning left and right, and in experiment 8, 3000 images of nodules were amplified by turning left and right and turning up and down. Comparing experiment 7 and 8 with experiment 4 and 5, to explore the influence of data generated by DCGAN and traditional data amplification method on the classification index of pulmonary nodules (Table 3).

Table 3. Lung nodules under different data amplification methods

	Precision (%)	Sensitivity (%)	F1 (%)
Amplification-1000	80.61	78.31	79.45
DCGAN-1000	86.49	84.32	85.17
Amplification-3000	88.43	84.95	86.60
DCGAN-3000	98.18	95.50	96.62

This figure is a comparison of the experimental results of using original image amplification and DCGAN to generate image amplification data. The same data is amplified, and the method of generating data using DCGAN is more effective. In the same amplification of 1000 data samples, the accuracy, sensitivity, and F1 score of the GAN-generated data compared to the graphic method of data augmentation classification were increased from 80.61% to 86.49%, 78.31% to 84.32%, and 79.45% Increased to 85.17%, and each index of classification results increased by about 6%. When 3,000 samples of data were also amplified, the accuracy, sensitivity, and F1 score of the data classification using GANs were increased from 88.43% to 98.18%,

84.95% to 95.50%, and 86.66 compared to the data from the graphics method %
Increased to 96.62%, and various indicators of classification results increased by about
10%.

**Explore the Impact of Quality of Generated Data on Classification of Lung
Nodules.** Experiment 9 used the original GAN network to generate 1000 pieces of
nodule data, and experiment 10 used the original GAN network to generate 3000 pieces
of nodule data. By comparing experiment 9 and 10 with experiment 4 and 5, the effect
of different quality data generated by GAN network and DCGAN network on the
classification results of pulmonary nodules was investigated (Table 4).

Table 4. Classification results of lung nodules under different generative networks

	Precision (%)	Sensitivity (%)	F1 (%)
GAN-1000	84.19	81.76	92.96
DCGAN-1000	86.49	84.32	85.17
GAN-3000	97.27	93.21	94.87
DCGAN-3000	98.18	95.50	96.62

Different GAN are used to generate nodule images, and the results obtained by using
these generated nodule images for training are different. Similarly, 1000 sample data
were generated. Compared with GAN, the accuracy of DCGAN increased from
84.19% to 86.94%, the sensitivity increased from 81.76% to 84.32%, the F1 score
increased from 82.96% to 85.17%, and the classification results improved by about
2.5%. Generated 3000 sample data. Compared with GAN, the accuracy of DCGAN
was increased from 97.27% to 98.18%, the sensitivity was increased from 93.21% to
95.50%, the F1 score was increased from 94.87% to 96.62%, and the classification
results were improved by about 2%.

**Comparison of the Method in This Paper with Other Related Research Methods
for Lung Nodule Detection.** With the development of machine learning, early
researchers used convolutional neural networks for the classification of lung nodules.
A deep neural network (DCNN) can better acquire the features of the image and
improve the accuracy of the detection results. However, the problem of low sensitivity
still cannot be solved, and researchers have begun to explore more features to reduce
sensitivity, such as DPN and MGI-CNN.

With the development of unsupervised learning, some people use intensive learning
to detect pulmonary nodules and achieve good results [21]. Some people use GAN to
generate medical images to achieve the effect of augmenting data. At the same time,
due to the characteristics of GAN, more features can be obtained, such as F&BGAN.

As shown in Table 5, we aims to improve the classification index of lung nodules by
generating nodule images using DCGAN. The accuracy of the method in this paper is
99.98%, which is much higher than the previous best performance of 98.83%, the
sensitivity is also higher than 96.20%, and the best result is 99.15% (Figs. 6 and 7).

Table 5. Classification results of lung nodules under different methods

	Precision (%)	Sensitivity (%)	F1 (%)
DCNN	98.83	78.90	87.75
DPN	90.44	94.60	92.47
MGI-CNN	94.20	96.20	95.19
F&BGAN	95.24	92.47	93.83
RL	99.20	99.10	99.15
DCGAN-20000	99.98	99.15	99.55

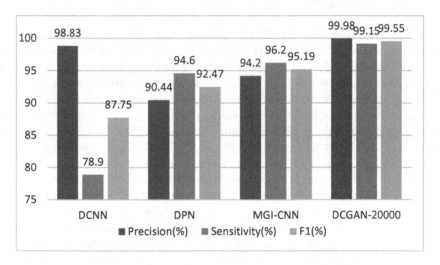

Fig. 6. The feature extraction method is compared with the results of this method

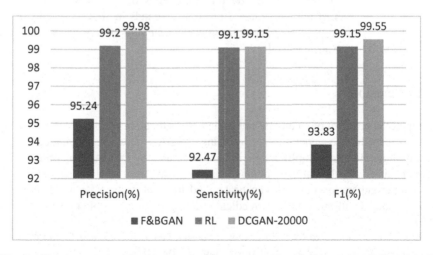

Fig. 7. The unsupervised learning method was compared with the results of this method

4 Discussion

First, it can be seen that as the gap between the positive and negative data samples decreases, the accuracy and sensitivity of the classification network have been greatly improved. When the gap between the sample ratios is larger, this improvement is more obvious. It is true that the imbalance of positive and negative sample data will have a great impact on the classification network. In the learning process, the discriminator only needs to judge most of the data as a class with few samples to obtain good results, so the final classification has low sensitivity and high false positive rate. As the sample ratio approaches equilibrium, the classification network needs to learn different features between different categories to achieve the classification effect.

The third set of experiments can clearly see the effects of two different data amplification methods. Because GAN finds the data distribution of the model by focusing on the potential probability density of the data, the traditional data amplification method only performs the original image. Some transformations did not generate new features, so the method of GAN generating data can better improve the performance of convolutional neural networks. And with the increase of the amplification data, the improvement effect is more obvious. The fourth group of experiments further verified the impact of the quality of the amplified data on the subsequent classification results. It can be clearly seen that the quality of the DCAGN generated image is higher than the quality of the GAN generated image, and the same data is amplified. The DCGAN indicators in all aspects are higher than the GAN indicators.

Early deep convolutional networks can learn more features due to the deeper layers and have a good accuracy rate. However, the deep convolutional network model cannot solve the problem of data imbalance, and the sensitivity obtained by classification is poor, such as DCNN. Since sensitivity is an important reference index for classification results, people began to use 3D networks to learn more features and improve the classification sensitivity as much as possible, such as DPN and MGI-CNN. However, the method in this paper uses the data amplification of DCGAN, which greatly increases the number of nodular images, not only expands the data, prevents overfitting, but also balances the ratio of nodular images and non-nodular images, so that the network can Better learning results.

5 Conclusion

Aiming at the problem of small amount of data in medical image processing, this paper proposes a method to generate data by using DCGAN network. Compared with the method of graphics data amplification, images generated by DCGAN can produce more nodule features, which is more conducive to the improvement of classification network. The counter generation network can be applied to amplify the data in detecting pulmonary nodules and other areas of medical imaging with the problem of lacking data.

Funding. Project supported by beijing Natural Science Foundation (5182018) and the Fundamental Research Funds for the Central Universities (PYBZ1834).

References

1. Bray, F., Ferlay, J., Soerjomataram, I., et al.: Global cancer statistics 2018: GLOBOCAN estimates of incidence and mortality worldwide for 36 cancers in 185 countries. CA Cancer J. Clin. **68**(6), 394–424 (2018)
2. Kostis, W.J.: Survival of patients with stage I lung cancer detected on CT screening. N. Engl. J. Med. **355**(17), 1763 (2006)
3. National Lung Screening Trial Research Team: Reduced lung-cancer mortality with low-dose computed tomographic screening. N. Engl. J. Med. **365**(5), 395–409 (2011)
4. Shen, D., Wu, G., Suk, H.I.: Deep learning in medical image analysis. Ann. Rev. Biomed. Eng. **19**(1), 221–248 (2017). annurev-bioeng-071516-044442
5. Golan, R., Jacob, C., Denzinger, J.: Lung nodule detection in CT images using deep convolutional neural networks. In: 2016 International Joint Conference on Neural Networks (IJCNN). IEEE (2016)
6. de Carvalho Filho, A.O., Silva, A.C., de Paiva, A.C., Nunes, R.A., Gattass, M.: Classification of patterns of benignity and malignancy based on CT using topology-based phylogenetic diversity index and convolutional neural network. Pattern Recogn. **81**, 200–212 (2018)
7. Zhu, W., Liu, C., Fan, W., et al.: DeepLung: deep 3D dual path nets for automated pulmonary nodule detection and classification. In: 2018 IEEE Winter Conference on Applications of Computer Vision (WACV), pp. 673–681 (2018). (IEEE 2018 IEEE Winter Conference on Applications of Computer Vision (WACV), Lake Tahoe, NV, 12 March 2018–15 March 2018)
8. Kim, B.C., Yoon, J.S., Choi, J.S., et al.: Multi-scale gradual integration CNN for false positive reduction in pulmonary nodule detection. Neural Netw. **115**, 1–10 (2019)
9. Goodfellow, I., Pougetabadie, J., Mirza, M., et al.: Generative adversarial nets. In: Neural Information Processing Systems, pp. 2672–2680 (2014)
10. Wolterink, J.M., Dinkla, A.M., Savenije, M.H.F., Seevinck, P.R., van den Berg, C.A.T., Išgum, I.: Deep MR to CT synthesis using unpaired data. In: Tsaftaris, S.A., Gooya, A., Frangi, A.F., Prince, J.L. (eds.) SASHIMI 2017. LNCS, vol. 10557, pp. 14–23. Springer, Cham (2017). https://doi.org/10.1007/978-3-319-68127-6_2
11. Wolterink, J.M., Leiner, T., Isgum, I.: Blood vessel geometry synthesis using generative adversarial networks (2018)
12. Calimeri, F., Marzullo, A., Stamile, C., et al.: Biomedical data augmentation using generative adversarial neural networks (2017)
13. Zhao, D., Zhu, D., Lu, J., et al.: Synthetic medical images using F&BGAN for improved lung nodules classification by multi-scale VGG16. Symmetry **10**(10), 519 (2018)
14. Radford, A., Metz, L., Chintala, S.: Unsupervised representation learning with deep convolutional generative adversarial networks
15. Ioffe, S., Szegedy, C.: Batch normalization: accelerating deep network training by reducing internal covariate shift (2015)
16. Glorot, X., Bordes, A., Bengio, Y.: Deep sparse rectifier neural networks. In: Proceedings of the 14th International Conference on Artificial Intelligence and Statistics (AISTATS) (2010)
17. Armato, S., Mclennan, G., McNitt-Gray, M., et al.: WE-B-201B-02: the lung image database consortium (LIDC) and image database resource initiative (IDRI): a completed public database of CT scans for lung nodule analysis. Med. Phys. **37**(6Part6) (2010)
18. Setio, A.A.A., Traverso, A., De Bel, T., et al.: Validation, comparison, and combination of algorithms for automatic detection of lung nodule in computed tomography images: the LUNA16 challenge. Med. Image Anal. **42**, 1–13 (2017). (14141414)

19. Lowekamp, B.C., Chen, D., Ibanez, L., et al.: The design of SimpleITK. Front. Neuroinform. **4**, 45 (2013)
20. Monkam, P., Qi, S., Xu, M., et al.: CNN models discriminating between pulmonary micro-nodules and non-nodules from CT images. Biomed. Eng. Online **17**(1) (2018)
21. Issa, A., Hart, G.R., Gowthaman, G., et al.: Lung nodule detection via deep reinforcement learning. Front. Oncol. **8**, 108 (2018)

CP-Net: Channel Attention and Pixel Attention Network for Single Image Dehazing

Shunan Gao, Jinghua Zhu[✉], and Yan Yang

School of Computer Science and Technology, Heilongjiang University,
Harbin 150080, China
zhujinghua@hlju.edu.cn

Abstract. An end-to-end channel attention and pixel attention network (CP-Net) is proposed to produce dehazed image directly in the paper. The CP-Net structure contains three critical components. Firstly, the double attention (DA) module consisting of channel attention (CA) and pixel attention (PA). Different channel features contain different levels of important information, and CA can give more weight to relevant information, so the network can learn more useful information. Meanwhile, haze is unevenly distributed on different pixels, and PA is able to filter out haze with varying weights for different pixels. It sums the outputs of the two attention modules to improve further feature representation which contributes to better dehazing result. Secondly, local residual learning and DA module constitute another important component, namely basic block structure. Local residual learning can transfer the feature information in the shallow part of the network to the deep part of the network through multiple local residual connections and enhance the expressive ability of CP-Net. Thirdly, CP-Net mainly uses its core component, DA module, to automatically assign different weights to different features to achieve satisfactory dehazing effect. The experiment results on synthetic datasets and real hazy images indicate that many state-of-the-art single image dehazing methods have been surpassed by the CP-Net both quantitatively and qualitatively.

Keywords: Image dehazing · Channel attention and pixel attention · Residual learning

1 Introduction

In the past 20 years, the problem of image dehazing has aroused wide attention in the computer vision field. Haze is a common atmospheric phenomenon caused by small floating particles such as dust and smoke in the air. These floating particles greatly absorb and scatter light, resulting in reduced image quality. Under the influence of haze, many practical applications such as video surveillance, remote sensing, and autonomous driving are vulnerable to threat, and advanced computer vision jobs such as segmentation [23–25] and object detection [11,21,22] are difficult to complete. Therefore, image dehazing has become

© Springer Nature Singapore Pte Ltd. 2020
J. Zeng et al. (Eds.): ICPCSEE 2020, CCIS 1257, pp. 577–590, 2020.
https://doi.org/10.1007/978-981-15-7981-3_42

an increasingly important technique and its purpose is to restore a hazy image to a haze-free image (see Fig. 1). Most of the successful approaches depend on the physical scattering model [3–5], which can be expressed by the following formula

$$I(z) = J(z)t(z) + A(1 - t(z))$$ (1)

Where I is a hazy image, J is a haze-free image, t is the transmission map and A is the global atmosphere light.

(a) Hazy Image   (b) Our dehazed Image

Fig. 1. An example of image dehazing

When the atmosphere is uniform, the transmission map t can be expressed as

$$t(z) = e^{-\beta d(z)}$$ (2)

where β is the atmosphere scattering parameter and d is the scene depth.

Equation 1 can also be transformed into the following form

$$J(z) = \frac{I(z) - A}{t(z)} + A$$ (3)

We know from Eq. 3 that since A and t have an infinite number of solutions, it is a pathological problem to use the atmospheric scattering model to dehaze. If A and t can be appropriately evaluated by leveraging the captured hazy image, a clear dehazing image we can get. However, it is often challenging to complete that.

Many early dehazing methods, such as [8,18,26,27] were based on atmospheric scattering models. [9,20] discovered the effective dark channel prior (DCP). In a haze-free image, every local area is likely to have shadows, or something of pure color, or something of black. Therefore, it is very likely that each local area will have at least one color channel with a low value. This statistical law is called Dark Channel Prior. However, the dark channel prior is not very suitable for images with the sky. DCP points out some pixels always have at least one color channel with a very low value in most local areas. This also indirectly proves that different channel characteristics contain different degrees of important information. The representation of network will be greatly limited if all features are treated equally.

Convolutional neural network (CNN) are widely used in computer vision. Many methods based on convolutional neural networks have excellent results in the field of image processing. One of the advantages of deep CNN model compared to traditional image processing algorithms is that it avoids the complex pre-processing of the image, especially manual participation in the image pre-processing process. Up to now, CNN has been widely used in various image-related applications. In the field of image dehazing, almost all methods are based on CNN. For example, DCP, AOD-Net [2], MSCNN [9], EPDN [31]. These dehazing methods have achieved very remarkable results.

The attention mechanism [14–16] is often used in convolutional neural networks, because it can improve the performance of the network remarkably. Inspired by the work [12], a new double attention (DA) module is proposed. Our DA module combines channel attention and pixel attention. DA can give different channel features and pixel features different weights. For example, important feature will be given greater weight, less important feature will be given less weight.

The deeper (complex, more parameters) CNN network is, the more expressive it is. [10] proposed ResNet which is conducive to training deep models based on convolutional neural networks. We incorporate the attention mechanism and the skip connection into DA module. CP-Net utilizes multiple local residual connections to not only transmit the information in the shallow part of the network to the deep part of the network as completely as possible, but also deepen the depth of the network and improve the network performance.

Many dehazing methods nowadays utilize peak signal to noise ratio (PSNR) and structure similarity (SSIM) indicators to measure the quality of dehazed image recovery. For human subjective evaluation, we also provide a large number of network outputs from corrupted inputs. Experimental results demonstrate that our network exceeds the previous state-of-the-art methods in terms of both the PSNR and SSIM metrics and qualitative comparisons. Not only that, we also made an ablation analysis to prove the effectiveness of our DA module.

The main contributions of our work can include the following:

1. We design a fully end-to-end single image dehazing algorithm CP-Net, which can directly output dehazing image without relying on the atmosphere scattering model in Eq. 1. It achieves state-of-the-art performance on both synthetic and real hazy images;
2. We combine the channel attention and pixel attention mechanism to design a new double attention (DA) module. DA module can focus more attention on important channel information and pixel information. The sum of the outputs of two attention module can further improve the capacity of CP-Net;
3. We combine double attention (DA) module and local residual learning to design a new basic block. Local residual learning can transfer the feature

information in the shallow part of the network to the deep part of the network through multiple local residual connections, DA module can give different channel features and pixel features different weights;

4. Our CP-Net contains multiple basic block connections, which not only reduces the loss of information during the flow, but also increases the depth of the network. In addition, CP-Net can also automatically learn different weights for different features.

2   Related Works

Most previous dehazing methods rely on Eq. 1. As we mentioned above, the atmosphere scattering model is ill-conditioned because global atmospheric light and transmission map can not be accurately estimated. No matter which method is used, we can not escape obtaining accurate transmission map and global atmospheric light. Traditional methods can only use different image statistical priors as constraints to minimize the information loss caused by corruption procedure. Modern methods can only use a large number of constraints to learn useful information in image continuously, but the structural design of dehazing network limits the performance.

The dehazing method introduced in [6] makes use of the Dark Channel Prior (DCP), which estimates the transmission map. However, this prior is proved to be unreliable when the scene objects are similar to the atmospheric light. Color attenuation prior was proposed by [7]. It points out that there is a linear relationship between the depth, brightness, and saturation of the hazy image, which can be used to form a function formula. The measure of calculating the optical transmission of foggy scene based on a single input image is proposed by [28], which assumed a premise that the surface shadow and the transfer function are statistically independent. [19] proposed a method (NLD) for estimating transmission map based on global priors to recover the clean image, and the algorithm assumes that the color of pixel points in a haze-free image can be clustered into hundreds of compact clusters in RGB space. Because the prior is based on the assumption under ideal conditions, the prior-based dehazing methods will fail in certain scenarios, such as bad natural weather.

In recent years, deep learning technology has been greatly developed. The emergence of large synthetic datasets [6] have solved the problem of data scarcity, which has also directly promoted the widespread development of data-driven image dehazing algorithms. Although these algorithms reduce the reliance on handmade priors, they still rely on the traditional strategies mentioned above. For example, DehazeNet [1] is an end-to-end system that directly learns and estimates the mapping relationship between a hazy image and its transmission map. [9] leverages a Multi-Scale CNN (MSCNN) that can estimate the transmission very well.

By transforming Eq. 1, the AOD-Net [2] no longer needs to estimate transmission map and atmospheric light. [13] propose a hazy image restoration method based on threshold fusion network (GFN) which consists of an encoding and decoding network. EPDN [31] simplifies the image dehazing problem to the image-to-image translation problem, which is embedded by the generative adversarial network (GAN) and does not depend on the physical scattering model.

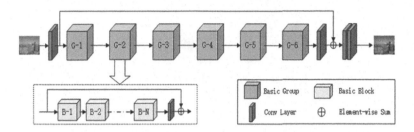

Fig. 2. Channel Attention and Pixel Attention Network (CP-Net) architecture

3 Channel Attention and Pixel Attention Network (CP-Net)

The detailed structure of CP-Net is described in Fig. 2. The input of CP-Net is a hazy image, which first goes through a convolutional layer, then is sent to six Group Architectures with multiple long skip connections and a convolutional layer. The output of convolutional layer is then fused with the output of shallow feature extraction part via element-wise addition. Finally, the features will flow into the reconstruction part, and then a dehazed image will be obtained.

In addition, local residual learning and B Basic Block structures constitute a Group Structure; Double Attention (DA) module and the skip connection constitute a Basic Block. Channel Attention and Pixel Attention constitute DA module.

We will introduce DA module and Basic Block structure in detail in Sect. 3.1 and 3.2 respectively. Finally, we will introduce Group Architecture and Global Residual Learning in detail in Sect. 3.3.

3.1 Double Attention (DA)

Because the distribution of haze on the image is not uniform, the network we designed can deal with different channel features and pixel features differently. Double Attention (see Fig. 3) is comprised of pixel attention and channel attention. DA can help our CP-Net assign different weights to each Channel of input features, extract more critical and important information, and make more accurate judgments of CP-Net. Next, we will elaborate on how CP-Net assigns weights to each channel feature and pixel feature.

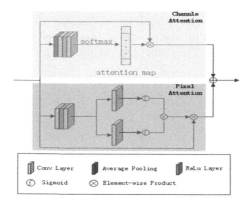

Fig. 3. Double Attention module

Channel Attention (CA). The main task of channel attention is to assign different weights to different channel features. By using the global average pooling, we transform the global spatial information on the channel into a channel descriptor.

$$g_c = G_p(F) = \frac{1}{H \times W} \sum_{i=1}^{H} \sum_{j=1}^{W} F_c(i,j) \tag{4}$$

Where G_p is the global pool function, $F_c(i,j)$ represents the pixel value of the position (i,j) on the c-th channel F_c, and the feature map with the shape of $C \times H \times W$ turns into the attention map of $C \times 1 \times 1$. The features are processed by two convolutional layers, a ReLu activation function and a latter softmax function, and then the attention map is obtained, that is, the weights of different channels.

$$CA_c = S\left(Conv\left(\delta\left(Conv\left(g_c\right)\right)\right)\right) \tag{5}$$

Where the S and δ represent the softmax and ReLu functions, respectively.

By merging the weights of the channel CA_c and the input F by element-wise multiplication, we can get the output of the CA module.

$$F_c = CA_c \otimes F \tag{6}$$

Pixel Attention (PA). Since the haze is distributed differently in different pixels of hazy image, the pixel attention (PA) module we proposed can learn the informative contextual feature of each pixel. Each pixel in the hazy image is treated differently by PA module which can pay more attention to those critical pixels.

Firstly, the input F are processed by two convolutional layers and a ReLu activation function.

$$F' = Conv(\delta(Conv(F))) \tag{7}$$

Then the feature map F' passes through a convolution layer and a sigmoid activation function, respectively.

$$F_1 = \sigma\left(\text{Conv}\left(F'\right)\right) \tag{8}$$

$$F_2 = \sigma\left(\text{Conv}\left(F'\right)\right) \tag{9}$$

Where the σ represents the sigmoid function. The shape of the feature map F', F_1 and F_2 remain unchanged as C × H × W.

In order to get pixel attention weight map, we merge F_1 and F_2 by element-wise multiplication.

$$PA = F_1 \otimes F_2 \tag{10}$$

At the end, we get the output of PA module by fusing PA and F by element-wise multiplication.

$$F_p = PA \otimes F \tag{11}$$

As shown in Fig. 3, we finally merge F_c and F_p by element-wise sum to further improve the performance of the network. The final output of DA module is F^*.

$$F^* = F_c + F_p \tag{12}$$

3.2   Basic Block Structure

As is shown in Fig. 4, the basic block structure is composed of double attention (DA) module and local residual learning. Local residual learning can transfer the feature information in the shallow part of the network to the deep part of the network through multiple local residual connections, DA module can give different channel features and pixel features different weights.

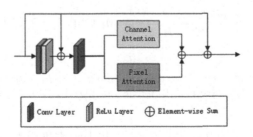

Fig. 4. Basic block structure

3.3   Group Architecture and Global Residual Learning

Our CP-Net contains six Group Architectures. A Group Architecture contains a skip connection and B basic blocks. Our B basic blocks are connected in sequence, followed by a convolutional layer. We fuse the input of Group Architecture and the output of the last convolutional layer by element-wise sum, which not only helps to reduce the loss of information in the flow process, but also helps to deepen the depth of the network. We added three additional convolutional layers and a global learning module after the last Group Architecture. Combining the features of the shallow part of the CP-Net and the features of the deep part through element-wise sum can significantly improve the dehazing effect of our network.

3.4   Loss Function

To train the proposed CP-Net, L1 loss, perceptual loss and SSIM loss are employed.

L1 Loss. Through L2 loss is widely used in many image dehazing networks, [32] proved that L1 loss can achieve higher PSNR and SSIM than L2 loss. Given an input hazy image I, the output of CP-Net is $CP(I)$ and the ground truth is J. Then the L1 loss over N samples can be written as

$$L_1 = \sum_{i=1}^{N} \|CP(I_i) - J_i\|_1 \tag{13}$$

Perceptual Loss. The perceptual loss leverages multi-scale features extracted from a pre-trained deep neural network to quantify the visual difference between the estimated image and the ground truth. In this paper, we adopt the VGG16 [30] pre-trained on ImageNet [29] and the three stages (ReLu1-2, ReLu2-2 and ReLu3-3). The perceptual loss is defined as

$$L_p = \sum_{j=1}^{3} \frac{1}{C_j \times H_j \times W_j} \|\emptyset_j(\mathrm{CP}(I)) - \emptyset_j(J)\|_2^2 \tag{14}$$

where $\emptyset_j(\mathrm{CP}(I))\,(\emptyset_j(J))$, $j = 1, 2, 3$, denote the aforementioned three VGG16 feature maps associated with the dehazed image $\mathrm{CP}(I)$ and the ground truth J, and C_j, H_j and W_j specify the dimension of $\emptyset_j(\mathrm{CP}(I))\,(\emptyset_j(J))$, $j = 1, 2, 3$.

SSIM Loss. SSIM is proposed to measure the structural similarity between two images. It can be written as

$$SSIM(CP(I), J) = \frac{2\mu_{CP(I)}\mu_J + C_1}{\mu_{CP(I)}^2 + \mu_J^2 + C_1} \cdot \frac{2\sigma_{CP(I)J} + C_2}{\sigma_{CP(I)}^2 + \sigma_J^2 + C_2} \tag{15}$$

Where μ_x, σ_x^2 are the average value and the variance of x, respectively. σ_{xy} is the covariance of x and y. C_1, C_2 are constants used to maintain stability. SSIM ranges from 0 to 1. SSIM Loss can be expressed by the following formula

$$L_s = 1 - SSIM(CP(I), J) \tag{16}$$

Total Loss. We take the sum of the L1 loss, the perceptual loss and the SSIM loss as the total loss L

$$L = L_1 + \alpha L_p + \beta L_s \tag{17}$$

Where α and β are positive weights. We set α and β to be 0.04 and 0.5, respectively.

3.5 Detailed Implementation of CP-Net

In this section, we specify the implementation details of our proposed CP-Net. We set up 6 Group Structures. Each Group Structure contains B = 14 Basic Blocks. The convolution layers kernel size of Channel Attention and Pixel Attention is set to 1×1, but other convolution layers kernel size is 3×3. Every Group Structure and every Basic Block Structure output 64 features which keep size fixed.

4 Experiments

4.1 Datasets and Metrics

During experiment, we used a synthetic dataset RESIDE [11] containing indoor and outdoor scenes. The indoor dataset contains 28850 hazy images and 2885 clear images for training. These hazy images are generated by corresponding clean images. The outdoor dataset contains 31430 hazy images and 898 clear images. The scatter parameters range from 0.04 to 0.2; the global atmosphere light changes from 0.8 to 1.0. Synthetic Objective Testing Set (SOTS) is used as the test dataset which contains 500 outdoor and 500 indoor images. We compare PSNR, SSIM and visualized dehazing results with previous state-of-the-art dehazing methods on our test dataset. In addition, we also conducted comparison experiments on Realistic Hazy Images.

4.2 Training Settings

In order to get a fine dehazing effect, random rotation and horizontal flip strategies are adopted to augment the training dataset. Two patches with a size of 240×240 are randomly cropped on the hazy image and its corresponding haze-free image as the inputs of CP-Net. CP-Net is trained for 1×10^6, 1×10^5 steps on indoor and outdoor datasets, respectively. CP-Net is optimized by Adam Optimizer, where the number of $\beta 1$ and $\beta 2$ is set up 0.9 and 0.999, respectively.

We set the initial learning rate as 1×10^{-4}. We leverage cosine annealing strategy [17] to adjust the learning rate from the initial value to 0. We define T as the total number of training steps, and η as the initial learning rate. The learning rate η_t is shown below at step t.

$$\eta_t = \frac{1}{2}\left(1 + \cos\left(\frac{t\pi}{T}\right)\right)\eta \tag{18}$$

All experiments are carried out with a Tesla V100 GPU.

4.3   Results on RESIDE Dataset

We compare our CP-Net both quantitatively and qualitatively with the previous state-of-the-art dehazing methods that include DCP, NLD, AOD-Net and EPDN. Table 1 shows the quantitative comparisons of our CP-Net and other networks in terms of PSNR and SSIM. It is clear that our experimental results are superior to previous advanced dehazing methods in terms of PSNR and SSIM. Not only that, but we also made a comparison of visualization in Fig. 5 for qualitative comparisons.

Table 1. Quantitative comparisons on SOTS for different methods.

Methods	Indoor		Outdoor	
	PSNR	SSIM	PSNR	SSIM
DCP	16.52	0.7433	17.48	0.7081
NLD	19.73	0.8043	17.72	0.8413
AOD-Net	21.15	0.8654	22.98	0.8982
EPDN	26.89	0.9401	22.05	0.9186
Ours	31.72	0.9858	30.47	0.9815

As is shown in Fig. 5, the top two lines are the results of indoor comparison, and the bottom two lines are the results of outdoor comparison. We can clearly discover that the dehazing images produced by DCP are very different from the real clear images (GT) in color, and the image details are seriously lost. This is because DCP uses prior assumptions. Images recovered by The NLD have a lot of black spots and the sky is highlighted. The dehazed images produced by AOD-Net have a little color distortion and some residual haze. However, our network can well preserve the true details of the images whether it is processing indoor or outdoor scenes. We can hardly see residual haze on our restored images.

We also evaluate the results on real hazy images and observe that although all models are trained with outdoor synthetic dataset RESIDE, our model can still produce better dehazed images. To a certain extent, our network can effectively remove haze, while maximizing the retention of image details. However, the images recovered from other networks not only have a lot of haze, but also are not as good as our network in terms of color fidelity and image detail.

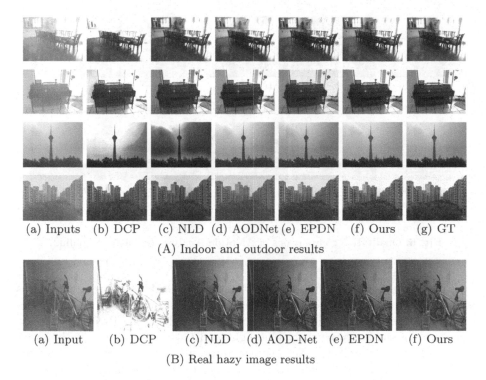

(a) Inputs   (b) DCP   (c) NLD   (d) AODNet   (e) EPDN   (f) Ours   (g) GT

(A) Indoor and outdoor results

(a) Input   (b) DCP   (c) NLD   (d) AOD-Net   (e) EPDN   (f) Ours

(B) Real hazy image results

Fig. 5. Qualitative comparisons on SOTS and Realistic Hazy Images testset

5   Ablation Analysis

In order to prove the rationality of our proposed CP-Net structure, we also designed two other networks defined as C-Net and P-Net. The only difference with CP-Net is that C-Net and P-Net contain only Channel Attention and Pixel Attention, respectively.

We use the same method which is used to train CP-Net to train the C-Net and P-Net. The final experimental results are shown in Fig. 6 and Table 2.

Table 2. Quantitative comparisons of Ablation Analysis.

Methods	Indoor		Outdoor	
	PSNR	SSIM	PSNR	SSIM
C-Net	29.98	0.9748	29.00	0.9713
P-Net	28.53	0.9621	28.45	0.9589
CP-Net	31.72	0.9858	30.47	0.9815

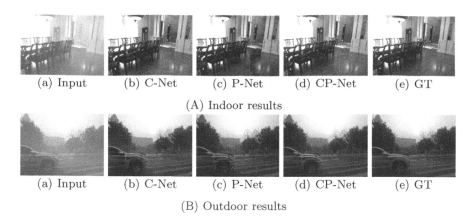

(a) Input   (b) C-Net   (c) P-Net   (d) CP-Net   (e) GT

(A) Indoor results

(a) Input   (b) C-Net   (c) P-Net   (d) CP-Net   (e) GT

(B) Outdoor results

Fig. 6. Qualitative comparisons of Ablation Analysis (color figure online)

Firstly, in the indoor results of Fig. 6, the area marked by the red box contains a large number of black spots in the image recovered by C-Net, but P-Net can restore the area well. In the area marked by the yellow box, P-Net produces a large area of shadow with navy blue light and C-Net can maintain the exact details of the area well. CP-Net can accurately restore these two areas at the same time and generate a dehazed image closer to the ground truth image.

Secondly, in the outdoor results of Fig. 6, the area marked by the red box contains a lot of haze in the image recovered by C-Net, but P-Net can remove the haze from the area cleanly. In the area marked by the yellow box, P-Net distorts the color, producing a thick black line but C-Net can restore this area well. CP-Net can accurately restore these two areas to produce a clearer image.

In order to further verify the superiority of CP-Net again, we compare CP-Net, C-Net and P-Net with respect to PSNR and SSIM. The comparative results are shown in Table 2. The results indicate that CP-Net is superior to C-Net and P-Net, and has achieved the highest PSNR and SSIM.

The above shows that Channel Attention and Pixel Attention are similar to a complementary relationship, and the fusion of the information captured by them can further enhance the effect of image dehazing.

6   Conclusion

We propose a new end-to-end single image dehazing network (CP-Net). Although our network structure is simple, it can surpass many previous state-of-the-art dehazing methods. CP-Net mainly leverages the combination of channel attention and pixel attention mechanisms. The combination of channel attention and pixel attention treats the information in the feature maps differently, filters out important information, and achieves excellent dehazing effect.

References

1. Cai, B., Xu, X., Jia, K., et al.: DehazeNet: an end-to-end system for single image haze removal. IEEE Trans. Image Process. **25**(11), 5187–5198 (2016)
2. Li, B., Peng, X., Wang, Z., et al.: AOD-net: all-in-one dehazing network. In: Proceedings of the IEEE International Conference on Computer Vision, pp. 4770–4778 (2017)
3. Hide, R.: Optics of the atmosphere: scattering by molecules and particles. Phys. Bull. **28**, 521 (1977)
4. Narasimhan, S.G., Nayar, S.K.: Chromatic framework for vision in bad weather. In: IEEE Computer Society Conference on Computer Vision and Pattern Recognition. IEEE (2000)
5. Narasimhan, S.G., Nayar, S.K.: Vision and the atmosphere. Int. J. Comput. Vis. **48**(3), 233–254 (2002)
6. He, K., Sun, J., Tang, X., et al.: Single image haze removal using dark channel prior. IEEE Trans. Pattern Anal. Mach. Intell. **33**(12), 2341–2353 (2011)
7. Zhu, Q., Mai, J., Shao, L.: A fast single image haze removal algorithm using color attenuation prior. IEEE Trans. Image Process. **24**(11), 3522–3533 (2015)
8. Ju, M., Gu, Z., Zhang, D.: Single image haze removal based on the improved atmospheric scattering model. Neurocomputing **260**, 180–191 (2017)
9. Ren, W., Liu, S., Zhang, H., Pan, J., Cao, X., Yang, M.-H.: Single image dehazing via multi-scale convolutional neural networks. In: Leibe, B., Matas, J., Sebe, N., Welling, M. (eds.) ECCV 2016. LNCS, vol. 9906, pp. 154–169. Springer, Cham (2016). https://doi.org/10.1007/978-3-319-46475-6_10
10. He, K., Zhang, X., Ren, S., et al.: Deep residual learning for image recognition (2015)
11. Li, B., Ren, W., et al.: Benchmarking single-image dehazing and beyond. IEEE Trans. Image Process. **28**, 492–505 (2018)
12. Zhang, Y., Li, K., Li, K., et al.: Image super-resolution using very deep residual channel attention networks (2018)
13. Ren, W., Ma, L., Zhang, J., et al.: Gated fusion network for single image dehazing. In: Proceedings of the IEEE Conference on Computer Vision and Pattern Recognition, pp. 3253–3261 (2018)
14. Xu, K., Ba, J., Kiros, R., et al.: Show, attend and tell: neural image caption generation with visual attention. Computer Science (2015)
15. Vaswani, A., Shazeer, N., Parmar, N., et al.: Attention is all you need. In: Advances in Neural Information Processing Systems, pp. 5998–6008 (2017)
16. Wang, X., Girshick, R., Gupta, A., et al.: Non-local neural networks (2017)
17. He, T., Zhang, Z., Zhang, H., et al.: Bag of tricks for image classification with convolutional neural networks. In: Proceedings of the IEEE Conference on Computer Vision and Pattern Recognition, pp. 558–567 (2019)
18. Meng, G., Wang, Y., Duan, J., et al.: Efficient image dehazing with boundary constraint and contextual regularization (2013)
19. Berman, D., Avidan, S.: Non-local image dehazing. In: Proceedings of the IEEE Conference on Computer Vision and Pattern Recognition, pp. 1674–1682 (2016)
20. Guo, T., Li, X., Cherukuri, V., et al.: Dense scene information estimation network for dehazing. In: Proceedings of the IEEE Conference on Computer Vision and Pattern Recognition Workshops (2019)
21. Li, B., Peng, X., Wang, Z., et al.: End-to-end united video dehazing and detection (2017)

22. Liu, Y., Zhao, G., Gong, B., et al.: Improved techniques for learning to dehaze and beyond: a collective study (2018)
23. Tu, Z., Chen, X., Yuille, A.L., et al.: Image parsing: unifying segmentation, detection, and recognition. Int. J. Comput. Vis. **63**(2), 113–140 (2005)
24. Tarel, J.-P., Hautière, N., Cord, A., et al.: Improved visibility of road scene images under heterogeneous fog. In: IEEE Intelligent Vehicles Symposium. IEEE (2010)
25. Sakaridis, C., Dai, D., Van Gool, L.: Semantic foggy scene understanding with synthetic data. Int. J. Comput. Vis. **126**(9), 973–992 (2018)
26. Fattal, R.: Dehazing using color-lines. ACM Trans. Graph. **34**, 1–14 (2014)
27. Jiang, Y., Sun, C., Zhao, Y., et al.: Image dehazing using adaptive bi-channel priors on superpixels. Comput. Vis. Image Underst. **165**, 17–32 (2017)
28. Fattal, R.: Single image dehazing. ACM Trans. Graph. **27**(3), 1–9 (2008)
29. Russakovsky, O., Deng, J., Su, H., et al.: Imagenet large scale visual recognition challenge. Int. J. Comput. Vis. **115**(3), 211–252 (2015)
30. Simonyan, K., Zisserman, A.: Very deep convolutional networks for large-scale image recognition. Computer Science (2014)
31. Qu, Y., Chen, Y., Huang, J., et al.: Enhanced pix2pix dehazing network. In: Proceedings of the IEEE Conference on Computer Vision and Pattern Recognition, pp. 8160–8168 (2019)
32. Lim, B., Son, S., Kim, H., et al.: Enhanced deep residual networks for single image super-resolution. In: Proceedings of the IEEE Conference on Computer Vision and Pattern Recognition Workshops, pp. 136–144 (2017)

Graphic Processor Unit Acceleration of Multi-exposure Image Fusion with Median Filter

Shijie Li$^{(\boxtimes)}$, Yuan Yuan, Qiong Li, and Xuchao Xie

National University of Defense Technology, Changsha, Hunan, China
lishijienudt@163.com

Abstract. A fast Graphic Processor Unit (GPU) accelerate algorithm of multi-exposure image fusion with median filter is presented in this paper. The proposed algorithm fuses images in YUV space instead of RGB space compared to traditional image fusion method. Furthermore, in YUV space the brightness components and the chromatism components were weighted fused separately with median filter. At last the filtered images were transferred to RGB and merged to the final fusion image. In the GPU acceleration part, three parallel methods were proposed, including sequence images concurrent execution, adjacent kernels merge, and parallel median filter techniques, to expand the concurrency of the algorithm on the GPU platform. In the experimental results, a 16–21 times speedup was obtained compared to the CPU implementation and up to 60 fps performance was achieved in a 1000 * 1000 * 6 multi-exposure sequence image fusion case. The results in the experiment demonstrate the high efficiency and high availability of our proposed method.

Keywords: GPU · Image fusion · Median filter · Acceleration

1 Introduction

Recently, many multi-exposure image fusion technologies are presented and widely used in image-related fields. The core idea of multi-exposure image fusion method is the image merge. More specifically, multiple exposure level images are merged into a fused output image. This method grows quickly in the last few years because it solves the image details lack problem caused by modern digital cameras' improper exposure setting. And compared to tone mapping based methods, image fusion-based method has a faster speed and equal image quality.

The application field of image fusion-based method is wide, such as remote sensing [1], medical imaging [2], computer vision [3] and some other areas. Although the image fusion method has been widely used, the execution speed still has much space to be further improve. Since the multi-exposure image fusion-based method is a pixel-level method, the pixel-level data calculation among images are computation intensity and time consuming. This computation workload becomes heavier with the fusion image size increasing. And the fact is that, in remote sensing area, medical imaging area, computer vision area, the images to be handled are usually large scale, the big size original multi-exposure images sequence brings a very heavy computation workload.

J. Zeng et al. (Eds.): ICPCSEE 2020, CCIS 1257, pp. 591–601, 2020.
https://doi.org/10.1007/978-981-15-7981-3_43

In some cases, for example, to merge a 1000 * 1000 * 6 resolution image sequence to a single fused image with the state of art image fusion methods costs almost 1 s, which is too long to endure in some real-time-limited applications.

Both the GPU and the CPU can process data, CPU has better versatility and can process different types of data. However, GPU is better at processing the same type of data, so GPU can be better suited for processing image fusion. In some other computation intensity applications, to improve the performance, Graphic Processing Unit (GPU) is a usually used accelerator and worked well in both stability and efficiency because of the thorough easy-to-use programming toolkit CUDA. Furthermore, the other reason of GPUs' wide use is that the GPU hardware is easy to obtain and deployed in almost any kind of computation facilities, down to mobile phones, embedded systems, up to workstations, supercomputers. So it is a wise choice to accelerate the image fusion algorithm with GPU in real life applications. However, as far as we know, there are very few GPU acceleration research in multi-exposure image fusion algorithm field, our work is meaningful and challenging at the same time.

In this work we present a faster than real-time GPU implemented based on multi-exposure image fusion algorithm with YUV space transferred median filters. Our method introduces three core parallel techniques including sequence images concurrent execution, adjacent kernels merge, and parallel median filters into original multi-exposure image fusion algorithm. And we further optimize the whole execution process to obtain a high performance.

The rest of our paper is organized as follows: in Sect. 2, we introduce the related work in multi-exposure image fusion algorithm. In Sect. 3, our proposed fast GPU algorithm's details are explained. In Sect. 4, the experimental results and analysis are given. At last, Sect. 5 gives the conclusions.

2   Related Works

Unlike tone mapping based methods [4–6], image fusion-based algorithms, specifically multi-exposure image fusion, skip the HDR image construction procedure to generate a tone-mapping-like fused image. This feature saves a lot of computation time which exists in tone mapping based algorithm. As a result, image fusion-based methods [24] have become more and more popular in consumer electronic device field. There are some kind of popular and widely used representative image fusion-based methods, such as pyramid-based method, the representative algorithm is Laplacian pyramid [7], the other one is wavelet transform based method the representative algorithms include discrete wavelet decomposition [8, 9] and stationary wavelet decomposition [10]. However, there are many challenges in these traditional presented algorithms. For example, in a pyramid-based method, the challenge is heavy computation workload. More specifically, the original images are reformed as Laplacian pyramids, then in each pyramid level from bottom to top, the images are fused and constructed the final fusion image. The extra pyramid generation process makes the overhead of image resize calculation and extra pyramid level image fusion calculation. This overhead consumes lots of hardware resources and execution time. On the other hand, the wavelet transform based method has many problems in shift variance and aliasing. Nowadays many new pixel-level image fusion methods are proposed, such as dual-tree complex wavelet [11], curvelet transform [12], contourlet transform [13, 14], and shearlet transform [15, 16] have been presented to solve the wavelet transform-based method's problems, and some fast designed pyramid methods [17, 18] are given to solve the performance problem in Laplacian pyramid algorithm.

As far as we know, there are very few GPU acceleration for image fusion algorithms, except for some remote sensing image fusion applications. For handling large scale remote sensing images captured by satellites, some GPU implemented methods have been proposed [19]. However, in that work, the GPU algorithm is not well designed, the speedup is ranging from 2 to 18, which is not stable, and considering the time consumption, the implementation still unsatisfies the real-time demand.

3 Faster Than Real-Time GPU Implementation of Multi-exposure Image Fusion with Recursive Filter

3.1 Algorithm Detail Design

In this subsection, we first explain the detail design of our original CPU based image fusion. The innovation of our original algorithm is the introduction of transformation between YUV and RGB space with the right median filters. By this way, we keep the fusion image quality better compared with the traditional methods.

Our proposed algorithm first transfers the original images sequence into YUV space from RGB space. Then contrast the brightness component and saturation the chromatism component separately. After getting the contrast ratio and exposure ratio from Y channel, we filter the data with weights and obtain the fused brightness component with the median filters. At the same time, we weighted filter the chromatism component with the saturation ratio and obtain the fused chromatism component. In the end, we transfer these two components into RGB space from YUV again and generate the final fusion image. The overall algorithm is described in Fig. 1.

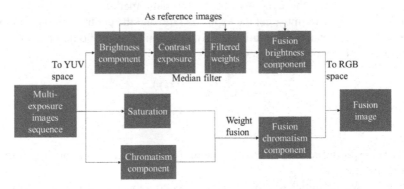

Fig. 1. The original CPU-based image fusion algorithm with median filters.

The transformation relationship between YUV space and RGB space is shown in (1):

$$Y = \frac{1}{3}(R+G+B)$$
$$U = \frac{1}{2}R - \frac{1}{2}B$$
$$V = \frac{1}{4}R - \frac{1}{2}G + \frac{1}{4}B$$

(1)

Y, U, V and R, G, B represent the related channels separately. At the same time, we can also obtain the matrix of image transformation from YUV space to RGB space from (1).

Before the weight filter operation, we need to get the fusion weight of i-th image's chromatism component. It can be calculated as:

$$ws_i(x,y) = \frac{S_i(x,y)}{\sum\limits_{j=1}^{N} S_j(x,y) + \varepsilon}$$

(2)

In Eq. (2), ws is the chromatism component, N is the number of images we need to handle in our image fusion application's input sequence, ε is a little value which can be any constant, in order to make the denominator above 0.

After working out the saturation weights, we can weight fuse the chromatism component to get the final fusion image, in the component, V component can be calculated as:

$$V_F(x,y) = \sum\limits_{i=1}^{N} V_i(x,y) \times ws_i(x,y)$$

(3)

And we can obtain the U component by the same method.

In our algorithm, we must get the exposure ratio and the contrast ratio of brightness component. After obtaining the ratios, a normalization to them should be made and the exposure weights can be calculated as (4):

$$we_i(x,y) = \frac{E_i(x,y)}{\sum\limits_{j=1}^{N} E_j(x,y)}$$

(4)

E is the exposure ratio, we is the exposure weights. We can get contrast weights by the same method in (4).

By the way, the initial weights of brightness component can be obtained by multiplying the exposure weights with contrast weights which is shown in (5).

$$wy_i(x,y) = we_i(x,y) \bullet wc_i(x,y) \tag{5}$$

The meanings of introducing the median filters is that brightness component weights in (5) vary rapidly in weights feature map. It will cause serious fusion image quality reduction. We expect the fusion weights to be smooth and have some abilities to overcome the noise in the weights feature maps. Motivated by it, we filter the weights which have cracks with median filters and treat it as a noise reduction process, as a result we hope to remove these cracks in weights map.

The formula of median filter is:

$$g(x,y) = med\{f(x-k,y-l),(k,l \in W)\} \tag{6}$$

Where *med* is the median operation to get the median value from the given set. f is the input function, g is the output function, x and y are pixel indexes from horizontal and vertical directions. W is a windows area usually set to 3 * 3 or 5 * 5.

Then we weight fuse the Y channel data after the median filter operation and obtain the brightness component of fusion image by the method described in (7):

$$Y_F(x,y) = \sum_{i=1}^{N} Y_i(x,y) \times wy_i'(x,y) \tag{7}$$

At last, we make a transformation from Y, U, V component to RGB space again to obtain the final fused image.

3.2   Sequence Images Concurrent Execution

In our GPU implemented design, the first novel and innovative parallel technology is sequence images concurrent execution. In the original CPU-based algorithm, the dataflow is shown in upper half of Fig. 2. For example, we intend to fuse a multi-exposure sequence with 4 images, the first image 1 is handled by CPU, after a series of

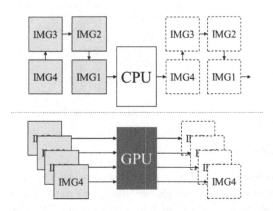

Fig. 2. Sequence images concurrent execution in GPU implementation compared with CPU.

operations including color space transformation, median filter and weight fusion, the first component of final fused image is ready. Then the second image 2 is sent to CPU after image 1, runs in the same way to image 1. Image 3 and image 4 are the same, all the images in the same sequence are executed in serial.

However, in our GPU design, we launch all the images at one time, pixels data in all images are mapped in GPU's global memory and be binding to threads in stream processors of the GPU. This method is from the task-level parallelism view, images are independent from each other in the same sequence.

The different operation schedule between CPU and GPU is illustrated in Fig. 2. The schedule we design will use much fuller of GPU's resources.

3.3   Adjacent Kernels Merge

The second technology used in our GPU parallel design is the adjacent kernels merge. It is a data-level parallelism technology. In a traditional GPU design, one kernel handles one function in the program, parallel data or pixels are sent to GPU's kernel to do the same functional operation. For example, we implement three kernels named Exposure, Contrast and Saturation. Since the kernels of GPU usually are launched one by one in common case, IMG1, IMG2 and IMG3 are sent to one kernel. This schedule develops the task-level parallelism but not data-level parallelism.

In our design, we improve the original kernel launch schedule to further develop the data-level parallelism. More specifically, we merge three kernels into one big kernel, and rewrite the function for each original step. Now in our new GPU kernel design, IMGs are calculated three original functions in one kernel in parallel. The GPU resources are fuller used compared with the original design. The new kernel launch schedule is shown in Fig. 3.

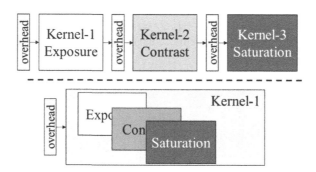

Fig. 3. Adjacent kernels merge technique

3.4   Parallel Median Filter

The last parallel technology used in our GPU implementation is parallel median filter. In original CPU-based algorithm implementation, this module consumes lots of time. For a n * n pixels image, with a m * m windows median filter, the computation complex is up to n * n * m * m, which will greatly increase the execution time in CPU version. Because the filter window moves in the image, the data in the window are calculated in serial. Another kind of filter is recursive filter, the recursive filter has a little better image fusion quality but much more time consumption, harder to be accelerated. So in our work, we choose median filter to accelerate our work.

In our GPU design, we duplicate the data sets generated by sliding window, and calculate them in parallel. We can find a fact that in each data sets, the pixels to be handle are independent from each other, it is the base reason why we can parallel our sliding windows data. Our method is illustrated in Fig. 4.

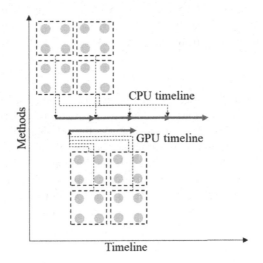

Fig. 4. Parallel median filter technique

4   Experiment and Analysis

4.1   Experimental Setup

In this subsection we introduce our experimental environment. The software platform can be listed as follows: Windows 10 professional operating system, Visual Studio 2019, OpenCV 4.1.0 [20], NVIDIA CUDA v10.1 [21], the hardware platform includes: gigabyte laptop computer with CPU i7 9750H@2.6 Ghz, RAM 16 GB, GPU GTX 1660Ti mobile.

The multi-exposure image sequences in our experiment are from the CAVE lab. In our experiment we choose 15 multi-exposure image sequences from the whole database. The related works are Reinhard [22], iCAM06 [23] and Li [17]. All the results in our experiment are executed for 20 times and we take the average values as the experiment final results.

4.2 The Effectiveness of Our Method

We use several quality metrics which are widely used in image fusion field to evaluate the effectiveness of our fusion method objectively, they are Q_0, $Q^{AB/F}$ and visual information fidelity (VIF). The results we get from related work and our own method are shown in Table 1, the results show that among all the methods, ours has the best performance in the evaluation of three quality metrics.

Table 1. The effectiveness of different image fusion methods.

Images	Indexes	Reinhard [22]	iCAM06 [23]	Li [17]	Ours
Garage	Q_0	0.555	0.576	0.675	0.701
	$Q^{AB/F}$	0.520	0.561	0.674	0.694
	VIF	0.467	0.443	0.522	0.602

Besides the image quality metrics, we also test the subjective image quality of our method. So we handle some image in database through our algorithm and show the fusion image results in Fig. 5, in these images, the left half part including 6 little images is a set of original multi-exposure image sequence, and the right part including two bigger images is a set of fusion image results generated by Li's method [17] and our method. From the subjective comparison between left bigger fused image and the right bigger fused image, we can find that our method almost keeps details well with Li's method in abnormal-exposed area, furthermore, our fusion image is more colorful than Li's.

Fig. 5. Subjective evaluation in different methods

4.3 Speed Performance Analysis

In this subsection, we evaluate the speedup performance of our proposed GPU method. First, we research the detail time consumption of different module and step in our proposed original algorithm. And the result shows that the exposure-contrast-saturation module and the median filter module cost the major part of the whole time consumption. After our GPU optimization, the time ratios of exposure-contrast-saturation module and the median filter module both decrease by the benefit of our well design. Which is shown in Fig. 6.

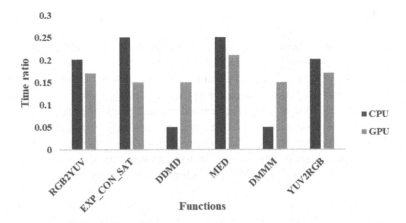

Fig. 6. Time ratio of different modules in GPU optimized implementation

At last, we evaluate the overall speedup of our proposed GPU implementation and some outstanding other fast image fusion methods. For the CPU version part, our original algorithm runs nearly the same but a little lower performance to Li's related work. Because in our original algorithm, the color space transformation is introduced in order to increase the subjective fused image quality, but our original CPU algorithm is much easier to be accelerate by GPU. It can be found that in small image size, for example, 100 * 100, the speedup is 21x, in large image size, 1000 * 1000, the speedup still has 16x. Our GPU implementation is efficient and stable in the application (Fig. 7).

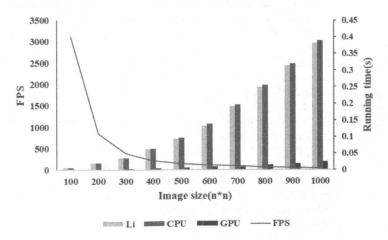

Fig. 7. Time ratio of CPU and GPU implementations

In small size level, the fps is up to 3119, even in large scale case, the fps still reaches 60, all the results demonstrate our method a faster than real-time speed and stable performance and fine image quality.

5 Conclusion

In this paper, an efficient, easy-to-use and fast GPU implemented multi-exposure image fusion approach is presented. In algorithm design aspect, we introduce a color space transformation into traditional image fusion algorithm to obtain brightness information in original multi-exposure image sequence to increase the subjective image quality. In the GPU acceleration design, we propose three parallel techniques from task-level to data-level named sequence images concurrent execution, adjacent kernels merge, and parallel median filter techniques. These techniques greatly improve the overall performance. Experiment results demonstrate that the proposed method can generate high visual quality fusion images, and keep high efficiency and speed at the same time.

Acknowledgment. This work is supported by National Key Research and Development Program of China (No.2018YFB0204301), the Advanced Research Project of China under grant 31511010202, and the National Natural Science Foundation of China under Grants (No. 61906207)

References

1. Ghassemian, H.: A review of remote sensing image fusion methods. Inf. Fusion **32**, 75–89 (2016)
2. El-Gamal, F.E.-Z., Ahmed, M.E., Atwan, A.: Current trends in medical image registration and fusion. Egypt. Inform. J. **17**(1), 99–124 (2016)
3. Xing, C., et al.: Image fusion method based on spatially masked convolutional sparse representation. Image Vis. Comput. **90**, 103806 (2019)
4. Eilertsen, G., Unger, J., Mantiuk, R.K.: Evaluation of tone mapping operators for HDR video. In: High Dynamic Range Video, pp. 185–207. Academic Press (2016)
5. Endo, Y., Kanamori, Y., Mitani, J.: Deep reverse tone mapping. ACM Trans. Graph. **36**(6), 177-1 (2017)
6. Eilertsen, G., Mantiuk, R.K., Unger, J.: A comparative review of tone-mapping algorithms for high dynamic range video. Comput. Graph. Forum **36**(2), 565–592 (2017)
7. Du, J., et al.: Union Laplacian pyramid with multiple features for medical image fusion. Neurocomputing **194**, 326–339 (2016)
8. Udhaya Suriya, T.S., Rangarajan, P.: Brain tumour detection using discrete wavelet transform based medical image fusion (2017)
9. Singh, D., Garg, D., Pannu, H.S.: Efficient landsat image fusion using fuzzy and stationary discrete wavelet transform. Imaging Sci. J. **65**(2), 108–114 (2017)
10. Jiang, Q., et al.: A novel multi-focus image fusion method based on stationary wavelet transform and local features of fuzzy sets. IEEE Access **5**, 20286–20302 (2017)
11. Yu, B., et al.: Hybrid dual-tree complex wavelet transform and support vector machine for digital multi-focus image fusion. Neurocomputing **182**, 1–9 (2016)

12. Ji, X., Zhang, G.: Image fusion method of SAR and infrared image based on Curvelet transform with adaptive weighting. Multimed. Tools Appl. **76**(17), 17633–17649 (2015). https://doi.org/10.1007/s11042-015-2879-8
13. Cai, J., et al.: Fusion of infrared and visible images based on nonsubsampled contourlet transform and sparse K-SVD dictionary learning. Infrared Phys. Technol. **82**, 85–95 (2017)
14. Meng, F., et al.: Image fusion based on object region detection and non-subsampled contourlet transform. Comput. Electr. Eng. **62**, 375–383 (2017)
15. Liu, X., Mei, W., Huiqian, D.: Structure tensor and nonsubsampled shearlet transform based algorithm for CT and MRI image fusion. Neurocomputing **235**, 131–139 (2017)
16. Yin, M., et al.: Medical image fusion with parameter-adaptive pulse coupled neural network in nonsubsampled shearlet transform domain. IEEE Trans. Instr. Meas. **68**(1), 49–64 (2018)
17. Li, S., Kang, X.: Fast multi-exposure image fusion with median filter and recursive filter. IEEE Trans. Consum. Electron. **58**(2), 626–632 (2012)
18. Gastal, E.S.L., Oliveira, M.M.: Domain transform for edge-aware image and video processing. ACM Trans. Graph. **30**(4), 69:1–69:11 (2011)
19. Al-Oraiqat, A.M., Bashkov, E.A., Babkov, V., Titarenko, C.: Fusion of multispectral satellite imagery using a cluster of graphics processing unit. arXiv preprint arXiv:1803.00737 (2018)
20. Kaehler, A., Bradski, G.: Learning OpenCV 3: Computer Vision in C ++ with the OpenCV Library. O'Reilly Media, Inc., Newton (2016)
21. Armstrong, D.E.: CUDA GPU Programming Applied to HSI Exploitation. No. LA-UR-17–20565. Los Alamos National Lab. (LANL), Los Alamos, NM, United States (2017)
22. Reinhard, E., Stark, M., Shirley, P., Ferwerda, J.: Photographic tone reproduction for digital images. In: Proceedings of the ACM SIGGRAPH, pp. 267–276, July 2002
23. Wang, Z., Bovik, A.: A universal image quality index. IEEE Signal Process. Lett. **9**(3), 81–84 (2002)
24. Qi, G., Chang, L., Luo, Y., et al.: A precise multi-exposure image fusion method based on low-level features. Sensors **20**(6), 1597 (2020)

Improved PCA Face Recognition Algorithm

Yang Tao[1]([✉])[iD] and Yuanzi He[2]

[1] Nanchang Institute of Science and Technology, Nanchang, China
taoyangxp@163.com
[2] Guangdong University of Science and Technology, Dongguan, China

Abstract. Face recognition technology, as a biometric recognition technology, is very mature and has many applications. It achieves in-depth applications in smart campus systems, such as classroom attendance, classroom behavior analysis, and smart restaurants. Using human faces as the face data foundation, computer vision and image processing technologies are applied to research and implement face recognition. Based on the principal component analysis (PCA) theory, this paper analyzed the characteristics of face data, studied the face recognition algorithm. Considering the LBP and SVM algorithm, an improved PCA face recognition algorithm was proposed. Through comparative experiments, the results show that the proposed algorithm can improve the accuracy of face recognition.

Keywords: LBP · PCA · SVM · Face recognition

1 Introduction

"Face recognition" and "big data" are the two most widely used methods in Internet Finance in recent years. Google, apple, Baidu and other well-known enterprises at home and abroad, as well as the Internet financial enterprises represented by Micro bank, E-commerce bank and Zhongke loan, are accelerating the layout of "face recognition" and "big data".

Face recognition technology, as a biometric recognition technology, is very mature and has many applications. Face recognition applications in China include public security, finance, access control, classroom attendance, and smart campuses [1]. It has also achieved in-depth applications in smart campus systems, such as classroom attendance, classroom behavior analysis, and smart cafeterias. These systems use the human face as the data foundation, and rely on computer vision and image processing technologies to research and implement face recognition. The reasons for the rapid development of face recognition include multiple aspects. For example, first, face recognition only requires people to complete the captured image capture, and it can automatically track, monitor, and recognize alarms in real time. Second, computer vision research has accelerated the development of face recognition technology. Machine vision is to allow computers to effectively understand pictures by analyzing pictures and video content, and sometimes it is necessary to recognize faces in pictures. Third, the demand for face recognition in information security, access control, human-computer interaction, visual monitoring, bayonet control, person identification and

© Springer Nature Singapore Pte Ltd. 2020
J. Zeng et al. (Eds.): ICPCSEE 2020, CCIS 1257, pp. 602–611, 2020.
https://doi.org/10.1007/978-981-15-7981-3_44

search is increasing, and the demand has accelerated the research speed of face recognition algorithms, such as in school classrooms. Automatic behavior analysis in China also requires face recognition technology to complete student identification and localization. It can be seen that the research on face recognition algorithms has very common practical value. However, many algorithms are protected by patents and cannot be widely and freely applied. This requires us to research or improve the face recognition algorithm according to the actual situation of specific applications, and apply it to application systems with urgent needs.

2 Related Works

Face recognition technology includes Geometrical Features Based (GFB) [2], Local Face Analysis (LFA) [3, 4], Convolutional Neural Networks (CNN) [5–7], Principal Component Analysis (PCA) [8] algorithm. PCA is one of the most popular algorithms at present.

At present, among the existing face recognition algorithms, the face recognition method based on Principal Component Analysis (PCA) has become mature, but the face recognition algorithm based on PCA often needs to convert the original image data to quantization, that is, to transform the image matrix into one-dimensional column vector. As a result, the dimension of the vectorized sample data is very high, which is easy to cause the curse of dimensionality and reduce the calculation speed of the algorithm. Moreover, the vectorization will destroy the spatial structure of the face image (for example, two pixels close to each other in the four neighborhoods will be separated after vectorization). In order to overcome the above shortcomings, Yang et al. [9] proposed a face recognition algorithm based on two-dimensional principal component analysis (2D-PCA), that is, the feature extraction is based on two-dimensional matrix, and the original image matrix does not need to be converted into vector. Therefore, 2D-PCA algorithm is better than PCA algorithm in running speed and recognition rate.

Although 2D-PCA algorithm has made some achievements in the research of face recognition, most of them focus on gray-scale image, and do not use the color information of face image. At present, some researches have shown that color information is also an important feature of human face, which can provide a lot of useful feature information for face recognition and improve the effect of face recognition. Color LBP [10] extends the texture feature extraction method LBP based on gray-scale image to color RGB space, and then uses PCA and LDA to classify the extracted face image. The classification accuracy is obviously improved. QPCA [11] uses quaternion to represent color information, which is better than PCA based on gray image. In order to use the color information of image more effectively, some scholars have realized the method of color transformation space, which is to transform the general RGB space into a more suitable color space for different face databases. In reference [12], BWDCS method is proposed and the classification accuracy of BWDCS space, RGB space and gray space is compared in different face databases. Finally, based on PCA classification Class results show that BWDCS space and RGB space are better than gray space and BWDCS space is the best.

In order to realize color face recognition based on 2D-PCA, Wang et al. In literature [13], represented an image of color RGB face with $I_1 \times I_2$ as a color information matrix with the size of $3 \times I$ (where $I = I_1 \times I_2$), and then directly used 2D-PCA to recognize the matrix. Experimental results show that compared with 2D-PCA algorithm based on gray image, the recognition rate is improved by 2%–3%. Wan [14] et al. first extract the features of R, G and B channels of a color image, then reconstruct the three feature matrices into a color information matrix, and finally use mdnib2dpca to improve the recognition rate. Although the 2D-PCA method in reference [13, 14] integrates RGB color information with color information matrix, which improves the recognition rate of the algorithm, the spatial relationship of color information is not considered. Therefore, because a two-dimensional matrix cannot keep the correlation between color information components and the spatial relationship of image information at the same time, some scholars propose tensor representation for face recognition [15]. The accuracy of tensor based algorithm is improved and the training time is reduced.

Based on the principal component analysis (PCA) theory, this paper analyzes the characteristics of face data, studies the face recognition algorithm, and then proposes an improved PCA face recognition algorithm.

3   Face Recognition Algorithms Based on PCA

3.1   PCA

PCA algorithm projects high-dimensional space vector into low-dimensional space through linear transformation. The linear transformation here generally uses K-L transformation to transform the sample matrix into one-dimensional vector, and then extracts features from the matrix composed of many sample vectors. K-L transform (Karhunen-Loeve transform) is a kind of orthogonal transform. Its essence is to project the image signal into the signal subspace. The obtained covariance matrix is a diagonal matrix, which can remove the correlation between signal components to the maximum extent. It is the best transform in the sense of mean square error (MSE), and plays an important role in data compression.

Given a set of decentralized picture vectors:

$$\Omega = \{\Phi_1 \Phi_2, \ldots, \Phi_m\} \qquad (1)$$

PCA is to reduce the dimension of the sample set in the new coordinate system is the largest sum of projections. According to the characteristics of face data, it can be converted into solving equation that the projection of data set in unit vector of u is the largest:

$$\max\left(\frac{1}{m}\sum_{i=1}^{m}\left(\phi_i^{\mathrm{T}}u\right)^2\right) = \max\left(u^{\mathrm{T}}\left(\sum_{i=1}^{m}\frac{1}{m}\phi_i\phi_i^{T}\right)u\right) \qquad (2)$$

where the covariance matrix of the sample is used.

3.2 SVM

SVM is a binary classification model. Its basic model is the linear classifier with the largest interval defined in the feature space [16]. Support vector machine model adopts nonlinear kernel function, which can project the input data to nonlinear space, so that it can become a nonlinear classifier. The learning strategy of Support Vector Machine is to maximize the interval, which can be formalized as a problem of solving convex quadratic programming, and its learning algorithm is the optimization algorithm of tangled convex quadratic programming [17]. SVM adopts the concept of kernel function, which transforms the original linear non separable data set into linear separable data set in high-dimensional space by mapping in high-dimensional space, so that the linear classification surface can be well used for classification. SVM commonly used kernel functions are: linear kernel function, polynomial kernel function, Gaussian kernel function, etc. In general, the face recognition process cannot be completed with a single kernel function, so we use the fusion kernel function selection rule, using the weight ρ to express the fusion relationship between the global kernel function and the local kernel function, that is

$$\varphi_m = \rho * \varphi_r + (1 - \rho) * \varphi_p \tag{3}$$

where φ_r and φ_p are global and local kernel functions, respectively.

3.3 LBP

LBP (Local Binary Pattern) is an operator used to describe the local texture features of an image. It has the advantages of rotation invariance and gray invariance. It was first proposed by T. Ojala, M. pietik ä inen, and D. Harwood in 1994 for texture feature extraction. Moreover, the extracted feature is the local texture feature of the image. LBP histogram sequence belongs to the first-order statistical feature, which can better describe the texture characteristics of image. However, due to the large difference of local features in each region of the image, the whole image is divided into several sub images in order to keep the texture characteristics and increase the microstructure information of the image. The detected face can be divided into 4 × 4 non overlapping regions [18], and the LBP histograms of each image region can be counted respectively, and then each histogram can be cascaded in the order of first column and then column, and the feature of the cascading is the LBP histogram of the whole image.

The classical LBP algorithm assigns the gray value of face pixels to a fixed nine palace lattice of 3 × 3, sets the gray value of the central pixel as the threshold value, and describes the local texture characteristics of the face image by comparing the gray value of the threshold value and its adjacent eight pixels. If the threshold value is larger than the gray value of its neighboring pixels, it will be encoded and marked as 1. If the threshold value is smaller than the gray value of its neighboring pixels, it will be encoded and marked as 0. All the encoded marks will be concatenated into an 8-bit binary number from a certain position in clockwise or counterclockwise order. Finally, the binary number will be given corresponding weight and converted into an 8-bit binary decimal number; the decimal value obtained is the LBP feature of the image.

Formally, for a given pixel at (x_o, y_o), it is obtained that LBP is represented in decimal form, with the formula of LBP defined as follows:

$$\text{LBP} = \sum_{i=0}^{p-1} s(p_i - p_0) * 2^i \tag{4}$$

Where p is the number of adjacent, then p = 8. p_o represents the pixel value at (x_o, y_o) and pi is the gray value of the neighboring pixels around the central pixel value p_o, where i is from 0 to 7. The function of S is defined by

$$S(x) = \begin{cases} 1 & x \geq 0 \\ 0 & x < 0 \end{cases} \tag{5}$$

If the pixel gray value of p_i is greater than that of p_o, the value of the function is set to 1, otherwise it is set to 0. LBP is applied to every pixel of the image, and an 8-bit binary code is obtained. The gray value of the pixel is replaced by the corresponding LBP code. The range of the gray value of the pixel after coding is 0 to 255 and only related to the binary code, the researcher concludes that LBP has the characteristics of gray invariance, and the image will treat the face under different lighting conditions, so LBP has the characteristics of gray monotone invariance, so it is not sensitive to the light in the face image.

3.4   Proposed Algorithm

According to the basic flow of the face recognition system, based on the specific face database, comprehensively considering the algorithms of LBP, SVM and PCA, an improved PCA face recognition algorithm is proposed, as follows:

STEP1:   load the face image library;
STEP2:   perform LBP feature extraction on the test sample image set;
STEP3:   perform PCA feature extraction on the LBP feature face image set;
STEP4:   use SVM to train the classification model and save the image features;
STEP5:   perform secondary feature extraction on the verified face image and send it to the classifier.
STEP6:   the verification image features are compared with the database; the classifier outputs the classification result.

Finally the recognition result will be obtained.

4   Experiments and Analysis

In our experiments, in order to test the performance of the proposed algorithm, we choose ORL database and Yale database which are representative of the experimental face database. The ORL face database was created by the University of Cambridge in the United Kingdom. The library contains a total of 400 facial images of 40 persons, and each person has 10 facial images with different expressions, postures, and face decorations. Each picture has a size of 92 × 112, and a single picture has a total of 10304 pixels. Some samples of ORL face database are shown in Fig. 1.

Fig. 1. Some samples of ORL face database.

The number of face objects in Yale database is slightly more than that in ORL database. There are 15 objects in total, 11 for each object, covering the front face images with different light intensity, expression and covering angle. Some samples of Yale face database are shown in Fig. 2.

Fig. 2. Some samples of Yale face database.

4.1 LBP Feature Extraction

We extract the LBP feature of a picture, and then we can get a feature map (the outermost circle is filled with zeros before operation). The value of the feature map is determined by the defined radius and the number of neighbors. The steps of LBP feature vector extraction are as follows:

STEP1: The detection window is divided into 16 × 16 small areas (cell).

STEP2: For one pixel in each cell, the gray value of the adjacent 8 pixels is compared with it. If the surrounding pixel value is greater than the central pixel value, the position of the pixel point is marked as 1, otherwise it is 0. In this way, 8 points in the 3 * 3 domain can generate 8-bit binary number after comparison, that is, the LBP value of the central pixel point of the window is obtained.

STEP3: Then calculate the histogram of each cell, that is, the frequency of each number, and then normalize the histogram;

Finally, the statistical histogram of each cell is connected into a feature vector.

Through the above algorithm, the extracted LBP operator can get an LBP code at each pixel, and the original LBP feature obtained is still "a picture". For example, a partial picture obtained by performing LBP feature extraction on the ORL dataset is shown in Fig. 3, and a partial picture obtained by performing LBP feature extraction on the Yale dataset is shown in Fig. 4.

Fig. 3. Some samples of Eigen faces extracted by LBP feature extraction from ORL database

Fig. 4. Some samples of Eigen faces extracted by LBP feature extraction from Yale database

4.2   PCA Feature Extraction

We already have the feature face image set extracted by LBP. On this basis, we implement PCA feature extraction. The steps of this method are as follows:

Step1:   Obtain a set of M feature face images and smooth them.

Step2:   Calculate the average image and obtain the deviation matrix.

Step3:   Calculate covariance matrix, calculate new eigenvalues and eigenvectors, which is the standard PCA algorithm flow.

Step4:   Principal component analysis.

In the new eigenvectors and eigenvalues, the larger the eigenvalues are, the more important it is for us to distinguish, that is, the principal component. We only need the eigenvectors corresponding to those large eigenvalues, while those very small or even zero eigenvalues are of little significance to us.

Through the above algorithm, with the ORL database, let's compare the different average images obtained by the PCA algorithm without the LBP feature extraction (original image) and the LBP feature extraction (LBP Eigen face), that is, the average face, as shown in Fig. 5. And then the new Eigen faces are shown in Fig. 6.

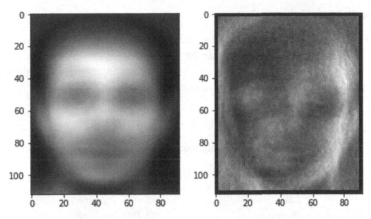

Fig. 5. The average faces, left one is for original image, right is for LBP Eigen face

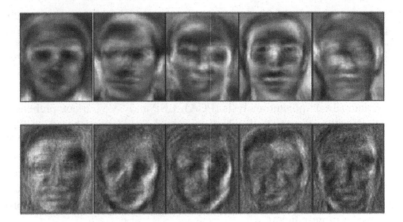

Fig. 6. PCA Eigen faces, the top line is a list of some for original image, and the bottom line is a list of some for LBP Eigen face

The same algorithmic process will be done with the Yale database, which will not be repeated here.

4.3 SVM Classification

Our experimental process is mainly divided into three stages, including image pre-processing (LBP feature extraction, PCA dimensionality reduction and normalization), image training process and image recognition process. Earlier we completed most of the data preprocessing, and then we normalize the preprocessing results. In SVM, we use Gaussian kernel to classify. In order to deal with the problem conveniently, in Yale database, we use 10 pictures of each person as the training set and another one as the test set to verify the algorithm. In ORL database, we use 9 pictures of each person as training set and another one as test set to verify the algorithm.

610   Y. Tao and Y. He

4.4   Result Analysis

Our proposed algorithm is tested on two datasets, including the ORL and Yale face databases. The recognition accuracy can be obtained from the above process. In order to facilitate performance comparison, we have implemented the PCA algorithm and LBP algorithm, and the experimental results are shown in Table 1.

Table 1. Experimental results

Face recognition algorithms	Face databases	
	ORL	Yale
PCA	94.50%	86.80%
LBP	95.39%	92.66%
Proposed	97.78%	98.81%

As can be seen from Table 1, the recognition accuracy of our proposed method has improved to some extent on different data sets.

5   Conclusion

Based on principal component analysis (PCA), LBP and SVM, a comprehensive face recognition algorithm is proposed, which is trained and verified based on ORL face database and Yale database respectively. The recognition accuracy is 97.78% and 98.81%, and the performance of the algorithm is verified. But the disadvantage is that the training time is too long, which is mainly due to the LBP operation. Of course, the running time can be allowed during the training. In the long run, it is still necessary to improve the complexity of the algorithm. In the future, the problem will be improved to make the system more efficient.

The further research is to apply this method to the intelligent classroom. Based on this face recognition algorithm, students' classroom behavior analysis can be carried out.

Acknowledgment. This paper is supported by the Ph.D. Research Initiation Fund of Nanchang Institute of Science and Technology with the Project (No. NGRCZX-18-01). It is also supported by the Science and Technology Project of Jiangxi Provincial Department of Education with the Project (No. GJJ191105).

References

1. Ke, P.: Study on face detection algorithm in subway security check. Dalian Jiaotong University (2017)
2. Zhao, D.: Research on face recognition method based on geometric features. Hebei University of Technology (2015)

3. Song, T.: Research on facial image analysis and recognition method based on local features. Zhejiang University (2015)
4. Jiao, F., Shan, S., Cui, G., Gao, W., Li, J.: Face recognition method based on local feature analysis. J. Comput.-Aided Des. Comput. Graph. (01), 53–58 (2003)
5. Wang, J., Meng, L.: Face detection algorithm based on convolutional neural network. Appl. Electron. Techn. **46**(01), 34–38 (2020)
6. Xu, T.: Research on face detection algorithm based on convolutional neural network. Nanjing University of Posts and Telecommunications (2019)
7. Liu, S., Liu, C., Zhang, A.: Real-time facial expression and gender recognition based on deep separable convolutional neural network. Comput. Appl., 1–8 (2020)
8. Turk, M.A., Pentland, A.P.: Recognition in faces pace. In: Proceedings of SPIE - The International Society for Optical Engineering, p. 1381 (1991)
9. Yang, J., Zhang, D., Frangi, A.F., et al.: Two-dimensional PCA: a new approach to appearance-based face representation and recognition. IEEE Trans. Pattern Anal. Mach. Intell. **26**(1), 131–137 (2004)
10. Choi, J.Y., Ro, Y.M., Plataniotis, K.N.: Color local texture features for color face recognition. IEEE Trans. Image Process. **21**(3), 1366–1380 (2012)
11. Jaha, E.S., Ghouti, L.: Color face recognition using quaternion PCA. In: International Conference on Imaging for Crime Detection and Prevention, pp. 497–500 (2011)
12. Li, B.Y.L., Liu, W., An, S., et al.: Face recognition using various scales of discriminant color space transform. Neurocomputing **94**(3), 68–76 (2012)
13. Wang, C., Yin, B., Bai, X., et al.: Color face recognition based on 2DPCA. In: 19th International Conference on Pattern Recognition, ICPR 2008, pp. 1–4 (2009)
14. Wan, M., Zhu, J., Lu, X.: MDNIB2DPCA method for image feature extraction. Comput. Eng. Appl. **52**(9), 177–183 (2016)
15. Wang, S.J., Yang, J., Zhang, N., et al.: Tensor discriminant color space for face recognition. IEEE Trans. Image Process. **20**(9), 2490–2501 (2011). A Publication of the IEEE Signal Processing Society
16. Bottou, L., Cortesm C., Denker, J.S., et al.: Comparison of classifier methods: a case study in handwritten digit recognition. In: International Conference on Pattern Recognition. IEEE Computer Society (1994)
17. Fadi, S., Nassifi, A.B., Aleksander, E.: Dimensionality reduction with IG-PCA and ensemble classifier for network intrusion detection. Comput. Netw. **148**(164), 175 (2019)
18. He, Y., Wu, H., Zhong, R.: Face recognition based on integrated learning of multiple LBP features. Comput. Appl. Res. **35**(01), 292–295 (2018)

Classification of Remote Sensing Images Based on Band Selection and Multi-mode Feature Fusion

Xiaodong Yu[1,2], Hongbin Dong[2], Zihe Mu[3], and Yu Sun[4(✉)]

[1] College of Computer Science and Technology, Harbin Normal University,
Harbin, China
[2] College of Computer Science and Technology, Harbin Engineering University,
Harbin, China
{yuxiaodong, donghongbin}@hrbeu.edu.cn
[3] College of Mathematical Sciences, Harbin Normal University, Harbin, China
mzh19961102@126.com
[4] Heilongjiang Institute of Construction Technology, Harbin, China
donglin_2016@163.com

Abstract. As feature data in multimodal remote sensing images belong to multiple modes and are complementary to each other, the traditional method of single-mode data analysis and processing cannot effectively fuse the data of different modes and express the correlation between different modes. In order to solve this problem, make better fusion of different modal data and the relationship between the said features, this paper proposes a fusion method of multiple modal spectral characteristics and radar remote sensing imageaccording to the spatial dimension in the form of a vector or matrix for effective integration, by training the SVM model. Experimental results show that the method based on band selection and multi-mode feature fusion can effectively improve the robustness of remote sensing image features. Compared with other methods, the fusion method can achieve higher classification accuracy and better classification effect.

Keywords: Remote sensing classification · Classification of features · Band selection · Multimodal feature fusion · SVM

1 Introduction

Remote sensing images can accurately and comprehensively describe the features of ground objects and are widely used in agriculture, geological exploration, environmental monitoring and military reconnaissance [1–4]. At present, many algorithms are used in identification classification, such as maximum likelihood method [5], minimum distance method [6], artificial neural network [7] (ANN), k-means clustering [8], and so on. Theoretically, the sample size of these methods must be large enough to ensure high-level, but in practice, it is not guaranteed to select enough classification samples. Support vector machine (SVM) method is a new type of data mining method [9], which

© Springer Nature Singapore Pte Ltd. 2020
J. Zeng et al. (Eds.): ICPCSEE 2020, CCIS 1257, pp. 612–620, 2020.
https://doi.org/10.1007/978-981-15-7981-3_45

has outstanding advantages in small sample, nonlinear and high-dimensional pattern recognition. Therefore, it can be widely used in distributed classification research and has achieved good results.

At present, the rapid development of satellite technology and the increasing number of sensors not only improve the quality of remote sensing images, but also result in the coexistence of multiple types and resolutions of remote sensing images. Optical sensor and synthetic aperture radar image, as the most important two kinds of sensor data, have their own advantages and disadvantages. Although they both have excellent performance in ground object identification, they rarely cooperate with each other and fail to make full use of the existing surface information. The fusion technology of remote sensing image came into being. The image fusion technology is classified according to different standards. According to the level of fusion, the fusion method is divided into three categories: pixel level, feature level and decision level. In the research of multi-source remote sensing image fusion method, the pixel-level fusion is widely used in the field of remote sensing image fusion due to its advantages of high computing accuracy, small data change and less information loss.

In the past 20 years, many researchers at home and abroad have proposed many fusion algorithms of different strategies. Ramakrishnan et al. fused panchromatic images with multispectral images using the improved IHS transform, and also discussed the application of fused images in earth science [10]. According to the spectral characteristics and application prospects of multi-spectral bands in Ref. [11], PCA, IHS and HPF methods were compared in maintaining the spectral quality of fusion images. Dupas proposed using Brovery transform and IHS method to fuse SAR images and multi-spectral images for land cover classification [12]. Single sensor data and fusion data are used for classification, and the overall classification accuracy of fusion data is higher than that of pure multi-spectral data. Chen et al. used IHS transformation to merge hyperspectral images with SAR images at hyperspectral resolution to obtain enhanced features of urban areas [13]. The fused image was superimposed on the digital elevation model (DEM) to obtain a three-dimensional image. This integration helps to resolve the ambiguity of different types of land cover. Pal et al. proposed a PCA-based fusion method to improve geological interpretation [14]. FFT method was used to filter SAR images, PCA was used to fuse multi-spectral data, and then the principal component selection method based on features was used to conduct false color synthesis of the fused data. Chandrakanth et al. applied HPF to the frequency domain and fused high-resolution panchromatic and high-resolution SAR images [15]. This method extracted low-frequency details from panchromatic images and high-frequency details from SAR images. Battsengel et al. proposed the fusion of high-resolution optical images and SAR images to improve the accuracy of land cover classification [16]. The method of multiplication was used to fuse the source images, and the fusion method based on BT, PCA and IHS was adopted. The results were compared visually on the basis of classification accuracy. Yang et al. used GS orthogonalization method to integrate optical images and SAR images to improve the classification of coastal wetlands [17]. The above methods are not computationally complex and can produce a fusion product with rich spatial information. However, these methods are highly dependent on the correlation of image fusion, so that the image after fusion has obvious spectral distortion.

In view of the above problems, this paper first selected the bands of Landsat8 images, selected the optimal three bands, and used HSV transform to effectively integrate the multi-spectral image features and sentinel-1 SAR image features in the form of vector or matrix according to the spatial dimension, so as to train the SVM model and obtain better ground object recognition effect.

2 Classification Principle of Support Vector Machines

The mechanism of support vector machine (SVM) is to find an optimal classification hyperplane that meets the classification requirements, so that the hyperplane can maximize the empty space on both sides of the hyperplane while ensuring the classification accuracy. Taking two types of data as examples, the given training set (x_i, y_i), $i = 1, \ldots, l$; $x \in R^n$, $y \in \{+1, -1\}$ and hyperplane are denoted as $(w \cdot x) + b = 0$. In order for the classifier to correctly classify all samples and have classification intervals, the following constraints need to be met:

$$y_i[(w \cdot x_i) + b] \geq 1 \quad i = 1, 2, \cdots, 1 \tag{1}$$

We can calculate the classification interval $2/\|w\|$, so the problem of constructing the optimal hyperplane is transformed into:

$$\min \phi(w) = \frac{1}{2}\|w\|^2 = \frac{1}{2}(w' \cdot w) \tag{2}$$

In high-dimensional space, if the training sample is indivisible or it is not known whether it is linearly separable, it will allow the introduction of a certain number of misclassified samples, and a non-negative relaxation variable ζ_i will be introduced. The above problem is transformed into a quadratic programming problem with linear constraints:

$$\min \frac{1}{2}\left(w' \cdot w + C \sum_i^n \zeta_i\right) \quad i = 1, 2, \ldots, n \tag{3}$$

However, in actual classification, it is difficult to ensure linear separability between categories. For the case of linear inseparability, SVM introduces a kernel function, maps the input vector to a high-dimensional feature vector space, and constructs the best classification surface in the feature space. Due to the good performance of the RBF kernel function, the radial kernel function is selected as the kernel function of the SVM in practical applications.

$$K(x, x_i) = \exp\left(-\frac{\|x - x_i\|^2}{2\sigma^2}\right) \tag{4}$$

3 Classification of Remote Sensing Images Based on Band Selection and Multi-mode Feature Fusion

3.1 Correlation-Based Feature Selection (CFS)

CFS [18] is a classic filtering type feature selection algorithm. It evaluates the influence of each feature on each classification separately by heuristics, so as to obtain the final feature subset. The evaluation method is shown in formula (5):

$$M_s = \frac{k r_{cf}}{\sqrt{k + k(k-1) r_{ff}}} \tag{5}$$

Among them, M_s represents the evaluation of the feature subset S containing k features; r_{cf} represents the average correlation between attributes and classification; r_{ff} represents the average correlation between attributes. The higher the computational correlation between each attribute and the classification attribute, or the lower the redundancy between the attributes, the more positive the evaluation. In CFS, information augmentation is used to calculate correlations between attributes. The method of computing information enhancement will be described below.

Assuming that the attribute is Y, γ represents each possible Y value, the entropy of Y is calculated from Eq. (6).

$$H(Y) = -\sum_{\gamma \in Y} P(\gamma) \log_2(P(\gamma)) \tag{6}$$

For the known attribute X, Y's entropy is calculated by Eq. (7)

$$H(Y \backslash X) = -\sum_{x \in X} P(x) \sum_{\gamma \in Y} P(\gamma \backslash x) \log_2(P(\gamma \backslash x)) \tag{7}$$

The difference $H(Y) - H(Y \backslash X)$ (namely the reduction of feature Y's entropy) reflects the additional information attribute X provides to attribute Y, also known as information enhancement. The information increment reflects the amount of information provided by X to Y, so the larger the information increment, the stronger the correlation. Since information enhancement is a symmetric measure, its disadvantage is that it tends to select those attributes with more potential values. Therefore, each attribute needs to be information enhanced and normalized to ensure that each attribute can be compared with other attributes so that different attribute selections can get the same result. The method of symmetric uncertainty is used to normalize it into [0, 1].

$$U_{XY} = 2.0 \times \frac{H(Y) - H(Y \backslash X)}{H(Y) + H(X)} \tag{8}$$

3.2   HSV Transform Fusion Method

HSV transform is an inverted cone color space composed of Hue, Saturation and Value, which transforms the color space composed of RGB three colors in multi-spectral images [19]. The basic idea of HSV transformation is to replace the brightness Value in the original image with a high-resolution full-color remote sensing image, and then re-sample the Hue and Saturation into the high-resolution size image through interpolation (nearest neighbor method, bilinear interpolation method and cubic con-volution interpolation method), and finally convert the image back to the RGB color space.

$$H = \begin{cases} \theta & B \le G \\ 360 - \theta & B > G \end{cases} \tag{9}$$

$$S = 1 - \frac{3}{(R+G+B)}\min(R, G, B) \tag{10}$$

$$V = \frac{1}{3}(R+G+B) \tag{11}$$

$$\text{Among them}: \theta = \arccos\left\{ \frac{\frac{1}{2}[(R-G)+(R-B)]}{\left[(R-G)^2 + (R-B)(G-B)^{\frac{1}{2}}\right]} \right\}$$

4   Experiments and Analysis

4.1   Data Source and Preprocessing

Landsat8 remote sensing data from September 10, 2015 and sentinel-1 standard polarization model data were mainly used in this paper. The Sentinel-1 standard polarization model data included a scene on December 23, 2015, with a VV spatial resolution of 15 m and an incidence Angle of 37°. The data processing of this paper mainly includes the following parts:

Landsat8 remote sensing data were imported into ENVI5.3 software, and on the basis of 1:50,000 topographic map, the intersection points of rivers and roads were selected as control points for geometric correction of Landsat8 remote sensing data. Afterwards, the 1–7 bands of Landsat8 remote sensing data were fused with panchromatic bands, and the required study areas were intercepted according to the research needs.

Using the ENVI extension package SAR scape, the Sentinel-1 standard polarization pattern data was radiative scaled, and the amplitude intensity image was converted into the backscattering coefficient image. Then the Lee filtering algorithm of 7 × 7 filter window is used to remove the speckle noise of radar image. The filtered image was converted into a transverse Mercator (UTM) projection and resampled to 30 m. Due to the different sensor of acquired radar image and optical image, the imaging mode of each satellite is different, so the space registration should be carried out. Using ENVI5.3 software, the radar images were registered based on Landsat8 image data dated 10 September 2015.

4.2 Experimental Data and Sample Selection

Study area selection, western songnen plains in western jilin province county territory of adlai was in charge of the town, momo, national nature reserve (" 4 "5" ^"°   " 4 ", "2" ^ ''' N ~ 4 "6" ^ ''° " 1 ". "8" ^' N ^ 123 °   "2" 7 "^" 'E ~ "12" "4" ^ 4 °^' E). According to the distribution of land features in this area, it is divided into 6 categories: unused land, grassland, marsh, wetland, water area, residential land and cultivated land. In this paper, a total of 16 160 pixel-level training samples were selected, among which 30% of each ground object was randomly selected as the verification set, and the remaining data samples were selected as the training set.

4.3 Classification and Accuracy Verification

SVM classifier using ENVI5.3 remote sensing software. The spectral characteristics of 1–7 bands of Landsat8 remote sensing data, spectral characteristics after band selection (CFS), and spectral characteristics after band selection (CFS) and radar fusion were used respectively to extract wetland information by integrating topographic assistance data and radar image backscattering. The results are shown in Fig. 1 (a), Fig. 1 (b) and Fig. 1 (c). The classification results of 578 measured sample points were selected for accuracy verification, and the error confusion Matrix was established by using the tool ENVI5.3 confusion Matrix. The accuracy evaluation results obtained are shown in Table 1, Table 2 and Table 3 respectively.

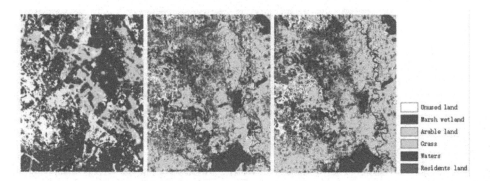

Fig. 1. (a), (b), (c) Results of extracting wetland information using SVM with three different features

Table 1. Confusion matrix of SVM classification results for spectral characteristics of 1–7 bands of Landsat8 remote sensing data

The sample points	Arable land	Grass	Marsh wetland	Waters	Residents land	Unused land	Total	User accuracy %
Arable land	82	10	25	4	3	0	124	66.13
Grass	0	39	21	14	0	0	74	52.70
Marsh wetland	4	9	93	0	0	0	106	87.74
Waters	17	3	31	60	14	10	135	44.44
Residents land	3	1	0	0	85	0	88	96.59
Unused land	0	0	0	0	0	51	51	100.00
Total	106	61	170	78	102	642	578	
Mapping accuracy %	77.36	63.93	54.71	76.92	83.33	83.61		

The overall accuracy = 70.93% Kappa = 0.64

Table 2. Confusion matrix of SVM classification results for spectral characteristics after band selection (CFS)

The sample points	Arable land	Grass	Marsh wetland	Waters	Residents land	Unused land	Total	User accuracy %
Arable land	94	0	14	0	12	2	122	77.05
Grass	0	70	14	0	0	0	84	83.33
Marsh wetland	6	19	89	0	1	0	115	77.39
Waters	0	0	0	107	0	0	107	100.00
Residents land	0	0	0	0	48	39	87	55.17
Unused land	0	0	0	0	19	44	63	69.84
Total	100	89	117	107	80	85	578	
Mapping accuracy %	94.0	78.65	76.07	100.00	60.00	51.76		

The overall accuracy = 78.20% Kappa = 0.73

Table 3. Confusion matrix of SVM classification results for band selection (CFS) after the spectrum and radar fusion characteristics

The sample points	Arable land	Grass	Marsh wetland	Waters	Residents land	Unused land	Total	User accuracy %
Arable land	102	0	13	0	0	4	125	81.60
Grass	0	62	13	0	6	0	75	82.67
Marsh wetland	1	14	101	0	0	0	116	87.07
Waters	0	0	0	106	0	0	106	100.00

(continued)

Table 3. (*continued*)

The sample points	Arable land	Grass	Marsh wetland	Waters	Residents land	Unused land	Total	User accuracy %
Residents land	6	0	1	0	78	4	88	88.64
Unused land	0	0	0	0	14	54	68	79.41
Total	109	76	127	106	98	62	578	
Mapping accuracy %	93.58	81.58	79.53	100.00	79.59	87.10		

The overall accuracy = 87.02% Kappa = 0.84

5 Conclusions

Based on the principle of image fusion, from a practical perspective, this paper puts forward a fusion method of multiple modal spectral characteristics and radar remote sensing image. It implemented the image Landsat8 band selection, chose the best three bands, and reused of HSV transform the multispectral image and effective Sentinel-1 SAR radar image fusion, and trained the SVM model, extract the feature information of wetland. Experimental results show that the method based on band selection and multi-mode feature fusion can effectively improve the robustness of remote sensing image features. Compared with other methods, the fusion method can achieve higher classification accuracy and better classification effect.

References

1. Zhao, M., et al.: A robust delaunay triangulation matching for multispectral/multidate remote sensing image registration. IEEE Geosci. Remote Sens. Lett. **12**(4), 711–715 (2015)
2. Izadi, M., Saeedi, P.: Robust weighted graph transformation matching for rigid and nonrigid image registration. IEEE Trans. Image Process. **21**(10), 4369–4382 (2012). A Publication of the IEEE Signal Processing Society
3. Shahdoosti, H.R., Ghassemian, H.: Fusion of MS and PAN images preserving spectral quality. IEEE Geosci. Remote Sens. Lett. **12**(3), 611–615 (2014)
4. Akhavan-Niaki, H., et al.: Evaluation of spatial and spectral effectiveness of pixel-level fusion techniques. IEEE Geosci. Remote Sens. Lett. **10**(3), 432–436 (2013)
5. Chen, F., Wang, C., Zhang, H.: Remote sensing image classification based on an improved maximum-likelihood method: with SAR images as an example. Remote Sens. Land Resour. **28**(1), 75–78 (2008)
6. Liu, J., Zhang, C., Wan, S.: The classification method of multi-spectral remote sensing images based on self-adaptive minimum distance adjustment. In: Li, D., Chen, Y. (eds.) CCTA 2012. IAICT, vol. 393, pp. 430–437. Springer, Heidelberg (2013). https://doi.org/10.1007/978-3-642-36137-1_50
7. Yu, X., Dong, H.: PTL-CFS based deep convolutional neural network model for remote sensing classification. Computing **100**(8), 773–785 (2018). https://doi.org/10.1007/s00607-018-0609-6

8. Wu, T., Chen, X., Xie, L.: An optimized K-means clustering algorithm based on BC-QPSO for remote sensing image. In: IGARSS 2017 - 2017 IEEE International Geoscience and Remote Sensing Symposium. IEEE (2017)
9. Yu, X., Dong, H., Patnaik, S.: Remote sensing image classification based on dynamic co-evolutionary parameter optimization of SVM. J. Intell. Fuzzy Syst. **35**(1), 343–351 (2018)
10. Ramakrishnan, N.K., Simon, P.: A bi-level IHS transform for fusing panchromatic and multispectral images. In: Proceedings of 5th International Conference on Pattern Recognition and Machine Intelligence, pp. 367–372 (2011)
11. Rokhmatuloh, R., Tateishi, R., Wikantika, K., et al.: Study on the spectral quality preservation derived from multisensor image fusion techniques between JERS-1 SAR and landsat TM data. In: Proceedings of International Geoscience and Remote Sensing Symposium (IGARSS), pp. 3656–3658 (2003)
12. Dupas, C.A.: SAR and LANDSAT TM image fusion for land cover classification in the Brazilian atlantic forest domain. In: Proceedings of 19th International Congress for Photogrammetry and Remote Sensing, pp. 96–103 (2000)
13. Chen, C.M., Hepner, G.F., Forster, R.R.: Fusion of hyperspectral and radar data using the IHS transformation to enhance urban surface features. ISPRS J. Photogram. Remote Sens. **58**(1), 19–30 (2015)
14. Pal, S.K., Majumdar, T.J., Amit, K.: ERS-2 SAR and IRS-1C LISS III data fusion: a PCA approach to improve remote sensing based geological interpretation. ISPRS J. Photogram. Remote Sens. **61**(5), 281–297 (2007)
15. Chandrakanth, R., Saibaba, J., Varadan, G., et al.: Fusion of high resolution satellite SAR and optical images. In: International Workshop on Multi-Platform/Multi-Sensor Remote Sensing and Mapping, pp. 1–6 (2011)
16. Battsengel, V., Amarsaikhan, D., Bat-erdene, T., et al.: Advanced classification of lands at TM and Envisat images of Mongolia. Adv. Remote Sens. **2**(2), 102 110 (2013)
17. Yang, J.F., Ren, G.B., Ma, Y., et al.: Coastal wetland classification based on high resolution SAR and optical image fusion. In: International Geoscience and Remote Sensing Symposium, pp. 886–889 (2016)
18. Yu, L., Liu, H.: Feature selection for high-dimensional data: a fast correlation-based filter solution. In: Proceedings of the Twentieth International Conference Machine Learning (ICML 2003), Washington, DC, USA, 21–24 August 2003. AAAI Press (2003)
19. Zhu, Q., Liu, B.: Multispectral image fusion based on HSV and red-black wavelet transform. Comput. Eng. (2012)

System

Superpage-Friendly Page Table Design for Hybrid Memory Systems

Xiaoyuan Wang[1,2,3,4], Haikun Liu[1,2,3,4]([✉]), Xiaofei Liao[1,2,3,4], and Hai Jin[1,2,3,4]

[1] National Engineering Research Center for Big Data Technology and System, Huazhong University of Science and Technology, Wuhan 430074, China
{xiaoyuanw,hkliu,xfliao,hjin}@hust.edu.cn
[2] Service Computing Technology and System Lab, Huazhong University of Science and Technology, Wuhan 430074, China
[3] Cluster and Grid Computing Lab, Huazhong University of Science and Technology, Wuhan 430074, China
[4] School of Computer Science and Technology, Huazhong University of Science and Technology, Wuhan 430074, China

Abstract. Page migration has long been adopted in hybrid memory systems comprising *dynamic random access memory* (DRAM) and *non-volatile memories* (NVMs), to improve the system performance and energy efficiency. However, page migration introduces some side effects, such as more *translation lookaside buffer* (TLB) misses, breaking memory contiguity, and extra memory accesses due to page table updating. In this paper, we propose superpage-friendly page table called SuperPT to reduce the performance overhead of serving TLB misses. By leveraging a virtual hashed page table and a hybrid DRAM allocator, SuperPT performs address translations in a flexible and efficient way while still remaining the contiguity within the migrated pages.

Keywords: Page table · Hybrid memory system · Page migration · Multiple page sizes · Address translation

1 Introduction

Recent years have witnessed many large-footprint applications. Traditional DRAM-based memory systems are unable to meet the ever-increasing memory demand due to the limited DRAM scaling in terms of memory density and power efficiency. The advent of *non-volatile memory* (NVM) technologies has attracted a lot of interests in constructing large-capacity and energy-efficient main memory systems with NVMs. However, since NVM cannot directly replace DRAM due to its shortcomings, such as lower performance and limited write endurance, hybrid memory systems composed of DRAM and NVM have been widely studied [1–4]. Most of these studies make efforts to improve system performance and save energy by using page migration [4, 5].

As the amount of memory required by applications increase significantly, the number of *page table entries* (PDEs) also grows rapidly. However, the capacity of *Translation Lookaside Buffer* (TLB) which is used to cache virtual-to-physical address

© Springer Nature Singapore Pte Ltd. 2020
J. Zeng et al. (Eds.): ICPCSEE 2020, CCIS 1257, pp. 623–641, 2020.
https://doi.org/10.1007/978-981-15-7981-3_46

translations cannot keep pace with the ever-increasing memory capacity due to access latency, energy consumption, and space constraints. The total address space that the TLBs can map directly, also known as TLB coverage, is far smaller than applications' footprint. In this scenario, the system performance is significantly degraded due to TLB misses. Previous studies show that the performance overhead is even up to 50% when running memory-hungry applications, i.e. applications with large footprints [6].

There have been a large body of studies on reducing the overheads serving TLB misses. These studies can be classified into two categories: one is to increase the TLB coverage (such as superpages, TLB coalescing, and range mapping), and another is to reduce the serving time of *page table walking* (PTW) which retrieves the *page table entries* (PTEs) to fill the TLBs upon TLB misses. Superpage [7–9] has long been used to reduce TLB misses. It can significantly improve the TLB coverage, e.g., a 2 MB superpage can enlarge the TLB coverage by 512 times when compared to a system in which memory pages are both managed and aligned in 4 KB. However, the using of superpage hinders lightweight page managements (such as page migration, page sharing) in hybrid memory systems [5]). TLB coalescing [10–12] and range mapping [13–15] are both practical and efficient schemes to improve TLB coverage. However, the frequent TLB updates due to page migrations limit the performance gain of these schemes.

Another direct and effective approach is to reduce the performance overhead of retrieving page tables. Since traditional binary-tree-based page table is very costly. For example, a TLB miss leads to four memory references in x86-64 architecture, and 24 memory references in virtualization environments [16–18]. Thus, it is essential to redesign the structure of page tables and the corresponding retrieving mechanism to reduce the cost of page table walking. *Hash page table* (HPT) leverages a hash function to map virtual addresses to physical ones in a constant time period. HPT significantly reduces the overhead of extra memory accesses caused by page table walking, at the expense of several undesirable advantages. For example, it is unable to support mixed page size and region mapping, and the cost of large page management is also extremely high. *Inverted page table* (IPT) is also developed to improve the physical-to-virtual address translation. By arranging an entry per memory page, IPT can significantly reduce the storage and the runtime overhead (such as searching), at the expense of lower performance of virtual-to-physical address translations. As a result, a hash function is usually used in IPT to speed up virtual-to-physical address translations, however, it is hard to support memory management at multiple page sizes.

We find that there is a remarkable contiguity in hot pages, which can be identified and migrated within a given monitoring period (10^8 cycles in our experiment) in hybrid memory systems. Previous works show that these contiguous pages can be leveraged by TLB coalescing [10–12]. In this paper, we study how to support fast virtual-to-physical and the reverse address translations while still remaining page contiguity between migrated pages. To accomplish this goal, several challenges should be addressed: 1) *identify page contiguity*, 2) *record page contiguity information in page tables, and* 3) *efficient virtual-to-physical and reserve address translations.*

To solve the above challenges, we propose superpage-friendly page table (called SuperPT), a novel page table design to support multi-grained page migrations in hybrid memory systems. SuperPT detects page contiguity within migrated pages and records them in a hash-based virtual page table. By leveraging this contiguity, multi-grained TLBs can deliver higher performance [10–12, 15]. What's more, even in systems without multi-grained TLB support, the system performance can also be improved by TLB prefetching [19, 20].

The remainder of this paper is organized as follows. Section 2 depicts the background and motivates of our design for multi-grained page migration in hybrid memory systems. Section 3 describes SuperPT designs in detail. Experimental results are presented in Sect. 4. We discuss related work in Sect. 5 and conclude in Sect. 6.

2 Background and Motivation

We first introduce virtual memory and page tables. Next, we experimentally study memory access statistics of typical applications to motivate the design of SuperPT.

2.1 Virtual Memory and Page Table

To extend the use of physical memory and enable memory protection, virtual memory is widely used in modern systems. There are two kinds of addresses in these systems, one for the virtual address space, and another for the physical or real address space. The virtual address is constructed by CPUs and used by processes, and the physical one is the real address space in memory systems. In order to accurately and conveniently perform the translation between virtual addresses and physical addresses, page tables are used to store and manage the virtual-to-physical address translations. To be specific, we classify most representative page table structures into the following categories.

Binary-Tree-Based Page Table. Figure 1 (a) shows the overview of binary-tree-based page table. Once a virtual-to-physical translation is required, the system looks up the corresponding page table entries layer by layer, thus four memory references are needed. In a virtualization environment, it even leads to 24 memory accesses per page table walking [16, 17, 21]. Therefore, this kind of page tables usually cause significant performance overhead [21]. Despite its high cost, it is widely adopted in modern computer systems, the reason is that this kind of page table is naturally friendly to cache locality because this mapping mechanism stores PTEs of adjacent pages in an adjacent manner.

Inverted Page Table. Figure 1 (b) shows the structure of *inverted page table* (IPT) [22]. IPT provides one-page table entry for each physical memory page. Each entry stores the *information of virtual address number (VPN)* and the corresponding *process ID* (PID). Thus, IPT is able to reduce the memory required to store the page tables since the number of IPT is equal to the number of physical memory pages. However, even when application memory requirement is low, the overhead of searching page tables is still very high upon a memory reference. Moreover, it also poses other problems: ① Mixed page sizes are hard to support. ② Due to the lack of tree structure, operations on regions are extremely expensive.

Hashed Page Table. As shown in Fig. 1 (c), *VPN* is the input of the hashing function. Through hashing, an entry in the hash table is related to the VPN. If the first field of the three entries matches the *virtual page number* (*VPN*) of the desired page, we get the *physical page number* (*PPN*). If the *VPN* is not hit, we access the next entry in the linked list. This hash-function based page table design improves the efficiency of searching page tables, but it also brings some side effects. ① They are not able to support multiple page sizes. Since different pages are assigned to different table entries by the hash function, the page continuity is not guaranteed. ② They lead to poor cache locality and low performance of TLB prefetching mechanism. Because the address space is fragmented by hash function, the system cannot guarantee continuous response to the adjacent address space. In this case, both the page table cache and the TLB prefetching mechanism are inefficient. ③ Hash collisions are expensive. Since page table walking tends to be on the critical path of applications, page tables are more desired to be optimized for speed. Therefore, the system may suffer from high performance overhead due to hash collisions. If the *physical page number* (PPN) is not hit, the entries on the collision chain will be checked one by one, causing extremely high latency of page table retrieving.

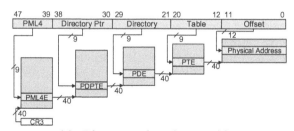

(a) Binary-tree based page tables

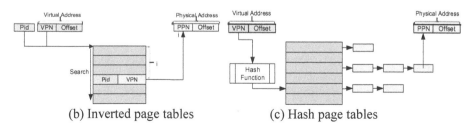

(b) Inverted page tables   (c) Hash page tables

Fig. 1. Layout of three kinds of page tables

2.2 Page Migration in Hybrid Memory Systems

Because NVM shows much lower energy consumption and higher density than DRAM, it has been studied by many works that prefer to use it to replace DRAM. However, NVM cannot directly replace DRAM for its shortcomings, such as limited write endurance, high write energy consumption, and high access latency, especially for write operations [23]. Therefore, the heterogeneous memory system composed of DRAM and NVM has become a practical approach to the current dilemma. In order to improve the system performance of these systems, previous works generally leverage page migration to take advantages of the two storage medium, and overcome their shortcomings [2–4, 24, 25]. However, the use of page migration in heterogeneous memory systems brings additional performance overhead, because extra update operations are required for every page migration operation.

To evaluate the extra memory access times caused by page migration, we run several representative applications and profile their memory usage in an interval of 10^8 cycles. These applications are selected from SPEC CPU2006 [26], Parsec3.0 [27], NAS Parallel Benchmarks [28], and Graph500 [29]. CactusADM, Mcf, and Omnetpp are all chosen from SPEC CPU2006. Canneal, X264, Facesim, and streamcluster are all multi-thread applications that chosen from Parsec3.0. MG, UA, and SP are selected from NAS Parallel Benchmarks. Graph500 is designed to evaluate the performance of supercomputer by using large scale memory-intensive graph processing algorithms. All experiments are conducted in a simulated platform, as presented in Sect. 4.1. We have the following observations.

Observation 1: There is a large number of update operations on page tables caused by page migration in heterogeneous memory systems.

For each selected application, the extra memory access times caused by page migration is over 83% on average compare to the no-migration scenarios, as shown in Fig. 2. On one hand, for applications whose page table operations account for less memory access, the increment is more pronounced. For example, the memory access times in Canneal increases by more than two times. On the other hand, for applications whose page table operations account for most memory access, memory access growth is also considerable. For example, the memory access times in Omnetpp increases about five percent. Note that, page migration not only causes increase of page table operations, but also causes increase of total memory access times.

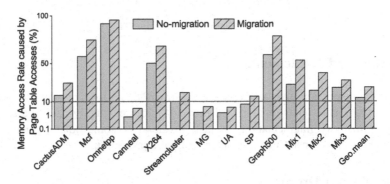

Fig. 2. Extra accesses caused by page migration in hybrid memory systems

Observation 2: There is a considerable continuity in migrated pages.

Figure 3 shows the *cumulative distribution function* (CDF) of the proportion of contiguous pages in the total number of migrated pages. For most applications, we find that almost 40% of the hot pages in the system are contiguous. For Graph500, the proportion of contiguous hot pages is over 90%. This implies that taking full advantage of the continuity of migrated pages can improve the efficiency of page table walking.

These above findings inspire us to conceive and model a new flexible and efficient memory management mechanism for hybrid memory systems that supporting multi-grained page sizes.

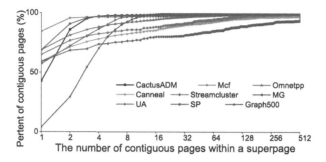

Fig. 3. Cumulative distribution function of contiguous hot pages within a superpage

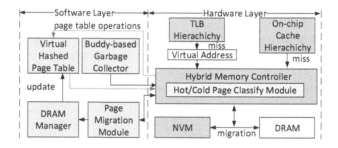

Fig. 4. Architecture of SuperPT

3   Design and Implementation

In this section, we first give an overview of SuperPT and then present the technical details of page table operations, hybrid memory allocator, buddy-based garbage collector, and memory fragmentation. At last, we describe some other implementation issues such as data consistency guarantee and page protection.

3.1   Architecture Overview

Figure 4 depicts the architecture of SuperPT. The hybrid memory controller counts and records the access information of each page in a given interval (10^8 cycles in our experiment). The hot/cold page classify module is added in the hybrid memory controller, which analyses access counts of each page in the memory and selects the hot pages among them.

In the *operating system* (OS) level, we design a novel page table. Like HSCC [4], a DRAM cache filter with utility-based migration mechanism is adopted to improve the performance of DRAM cache. The DRAM cache manager module is responsible for page allocation and replacement. The migration module moves hot and cold pages between slow NVM and fast DRAM. To reduce the impact on system performance, SuperPT performs these modules periodically in the background.

Similar to CHOP [30], active pages in SuperPT are ranked according to the number of page accesses. We identify the top-N hot pages if the total accesses of them contribute 70% of the application's memory accesses in every period. When a NVM page is identified to be a hot page, it would be migrated from NVM medium to DRAM medium by the migration module. When the migration operation completes, the mapping of the page should be also updated.

3.2 Virtual Hashed Page Table

Figure 5 shows the structure of virtual hashed page table. The following terms are used: *virtual page number* (*VPN*), *virtual superpage number* (*VSN*), *index*, and *offset*. *Index* indicates the normal page number within a superpage, while *offset* indicates the real address within the normal page. When a virtual address comes, *VSN* is used as the key to find the location of superpage that the required page stays in. Note that, since the *virtual page number* is used as the key in the *Hash Function*, when a group of applications with the same or similar access patterns run together, the collision rate of the hash function increases significantly. To be more specifically, when two identical workloads running on the same machine, the hot pages they migrate will have the same virtual address and then be assigned to the same superpage by *Hash Function*, resulting in an increased collision rate.

To solve this challenge, the *process ID* (*PID*) is used as the key of the hash function along with the *virtual superpage number* (*VSN*). In this way, even the same application is allocated to different large pages, the use of PID avoids unnecessary hash conflicts.

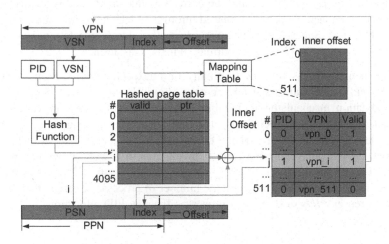

Fig. 5. Virtual hashed page table in SuperPT

3.3   Page Table Operations

Since physical address indexed cache is commonly used in modern computer systems, mapping virtual addresses to physical ones is on the critical path of applications. At this time, the system will first access TLBs. Upon a TLB miss, the page table walk operation is performed for virtual-to-physical address translation.

Lookup: ① Virtual-to-physical address translations. Upon a TLB miss, a hash query is done to find the corresponding superpage number according to the *virtual superpage number* (VSN) and *process ID* (PID) by the *memory manage unit* (MMU). Meanwhile, the *index* is used as the key to lookup the *mapping tables* along with the value that the *hash function* gives. Note that, since *mapping table* and *hashed page table* are searched in parallel, the lookup overhead is acceptable. If the found hash table entry is valid, the *ptr* points to the starting address of the table that consists of 512 normal page table entries. If the entry is hit (the *valid* is true), the physical superpage address (i), the inner offset (j), and *offset* are used together to form the physical address. ② Physical-to-virtual address translation. Similarly, the physical address is also divided into three sections: *PSN, index,* and *offset*. By using *PSN* to search *hashed page table*, the starting address of normal page table is obtained, and then *index* is used to find the related internal normal page table of the large page to find the required *virtual page number* (VPN). At last, the complete virtual address is formed by combining it with the *offset*.

Insert: When a page or a set of pages are migrated from NVM to DRAM, the corresponding new page table entry is inserted into the *hash page table*. To be more particular, according to the virtual address information, the hash function is first used to find the physical superpage address. Second, the corresponding small page within the superpage is allocated to store the corresponding data. Finally, the corresponding hash page table, internal page table, and mapping table are updated.

Update: When there is insufficient free small pages within a superpage, or when the available free DRAM size is too small, the page write-back operation is triggered. Correspondingly, the content of *hash page table* also needs to be updated along with the page write-back. Specifically, SuperPT will find the corresponding physical superpage according to the *VSN*, and the corresponding small page is located through the *index* field. At last, the page is replaced through a LRU algorithm, and written back to NVM. Note that, SuperPT significantly reduces the address translation latency due to its good performance in physical-to-virtual address translation.

Delete: When a process finishes (normally or unexpectedly), its memory space is reclaimed and the relevant page table entries should be invalidated. SuperPT uses a lazy invalidation mechanism—reclaiming those pages when they are reallocated or during garbage collection. Specifically, the *PID* of the relevant page is checked at the time of memory allocation or garbage collection, and if its *PID* exists in the invalidation list, the relevant page table entry is invalidated. It is well known that in x86-64 platform, Linux uses 22 bits to identify process numbers (up to 2^{22} processes can be identified at the same time), and when the number of processes exceeds this threshold, the system reuses *PIDs* of destroyed processes for new PID allocation.

Page Sharing: Sometimes there may be a part of main memory that shared by more than one process. In this case, a single page table entry can be mapped to at least two virtual pages, and then SuperPT leverages a link pointer to bind the information of those virtual pages to the root page table.

Page Protection: Like traditional page table entry, the PTEs in SuperPT contain physical address and other various flags. The *present* bit reflects whether the page is already in memory or not. The *writable* bit reflects whether the page is allowed to write. The *user accessible* bit reflects whether the page can be accessed by the user mode code. The *write through caching* bit reflects whether the writes can be directly passed to the main memory. The *disable cache* bit reflects whether the page can be cached. The *accessed* bit reflects whether the page has been accessed yet, i.e., when the page is used, the CPU sets this bit. When a page is written, the *dirty* bit is set by the CPU. The *global* bit reflects whether the page can be flushed from caches or not when the content is switched. The *available* bit reflects if the page can be used freely by the OS or not. If the *no execute* bit is set, the system forbids executing code on this page. Since SuperPT supports range mapping, the *huge page/null* is overridden to reflect this is a regular page or a range page mapping.

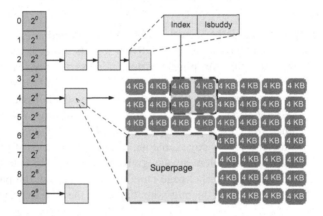

Fig. 6. Buddy-based garbage collator in SuperPT

3.4 DRAM Allocator

When a group of pages are identified as hot pages, the page migration module will migrate them to DRAM. With the virtual address and process number, the system first uses *Hash Function* to find the starting address of the corresponding superpage, and then carries out the corresponding normal page allocation operation. Finally, the system updates the corresponding page table entries.

Since SuperPT supports multi-granularity page migration and multi-granularity page mapping, at the end of each page monitoring period, a simple sort is made to merge the migration requests of small pages with adjacent addresses, making sure that the adjacent pages remain adjacently after migration as much as possible. However,

632   X. Wang et al.

when a set of consecutive pages are migrated together, there may not be enough free pages for them. At this point, SuperPT splits these pages to fit the available memory space. This sacrifices some page continuity but has less impact on system performance.

3.5 Buddy Based Garbage Collection

To reduce external fragmentation, we leverage a buddy-system based garbage collector. It merges adjacent free blocks by keeping track of its neighbors. In particular, a bitmap is used to track whether neighbors are in use. What's more, since the memory blocks are aligned by power-of-two, they could be merged to construct a double-sized block.

As shown in Fig. 6, the space within a superpage is divided into 10 groups, each of which is a collection of exponential successive pages. For example, each element in group i represents a set of 2^i consecutive pages. In addition, for the convenience of management and statistics, any element in each group only records the first page address of the contiguous page addresses within the superpage.

3.6 Data Consistency

Page migrations may raise data inconsistency problems. We address those issues as follows.

Data Consistency Between DRAM and NVM. As mentioned before, there may be two replicas of the migrate pages, one is in DRAM and the other is resided in NVM. Conversely, the operation that mapping a virtual address to physical ones may be performed by both normal page table used for NVM and *Hashed page table* used for DRAM. To guarantee the consistency of the data, we extend the normal page table with a migration flag (M) by using reserved bit of PTE, identifying whether a page is migrated or not. Since the normal page table and *hashed page table* are searched in parallel, the more efficient hashed page table is always returned first if the page being searched is in DRAM, so the system always gives high priority to data that has been cached in DRAM. Even if the normal page table returns data first, the page migration status can be determined based on the migration flag bit in the PTE to prevent accessing to the wrong data.

Cache Consistency. To achieve higher performance, a large number of modern processors use write-back cache solutions, in which the data modify operations (i.e. writes) are directed to the cache without informing the main memory about these modifications. In this way, the memory is finally modified only when the cache is evicted. Since a set of cache lines may be mapped by a single page, once page migration occurs before all the cache lines are written back to memory, the stale data may be chosen for migration, resulting in data inconsistence problems. *Clflush* instructions is used in SuperPT to solve this problem. When pages are migrated, all cache lines corresponding to these pages are invalidated. Meanwhile, the invalidation operation would be broadcast to other parts in the same cache consistency domain. As a result, the relevant cache lines throughout the cache hierarchy is either be invalidated (clean pages) or written back (dirty pages). Finally, before a page is migrated, SuperPT writes back all the related dirty cache lines to the memory and invalidates all the corresponding clean cache lines.

4 Evaluation

4.1 Experimental Methodology

We implement SuperPT in a full-system simulator by integrating Zsim [31] and NVMain [33]. Zsim is based on Pin tools [32], and we leverage it to simulate on-chip systems, because it is fast and supports x86-64 multi-core and many-core architectures well. Meanwhile, we also add many OS-level functions to Zsim, such as memory allocator, *memory management unit* (MMU) for TLB, and page table simulation. As a widely studied cycle-accurate memory simulator, NVMain is used in SuperPT to simulate the hybrid memory system composed of DRAM and NVM in detail.

Table 1. System configuration of simulated platform

CPU	8 cores, 3.2 GHz, out-of-order	
TLB hierarchy	L1 DTLB	32 entries for 2 MB superpages, 64 entries for 4 KB small pages in each core, 4-way set-associate, 1 cycle per access
	L2 DTLB	512 entries for 2 MB superpages, 1024 entries for 4 KB pages, both can be used for Data and Instruction. 8-way set-associate, 8 cycles per access
Cache hierarchy	L1 Cache	Private 64 KB in each core, 4-way, split Data and Instruction, 3-cycles per access
	L2 Cache	Private 256 KB in each core, 8-way, set associate, 10-cycles per access
	L3 Cache	Shared 8 MB, 16-way set-associate, 34 cycles per access
DRAM	4 GB, channel-rank-bank-row-col: 1-4-32-32768-64, FR-FCFS, Bandwidth (GB/Sec): 10.7, Timing (cycles): (cas-red-rp-ras: 7-7-7-18), Read Delay (ns): 13.5, Write Delay (ns): 28.5	
PCM	32 GB, channel-rank-bank-row-col: 4-8-64-65536-64, FR-FCFS, Bandwidth (GB/Sec): 10.7, Timing (cycles): (cas-red-rp-ras: 9-37-100-53), Read Delay (ns): 13.5, Write Delay (ns): 171	

Configuration. Table 1 depicted the experimental platform and detailed configuration. We choose PCM as the representative storage medium of memory for it has been widely studied. The timing parameters are referred to previous works [4, 5, 34]. In addition, the latencies of manage mechanisms that associated with data consistency, such as *Clfulsh* and data migration are modeled in detail based on the timing parameters of on-chip systems and memory.

Table 2. Workloads for evaluation

Workloads	Applications
SPEC CPU 2006	CactusADM x8, Mcf x8, Omnetpp x8
Parsec 3.0	Canneal x8, X264 x8, Streamcluster x8
NPB	MG x8, UA x8, SP x8
Large footprints	Graph500 x8, GUPS x8
Mix1	CactusADM x2 + Mcf x2 + Canneal x2 + Omnetpp x2
Mix2	Canneal x2 + X264 x2 + SP x2 + GUPS x2
Mix3	Mcf x2 + X264 x2 + SP x2 + UA x2

Alternative Policies. To better evaluate the system performance, SuperPT is compared with several alternative page migration mechanisms for heterogeneous memories as follows.

①**Flat-static:** 4 GB DRAM and 32 GB NVM are both used as main memory in the same address space [4], and are managed at 4 KB granularity. Based on the capacity ratio of DRAM to NVM, the data is evenly distributed in the address space. Since the fast DRAM and slow NVM are used indiscriminately in this system, there is no page migration between DRAM and NVM. This system is used as a baseline for comparison.

②**HSCC:** This is a state-of-the-art hybrid main memory system with traditional four-level page tables [4]. HSCC leverages a utility-based page migration strategy to migrate hot pages between fast DRAM and slow NVM.

③**DRAM:** This is a memory system consisting of only 32 GB DRAM, which is used as the applications' performance upper bound.

Benchmarks. As shown in Table 2, we choose several representative applications from SPEC CPU2006 [26], Parsec [27], NAS Parallel Benchmarks [28], and Graph500 [29]. CactusADM, Mcf, and Omnetpp are chosen from SPEC CPU2006. Among them, CactusADM is designed as a computational kernel to represent many programs in numerical relativity. Mcf is designed to solve a scheduling problem in public mass transportation. Omnetpp is implemented to simulate discrete event of large Ethernet networks. Canneal, X264, and Streamcluster are all multiple thread workloads chosen from Parsec3.0. MG, UA, and SP are chosen from NAS Parallel Benchmarks. MG is a simple multiple-grid kernel that needs highly structured communication at long distance and used to evaluate data communication both for short and long distance. UA solves heat equation with convection and diffusion from moving ball. SP leverages scalar penta-diagonal to solve nonlinear PDEs problems. Graph500 is designed to evaluate the performance of supercomputer by using large scale memory-intensive graph processing algorithms. To verify the effectiveness of the system in running the same application, we ran eight instances for each application at the same time.

4.2   Extra Memory Accesses Time

Figure 7 shows the *memory access times* (MAT) caused by page migration of each workload in our experiment, and all the result are normalized to the baseline system (Flat-static). SuperPT significantly reduces the MAT of those applications with lower data locality and large footprint. For example, for Canneal and GUPS, SuperPT can reduce 24.3% and 23.2% access times, respectively. Meanwhile, the access counts of mixed workloads with different applications, such as Mix1, Mix2, and Mix3, are also significantly reduced. Compared to HSCC, SuperPT reduces 19.3% memory accesses on average. This indicates that SuperPT significantly reduces the extra memory accesses due to page migrations. Moreover, the well page contiguity deliver higher performance to the page table walker cache.

4.3   Application Performance

Figure 8 shows the *instructions per cycle* (IPC) of every workload, all normalized to the Flat-static system. For applications with poor locality and large footprint (such as Canneal and GUPS), SuperPT can significantly improve system performance. We also notice that for highly parallel applications, such as MG, UA, and SP, although SuperPT reduced the proportion of memory access times by a considerable amount, the performance improvement of SuperPT on such applications is not very significant due to the small proportion of memory access caused by page table access (Fig. 2). Overall, SuperPT improves system performance by 77.9% and 9.5% on average, compared to Flat-static and HSCC, respectively. The performance gap between SuperPT and the upper bound (DRAM) is only 6.8% on average.

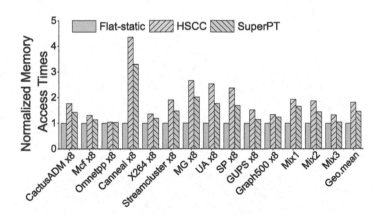

Fig. 7. Normalized memory access times relative to the flat system

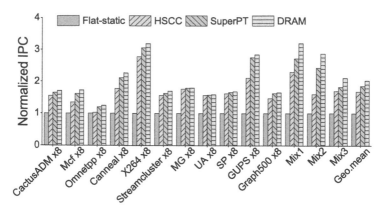

Fig. 8. Normalized IPC relative to the flat system

4.4 DRAM Allocation Collision

Figure 9 shows the hash conflict rate caused by DRAM allocation in SuperPT. To evaluate the performance of hash function in SuperPT, we compare SuperPT with several open-source hash functions, such as VHPT-remainder[1], VHPT-wyhash[2]. We have the following observations: ① SuperPT brings a very low conflict rate (less than 0.1% on average), especially for applications with small memory footprint, such as CactusADM. ② More efficient hash algorithms can significantly reduce the conflict rate, such as SuperPT-wyhash, which reduces the conflict rate by 84.4% on average, compared to SuperPT-remainder. Since a higher conflict rate is also associated with higher performance overhead, SuperPT adopts the hash function of SuperPT-wyhash in order to reduce system overhead.

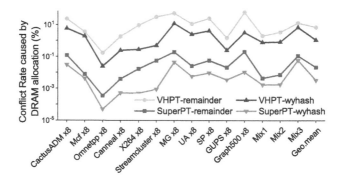

Fig. 9. Conflict rate of DRAM allocation

---

[1] Remainder is a simple hash function design with remainder operation.

[2] Wyhash [35] is a fast hash function on x86-64 without quality problems.

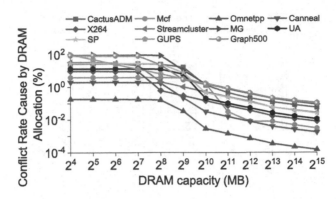

Fig. 10. DRAM collision rate sensitive to different DRAM size

To further study how the hash conflicting rate in SuperPT is sensitive to DRAM size, we run selected workloads with different DRAM capacity. As shown in Fig. 10, the conflict rate is calculated as the ratio of the number of conflicts to the total number of allocated pages. We find that, the conflict rate in SuperPT is less than 1% for most evaluated applications as long as the DRAM capacity is large than 4 GB. Moreover, the larger memory capacity leads to a lower collision rate.

4.5   Storage and Runtime Overheads

In common x86-64 systems, 46 bits are used for physical address and 48 bits for virtual address. Since 20 bits are needed for identifying the offset inner a superpage, the *virtual superpage number* (VSN) uses 28 bits. Similarly, 26 bits are needed for *physical superpage number* (PSN). Thus, to store the *virtual hashed page table* (VHPT), SuperPT consumes $\frac{4096 \times (26 + 16)}{8 \times 1024} = 13.5$ KB. The mapping table uses $\frac{9 \times 512}{8 \times 1024} = 0.56$ KB. Meanwhile, to index the normal pages in the superpage, $\frac{(22 + 34 + 1) \times 512 \times 4096}{8 \times 1024 \times 1024} = 14.25$ MB are needed to store the mappings between inner small pages and superpages. In all, the total memory usage is 14.25 MB. This storage overhead is negligible in modern computers which usually have several terabytes memory.

Figure 11 shows the breakdown of performance overhead caused by *page table walking* (PTW), DRAM mapping, Allocate pages, Confliction, Clflush, and TLB shootdown. All these operations are modeled by adding reasonable latencies in our simulator. We find that the runtime overhead of these selected programs varied greatly. For applications with large footprint and good hot page contiguity, such as MG, SP, mix2, and mix3, conflicts accounts for a larger partition of runtime overhead. In summary, the runtime performance overhead of SuperPT is 3.7% on average. This overhead is acceptable given the large performance benefits of employing multi-grained pages and efficient page table policy.

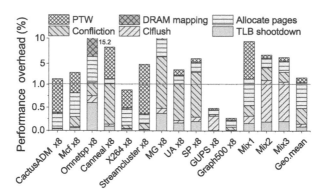

Fig. 11. Breakdown of running time overhead in SuperPT

5   Related Work

Page Contiguity. There has been a number of studies on exploring page contiguity to improve the TLB performance. Gorman *et al.* [36] design a new memory allocator in the operating system to mitigate memory fragmentation and promotes contiguity by aggregating pages based on the relocation information of them. Libhugetlbfs [37] exposes a user interface to programs to explicitly use huge pages. GLUE [12] leverages a single speculative superpage translation in the TLBs to map contiguous small pages. To reduce the overhead of page table walk, previous works [38, 39] allow the verification of speculative translations off the critical path of program execution. GTSM [40] leverages page contiguity at the hardware layer to construct superpages, even in system with retired bits. A large number of previous works focus on leveraging page contiguity to reduce TLB misses and to mitigate memory fragmentation (e.g., internal and external memory fragmentations). By using memory compaction, *contiguous memory allocators* (CMAs) [41] migrate memory fragmentation and offer a large contiguous memory space. By aggressively merging discrete physical frames into contiguous regions, Translation Ranger [14] enlarges the reach of contiguity-aware TLBs.

Contiguity-Aware TLBs. To leverage the page contiguity, TLB coalescing [10, 11, 42] and *memory management unit* (MMU) cache coalescing [6] are designed to increase the coverage of TLBs and MMU caches. To further enlarge TLB reach to cover the modern gigabyte-to-terabyte physical memory, direct segments [43] leverage a large segment to do fast address mapping between contiguous virtual address region and contiguous physical address region. *Redundant memory mappings* (RMM) [15] significantly enlarge TLB coverage by using range TLBs to map regions that both contiguous in virtual and physical addresses.

6   Conclusion

Energy efficiency in heterogeneous memory systems made up of DRAM and NVMs. However, page migration brings some side effects to the system, such as extra memory access due to page table modification and being hard to remain the contiguity of pages. In this paper, we propose a Superpage-friendly Page Table called SuperPT to reduce the overheads serving TLB misses. By leveraging a virtual hashed page table and a hybrid DRAM allocator, SuperPT conducts the address translation in a flexible and efficient way while rarely destroy the contiguity within the migrate pages. Experimental results show that SuperPT significantly reduces memory access times by 19.3% on average and thus improves system performance by 9.5% on average.

References

1. Dhiman, G., Ayoub, R., Rosing, T.: PDRAM: a hybrid pram and dram main memory system. In: Proceedings of the 46th Annual Design Automation Conference, pp. 664–469. ACM, New York (2009)
2. Qureshi, M.K., Srinivasan, V., Rivers, J.A.: Scalable high performance main memory system using phase-change memory technology. In: Proceedings of the 36th Annual International Symposium on Computer Architecture, pp. 24–33. ACM, New York (2009)
3. Ramos, L.E., Gorbatov, E., Bianchini, R.: Page placement in hybrid memory systems. In: Proceedings of the International Conference on Supercomputing, pp. 85–95. ACM, New York (2011)
4. Liu, H., et al.: Hardware/software cooperative caching for hybrid DRAM/NVM memory architectures. In: Proceedings of the International Conference on Supercomputing, pp. 26:1–26:10. ACM, New York (2017)
5. Wang, X., et al.: Supporting superpages and lightweight page migration in hybrid memory systems. ACM Trans. Archit. Code Optim. **16**(2), 11:1–11:26 (2019)
6. Bhattacharjee, A.: Large-reach memory management unit caches. In: Proceedings of the 46th Annual IEEE/ACM International Symposium on Microarchitecture, pp. 383–394. ACM, New York (2013)
7. Romer, T.H., Ohlrich, W.H., Karlin, A.R., Bershad, B.N.: Reducing TLB and memory overhead using online superpage promotion. In: Proceedings of the 22nd Annual International Symposium on Computer Architecture, pp. 176–187. ACM, New York (1995)
8. Talluri, M., Hill, M.D.: Surpassing the TLB performance of superpages with less operating system support. In: Proceedings of the Sixth International Conference on Architectural Support for Programming Languages and Operating Systems, pp. 171–182. ACM, New York (1994)
9. Swanson, M., Stoller, L., Carter, J.: Increasing TLB reach using superpages backed by shadow memory. In: Proceedings of the 25th Annual International Symposium on Computer Architecture, pp. 204–213. IEEE Computer Society, Washington, DC (1998)
10. Pham, B., Vaidyanathan, V., Jaleel, A., Bhattacharjee, A.: Colt: coalesced large-reach TLBs. In: Proceedings of the 2012 45th Annual IEEE/ACM International Symposium on Microarchitecture, pp. 258–269. IEEE Computer Society, Washington, DC (2012)

11. Pham, B., Bhattacharjee, A., Eckert, Y., Loh, G.H.: Increasing TLB reach by exploiting clustering in page translations. In: Proceedings of the 2014 IEEE 20th International Symposium on High Performance Computer Architecture, pp. 558–567. IEEE Computer Society, Washington, DC (2014)
12. Pham, B., Veselý, J., Loh, G.H., Bhattacharjee, A.: Large pages and lightweight memory management in virtualized environments: can you have it both ways? In: Proceedings of the 48th International Symposium on Microarchitecture, pp. 1–12. ACM, New York (2015)
13. Gandhi, J., et al.: Range translations for fast virtual memory. IEEE Micro **36**(3), 118–126 (2016)
14. Yan, Z., Lustig, D., Nellans, D., Bhattacharjee, A.: Translation ranger: operating system support for contiguity-aware TLBs. In: Proceedings of the 46th International Symposium on Computer Architecture, pp. 698–710. ACM, New York (2019)
15. Karakostas, V., et al.: Redundant memory mappings for fast access to large memories. In: Proceedings of the 42nd Annual International Symposium on Computer Architecture, pp. 66–78. ACM, New York (2015)
16. Bhargava, R., Serebrin, B., Spadini, F., Manne, S.: Accelerating two-dimensional page walks for virtualized systems. In: Proceedings of the 13th International Conference on Architectural Support for Programming Languages and Operating Systems, pp. 26–35. ACM, New York (2008)
17. Gandhi, J., Basu, A., Hill, M.D., Swift, M.M.: Efficient memory virtualization: reducing dimensionality of nested page walks. In: Proceedings of the 47th Annual IEEE/ACM International Symposium on Microarchitecture, pp. 178–189. IEEE Computer Society, Washington, DC (2014)
18. Yan, Z., Veselý, J., Cox, G., Bhattacharjee, A.: Hardware translation coherence for virtualized systems. In: Proceedings of the 2017 ACM/IEEE 44th Annual International Symposium on Computer Architecture, pp. 430–443. ACM, New York (2017)
19. Kandiraju, G.B., Sivasubramaniam, A.: Going the distance for TLB prefetching: an application-driven study. In: Proceedings of the 29th Annual International Symposium on Computer Architecture, pp. 195–206. IEEE, Anchorage (2002)
20. Saulsbury, A., Dahlgren, F., Stenström, P.: Recency-based TLB preloading, In: Proceedings of the 27th Annual International Symposium on Computer Architecture, pp. 117–127. ACM, New York (2000)
21. Yaniv, I., Tsafrir, D.: Hash, don't cache (the page table). In: Proceedings of the 2016 ACM SIGMETRICS International Conference on Measurement and Modeling of Computer Science, pp. 337–350. ACM, New York (2016)
22. Stallings, W.: Operating Systems: Internals and Design Principles, 7th edn. Pearson/Prentice Hall, Upper Saddle River (2011)
23. Raoux, S., et al.: Phase-change random access memory: a scalable technology. IBM J. Res. Dev. **52**(4.5), 465–479 (2008)
24. Park, H., Yoo, S., Lee, S.: Power management of hybrid DRAM/PRAM-based main memory. In: Proceedings of the 48th Design Automation Conference, pp. 59–64. ACM, New York (2011)
25. Wei, W., Jiang, D., McKee, S.A., Xiong, J., Chen, M.: Exploiting program semantics to place data in hybrid memory. In: Proceedings of the 2015 International Conference on Parallel Architecture and Compilation, pp. 163–173. IEEE Computer Society, Washington, DC (2015)
26. SPEC CPU2006. https://www.spec.org/cpu2006. Last Accessed 21 Nov 2019
27. Parsec. http://parsec.cs.princeton.edu/index.htm. Last Accessed 21 Nov 2019
28. Bailey, D., et al.: The NAS parallel benchmarks. Int. J. Supercomput. Appl. **5**(3), 63–73 (1991)

29. Graph500. http://graph500.org/. Last Accessed 21 Nov 2019
30. Jiang, X., et al.: CHOP: adaptive filter-based DRAM caching for CMP server platforms. In: Proceedings of the Sixteenth International Symposium on High-Performance Computer Architecture, pp. 1–12. IEEE Computer Society, Washington, DC (2010)
31. Sanchez, D., Kozyrakis, C.: ZSim: fast and accurate microarchitectural simulation of thousand-core systems. In: Proceedings of the 40th Annual International Symposium on Computer Architecture, pp. 475–486. ACM, New York (2013)
32. Luk, C.K., et al.: Pin: Building customized program analysis tools with dynamic instrumentation. In: Proceedings of the 2005 ACM SIGPLAN Conference on Programming Language Design and Implementation, pp. 190–200. ACM, New York (2005)
33. Poremba, M., Zhang, T., Xie, Y.: NVMain 2.0: a user-friendly memory simulator to model (non-)volatile memory systems. IEEE Comput. Archit. Lett. **14**(2), 140–143 (2015)
34. Lee, B.C., Ipek, E., Mutlu, O., Burger, D.: Architecting phase change memory as a scalable DRAM alternative. In: Proceedings of the 36th Annual International Symposium on Computer Architecture, pp. 2–13. ACM, New York (2009)
35. Wyhash. https://github.com/rurban/smhasher. Last Accessed 21 Nov 2019
36. Gorman, M., Healy, P.: Supporting superpage allocation without additional hardware support. In: Proceedings of the 7th International Symposium on Memory Management, pp. 41–50. ACM, New York (2008)
37. Huge Pages Part 2 (Interfaces). https://lwn.net/Articles/375096/. Last Accessed 21 Nov 2019
38. Barr, T.W., Cox, A.L., Rixner, S.: SpecTLB: a mechanism for speculative address translation. In: Proceedings of the 38th Annual International Symposium on Computer Architecture, pp. 307–318. ACM, New York (2011)
39. Papadopoulou, M.M., Tong, X., Seznec, A., Moshovos, A.: Prediction-based superpage-friendly TLB designs. In: Proceedings of the 2015 IEEE 21st International Symposium on High Performance Computer Architecture, pp. 210–222. IEEE Computer Society, Washington, DC (2015)
40. Du, Y., Zhou, M., Childers, B.R., Mossé, D., Melhem, R.: Supporting superpages in non-contiguous physical memory. In: Proceedings of the 2015 IEEE 21st International Symposium on High Performance Computer Architecture, pp. 223–234. IEEE Computer Society, Washington, DC (2015)
41. Corbet, J., Rubini, A., Kroah-Hartman, G.: Linux Device Drivers: Where the Kernel Meets the Hardware. 3rd edn. O'Reilly Media, Sebastopol (2005)
42. Wang, X., Liu, H., Liao, X., Jin, H., Zhang, Y.: TLB coalescing for multi-grained page migration in hybrid memory systems. IEEE Access **8**, 66304–66313 (2020)
43. Basu, A., Gandhi, J., Chang, J., Hill, M.D., Swift, M.M.: Efficient virtual memory for big memory servers. In: Proceedings of the 40th Annual International Symposium on Computer Architecture, pp. 237–248. ACM, New York (2013)

Recommendation Algorithm Based on Improved Convolutional Neural Network and Matrix Factorization

Shengbin Liang$^{(\boxtimes)}$ (iD), Lulu Bai (iD), and Hengming Zhang (iD)

School of Software, Henan University,
Kaifeng 475004, People's Republic of China
liangsbin@henu.edu.cn

Abstract. The traditional collaborative filtering algorithm uses the user rating information as a recommendation basis, but the ratings matrices are usually sparse and cannot reflect users' preference exactly, so the recommendation results are not very accurate. Therefore, this paper proposes an improved convolutional neural network for collaborative filtering (CNNCF), using the deep learning model to deeply mine the hidden feature information. then implicit the semantic model, Then the extracted explicit feature information was replaced by the implicit feature information in the LFM to further improve the prediction accuracy, and finally personalized recommendation through the user-item preference matrix. Experimental results on the MovieLens dataset show that the model can overcome data sparse, and recommendation accuracy is better than the traditional collaborative filtering model.

Keywords: Collaborative filtering · Latent factor model · Convolutional neural network · Recommender System · Hybrid recommendation

1 Introduction

With the rapid development of information technology, Internet information resources are growing rapidly. Recommendation System (RS) plays an important role in solving the problem of information overload and providing personalized services for users. It has been widely used in news, music, e-commerce and other fields.

The recommendation algorithms are mainly divided into content-based recommendation algorithm [1], collaborative filtering algorithm [2] and hybrid recommendation algorithm [3]. The content-based recommendation algorithm mainly uses the user's personal features and content information for recommendation. While CF algorithm based on user item ratings matrix is the most widely used recommendation algorithm. But the ratings matrix is usually very sparse, CF algorithm faces data sparsity, cold start and other problems. In order to solve the problems, a content-based algorithm and a collaborative filtering algorithm is combined to form a hybrid recommendation algorithm [4]. For example, the hidden feature information extracted from the content information such as item and user portrait or the combination of auxiliary information and collaborative filtering can improve the recommendation accuracy [5].

© Springer Nature Singapore Pte Ltd. 2020
J. Zeng et al. (Eds.): ICPCSEE 2020, CCIS 1257, pp. 642–654, 2020.
https://doi.org/10.1007/978-981-15-7981-3_47

At present, deep learning has made breakthroughs in image recognition, machine translation, speech recognition and other fields, and has achieved good results in feature extraction [6]. Therefore, deep learning is introduced into the recommendation system for feature learning. In order to strengthen the model prediction sequence evolution ability, Li proposed a sequence stream recommendation algorithm based on circular time convolutional network, and used time-series convolutional neural network for feature extraction [7]. At the same time, a threshold cycle unit is introduced to capture the logical relationship between feature vectors. In order to solve the problem of data sparseness and cold start, Zhang proposed a standardized matrix decomposition model based on attention mechanism to build a user-item heterogeneous network based on user trust network, score records and based on the heterogeneous network mining similarity between users to solve the problem caused by data sparsity [8]. Jiang et al. proposed a combined recommendation algorithm, which introduced the interception factor into the content-based recommendation algorithm and the user-based collaborative filtering algorithm, and then generated a hybrid recommendation algorithm [9]. Wang use potential group recommendation based on dynamic convolution probability matrix decomposition to realize group recommendation by analyzing the relationship between users, groups and services, but many parameters in the model need to be adjusted manually [10]. Kim put forward a kind of recommendation model based on tags, but the model only uses the auxiliary information provided, cannot fully mine the deep information implied in the data, and the effect is not ideal [11].

This paper proposes an improved matrix factorization model integrated with convolutional neural network (CNNCF). The improved CNN is used to extract implicit feature information from users and items, and then compounded to matrix decomposition. It reduces the sparseness of the data, improves the accuracy of prediction scores, and also improves the accuracy of user recommendations.

2 Related Work

2.1 CNN

CNN is a deep neural network model, which includes Embedding layer, convolutional layer, pooling layer and output layer.

Embedding Layer. The embedding layer mainly converts the user's description of items into an embedding matrix. Each user's description of items can be regarded as a sentence consisting of n words, and then the sentence consisting of n words is processed into a description information matrix. For the i-th word containing the description information D of a finite number of words, each word of the document is processed into a p-dimensional vector $x_i \in R^p$ after passing through the embedding layer, however $D \in R^{P*|D|}$, The result information matrix is:

$$X = [x_1 \ldots x_i, x_{i+1}, x_{i+2}] \tag{1}$$

Convolution Layer. The main task of the convolutional layer is to extract different features [12]. Usually, the first layer of convolution can only extract some low-level features such as edges and lines in the input data, and then learn more deeply through multi-layer convolution. The analysis yields more abstract features. However, the traditional CNN does not perform well on local feature attribute information, so this paper uses the literature [13] put forward the improved convolution model proposed in to extract user attribute and item description information, and adopts a multi-channel method by setting different sizes. The convolution kernel and the weight matrix and the bias are spliced into a deeper matrix by using different filters, resulting in extraction of different forms of convolution mapping attribute vectors. And processed by the non-linear activation function ReLU to get the final feature vector.

Pooling Layer. The pooling layer mainly down-samples the feature maps extracted by the convolutional layer [14], which can effectively reduce the size of the matrix and amount of the training parameters of the last fully connected layer. Moreover, using the pooling layer can speed up the calculation speed and avoid overfitting. Since the output results of the convolutional layers are unequal, each convolutional mapping attribute vector contains all the mapping attributes [15], and cannot represent the obvious features of the context. This paper uses maximum pooling. Assuming that the size of the sampling window is w and the height is h, the process of pooling is to first take the window size as the step size, divide the feature result of the convolution layer into several sub-areas of w_h size, and then use the maximum sampling method to obtain The maximum eigenvalue of the region.

Fully Connected Layer. In convolutional neural networks, there are generally 1 to 2 layers of fully connected layer at the end to classify. After processing the convolutional layer and pooling layer in the previous several rounds, the original data is mapped to the hidden feature vector. In the space, the role of the fully connected layer is to map these hidden features into k-dimensional space. However, in order to prevent the problem of overfitting, a dropout layer needs to be added in front of the fully connected layer.

2.2   Hidden Factor Matrix Factorization

The hidden factor matrix decomposition model has been widely used in the field of recommendation systems, decomposes the user-item ratings matrix $R_{m \times n}$ into two low-dimensional matrices and multiplies them to obtain the user feature matrix $U_{m \times k}$ and the item feature matrix $V_{k \times n}$.

$$R \approx U_{m \times k} V_{k \times n} \tag{2}$$

$U_{m \times k}$ and $V_{k \times n}$ are matrices after dimensionality reduction, then the predicted value $R(u, i) = r_{ui}$, which user u on item i can be calculated by formula (3):

$$\hat{r}_{ui} = \sum_f p_{uf} q_{if} \tag{3}$$

where $p_{uf} = P(u,f)$, $q_{if} = Q(i,f)$. Matrix P and Q are obtained by minimizing RMSE learning from the observations of the training set. So, the loss function of LFM is:

$$C(p,q) = \sum_{(u,i) \in Train} \left(r_{ui} - \sum_f^F p_{uf} q_{if} \right)^2 \tag{4}$$

However, the implicit semantic model is easy to overfit when meet a sparse rating matrix. Therefore, formula (4) is prone to overfitting. To prevent overfitting, it is necessary to introduce a parameter attenuation term $\lambda \left(\|p_u\|^2 + \|q_i\|^2 \right)$, which λ is the regularization parameter, so we get:

$$C(p,q) = \sum_{(u,i) \in Train} \left(r_{ui} - \sum_f^F p_{uf} q_{if} \right)^2 + \lambda \left(\|p_u\|^2 + \|q_i\|^2 \right) \tag{5}$$

3 Improved Convolutional Neural Network and Matrix Factorization Algorithm Model

3.1 Improved Convolutional Neural Network Model

Since the convolutional layer of the traditional convolutional neural network generally obtains the feature mapping through a linear convolution filter [16], and then activates it non-linearly, the calculated feature mapping is as formula (6):

$$f_{i,j,k} = \max \left(w_k^T x_{i,j} + b, 0 \right) \tag{6}$$

where (i,j) is the coordinates of the feature mapping, is the input with (i,j) as the center position in the sliding window, k is the channel index of the feature mapping, w is the weight, and max is the nonlinear excitation function.

This linear convolution filter has a better effect in the case of linearly separable features, while the feature vectors extracted are commonly non-linear, so in order to extract more abstract and effective feature information, this paper uses an improved CNN which uses multiple convolution kernels in parallel, and multiple different convolution kernels are used to learn features to achieve local to global correlation learning ability.

Traditional CNNs have multiple convolutional layers and pooling layers, and each convolutional unit is a single convolution kernel for feature learning. Figure 1 indicates a convolutional layer, and the convolutional layer has multiple different convolution kernels. The convolution kernel slides down with a step size of 1 and performs a convolution operation every time a text vector is passed to generate a new feature vector. The calculation method of the eigenvector is shown in formula (7).

$$c_i = f(w \times W_{i:i+h} + b) \tag{7}$$

Where w represents different convolution kernel matrix weight parameters, h represents the height of the convolution kernel, b is the offset, and f is the activation function.

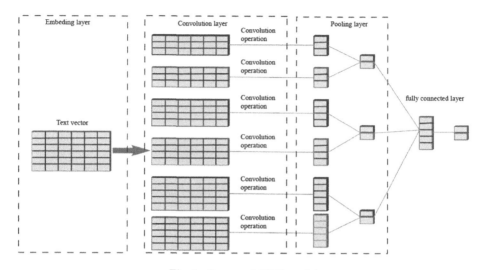

Fig. 1. Improved CNN model

After the user description attributes and the item description attributes are processed by the embedding layer, the user and item attributes can be represented by a matrix with word vectors, and then convolved to obtain one after each type of convolution check text processing feature map:

$$c = (c_1, c_2, c_3, \ldots, c_{n-h+1}) \tag{8}$$

where n is the number of words in the text. In the pooling layer, we use the maximum pooling layer (max-pooling) to perform feature extraction on the feature mapping output by the convolution layer, that is, $\hat{c} = \{c\}$. Because the length of the feature mappings obtained by the convolution layer is different, after the pooling layer processing, texts of different lengths have become feature vectors of the same length. Full connection is to map the hidden features processed by the pooling layer to k-dimensional space. In this paper, k is selected as 10-dimensional.

3.2 Convolutional Neural Network and Matrix Factorization Fusion Model

At present, matrix factorization technology can be effectively used in scoring prediction in recommendation systems, but as the amount of data increases, the problem of sparse data leads to matrix factorization technology only using user-item rating matrix learned user and item feature vector, Ignoring the user's attention to different attribute features, cannot accurately reflect the user's interest preferences, thus greatly affecting the model's score prediction performance. Therefore, in the face of sparse data, this paper analyzes the user's preference for each attribute of the project to mine the user's interest preferences, thereby recommending items that are more in line with the user's interests. At the same time, this paper uses an improved convolutional neural network model to perform feature extraction on the processed user and item feature attribute information, and then replaces the extracted user and item feature vectors with implicit feature information in matrix decomposition. The inner product operation is used as the target user's rating of the item. Thus, the user-item feature matrix is converted from the hidden factor feature matrix to the explicit factor feature matrix, the hidden feature information mined based on the content is applied to the collaborative filtering recommendation algorithm, and the final learned user and item implicit representation pairs are used. The user makes a recommendation.

The framework of the convolutional neural network and matrix factorization fusion model is shown in Fig. 2. The framework consists of two processes: first, feature extraction of the description attribute information of the user-item matrix through an improved CNN model; then, the LFM algorithm is used to replace the explicit feature information extracted in the previous process with the hidden features in the LFM Feature information to improve prediction accuracy.

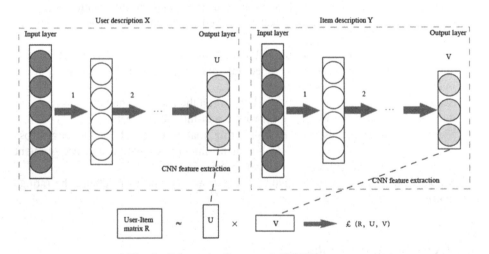

Fig. 2. Schematic diagram of CNNCF structure

In order to make the scores obtained by the inner product operation of the user and item feature vectors U and V closer to the true value and more accurately reflect the user's interest preferences, a new loss function is finally constructed:

$$L(P, Q, W) = \sum_{(u,i) \in R} (r_{ui} - (\mu + b_u + b_i + p_u^T q_i))^2$$
$$+ \left[\lambda_1 \sum_u \|p_u - cnn(W, x_i)\|^2 + \lambda_2 \sum_u \|b_u\|^2 + \lambda_3 \sum_i \|b_i\|^2 + \lambda_4 \sum_i \|q_i\|^2 \right]$$
$$(9)$$

Among them, $L(P, Q, W)$ represents the loss function, r_{ui} represents the value of the original rating matrix, μ is the average value of the user's rating items in the training set, b_u represents the user bias term, b_i represents the item bias term, $cnn(W, x_i)$ represents the scoring value constructed by the improved convolutional neural network of the user and item description attribute information λ_1, λ_2, λ_3 and λ_4 represent the regularization equilibrium coefficients.

The overall steps of the fusion recommendation model of the improved convolutional neural network and matrix factorization model proposed in this paper are as follows:

Algorithm: CNNCF algorithm description.

Input. user's characteristic information, item's characteristic information, user-item scoring record.

Output. User-item prediction score.

Step1. Extract user-item scoring records, user characteristic information, and item characteristic information.

Step2. Input user feature information and item feature information into the embedding layer to obtain user feature embedding matrix $\{e_u\}$ and item feature embedding matrix $\{e_i\}$.

Step3. The traditional convolutional neural network is improved into a multi-channel network, and the user embedding matrix $\{e_u\}$ and the item embedding matrix $\{e_i\}$ are input respectively.

Step4. Obtain the relationship matrix U between the user and the hidden factor, and the relationship matrix V between the item and the hidden factor.

Step5. Construct the user interaction rating matrix R, construct the target loss function by fitting the error (R, U, V) of the interaction matrix, and use gradient descent to train and optimize the target function.

Step6. After training, get the final matrices P and Q, and then rebuild the ratings matrices to fill in the vacant scores.

4   Experimental Results

4.1   Experimental Design

We use the movie ratings data set MovieLens1 M, which contains more than 1 million rating records of more than 6,000 users for nearly 4,000 movies, and the ratings matrix is very sparse. The user information includes descriptive information such as user ID, gender, age, occupation, zip code, and the movie information includes descriptive information such as movie ID, name, movie classification, label, and release time. This paper divides the data set into 80% for training set, the remaining 20% are test set (Table 1).

Table 1. Data statistics for the MovieLens1M data set

Filed	Value
#users	6 040
#movies	3 883
#ratings	993 482
#timestamps	993 482
Sparsity	0.95765

Experimental Environment and Specific Parameter Settings. The setting value of the network model parameters greatly affects the effect of the experiment. In order to improve the accuracy of prediction, this paper uses mini-batch gradient decent when training CNN. The method divides the data into several batches, update parameters by batch. The specific parameter settings are shown in Table 2.

Table 2. CNNCF parameters setting

Parameter name	Parameter value
Embedded layer length	64
Convolution kernels	16
Convolution kernel window height	(3,4,5)
Pooling method	Max-pooling
batch-size	200
Iterative rounds	10
Learning-rate	0.001
Dropout rate	0.5
$\lambda_1, \lambda_2, \lambda_3, \lambda_4$	0.1
Hidden factor feature space dimension (k)	10

Evaluation Indicators and Comparison Methods. This paper chooses the metric commonly used in recommendation systems: root mean square error (RMSE) to measure the performance of the proposed algorithm model.

$$RMSE = \sqrt{\frac{\sum_{i=1}^{S}(r_i - \hat{r}_i)^2}{S}} \tag{10}$$

where r_i is the real score of user i on the item, $\hat{r}_i$ is the predicted score, and S is the number of ratings as the test set. It can be seen from formula (10) that the smaller value of RMSE, the higher the recommended accuracy.

Based on the pre-processed MovieLens1M data set, this paper selects the traditional hidden factor model, neural collaborative filtering model, conditional convolutional hidden factor algorithm model proposed by Li and other common convolutional neural network and matrix decomposition model (FM & CNN) Compared with the fusion model of improved CNN and matrix factorization model proposed in this paper.

1. The neural collaborative filtering model (NCF) proposed in [17] is a combination model based on a generalized matrix factorization model and a multilayer perceptron. Multilayer perceptron extracts user and item features by weighted summation.
2. The conditional convolution hidden factor model proposed in [18] is a recommendation model that uses conditional convolution to extract user and item features and compound with matrix decomposition.
3. FM & CNN is a recommendation model that compound ordinary convolutional neural networks with matrix decomposition models.

4.2   Experimental Results and Analysis

Influence of Model Parameters. The different parameters of the improved convolutional neural network model will also affect the experimental results. The key factor affecting the performance of the model is the length of the hidden factor space dimension embedding layer. This paper compares RMSE under different hidden factor space dimensions and embedding layer length.

Since the hyperparametric hidden factor space dimension k is not only the effective dimension of the features extracted by the CNNCF from users' description information and items' description information, but also the dimension of the hidden feature vector of matrix decomposition, the value of k in Eq. (3) directly affects accuracy of the algorithm. Figure 3 shows the effect of different k values on the performance of the CNNCF model and it can be seen that when k is very small, CNNCF cannot extract valid user and item features; when k is larger than 10, more and more interference information will be introduced and the accuracy of the recommendation will be reduced. Therefore, we set the value of k is 10, the performance of the model is optimal.

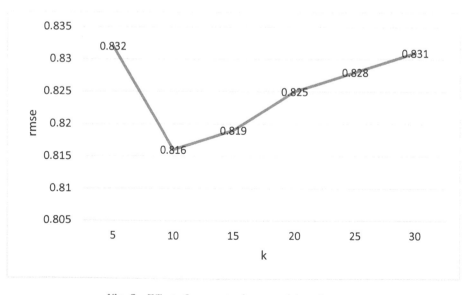

Fig. 3. Effect of parameter k on model performance

Because the length of the embedded layer has a great influence on the model, it will not represent the word well if it is too short, otherwise it will make overfitting due to insufficient data. It can be seen from Fig. 4 that when the length of the embedded layer is greater than 64, the overly long feature vector is prone to overfitting in the case of insufficient data, and when the length of the embedded layer is 64, the prediction error of the model algorithm is: 0.8176. It can be seen that the model algorithm can better learn the potential characteristics of users and items when the length of the embedded layer is 64, and then build a user's rating prediction model for the item based on these potential features, improving the model's recognition of users and movies degree.

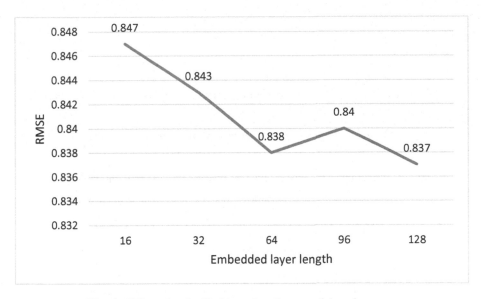

Fig. 4. Effect of embedded layer length on model performance

Performance Comparison of Different Models. By combining the traditional hidden factor model, FM & CNN model, neural collaborative filtering model (NCF) and conditional convolution hidden factor model with the proposed fusion algorithm (CNNCF) based on improved convolutional neural network and matrix factorization proposed in this paper on the public dataset For MovieLens1M, the root mean square error of multiple experiments is shown in Fig. 5 and Table 3.

As can be seen from Table 3, compared with the LFM model, the model proposed in this paper can improve the score prediction ability by 18.36%, indicating that the deep recommendation model is significantly better than the traditional CF model. Compared with the FM & CNN model, the score prediction is improved by about 5.17%, indicating that it has better feature extraction capabilities, a higher recognition

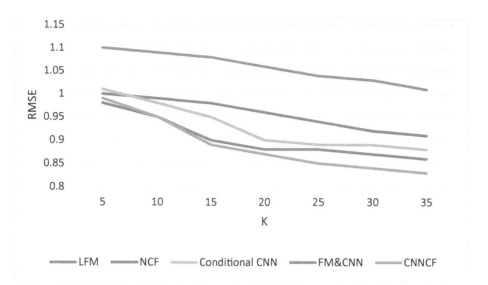

Fig. 5. Comparison of experimental results with other models

Table 3. Experimental results of different model on the MovieLens 1M dataset

Model	RMSE	CNNCF performance improvement ratio%
LFM	1.0015	18.36
Neural Cooperative Filtering (NCF)	0.9106	10.21
Conditional convolution hidden factor	0.8881	7.93
FM&CNN	0.8622	5.17
CNNCF	0.8176	—

rate for user and item description information, and also illustrates convolutions of different sizes. The kernel is significantly superior to traditional CNN in nonlinear feature extraction. Compared with the conditional convolutional hidden factor model, the CNNCF model score prediction performance is improved by 7.93%. The conditional convolutional hidden factor model performs excellently in extracting user features and item features, and even without using additional parameters, even using a shallow network achieves better results than the traditional deep network, but the performance of score prediction is not as good as the CNNCF model. Compared with the neural collaborative filtering model, the score prediction performance is improved by 10.21%, so it also shows that the improved convolutional neural network can be well integrated into the matrix decomposition model, which is conducive to improving the prediction of the score.

5 Conclusions

The traditional single collaborative filtering model is easily constrained by data sparseness and other issues. In order to improve the accuracy of the recommendation system, this article makes improvements on the recommendation model algorithm, and compounds the improved convolutional neural network with the matrix decomposition model. And then it compared FM & CNN model, neural collaborative filtering model and conditional convolution hidden factor model respectively. The experimental results show that the hybrid deep learning-based recommendation model is better than the single, traditional collaborative filtering recommendation model. The improved feature of the convolutional neural network model is significantly better than the traditional convolutional neural network model. It also fully proves that the improved convolutional neural network and matrix decomposition model fusion model proposed in this paper has high score prediction accuracy and outstanding recommendation effect.

Acknowledgements. This work has been financially supported by the Key Scientific Research Projects of Education Department of Henan province, China (No. 20A520008).

References

1. Lang, K.: NewsWeeder: learning to filter Netnews. In: The 12th Machine Learning Proceedings, pp. 331–339 (1995)
2. Huang, L., Jiang, B., Lu, S.: Review of research on deep learning-based recommendation systems. Chin. J. Comput. **41**(7), 1619–1647 (2018)
3. Yang, B., Zhao, P.: A summary of recommended algorithms. J. Shanxi Univ. (Nat. Sci. Edn.) **34**(3), 337–350 (2011)
4. Abhishek, K.: A review on personalized information recommendation system using collaborative filtering. Int. J. Comput. Sci. Inf. Technol. **2**(3), 1272–1278 (2011)
5. Zhang, F., Yuan, N., Lian, D.: Collaborative knowledge base embedding for recommender systems. In: Proceedings of the 22nd ACM SIGKDD International Conference on Knowledge Discovery and Data Mining, pp. 353–362. ACM, San Francisco (2016)
6. Wang, X., Wang, Y.: Improving content-based and hybrid music recommendation using deep learning. In: Proceedings of the 22nd ACM International Conference on Multimedia, pp. 627–636. ACM (2014)
7. Li, T., He, Z., Wang, B., Yan, Y., Tang, X.: Sequence stream recommendation algorithm based on cyclic time convolution network. Comput. Sci. **7**(03), 103–109 (2020)
8. Zhang, Q., Wang, B., Cui, N., Song, X., Qin, J.: Normalized matrix factorization recommendation algorithm based on attention mechanism. J. Softw. **31**(03), 778–793 (2020)
9. Jiang, X., Qi, X., Liu, L.: Research on personalized information recommendation method. J. Intell. Syst. **13**(02), 189–195 (2018)
10. Wang, H., Dong, M.: Potential group recommendation based on dynamic convolution probability matrix factorization. J. Comput. Res. Dev. **54**(8), 1853–1863 (2017)
11. Kim, B. S., Kim, H., Lee, J.: Improving a recommender system by collective matrix factorization with tag information. In: 2014 Joint 7th International Conference on Soft Computing and Intelligent Systems (SCIS) and 15th International Symposium on Advanced Intelligent Systems (ISIS), pp. 980–984. IEEE (2014)

12. Taigman, Y., Yang, M., Ranzato, M.: Closing the gap to human-level performance in face verification. In: IEEE Computer Vision and Pattern Recognition (CVPR), pp. 5–6. IEEE (2014)
13. Wang, G., Huang, X.: Convolutional neural network text classification model based on Word2vec and improved TF-IDF. Small Microcomput. Syst. **40**(05), 1120–1126 (2019)
14. Lu, Q., Wang, R., Zhang, J.: Recommendation algorithm combining neighborhood model and implicit semantic model. CEA **49**(19), 100–103 (2013)
15. Mikolov, T., Sutskever, I., Chen, K.: Distributed representations of words and phrases and their compositionality. In: Advances in Neural Information Processing Systems, pp. 3111–3119 (2013)
16. Zhang, Q., Zhang, S., Lei, Z.: Chinese sentiment classification based on improved convolutional neural network. Comput. Eng. Appl. **53**(22), 111–115 (2017)
17. He, X., Liao, L., Zhang, H.: Neural collaborative filtering. In: Proceedings of the 26th International Conference on World Wide Web, pp. 173–182 (2017)
18. Li, N., Sheng, Y., Ni, H.: Conditional convolutional hidden factor model for personalized recommendation. Comput. Eng. **46**(04), 85 − 90 + 96 (2020)

Research on Service Recommendation Method Based on Cloud Model Time Series Analysis

Zhiwu Zheng[1](✉), Jing Yao[1,2], and Hua Zhang[1]

[1] School of Software, Hunan University of Information Technology,
Changsha, China
neroe2005@126.com
[2] School of Computer, Central South University, Changsha, China

Abstract. The problem of information overload is becoming increasingly prominent, and recommendation systems are developing rapidly in various fields. How to find the most user-friendly services has become the focus. Service recommendation based on QoS is an important technology to select appropriate services for users. In this paper, a service selection method based on time series analysis of cloud model is proposed. Firstly, the noise was removed by clustering algorithm, clustering was divided, and similar user sets were obtained. Then, the cloud model was established by using similar user history data in different periods, and the comprehensive cloud model was obtained by combining time decay function. Finally, the recommended service was obtained by comparing TOPSIS method with ideal cloud model. The experimental results on WS-Dream dataset show that the accuracy of recommendation is improved compared with the existing recommendation algorithms.

Keywords: Cloud model · Time series · Clustering algorithm · Service recommendation

1 Introduction

In the era of rapid development of Internet technology, the number of Internet users has increased exponentially, and user experience has become an important aspect of developers'attention. At the same time, displaying too much information makes users unable to get useful information from some parts, which reduces the efficiency of use. Recommendation system is not only an important means of information filtering, but also a very promising way to solve the problem of information overload. It can find suitable user services from large data sets and recommend them to users. It has a good application in many scenarios [1].

Recommendation system mainly focuses on content information, at the same time, continuous information also reflects a lot of information about user behavior. By predicting the ratings of different projects to provide personalized recommendations for specific users, to help people find products that they are interested in and save time in the search process. For users, such a system can learn from the user's record experience and recommend the list of products that best suit them.

© Springer Nature Singapore Pte Ltd. 2020
J. Zeng et al. (Eds.): ICPCSEE 2020, CCIS 1257, pp. 655–665, 2020.
https://doi.org/10.1007/978-981-15-7981-3_48

In the past few years, recommendation technology has developed rapidly, and many recommendation systems have been developed. Clustering algorithm is also used in recommendation system, which divides users (items) into different clusters [2]. Document [3] Users are divided into several clusters by user item scoring matrix, and target users are clustered as nearest neighbor search space, which reduces the scope of nearest neighbor search, improves scalability and improves prediction performance. Literature [4] proposes an algorithm for clustering project types. Different clusters are divided according to different project types. Each cluster calculates its own rating prediction, and then uses weighting strategy to merge the final prediction results. Document [5] applies fuzzy C-means clustering to user-based collaborative filtering, and shows that fuzzy clustering is superior to other clustering techniques. [6] A recommendation algorithm based on multivariate clustering is proposed. Users are clustered from two aspects of rating mode and trust relationship. Support vector machine (SVM) is introduced to determine the prediction score according to the characteristics of users and items. Literature [7] Clustering users by K-means algorithm based on user behavior statistics of different projects improves the quality of recommendation. Users' behavior statistics of different project types perform well in measuring user interest and clustering. The traditional collaborative filtering recommendation algorithm has sparse user data. It needs to consider the cold start problem, but also does not consider the changes in user time. It can not accurately predict user interest and the range of nearest neighbors, which is not conducive to real-time recommendation. In order to solve the above problems, this paper introduces cloud model into service recommendation process and proposes a service recommendation method based on time series of cloud model combined with the characteristics of multi-period user differences. In different application scenarios, the service recommendation accuracy is conquered through simulation experiments based on real data sets, and satisfactory recommendation results are achieved.

This paper describes the current research situation and main problems of service selection, proposes a service selection method based on time series analysis of cloud model, and then briefly introduces clustering algorithm and cloud model. Through simulation experiments, combined with the effect of service recommendation of time series, the calculation accuracy in different application scenarios is synthetically determined. Finally, the full text is summarized.

2   Related Work

2.1   Clustering Algorithm

Clustering is the process of synthesizing objects into similar sets. There are as many objects as possible in the set, and there is no intersection between the aggregated set classes. By introducing clustering algorithm into recommendation algorithm, users or items can be clustered. Similar users or items can be clustered into the same class without calculating the similarity between different users, which greatly reduces the amount of calculation.

Clustering algorithm can be divided into: partition method, hierarchical method, density-based method, grid-based method and model-based method [8–10]. Clustering algorithm is widely used at present, but it has certain pertinence. No clustering algorithm can complete data clustering in different application scenarios.

K-means clustering algorithm is one of the classical clustering algorithms. The main idea is to find K clustering centers {c1, c2, c3,..., ck} so as to minimize the sum of the square distance of each data point Xi to the nearest clustering center ck. This process allows objects with high similarities to be clustered in a class. Because K-means clustering algorithm is simple and effective, it is suitable for large data sets and can meet the needs of collaborative filtering algorithm. This paper chooses K-means clustering algorithm as clustering algorithm. The cost function is as follows:

$$E = \sum_{i=1}^{k} \sum_{x_i \in S_i} |x_i - \text{avg}_k|^2 \tag{1}$$

The sum of the square errors of the objects in the data set is E, Xi is any point in the data set, and AVG is the average value of cluster si. The objective function makes the clusters as compact and independent as possible, and is measured by Euclidean distance. The algorithm flow of K-means clustering algorithm is as follows:

Step1: Initialize the data and randomly determine the cluster centers {c1, c2, c3,..., ck} according to the number of clusters specified.
Step2: Find the nearest cluster center CK and assign each sample Xi to this class.
Step3: Recalculate the average value of each class and repeat step 2.
Step4: When the average value of the class and the clustering center {c1, c2, c3,..., ck} no longer change, stop the calculation and get the final clustering.

Clustering algorithm is efficient and widely used in clustering large-scale data. Many algorithms are extended and improved around this algorithm.

2.2 Cloud Model

There are many uncertain phenomena and things in the real world. Representation of uncertainties is a difficult problem. At present, the research mainly focuses on Fuzziness and randomness. In recent years, the theory system of fuzzy sets with membership degree as the criterion of fuzziness is becoming more and more mature. But membership degree is obtained by membership function, which makes it become an accurate mathematical field. To solve this problem, Academician Li Deyi put forward cloud model theory [12]. Cloud model combines fuzziness and randomness to transform qualitative concepts and quantitative values.

Academician Li Deyi's cloud model [13] mainly reflects the fuzziness and randomness of things. Qualitative concepts and quantitative data can be transformed into each other through these two uncertainties.

Any element X in the universe U has a number A(x) < [0,1], which corresponds to it. A(x) is called a fuzzy set on U, and A(x) is called the membership degree of X to A.

This concept is extended to cloud models where x A (x), x < U, and order pairs (x, A (x)) are called cloud droplets [11].

A cloud droplet is a qualitative concept. A large number of cloud droplets make up a cloud. The numerical characteristics of clouds are represented by three numerical values: Expected Ex, En and Hyper-En He. Qualitative and quantitative features are usually represented by triples. The digital characteristics of the cloud are shown in Fig. 1.

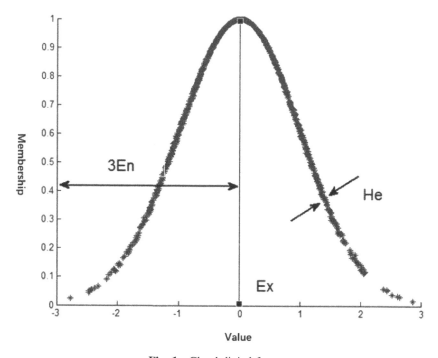

Fig. 1. Cloud digital features

Expectation Ex: The central point of the data is also the average in the mathematical sense.

Entropy En: The measurable granularity that represents a qualitative concept, usually the larger the concept of entropy is, the more macroscopic it is. Entropy also reflects the uncertainties of qualitative concepts, indicating the range of values acceptable to qualitative concepts in the domain space, i.e. ambiguity, which is the measurement of qualitative concepts and the same.

Hyperentropy He: The measure of uncertainty of entropy, which reflects the randomness of samples representing qualitative conceptual values and reveals the correlation between fuzziness and randomness [14].

The formulas for calculating the three digital characteristics of clouds are shown in Formula (2)–Formula (4).

$$Ex = \overline{X} = \frac{1}{N}\sum_{i=1}^{N} x_i \qquad (2)$$

$$En = \sqrt{S^2 - \frac{1}{3}He} \qquad (3)$$

$$He = \sqrt{\frac{\Pi}{2} \times \frac{1}{N}\sum_{i=1}^{N} |x_i - Ex|} \qquad (4)$$

Among them, S2 is the sample variance, expressed as follows:

$$S^2 = \frac{1}{N-1}\sum_{i=1}^{N}(x_i - \overline{X})^2 \qquad (5)$$

3   Service Selection Method Based on Cloud Model Time Series Analysis

The main idea of service selection method combined with time series analysis of cloud model is as follows: Firstly, the user's historical data is divided into initial clustering using clustering algorithm. Each cluster can be regarded as a set of similar users. Users in the same cluster can be regarded as similar users. Each user can be regarded as a "cloud", and each user's historical data as a "cloud droplet". Combined with service data with time series, according to different time periods, the data can be divided into several parts, which are represented by cloud model, and then combined with time decay factor. To the integrated cloud model. Then the optimal service is obtained by comparing with the ideal cloud model obtained by TOPSIS method.

The specific steps of the algorithm are as follows:

Input: User history data, clustering number k, time attenuation factor, time modeling number M.

Output: Predicted top-N recommendation set for target user Ru [S1, S2, S3,... Sn].

Step1: Cluster the user's historical data and eliminate the noise data.

Step2: Set the corresponding threshold and integrate the clustered users into similar user sets.

Step3: According to different time periods, the service data is divided and modeled with cloud model.

Step4: The comprehensive cloud model is obtained by combining the cloud model of each period with the time attenuation factor.

Step5: Using the historical data of similar users, we can get positive and negative ideal cloud A + , A-.

Step6: Calculate the distance from each alternative to positive and negative ideal solutions, and get the top-N recommendation set according to the distance.

Step7: The average absolute error between the calculation and the actual potential user's QoS value is analyzed, and the recommendation accuracy is analyzed.

Select user's historical service data, which is represented by user's response time. The goal is to select data that different users have common service. Although there are many data, there are few data that meet the requirements. Place the selected data in matrix Tu, j, which is sparse. Where UI represents user i, jn represents the nth service, tui, jn represents the response time of user I under N service. Its representation is shown in (6).

$$T_{u,j} = \begin{bmatrix} t_{u1,j1} & \cdots & t_{u1,jn} \\ \vdots & \ddots & \vdots \\ t_{ui,j1} & \cdots & t_{ui,jn} \end{bmatrix} \tag{6}$$

After data pre-processing, the data in Tu and j are clustered by columns using k-means clustering algorithm. To avoid the deviation of recommendation effect and improve the similarity accuracy, a threshold is set. If the calculation distance is larger than the threshold, the threshold is discarded, so as to ensure the accuracy of the data. The data from the end of clustering is stored in the matrix (e.g. 7), where each column represents all the users who have eliminated the noise data under a certain service and have completed the classification. The column is composed of a small Ui matrix, i.e. Ui classes, such as the column vectors shown in (8).

$$H_{u,j} = \begin{bmatrix} B_{U1,j1} & \cdots & B_{U1,jm} \\ \vdots & \ddots & \vdots \\ B_{Ui,j1} & \cdots & B_{Ui,jm} \end{bmatrix} \tag{7}$$

$$B_{Ui,j} = \begin{bmatrix} t_{u1,j}, t_{u2,j}, \ldots, t_{ui,j} \end{bmatrix}^T \tag{8}$$

Set the number of users in the N = Ui matrix and traverse them. After encountering Ui, increase the value of A [Ui] by 1. After traversing the matrix, the final clustering is obtained. Choose user history service data with time nodes and build cloud model in different time periods, then get cloud model C (t) in different time periods. Combining with time attenuation factor, this paper uses formula (9) to combine time attenuation factor and negative exponential function to get the final comprehensive cloud model.

$$C'(t) = C(t) * e^{-\lambda t} \tag{9}$$

Calculate positive and negative ideal clouds. For standardized data, processing formulas such as (10), introducing decision factor wi, get positive and negative ideal clouds A + {a1 + , a2 +,... An +}, A-{a1-, a2-,...} Then calculate the distance from each user to the positive and negative ideal cloud D + {d1 + , d2 +,... Dn +}, D-{d1-, d2-,... Dn-}. The top-N recommendation set [15] is obtained by calculating the relative closeness of D + , D-and Cn.

$$x'_{ij} = \frac{x_{ij}}{\sqrt{\sum_{k=1}^{m} x_{kj}^2}}, i = 1, \ldots, m; \, j = 1, \ldots, n \tag{10}$$

$$a^+ = w_i * x'_{ij} \tag{11}$$

$$C_n = \frac{d_n^-}{d_n^- + d_n^+}, n \in [1, n] \tag{12}$$

By comparing the comprehensive cloud model with the positive and negative ideal cloud, the best service for users is obtained and recommended. The error of the recommended results is calculated, and the average absolute error between the recommended results and the actual potential users' QoS values is calculated, and the recommendation accuracy is analyzed.

4 Experiment

We developed a test program on MATLAB, and compared the effectiveness of this multi-period clustering recommendation method with the experimental data of WS-Dream #2 [16, 17] dataset. The data set describes the results of real QoS assessment from 142 users, providing 4,500 Web services on 64 different time slices, with 15 min interval. Combined with clustering algorithm, similar user sets are partitioned. By calculating the QoS data of some similar users, the cloud model and time attenuation factor are combined to make a comparative analysis.

In this paper, we randomly select several services to cluster the users who have invoked data for each service, and then compare and analyze the calculated data, and extract a group of data from them.

It defines the corresponding time of users of User ID 0 for 12 different services in the first time slice. User ID denotes user ID, representing different users; Service ID denotes service ID, representing different services; Time Slice ID denotes different time slices, totaling 64 time slices, each time slice is 15 min apart; Response Time denotes response time.

In WS-Dream #2 data set, data preprocessing is carried out first, and services involving more users are selected. These data are divided into two parts. One part of the data is selected for user training and the other part is tested. For training data, firstly, the noise data is removed by clustering algorithm, and the similar user set after clustering is obtained. Combined with the service data with time series, the data is divided into several parts according to different time periods, which are represented by cloud model, and then integrated cloud model is obtained by combining time attenuation factor. Then the optimal service is obtained by comparing with the ideal cloud model obtained by TOPSIS method. Finally, the average absolute error MAE (Mean Absolute Error) [18] is used as the measurement index to measure the accuracy of the recommendation algorithm. This method mainly measures the accuracy of the prediction by calculating the deviation between the predicted user rating and the actual user rating. Assuming that the user's predictive score set is {s1, s2,, sN} and the corresponding actual score set is {S1, S2, SN}, the MAE can be calculated by the following expression:

$$MAE = \frac{\sum_{i=1}^{N}|s_i - S_i|}{N} \tag{11}$$

Figure 2 is a multi group data service to predict differences for the number of clustering after MAE value graph set 2,3,4,5,7,9 initial clustering number simulation. It can be seen that with the increase of the number of clusters, the accuracy of the final recommendation first increased and then decreased, when the selection of cluster number is 4, the best effect of the highest precision. Therefore, proper number of clustering, is beneficial to improve the accuracy of recommendation later.

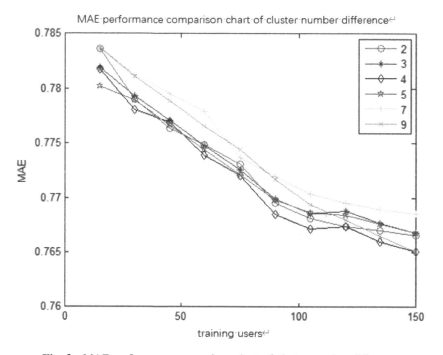

Fig. 2. MAE performance comparison chart of cluster number difference

Time-segment modeling is to introduce time factors into the model. By comparing the number of different segments, it can show the recommendation effect of the proposed algorithm in different time periods. Figure 3 is a MAE performance comparison chart of the recommended results under different segments. 2,4,8,16 segments are selected to simulate. From the graph, it can be seen that when the time period is divided into 8 segments, the recommendation effect is better.

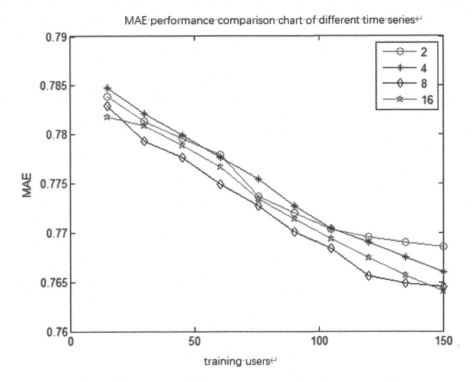

Fig. 3. MAE performance comparison chart of different time series

Based on the above two experiments, it can be seen that when the number of clusters is 4 and the time period is divided into eight segments, the effect is the best. In this scenario, compared with other recommended models based on clustering algorithm, the traditional collaborative filtering algorithm CF, the FCM proposed in document [5], and the IF-CF proposed in document [7], the results are fed back to Fig. 4.

As can be seen from the figure above, this paper takes the user's time factor into account in the recommendation process, and combines the time attenuation factor. Compared with other algorithms, the overall performance of several mainstream algorithms improves with the increase of the number of users. When the number of users is small, the recommendation accuracy of this algorithm is not ideal. It may be that the amount of user data is small, the number of users in each cluster is small, and the deviation from users is large. In the range of 50–100 neighbor users, several algorithms have the same effect. With the increase of the number of users, the recommendation accuracy of this algorithm has a slight advantage. The goal of this paper is to improve the accuracy of recommendation by effectively introducing time factors.

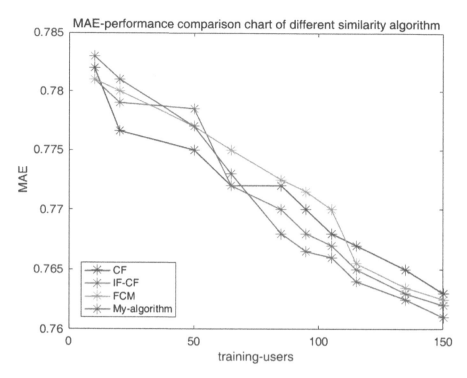

Fig. 4. MAE performance comparison chart of different similarity algorithm

5 Conclusion

Collaborative filtering algorithms recommend items to active users based on past records of neighbor users. Because the collaborative filtering algorithm cannot accurately find similar neighbors when user-provided scores are very sparse, the prediction accuracy in user's project preferences will be very low. At the same time, many collaborative filtering algorithms do not consider the user's own feature attributes or only consider single feature attributes, but also ignore the time-based factors, resulting in poor recommendation quality. In this paper, a service selection method based on time series analysis of cloud model is proposed, which combines clustering algorithm and multi-period user difference characteristics. Experiments show that this method improves the accuracy of recommendation. By using clustering algorithm, the steps of calculating similar users are omitted and the amount of calculation is reduced. However, the use of single feature attributes in this paper cannot well characterize the personalized characteristics of users. Therefore, in the future work, we will try to introduce multi-attribute user feature data, and better integrate user characteristics and interests into it.

Acknowledgment. The project is provided with Hunan Provincial Department of Education Outstanding Scientific Research Youth Project "Research on Weibo Information Mining and Precision Recommendation Technology" Fund Number: 18B570.

References

1. Artemenko, O., Pasichnyk, V., Korz, H., et al.: Using big data in e-tourism mobile recommender systems: a project approach. In: ITPM 2020, pp. 194–204 (2020)
2. Cai, Z., Wang, J., He, K.: Adaptive density-based spatial clustering for massive data analysis. IEEE Access **8**, 23346–23358 (2020)
3. Shen, J., Zhou, T., Chen, L.: Collaborative filtering-based recommendation system for big data. IJCSE **21**(2), 219–225 (2020)
4. Frémal, S., Lecron, F.: Weighting strategies for a recommender system using item clustering based on genres. Expert Syst. Appl. **77**, 105–113 (2017)
5. Koohi, H., Kiani, K.: User based collaborative filtering using fuzzy c-means. Measurement **91**, 134–139 (2016)
6. Guo, G., Zhang, J., Yorke-Smith, N.: Leveraging multiviews of trust and similarity to enhance clustering-based recommender systems. Knowl. Based Syst. **74**(1), 14–27 (2015)
7. Chen, J., Wang, X., Zhao, S., Qian, F., Zhang, Y.: Deep attention user-based collaborative filtering for recommendation. Neurocomputing **383**, 57–68 (2020)
8. Talasbek, A., Serek, A., Zhaparov, M., Yoo, S.M., et al.: Personality classification by applying k-means clustering. In: ICAIIC 2020, pp. 421–426 (2020)
9. Taoying Li, X., Zhang, J.: Time series clustering model based on DTW for classifying car parks. Algorithms **13**(3), 57 (2020)
10. Rehioui, H., Idrissi, A.: New clustering algorithms for twitter sentiment analysis. IEEE Syst. J. **14**(1), 530–537 (2020)
11. Yan, G., Jia, S., Ding, J., Xu, X., Pang, Y.: A time series forecasting based on cloud model similarity measurement. Soft. Comput. **23**(14), 5443–5454 (2018). https://doi.org/10.1007/s00500-018-3190-1
12. Deyi, L.: Uncertainty in knowledge representation. Eng. Sci. **10**, 73–79 (2000)
13. Deyi, L., Changyu, L.: Universality of normal cloud model. Eng. Sci. **08**, 28–34 (2004)
14. Li, D.Y., Liu, C.Y., Gan, W.Y.: A new cognitive model: cloud model. Int. J. Intell. Syst. **3**(24), 357–375 (2009)
15. Krohling, R.A., Pacheco, A.G.C.: A-TOPSIS – an approach based on TOPSIS for ranking evolutionary algorithms. Procedia Comput. Sci. **55**, 308–317 (2015)
16. Zheng, Z., Zhang, Y., Lyu, M.R., et al.: Investigating QoS of real-world web services. IEEE Trans. Serv. Comput. **7**(1), 32–39 (2014)
17. Zibin, Z., Yilei, Z., Michael, R.: Distributed QoS evaluation for real-world web services. In: Proceedings of the 8th International Conference on Web Services (ICWS 2010), Miami, Florida, USA, 5–10 July, pp. 83–90 (2010)
18. Guisselle, A., Garcia, L., Rakesh, N.: Network and QoS-based selection of complementary services. IEEE Trans. Serv. Comput. **9**(1), 79–91 (2015)

Cloud Resource Allocation Based on Deep Q-Learning Network

Zuocong Chen$^{(\boxtimes)}$

College of Computer Science and Technology, Hainan Tropical Ocean
University, Sanya 572022, China
Twsf2005@163.com

Abstract. The goal of resource allocation is to allocate the optimal resource to the candidate tasks, so that all the tasks can be finished in less time and the users' demands can be satisfied. To have better performance on the time span, CPU usage ratio and the load balance compared with existed methods, it proposes an allocation method that can map the tasks to the resources effectively, where an optimal allocation program will be generated. Firstly, the resource allocation model for tasks was proposed and the goal function was designed. Afterward, the deep Q-learning algorithm was defined to get an optimal allocation program, and the algorithm was analyzed in detail. The experiment was implemented to verify the proposed method. The simulation experiments prove that the method in this paper can effectively implement task scheduling, which has the advantages of high CPU utilization, short scheduling time and strong load balancing ability.

Keywords: Deep Q learning · Resource allocation · Reinforcement learning · Cloud computing

1 Introduction

Cloud computing (Cloud Computing) is a computing model that uses computing resources as a service and then provides them to users through the network, so that users can use computing resources on demand [1]. These resources mainly include data, software, hardware, and bandwidth. The services provided by cloud computing mainly include three categories: software as a service, platform as a service, and infrastructure as a service. Resources in cloud computing have the characteristics of geographical dispersion, heterogeneous resources, and dynamic changes [2]. Therefore, it is necessary to comprehensively consider these heterogeneous communication and computing resources when making offloading decisions for users' computing tasks, and optimize resource scheduling to ensure the user experience.

In recent years, some research work on tasks for resource allocation in cloud computing environment. There are many research results on offloading strategies and resource scheduling issues. Some studies have proposed a distributed task off-loading scheme based on game theory, which optimizes energy consumption and maximizes the game among users.

© Springer Nature Singapore Pte Ltd. 2020
J. Zeng et al. (Eds.): ICPCSEE 2020, CCIS 1257, pp. 666–675, 2020.
https://doi.org/10.1007/978-981-15-7981-3_49

Resource scheduling is a core issue of cloud computing, which is directly related to the stability of cloud computing services, the efficiency of resource use, and user satisfaction. The research of cloud computing resource scheduling has very important meaning from the theory and technology itself. There have been many researches about how to achieve resource allocation in the cloudy environment.

Li et al. [3] apply the successive approximation method to study the generalization-form model and present a gradient-based resource allocation scheme to solve the approximation problems, to solve the problems of non-strict convexity and non-separation in generalization-form optimization problem. Jin et al. [4] propose an incentive-compatible auction mechanism (ICAM) for the resource trading between the mobile devices as service users (buyers) and cloudlets as service providers (sellers), so that an auction mechanism that holds certain desirable properties for the cloudlet scenario can be designed. Zhang et al. [5] combined energy-efficient user scheduling and power allocation schemes for the downlink NOMA heterogeneous network for perfect and imperfect CSI to study the trade-off between data rate performance and energy consumption in NOMA. Cánovas et al. [6] advocate an intelligent media distribution system architecture for delivering video streaming, where the network parameters such as bandwidth, jitter, delay and packet loss that influence the QoE of the end-users and the other parameters of the energy consumption such as CPU, RAM, temperature and number connected users that impact the result of the QoE are considered. Rezaee et al. [7] propose three quality management resources (human, organizational and technological) and eight different strategies related to quality, to explore the relationship between effective strategies and improve the quality and quality management of allocated resources for the successful implementation of the strategies. Lei et al. [8] propose a heterogeneous resource allocation approach, called skewness-avoidance multi-resource allocation (SAMR), to allocate resource according to diversified requirements on different types of resources, so alieving the problem that the existing homogeneous resource allocation mechanisms cause resource starvation. Ghribi et al. [9] mapped the allocation problem to the combination of an optimal allocation algorithm with a consolidation algorithm relying on migration of VMs at service departure, where the optimal allocation algorithm is solved as a problem with a minimum power consumption objective.

To have a better performance on the time span, CPU usage ratio and the load balance, we proposed an allocation method that can map the tasks to the resources effectively by deep learning, where an optimal allocation program can be generated. The combination of deep learning and resource allocation can provide a vivid prospect to improve the efficiency and generality.

2   The Goal for Resource Allocation

The goal of resource allocation is to allocate the optimal node to the task, so that all the tasks can be finished in less time. The factors considered in resource allocation in cloud computing mainly include the estimated execution time $time\_\cos t(r)$, the network bandwidth $band\_width(r)$ and the network delay $delay(r)$.

The estimated execution time $time\_\cos t(r)$ defined as the earliest finished time for t_i at the node tn_j, and it is denoted as Eq. (1)

$$time\_\cos t(r) = ECT_{ij} + start(tn_j) \tag{1}$$

where $start(tn_j)$ is represented as the earliest available time for the node, ECT_{ij} is the required execution time of task t_i on node tn_j.

The network bandwidth $band\_width(r)$ is the maximum bandwidth provided by the route r, and the maximum network delay generate in the route r can be denoted as $delay(r)$.

By combining the above, the goal function of the resource allocation can be denoted as Eq. (2).

$$F = \min \frac{atime\_\cos t(r) + bdelay(r)}{cband\_width(r)} \tag{2}$$

Satisfying:

$$\begin{cases} time\_\cos t(r) < TLT \\ band\_width(r) < ELT \\ delay(r) < DLT \end{cases} \tag{3}$$

where a, b and c are the weights for $time\_\cos t(r)$, $band\_width(r)$ and $delay(r)$ with the corresponding boundaries for them are TLT, ELT and DLT.

3   Allocation Algorithm Based on Improved Deep Q-Learning Network

3.1   MDP Model

Reinforcement learning [10] is an unsupervised machine learning method. It can get optimal policies by maximizing the long-term return. The interaction between agent and the environment in reinforcement learning can be denoted a Markov process. The MDP can be defined a four-tuple, namely, $M = \;<X,\;U,\;f,\;u>$, where $X \in \mathbb{R}^k$ is the state space, $U \in \mathbb{R}^n$ denotes the action space, $f : X \times U \times X \to [0,\;+\infty]$ is defined as the state transition function, $r : X \times U \times X \to R$ is the reward function. At the time t, the state is denoted as x_t, according to the environment dynamics $f(x_t,\;u_t,\;x_{t+1})$, the immediate reward is $r(x_t,\;u_t,\;x_{t+1})$. The Agent attempt to find the optimal policy though maximizing the accumulative rewards by maximizing the expected rewards.

Policy $h : X \to U$ is the mapping from the state space X to the action space U, where the mathematical set of h depends on specific domains. The goal of the agent is to find the optimal policy h^* that can maximize the cumulative rewards. The cumulative rewards are the sum or discounted sum of the received rewards, and here, we use the latter case.

Under the policy h, the value function $V^h : X \rightarrow \mathbb{R}$ denotes the expected cumulative rewards, which is shown as:

$$V^h(x) = \mathrm{E}_h \left\{ \sum_{k=0}^{\infty} \gamma^k r_{t+k+1} | x = x_t \right\}, \tag{4}$$

where $\gamma \in [0, 1]$ represents the discount factor and x_t is the current state.

The optimal state-value function $V^*(x)$ is computed as:

$$V^*(x) = \max_h V(x), \forall x \in X. \tag{5}$$

Therefore, the optimal policy h^* in state x can be obtained by

$$h^*(x) = \arg\max_h V^h(x), \forall x \in X. \tag{6}$$

3.2   Deep Q-Learning Network

Q-Learning is a temporal difference algorithm proposed by Watkins in 1989. It is characteristic of the different of the applied policy in exploration and in evaluation. The exploration policy generally adopts $\varepsilon - greedy$, while the evaluation takes the *greedy* policy, the update equation for action value function can be denoted as:

$$Q(x_t) = Q(x_t) + \alpha(r_{t+1} + \gamma \max_{u_{t+1}} Q(x_{t+1}) - Q(x_t)) \tag{7}$$

The deep Q-Learning network (DQN) use the deep neural network to approximate the action value function. Directing using the samples collected from on-line are sequential to learn the value function will make the learning process unsteady. Therefore, the samples are firstly stored in an experience pool, and then randomly sample it in the learning process. There are also two networks, the one is the value function network and the other is the goal value function network. The goal value function network can be a copy from the value function network, and it is changed after several batches of samples are learned in the learning process.

In the cloud environment, there is no determined goal terminal node. Therefore, we need to get an episode of samples, where the number of the numbers is equal to the tasks. The immediate task in the allocation process will not get any reward until the final task. Therefore, the former allocated samples will get a reward of 0 and the final task will get a reward shown as Eq. (2).

At every time step, the agent tries to select the optimal action to execute, but it also with some probability to explore the action space. Namely, the candidate node. We take the $\varepsilon - greedy$ exploration policy, there the exploration probability of ε is set as the following equation:

$$\varepsilon = \varepsilon_0 * \left(1 - \left(\frac{i}{n}\right)^2\right) \tag{8}$$

where the initial value of ε is set to 0.9, n is the sum of the iterations and i is the current iteration.

The allocation process of task to the node by using improve DQN can be describes as:

(1) Initialize the action space and state space. The state space X is the history allocated tasks, and the action space U is the set of available nodes. The current episode is set to $t = 0$, and the number of the episodes is T;

(2) Initialize the action value network and the goal action value network. For the state-action pair $\forall (x, u) \in X \times U$ in the state action space, the state action value for any state action pair can be initialized as $Q_{0e}(x, u) = 0$, and the goal action value function is set as $Q_{0g}(x, u) = Q_{0e}(x, u) = 0$;

(3) Set the current state as $x = t_0$, x will be the first task in the task queue $T = t_0, t_1, \ldots t_n$;

(4) According to $\varepsilon - greedy$ exploration policy, the optimal action $a \in \{node_1, \ldots node_m\}$ will be selected as the probability $1 - \varepsilon$, and the other actions will be executed as the exploration probability ε;

(5) If the current task is the final task in the task queue, the reward r will be computed according to Eq. (2), else the reward will be $r = 0$. The next state is represented as $x = t_0 t_1$;

(6) Construct the current sample $(t_0, node_0, t_0 t_1, r)$;

(7) Update the value of ε according to $\varepsilon = \varepsilon_0 * \left(1 - \left(\frac{i}{n}\right)^2\right)$

(8) Transfer to the step (4), until the current state $x = t_0 t_1, \ldots, t_n$ and the number of the episodes equal to T;

(9) Put all the episodes to the experience pool.

(10) Randomly sample an episode e from the experience pool. From the final state $x = t_0 t_1, \ldots, t_n$ to the initial state of this episode e, all the samples are used to update the action value network. The action value network for evaluation can be updated as:

$$Q_e(x, u) = Q_e(x, u) + \alpha(r + \gamma \max_{u'} Q_g(x', u') - Q_e(x, u)) \tag{9}$$

(11) The network of goal action value network is copied form the evaluation action value network $Q_g(x, u) = Q_e(x, u)$

(12) Transfer to the step (9) until all the episodes are iterated.

After the follow algorithm has been implemented, the action value function can be learned. The action to execute at each time step can be shown as:

$$h^*(x) = \arg\max_a Q^h(x, a), \forall x \in X. \tag{10}$$

Therefore, the whole algorithm is defined:

Algorithm 1: DQN-based method for resource allocation

Initialize: $t=0$, $\forall (x, u) \in X \times U$, $Q_{0e}(x, u) = 0$,

$Q_{0g}(x, u) = Q_{0e}(x, u) = 0$

Input: X, U, T

Step1: Repeat T times:

Step2:   set $x=t_0$, $T=t_0, t_1, ...t_n$;

Step3:   select $a \in \{node_1, ...node_m\}$ with $1-\varepsilon$, and the others with $1-\varepsilon$;

Step4:   **if** current task is the last:

compute reward $F = \min \dfrac{atime\_cost(r) + bdelay(r)}{cband\_width(r)}$

     else:

reward=0

Step5:   update the next state $x=t_0 t_1$

Step6:   update the current sample: $t_0, node_0, t_0 t_1, r$ to pool;

Step7:   update $\varepsilon = \varepsilon_0 * (1 - (\dfrac{i}{n})^2)$

Step8:   update $Q_e(x, u) = Q_e(x, u) + \alpha(r + \gamma \max_{u'} Q_g(x', u') - Q_e(x, u))$;

Step9:   update the goal network $Q_g(x, u) = Q_e(x, u)$

Step10:end repeat

Output: obtain the optimal action $h^*(x) = \arg\max_a Q^h(x, a), \forall x \in X.$

---

4   Experiment Result

4.1   Experiment and Parameter Setting

The grid computing simulation tool Cloudsim is used to simulate the proposed method. There is three node clusters in the cloud environment are Cluster1, Cluster2 and Cluster

3. The number of initial nodes in the three clusters are 10, 12 and 15. The number of CPU is of 4. The numbers of the resources in three clusters are 40, 48and 60. The numbers of the users are 10, 8 and 9, and the corresponding tasks are 100, 300, 500.

The setting of the parameters are set as the exploration factor $\varepsilon = 0.01$, the discount factor is $\gamma = 0.8$, the learning rate $\alpha = 0.1$, the maximum number of episodes is T.

We will compare with the representative methods, such as Literature [8] and Literature [9] in the time for task scheduling, CPU usage ratio and the load balance.

4.2   The Time for Task Scheduling

The time span for executing all the tasks are simulated, and the compared result of the time for task scheduling is shown as Fig. 1.

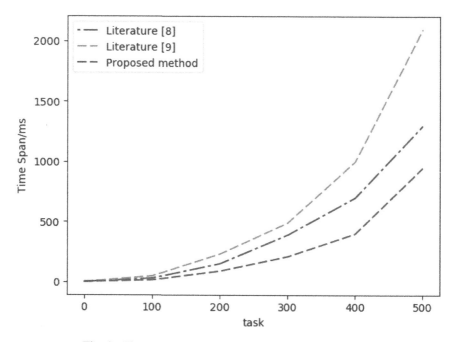

Fig. 1. The compared result of the time for task scheduling

It is easy to see from Fig. 1, the time span for the three methods are increasing in the whole simulation. However, the method of Literature [9] requires most time to execute all the tasks. Our method get the best performance in the whole experiment. The time span for Literature [8], Literature [9] and our method are about 1300 ms, 2100 ms and 850 ms, respectively. The deep reinforcement learning method can optimize the allocation program, so that it can get the far better performance on time span.

4.3   CPU Usage Ratio

The CPU usage ratio is a very import index to evaluate the reasonability of the resource allocation program. The results of the three methods after allocating all the tasks are shown in Fig. 2.

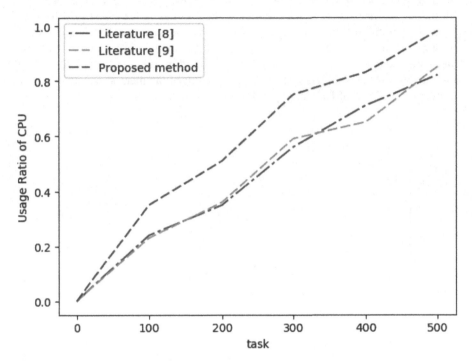

Fig. 2. Comparison of CPU usage ratio

From Fig. 2, we can evidently see that the CPU usage ratio for Literature [9] and Literature [10] are almost the same in the former 200 tasks. Afterward, the method in Literature [8] behaves a litter better than that of Literature [9]. But after 450 tasks are executed, the method in Literature [9] outperforms that of Literature [8] again. However, the CPU usage ratio is still higher than of those in Literature [8] and Literature [9], with a value about 0.98 after all tasks are finished.

4.4   Load Balance

The load balancing dispersion reflects the load balancing degree of the system, and its value can be calculated by the following formula:

$$\phi = \sqrt{\frac{\sum_{j=1}^{m}(LB_j - \bar{LB}_j)^2}{m-1}} \tag{11}$$

where LB_j is the load balancing factor, and its value is the number of the tasks allocated to the node j, and $\overline{LB}_j$ is the average value of load balancing factor. The simulated result of the load balancing dispersion is shown as:

Though the load balancing dispersion obtained from three methods are increasing with the tasks, but it is clear that the prosed method has the smallest load balancing dispersion with the increase of the tasks. The method in Literature [8] is higher that of Literature [9] in the former 140 tasks. Afterward, the method in Literature [8] behave better until 350 tasks have been executed. The method in Literature [8] performs better between 350 tasks and 470 tasks. However, the result of the method in Literature [9] is still outperforms that of Literature [8] (Fig. 3).

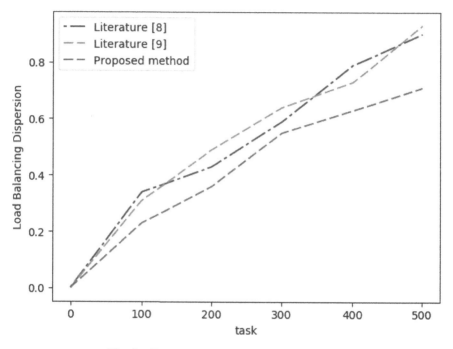

Fig. 3. Comparison of load balancing dispersion

5   Conclusion

Aiming at the shortcomings of long scheduling time and low CPU utilization of traditional scheduling algorithms in complex cloud environments, a cloud resource allocation algorithm based on deep reinforcement learning is proposed. Firstly, the resource allocation model for tasks is proposed. Afterward, the deep Q-learning algorithm is designed to get an optimal allocation program, and the algorithm is analyzed. Simulation experiments prove that the method in this paper can effectively implement task scheduling, which has the advantages of high CPU utilization, short scheduling time and strong load balancing ability, and strong feasibility.

Acknowledgments. This work was financially supported by Hainan Provincial Department of Science and Technology under Grant No. ZDKJ2016021.

References

1. Chen, X., Lin, J., Ma, Y., et al.: Self-adaptive resource allocation for cloud-based software services based on progressive QoS prediction model. Sci. China Inf. Sci. **62**, 1–3 (2019)
2. Sharma, V., Choudhary, G., You, I., et al.: Self-enforcing game theory-based resource allocation for LoRaWAN assisted public safety communications. J. Internet Technol. **19**, 515–530 (2018)
3. Li, S., Sun, W.: Utility maximization for resource allocation of migrating enterprise applications into the cloud. Enterp. Inf. Syst., 1–33 (2020)
4. Jin, A., Song, W., Zhuang, W.: Auction-based resource allocation for sharing cloudlets in mobile cloud computing. IEEE Trans. Emerg. Top. Comput. **6**, 45–57 (2018)
5. Zhang, H., Fang, F., Cheng, J.: Energy-efficient resource allocation in NOMA heterogeneous networks. IEEE Wireless Commun. **25**, 48–53 (2018)
6. Cánovas, A., Taha, M., Lloret, J., et al.: Smart resource allocation for improving QoE in IP Multimedia Subsystems. J. Netw. Comput. Appl. **104**, 107–116 (2018)
7. Rezaee, E., Pooya, A.: Resource allocation to strategies of quality management with FANP and Goal Programming approach. TQM J. **31**, 850–870 (2019)
8. Wei, L., Foh, C.H., He, B., et al.: Towards efficient resource allocation for heterogeneous workloads in IaaS clouds. IEEE Trans. Cloud Comput. **6**, 264–275 (2018)
9. Ghribi, C., Hadji, M., Zeghlache, D.: Energy efficient VM scheduling for cloud data centers: exact allocation and migration algorithms. In: IEEE/ACM International Symposium on Cluster, Cloud & Grid Computing. ACM (2013)
10. Zhong, S., Liu, Q., Zhang, Z., Fu, Q.: Efficient reinforcement learning in continuous state and action spaces with Dyna and policy approximation. Front. Comput. Sci. **13**(1), 106–126 (2019). https://doi.org/10.1007/s11704-017-6222-6
11. Chen, Z.: Method for extraction and fusion based on KL measure. In: Cheng, X., Jing, W., Song, X., Lu, Z. (eds.) ICPCSEE 2019. CCIS, vol. 1058, pp. 42–51. Springer, Singapore (2019). https://doi.org/10.1007/978-981-15-0118-0_4
12. Yang, L., Wen, K., Gao, Q., et al.: SVM based multi-label learning with missing labels forimage annotation. Pattern Recogn. **78**, 307–317 (2018)
13. Chen, Z.-c.: Task scheduling algorithm based on campus cloud platform. In: Sun, X., Pan, Z., Bertino, E. (eds.) ICAIS 2019. LNCS, vol. 11633, pp. 299–308. Springer, Cham (2019). https://doi.org/10.1007/978-3-030-24265-7_26
14. Cheng, J., Xu, R., Tang, X., et al.: An abnormal network flow feature sequence prediction approach for DDoS attacks detection in big data environment. Comput. Mater. Continua **55**, 95–119 (2018)
15. Ebrahimi, M.A., Khoshtaghaza, M.H., Minaei, S., et al.: Vision-based pest detection based on SVM classification method. Comput. Electron. Agric. **137**, 52–58 (2017)

Application of Polar Code-Based Scheme in Cloud Secure Storage

Zhe Li, Yiliang Han$^{(\boxtimes)}$, and Yu Li

College of Cryptographic Engineering, Engineering University of PAP,
Xi'an 710086, Shaanxi, China
hanyil@163.com

Abstract. In view of the fact that the quantum computer attack is not considered in the cloud storage environment, this paper selects the code-based public key encryption scheme as the security protection measure in the cloud storage. Based on random linear code encryption scheme, it employs the structure of the RLCE scheme and Polar code polarization properties, using the Polar code as underlying encoding scheme, through the method of RLCEspad, putting forward a kind of improved public key encryption scheme which considers semantic security and is resistant to adaptively chosen ciphertext attacks. The improved scheme is applied to cloud storage to ensure that the storage environment will not be attacked by quantum computer while ensuring the confidentiality, availability and reliability.

Keywords: Post-quantum cryptography · Polar code · Random linear code encryption · Code-based scheme · Cloud storage

1 Introduction

At the beginning of the New Year in 2020, COVID-19 outbreak broke out in Wuhan. China is experiencing a major public health emergency with the fastest transmission rate, the widest infection range and the greatest difficulty in prevention and control since the founding of new China. While the whole country is doing its best to fight the epidemic, foreign hostile forces and domestic criminals have not stopped cyber-attacks and sabotage. According to relevant reports, some overseas hacker groups have taken advantage of the situation to carry out network penetration attacks on key areas of China with pneumonia and other relevant information [1]. We will integrate cyber security with political security, military security, economic security and biological security, systematically plan the development of a cybersecurity risk prevention and control and governance system, and comprehensively improve the country's ability to comprehensively manage cyber society. Cryptography will play a crucial role.

Cloud storage system should ensure availability (for, any legal customers can access the uploaded data from some networking equipment), reliability (outsource to the cloud user data) and efficient retrieval (outsource to the cloud user data), data sharing (between authorized users), security (including confidentiality and integrity), and other functional requirements/regulations [2].

© Springer Nature Singapore Pte Ltd. 2020
J. Zeng et al. (Eds.): ICPCSEE 2020, CCIS 1257, pp. 676–689, 2020.
https://doi.org/10.1007/978-981-15-7981-3_50

The rapid development of quantum computer technology has a profound impact on the present and future society. At present, 5G communication, transportation, cloud storage, public opinion control and other fields are closely related to cryptography. Key cryptography techniques used in various fields are still relatively safe until the arrival of quantum computers. The proposed Shor algorithm and Grover [3] algorithm pose severe challenges to the security of classical cryptography schemes.

In order to deal with the threat of quantum computer to cryptography schemes based on classical mathematical difficulties, countries all over the world are seeking new cryptography schemes that can resist the attack of quantum computer, namely Post Quantum Cryptography [4]. The National Institute of Standards and Technology (NIST) of the United States launched the standardization project of post quantum algorithm in 2012, and the post quantum cryptography was recruited globally in 2016, and the second round of selection was completed in 2019 [5]. In the time when countries formulate post quantum cryptography algorithm standards, China has also promoted post quantum cryptography design competition and formulated post quantum cryptography standards. So far, post quantum cryptography standardization has entered the second round of selection [6]. Post quantum cryptography schemes mainly include 5 kinds [7], each of which has its unique application range. At present, code- based scheme and lattice - based scheme [8] have become the focus of research. Compared with other post quantum cryptography schemes, the code-based cryptography scheme is more suitable for the construction of encryption schemes. Among the post quantum cryptography, the code-based encryption scheme has a good research prospect.

The code-based cryptography scheme can be traced back to the McEliece scheme in 1978 [9]. Up to now, the scheme is still safe under the appropriate parameter selection, and it has entered the second round of PQC collection algorithm. Because the security of the Goppa code-based McEliece scheme is based on the decoding difficulty of general linear code, it can be specified to the NPC problem and has the characteristics of resisting the attack of quantum computer, many experts began to focus on this scheme. In 1986, Niederreiter [10] constructed the Niederreiter scheme using the dual form of the McEliece scheme. The security of the two schemes is equivalent. Both the McEliece scheme and Niederreiter scheme have the shortcoming of too large key size, which is difficult to be applied in the actual scenario. In view of the characteristics of poor utility of McEliece schemes, experts used other codes with more compact structures as the underlying codes to reduce the key size, and the constructed deformation schemes had security problems and were vulnerable to structural attacks [11].

In 2016, Wang et al. [12] constructed a random linear code encryption scheme, namely RLCE public key encryption scheme, by inserting random columns in each column of the generated matrix, which was selected by the first round of PQC collection algorithm. Unlike other cryptography schemes that make use of compact codes, this scheme does not depend on the structure of any underlying code. The RLCE scheme achieves randomness by inserting random columns, and the security of the scheme depends on the NPC problem of linear random code decoding, which avoids the structural attack introduced by the underlying coding structure. In 2017, Wang [13] further studied the padding method of RLCE scheme, aiming at the potential attack of RLCE scheme. The improved RLCE scheme reduced the key size, improved encryption and decryption performance, and the IND-CCA2 (adaptive selective ciphertext attack) security of the padding scheme was achieved. Matthews [14] at the 2019 CBC (code-based Cryptography) conference, it was proposed to use Hermitian code as the underlying code of RLCE scheme. At the 2019 A2C (Algebra, Codes and Cryptology) conference, Liu et al. [15] proposed to use Polar code as the underlying code of RLCE scheme. By taking advantage of the polarization nature of Polar code, the encryption and decryption complexity was reduced. The scheme proposed by Liu in literature [15] does not consider semantic security and is vulnerable to IND-CCA2.

Based on RLCE scheme and PolarRLCE scheme, this paper proposes an IND-CCA2 secure RLCE Public key encryption scheme based on Polar code, which is based on Polar code polarization property and improves key storage method. And the improved scheme will be applied to the cloud storage.

2   Basic Knowledge

2.1   Relevant Definitions

Definition 1. A Binary Input Discrete Memoryless Channel [16].
It can be expressed as $W: X \rightarrow Y$, X is the set of input symbols, Y is the set of output symbols, and the transition probability is $W(Y|X), x \in X, y \in Y$. For channel W, the channel after N times polarization can be expressed as W^N, then the transition probability of channel W^N is: $X^N \rightarrow Y^N$

$$W^N\left(y_1^N | x_1^N\right) = \prod_{i=1}^{N} W(y|x) \tag{1}$$

For a B-DMC, there are two important channel capacity parameters:
Symmetric Capacity:

$$I(W) \triangleq \sum_{y \in Y} \sum_{x \in X} \frac{1}{2} W(y|x) \log \frac{W(y|x)}{\frac{1}{2}W(y|0) + \frac{1}{2}W(y|1)} \tag{2}$$

Bhattacharyya Parameter:

$$Z(W) \triangleq \sum_{y \in Y} \sqrt{W(y|0)W(y|1)} \tag{3}$$

$I(W)$ is a measure of channel rate and $Z(W)$ is a measure of channel reliability. Under the condition of equal probability input, the maximum rate of channel W in reliable transmission is $I(W)$. In the case that channel W only transmits 0 or 1, the upper limit of maximum likelihood judgment error probability is $Z(W)$. The value range of $I(W)$ and $Z(W)$ is both $[0, 1]$. $I(W)$ and $Z(W)$ satisfy the following relationship: if and only if $Z(W) \approx 0, I(W) \approx 1$; If and only if $Z(W) \approx 1, I(W) \approx 0$.

Definition 2. Channel Polarization.

Channel polarization is divided into two stages: Channel combination and Channel Splitting.

(1) Channel Combination

Combine N independent channels of B-DMC W and generate a vector channel W^N through recursion: $X^N \rightarrow Y^N$, where N is the power of 2, $N = 2^n, N \geq 0$. $u_1^N \rightarrow x_1^N$ is the mapping from the input of the complex channel W_N to the input of the original channel W^N. So, get $u_1^N = x_1^N G_N . u_1^N$ is the original bit sequence, x_1^N is the encoded bit sequence, G^N is a N dimension generated matrix, code length is $N = 2^n$.

The transition probability of channel W_N and W^N has the following relationship:

$$W_N\left(y_1^N | u_1^N\right) = W^N\left(y_1^N | u_1^N G_N\right), y_1^N \in Y^N, u_1^N \in X^N. \tag{4}$$

(2) Channel Splitting

The complex channel W^N formed by the combination of channels splits into N coordinate channels of binary input. $W_N^{(i)} : X \rightarrow Y^N \times X^{i-1}, 1 \leq i \leq N$,

$$W_N^{(i)}\left(y_1^N, u_1^{i-1} | u_i\right) \triangleq \sum_{u_{i+1}^N \in X^{N-i}} \frac{1}{2^{N-1}} W_N\left(y_1^N | u_1^N\right) \tag{5}$$

Definition 3. Polarization Coding Principle.

The basic idea of polarization coding is to send data bits only on the coordinate channel $W_N^{(i)}$, where $Z\left(W_N^{(i)}\right)$ approaches 0 $(I\left(W_N^{(i)}\right)$ approaches 1).

2.2 RLCE Encryption Scheme

Table 1, 2 and 3.

Table 1. RLCE key generation

RLCE Key generation

Input: (n, k, d, t, w), $n, k, d, t > 0$, $w \in \{1, 2 \cdots n\}$, $k + 1 \geq d \geq 2t + 1$.
Output: G^{pub}, (S, G_1, P, A).
1. G: $k \times n$ order generator matrix of code C whose dimension on domain F is k.
2. Generate w random column vectors $r_1, r_2, \cdots, r_w$, and by inserting w random $k \times 1$ column vectors into generator matrix G, obtained the $k \times (n + w)$ matrix G_1,
$G_1 = (g_1, \cdots, g_{n-w}, g_{n-w+1}, r_1, \cdots g_n, r_w)$.
3. To mix the columns, choose w random non-singular binary 2×2 non-singular matrices $A_1, A_2, \cdots, A_W$.
4. Denote $A = [1, \cdots 1, A_1, A_2, \cdots, A_W]$ the $(n + w) \times (n + w)$ nonsingular invertible matrix.
5. Let S be a randomly chosen $k \times k$ non-singular matrix, P be the $(n + w) \times (n + w)$ permutation matrix.
6. Output $k \times (n + w)$ public key $G^{pub} = SG_1AP$, Private key (S, G_1, A, P).

Table 2. RLCE encryption

RLCE encryption

Input: Public key G^{pub}, message $m \in F_2^k$, error vector $e \in F_2^{n+w}$.
Output: Ciphertext $c \in F_2^{n+w}$.
1. $c = mG^{pub} \oplus e$, $w_H(e) \leq t$.

Table 3. RLCE decryption

RLCE decryption

Input: Ciphertext $c \in F_2^{n+w}$, Private key (S, G_1, A, P).
Output: Message $m \in F_2^k$, or decoding error identification $\perp$.
1. $cP^{-1}A^{-1} = mSG_1 \oplus eP^{-1}A^{-1} = (c_1', c_2', \cdots c_{n+w}')$.
2. From length of $n + w$ vector $cP^{-1}A^{-1}$ delete w column vectors, get length of n
$c' = (c_1', c_2', \cdots, c_{n-w+1}', c_{n-w+3}', c_{n-w+5}', \cdots, c_{n+w-1}')$.
3. $c' = mSG_1 \oplus e'$, $e' \in F_2^n$, $w_H(e') \leq t$.
4. Using a decoding algorithm, to calculate $m' = mS$, $m = m'S^{-1}$.
5. Calculate the hamming weight $w = wt(c - mG_1)$, if $w \leq t$, output message m; else, output $\perp$.

2.3   Message Encoding

In order to resist the known attacks of code-based schemes, the security of the code-based schemes can be guaranteed by means of message encoding (message padding). In general, there are three ways to encode messages in ciphertext [13]:

(1)   basicEncoding: Encode information within the vector $m \in GF_2^k$ and the ciphertext is $c = mG + e$. In this case, we can encode $mLen = mk$ bits information within each ciphertext.

(2) mediumEncoding: In addition to basicEncoding, further information is encoded in the non-zero. In this case, we can encode $mLen = m(k+t)$ bits information within each ciphertext.

(3) advancedEncoding: In addition to mediumEncoding, further information are encoded within the choice of non-zero entries within e. Since there are $\binom{n+w}{t}$ candidates for the choice of non-zero entries within e, we can encode $mLen = m(k+t) + \left\lfloor \log 2 \binom{n+w}{t} \right\rfloor$ bits information within each ciphertext.

Pointcheval padding [17]:

$$c = Enc(G, r_1, H_1(m\|r_2))\|(H_2(r_1) \oplus (m\|r_2)) \tag{6}$$

Fujisak-Okamato padding [18]:

$$c = Enc(G, r_1, H_1(m\|r_1))\|(H_2(r_1) \oplus m) \tag{7}$$

Kobara-Imai's α-padding [19]:

$$c = Enc(G, y_1, H_1(m\|r_1))\|y_2(H_2(r_1) \oplus m), \; y_1\|y_2 = H_2(H_1(m\|r_1) \oplus (m\|r_1) \tag{8}$$

Kobara-Imai's β-padding:

$$c = y_1\|Enc(G, y_2, H_1(r_1)), \; y_1\|y_2 = (r \oplus H_1(H_2(r) \oplus m)) \oplus (H_2(r) \oplus m) \tag{9}$$

Kobara-Imai's γ-padding:

$$c = y_3\|Enc(G, y_1, y_2) \tag{10}$$

$$y_3\|y_2\|y_1 = (r \oplus H_1(H_2(r) \oplus (m\|const)))\|(H_2(r) \oplus (m\|const)) \tag{11}$$

Let H_1; H_2 be random oracles (they could be pseudo-random-bits generators or hash functions) that output random strings of appropriate lengths and let r_1; r_2 be randomly selected strings with appropriate length.

2.4   Pre-computation for Private Key

Table 4.

Table 4. Pre-computation for private key

Pre-computation for Private key [20]
Input:
1. $(n+w) \times (n+w)$ permutation matrix P.
2. $k \times (n+w)$ generator matrix G.
3. integer u_0 $(u_0 < k)$.
Output: $k \times (u_0 + 1)$ matrix X.
1. $0 \leq I_1 \leq I_2 \leq \cdots I_u \leq k$ be a list of all integers in the interval $[0, k-1)$ with $P[I_i] \geq n - w$ for all $1 \leq i \leq u$. If $u > u_1$ return an error.
2. $0 \leq J_1 < J_2 < \cdots J_{k-u} < k$ be a list of all integers in the interval $[0, k-1)$ with $P[J_i] < n - w$ for all $1 \leq i \leq k - u$.
3. $k \leq T_1 < T_2 < \cdots T_u < n + w$ be the first u integers such that $P[T_i] < n - w$ for all $1 \leq i \leq u$.
4. W be a $u \times u$ matrix such that $W[i][j] = G[i][j]$ for all $1 \leq i,j \leq u$.
5. V be a $(k - u) \times u$ matrix such that $V[i][j] = G[i][j]$ for all $1 \leq i \leq k - u$ and $1 \leq j \leq u$.
6. $U = (P_2[T_0], \cdots, P_2[T_u])$ be a $1 \times u$ matrix.

3   Improved RLCE Public Key Encryption Scheme Based on Polar Code

3.1   Improved RLCE Public Key Encryption Scheme

The improved RLCE Public key encryption scheme based on Polar code proposed in this paper is based on the RLCE encryption scheme proposed by Wang and the RLCE encryption scheme based on Polar code proposed by Liu. Aiming at the shortcomings of the Polar RLCE encryption scheme proposed by Liu, which is larger in key size and has no IND-CCA2 security, an improved scheme is proposed. In this paper, Polar code is used as the underlying code of RLCE encryption scheme by virtue of its polarization property and low decoding complexity. The improved scheme still adopts the structure of Wang's RLCE scheme, without changing the form of RLCE encryption scheme. The IND-CCA2 formal security proof of Wang's RLCE encryption scheme was applied to the encryption scheme based on Polar code, the Polar RLCE encryption scheme was padded with messages, each ciphertext was semantically secure mediumEncoding, and the public key matrix was converted into the system matrix, part of the private keys were estimated to reduce the key storage space. This scheme is similar to the RLCE encryption scheme, including key generation (polarRLCE.keysetup), encryption (polarRLCE.enc), and decryption (polarRLCE.dec).

3.1.1   PolarRLCE.KeySetup

(1) Parameter   selection: (n, k, d, t, w), $n, k, d, t > 0$, $w \in \{1, 2 \cdots n\}$, $k + 1 \geq d \geq 2t + 1$.

(2) G: $k \times n$ order generating matrix of Polar code C whose dimension on domain F is k and whose minimum distance d $\geq$ 2t + 1.

(3) G_1: Generate w random column vectors $r_1, r_2, \cdots, r_w$, and by inserting w random $k \times 1$ column vectors into generator matrix G, obtained the $k \times (n + w)$ matrix G_1, $G_1 = (g_1, \cdots, g_{n-w}, g_{n-w+1}, r_1, \cdots g_n, r_w)$.

(4) A: $(n + w) \times (n + w)$ nonsingular invertible matrix,

$$A = \begin{pmatrix} I_{n-w} & \cdots & 0 \\ \vdots & A_1 & \vdots \\ \vdots & & \ddots & \vdots \\ 0 & \cdots & A_w \end{pmatrix},$$

$$A_1 = \begin{pmatrix} a_{1,11} & a_{1,12} \\ a_{1,21} & a_{1,22} \end{pmatrix}, \cdots A_w = \begin{pmatrix} a_{w,11} & a_{w,12} \\ a_{w,21} & a_{w,22} \end{pmatrix} \in F_2^{2 \times 2}, \ A_1, A_2, \cdots, A_w \text{ be } 2 \times 2$$

nonsingular invertible matrix.

(5) $S : k \times k$ nonsingular invertible matrix.

(6) $P : (n + w) \times (n + w)$ permutation matrix.

(7) $G^{pub} : k \times (n + w)$, Public key $G^{pub} = SG_1AP$.

Public key $G^{pub} = SG_1AP$, Private key (S, G_1, A, P).

3.1.2   PolarRLCE.Enc

(1) The sender selects the plaintext $m \in F_2^k$, randomly pick the error vector $e = [e_1, \cdots, e_{n+w}] \in F_2^{n+w}$, the hamming weight of e the error vector is at most t, $wt_H(e) \leq t$.

(2) The sender uses the public key G^{pub} of the receiver to encrypt the plaintext m, get the ciphertext $c \in F_2^{n+w}$.

(3) $c = mG^{pub} \oplus e$.

3.1.3   PolarRLCE. Dec

(1) The receivers receive the ciphertext c, use own private key, get

$$cP^{-1}A^{-1} = mSG_1 \oplus eP^{-1}A^{-1} = (c_1', c_2', \cdots c_{n+w}') \tag{12}$$

(2) From length of $n + w$ vector $(c_1', c_2', \cdots c_{n+w}')$ delete frozen bit column vectors, get   length   of $n, c' = (c_1', c_2', \cdots, c_{n-w+1}', c_{n-w+3}', c_{n-w+5}', \cdots, c_{n+w-1}')$, $c' = mSG_1 \oplus e'$.

(3) The receiver uses the SC decoding algorithm of Polar code, decode c' to obtain plaintext m.

(4) The receiver calculates hamming weight of $wt = wt(c - mG_1)$, if $wt \leq t$, Output message m; else, output $\perp$.

The key storage space is reduced by transforming the public key matrix into the system matrix and no longer storing the private key S.

3.2 Message Padding

The PolarRLCE schemes proposed by Liu has no IND-CCA2 security. This paper adopts the method of message padding to make the improved scheme IND-CCA2 security. This paper adopts the RLCEspad message padding scheme designed by Wang, which is suitable for encrypting short length information. After the information is filled in, the information is converted into information bits or other information in the RLCE scheme. If mediumEncoding or advancedEncoding is used, the error vector e is also partially encoded. The general encoding used in this paper is that the populated information can be encoded in non-zero terms (Table 5).

RLCEspad $(mLen, k_1, k_2, k_3, t, m, r)$:

a. $k_1 + k_2 + k_3 = \lceil \frac{mLen}{8} \rceil$, $v = 8(k_1 + k_2 + k_3) - mLen$.

b. H_1, H_2, H_3 be a random oracle, takes any-length inputs and outputs k_2, $k_1 + k_2$, k_3 bytes Output bit string.

c. $m \in \{0,1\}^{8k_1}$ be a message to be encrypted, $r_1 \in \{0,1\}^{8k_3-v}$, $r = r_1 \| 1^v$ be a randomly selected binary bit string.

RLCE padding process is as follows:

Random select $1 \leq l_1 < l_2 < \cdots l_t \leq n + w$, $e_1 = l_1 \| l_2 \cdots \| l_t \in \{0,1\}^{16t}$, By calculation r to get e_1.

$$y = ((m \| H_1(m, r, e_1)) \oplus H_2(r, e_1)) \| H_3(m \| H_1(m, r, e_1)) \oplus H_2(r, e_1))) \tag{13}$$

Convert y to $(y_1, e_1) \in GF_2^{k+t}$, $y_1 \in GF_2^k$, $e_1 \in GF_2^t \cdot e \in GF_2^{n+w}$, for $0 \leq i \leq t$, $e[l_i] = e_1[i]$; for $j \neq l_i$, $e[j] = 0$. output (y_1, e_1). Ciphertext $c = y_1 G + e$.

Table 5. RLCEspad

RLCEspad
Input: $mLen, t, m, r, H_1, H_2, H_3$.
Output: (y_1, e_1), c.
1. Calculate $v = 8(k_1 + k_2 + k_3) - mLen$.
2. Calculate $e_1 = l_1 \| l_2 \cdots \| l_t \in \{0,1\}^{16t}$, for $m \in \{0,1\}^{8k_1}$, $r_1 \in \{0,1\}^{8k_3-v}$, $r = r_1 \| 1^v$.
3. (mediumEncoding): $y = ((m \| H_1(m, r, e_1)) \oplus H_2(r, e_1)) \| H_3(m \| H_1(m, r, e_1)) \oplus H_2(r, e_1)))$.
4. Convert y to $(y_1, e_1) \in GF_2^{k+t}$, output (y_1, e_1), get ciphertext $c = y_1 G + e$.

4   Performance Analysis

Public key size:

$$G^{pub} = SG_1AP = [I|Q] \tag{14}$$

The size of G^{pub} is $k \times (n+w)$, by using the system matrix, the improved G^{pub} size is $k \times (n+w-k)$.

Private key size:
Private key (V, W^{-1}, G_1, A, P), the size of V is $(k-u) \times u$, the size of W^{-1} is $u \times u$, the size of G_1 is $k \times (n+w)$, the size of A is $(n+w) \times (n+w)$, the size of P is $(n+w) \times (n+w)$. Pre-computation for private key W^{-1}, the improved scheme is no longer stored $k \times k$ matrix S, reduced private key storage space (Table 6).

Table 6. Comparison of public key size of codes-based schemes (Kbytes)

ISD	Our	PolarRLCE [15]	HermitianRLCE [14]	GRSRLCE [12]	GoppaMcEliece [9]
128	98	98	103	183	188
192	256	256	198	440	490
256	380	380	313	1203	900

From the table, get the analysis:

(1)   At the same bit security level, the public key size of the proposed scheme is the same as that of the Liu scheme, but the proposed scheme has IND-CCA2 security, and part of the private keys are estimated in this paper, which can significantly reduce the storage space of the private keys.

(2)   At the same bit security level, the public key size of the cryptography scheme based on Polar code is the smallest in this paper.

5   Security Analysis

5.1   Brute Force Attack

A brute force attack is a trial-and error method used to obtain the correct keys. By taking the form of a system matrix, the private key are (V, W^{-1}, G_1, A, P). In this paper, Polar code is used as the underlying code of the scheme, and its equivalence class code family of permutation matrix P is large. It is difficult to find the correct one in polynomial time through exhaustive attack, and it is difficult to operate the ciphertext. $cP^{-1}A^{-1} = mSG_1 \oplus eP^{-1}A^{-1} = (c_1', c_2', \cdots c_{n+w}')$, unable to recover the correct plaintext. In addition, according to the parameters selected by Liu scheme, it is not feasible for the attacker to find the other three types of keys by exhaustive method. Therefore, exhaustive attacks do not affect the security of the scheme in this paper.

5.2   Information Set Decoding Attack

Information set decoding attacks are by far the most effective attacks against McEliece encoding schemes. Among all known attacks, the attacks based on information set decoding have the lowest computational complexity. Stern [21] firstly proposed an information set decoding attack on the McEliece scheme. Since then, in order to improve the effectiveness of the attack on the code-based scheme, there have been many improved information set decoding attacks [22]. Information set decoding attack does not attack the cryptography scheme by using the underlying structure of codes, but by searching information set. The working factor of information set decoding attack increases with the increase of error vector. For RLCE encryption schemes, the information set decoding attack looks for the number of columns of the public key G^{pub}, not the number of columns of the private key G_1. A random selection of k bits from the $n + w$ bit ciphertext containing t errors constitutes, a bit composition of the corresponding position is selected from the error vector, and a matrix is formed by selecting the corresponding column. If the selected bits do not contain the errors bits, that is $e_k = 0$, the attacker can easily recover the plaintext m.

For randomly selected k columns from the public key of the RLCE encryption scheme, the probability that the ciphertext contains no errors at these locations is $\dfrac{\binom{n+w-t}{k}}{\binom{n+w}{k}}$. In this paper, the difficulty of finding the error-free position of the selected error vector is e_k increased by inserting a random w column. Given the appropriate parameters, it is extremely difficult for an attacker to decode the information set to obtain the plaintext m. Therefore, by inserting random columns, this paper guarantees that the cryptography scheme is not affected by the information set decoding attack.

5.3   Reaction Attack

Reaction attack is that the attacker modifies a small amount of ciphertext c and sends the modified ciphertext c to the receiver to observe whether the receiver can correctly decode. For a given ciphertext c, the attacker randomly selects the location i, adds errors in the location i, sends the added wrong ciphertext c to the receiver, and observes whether the receiver can decode correctly. If the receiver can decode correctly, it means that the attacker has added an error in the position i.If the receiver decoding fails, the attacker did not add an error to the location i. Repeat the operation until the attacker gains an k error-free bit. This article resists response attacks by inserting a random w column and then padding it with a message. The Polar code-based RLCE encryption scheme proposed by Liu did not consider the response attack. This paper analyzed the possible response attacks and found that, unlike the scheme with the underlying code of hamming cyclic code, this scheme could resist the possible response attacks through the method of message padding.

5.4   Key Recovery Attack

The basic idea of key recovery attack is to recover the correct private key from the public key G^{pub}. A key recovery attack is a structural attack, usually against a specific code, for example QC-LDPC codes, GRS codes etc. The scheme proposed in this paper inserts random columns. At the same time, random nonsingular matrix A is used to mix the inserted random columns, the structure of the private key is scrambled. After mixing, the frozen bit column is randomly deleted from the ciphertext, since the attacker cannot know the polarization property of the Polar code, the complexity of the private key structure is further increased, and it is difficult for the attacker to recover the correct private key structure through the public key in polynomial time. Bardet et al. [23] proposed an attack to determine the minimum weight of Polar code, and solved the equivalence problem of polarized code relative to decreasing monomial code. The random columns inserted into the improved scheme in this paper disturb the original structure of the generated matrix, so the security of the scheme in this paper will not be affected by reference [23]. Since the structure of Polar code is different from that of hamming cyclic code, GRS code and RM code, the structural attack against the underlying codes of cyclic code, GRS code and RM code will not threaten the security of the scheme in this paper.

5.5   Adaptively Chosen Ciphertext Attacks

In this paper, the improved system matrix is used to encrypt the plaintext m, which may lead to the IND-CCA2 security reduction. Liu's Polar code-based RLCE encryption scheme does not have semantic security, and an example of a correlation attack is illustrated: two different ciphertexts can be obtained by encrypting the same plaintext twice, and two different ciphertexts can be compared and analyzed to find the original message.

$y = ((m\|H_1(m, r, e_1)) \oplus H_2(r, e_1))\|H_3(m\|H_1(m, r, e_1)) \oplus H_2(r, e_1)))$. According to the message padding method described above, this paper adopts the RLCEspad padding method to perform mediumEncoding for each ciphertext, so that some messages are encoded in the non-zero term of the error vector e. Before sending the plaintext m, the sender scrambles the information through three random oracle models H_1, H_2, H_3, encodes part of the message in the non-zero term of the error vector e, and then sends the encoded message to the sender. In order to obtain the plaintext, the difficulty of attacking the populated cryptography scheme is equivalent to that of the original McEliece scheme based on Goppa code. Therefore, the scheme after message padding by RLCEspad method in this paper has semantic security and can achieve IND-CCA2 security.

6 Application of Polar Code-Based Scheme in Cloud Secure Storage

We apply the improved code-based scheme to cloud storage. In our cloud storage system, we adopt public-key encryption and symmetric encryption based on encoding. Specifically including the message sender, message receiver, a number of cloud storage servers. Specific details of cloud storage can be referred to the literature [24].

7 Conclusion

Based on RLCE encryption scheme and PolarRLCE encryption scheme, this paper takes advantages of Polar with more equivalence classes and low complexity of encryption and decryption, proposing an IND-CCA2 secure RLCE public key encryption scheme. After security analysis, it can resist the known information set decoding and other attacks. Compared with the scheme based on hamming code, this scheme can resist the known structural attacks.

Polar has become a research hotspot due to its unique polarization and low decoding complexity. The application of Polar code to the code-based cryptography scheme has a good prospect in the future. In this paper, the public key encryption scheme based on Polar code is applied to cloud storage to ensure that the data stored in the cloud environment is safe enough under the attack of quantum computer. In the following research, the application prospect of public key encryption scheme based on Polar code can be further broadened, and the code-based scheme can be applied to more practical scenarios. Applying code-based cryptography schemes to blockchain is an interesting work.

Acknowledgment. This work was supported the National Natural Science Foundation of China (No.61572521); The Scientific Foundation of the Scientific Research and Innovation Team of Engineering University of PAP (No.KYTD201805).

References

1. Hongzhe, D., Yongfang, Z.: Fire prevention and extinguishing: generation and management of secondary public opinions in COVID 19 public crisis events. J. Univ. Electron. Sci. Technol. **22**(2), 1–7 (2020)
2. Zhang, L., Xiong, H., Huang, Q., et al.: Cryptographic solutions for cloud storage: challenges and research opportunities. IEEE Trans. Serv. Comput. (2019)
3. Jordan, S.P., Liu, Y.K.: Quantum cryptanalysis: Shor, Grover, and Beyond. IEEE Secur. Priv. **16**(5), 14–21 (2018)
4. Bernstein, D.J., Lange, T.: Post-quantum cryptography. Nature **549**(7671), 188–194 (2017)
5. Ding, J., Steinwandt, R. (eds.): PQCrypto 2019. LNCS, vol. 11505. Springer, Cham (2019). https://doi.org/10.1007/978-3-030-25510-7
6. Wu, W.L.: Preface of special issue on block cipher. J. Cryptologic Res. **6**(6), 687–689 (2019)

7. Alagic, G., Alagic, G., Alperin-Sheriff, J., et al.: Status report on the first round of the NIST post-quantum cryptography standardization process. US Department of Commerce, National Institute of Standards and Technology (2019)
8. Nejatollahi, H., Dutt, N., Ray, S., et al.: Post-quantum lattice-based cryptography implementations: a survey. ACM Comput. Surv. (CSUR) 51(6), 1–41 (2019)
9. Mceliece, R.J.: A public-key cryptosystem based on algebraic coding theory. DSN Prog. Rep. 42(44), 114–116 (1978)
10. Niederreiter, H.: Knapsack-type cryptosystems and algebraic coding theory. Probl. Control Inf. Theory 15(2), 159–166 (1986)
11. Faugère, J.C., Otmani, A., Perret, L., et al.: Structural cryptanalysis of McEliece schemes with compact keys. Des. Codes Crypt. 79(1), 87–112 (2016)
12. Wang, Y.: Quantum resistant random linear code based Public key encryption scheme RLCE. In: 2016 IEEE International Symposium on Information Theory (ISIT), pp. 2519–2523. IEEE (2016)
13. Wang, Y.: Revised quantum resistant public key encryption scheme RLCE and IND-CCA2 security for McEliece schemes. IACR Cryptology ePrint Arch. 2017, 206 (2017)
14. Matthews, Gretchen L., Wang, Y.: Quantum resistant public key encryption scheme HermitianRLCE. In: Baldi, M., Persichetti, E., Santini, P. (eds.) CBC 2019. LNCS, vol. 11666, pp. 1–10. Springer, Cham (2019). https://doi.org/10.1007/978-3-030-25922-8_1
15. Liu, J., Wang, Y., Yi, Z., Pei, D.: Quantum resistant public key encryption scheme polarRLCE. In: Gueye, C.T., Persichetti, E., Cayrel, P.-L., Buchmann, J. (eds.) A2C 2019. CCIS, vol. 1133, pp. 114–128. Springer, Cham (2019). https://doi.org/10.1007/978-3-030-36237-9_7
16. Arikan, E.: Channel polarization: a method for constructing capacity-achieving codes for symmetric binary-input memoryless channels. IEEE Trans. Inf. Theory 7(55), 3051–3073 (2009)
17. Fouque, P.-A., Pointcheval, D.: Threshold cryptosystems secure against chosen-ciphertext attacks. In: Boyd, C. (ed.) ASIACRYPT 2001. LNCS, vol. 2248, pp. 351–368. Springer, Heidelberg (2001). https://doi.org/10.1007/3-540-45682-1_21
18. Fujisaki, E., Okamoto, T.: Secure integration of asymmetric and symmetric encryption schemes. J. Cryptology 26(1), 80–101 (2013)
19. Kobara, K., Imai, H.: Semantically secure McEliece public-key cryptosystems -conversions for McEliece PKC. In: Kim, K. (ed.) PKC 2001. LNCS, vol. 1992, pp. 19–35. Springer, Heidelberg (2001). https://doi.org/10.1007/3-540-44586-2_2
20. Wang, Y.: RLCE Key Encapsulation Mechanism (RLCE-KEM) Specification (2019)
21. Stern, J.: A method for finding codewords of small weight. In: Cohen, G., Wolfmann, J. (eds.) Coding Theory 1988. LNCS, vol. 388, pp. 106–113. Springer, Heidelberg (1989). https://doi.org/10.1007/BFb0019850
22. Welch, Z.D.: An Analysis of Potential Standards for Post-Quantum Cryptosystems. Carleton University (2019)
23. Bardet, M., Chaulet, J., Dragoi, V., Otmani, A., Tillich, J.-P.: Cryptanalysis of the McEliece public key cryptosystem based on polar codes. In: Takagi, T. (ed.) PQCrypto 2016. LNCS, vol. 9606, pp. 118–143. Springer, Cham (2016). https://doi.org/10.1007/978-3-319-29360-8_9
24. Zeng, P., Chen, S., Choo, K.K.R.: An IND-CCA2 secure post-quantum encryption scheme and a secure cloud storage use case. Hum.-Centric Comput. Inf. Sci. 9(1), 1–15 (2019)

Container Cluster Scheduling Strategy Based on Delay Decision Under Multidimensional Constraints

Yijun Xue[1], Ningjiang Chen[1,2(✉)], and Yongsheng Xie[1]

[1] School of Computer and Electronic Information, Guangxi University,
Nanning 53004, China
chnj@gxu.edu.cn

[2] Guangxi Key Laboratory of Multimedia Communications and Network
Technology, Nanning 53004, China

Abstract. With the rise of online applications such as machine learning, stream processing, and interactive data-intensive applications in shared clusters, container cluster scheduling in data centers is facing new challenges. In order to solve the problem that application performance and economic cost cannot be balanced in a container cluster deploying a hybrid application, this paper proposes a container cluster scheduling strategy based on delay decision under multi-dimensional constraints. Formal language-based application placement constraints were introduced, and a task reorder model was established based on delayed decision-making. The experiments show that this strategy improves application performance and cluster utilization.

Keywords: Container cluster · Multi-dimensional constraint · Delay decision · Application performance · Cluster utilization

1 Introduction

With the extensive applications of Hadoop (processing analysis) [1], TensorFlow (deep learning) [2], core e-commerce [3] and other long-term online running applications, according to the principle that application resource demand is less than supplied physical resource, the application and resource are dynamically adjusted at runtime. It is difficult to apply a cluster resource sharing management model that optimizes supply and demand management.

Modern large cloud data center container clusters usually run many different types of applications. In addition to traditional batch processing applications, they also include stream processing [4], iterative computing [5], data-intensive interactions [6], and delay-sensitive Online application [7]. Studies [8–10] show that in the production environment of actual data center clusters, the operational utilization of global cloud facilities and commercial clusters is only 6% to 12%. We analyzed the set of newly released tracking data [11] by Alibaba. The statistical results are shown in Fig. 1. There are space imbalances (heterogeneous resource utilization across machines) and time imbalances when the cluster is running. (The resource usage time of each machine

© Springer Nature Singapore Pte Ltd. 2020
J. Zeng et al. (Eds.): ICPCSEE 2020, CCIS 1257, pp. 690–704, 2020.
https://doi.org/10.1007/978-981-15-7981-3_51

varies). Alibaba reserves fixed resources for online applications. Unlike batch jobs for short-term containers, these applications take longer to run, so called long-running application (LRA). Containers of LRA have a relatively long service life, avoiding repeated container initialization costs and reducing scheduling load. The overly simple scheduling strategy will result in poor placement of LRAs tasks, exacerbate these two imbalances in the cluster, and cause waste of resources.

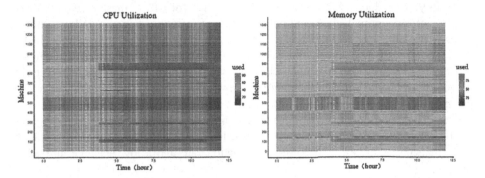

Fig. 1. Current status of cluster resource utilization

This paper proposes a cluster scheduling method based on delayed decision and LRAs. This method uses a formal language to dynamically describe multi-dimensional constraints and construct a constraint management model. At the same time, the optimal task scheduling queue is obtained by using the node matching optimization strategy and multi-attribute decision making. The contributions of this article are mainly three points:

- A multi-dimensional labeling constraint characterization model is proposed, which uses the flexibility of formal language to realize the dynamic expression of multi-dimensional constraints and reduce the constraint violation rate.
- A scheduling management mechanism based on delay decision is proposed, which includes a node matching optimization strategy and a task reorderer. Using the idea of delay, tasks are optimally placed, which improves throughput and data utilization.
- Experimental results show that the average utilization rate of this method is improved by about 10% compared with the existing cluster scheduling methods, and the constraint violation rate does not exceed 5%, which effectively ensures the application performance.

2   Related Work

Some existing scheduling systems for LRA support [12–14] are still in the exploratory stage for the constraints' expression between deployed application containers. Simple affinity and anti-affinity constraints have been partially implemented in a few scheduling systems, such as Mesos [15], YARN [16], Borg [17], but their constraints are implicitly supported by static machine attributes, lacking some flexibility.

In [16] and [17], YARN and Borg act as matchers between the resource requirements of various applications and the resources available on the machine nodes, but which will cause blocking situations. References [18–20] used a scheduling delay mechanism to alleviate the head-of-line blocking situation caused by FIFO ordering. Apollo [21] and Sparrow [22] independently decide where to run tasks to improve scalability and reduce allocation delays. But for tasks of LRAs, these methods cannot achieve global optimal allocation. Choosy [23], a resource fair scheduling method, realizes resource sharing under placement constraints, but lacks research on better convergence time and constraints between tasks. Paper [24] proposed the Quincy scheduling strategy to instantly calculate and optimize the global matching of scheduling decisions through the minimum flow graph. Due to the high complexity of the graph, there will be a huge delay in the cluster scheduling of large-scale data centers. To sum up, the existing methods are difficult to achieve flexible expression of multi-dimensional constraints, and cannot balance scheduling resources and application performance. There is a high constraint violation rate and a low resource utilization rate.

3   General Framework of the Method

The approach overview is shown in Fig. 2. In this paper, constructing a constraint management model with multi-dimensional constraints based on the formal language of each application. The node matching optimizer and task reorder are used to determine the scheduling order of the tasks, so as to achieve the optimal placement of tasks.

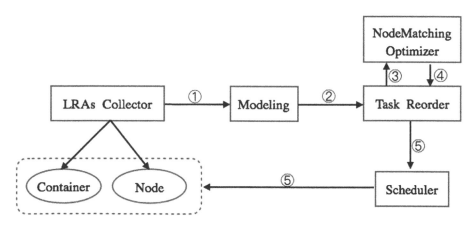

Fig. 2. Approach overview (The flow of the system approach can be described as follows)

① The LRAs collector collects the load and CPU, memory and other resource usage of each container and node.

② A performance modeler that builds a constraint management model and characterizes the constraint relationships between containers and nodes.

③ The Task Reorder analyzes multiple characteristics of tasks through multi-attribute decision-making, establishes a decision matrix, obtains the optimal task queue scheduling order, and achieves optimal task placement.

④ The node matching optimizer uses the resource utilization rate in the step ① as a benchmark, uses a set of priority functions to process the nodes in the cluster, and passes the processing results to the task reorder in real time.

⑤ The container scheduler performs a certain delay scheduling according to the task queue scheduling order and node processing results.

⑥ After performing container scheduling, continue to execute the steps to form a method closed loop.

4   Dynamic Characterization of Multidimensional Label Constraints

In the actual production environment, cluster scheduling needs to meet various constraints, but there will always be conflicts, so it is necessary to ensure a low constraint violation rate. It has an important impact on the performance of the application, which is possible to further optimize the core of the cluster. Label is a simple but powerful constraint mechanism for referencing containers of the same or different (possibly not yet deployed) applications. For example, using the label 'hb' to reference the current and future containers of HBase application. This paper uses a formal language (label) to specify multi-dimensional constraints for long-running application containers, build a constraint management model, as shown in Fig. 3.

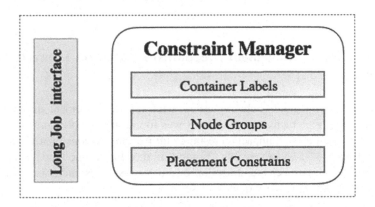

Fig. 3. Constrains Manager Model, and simply record container information generated by the application, including application ID, application type, deployment node, and resource specifications. The node label set is updated in real time. When a container is assigned to a node, the container label is added to the node label set. Only when the container has completed execution and is not in a running state, the label associated with the node is removed.

(1) Establish the set of tags: Setting the label set in the unit of node, each label can be associated with multiple containers on the node;

(2) Specify node group: The predefined node group is a node and a rack. A node corresponds to a single element of a cluster node. The rack contains all nodes of a physical machine.

(3) Define constraint forms: Allow application owners and cluster operators to use labels to specify container placement constraints that reference containers of the same or different applications, and point to a specific node set of node groups. The form of the constraint defined in this paper is as follows:

$$P = \{subject\_label, label\_constraint, node\_group\}$$

subject_label is a label or label association that identifies the container subjected to constraints; *label_constraint* is a constraint of the form $\{p\_label, pmin, pmax\}$, p_label is the container label (or label association), $pmin$ and $pmax$ are positive integers and represent the number of containers; *node_group* represents a node group.

At this paper, there are the following constraints:

Definition 1: Affinity constraints. Coordinating some LRA containers on the same node or node group can bring benefits to the cluster. Use $pmin = 1$ and $pmax = \infty$ to express affinity constraints.

Definition 2: Anti-affinity constraint. Minimize resource interference between long-running applications by placing containers on different machines through anti-affinity constraints inside and between applications, and use $pmin = 0$ and $pmax = 0$ to express anti-affinity (anti-affinity) constraints.

Definition 3: Cardinality constraint. Affinity and anti-affinity constraints represent the two extremes of container placement. In order to achieve a balance between the two, a more flexible cardinality constraint is used, which limits the number of juxta-posed containers. Cardinality constraints are expressed for other values of $pmin$ and $pmax$.

5   Scheduling Management Mechanism Based on Delayed Decision

Considering the constraints between tasks, a reasonable placement decision can be made, but in the implementation process, it was found that the placement order of the tasks also has an important impact on resource utilization and constraint violation rates. As shown in Fig. 4. (a), there are idle resources in the cluster, but there are still waiting task queues, which reduces the utilization of cluster resources and affects the running time of jobs, resulting in poor quality of service.

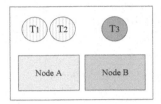 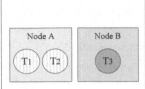 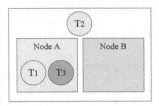

Fig. 4. Task deployment results with different queue order: it is assumed that there are two tasks (T1, T2) have affinity constraints and can only be deployed on node A; task T3 is unconstrained; each node can only deploy two tasks. The ideal deployment result is Fig. 4. (b), and the actual deployment result is Fig. 4. (c).

Compared with the queue management functions of existing scheduling systems, most of them support queuing on nodes, but do not support global task queuing. This paper combines the multi-dimensional constraints between tasks and considers multiple task requests at once, and proposes a scheduling management mechanism based on delay decision, including node matching optimization strategy and task queue reordering model.

5.1 Strategy of Node Matching Optimization

In practical application scenarios, it is often the case that the working nodes to be scheduled are "picked", that is, certain containers are required to be scheduled to run only on specific hosts, so a node matching optimization strategy is designed in this paper. The nodes are matched to achieve container scheduling to eligible nodes, and the nodes are preferably used to implement container scheduling to appropriate nodes.

Considering multiple task requests within a certain time interval π, define a task set $T = \{T1, T2, \ldots, Tn\}$. For each task Ti, the basic filtering generally evaluates some common factors such as node port availability, whether resources are satisfied, and whether the mounted disks conflict. The constraint filtering generally considers the anti-affinity constraint relationship between tasks, and takes the difference set. Find the node set Ni matched by each task.

Use a set of priority functions to process each node in the node set Ni, AffinityPriority, AntiAffinityPriority (there is a constraint relationship between Ti and the task running in node Nij. Return a function value of 1, otherwise Returns 0). This article uses a data collector to obtain the total CPU, memory (Nodeij.capacityCPU, Nodeij.capacityMemory) on the node, and the sum of the requested CPU and memory (Nodeij.requestCPU) of the container that has been scheduled on this node and the Ti to be scheduled, Nodeij.requestMemory). The specific formula is as follows:

$$Node_{ij}.restCPU = \left(\frac{Node_{ij}.capacityCPU - Node_{ij}.requestCPU}{Node_{ij}.capacityCPU} \right) \qquad (1)$$

$$Node_{ij}.restMemory = \left(\frac{Node_{ij}.capacityMemory - Node_{ij}.requestMemory}{Node_{ij}.capacityMemory} \right) \qquad (2)$$

$$priorityFunc4 = \frac{1}{2}(Node_{ij}.restCPU + Node_{ij}.restMemory) * 10 \tag{3}$$

$$priorityFunc5 = 10 - \left| \left(\frac{Node_{ij}.requestCPU}{Node_{ij}.capacityCPU} \right) - \left(\frac{Node_{ij}.requestMemory}{Node_{ij}.capacityMemory} \right) \right| * 10 \tag{4}$$

Among them, Nodeij.restCPU and Nodeij.restMemory represent the remaining rates of CPU and memory resources in Nodeij respectively; the priority function priorityFunc4 is used to evaluate the resource consumption of the node; the priority function priorityFunc5 is used to evaluate the resource balance of the node. The node's final score (anti-affinity constraint) is obtained by adding the values returned by multiple priority functions. The larger the score, the better the quality of the node. According to the order of FianlScoreNodeij, the Ti queue of task Ti candidate nodes is formed.

$$FianlScoreNode_{ij} = \sum_{t}^{m} wt * priorityFunct \tag{5}$$

5.2   Task Priority Multi-attribute Ranking Model

Tasks have multiple characteristics, which lead to different scheduling orders affecting the optimal placement of tasks. This paper analyzes the multiple characteristics of tasks and introduces multi-attribute decision theory to transform the reordering of task queues into multi-attribute decision problems. The number of nodes matched by the task, the amount of requested resources of the task, and the starvation status of the task are of great significance to the task, and their corresponding attribute values are all expressed in the form of real numbers. This article first makes relevant definitions and assumptions, and then builds a Task Reorder Model (**TRM**).

Definition 4:   Starvation time STi (Starved Time) of task Ti.

$$STi = (t - ai) - \pi \tag{6}$$

STi represents the length of the starvation time of the task Ti, which reflects the starvation state of the task. It is calculated and determined by the current time t, the task's arrival time ai, and the scheduling interval π, and they have the same unit of measurement.

Definition 5:   Number of Matched Nodes (NMNi) for task Ti.

$$NMNi = NiQueue\_Size \tag{7}$$

NMNi represents the number of matching nodes of task Ti, and the queue length of the queue NiQueue of task Ti candidate nodes.

Definition 6: Requested Resource (RRi) of task Ti.

$$RRi = Resquest.CPU + Resquest.Memory \qquad (8)$$

RRi represents the sum of the total request amount of CPU resources and memory resources of task Ti. Under certain node resources, deploying more tasks can improve the system throughput.

Multi-attribute decision-making needs to evaluate any task Ti from different attributes to get m evaluation results, which corresponds to the attribute vector of the task Ti (Ai1, Ai2, Ai3, ..., Aim). Collect the task set T according to the evaluation attribute set Attr Attributes of each task in Ti, thus establishing a decision matrix. The existing evaluation attributes include the three characteristics of task starvation time STi, the number of matching nodes NMNi, and the resource request amount RRi. Therefore, the decision matrix corresponding to this article is shown in the following formula.

$$E = \begin{pmatrix} ST_1 & NMN_1 & RR_1 \\ ST_2 & NMN_2 & RR_2 \\ \vdots & \vdots & \vdots \\ ST_n & NMN_n & RR_n \end{pmatrix} \qquad (9)$$

This paper proposes a task queue reordering model (TRM) based on multi-attribute decision making, which is expressed as {**Attr, T, ρ, E, β, w, ψ, RANK**}. Each element in the model is:

Attr is the task evaluation attribute set. In this paper, Attr = {starvation time (STi), number of matching nodes (NMNi), resource request amount (RRi)}; **T** is the task set that arrives within the scheduling interval π, T = {T1, T2, ..., Tn}; **ρ** represents the mapping relationship: Ti → (Ai1, Ati2, Ai3, ..., Aim), which represents the evaluation of multi-dimensional features of task Ti according to the task evaluation attribute set Attr; **E** means according to the task set Decision matrix $E = \begin{pmatrix} E_{11} & \cdots & E_{1m} \\ \vdots & \ddots & \vdots \\ E_{n1} & \cdots & E_{nm} \end{pmatrix}$ based on attributes of each task, among them, i in Eij represents the i-th task in the task set T, and j represents the j-th task evaluation attribute value of the i-th task, i {1, 2, ..., n}, j {1, 2, ..., m}; **β** indicates the mapping relationship Attr → w, and the subjective setting of the weight vector w according to the evaluation attributes; **w** represents the task attribute evaluation attribute weight vector, w = (w1, w2, ..., wn) T and the matrix E in the task priority decision process Decide the results together. **ψ** represents the mapping relationship (w, E) → RESULT of the decision process. Task priority selection is performed based on the attribute evaluation attribute weight vector w and decision matrix E, and the ranking result of the T task set is calculated. RESULT represents the ranking result of the task. **RANK** = {r1, r2, ..., rn}, where rk = {Tk, rNbk}, k {1, 2, ..., n}.

Using this model to analyze task evaluation attributes, starvation time (STi) and resource request volume (RRi) are all benefit attributes. The number of matching nodes (NMNi) is a cost attribute. Because the above three task attribute values have different dimensions, in order to eliminate the influence on the ranking decision result, these task attribute values are processed as follows respectively.

The normalization processing of task evaluation attribute starvation time (STi) and resource request amount (RRi) is as follows:

$$rij = \frac{aij - \min_i aij}{\max_i aij - \min_i aij} \tag{10}$$

The normalized processing of the task evaluation attribute matching node number (NMNi) is as follows:

$$rij = \frac{\max_i aij - aij}{\max_i aij - \min_i aij} \tag{11}$$

$\max_i aij$ and $\min_i aij$ represent the maximum and minimum values of the element aij in the j-th feature attribute Aj in the i-th task, rij $\in$ [0, 1]. The processed decision matrix is D = (rij) n * m, and then a weighted normalization decision matrix C = wD is reconstructed by considering the weight w.

$$C = \begin{pmatrix} w_1 r_{11} & w_2 r_{12} & \cdots & w_m r_{1m} \\ w_1 r_{21} & w_2 r_{22} & \cdots & w_m r_{2m} \\ \vdots & \vdots & \cdots & \vdots \\ w_1 r_{n1} & w_2 r_{n2} & \cdots & w_m r_{nm} \end{pmatrix} \tag{12}$$

Finally, the priority multi-attribute decision evaluation value of task Ti can be expressed as:

$$C(Ti) = \sum_{j=1}^{m} w_j r_{ij}, i = 1, 2, \ldots, n \tag{13}$$

Each task in the task set is sorted according to the value of C (Ti), and the reordering result RANK is obtained, and finally an optimal task scheduling queue is formed. The TRM algorithm is shown below.

Algorithm: Task queue reordering algorithm

Input : T,   The set of tasks that arrive within the scheduling interval π ;

      N_iQueue,   Candidate node queue of T_i ;

      w, The set of task attribute weight

Output : *TQueue*,   Task queue

    1 T ={ T1,T2,...,Tn }, N_iQueue =$\varnothing$

    2 **while** (each $T_i \in$ T)

    3    $STi= (t - ai) - \pi$

    4    $NMNi = N_iQueue\_Size$

    5    $RRi = Resquest.CPU + Resquest.Memory$

    6 $E=build\_matrix(STi, NMNi, RRi)$

    7 **for**($i = 1 , i \leq n , i++$)

    8   **if**($j < m$)**then**

    9   r_{ij}=function_Normalized(*STi, NMNi, RRi*)

    10   **end if**

    11   D= $build\_matrix(r_{ij})$

    12   C= $build\_matrix(w,r_{ij})$

    13 $$C(T_i) = \sum_{j=1}^{m} w_j r_{ij}$$

    14 *Rank*=get_set(C(T$_i$))

    15 *TQueue=bulid_queue(RANK)*

    16 **end**

---

6   Experiment

The implementation of the cluster scheduling prototype system MD-Kubernetes designed in this paper is based on Kubernete. Kubernete is a relatively comprehensive container scheduling system so far. The overall design is shown in Fig. 5, which mainly includes Client, Kube-API Server, Resource Manager, Gateway, Node Manager and other modules.

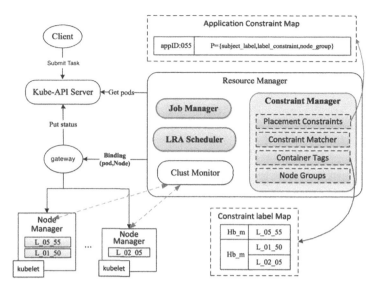

Fig. 5. Architecture of the Prototypal System

The experiment is compared with the existing scheduling system to verify the effectiveness of the system model in the container cluster scenario of hybrid applications. In the experiment, 10 blade server machines were selected to build a 400 virtual node cluster simulation system, configured with Intel Core i7, CPU3.40 GHz, 8 GB memory and Gigabit network card, and these machines were divided into 10 racks.

Software Configuration: In order to realize various configurations in the experiment, this article extends the workload generator GridMix, which can generate long-term running application examples of custom constraints. This article mainly deployed the following applications in the experimental cluster:

(1) HBase instance. Each instance is set up with 10 workers, including simple operations such as adding, deleting, modifying, and checking. Through the API, benchmark tests are performed using YCSB and 0.5 TB data.
(2) TensorFlow instance. Each instance is set up with 8 workers and 2 parameter servers, and runs a machine learning workload involving more than 1,200 iterations.

Each worker is set up with the corresponding configuration, where HBase and TensorFlow workers use containers configured with <2 GB, 1 CPU>, and the main workers of TensorFlow instances use containers configured with <4 GB, 1 CPU>.

Placement Constraints: When deploying HBase and TensorFlow instances, we use the following placement constraints:

(1) Set **affinity constraints** within the application to minimize network traffic. All workers deploying the same HBase instance or TensorFlow instance should be on the same rack;

(2) **Anti-affinity constraints** for different applications. The containers generated by setting HBase instances and TensorFlow instances should be deployed in different racks, thereby improving the stability of the service itself.
(3) This paper implements the **cardinality constraints** between applications. No more than two HBase workers or four TensorFlow workers are placed on the same node, which can minimize resource interference.

This article uses the following three scheduling systems for comparison:

MD-Kubernetes: This paper expresses constraints dynamically. Considering multiple task requests at a time, there is a certain delay. Compared with long-term running applications, the running time is negligible.

YARN: only supports the affinity for specific nodes/racks, and lacks support for constraints between applications. Through comparison, you can intuitively see the impact of constraints on task placement.

Kubernetes: The Kubernetes scheduling system is by far the most complete system that supports placement constraints, but Kubernetes considers one container request at a time during scheduling and does not support cardinality constraints.

6.1   Comparison of Application Performance

Deploy 45 TensorFlow instances and 50 HBase instances in the above 400-node cluster, and submit a job request that uses 50% of the cluster's memory through the GridMix workload generator. The running time of the application indicates different application performance. Figure 6. (a) describes the running time of the deployed machine learning workflow on TensorFlow, and Fig. 6. (b) describes the running time of data insertion on the Hbase instance. Box plots to show their runtimes. It can be seen from the comparison that compared with YARN that does not support constraints between tasks, the running time of the MD-Kubernetes scheduling instance is reduced by 2.1 times and 2.4 times from the median and maximum values of YARN. Unpredictability, because the satisfaction of some constraints is random. Compared with Kubernetes with simple constraints, the running time of TensorFlow instances scheduled by MD-Kubernetes is reduced by 32% at the median, and the running time of HBase instances is reduced by 23% at the median.

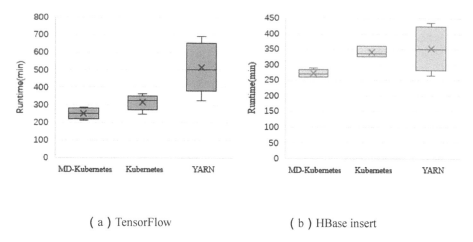

(a) TensorFlow   (b) HBase insert

Fig. 6. Comparison of application performance

6.2   Comparison of Resource Utilization

In order to simulate the real generation of the cluster, GridMix was used to generate the same load in a 400 virtual node cluster simulation system, and the CPU and memory resource usage of the cluster was continuously monitored to calculate the average CPU and memory utilization efficiency of all open containers. By analyzing the average CPU and memory usage efficiency of a container that is open, you can measure the proportion of pods (containers) that are open but not working (that is, the idle rate of container use). When using Kubernetes for scheduling in a cluster, the average CPU resource utilization of a node is mostly between 20% and 40%, and the average resource utilization of a node's memory is mostly between 40% and 60%. Pass this verification MD-Kubernetes system has the advantage of high resource utilization in terms of task placement strategy (Fig. 7).

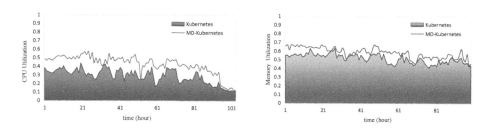

Fig. 7. Average resource utilization of cluster nodes

7 Conclusions

This paper proposes a constraint expression model based on label to support the dynamic expression of multi-dimensional constraints for long-term running tasks in a cluster. A cluster-oriented delay decision scheduling management mechanism guarantees the placement quality of long-term running tasks under multi-dimensional constraints and improves cluster efficiency. The designed MD-Kubernetes prototype system is implemented on Kubernetes in the form of plug-ins. The constraint expression model in this paper also has limitations, and further research on the adaptiveness of cluster scheduling is needed.

Acknowledgment. This work is supported by the Natural Science Foundation of China (No. 61762008), the Guangxi Natural Science Foundation Project (No. 2017GXNSFAA198141), and the National Key Research and Development Project of China (No. 2018YFB1404404).

References

1. Vavilapalli, V.K., Murthy, A.C., Douglas, C.: Apache Hadoop YARN: yet another resource negotiator. In: Symposium on Cloud Computing, pp. 1–16. ACM (2013)
2. Martín, A.: TensorFlow: learning functions at scale. In: ACM Sigplan International Conference on Functional Programming, p. 1. ACM (2016)
3. Verma, A., Pedrosa, L., Korupolu, M.: Large-scale cluster management at Google with Borg. In: Tenth European Conference on Computer Systems, pp. 1–17. ACM (2015)
4. Xingcan, C., Xiaohui, Y., Yang, L.: Overview of distributed stream processing technology. Comput. Res. Develop. **52**(2), 318–332 (2015)
5. Abadi, M., Barham, P., Chen, J.: TensorFlow: a system for large-scale machine learning. In: Usenix Conference on Operating Systems Design and Implementation, pp. 265–283. USENIX Association (2016)
6. Zaharia, M., Chowdhury, M., Franklin, M.J.: Spark: cluster computing with working sets. In: Usenix Conference on Hot Topics in Cloud Computing, p. 10. USENIX Association (2010)
7. Apache HBase[EB/OL] (2018). http://hbase.apache.org
8. Jyothi, S.A., Curino, C., Menache, I.: Morpheus: towards automated SLOs for enterprise clusters. In: Usenix Conference on Operating Systems Design and Implementation, pp. 117–134. USENIX Association (2016)
9. Rajan, K., Kakadia, D., Curino, C.: PerfOrator: eloquent performance models for Resource Optimization. In: ACM Symposium on Cloud Computing, pp. 415–427. ACM (2016)
10. Xu, G., Xu, C.-Z.: Prometheus: online estimation of optimal memory demands for workers in in-memory distributed computation. In: ACM Symposium on Cloud Computing, pp. 655–667. ACM (2017)
11. Alibaba trace [DB/OL] (2018). https://github.com/alibaba/clusterdata
12. Nathuji, R., Kansal, A., Ghaffarkhah, A.: Q-clouds: managing performance interference effects for QoS-aware clouds. In: European Conference on Computer Systems, Proceedings of the, European Conference on Computer Systems, EUROSYS 2010, Paris, France, April, pp. 237–250. DBLP (2010)
13. Avadi, B., Abawajy, J., Buyya, R.: Failure-aware resource provisioning for hybrid Cloud infrastructure. J. Parallel Distrib. Comput. **72**(10), 1318–1331 (2012)

14. Tumanov, A., Zhu, T., Park, J.W.: TetriSched: global rescheduling with adaptive plan-ahead in dynamic heterogeneous clusters. In: Eleventh European Conference on Computer Systems, pp. 35–36. ACM (2016)
15. Hindman, B., Konwinski, A., Zaharia, M.: Mesos: a platform for fine-grained resource sharing in the data center. In: Proceedings of the 8th USENIX Conference on Networked Systems Design and Implementation, pp. 429–483. USENIX Association (2010)
16. Karanasos, K., Suresh, A., Douglas, C.: Advancements in YARN resource manager 43(3), 51–60 (2018)
17. Verma, A., Pedrosa, L., Korupolu, M.: Large-scale cluster management at Google with Borg. In: Tenth European Conference on Computer Systems, pp. 1–17. ACM (2015)
18. Ananthanarayanan, G., Kandula, S., Greenberg, A.: Reining in the outliers in map-reduce clusters using Mantri. In: Usenix Conference on Operating Systems Design and Implementation, pp. 265–278. USENIX Association (2010)
19. Ferguson, A.D., Bodik, P., Kandula, S.: Jockey: guaranteed job latency in data parallel clusters. In: European Conference on Computer Systems, EUROSYS, pp. 99–112 (2012)
20. Zaharia, M., Konwinski, A., Joseph, A.D.: Improving MapReduce performance in heterogeneous environments. In: Usenix Conference on Operating Systems Design and Implementation, pp. 29–42. USENIX Association (2008)
21. Boutin, E., Ekanayake, J., Lin, W.: Apollo: scalable and coordinated scheduling for cloud-scale computing. In: Usenix Conference on Operating Systems Design and Implementation, pp. 285–300. USENIX Association (2014)
22. Ousterhout, K., Wendell, P., Zaharia, M.: Sparrow: distributed, low latency scheduling. In: Twenty-Fourth ACM Symposium on Operating Systems Principles, pp. 69–84 (2013)
23. Ghodsi, A., Zaharia, M., Shenker, S.: Choosy: max-min fair sharing for datacenter jobs with constraints. In: ACM European Conference on Computer Systems, pp. 365–378 (2013)
24. Isard, M., Prabhakaran, V., Currey, J.: Quincy: fair scheduling for distributed computing clusters. In: IEEE International Conference on Recent Trends in Information Systems, pp. 261–276 (2009)

Trusted Virtual Machine Model Based on High-Performance Cipher Coprocessor

Haidong Liu$^{(\boxtimes)}$, Songhui Guo, Lei Sun, and Song Guo

Information Engineering University, Zhengzhou 450000, China
2498779925@qq.com

Abstract. In the cloud computing environment, with the complex network environment, the virtualization platform faces many security problems. At the same time, trusted computing can greatly enhance the architecture security of virtualization platform systems, but there are many problems when trusted computing is deployed directly in the cloud environment. Therefore, this paper proposes a trusted virtual machine model based on high-performance cipher coprocessor to solve the security problems such as the isolation and insufficient performance of virtual TPM (vTPM) on the existing virtual platform. In this model, virtio technology was used to realize the virtualization of TPM, and a management architecture was designed to manage the life cycle of vTPM. The analysis shows that the model can complete the isolation of vTPM, and protect the security of vTPM during the migration process through the migration control server, and can strengthen the security of the virtualization platform. Finally, the simulation results show that the model is more feasible and suitable for cloud platform than hardware TPM.

Keywords: VTPM · Trusted computing · Cloud computing · Virtualization

1 Introduction

In the era of data and information, the development of emerging industries as the Internet, cloud computing, information technology, and so on, rapidly promotes the in-depth development of the entire information industry [1]. Cloud computing technology has been widely deployed and applied because it can provide economical, portable and dynamic on-demand computing services. However, cloud computing uses virtual machines to provide tenant with the environment of computing service, which results in tenant losing direct control of physical hardware. As a result, it is difficult for ordinary people to ensure the security of the data that they stored in the cloud and the services provided by cloud service providers, or to fully trust the security commitments of service providers [2]. At the same time, most of the current network security systems are mainly composed of traditional security facilities such as firewalls [3]. These traditional methods are difficult to adapt to the complex network environment of cloud computing environment. In summary, how to solve the security issues of cloud computing has become an important factor to restrict the further development of cloud computing [4].

© Springer Nature Singapore Pte Ltd. 2020
J. Zeng et al. (Eds.): ICPCSEE 2020, CCIS 1257, pp. 705–715, 2020.
https://doi.org/10.1007/978-981-15-7981-3_52

The structural characteristics of cloud computing environment are the main reasons for security problems [5]. Only through the comprehensive measures from hardware chip to network service, the security of the whole information system can be guaranteed [1]. This provides a basic idea to solve the security problems of cloud computing. In traditional information systems, trusted computing organization (TCG) provides an effective technology program for trusted platform through using trusted platform module (TPM).

TPM is successful for PC systems, but there are three issues that need to be processed, when using it in a cloud computing environment. First, the TPM that is mounted on the computer LPC bus is not suitable for processing large amounts of data. Second, Large amounts of data cannot be saved on the TPM, but this is critical for the cloud computing environment. Third, In the design of TPM chip, TCG considers that the chip is an extra cost in the computer system, which leads to more attention to the cost of the chip. Therefore, the computing performance of TPM is very limited. and TPM mainly achieves hardware acceleration of asymmetric cipher algorithms. Without Symmetric encryption algorithm, it is difficult to deal with the cipher operations in cloud computing environment. These issues lead to TPM not adapting to the cloud environment. In addition, it is necessary to realize the migration of vTPM in the cloud computing environment. In conclusion, the virtual trusted platform, that directly rely on hardware TPM virtualization in the cloud environment, will face various issues such as vTPM security isolation, TPM speed, and so on.

Therefore, in the cloud environment characterized by virtualization technology, a hardware security anchor is required to deal with the trustworthy problems of cloud platforms. In this case, a trusted virtual machine model based on high-performance cipher coprocessor is proposed to solve various problems that are caused by insufficient performance of TPM hardware.

This paper organized as follows. Section 2 introduces the research of trusted virtualization platform. Section 3 introduces Virtio technology. Section 4 designs the whole architecture of trusted virtual machine model based on high performance cipher coprocessor virtualization. Section 5 demonstrates that the trusted virtual machine model is more suitable for cloud environments than hardware TPM through simulation experiments. Finally, we summarize the paper in Sect. 6.

2   Research on Trusted Virtualization Platform

In this section, through the analysis of the existing trusted virtualization platform, the corresponding design objectives are proposed.

2.1   Existing TPM Virtualization Technology

Trusted computing technology is based on TPM. In order to realize the trusted virtualization platform in cloud computing environment, TPM must be virtualized. Trusted computing group (TCG) has proposed the concept of virtual trusted platform module (vTPM) [6] This provides two technical ideas for the implementation of vTPM, one is based on the virtualization of hardware TPM, the other is based on the virtualization of high-performance cipher coprocessor. At present, the virtualization technologies of TPM mainly include TPM passthrough, simulation based on function library and character device in user space (CUSE) [7]. The latter two belong to software simulation of TPM virtualization. Passthrough technology allocates a hardware TPM device directly to a virtual machine, which is characterized by simple architecture, high efficiency and strong security. Because each virtual machine has to prepare a hardware TPM in passthrough, it is not suitable for cloud computing environment. The virtualization technology based on function library has high performance and does not need TPM hardware, but its sensitive information is directly exposed in user space. CUSE is a general virtualization technology, which can provide good support for the upper interface of virtual machine, but there are also security problems.

At present, the virtualization program based on hardware TPM includes Xen, which realizes the protection of vTPM through hardware TPM. Its vTPM instance and management are in the privileged virtual machine. In addition, hardware isolation technologies such as Intel SGX and ARM Trust Zone are used to deal with the isolation of VTPM [8]. For example, eTPM uses Intel SGX technology to achieve the isolation of vTPM, and protect the confidentiality of information [9]. But hardware isolation technology also has some security problems, it is difficult to achieve the pure physical isolation effect like TPM. For example, foreshadow attack can obtain sensitive information protected by SGX, which destroys the confidentiality of vTPM [10].

2.2   Design Goal of Trusted Virtual Machine Model

According to the problems of the existing trusted virtualization model and the security requirements of cloud computing environment. In this paper, the design goal of trusted virtualization model includes the following objectives.

- This model improves the availability of the system by solving the TPM performance problems that is caused by LPC bus and hardware design in cloud computing.
- The model solves the problem of isolation between instances of vTPM, and protects the confidentiality, integrity and system security of instances of vTPM.
- This model realizes the binding with virtual machine and security migration of vTPM in the cloud computing environment.

3   Related Work

Virtio is a series of efficient, well-maintained Linux drivers which can be adapted for various different hypervisor implementations using a shim layer [11]. In full virtualization, privileged instructions of VM need to be translated, which can lead to complex processing of instructions. In para-virtualization, virtio uses the virtio-vring to achieve efficient messaging between front-end and back-end drivers. And It can achieve almost the same I/O performance as native devices (Fig. 1).

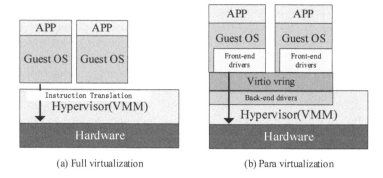

(a) Full virtualization   (b) Para virtualization

Fig. 1. Full virtualization and para-virtualization of virtio

Meanwhile, the virtio is widely implemented at present. For example, CentOS 7.6 has set virtio driver as the default compilation option. Therefore, we use Virtio technology to realize the virtualization of high-performance cipher coprocessor, as shown in Fig. 2.

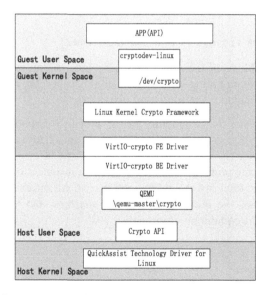

Fig. 2. Framework of virtual cipher coprocessor based on virtio

This framework can achieve the virtualization of vTPM, and provide the same interface with the hardware TPM in virtual machine. At the same time, it ensures that both the management of vTPM and the instance of vTPM are in the cipher coprocessor, which eliminates the possibility of being attacked by malicious software. And the framework provides the basic hardware support for the secure migration of vTPM. Finally, the vTPM based on this architecture can realize the security of cloud environment. Meanwhile, it also can deal with the security isolation of virtual machine and ensure the security of tenant information by using the whole disk encryption technology based on the key protected in vTPM.

4 Design of Trusted Virtual Machine Model

4.1 Framework Design

As shown in Fig. 3, the system architecture is divided into five levels: virtual machine user space, virtual machine kernel space, host user space, host kernel space and hardware. The management and instantiation of the whole system are realized in the cipher coprocessor. And it provides vTPM service for virtual machine through virtio technology. The application for a virtual machine calls the API to perform the operation process through TSS. First, the application of a virtual machine accesses the device file /dev/vtpm. Second, the driver in QEMU accesses the API interface of cipher coprocessor. Third, the cipher API calls the cipher coprocessor driver. Finally, the final operation is completed in the cipher coprocessor.

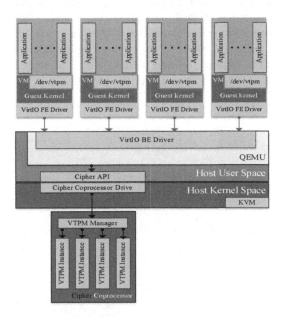

Fig. 3. Trusted virtual machine model architecture

In this architecture, the management of vTPM and the storage of sensitive information need to be completed in the cipher coprocessor. First, the manager of vTPM needs to complete the unique binding of vTPM to VM and the persistence of the vTPM instance, and deal with the migration mechanism of vTPM. That is to say, in the binding of vTPM and VM, vTPM manager of the cipher coprocessor needs to distinguish the corresponding relationship between VM and vTPM, and complete the isolation of vTPM. Secondly, in the life cycle of vTPM, as RTM and RTR of trusted computing platform, the architecture of vTPM needs to complete the integrity authentication and trust chain transmission for the trusted cloud platform. Therefore, it needs to ensure that vTPM early starts before VM and remains stable in the life cycle of VM. Finally, because there is a migration of virtual machine in cloud computing environment, it is necessary to implement the migration and recovery of vTPM in the cipher coprocessor. In the process of migration, vTPM is separated from hardware, so it is necessary to solve the problems of protecting sensitive information stored in vTPM, such as endorsement key (EK), identity authentication key (AIK) and storage key (SK).

In Fig. 4 the vTPM management model in cipher coprocessor is as follows.

- The vTPM scheduling and access control module is designed to manage the vTPM. It is mainly divided into two parts. First, the access control function is to control the VM access according to the access policy to ensure the security of the system. Secondly, vTPM scheduling is to identify the vTPM bound to the VM for the access allowed by the access policy, complete the corresponding operation, and ensure the quality of service (QoS) of the system.
- The vTPM management control module mainly provides management interface and vTPM life cycle management such as creation, registration, change and destruction.
- The migration and information management module is to complete the migration and information storage of vTPM. In migration, the module needs to complete the packaging of vTPM information and binding with VM image. In addition, some context information needs to be saved in the process.

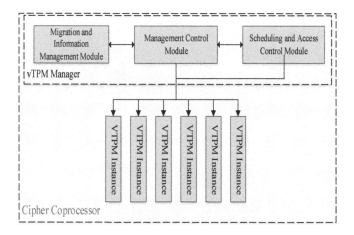

Fig. 4. Cipher coprocessor architecture

4.2 Analysis of Single Node Trusted Virtual Machine Model

In the framework of trusted cloud platform, the state of vTPM needs to be consistent with VM, just like the one-to-one correspondence between hardware TPM and computer. Only in this way can we ensure the credibility of the cloud platform. In other words, on a single cloud node, the state of VM, such as create, start, run, shut down, suspend, destroy, should be consistent with VM [12]. In addition, it is also necessary to consider the measurement about various states of host, such as startup, operation, shutdown, etc. In this way, the cloud computing environment can be trusted through the trust chain from the host to the VM.

The workflow of the system is as follows. In the phase of the host, the high-performance cipher coprocessor needs to generate a new TPM instance for the host, and the trusted model uses trusted computing to ensure the security of the system by completing the measurement from the core root of trust for measurement (CRTM) through BIOS, system boot, operating system to application. Meanwhile, it provides the basic support for the vTPM migration. In the virtual machine creation phase, it is necessary to assign a vTPM instance to the virtual machine and complete the configuration of EK, AIK, SK and PCR. In this way, we can ensure the long-term effectiveness and stability of vTPM. In the migration of virtual machine, the trusted recovery of virtual machine is ensured by saving the relational information in the running environment. In the destruction of virtual machine, vTPM must be destroyed to maintain the consistency with VM, so as to protect the system security.

4.3 VTPM Migration Process

Virtual machine migration triggered by load balancing and other reasons often occurs in cloud computing, and its purpose is to ensure high availability and stability of the system. At the same time, in trusted cloud computing, vTPM binds to virtual machines one by one. Therefore, during migration, vTPM needs to follow the virtual machine. Normally, vTPM contains virtual machine sensitive information, which is hard to obtain because of hardware protection of cipher coprocessor in a single cloud computing node. However, during the migration process, vTPM is not protected by hardware, so the confidentiality, integrity and binding relationship of vTPM must be protected by security protocol.

Therefore, the migration process in the trusted virtual machine model is divided into four steps as shown in Fig. 5.

- The migration control server receives the migration requirement, and then sends a migration command to the trusted virtualization environment A. After the high-performance Cipher coprocessor in environment A receives the migration command, it begins preparing for the vTPM migration, which includes registering the information for the vTPM that needs to be migrated and packaging the information.

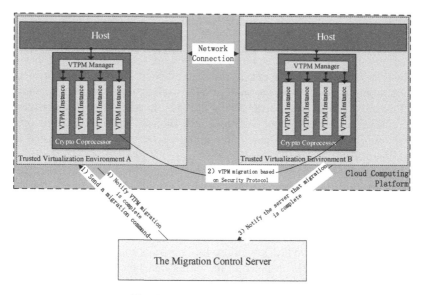

Fig. 5. VTPM migration process

- The cloud computing migration control server uses TNC technology to ensure that the migrated objects are trusted. When the server completed the authentication, the server achieves the secure transmission of vTPM from A to B by distributing keys.
- After receiving the encrypted vTPM package, the trusted virtualization environment B completes the decryption by requesting a key from the cloud computing migration control server. Then the server can confirm that B has received the vTPM package by waiting for B to return the confirmation information, and then notify A to destroy the migrated vTPM information. This process ensures the integrity, confidentiality, and uniqueness of the vTPM during the migration.
- The server informs the trusted virtualization environment B that the vTPM migration is complete. After receiving the message, the high-performance cipher coprocessor deletes the migrated vTPM to ensure uniqueness.

5   Simulation Experiment of Trusted Virtual Machine Model

According to the above, a simulation experiment is designed to verify the feasibility of the model. The operating systems of the host and VM in this experimental environment are Centos 7.6, the Linux kernel version is 3.10.0-862.el7.x86_64, and the CPU of the server is Intel Xeon CPU 6238.

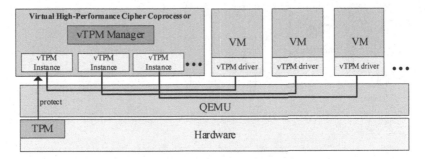

Fig. 6. VTPM experimental architecture

In the experiment, as shown in Fig. 6, the high-performance cipher coprocessor is simulated by using simulation software to realize vTPM instance and vTPM manager on the host. We created 10 QEMU virtual machines on the host and modified the code of QEMU to support vTPM. The security of the platform is ensured by using Trusted Grub2 to implement the trusted boot of VM. Finally, TSS is tested in VM, and the results are shown in Table 1.

Table 1. TPM execution time.

Command	Time
TPM_GetRandom	140 ms
TPM_Sign	190 ms
TPM_Quote	200 ms
TPM_Seal	160 ms

In Table 1, the execution time of vTPM is at millisecond level for the basic operation, which shows that the vTPM in trusted virtual machine model has the ability to support the security of cloud computing platform. At the same time, due to the performance of TPM, the trusted virtual machine model based on high-performance cipher coprocessing is more suitable for the cloud computing platform with a large number of virtual machines.

As shown in Fig. 7, the demo of vTPM migration is implemented by modifying the code of QEMU. The process is as follows. First, "migrate exec:cat > testvm.bin" command completes the packaging of the binary for the vTPM. Secondly, "incoming exec:cat < testvm.bin" command implements the migration of vTPM to a new virtual machine.

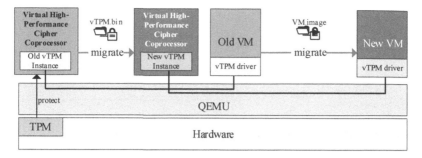

Fig. 7. Demo of vTPM migration

6 Conclusion

This paper presents a trusted virtual machine model based on high-performance cipher coprocessor. This model uses the cipher coprocessor to enhance the security of the virtualization platform, and proposes a framework to implement vTPM by virtio technology. Based on this framework, the whole security architecture is designed. Then a secure vTPM migration process is designed on this security architecture. Finally, the feasibility and validity of the model are proved by these experiments.

This paper provides a general framework, but only completes a simple simulation test. Therefore, a lot of work will be needed to implement the trusted virtual machine model on the cipher coprocessor.

In addition, the following work will be carried out in the future

- The study of security protocols during migration, as it will affect the security of the entire model.
- Research on the management methods for the entire life cycle of vTPM instances. It involves scheduling optimization based on trusted virtual machine models, virtual machine security, and so on.

References

1. Ren, C., Wang, Z., Zhao, B.: Industrialization application of trusted computing structure specification. Inf. Secur. Commun. Priv. **1**, 43–49 (2019)
2. Liu, C.Y., Wang, G.F., Lin, J.: Practical construction and audit for trusted cloud execution environment. Chin. J. Comput. **39**(2), 339–350 (2016). https://doi.org/10.11897/SP.J.1016.2016.00339
3. Shen, C., Shi, L., Zhang, H., Liu, C., Shang, Z.: Trusted computing and trusted cloud security framework. Sci. Manage. **38**(02), 1–6 (2018). https://doi.org/10.3969/j.issn.1003-8256.2018.02.001
4. Feng, D.G., Zhang, M., Zhang, Y., Xu, Z.: Study on cloud computing security. J. Softw. **22**(1), 71–83 (2011). https://doi.org/10.3724/SP.J.1001.2011.03958
5. Ali, M., Khan, S.U., Vasilakos, A.V.: Security in cloud computing: opportunities and challenges. Inf. Sci. **305**, 357–383 (2015). https://doi.org/10.1016/j.ins.2015.01.025

6. Stefan, B., Ramón, C., Kenneth, A.G.: vTPM: virtualizing the trusted platform module. Usenix Secur. **15**, 305–320 (2007)
7. Berger, S.: Scalable attestation: a step toward secure and trusted clouds. In: 2015 IEEE International Conference on Cloud Engineering. IEEE (2015). https://doi.org/10.1109/ic2e.2015.32
8. He, X., Tian, J., Liu, F.: Survey on trusted cloud platform technology. J. Commun. **40**(02), 154–163 (2019). https://doi.org/10.11959/j.issn.1000-436x.2019035
9. Sun, H.: eTPM: a trusted cloud platform enclave TPM scheme based on intel SGX technology. Sensors **18**(11), 3807 (2018). https://doi.org/10.3390/s18113807
10. Van Bulck, J., Minkin, M., Weisse, O.: Foreshadow: extracting the keys to the intel SGX kingdom with transient out-of-order execution. In: 27th USENIX Security Symposium, USENIX Security 18 (2018)
11. Russell, R.: virtio: towards a de-facto standard for virtual I/O devices. ACM SIGOPS Oper. Syst. Rev. **42**(5), 95–103 (2008). https://doi.org/10.1145/1400097.1400108
12. Zhang, J., Yang, S., Tu, S., Wang, X.: Research on vTPCM trust management technology for cloud computing environment. NETINFO Secur. **208**(04), 15–20 (2018). https://doi.org/10.3969/j.issn.1671-1122.2018.04.002

Author Index

Al-Neshmi, Hamzah Murad Mohammed
  II-216

Bai, Hongpeng I-428
Bai, Lulu I-642
Bai, Xin II-423
Bai, Yulin II-274
Bian, Qinyu II-3
Bian, Xuanyi II-384

Cai, Lei I-563
Cao, Wenlong II-138
Chai, Rui I-230
Chang, Qi II-237
Che, Chao I-476
Chen, Bing II-521
Chen, Dehong II-237
Chen, Jingyi I-327
Chen, Liang I-534
Chen, Ningjiang I-690, II-445
Chen, Yan II-445
Chen, Zuocong I-666
Cui, Qianna I-327
Cui, Wei II-423

Dai, Huanhuan I-72
Dai, Mingzhi II-197
Deng, Liping I-463
Deng, Quanxin I-292
Di, Xiaoqiang I-428
Dong, He I-341
Dong, Hongbin I-518, I-612
Dong, Liang I-447
Dong, Yang I-327
Dong, Yu II-354
Dong, Zongran II-384
Dou, Kangkang II-151
Du, Qi I-268
Du, Xiaochen I-488
Du, Xiujuan II-129
Du, Xuehui I-130

Fan, Lilin I-72
Fan, Yaqing I-389
Fan, Youping I-259
Feng, Fan I-61
Feng, Guangsheng I-44, I-341
Feng, Hailin I-488
Feng, Xiang II-197
Fu, Qiang I-518

Gan, Qinghai II-237
Gao, Guannan I-447
Gao, Jingyang I-563
Gao, Liwei I-563
Gao, Qingji I-463
Gao, Shunan I-577
Gu, Ruizhe II-138
Guo, Song I-705
Guo, Songhui I-705
Guo, Weibin II-197
Guo, Weiyu II-521

Han, Haiyun II-458
Han, Yiliang I-676
Han, Zhongyuan II-176
He, Jiakang II-185
He, Shuning I-327
He, Yuanzi I-602
He, Ziqi II-445
Hong, Chaoqun I-534
Hu, Honggang II-138
Hu, Xianlang I-44, I-341
Huang, Chengzhe II-176
Huang, Chun I-268
Huang, Fan II-138
Huang, Guili II-237
Huang, Guimin I-389
Huang, Heyan I-283
Huang, Hui I-268
Huang, Kai I-389
Huang, Kangzheng I-145
Huang, Lin I-380

Huang, Song II-638
Huang, Weishan II-445
Huang, Xia I-145
Huang, Xiaoyan I-368
Huang, Yong II-36
Huo, Jiuyuan II-216

Ji, Haipeng II-475
Jia, Juan II-101
Jin, Hai I-623
Ju, Tao II-216

Kong, Guangqian II-506
Kong, Leilei II-176
Kong, Linghong II-393
Kong, Xiangrui I-341

Li, Bingjie I-72
Li, Caimao I-440
Li, Guoliang I-447
Li, Lin II-237
Li, Qiong I-591
Li, Ruibing I-341
Li, Shaozhuo I-130
Li, Shijie I-591
Li, TianCai II-344
Li, Wenhan I-299, I-550
Li, Wujing II-423
Li, Xiaoguang II-90
Li, Xichen I-166
Li, Xuan I-130
Li, Yan II-393
Li, Yang I-247
Li, Yongsheng II-36
Li, Yu I-676
Li, Yuanhui II-458, II-574
Li, Yuhang I-16
Li, Yunkai II-354
Li, Zequn II-393
Li, Zhe I-676
Liang, Boxiang II-101
Liang, Shengbin I-642
Liang, Yongyu I-416
Liao, Liang II-574
Liao, Xianglin II-638
Liao, Xiaofei I-623
Lin, Changhai II-101

Lin, Chengrong I-440
Lin, Jing II-409
Lin, Junyu II-369
Lin, Li I-354
Lin, Xianmin II-246
Liu, Dianting I-145
Liu, Dong I-72
Liu, Haidong I-705
Liu, Haikun I-623
Liu, Jian II-151
Liu, Jiao II-638
Liu, Jishun II-167
Liu, Jun II-601
Liu, Lu I-310
Liu, Minghao I-310
Liu, Pingshan I-389
Liu, Rui I-118
Liu, Shaohua II-101
Liu, Siwei II-36
Liu, Xia II-574
Liu, Xiao I-380
Liu, Xinduo II-475
Liu, Yi II-237
Liu, Yu II-258, II-274
Lu, Peng II-113
Lu, Yao II-290
Luo, Liming II-620
Luo, Yu I-259
Lv, Hongwu I-44, I-341
Lv, Ye I-101

Ma, Xinxing II-21, II-62
Ma, Zhiqi II-246
Mai, Feng I-292
Mi, Chunqiao II-409
Mu, Lin II-216
Mu, Zihe I-612

Nima, Zhaxi I-292

Pan, Haiwei I-327
Pan, Lihu II-354
Pang, Hong I-259
Pang, Jinhui I-61
Pang, Shengnan II-588
Peng, Chunyan II-129
Peng, Kai I-440

Peng, Shengwang   II-246
Peng, Wanwan   I-259
Pu, Chunfen   I-405

Qi, Chunhong   II-543
Qi, Chunqi   II-329
Qi, Haoliang   II-176
Qi, Minhui   I-292
Qi, Xiaoyao   I-222
Qi, Yanming   II-101
Qian, Kaiguo   I-354
Qian, KaiGuo   I-405
Qian, Zheng   I-488
Qiao, Gangzhu   I-230
Qiao, Lei   II-393
Qin, Jing   II-113
Qin, Pinle   I-230
Qin, Xuan   II-216
Qu, Cong   I-440
Qu, Fang   I-440
Qu, Liangdong   II-36

Rao, Yuan   II-3
Ren, Dan   II-574
Ren, Leiming   II-258
Ren, Long   II-369
Ru, Yan   I-72

Shan, Lin   I-476
Shan, Shimin   II-258
Shang, Yimeng   II-467
She, Chundong   II-101
Shen, Qing   II-185
Shen, Shikai   I-354, I-405
Shi, Jianting   I-506
Shi, Shouchuang   II-369
Shi, Shumin   I-283
Shi, Zhengguang   II-138
Shi, Zhicai   II-393
Song, Jinyu   II-638
Song, Wei   I-310
Song, Yu   I-310
Sun, Bin   I-222
Sun, Le   II-90
Sun, Lei   I-705
Sun, Rui   I-428

Sun, Xu   II-176
Sun, Yu   I-612
Sun, Yuhong   II-588

Tan, Aoqi   II-475
Tan, Chengming   I-447
Tan, Zheng   II-344
Tang, Huabin   II-167
Tang, Yifan   I-292
Tao, Yang   I-89, I-602
Tian, Huan   II-506
Tian, Yuan   I-189
Tong, Heng   II-490

Wang, Cong   II-423
Wang, Diangang   I-380
Wang, Futian   II-490
Wang, Hongzhi   I-3, I-33
Wang, Ke   II-445
Wang, Lei   II-290
Wang, Leipeng   II-3
Wang, Linteng   II-90
Wang, Ming   I-447
Wang, Na   I-130
Wang, Qian   I-476
Wang, Shuo   II-3
Wang, Wenxiang   I-310
Wang, Xiao   I-310
Wang, Xiaoyuan   I-623
Wang, Xinsheng   I-222
Wang, Xiuli   I-247, II-309, II-329, II-521
Wang, Yaling   II-77
Wang, Yanbo   I-230
Wang, Yishu   I-101
Wang, Yonglu   I-118
Wang, Yue   II-77, II-309, II-329, II-467,
    II-521
Wang, Yujian   I-354, I-405
Wang, Zhanghui   II-90
Wang, Zhifang   I-299, I-550
Wang, Zhijie   II-475
Wang, Zhiwen   II-369
Wei, Qiang   II-506
Wei, Xingchen   II-620
Wei, Yu   II-393
Wen, Bin   II-47

Wen, Yiping II-344
Wu, Hao I-447
Wu, Maochuan II-369
Wu, Minhua II-620
Wu, Wenjun II-309, II-329
Wu, Xiaohua I-259
Wu, Yanqiu I-534
Wu, Yue I-476
Wu, Yun II-506

Xia, Kai I-488
Xia, Shuang II-436
Xie, Nannan I-428
Xie, Wenqing I-299, I-550
Xie, Xuchao I-591
Xie, Yongsheng I-690
Xu, Da I-463
Xu, Jin I-518
Xu, Lu II-274
Xu, Xiujuan II-258, II-274
Xu, Zhenzhen II-274
Xue, Chen II-423
Xue, Yijun I-690

Yan, Fengting II-393
Yan, Guanghui I-101
Yan, Huimin II-354
Yan, Yu I-3
Yang, Guozheng I-416
Yang, Nan II-423
Yang, Song II-384
Yang, Yan I-16, I-577
Yang, Yinhui I-488
Yao, Jing I-655
Yi, Yaling I-33
Yin, Baihui II-588
Yin, Shengjun I-33
You, Huanying II-101
Yu, Bing II-475
Yu, Chenshuo I-247
Yu, Haihao II-176
Yu, Minghao I-563
Yu, Nenghai II-138
Yu, Shaojun I-354
Yu, ShaoJun I-405
Yu, Xiaodong I-612
Yu, Xiaohan II-638
Yu, Xiaoqing I-230

Yu, Yongbin I-292
Yuan, Guowu I-447
Yuan, Jie I-506
Yuan, Peiyan I-368
Yuan, Yuan I-591
Yue, Chenbo II-62

Zeng, Yong I-380
Zeng, Zhiqiang I-534
Zhang, Cheng II-490
Zhang, Dafang I-44
Zhang, Ding I-61
Zhang, Guiqiang I-506
Zhang, Hang II-47
Zhang, Hao I-368
Zhang, Hengbo I-476
Zhang, Hengming I-642
Zhang, Hua I-655
Zhang, Jianxin I-476
Zhang, Meng I-101
Zhang, Shuo II-21, II-62
Zhang, Suojuan II-638
Zhang, Xiao II-543
Zhang, Xiaoming II-185
Zhang, Xinglei II-101
Zhang, Yan II-167
Zhang, Yingtao I-506
Zhao, Jun I-118
Zhao, Shizhao I-44
Zhao, Xiaowei II-274
Zhao, Yang II-561
Zhao, Yuechen II-309
Zheng, Aihua II-246
Zheng, Jinghua I-416
Zheng, Mengce II-138
Zheng, Zhiwu I-655
Zhong, Xiaofang II-588
Zhong, Yingli I-16, II-21
Zhou, Hao I-447
Zhou, Weikang I-61
Zhou, Xiaoyi II-246
Zhou, Yueyang I-283
Zhou, Zhengda II-167
Zhu, Jinghua I-577, II-21, II-62
Zhu, Tiejun II-601
Zhu, Yunlong II-423
Zhuo, Yongning I-380
Zuo, Kaizhong I-118

Printed in the United States
By Bookmasters